A Fair Share: Doing the Math on Individual Consumption and Global Warming

Steffen E. Eikenberry, MD, PhD

May, 2018

Contents

Part I

Introduction and Background

Chapter 1

Introduction, or, the banality of conservation

> Unprovided with original learning,
> unformed in the habits of thinking,
> unskilled in the arts of composition, I
> resolved to write a book
>
> *Edward Gibbon, Memoirs of My Life and*
> *Writings*

Global warming, and widespread environmental damage have weighed heavily on my mind, but typically in the abstract. As a member of the Western consumer class, one is faced daily with decisions that one vaguely believes have a broader impact on the environment, and a confusing array of news reports send mixed messages daily. Is it worthwhile to even try consuming ethically? Does it have even a marginal impact, or is it mere vanity or worse: are hybrid cars and green-labeled products merely indulgences for the rich that fail to address or even exacerbate the underlying drivers of environmental degradation? The notion that consumer choices are meaningful often faces derision from both the political right and left. Could it be that such banal acts as turning down the heater/AC or driving less actually have global significance? Doesn't the answer, "drive less," make a mockery of the enormity of global warming? Hence this book.

This book is one I wrote for myself. I wished to answer the question, to my own satisfaction, of whether one can consume, in the broad sense of the word, in a manner that is environmentally responsible, or whether any such effort is mere vanity. To answer this question, we must first answer a prerequisite inquiry: is consumption at the level of individual (Western) consumers a meaningful driver of environmental damage in general, and global warming in particular? The answer, as I will show, is decidedly yes. And, answering the natural follow-up question, yes, this consumption can be readily modified at the individual scale to dramatically reduce environmental impact.

To answer my previous inquires, it is indeed true that those acts most necessary are also most banal. Personal transportation (mainly in private vehicles), direct energy use within residential and commercial spaces, and food production collectively account for the greater share of America's environmental impact. Upstream energy and resource consumption for other goods and services account for most of the remainder. The Western consumer is the final common driver of the greater part of environmental degradation, and his day-to-day choices collectively drive the global carbon emissions and land use changes that chiefly underly global warming, mass extinction, and the emptying of the wilds.

It is my purpose to quantify the impact of Western, and in particular American consumption. It is my opinion that we are often distracted by red herrings, with stories in the popular environmental press expounding on the evils of everything from straws to toilet paper. While not all necessarily false, such reports can detract from the factors of major importance, and, I believe, induce defeatism. After all, if literally everything is killing the planet in equal measure, why even bother trying? Clearly then, one would be forced to the conclusion that there is simply no reasonable way to live in modern society without an enormous impact, short of renouncing the world? Hence the importance of quantifying the impact of our behaviors in a rigorous way, and determining those of true import.

Greenhouse gas (GHG) emissions, measured in carbon dioxide-equivalents (CO_2e), are ultimately generated, in large part, by acts of consumption, and so it is imperative to quantify them with respect to such acts. However, they are most typically inventoried at national, regional (e.g. North America, Europe, etc.), and global territorial scales, and are attributed to different *economic* sectors. At the coarsest scale, the EPA divides these sectors into (1) transportation, (2) industrial, (3) residential and commercial, and (4) agriculture; electricity generation, which supports all other economic sectors, may also be disaggregated as its own sector. This division is useful, but it masks how household level consumption acts across all sectors, and ultimately drives emissions across all sectors as well. Therefore, to truly understand individual emissions, it is far more instructive to make our unit analysis not *economic sector*, but *consumption category*, and many authors have undertaken just such an analysis, relying on a variety of data sources. In general, these authors (e.g. [1, 2]), and my own analysis (presented at length in the book before you), find that each American household generates on the order of 48–57 metric tons (or "tonne") of CO_2e annually, and, since the average household consists of just over 2.5 persons, this equals about 19–23 tonnes CO_2e per person. The majority of this impact is attributable to those three basic things everyone does every day: get around (personal transportation), directly use energy (heating, cooling, other residential energy uses, etc.), and eating, all things one has direct control over day by day, and the principle focus of this text.

American household consumption CO_2e emissions estimates equate, in the aggregate, to about 85% to *over* 100% of US territorial emissions. The latter possibility arises from the fact that emissions are widely "imported" and "exported" across national borders. To see this, consider some product manufactured in a Chinese factory powered by coal and shipped via international freight to the US for consumption. Under territorial accounting, the resulting CO_2 is assigned to China (and international freight CO_2 is assigned not at all), despite the fact that the US consumer is more ultimately responsible for its generation, and consumption-based accounting allocates this CO_2 to the American consumer. Perhaps 10–30% of US emissions are similarly imported: Weber and Matthews [1], at the high end, calculated that about 30% of US household emissions are imported from outside the country; other estimates are somewhat lower, with Hertwich and Peters [4] giving 18% of all US consumption emissions imported, while Davis and Caldeira [5] estimated a net 10.8% of total US consumption emissions were imported from outside the country. Many other developed, mainly European countries, import even larger fractions of their consumption footprint [5], and it is especially noteworthy that while territorial emissions have fallen in many such countries over the last couple decades, this is likely at least partially due to "outsourcing" emissions generation to developing countries, with net consumption emissions actually increasing for some of these nations [6].

Now, at least 70% of household CO_2e emissions can be attributed to, in descending order of importance, (1) personal transportation (25–40%), (2) household energy use and operations (25–30%), and (3) food consumption (15–20%), with the remaining 15–30% due to the emissions involved in conveying various goods and services. Thus, even at this coarse scale, it should be obvious that how much, and in what vehicles, Americans drive matters quite a bit. So do all the other banal and half-invisible everyday behaviors, from the choice of thermostat setting to one's daily bread. In the subsequent sections, I try to put these emissions numbers in a larger context, and look closer at the consumption that drives them. The overarching goal of the chapters that follow is to develop a comprehensive understanding of the major greenhouse gases, technologies, and industrial processes that all converge at the consumer level.

1.1 A fair share

In this section I perform some basic calculations to demonstrate that each US citizen is responsible for annual emissions of roughly 20 metric tons of CO_2e, and that, to meet near-term climate stabilization targets, one is entitled to, at *most* only 15 tonnes of CO_2e/year, while 10 $MgCO_2e$/year is a more reasonable short-term "fair share." In the longer-term, fair share per capita emissions are <4 MgCO2e/year.

It has been generally accepted that global atmospheric warming must be capped at a 2 °C (3.6 °F) rise to avoid the most dangerous consequences of global warming, and to achieve this, atmospheric CO_2 concentrations must be stabilized at the 450 ppm level. In its fourth assessment report, the International Panel on Climate Change (IPCC) estimated that this stabilization target would require developed countries (i.e. "Annex I" countries) to reduce their emissions by 25–40% by 2020, and by 80–95% by 2050 relative to a 1990 baseline [9]. In 2015, the US EPA inventory [10] estimated 1990 US territorial emissions at 6.301 billion metric tons of CO_2-equivalents (CO_2e). Using the January 1, 2016 Census estimate of the US population, 322.3 million persons, this amounts to about 19.5 metric tons of CO_2e ($MgCO_2e$) per capita, as the baseline from which reductions must be made. Note that I am using 1990 baseline emissions data, but the current population estimate. Therefore, if we assume all US citizens are equally responsible for meeting IPCC reduction targets, we may consider the maximum per capita CO_2e "fair share" for the year 2020 to be 14.6 $MgCO_2e$, a 25% reduction from the 1990 baseline. We get a more conservative 11.7 $MgCO_2e$ using a 40% decrease from the baseline. However, the maximum longer-term fair share is just 3.9 $MgCO_2e$ per capita (80% decrease from 1990 baseline).

Consider now that current per-capita emissions in the US now amount to about 21 $MgCO_2e$/person/year, based upon territorial accounting, and using 2013 population and emissions figures. Each US household, via direct consumption, likely generates a very similar amount of CO_2e on average (see below). Therefore, a 25% reduction from the 1990 baseline is actually a slightly larger 30% reduction from current emissions rates. Nevertheless, the difference in figures is small, and the essential fact to remember is that the average US citizen is directly responsible for just over 20 metric tons of CO_2e per year, and this is the baseline that all other emissions numbers are (at least implicitly) always being compared to.

Now, these initial calculations ignore projected population growth, nor do they take into account the marked heterogeneity in per capita emissions both between developed and developing nations, and among developed nations. A better approach might be to assume all citizens of Annex I countries (essentially all developed countries) are entitled to the same carbon budget. United Nations Framework Convention on Climate Change (UNFCCC) estimates for total Annex I emissions in 1990 are 19.9 and 18.9 billion $MgCO_2e$, with and without CO_2e from land use changes, respectively. Using 2013 World Bank population estimates, the sum population of Annex I countries is just shy of 1.3 billion persons (excluding Liechtenstein and Monaco), and thus we arrive at about 15 $MgCO_2e$ per capita, as our baseline for emissions reductions, across most developed countries. Twenty-five, 40, and 80% reductions from this baseline are 11.25, 9, and 3 $MgCO_2e$ per capita, respectively.

Thus, "reasonable" upper limits to what can be considered fair levels of carbon consumption are about 10–15 MgCO$_2$e per capita in 2020, just three years from the time of this writing, and just 3–4 MgCO$_2$e per capita in 2050, 33 years from the time of writing. These are still somewhat inflated, for if we believe that emissions should be shared equally among all citizens of the globe, and therefore our 25, 40, and 80% reduction targets should apply equally, dividing 38 GtCO$_2$e (approximate 1990 global emissions) among 7.12 billion people (2013 World Bank estimate), baseline global per capita emissions are a mere 5.3 MgCO$_2$e, with corresponding reduction targets of 4.00, 3.20, and 1.07 MgCO$_2$e/capita. Therefore, *truly* fair per capita carbon emissions can only be considered to amount to 3–4 MgCO$_2$e in the short-term, with a medium- to long-term share of only 1 MgCO$_2$e.

These calculations have worked out rather conveniently because, as already mentioned, current US per capita emissions are roughly 21 MgCO$_2$e, on both territorial and consumption bases. We may round down to 20 MgCO$_2$e per capita as a rule of thumb and for numerical convenience, and then depending on how we do our calculations, 25% (to 15 MgCO$_2$e), 40–50% (to 10–12 MtCO$_2$e), 80% (to 4 MgCO$_2$e), and 95% (to 1 MgCO$_2$e) reductions from this baseline all represent a different "level of fairness." Is it possible to achieve such emissions levels as a US citizen without retiring to a cave? I will argue that the first two are, in fact, eminently achievable for the majority of households and individuals, while one may even approach the 80% reduction target through ambitious changes in one's own lifestyle. The final goal requires deeper society-wide decarbonization, but such an achievement would be greatly abetted by, and likely mandates as a component, lower carbon lifestyles across the consumer class.

1.1.1 Fair share based on cumulative emissions

The fifth assessment report of the IPCC has estimated that cumulative anthropogenic (Greek "born of man") carbon dioxide emissions must be limited to 3,670 GtCO$_2$ (1,000 GtC) for a 66% chance of limiting global warming to <2 °C. More than 50% of this global carbon budget had already been spent by 2011, at an estimated 1,890 GtCO$_2$ (1,630 to 2,150) (515 GtC [445 to 585]). Moreover, when including non-CO$_2$ forcers (e.g. methane), the cumulative CO$_2$ limit falls to just 3,300 GtCO$_2$ (900 GtC), and at current emissions rates (about 50 GtCO$_2$e/year globally), it will likely be less than 30 years before the carbon budget is exhausted [8].

Supposing about 135 million people are born per year until 2050[1] we have, from the year 2015, about 4.75 billion births added to the existing population of 7.2 billion; this amounts to roughly 12 billion lives. Extending to 2100 with the same annual birth number adds another 6.75 billion lives for a total of 18.75 billion individual human lives until 2100. Dividing the global carbon budget of about 1,778 billion MgCO$_2$ (485 billion MgC) among 12 billion humans yields a scant 148 MgCO$_2$ per person born up to 2050, or 95 MgCO$_2$ per person if we assume everyone born up to the year 2100 has an equal right to emissions.

Assuming everyone born up to 2050 has an equal claim on the global carbon budget, then we each are entitled to 148 MgCO$_2$e, and this yields about 1.9 MtCO$_2$e/year over a life expectancy of 78.74 years (US, 2012), or 2.5 MgCO$_2$e/year if the counter starts at age 18. Similarly, about 1.2–1.6 MtCO$_2$e/year is the fair allocation out to the 2100 window.

Put more simply, on the basis of a cumulative emissions cap, everyone has a roughly 150 MgCO$_2$e lifetime allocation, or about 2 MgCO$_2$e/year, and under the assumption that human activity is largely de-carbonized by the latter half of this century. The allocation falls to closer to 1 MgCO$_2$e/year if overall decarbonization takes longer, similar to the long-term, global-average fair share I derived above.

---

[1]Birth rate based upon CIA World Factbook fertility figures and, based upon UN projected fertility rates, population growth, and life expectancy through 2050 [11], the total *number* of annual global births is likely to remain roughly constant for the next few decades.

1.2 Typical household emissions patterns and reductions potential

In subsequent chapters, I examine and quantify the major consumer drivers of carbon emissions in detail, and the results of these investigations are summarized here. I conclude that consumptive household emissions are driven by four major classes of activity, with approximate overall annual per capita emissions in parentheses: (1) personal transportation (6 $MgCO_2e$) (2) residential shelter (5 $MgCO_2e$), (3) food consumption (3–5 $MgCO_2e$), (4) and all other consumer goods and services (5–6 $MgCO_2e$), for a grand total of about 21 $MgCO_2e$/capita (or 53 $MgCO_2e$ per average household), and within the range reported by existing analyses, e.g. [1, 2, 3]. Note again, that this individual carbon impact, when aggregated across the entire US population, roughly matches 100% of *all* US territorial CO_2e. However, these averages mask marked variations about the mean, and in any given category of consumption, the top and bottom 20% of consumers (whether our unit of consumption be household or individual) vary by at least a factor of two in terms of their carbon impact, and generally much more: transportation emissions likely vary by more than a factor of *ten* between the top and bottom quintiles, residential energy emissions vary over threefold, and food and goods/services emissions also both vary by at least a factor of two or three.

Given this, it is clear that marked reductions in one's carbon footprint relative to the mean are readily achievable, and that the greatest absolute benefits are realized when the heaviest consumers alter their habits. For example, an individual in the top 20% across consumption categories might easily generate 40–45 $MgCO_2e$ in total, and thus, just bringing this person's consumption down to the mean would yield carbon savings as great as those to be had if a more typical individual's footprint went completely to zero. For comparison, a bottom 20% individual might be responsible for just 8–12 $MgCO_2e$ yearly, and such a person has already met a reasonable short-term emissions reductions target.

Carbon emissions for the first two consumption categories above (transportation and residential shelter) are primarily direct (including electricity generation), and under an individual's immediate control. Direct consumption of gasoline via passenger automobile transport dominates the personal transportation category, with emissions due to both the direct combustion of fuel (about 75–80%), and to the energy-intensive processes of petroleum extraction and refining. Jet fuel for air travel and the amortized costs of automobile manufacture are the two other major contributors, while public transportation amounts to a rounding error. Direct consumption of electricity and heating fuels dominate emissions attributable to the residence, along with the amortized carbon cost of home construction. Space and water heating are the two single largest users of residential energy, while a variety of miscellaneous electric loads are also of major importance. These two categories, personal transportation, and residential shelter, account for over half of household-level emissions, and are, again, essentially under the direct control of the consumer.

Food consumption is the largest indirect driver of emissions, with the majority occurring at the production stage, and with methane produced via enteric fermentation in beef (primarily) and dairy (secondarily) cattle the single greatest CO_2e source in US agriculture. A variety of post-farm processing and packaging activities downstream of the farm gate account for around 20–33% of emissions, while transport *per se* is a relatively minor component of the carbon footprint, despite the prominent billing of "food miles" in public discourse; landfilling of food waste also results in warming methane emissions. In terms of actual food products, beef alone likely accounts for almost one-third of all dietary emissions, despite making up only about 4% of the American diet. Meat consumption overall accounts for as much as half of diet-related emissions (and even more for heavier meat eaters), while dairy also has a substantial impact, and

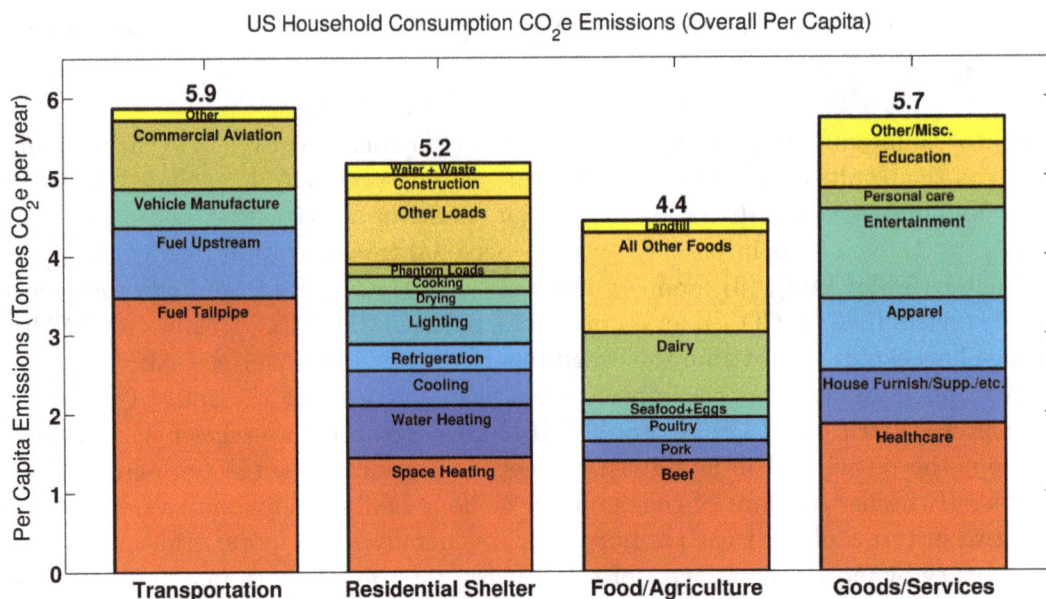

Figure 1.1: Approximate overall mean per capita consumption-based CO_2e emissions attributable to US households, disaggregated primarily by end-use or final product (e.g. space heating, beef), although note that gasoline for passenger vehicles is divided into the tailpipe emissions and those emissions due to upstream gasoline extraction and refining.

animal products all together may sum to almost 70% of the dietary carbon footprint. Finally, consumer-level food wastage (across all food types) amounts to up to one-third of the total impact of diet.

Overall per capita emissions attributable to household consumption, as calculated in the remainder of this book, are summarized in Figures 1.1 and 1.2 (following roughly the format introduced in [2]), with the two figures giving different disaggregations of residential and food emissions (note that food-related emissions differ slightly between the figures due to different methodologies for calculations).

Now, the above sums give household level consumption averaged across the entire US population, a population that includes many household types (single adults, large families with many children, poor and rich, etc.) spanning disparate geographic regions. Furthermore, nearly a quarter of US persons are under age 18, and they clearly have much less direct control over household consumption patterns in general. Therefore, a typical adult American will have an even larger carbon footprint than these numbers suggest; single adults especially, tend to have a relatively large impact, as there is little "sharing" of emissions across other household members. Tabulating the numbers for an average single-person household (and assuming such a person is also a typical driver), I arrive at a carbon footprint of around 31 $MgCO_2e$/year, roughly 50% higher than the overall per capita average. This sum disaggregates into about 8 $MgCO_2e$ for transportation, 10 $MgCO_2e$ from the residence, 4.5 $MgCO_2e$ for food, and 8.5 $MgCO_2e$ due to spending on general goods/services. Thus, both the potential for, and importance of, emissions reductions for single adults are appreciably greater than for the general US populace.

The above discussion gives us our baseline averages, but the next essential tasks are to examine (1) *existing* variation in consumption emissions, and (2) how basic conservation strategies might feasibly affect the consumption footprint, relative to baseline. Concerning the first point, I have already mentioned that the data presented in this book indicates dramatic variation, but it is instructive to consider transportation as an example, where differences in consumption emissions are especially profound, and indeed, in terms of gasoline use, the bottom and top 20%

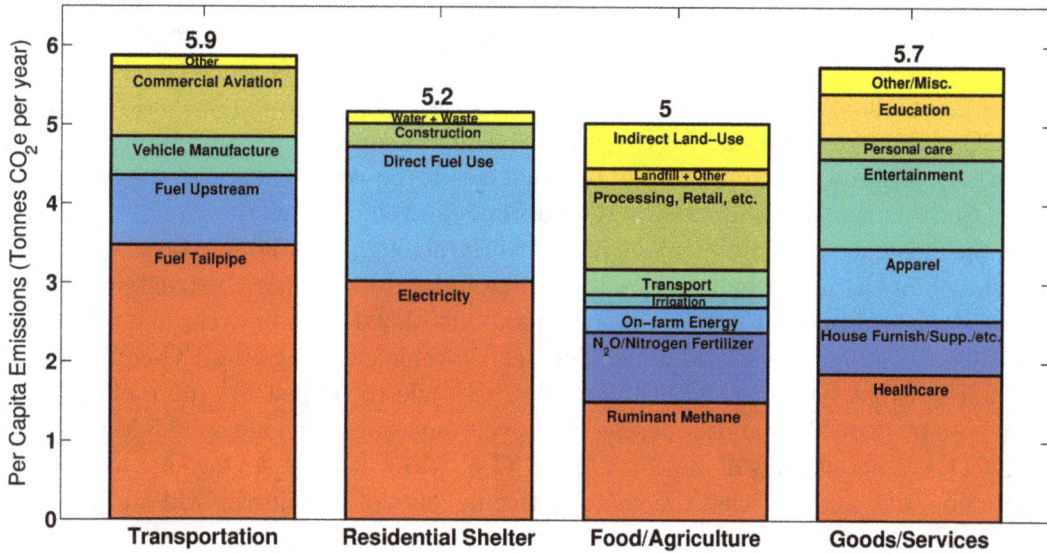

Figure 1.2: Approximate overall mean per capita consumption-based CO_2e emissions attributable to US households, but differing from Figure 1.1 in that residential emissions are given mainly in terms of energy source, and food emissions are divided by broad mechanism.

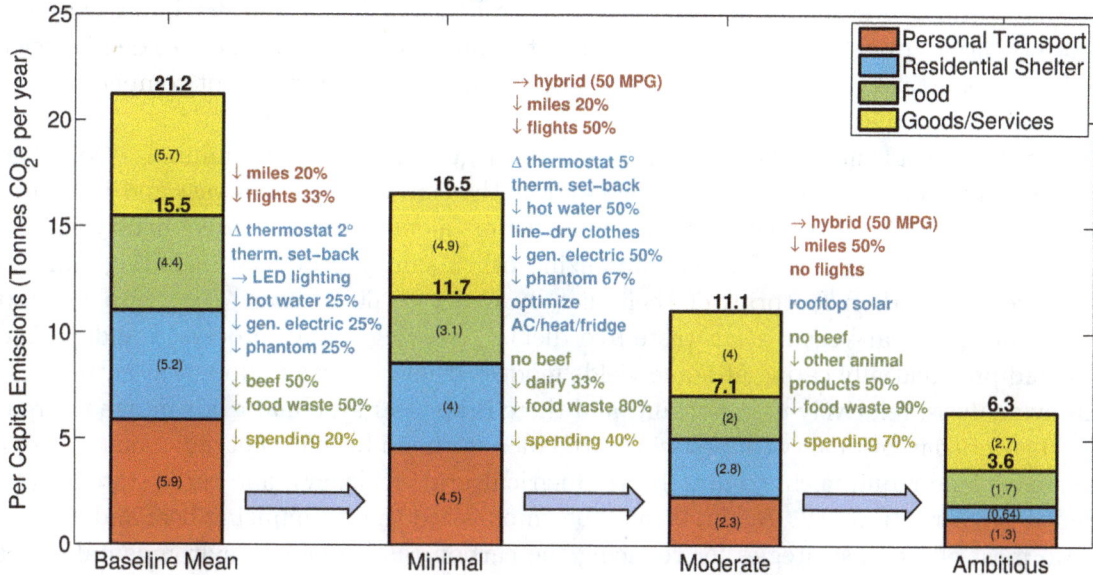

Figure 1.3: Carbon-equivalent emissions reductions from the mean US baseline under various collections of "minimal," "moderate," and "ambitious" changes in consumption patterns. Note that I consider healthcare spending relatively inflexible, and spending reductions in this category are only a fourth of the others (e.g. a 20% reduction in general spending gives just a 5% reduction in healthcare spending).

households vary over 11-fold, with mean gasoline-related emissions of 2.5 and 28 $MgCO_2e$ per household, for the bottom and top quintiles, respectively. In terms of vehicle ownership, the top 20% owned 3.5 cars to every one in the bottom 20% (See Chapter 6). Concerning aviation, I calculate that the top 20% of fliers travel roughly four times the air miles as the American average, while over 50% of US adults actually take *no* airplane trips in any given year. Thus, a "top-20% individual" might generate over 20 $MgCO_2e$ per year from personal transportation (15.4 $MgCO_2e$ from gasoline, 1.6 $MgCO_2e$ from vehicle manufacture, and about 3.5 $MgCO_2e$ from aviation), while a "bottom-20%" individual would be responsible for <2 $MgCO_2e$ (just 1.33 $MgCO_2e$ from gasoline and <0.5 $MgCO_2e$ from the vehicle itself).

While the existence of variation alone indicates the strong potential for conservation, a more bottom-up look at the major emissions sources may ultimately be more instructive. Relative to the baseline for a typical adult driver, who puts in about 12,500 miles per year in a vehicle getting just 22 MPG, upgrading to a hybrid or smaller electric vehicle getting 50 MPG-equivalent, while reducing miles driven by 20%, and changing driving style to be just 5% more efficient, would reduce fuel use by two-thirds and associated fuel-cycle emissions from almost 6.5 $MgCO_2e$ to just over 2 $MgCO_2e$, a net savings of almost 4.5 $MgCO_2e$. Most of these savings are attributable to the vehicle upgrade, and I discuss the comparative benefits of gasoline hybrid and pure electric vehicles extensively in Section 6.6.

Moving to residential energy, a variety of simple strategies that, while requiring conscientiousness, ultimately represent a minimal burden (in my view) and no true change in quality of life, can collectively reduce the emissions associated with household energy use by at least 33%, relative to baseline. The two most important are adjusting one's baseline thermostat (by, say 3–5 °F from baseline, e.g. 78° instead of 73° in the summer, and 66° instead of 70° in the winter) and setting the thermostat back when the dwelling is unoccupied (especially during the day in summer) or at night (in the winter), and avoiding excessive general electric loads (leaving TVs on, phantom loads, etc.). Minimizing hot water use by limiting shower water and general faucet use, turning over any incandescent lighting to LEDs or CFLs, and dry-lining clothes are also of potential. Investing in rooftop solar has the potential to reduce net residential energy emissions to effectively zero, although this is best pursued in *conjunction*, not competition, with other conservation measures.

A diet that avoids nearly all beef (*including* 100% grass-fed beef) and minimizes food waste (up to one-third of all food in the US is wasted at the consumer level, a vast and inexcusable waste of land and other resources) may, via these factors alone, have associated dietary emissions only 50% of the average. By further minimizing other animal products, including dairy, one may reduce the carbon footprint of their diet by closer to 60 or even 70% (with the latter possible under a near-vegan, near-waste-free diet). Avoiding highly processed and packaged foods, and preferentially eating at home yields modest benefits as well. Note that eating locally is, in my analysis, relatively unimportant, and there is no clear systematic advantage to organic products over conventional (with pesticide avoidance the main benefit of organic, but at the cost of increased land requirements due to lowered agricultural efficiency), and these statements are defended further in Part IV. Again, it must be emphasized that minimizing beef and waste are the two most effective strategies for reducing the carbon (and broader environmental) impact of diet, by far, and one need not be either strictly vegetarian or vegan to realize appreciable benefits in diet (and indeed, dairy is roughly comparable to poultry or pork, in its per-unit environmental impact).

In terms of general goods and services, nearly all direct consumer spending has an associated carbon cost on the order of 0.5–1.5 $kgCO_2e$ per dollar, due to the fundamental and wholly inescapable dependence of the modern industrial and commercial infrastructure upon fossil fuels. Thus, minimizing spending, but especially upon clothing and apparel, household goods in general, and entertainment is the only real way to reduce one's impact in this area. Healthcare is likely the most impactful category of goods and services overall, but I consider use of this resource relatively inflexible and often beyond a typical individual's control.

Not fitting neatly into any other category, I also note that maximal household recycling of all commonly recycled materials may annually offset around 0.35 $MgCO_2e$ per person, relative to a counterfactual with no recycling, a nontrivial but still quite modest sum. About two-thirds of this potential benefit is related to *paper* recycling (with the benefit mainly manifested in preserved forest carbon), with plastic and aluminum recycling of secondary importance. Otherwise, direct waste collection and management has only a very minimal carbon impact, with the partial exception of landfill gas emissions from food waste, which I attribute to diet in my accounting.

Potential emissions savings, relative to the average baseline, using several combinations of the above strategies are summarized graphically in Figure 1.3, which I lump together as "minimal," "moderate," and "ambitious," defined as:

1. **Minimal (20–25% CO_2e decrease from baseline):** (1) Reduce miles driven by 20% and omit one-third of flights; (2) change the thermostat by 2 °F and set it back by 3–5 °F when absent or at night in winter, change lighting to all LED (or CFL), decrease hot water, general electric, and phantom electric loads by 25%; (3) cut out 50% of beef and food waste; and (4) decrease general consumer spending by 20%.

2. **Moderate (45–50% CO_2e decrease):** (1) Upgrade vehicle to a hybrid or electric vehicle getting 50 MPG-equivalent, reduce miles driven by 20%, and omit half of flights; (2) change thermostat by 5 °F and employ the set-back strategy, decrease hot water and generic electric loads by 50%, avoid two-thirds of phantom loads, line-dry clothes, and upgrade old HVAC equipment and refrigerators; (3) Cut out beef entirely, decrease dairy 33%, and avoid 80% of all food waste; and (4) cut general spending by 40%.

3. **Ambitious (70–75% CO_2e decrease):** (1) Upgrade vehicle to a 50 MPG-equivalent vehicle, reduce miles driven by 20%, and take no flights; (2) add rooftop solar sufficient to cover 100% of residential energy use; (3) avoid beef, decrease all other animal products by 50%, and avoid 90% of food waste; and (4) decrease general spending by 70%.

We see that, even if all individuals and households undertook a *minimal* emissions-reduction strategy, net US household emissions (which in aggregate equal roughly 100% of the US's territorial emissions) would fall by over 20%, enough to very nearly meet near-term climate stabilization targets. This would also have the additional benefit of reducing pressure on fossil energy supplies, and reduce the amount of energy infrastructure that must ultimately turnover to renewable, near-zero carbon sources. More significant ("moderate") changes from the baseline can nearly halve the household consumption footprint, and although these do represent much more overt changes in lifestyle, they are by no means beyond the capabilities (financial or otherwise) of the greater part of Americans.

For the extremely ambitious individual, it is possible to push the cumulative impact of the three most direct categories of consumption, transportation, the residence, and diet, to under 4 $MgCO_2e$, a rather remarkable feat. It is at this point (and only at this point), that those various goods and services conveyed by the fossil economy become the dominant impact category, with the impact dominated by largely involuntary participation in the carbon-intensive US healthcare system. For one who truly reaches this point, further decreases in one's environmental individual footprint mandate deeper decarbonization throughout the economy and energy systems.

We see then, that collective individual action (if this is not a contradiction in terms) can eliminate the greater part of the US's carbon emissions (as well as dramatically reduce fossil fuel, water, and agricultural land use), and the collective lifestyles of Americans are indeed the fundamental driving force underlying the majority of emissions. To fully decarbonize the economy requires a deeper, infrastructure-wide shift to other energy sources, but this can be aided greatly by conservation measures, and individual conservation can synergize with, for example, the deployment of renewable electricity for a much more rapid decarbonization than could otherwise occur.

1.2.1 A note on household vs. individual

Energy use is typically reported using energy per household as the basic unit, and several works have examined household-level carbon footprints; the mean US household size was 2.54 persons in 2015, per the US Census. As already noted, then, using per capita emissions numbers on the basis of household-level consumption includes a significant number of children in the calculation, and lumps a variety of households. Therefore, I have, when especially salient, examined household consumption stratified by household size, and I have made a special focus upon single-person households vs. the overall mean household. Since it is probably most correct to attribute consumption emissions primarily to adults, single-person household emissions may be more salient for most readers, and may better represent the impact of typical adults compared to an overall per capita average. My primary metric, throughout the book, is CO_2e per capita (or individual), and not CO_2e per household, although the latter features prominently as well.

1.2.2 A final summary

In the near-term, every American is "entitled" to emit, directly and indirectly, no more than 10–15 $MgCO_2e$ at most, which is 25–50% less than current US per capita emissions. The 15 $MgCO_2e$/capita goal is likely *immediately* achievable for most households, and 10–12 $MgCO_2e$ is a reasonable five- or ten-year goal for an individual to set, particularly if they live in a 2+ person household. It is achievable through a low-waste low-meat diet, a high mileage vehicle, or alternatively, cutting vehicle-miles by 30–50%, and reductions in household energy consumption that are readily achievable. Furthermore, limiting consumer spending is, like it or not, also important. None of these changes are qualitatively drastic, but do represent significant departures from business-as-usual.

An 8 $MgCO_2e$/capita goal represents a roughly 60% reduction from the US baseline under territorial accounting, and a similar or even larger reduction on a consumption-based accounting. Longer-term, this 8 $MgCO_2e$ figure must itself be cut at least in half. This can be achieved through more spartan changes in lifestyle under current conditions, through investment in renewable energy at the household level, and/or via major changes in American infrastructure. For those with the financial resources, household-level emissions can be dramatically reduced by investing in rooftop solar and alternative drive-train vehicles (hybrid-electric or electric). Since, on average, household emissions increase markedly with income, it is not unreasonable, in my view, to ask this of higher earners. If one does make this investment, one must still be mindful of the embedded emissions that go into residential construction: even a fully solar-powered house still embodies significant carbon if it is very large.

Through individual and household behavior and consumption changes alone, getting to a short-term US/Annex I fair share is actually quite doable; getting to a short-term global fair share (e.g. an 80% reduction from the US baseline) is challenging but can be largely achieved with some sacrifice and care. Such reductions alone would amount for the greater part of global emissions, would be of profound benefit to the planet, and are urgently needed. Ultimately, however, those last few metric tons of emissions need to disappear as well, and achieving this at the individual level is much more difficult: it will require a larger society-wide investment in low-carbon infrastructure.

1.3 Goals

> The man of knowledge in our time is bowed down under a burden he never imagined he would ever have: the overproduction of truth that cannot be consumed...it is strewn all over the place, spoken in a thousand competitive voices. Its insignificant fragments are magnified all out of proportion, while its major and world-historical insights lie around begging for attention.
>
> *Ernest Becker, The Denial of Death*

My most basic goal is to present to the reader a thorough analysis that shows how human activity causes greenhouse gas emissions, quantifies the direct and indirect contributions of individual (or household) level activities to these emissions, and further quantifies how the magnitude of emissions may be reduced through generally achievable changes in lifestyle. An overarching theme is to then put these numbers into a simple framework that gives a sense of their ultimate significance. That framework is the typical and "fair-share" per-capita emissions that I have discussed in this introduction. It is not enough, I think, to tell someone in isolation that behavior x will reduce their emissions by, say, one metric ton of CO_2. But this can be useful when one has a framework for the actual magnitude, and importance, of household emissions.

The approach I have taken in this book is to review major categories of household-level activity and determine how emissions are ultimately generated from these activities. To fully understand this, requires, to give just a few examples, reviews of oil extraction and combustion, electricity-generating technologies, large-scale water provision projects, the thermodynamic principles of heat engines, various agricultural practices, and high altitude jet fuel combustion. Therefore, each category of consumption serves as a springboard for a variety of discussions much deeper than might be immediately apparent. While my major focus is climate change throughout, I do review other major environmental problems associated with certain technologies, for example, the ecological effects of pesticides, and the effects of fracking for natural gas upon water supplies.

In all chapters of this book I have relied as exclusively as possible upon the peer-reviewed scientific literature, as well as IPCC publications and several major governmental surveys and reports, but ultimately present my own analyses and conclusions concerning the problems at hand. While likely beyond the interest of many general readers, I have included a large amount of technical detail in this text, an exercise that is meant to show how I have arrived at my conclusions, and to point the interested reader to further resources. It is not, I think, enough simply to know the bottom line, but how one arrives there.

Furthermore, by providing a broader background on most issues that meaningfully affect global warming, I hope to equip the reader to put into context environmental news stories and to *critically* evaluate the wide variety of claims made by both the political right and left. Finally, I hope that those reading this take it upon themselves to, to the extent possible, honestly evaluate their own participation in the systems driving climate change and then alter their behaviors in a meaningful way, and moreover, understand that such acts are not mere vanity, but a potentially powerful counter to the fundamental drivers of the environmental crises we face today.

Finally, it is not necessary (or necessarily even advised) to read this book in sequential order, with the book divided into several largely independent parts, each focusing either on broad background information or a different category of consumption. It is hoped that the early sections in each major part will provide the general reader with the basic conclusions on how each partic-ular category of consumption generates carbon-equivalent emissions and the most meaningful individual acts to mitigate this; further reading may be tailored to each individual's interests, and I encourage each reader to focus upon those sections of most personal interest or salience (I would, for example, advise a closer reading of the comparisons of hybrid, plug-in hybrid, and pure electric vehicles for those contemplating a new car purchase). Furthermore, many longer sections begin with a box of bullet points highlighting that section's major conclusions, both to motivate the subsequent material and to aid the more casual reader.

1.3.1 The technical nature of this work

This book is open to the criticism that it does not address, certainly not explicitly, how the basic structure of an industrial civilization and an economic model that demands everlasting growth drive global climate change. The fundamental imperatives of corporations, and indeed most institutions, are growth and profit. Add to this that we live in a pervasively materialistic society, one where material consumption is a general proxy for status, and indeed, where most livelihoods now depend on the material consumption of others. Furthermore, the basic relationship between man and nature is generally viewed not even so much as one of two equal antagonists (nature is no longer so mighty), but as one of master and plantation.

The proper way to live, the proper relationship we should have with "nature," how to contend with the seeming all-consuming nature of the economic system, these are fundamental questions that must be answered, and as a society we probably must answer them differently than we (implicitly) have, if there is to be any hope for the global ecology as we know it. But having the right goals is not enough to guide one's hand, one needs the basic facts to know what course to take. That is why this book is mostly technical in nature. It is meant to be a rigorous guide for those interested in a particular path, and not, primarily, a meditation on what path to take.

Chapter 2

General principles

In this chapter, I introduce some generally qualitative principles that (I hope) can serve as a guide to thinking when I turn to the gritty quantitative work that dominates the rest of this book, preventing one from, too much, getting lost in the weeds of numbers.

2.1 The profound importance of transition

M. King Hubbert, a geophysicist most famous for his theory of "Peak Oil," (discussed at further length in Section 3.7.3), was an early and deep thinker on the subject of the energetic basis for civilization. He recognized that, by virtue of their finite nature and rapid exploitation, fossil fuel use must, on the timescale of human civilization, rapidly peak and decline to zero, the continuation of an industrial civilization depending on whether societies successfully develop alternative solar-based technologies in the energy-rich window provided by fossil fuels. It is worth quoting him at length, from the conclusions of his famous 1949 paper in Science [12] (emphasis added):

> These sharp breaks in all the foregoing curves [showing a rapid rise and fall in fossil energy over geologic time] can be ascribed quite definitely, directly or indirectly, to the tapping of the large supplies of energy stored up in the fossil fuels. The release of this energy is a unidirectional and irreversible process. It can only happen once, and the historical events associated with this release are necessarily without precedent, and are intrinsically incapable of repetition.
>
> It is clear, therefore, that our present position on the nearly vertical front slopes of these curves is a precarious one, and that the events which we are witnessing and experiencing, far from being "normal," are among the most abnormal and anomalous in the history of the world. *Yet we cannot turn back; neither can we consolidate our gains and remain where we are. In fact, we have no choice but to proceed into a future which we may be assured will differ markedly from anything we have experienced thus far.*
>
> Among the inevitable characteristics of this future will be the progressive exhaustion of the mineral fuels, and the accompanying transfer of the material elements of the earth from naturally occurring deposits of high concentration to states of low concentration dissemination. Yet despite this, *it will still be physically possible to stabilize the human population at some reasonable figure, and by means of the energy from sunshine alone to utilize low-grade concentrations of materials and still maintain a high-energy industrial civilization indefinitely.*

Whether this possibility shall be realized, or whether we shall continue as at present until a succession of crises develop—overpopulation, exhaustion of resources and eventual decline—depends largely upon whether a serious cultural lag can be overcome. In view of the rapidity with which the transition to our present state has occurred it is not surprising that such a cultural lag should exist, and that we should continue to react to the fundamentally simple physical, chemical, and biological needs of our social complex with the sacred-cow behavior patterns of our agrarian and prescientific past. *However, it is upon our ability to eliminate this lag and to evolve a culture more nearly in conformity with the limitations imposed upon us by the basic properties of matter and energy that the future of our civilization largely depends.*

The final point cannot be emphasized enough: the basic stoichiometry of human civilization and energy use cannot be denied. Fossil fuels represent 500 million years of sunlight trapped in minerals that may be burned away within scarcely more than the span of a few human lives. If this energy is squandered, rather than put at least in part towards a *transition* to a new infrastructure that more directly captures the sunlight falling upon earth, industrial civilization cannot persist in anything resembling its current form. Adding to the urgency of transition is the carbon budget that must not be exceeded to avoid dangerous climate change, and which will be spent far before fossil reserves are ultimately exhausted.

We must understand, then, that efficiency measures and use-reduction, while extremely important, *by themselves* will act only to push back somewhat those dates that the carbon budget is spent, and ultimately the fund of fossil energy depleted. They must be coupled to a larger energy transition to be of any long-term efficacy. Thus, we must divide our thinking into near- and medium-term time horizons, say 10–30 years, and the longer time-horizon. In the nearer-term, our goal should be to *reduce* the pressure on the climate system and energy resources, and the most efficacious actions must be evaluated with respect to the current and likely near-term energy mixes. So for example, under the *current* energy system, hybrid-electric vehicles powered by gasoline generate emissions comparable to, or even lower than, those of pure battery-electric vehicles (both yield around 40–60% the emissions of comparable gasoline-only vehicles). In the longer-term, gasoline-powered vehicles must disappear from the face of the earth, but hybrids are a rational *short-term* solution to relieve pressure.

These short-term measures can provide a window for the concurrent implementation of longer-term measures, which must include a near-complete transition of the energy system (not just electricity, but all energy) to near zero-carbon, non-fossil, sources, including wind, solar, and hydro renewables, and potentially fourth-generation nuclear technologies. Note that biomass, a leading renewable energy source, can only provide a relatively small fraction of the energy for modern civilization without devastating ecological effects. Further, existing biomass technologies are not zero- or even low-carbon over a timescale of decades to centuries, and are, in fact, probably more similar to fossil fuels than to other renewable technologies in this respect (see Sections 3.8 and 4.10).

Whether, as a global civilization, this transition is made in a way that allows a relatively high energy lifestyle for most, or even a fraction, of the globe's citizens has yet to be seen. In the meantime, the focus of this book is on understanding how the (Western, and in particular, the American) individual can help the world to meet short- and medium-term carbon reduction targets, providing precious breathing room for both the climate and industrial civilization.

2.2 Nothing is "zero-impact"

It is important to briefly emphasize that essentially every act of consumption carries a cost: there can be no zero-impact man. But the lesson is not to approach consumption fatalistically, but to rationally assess the costs and *relative* magnitude of impacts compared to reasonable alternatives. For example, it is quite true that it requires silica mining and a significant amount of energy to produce solar panels. They carry a cost, but the cost is orders of magnitude lower than that of fossil-based alternatives (natural gas or coal), and it is disingenuous to oppose renewable energy on the basis of such costs (biofuels being, in my opinion, an exception where the costs truly do often outweigh the benefits).

This principle should also warn one against a kind of rebound effect which might manifest, for example, when the owner of an electric vehicle considers his ride "zero-emissions." Yet, electric vehicles require large amounts of (surprisingly) electricity, the production and delivery of which is the single largest source of greenhouse gas emissions in the US. Compared to the alternative of a comparable gasoline vehicle, electric vehicles surely reduce carbon emissions, but by no means eliminate them.

2.3 Lifecycle assessment as a foundation

In determining the environmental impact of a particular product, fuel, or technology, etc., it is important, insofar as possible, to assess impact over the entire *lifecycle* of the item in question, and I attempt to do so throughout this book. A complete lifecycle assessment is often termed a "cradle-to-grave" analysis, while more limited assessments, e.g. a "cradle-to-factory gate" may be performed. For example, at the tailpipe, gasoline combustion generates about 8.887 $kgCO_2e$ per gallon of fuel, but the extraction and refining of gasoline generates about 2.26 $kgCO_2e$/gallon, and so our "well-to-wheel" emissions factor is a larger 11.146 $kgCO_2e$/gallon. To more completely assess the impact of personal vehicles, we should also consider the carbon footprint of vehicle manufacture, maintenance, and disposal, and over the expected vehicle lifetime, this increases the global warming impact of driving by about 10%. We may go even further, and estimate the carbon emissions embodied in the infrastructure that supports vehicle travel (roadways, parking, etc.) for an extended lifecycle analysis.

In general, for more or less direct energy uses, including transportation fuels (gasoline and diesel), electricity, and direct fuel use (e.g. residential natural gas), lifecycle emissions are dominated by the use-phase (or generation phase, for electricity). The environmental impact of food, on the other hand, is dominated by production-level emissions.

2.4 The discipline of the mind

> You don't use science to show that you're right, you use science to become right
>
> *xkcd.com (Randall Munroe)*

I should like to briefly emphasize that it is essential to approach any study dispassionately and with the discipline of mind to (at least sincerely attempt to) reach a conclusion based only on what the data and science show, whatever one's prior ideological leanings. This is especially important in environmentalism, where any number of controversies are animated by fierce passions, such as those concerning nuclear power, alternative energy, natural gas fracking, organic and sustainable agriculture, plastic bag bans, etc. In some cases, the conclusions of my own inquiries do not necessarily conform with what might be viewed as the "default" environmentalist position, but of course in many they do, while in still others wholly firm conclusions are difficult to arrive at. Regardless, I attempt a sincere inquiry throughout this text.

2.5 The promise and peril of efficiency

- Energy-efficiency, via such technologies as high-efficiency vehicles, LED lighting, etc., can markedly reduce one's carbon footprint, but one must have a care that improved efficiency is not used as a license to increase consumption in response, a phenomenon known as rebound.

- The rebound factor (RE) quantifies the fraction of expected energy savings from an efficiency measure that fails to be realized due to an induced increase in energy use, with RE = 100% implying no net energy savings, and RE > 100% a special case, also known as Jevons' Paradox, where increased efficiency leads to a net *increase* in energy consumption.

- Energy-efficiency governmental mandates, such as CAFE fuel economy standards, are probably effective in reducing net energy use and carbon footprint and have very dramatically increased the efficiency of many major appliances over the last couple decades, but it must be noted that economy-wide energy efficiency has been increasing since the early 1800s, in the face of massive net increases in both energy and carbon emissions.

- Efficiency advances can, in an ideal scenario, synergize with reductions in overall consumption, a case where RE < 0%, and it is within the power of the individual to realize such a scenario in their own habits (e.g. both drive less and take a hybrid or upgrade the AC while changing the thermostat).

With the "oil shocks" of 1973 and 1979, Western governments, including the US, enacted various energy-efficiency measures. In the US, major legislation was first introduced in 1975, with the Energy Policy and Conservation Act (EPCA), which included the Corporate Average Fuel Economy (CAFE) standards for passenger cars, probably the most effective piece of energy-efficiency legislation in this country by a wide margin [204], and largely responsible for a 65% increase in new car fuel economy from 1973 to 1987, as well as more recent increases in fleet-average economy. Thirty years later, the Energy Policy Act of 2005 was enacted, followed closely by the The Energy Independence and Security Act (EISA) of 2007. Geller et al. [204] estimated that various energy efficiency policies and programs saved 11% of the US's primary energy use in 2002, and the EISA has been projected to reduce energy consumption by 9% in 2030, relative to a business-as-usual scenario [18]. It seems obvious then, that efficiency standards are an effective means of combating climate change and conserving resources.

However, despite the fact that the conservation measures mentioned above have clearly markedly decreased US energy consumption and carbon emissions compared to a counterfactual where such measures were not enacted, *all else being equal*, the true effectiveness of efficiency standards/increases in reducing overall energy consumption (and hence carbon emissions) has been of some controversy: increased efficiency also reduces the cost of energy use, and thus consumers and/or manufacturers have an incentive to increase energy consumption, and the degree to which the expected efficiency offset is reduced by increasing consumption referred to as the rebound effect (RE).

The rebound effect may be simply quantified as the fraction (or percentage) of the expected energy offset of increasing an energy end-use that is not ultimately realized; this can be mathematically expressed as [13]:

$$RE = 1 - \frac{AES}{PES} \tag{2.1}$$

where PES is potential energy savings, and AES is the actual energy savings. As an example, suppose a new light duty vehicle consumes 50% less fuel per mile, but because of the reduced cost of driving the owner drives 20% more miles, and hence 20% of the energy offset that would have been achieved without this behavioral response is lost, and the RE is 20%. Thus, an RE of 0% implies the efficiency offset is exactly as expected (no behavioral or other downstream response), an RE > 0% means that at least some of the expected energy savings are lost, with an RE = 100% implying no net energy change at all (for example, a 50% reduction in fuel per mile with a concomitant doubling in miles driven). The prior example is one of a *direct* RE effect, where the expected reduction in energy use was undermined by a direct increase in the end-use made more efficient; a second example might be leaving a more efficient light bulb on for longer than one would otherwise.

The *indirect* rebound effect (potentially) arises when energy-efficiency improvements result in monetary savings on direct energy expenditures, which are then in turn put towards other goods or services that themselves embody large amounts of energy and/or emissions. However, direct energy sources, which for consumers are mainly gasoline, electricity, and natural gas, all have extremely high CO_2e per dollar emissions factors, amounting to about 4.5 kgCO_2e/\$ for gasoline, 5.2 kgCO_2e/\$ for electricity, and 7.5 kgCO_2e/\$ for natural gas[1] which are all about 5–10 times the CO_2e/\$ factors for most other consumer expenditures (except for certain food categories, such as beef; see Chapter 23 on goods and services), and this, very crudely, suggests that the indirect rebound effect, at least with respect to carbon emissions, is unlikely to exceed 10–20%, on average. Note that any direct RE also reduces the potential magnitude of the indirect RE by absorbing some of the monetary savings [17].

Finally, energy efficiency may result in an *economy-wide* rebound effect, whereby newly available energy resources may be put towards other ends, or changes in energy supply and demand promote overall energy consumption. Especially for efficiency improvements in the *production* of goods, it may be possible for RE to *exceed* 100%, the special case where total energy consumption actually increases with efficiency improvements. The notion that RE can be >100% was perhaps first elucidated (although in different terms) by the nineteenth century British economist William Stanley Jevons, whose famous paradox (which is referenced several times throughout this book), was formulated in reference to coal consumption, and can be summarized best in his own words [19]:

> *It is wholly a confusion of ideas to suppose that the economical use of fuel is equivalent to a diminished consumption. The very contrary is the truth....It is the very economy of its use which leads to its extensive consumption...It needs but little reflection to see that the whole of our present vast industrial system, and its consequent consumption of coal, has chiefly arisen from successive measures of economy...Civilization, says Baron Liebig, is the economy of power, and our power is coal. It is the very economy of the use of coal that makes our industry what it is; and the more we render it efficient and economical, the more will our industry thrive, and our works of civilization grow.*

Indeed, to partially reiterate the final point above, the whole industrial revolution may perhaps be regarded as a manifestation of Jevons' Paradox, with increasingly efficient uses of fossil energy leading to their global adoption across almost every sector of human activity.

---

[1]Factors derived using 11.146 kgCO_2e/gallon and \$2.50/gallon for gasoline, \$0.13/kWh and 0.682 kgCO_2e/kWh for grid-average US electricity, and 0.2613 kgCO_2e/kWh and \$0.035/kWh for natural gas.

However, we must caution that Jevons was explicitly referring to the use of coal in manufacturing and industry and *not* its domestic consumption, even stating

> I speak not here of the domestic consumption of coal. This is undoubtedly capable of being cut down without other harm than curtailing our home comforts, and somewhat altering our confirmed national habits. The coal thus saved would be, for the most part, laid up for the use of posterity.

Thus, it is entirely sensible that increasing the efficiency by which an energy source is exploited towards productive ends may lead to an economy-wide net expansion of its use and indeed, this is not always a negative: increased efficiency of solar cells and wind turbines helps drive the wider adoption of such technologies. However, most individual and domestic uses of energy are somewhat dissimilar, where energy demands are being used toward relatively inelastic demands: one only wants/needs to drive so many miles, and the most comfortable room temperature does not change with a heater's efficiency. How efficiency improvements at the consumer-level interact with the larger productive economy is an open question, and in this case modeling studies suggest that the economy-wide RE could range from $< 0\%$ (i.e. net energy decreases beyond those expected) to $> 100\%$ [13].

One problem with many studies on the rebound effect is that they consider only the use-phase cost of efficiency measures, failing to account for the substantive up-front capital costs often associated with more efficient technologies or other energy-conservation measures: upgrading HVAC equipment costs many thousands of dollars, high-efficiency lighting is relatively expensive, insulation does not come for free, the premium for an electric vehicle is on the order of \$10,000 or more, and solar panels can run into the tens of thousands as well. Such costs reduce one's spending potential (at least in the short-term), and clearly provide an incentive to maximize one's return on investment, thus quite plausibly driving both direct and indirect REs < 0. That is, there may be a greater than expected direct benefit from efficiency measures, because now the consumer is more invested in reducing use-phase energy costs, while the implementation costs counteract (short-term) any indirect rebound, possibly pushing it negative.

Furthermore, government-mandated efficiency standards for passenger vehicles and appliances, including major energy-users such as air conditioners and heaters, incur a cost to the manufacturer without any energy savings benefit at the manufacturing stage (explaining the general opposition of such mandates by industry), a fact that is typically disregarded [17], and thus we can expect a potentially negative indirect RE at the producer level, which may also act to contract the economy and force economy-wide RE < 0. Given this, we have little reason to think that such mandated efficiency advances will lead to a Jevons' Paradox. Another general problem with economy-wide studies is that they rely on general equilibrium models of the economy, and assume the existence of that fantastical beast, *homo economicus*, who always seeks to increase his utility via consumption, despite evidence that this corresponds but poorly with reality [13].

Suffice it to say that both indirect and economy-wide REs for efficiency improvements in consumer energy end-uses are highly uncertain, may even be negative once upfront capital and manufacturing costs are accounted for, and in any case, since most energy use and associated emissions in the American economy are ultimately driven by consumption, it is difficult for me to imagine that any program of collective conservation at the consumer scale could lead to anything like Jevons' Paradox, so long as our overall goal is to decrease our *total* individual energy and GHG footprint.

Returning now to the direct rebound effect, multiple studies have attempted to quantify the direct RE for automobile travel, heating and cooling, appliances, etc., and these tend to show relatively small, although positive direct REs [17], suggesting that efficiency standards and advances are indeed effective means for conservation, but with benefits slightly smaller than might be expected *prima facie*. There is a fairly extensive econometric literature devoted to quantifying the rebound effect for fuel efficiency and motor vehicles, and as reviewed in [14], the vast majority of works have concluded that the direct RE is small, in the 5–30% range (notwithstanding a few outliers); similarly, Gillingham et al. [17] found that most recent direct RE estimates for both gasoline and electricity use fall in the 5–25% range. Additionally, Small and and Van Dender [14] concluded that the RE has fallen over time, and moreover, that the RE tends to be smaller for wealthier households. As a general rule, direct energy use by the better-off is relatively independent of energy prices: the wealthy will drive what they will regardless of the price of gas or their vehicle's MPG, and condition their homes to comfort, rather than the price of heating fuel or electricity.

The final point I would like to emphasize is that, optimally, it is within the power of the consumer to approach efficiency and total consumption such that they act not to cancel (at least partially) each other out via rebound, but with synergy. For example, consider upgrading from a typical American passenger vehicle to the best hybrid on the market today, and one may increase their MPG from about 22 to 56, saving an impressive 60% of fuel emissions; reduce miles driven by 20%, and the overall fuel costs drop by nearly 70% from baseline. Upgrading HVAC equipment and insulation such that one's home uses 30% less energy for heating and cooling, *in conjunction* with reasonable thermostat strategies that also avoid 30% of heating/cooling energy from baseline yields a net 51% energy savings.

This general idea applies to other conservation measures as well. For example, in the Phoenix, AZ area I have noticed that it is common for many roof-top solar installers to promote, at least in advertising materials, the notion that solar gives a license to consume more: "Go ahead and turn down the AC, you've got solar." Now, to be sure on balance such an approach will still yield net fossil energy savings, but consider two hypothetical counterfactuals, where we begin with a household using 20,000 kWh of electricity per year, equivalent to about 13.64 $MgCO_2e$. In the first case, suppose a rooftop solar system is installed that offsets 10,000 kWh/year, and the family uses an additional 2,000 kWh (equivalent to a 20% RE), and so ultimately 12,000 kWh are drawn from the grid, yielding 8.18 $MgCO_2e$, a 40% total offset. In the second case, via efficiency improvements and behaviorial changes alone we first reduce energy consumption by 40%, to 12,000 kWh, an entirely plausible endeavor, and just as good as the final outcome in the first case. Now, adding our 10,000 kWh solar system drops final consumption to 2,000 kWh net, an impressive 90% total offset.

2.6   Is sustainability a luxury of the rich? (No)

It is common to see the idea of sustainability, especially by means of conscientious consumerism, to be derided as merely a luxury of the upper middle class. This is an absurd viewpoint almost on its face. At the national scale, carbon emissions increase strongly with GDP [4]; worldwide, per capita emissions are only about a third of US per capita emissions, and yet the US is the richest country. Within the US, household income is also strongly correlated with the household carbon footprint, which roughly doubles (on average) from the bottom to top income quintiles [2]. More wealthy households consume far more goods and services and tend to have larger houses that use more electricity and fuel. Wealthier individuals do not generally rely on public transportation, can drive more expensive cars with lower fuel efficiency, and are more free to engage in discretionary travel.

Luxury and performance vehicles tend to be very heavy, with very poor gas mileage, and in general, a higher base MSRP is correlated with a lower MPG for the top-selling vehicle brands in the US. Furthermore, lower income households have fewer vehicles per household, utilize transit more, and a greater proportion of car trips have multiple, rather than a single, occupants. Increases in income also are strongly correlated with increases in vehicle miles and more vehicles owned, with the richest households burning nearly thrice the fuel in about twice as many vehicles. Emissions from residential energy use increase by about 50% from bottom to top quintile, even when controlling for the larger size of wealthier households (see Section 11.2.4). When it comes to general goods and services, thanks to the profligate habits engendered by money, emissions of those in the top quintile of wealth are likely around 2.5 times those in the bottom quintile (again controlling for household size).

Curiously, the weakest association between wealth and consumption is found in the food category, at both the international level (although the association is still weakly positive at this scale) [4] and among US households [2]. Most food and agriculture emissions ultimately stem from animal products (organic or conventional), and the poor and rich alike eat similar amounts, with the wealthy just seeming to prefer more expensive versions of the same basic diet [2].

I will make this point again throughout the book, when appropriate, but the fundamental point is that, to a great degree, we must look to the poor as an example, and Americans must figure out how to live a lifestyle that is not so utterly dependent upon heavy energy and, indirectly, financial consumption. The American Way of Life must become negotiable.

Chapter 3

Essential background

In this chapter, I delve more deeply into several topics of major importance to the rest of the book, beginning with a brief primer on mathematical units and the differences between energy and various emissions metrics, with an especial focus on the global warming potential (GWP) metric and its mathematical description for various GHGs, a review of the properties and emissions sources of most major anthropogenic radiative forcers (i.e. substances released into the atmosphere that have a global warming *or* cooling effect, and include all GHGs), and major carbon stocks and sinks.

Of particular importance in the earlier part of this chapter is a discussion of the 1987 Montreal Protocol and the ozone-depleting halocarbons (e.g. CFCs and HCFCs) widely used as refrigerants, propellants, and for other industrial uses. The Montreal Protocol, by far the most successful international environmental treaty, has not only averted a crisis in the ozone layer by directing the ongoing phase-out of these substances (especially CFCs), but has also (incidentally) avoided much radiative forcing (i.e. global warming) that would otherwise have occurred, for the Montreal gases are also extremely potent greenhouse gases.

A key historical lesson also comes from the recognition that early consumer behavior was important to the CFC phase-out: the possible harms of CFCs were first described in 1974, and this "1974 warning" resulted in widespread voluntary consumer changes and some national-level restrictions (but no international action) that likely avoided around 50% of the CFC emissions that would have otherwise occurred until definitive international action occurred with Montreal in 1987. Thus, consumer changes markedly lowered the overall CFC burden on the atmosphere, and awareness and willingness to change among the consumer class was likely essential to Montreal being politically feasible. There is a clear parallel to climate change: while high-level international action and broad technological changes were necessary to ultimately phase-out CFCs, and the same is likely true for GHGs, the task was greatly abetted by early consumer action, which is almost certain to be true for climate change as well.

The latter part of this chapter focuses more on fuels, especially fossil fuels, beginning with an introduction to the process of combustion, by which various hydrocarbons are converted to heat, water, and CO_2, along with other byproducts. The intrinsic chemical makeup of different fossil fuels governs their energy (heat) content, as well as the amount of CO_2 generated per kWh in a predictable manner, with a higher hydrogen:carbon ratio giving more energy and lower CO_2/kWh. I then discuss the geologic processes by which fossil fuels come into being, focusing especially upon unconventional natural gas and oil that is increasingly extracted via hydrofracking, and conclude with an overview of M. King Hubbert's famous, and still highly relevant, "Peak Oil" theory. Next, various biofuels (corn ethanol, soy biodiesel, and wood) and their scalability as replacements for major fossil fuels is addressed, and we see that none are viable replacements at anything approaching current US energy use, although wind and solar clearly are. The chapter closes with an overview of historical primary energy consumption in the US, demonstrating the remarkable transition from a wood-fired world to one powered almost entirely by fossil fuels in just the last century and a half.

3.1 Some notes on units

There are two major systems of measurement, the metric system, known in its modern incarnation as the International System of Units, or SI, from the French Système International d'Unités, and the British (or "Imperial") system, familiar principally to Americans and member states of the Commonwealth. Essentially all base calculations in this book are performed using metric units, but because some British units are so prevalent, I sometimes present results converted to British units. Most prominently, I have chosen to use miles for distance and degrees Fahrenheit for temperature, as well as gallons as the base unit for fuel volumes. I occasionally use acres for land area as well, and horsepower for power. Most other units are given in metric; of special note, emissions factors are always given in metric terms. British (or "short") tons are never used, and the term "tonne" is synonymous with metric ton (1,000 kg).

Throughout this book, comfort with the metric system of units, scientific notation, and converting between different orders of magnitude is a necessity for the fullest comprehension. The most natural unit for household CO_2-equivalent emissions is the metric ton, or "tonne," equivalent to 1,000 kg or 10^6 grams, and thus can also be referred to as a *megagram* (Mg); this is precisely what I do. A Mg consists of 1,000 kilograms (kg), and a kg in turn is equal to 1,000 grams (g). Depending upon the context, any of these units may be most natural. For example, burning a gallon of gasoline generates about 11 $kgCO_2e$, while burning 500 gallons generates around 5,500 $kgCO_2e$, more concisely given as 5.5 $MgCO_2e$. With this, we dip but a toe in the sea of dimensional analysis, and far more treacherous mathematical waters lie ahead: reader, consider thyself warned.

3.2 Emissions, energy, and primary energy

I take this opportunity to emphasize the differences between several terms that appear throughout this book, both to improve understanding, and to avoid some common pitfalls in thinking. My primary focus in the book is CO_2-equivalent emissions (CO_2e), generally measured via the global warming potential (GWP) metric, which, as described below, measures the warming effect of a substance (or behavior) over 100 years and rescales this in terms of the warming effect of a pulse emission of CO_2 over 100 years. The 100-year horizon is ultimately arbitrary, and 20-year, 50-year, etc. GWPs may also be described, and when applicable I label such values GWP_x, where x is the time-horizon.

Sometimes carbon (C), *per se*, and carbon dioxide (CO_2) are confused. Carbon has a molecular weight of about 12, while CO_2 has a molecular weight of 44, and thus when emissions are given in units of C, they must be converted to CO_2 by a scaling factor of 44/12, e.g. 1 MgC = 44/12 $MgCO_2$. Adding to potential confusion, "carbon emissions" and related terminology are frequently used in a generic sense, referring to greenhouse gases in general. While in prose I sometimes use generic terminology, all numbers are accompanied by precise units, so, for example, 1 MgC, 1 $MgCO_2$, and 1 $MgCO_2e$ mean precisely 1 tonne of carbon (C), 1 tonne CO_2, and 1 tonne CO_2-equivalent (and in this last case the physical component gases may be anything, and further, unless otherwise stated this means CO_2-equivalents over a 100-year timeframe), respectively.

Carbon emissions are strongly linked to fossil energy use, and it is a common error in the popular literature to confuse emissions with energy. Energy is typically measured in either joules

(J), or, more commonly, watt-hours (Wh) or kilowatt-hours (kWh). Energy is the integral (or summation) of power, and so 1 kWh is the energy used when drawing 1 kW of power over 1 hour. Now in general, any use of energy has some associated carbon intensity, which we may quantify in $kgCO_2e/kWh$, and this quantity may vary over orders of magnitude, being, for example, about 0.01 $kgCO_2e/kWh$ for wind-derived electricity and 1 $kgCO_2e/kWh$ for coal-fired electricity, respectively. Further, emissions intensities tend to be lower for direct fuel use, around 0.2–0.3 $kgCO_2e/kWh$ for typical fossil fuels, but in the 0.6–1.1 $kgCO_2e/kWh$ range for fossil-derived electricity (including all lifecycle emissions). But note that we are here comparing two different forms of energy, thermal (heat) in the former case, and electrical in the latter; sometimes the notation kWh(t) and kWh(e) is used to distinguish between the two. Electrical energy is "higher quality," and this leads us to the concept of *primary energy*, which is the energy contained in raw fuel before any downstream transformation.

In typical fossil fuel generating plants, only 30–45% of the thermal energy contained in fuel is converted to electrical energy (the "thermal efficiency"). This electrical energy is then distributed at a loss and devoted towards various end uses, at varying efficiencies. For example, suppose we used a motor to convert electricity to motion with 85% efficiency, but if the electricity was derived from a gas generator at 40% thermal efficiency and 6.5% was lost in the distribution system, then our primary energy efficiency would be just 85% × 40% × 6.5% = 31.8%. Several renewable energy sources yield electricity at the primary level, e.g. hydropower, solar, and wind. Therefore, when we do calculations, as in Section 4.14, on the renewable energy resources needed to replace all global primary energy, renewables suffer a penalty in a straight one-to-one comparison, as much fossil primary energy is lost in transformation to electricity or other ends. The idea of primary energy is also important when comparing the efficiency of, e.g., electric and gasoline-based vehicles, or gas and electric heaters.

3.3   The GWP metric and mathematics, oh my

As already discussed, in order to assess and compare the impact of various activities, from the individual to global scale, we must use some common unit of measurement, or metric. The fundamental metric that I will use throughout this book, in line with the IPCC, is *metric tons of carbon dioxide equivalents* ($MgCO_2e$), defined on the basis of global warming potential (GWP). Note that we use CO_2 *equivalents*. That is, there are multiple greenhouse gases, each with unique properties, but we express the effect of each on the climate in terms of an equivalent amount of CO_2. But how to do this, you may ask, and while there are several commonly used methods, no one is necessarily more "correct" than another. To further complicate matters, because different gases act over different timescales, depending on how far into the future we look, a given gas will be "equivalent" to different amounts of CO_2 at different points in time.

The standard IPCC method is to compare gases in terms of GWP, with the GWP of gas x defined as, and bear with me dear reader, the integrated *radiative forcing* (RF) due to a pulse emission of one kg divided by the integrated radiative forcing of a single kg of CO_2. To understand this sentence first requires understanding radiative forcing. Consider a kg of gaseous CO_2 spread uniformly throughout the atmosphere. This CO_2 will continuously trap some heat and we quantify the amount of heat trapped per unit time in units $W\ m^{-2}$: this quantity is the *instantaneous* radiative forcing. Note that the term radiative forcing stems from the notion that greenhouse gases trap long-wave infrared *radiation* and thereby *forces* heating of the earth. The *radiative efficiency* (RE) of a gas is the per-kg radiative forcing ($W\ m^{-2}\ kg^{-1}$).

Now, when a pulse of CO_2, or any other gas, is emitted, not all of it stays in the atmosphere. For many gases, the pulse decays away according to a simple exponential equation, and the rate of decay is quantified by a single parameter, the "perturbation lifetime." Formally, for some gas x we have the fraction remaining in the atmosphere at time t, $R_x(t)$, as

$$R_x(t) = e^{-\frac{t}{\tau_x}}, \tag{3.1}$$

where τ_x is the perturbation lifetime, and the instantaneous radiative forcing at any time is

$$RF_x(t) = A_x R_x(t), \tag{3.2}$$

where A_x is the radiative efficiency. Finally, integrating RF out to some time horizon H gives us the absolute GWP (AGWP) for gas x,

$$\text{AGWP}_x(H) = \int_0^H RF_x(t)dt \tag{3.3}$$

and we then have the mathematical expression for GWP_x as the $\text{AGWP}_x/\text{AGWP}_{CO2}$ ratio,

$$\text{GWP}_x(H) = \frac{\text{AGWP}_x(H)}{\text{AGWP}_{CO2}(H)}. \tag{3.4}$$

Figure 3.1 shows the time-course of the remaining mass and integrated RF, i.e. AGWP, of a CO_2 bolus over 100 years, and further compares these time-courses to those of a methane (CH_4) bolus. Dividing the respective AGWPs at each time-point gives the CH_4 GWP at any given time horizon, also shown in Figure 3.1. We may also modify GWP values via consideration of climate-carbon feedbacks, as done in the most recent IPCC assessment [7], and while there is uncertainty, the GWP metrics including climate-carbon feedback are likely better estimates. For example, 100-year GWPs for methane are given as 28 and 30 for biogenic and fossil methane, respectively, without climate-carbon feedback, but 34 and 36 with such feedback. The GWP values for biogenic and fossil methane vary slightly because, as reviewed in Section 3.4.2, CH_4 is oxidized to CO_2 in the atmosphere. For biogenic CO_2, this is not *new* to the short-term carbon cycle, while for fossil methane this represents an additional anthropogenic warming source. I now take a moment to review the slightly more complex CO_2 pulse dynamics.

3.3.1 CO_2 pulse dynamics

In the case of a CO_2 pulse emission, over time some is absorbed into the ocean, and some is incorporated into various carbon pools such as forests, other biomass, soil organic carbon, etc., while a substantial portion will persist in the atmosphere for millennia. The method for deriving a relatively simple mathematical description of this process has been to run large-scale biophysical simulations that track the fraction of a large pulse (100 Gt) of CO_2 emitted into the atmosphere that remains over the course of a millennium. Then, a simple mathematical equation is fit to the result; the form of this equation is a sum of exponentials, each with a unique time constant. Roughly speaking, each exponential describes the incorporation into a different pool[1]. From Joos et al. [20], we have that the fraction of a CO_2 pulse remaining in the atmosphere, $R_{CO_2}(t)$, at time t (in years) is

$$R_{CO_2}(t) = a_0 + \sum_{i=1}^{3} a_i \exp\left(\frac{-t}{\tau_i}\right) \tag{3.5}$$

---

[1]This mathematical form is equivalent to a "multi-compartment" pharmacokinetic model describing how the plasma concentration of a drug changes over time following injection into a human (or some unfortunate animal). The compartments are thought to represent various tissues, and the drug moves into different peripheral compartments from the plasma at different rates, analogous to the atmospheric carbon dioxide being incorporated into various terrestrial or marine carbon pools at different rates.

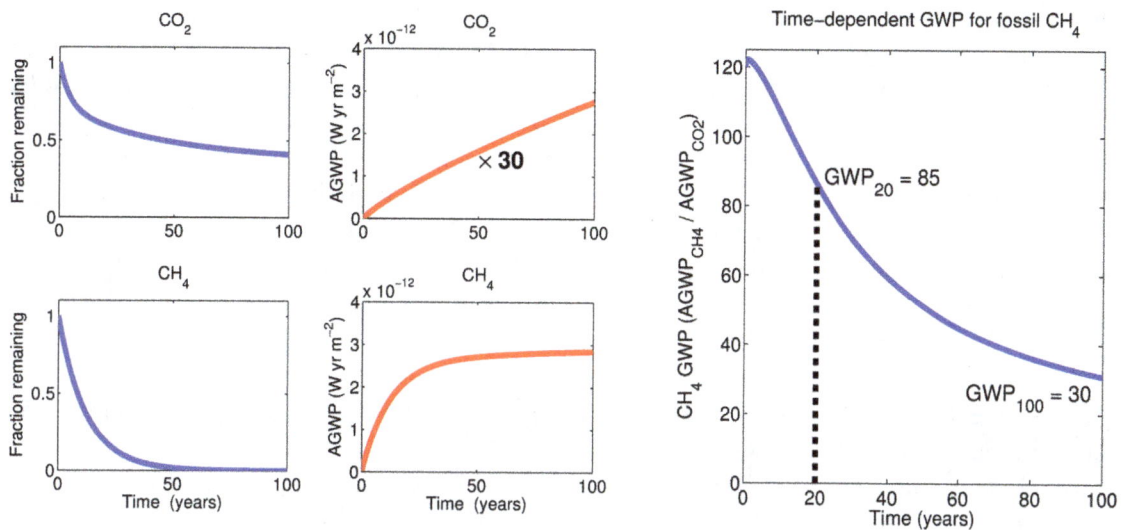

Figure 3.1: The leftmost panels show the time-courses of CO_2 (top) and fossil CH_4 (bottom) removal from the atmosphere, following a 1 kg pulse emission. We see that CO_2 is quite long-lived, while methane is cleared within a few decades. Moving to the right, we have the *integrated* RFs (AGWP) from these gases at each year, with the CO_2 AGWP inflated 30-fold to make it visually comparable to that of CH_4. Now, the AGWP for methane becomes very large initially, but, because the gas clears by the fifth decade, it then plateaus (i.e. no further warming effect), while the CO_2 AGWP steadily rises, reflecting the persistent warming effect. Dividing these two quantities shows that the relative warming effect of a CH_4 bolus, relative to CO_2, is very large in the first years and steadily declines, but is still 30 at the 100-year horizon. These calculations include the slight effect of methane oxidizing to CO_2, but omit climate-carbon feedbacks for CH_4.

The a_i parameters are given as 0.2173, 0.2240, 0.2824, and 0.2763 in ascending order, and time-constants, τ_i, are 394.4, 36.54, and 4.304 years, also in ascending order. The radiative efficiency of CO_2, A_{CO2}, is 1.37×10^{-5} W m^{-2} ppb^{-1}, or 1.7561×10^{-15} W m^{-2} kg^{-1}. This equation and RE value is used to inform Figure 3.1.

3.4 Major radiative forcers

Here I discuss major radiative forcers, i.e. those substances that exert either an atmospheric warming or cooling effect, and almost all anthropogenic alteration of the climate is thus manifested through such agents. The radiative forcing concept underlies the GWP metric, and there are several ways to define it. As above, we can consider the instantaneous radiative forcing of some species, in W m^{-2} kg^{-1}, but, in climate assessments, the more typical method is to compare the total radiative forcing attributable to a substance *relative* to a pre-industrial baseline, taken as the 1750th year of the Common Era by the IPCC. Since 1750, total anthropogenic forcing reached 2.3 W m^{-2}, on balance, in 2011 by the IPCC's estimate, driven principally by CO_2, CH_4, and several other lesser gases, and partially counteracted by cooling aerosols (mainly nitrate and sulfate) also released by the activities of industrial civilization.

Different future emissions scenarios have been defined by the IPCC on the basis of projected radiative forcing in 2100, referred to as representative concentration pathways (RCPs). A low-emissions scenario where emissions rapidly decrease after 2020 and global warming stabilizes at around a 1.5 °C increase in global mean air temperature is the RCP2.6 pathway (i.e. RF of 2.6 W m^{-2} in 2100), while a high-emissions scenario, the RCP8.5, corresponds to continuing emissions over the twenty-first century and dramatic warming; depressingly, humanity seems much more on track for the latter than the former.

Broadly speaking, radiative forcers are divided into the overlapping categories of well-mixed greenhouse gases (WMGHG) and near-term climate forcers (NTCF) [7]. The WMGHGs include CO_2, CH_4, nitrous oxide (N_2O), and some halocarbons, with the first three forming a canonical group most often considered in lifecycle assessments, while major NTCFs include methane (the only major overlapping species), ozone (O_3), aerosols (mainly sulfate (SO_4), nitrogen oxides (NO_x), and black and organic carbon), and other halocarbons. Stratospheric ozone is essential for blocking UV light, as discussed in Section 3.4.4, but O_3 is also a powerful greenhouse gas with a complex atmospheric chemistry coupled to many other radiatively active molecules, and many emissions, e.g. methane, can increase O_3 for a downstream warming effect. Note that water vapor (H_2O) is actually the largest contributor to the atmospheric greenhouse effect overall, but, in the lower atmosphere water concentration is essentially a feedback response to other greenhouse gases, and anthropogenic water emissions are trivial in magnitude compared to the background hydrologic cycle, and so it is *not* considered an anthropogenic GHG. However, statospheric H_2O emissions, e.g. from high-flying aircraft or methane oxidation, do remain in the stratosphere and exert a warming effect. Natural variation in solar irradiance (very minor), land use changes that alter surface albedo (i.e. reflectivity), and aircraft contrails and induced cloudiness also affect radiative forcing.

All the WMGHG and most of the NTCF have a net warming effect, with the major exception of several anthropogenic aerosols, which exert a powerful (if short-lived) cooling effect, partially counteracting the warming gases. It is also apropos to introduce the idea of *abundance-* and *emissions*-based radiative forcing. The former is a measure of warming strictly due to the concentration of any given substance in the atmosphere (relative to pre-industrial times), while the latter includes the downstream effects of a past emission stream. For example, methane emissions increase ozone, CO_2, and stratospheric water concentration, and so the warming attributable to methane *emissions* is greater than that due to its atmospheric abundance alone.

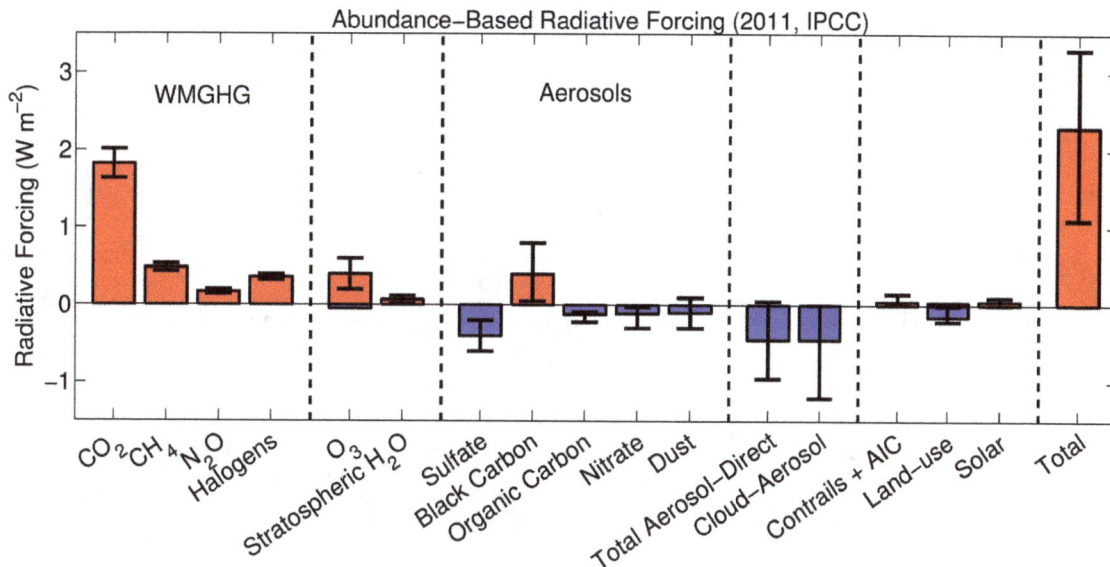

Figure 3.2: Abundance-based radiative forcing since 1750 attributable to major species, in 2011, per the fifth IPCC assessment [7].

Figures 3.2 and 3.3 summarize the IPCC's most recent estimates for abundance- and emissions-based radiative forcing. In either case, CO_2 is dominant with CH_4 second, but relative impact of CH_4 approaches 60% of CO_2's on an emissions-basis.

3.4.1 Carbon dioxide (CO_2)

Carbon dioxide is the canonical greenhouse gas, and is the dominant contributor to anthropogenic radiative forcing, on both an abundance and emissions basis. Globally, and in the US, most CO_2 emissions are attributable to fossil fuel combustion, either for energy, transportation, residential heat, or industrial activity, with CO_2 from deforestation and other land use changes contributing about 15% of CO_2 emissions in recent years [7] (but a much greater proportion during the early industrial era); cement production also directly yields 5–7% of global CO_2 [21]. There is little disagreement as to the sources and magnitude of CO_2 emissions [22], and the fundamental role of this gas in anthropogenic global warming is incontrovertible [7].

3.4.2 Methane (CH_4)

After CO_2, methane is the second most important GHG, and atmospheric concentrations have increased dramatically since preindustrial times, nearly tripling from 722±25 ppb to just over 1803±2 ppb in 2011 [7]. On a concentration basis, the IPCC estimated radiative forcing attributable to CH_4 at 0.48±.05 W m^{-2} in 2011, about a quarter of the 1.82±0.19 W m^{-2} from CO_2. However, atmospheric methane chemistry is complex, and CH_4, NO_x aerosol, hydroxyl radical (HO), and ozone chemistry are all coupled. Methane is a major ozone precursor (tropospheric ozone is a powerful greenhouse gas), oxidation by HO leads to stratospheric water and CO_2, and CH_4 enhances its own lifetime through a feedback effect upon the hydroxyl radical. After accounting for these indirect effects, on an *emissions* basis, CH_4 forcing is doubled relative to its abundance forcing, to 0.97 W m^{-2}, compared to 1.7 W m^{-2} for CO_2 (in 2011) [7, 23]. Most of this increase is attributable to ozone produced downstream of methane, and the IPCC quantifies CH_4 AGWP at year H as

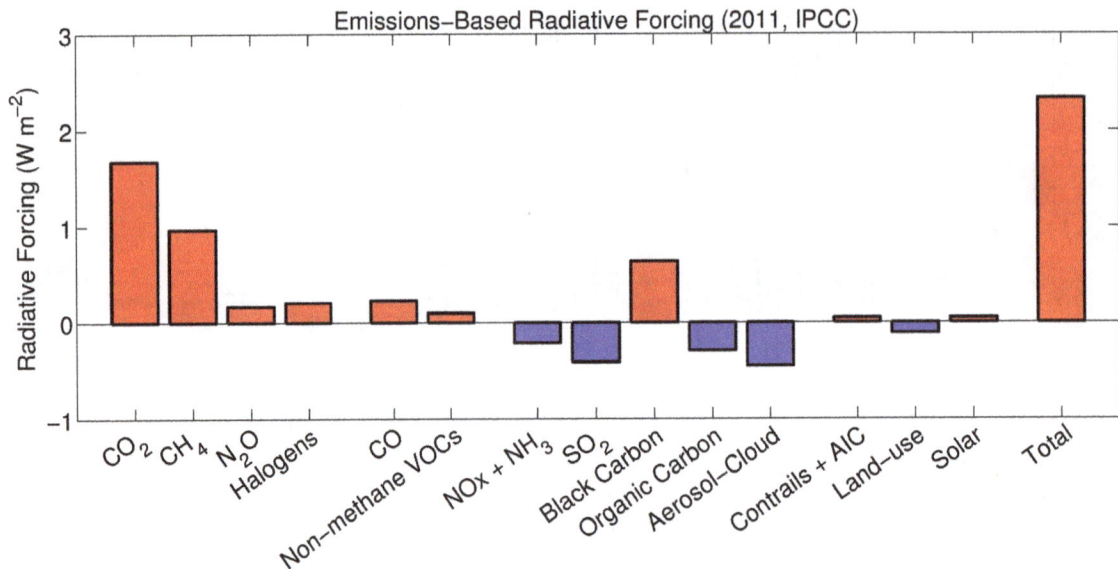

Figure 3.3: Approximate emissions-based radiative forcing for major species, per the fifth IPCC assessment [7]. Omitted from this figure is the significant uncertainty associated with each metric, especially for the aerosol precursor species.

$$\text{AGWP}_{CH4} = (1 + f_1 + f_2)A_{CH4}\int_0^H e^{-\frac{t}{\tau}}dt = (1 + f_1 + f_2)A_{CH4}\tau(1 - e^{-\frac{H}{\tau}}), \qquad (3.6)$$

where $\tau \approx 12.4$ years is the perturbation lifetime of methane, and $f_1 = 0.5$ and $f_2 = 0.15$ account for the indirect effects on ozone and stratospheric water vapor, respectively; the radiative efficiency, A_{CH4}, equals 3.36×10^{-4} W m^{-2} ppb^{-1}, or 1.2767×10^{-13} W m^{-2} kg^{-1} [7]. For fossil methane, new CO_2 from CH_4 oxidation must also be tracked. Methane GWP$_{100}$ estimates have steadily increased across IPCC assessments, from 21 initially, to 25 in the fourth assessment, and now 28/30 and 34/36 for biogenic/fossil methane with and without climate carbon feedbacks, respectively. Furthermore, these estimates do not include indirect aerosol effects, and their inclusion could further increase GWP values, especially in the short-term (i.e. 20 years) [24]. These progressive updates also complicate somewhat our assessment of most earlier literature, which tends to use smaller GWP$_{CH4}$ values, and when I feel it is especially appropriate, I adjust earlier GWP estimates.

Methane is mainly removed from the atmosphere via oxidation by the hydroxyl radical, and knowledge of this sink constrains global emissions to around 550 TgCH$_4$/year [95], but there is great uncertainty as to the exact origin of methane emissions, both on a national and sectorial basis. In the US, CH_4 emissions are due primarily to, in descending order of importance [25, 95, 10]: (1) livestock, via enteric fermentation and manure management (with enteric fermentation accounting for perhaps 75% of livestock CH_4, based on the EPA inventory [10]), (2) oil and natural gas exploration, with fugitive emissions from natural gas extraction by far dominant, (3) landfills (due to anaerobic breakdown of organic wastes), (4) coal mining emissions, and (5) a variety of minor sources, including wastewater treatment, rice cultivation, abandoned mines, combustion, and wildfires (wastewater is often lumped with the landfill category in top-down estimates). While the EPA's bottom-up inventory of CH_4 emissions has remained quite stable over the last decade, multiple top-down estimates suggest much higher and recently increasing emissions, especially due to increasing natural gas extraction over the southern-central US, and

42

I discuss this further in the context of unconventional natural gas extraction ("fracking") in Section 4.6.

Several recent studies based on satellite observations have shown very high and increasing methane emissions concentrated over the central and south-central US [95, 96], areas of both large-scale agriculture and livestock production, and oil and gas exploration, and Turner et al. [95] estimated US emissions to be about 50% higher than those tabulated in the EPA's bottom-up inventory.

3.4.3 Nitrous oxide (N₂O)

Nitrous oxide follows CO_2, CH_4, and, as discussed below, the halocarbons, in importance as a radiative forcer. Although halocarbons exert a stronger warming effect, they are being phased out under the Montreal Protocol, while N_2O is increasing in importance with increased global nitrogen fertilizer use and expanding livestock systems. As one might infer, N_2O is primarily associated with agricultural activity and anthropogenic alteration of the global nitrogen cycle, as discussed extensively in Section 20.1.3. Also noteworthy is that, with the phase-out of the ozone-depleting Montreal gases (discussed next), N_2O may become the major ozone-depleting substance. Nitrous oxide is relatively long-lived, with a lifetime of 121 years, and a 100-yr GWP of 298 (which actually slightly exceeds the 20-yr GWP of 268) [7].

3.4.4 The Montreal gases, halocarbons, and other fluorinated gases

Commercially useful halocarbons, broadly divided into *chloro*fluorocarbons (CFCs), *hydrochloro*-fluorocarbons (HCFCs), and *hydro*fluorocarbons (HFCs), have been widely used in aerosol propellants, foam products (e.g. styrofoam), fire retardants, and as industrial solvents, but they are mainly used as refrigerants (i.e. the working fluid for heat exchange in air conditioners and refrigerators). Halocarbons are not only potent GHGs that, as a class, are the third most important well-mixed GHG after methane [7], but those containing chlorine—CFCs and HCFCs, but *not* HFCs—are also powerful ozone depleting substances (ODS), a fact whose discovery led to, by far, the most successful international environmental treaty in history, the Montreal Protocol. Not only has this treaty stabilized the ozone hole, but it has (essentially incidentally) done far more than any other treaty to mitigate climate change [26].

Background for the Montreal Protocol and ozone depletion

In 1974, Molina and Rowland [27] determined that halogenated hydrocarbons containing chlorine (specifically CF_2Cl_2 and $CFCl_3$, better known as CFC-12 or Freon-12, and CFC-11 or Freon-11, respectively) could lead to stratospheric ozone depletion: these molecules are insoluble and inert within the lower atmosphere, thus allowing them to diffuse into the stratosphere where they persist for 50–100 years [7], and where UV radiation causes photolytic dissociation of the chlorine atom; free chlorine in turn catalytically destroys ozone (in a two-step reaction):

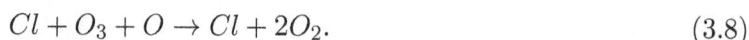

$$CF_xCl_y \xrightarrow{\ hv\ } CF_xCl_{y-1} + Cl, \tag{3.7}$$

$$Cl + O_3 + O \rightarrow Cl + 2O_2. \tag{3.8}$$

Further work confirmed the ozone threat, and the US, Canada, and several other countries banned the use of CFCs in aerosols (but not other uses) in the late 1970's, but many European countries opposed significant limits on CFCs [30]. Consumers also voluntarily reduced use of CFC-containing products, arresting the previously rapid growth in CFC consumption, and even leading to a transient decrease in CFC emissions [32].

Despite CFC stabilization resulting from the "1974 warning," CFC production began to tick up in the later 1980's with increased demand [30, 32]. Given that early models predicted the overall ozone-depleting effect of CFCs to be relatively small [28], concern over CFCs waned [28] and there was little further action until, in 1985, Farman and colleagues [31] published data revealing the existence of dramatic and worsening seasonal ozone depletion occurring above Antarctica, with October ozone levels about 30% lower than during previous decades, a completely unexpected phenomenon [28, 29] that turned out to be a consequence of chemical reactions in polar stratospheric clouds (PSCs) that liberate chlorine from chemical "reservoirs," and global meteorological patterns.

Briefly, the global atmospheric circulation moves chlorine to the Antarctic atmosphere. In the Antarctic winter (spring in the Northern Hemisphere), the Antarctic Vortex isolates the regional atmosphere such that, in the absence of sunlight and fresh warm air, the stratosphere drops to a bitterly cold -80 to -90 °C, in turn causing PSCs to form. Chemical reactions on the PSCs liberate chlorine from reservoirs such as HCl, and the chlorine then catalytically destroys ozone via several novel reactions [28, 29].

The striking demonstration of a large and growing ozone "hole" helped galvanize support for serious international political action, via the Vienna Convention of 1985 and then the Montreal Protocol, ratified in 1987, just 13 years after Molina and Rowland's seminal work and in the face of significant scientific uncertainty and industry opposition [30]. The Montreal Protocol imposed emissions limits and a gradual phaseout first of CFCs and then HCFCs (HCFCs were initially developed as CFC replacements with a lower, but non-zero, ozone-depleting potential). Limits were strengthened and the phaseout schedule accelerated over eight subsequent revisions to the protocol. Complete CFC phaseout was completed in 2010, while the less potent HCFCs, which are still widely used for refrigeration/air conditioning in developing countries and to a lesser extent in the West, are to be eliminated by 2030 [26].

In the absence of the Montreal Protocol, CFC emissions almost certainly would have continued to increase from the 1970's to cause eventual widespread global ozone depletion [29], along with severely exacerbated global warming. Instead, atmospheric chlorine concentrations are slowly decreasing and stratospheric ozone appears to be recovering, although long CFC lifetimes mean that these chemicals will continue to affect the atmosphere for many years.

Global warming and Montreal

All halocarbons are potent GHGs, and the IPCC [7] estimated that, as a class, these gases exerted a direct RF of 0.36 W m^{-2} in 2011, with CFC-12 and CFC-11 accounting for about two-thirds of this forcing; CFC-113 and HCFC-22 account for most of the rest. Note that this is despite CFC production being fully phased out and relates to the long lifetimes and very large radiative efficiencies of these substances. Subdividing, the Montreal gases (i.e. CFCs and HCFCs) have a direct RF of 0.33 W m^{-2}. However, ozone is also a strong GHG and so its depletion has a global cooling effect; this RF estimate adjusts downward to 0.18 W m^{-2} after accounting for the loss of ozone. The HFCs have an RF of 0.02 W m^{-2} (dominated by HFC-134a which is the principle refrigerant for mobile/automotive air conditioners), while other fluorinated gases add 0.01 W m^{-2}.

On the whole, the Montreal Protocol has been a boon for the climate: the direct RF from Montreal gases was 0.32 W m^{-2} in 2000 and has remained essentially stable since, but would likely have reached 0.60–0.65 W m^{-2} by 2010 without intervention [26]. Even accounting for the indirect cooling effect of ozone depletion (30–45% RF offset), this would have meant a 7–9% higher total anthropogenic RF in 2011. Even more dramatically, a 2007 work by Velders et al. [32] estimated that, had the 1987 Montreal Protocol not been put into place, in terms of GWP, ODS emissions could have reached 15–18 GtCO$_2$e/yr (equivalent to about half of

CO_2 emissions), while had the 1974 "early warning" not occurred, ODS may have reached a staggering 24–76 $GtCO_2e/yr$ (comparable to or even worse than CO_2 emissions); see also Figure 3.4 below. After adjusting for the cooling effect of ozone depletion, the Montreal Protocol still prevented emissions on the order of 10 $GtCO_2e/yr$ in 2010, or about 20% of all anthropogenic GHG emissions in that year [26]. Instead of these dire numbers, carbon equivalent halocarbon emissions were about 1 $GtCO_2e$ in 2011 [7].

Now, despite the clear climate benefits of Montreal, it has had one major weakness with respect to climate: HFCs (which again, lacking chlorine do not deplete ozone) used as CFC/HCFC replacements are also strong GHGs. The current mix of HFCs has an average atmospheric lifetime of 15 years and a GWP of 1600 [26], but low-GWP alternatives are available and in use. These alternatives include very short-lived HFCs, hydrocarbons such as propane, and CO_2. Without action on HFCs, given their rapid recent growth, HFC radiative forcing could increase from 0.012 W m^{-2} in 2010 to as much as 0.4 W m^{-2} in 2050 (or about 20% the current CO_2 RF) [34].

Fortunately, an amendment was very recently adopted (October 15, 2016) to largely phase HFCs out by 2050, under the Montreal Protocol. The plan has developed countries reducing HFC use beginning in 2019 with an 85% reduction by 2036, while most developing countries will reduce consumption 80% by 2045 [33]. This phaseout has the potential to avoid up to 0.5 °C of warming by 2100 and 80 billion tonnes CO_2e [33, 34].

Current halocarbons and use

HFC-134a is widely used in mobile (automotive) air conditioners (MACs), and it is the most important HFC. It is an unfortunate feature of these ACs that they leak slightly during operation, and it is eventually necessary to recharge the refrigerant, a process that itself can result in significant HFC-134a venting (especially for DIYers). For a typical passenger vehicle in the US, leak rates are on the order of 50–100 g/yr for older models [36], but this rate is trending down, with more recent model years leaking from <10 to about 25 g/yr [35]. Therefore, a typical driver adds anywhere from 16 to 156 kgCO_2e (at an HFC-134a GWP of 1550) from their onboard AC. While nontrivial in the aggregate, this is but a fraction of the CO_2e emissions resulting from the fuel used to power MACs (up to 10% of all automotive fuel, as discussed in Section 7.1.2), and less than 2% of all passenger vehicle CO_2e emissions. Note that summer HFC-134a emissions are two to three times those in winter [37], a fact largely attributable to higher use and service rates in summer (hence, further motivation to avoid unnecessary AC use).

A lesson for climate change?

It is clear that robust international cooperation was necessary to avoid catastrophic CFC emissions, a situation analogous to that of carbon today. There is, perhaps, a lesson to be learned from the two stages of the CFC phaseout. That is, following the 1974 warning, CFC emissions halted their rapid increase as a result of limited government restrictions (isolated to individual nations) and changing *consumer* behavior and demand. It took a high-level international treaty 13 years later, along with new technologies, i.e. HCFCs and HFCs with lowered or zero ozone-depleting potential, to definitively address the crisis. It is also unlikely that, without a willing and aware population, international action could have been effected.

I believe the carbon situation is somewhat similar to the 1974-1987 interim today: climate change is a global problem that will require international cooperation *and* alternative technologies (widespread renewable energy, etc.) to definitively solve. However, conservation measures, relatively isolated political actions (i.e. at the community, city, state, or national, but not international, scale), and an informed consumer class that takes emissions reductions seriously *can*

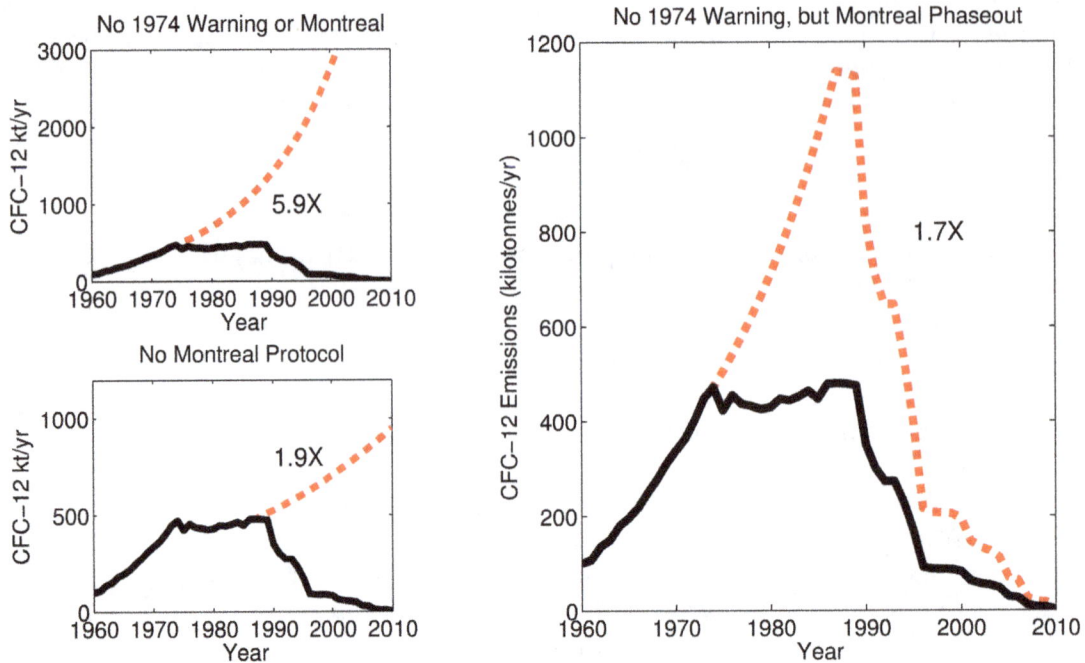

Figure 3.4: The left two panels show hypothetical CFC-12 emissions scenarios: (1) 7% annual growth in CFC production/emissions from 1974 onwards without the 1974 warning or Montreal Protocol, or (2) the 1974 warning but without the 1987 Montreal Protocol, leading to 3% annual CFC growth subsequent to 1987. The black line shows the actual CFC-12 production (the CFC-11 curve is similar); inset numbers show hypothetical cumulative emissions from 1960 to 2010 relative to truth. On the right, we have a third scenario: no consumer/national response to the 1974 warning, but definitive phaseout via international action in 1987. Although disaster is ultimately averted, cumulative CFC-12 emissions are 70% higher, and 100% higher between 1974 and 2010.

reduce CO_2 emissions, even without broader political action or technology change.

By checking carbon emissions until definite phase-out of the carbon economy, we can dramatically lower the overall burden on the climate system. To see this by way of comparison, suppose, as a thought experiment, that CFC emissions had continued unabated after 1974 (ODS emissions were likely to have continued growing by at least 7% annually without the 1974 warning [32]), but that the Montreal Protocol of 1987 yielded a comparable phaseout (but of course, from higher baseline CFC emissions). In this case, cumulative CFC-12 and CFC-11 emissions would have been almost 70% higher before complete phase-out in 2010 (and clearly a full phaseout by 2010 would have been less likely, with more infrastructure and political capital committed to CFCs). Looking at just 1974 on, cumulative emissions would have doubled without the 1974 consumer response. Thus, in a crude sense, consumer behavior accounted for about 50% of the avoided CFC emissions from 1974 through 2010. This, and alternative worlds where either the 1974 warning or Montreal protocol did not exist (adapted from [32]), are illustrated in Figure 3.4 for CFC-12.

Of course, fossil fuels are fundamental to industrial civilization in a way that CFCs never were, and one may challenge my analogy on such grounds, but the hard fact remains that fossil fuels *must* be phased out sooner or later, either via geologic or political constraints, and so, I think, the basic lesson of the CFCs still holds.

3.4.5 Aerosols

Sulfate

Sulfate (SO_4) aerosols are the major anthropogenic cooling aerosols, with the AR5 estimating a global radiative forcing of -0.40 W m^{-2} (-0.60 to -0.20) in 2011 due to these substances, which are primarily formed from sulfur dioxide (SO_2) emitted by fossil fuel burning, especially coal-fired power plants and fuel oils. To derive GWP values for SO_2, we first have that some fraction of an SO_2 pulse emission evolves to SO_4 (sulfate), which persists in the atmosphere for about 4 days and causes cooling by directly reflecting solar radiation. Now, by dividing the steady-state sulfate forcing by annual SO_2 emissions, we get an equivalent instantaneous RF for SO_2, estimated at -3.2×10^{-10} W m^{-2} kg^{-1} by [39]. Multiplying by the lifetime of 4 days, we get the total time-integrated radiative forcing from SO_2, and we then divide by the time-integrated forcing of a CO_2 bolus at any time. Via this process, Fuglestvedt et al. [39] derived GWP$_{20}$ and GWP$_{100}$ values of -140 and -40, respectively, for the direct radiation scattering effect of SO_2 emissions; several other similar estimates are also summarized in this work.

Sulfate aerosols also alter cloud structure for a further cooling effect, and including these indirect effects in our GWP estimates may increase their magnitude greatly, with Lauer et al. giving several 100-yr GWP estimates roughly ten times those above, for SO_2 emissions from shipping; Shindell et al. [24] somewhat similarly estimated GWPs about twice those above (GWP$_{20}$ -268, GWP$_{100}$ -76), but the uncertainty in these indirect effects is extreme.

Increased controls have reduced SO_2 emissions in the developed world, e.g. from scrubbers installed at coal power plants and the adoption of ultra-low sulfur diesel fuel. While valuable for improving local and regional air quality, absent concomitant CO_2 reductions, these controls can increase warming, and represent, to some degree, environmental problem shifting.

NO$_x$

Nitrogen oxides (NO, NO_2, and NO_3^-, collectively labelled NO$_x$) have competing warming and cooling effects that act over different timescales, and the balance depends strongly on the altitude of emissions, latitude, and local atmospheric conditions. In the troposphere (lower atmosphere), NO$_x$ emissions catalyze the oxidation of CH_4 and other VOCs to form ozone (O_3) by the hydroxyl radical (OH), increase levels of OH and thus decrease the lifetime of CH_4, and form nitrate aerosols that reflect solar radiation [7]. That is, somewhat confusingly, NO$_x$ has two separate effects on methane: (1) oxidation leading to methane destruction (cooling) and ozone formation (warming), and (2) decreased methane lifetime (cooling). Ozone has a strong, but very short-lived warming effect, as tropospheric ozone has a mean lifetime of 22.3 days [38], while the decrease in average methane lifetime has a longer-lived cooling effect. In general, NO$_x$ has a long-term overall negative radiative forcing, i.e. 'tis cooling.

Note that NO$_x$ emissions can also enter into the global nitrogen cycle (see Section 20.1.2) by depositing onto soils, increase plant CO_2 uptake via nitrogen fertilization, and evolve to N_2O with associated warming (see Figure 20.2 of Section 20.1.2), and these second-order influences are poorly quantified.

NO$_x$ emissions may be divided by source into aviation, surface, and shipping NO$_x$. Shipping NO$_x$ has a strong cooling effect over all time-horizons, with net GWP$_{20}$ between -76 and -31 (mid-estimate -47), and GWP$_{100}$ -36 to -25 (mid-estimate -32, and GWPs per kg N), based on several studies compiled by [39], and considering the effects of NO$_x$ on ozone formation, ozone-related methane destruction, and methane lifetime.

Aviation NO$_x$, on the other hand, likely has net warming effects, at least in the near-term, as NO$_x$ efficiently creates ozone at high altitude, and the warming effect of ozone is greater at altitude [263]; 100-GWP estimates for aviation NO_x range from -2.1 to +71 [39], and such emissions are discussed in further detail in Section 8.3.1.

3.4.6 Black carbon

Black carbon (BC), or soot, is a particulate product of combustion of some fossil fuels and biomass that has been increasingly recognized as an important near-term driver of climate change. Chemically, BC is an insoluble aggregate of small carbon spherules that are chemically inert, and it is lost from the atmosphere either by physically settling to the earth's surface or removed via precipitation ("dry" or "wet deposition"). The defining characteristic of BC is its strong absorption of light throughout the visible spectrum: it is quite literally black, and this broad spectrum absorbance translates into a massive heat gain compared to typical greenhouse gases. In addition to absorbing solar radiation, BC affects climate forcing by altering clouds and enhancing the melting of ice and snow upon which it settles. Significant uncertainty surrounds BC in the scientific literature, but I refer the motivated reader to a recent exhaustive (but relatively reader-friendly) review by Bond et al. [41].

Black carbon generally persists in the atmosphere for only a few days to a few weeks, and thus a pulse of BC exerts an extremely powerful but very brief (i.e. days) warming force, in great contrast to the long-lived CO_2. Unlike the well-mixed greenhouses gases, BC exerts its warming effects primarily at a local/regional scale, near the source of emissions. Additionally, BC aerosols have multiple indirect warming and cooling effects, mainly via interaction with clouds that may either increase or decrease cloud cover and through alteration of the regional hydrologic cycle. Black carbon deposited on snow and ice also increases the absorption of solar radiation to increase melting.

Broadly speaking, black carbon is emitted from a few major sources [41]: (1) diesel engines, (2) industrial fossil fuels, (3) residential solid fuels, e.g. wood or dried dung, and (4) "open biomass burning," i.e. open burning of forests, savannahs, and agricultural waste. Because various organic and sulfur-containing co-emissions that result from uncontrolled open biomass burning have a cooling effect on the climate, the net effect of open biomass burning is likely actually climactic cooling. Indeed, taken together, the net climate effect of aerosols from all black carbon-rich sources may be a cooling one. On the other hand, BC from diesel combustion and residential fuel use almost certainly has a strong warming effect, and these two areas represent attractive mitigation targets [41].

Global emissions and radiative forcing

As reviewed in AR4, estimates of global BC emissions vary between about 5.8 to 8.0 Tg BC [9]; recently Bond and colleagues [41] estimated global emissions of 7.5 Tg BC for the year 2000 from a bottom-up inventory, but the uncertainty of this estimate is great.

Several global-scale estimates of the radiative forcing attributable to BC implicate it as the second- or third-most important contributor to anthropogenic climate change. The most recent IPCC assessment (AR5) estimated the overall direct radiative forcing of aerosol black carbon (BC) to be 0.40 (0.05 to 0.80) W m^{-2}, a doubling over the prior IPCC estimate (point estimate 0.20 with the range 0.05 to 0.35 in AR4). Black carbon deposited on ice and snow was considered an additional minor contributor at 0.04 W m^{-2} (a reduction from the AR4 estimate).

Direct radiative forcing and GWP

While its indirect effects and disparities in time- and spatial-scales relative to CO_2 have led some authors to criticize application of the GWP framework to BC [42], there do exist several

48

published estimates of a GWP for BC, and indeed, the mathematical description of the *direct* radiative forcing of BC is quite straightforward, mimicking that of short-lived greenhouse gases.

Bond and Sun [43] estimated 100-yr and 20-yr GWPs of 680 (210–1500) and 2200 (690–4700), respectively. They assumed a simple exponential decay pattern for BC, and compiled several estimates for BC atmospheric lifetime which ranged from 2.4 to 8.4 days. Point estimates from seven modeling studies ranged from 4.4 to 7.3 days, with a mean of 5.5 days (or 0.0151 years), which was taken to be the exponential decay factor. Their median estimate for radiative efficiency was 1800 W/g (range 900–3200), which converts to 3.5287×10^{-9} W m^{-2} kg^{-1}, assuming a global surface area of 510.1 trillion m^2. This is about 2 million times the radiative efficiency of CO_2. Given the very brief atmospheric residence time, we can simply assume that all heat energy is deposited instantaneously, and we have the absolute GWP of some quantity of BC as:

$$\text{AGWP}_{\text{BC}} = A\tau R \approx A(5.3172 \times 10^{-11}) \tag{3.9}$$

where A is BC mass (in kg), τ is the mean lifetime (in years), R is the radiative efficiency given above, and the approximation assumes $\tau = 5.5$ days.

3.5 Carbon stocks and sinks

3.5.1 Soils

The soil is the largest terrestrial carbon store, and therefore human activities that perturb soil may have a profound influence on global warming. Indeed, Jobbgy and Jackson [44] estimated global soil stores, within the top three meters of soil, at 2,344 GtC, about five times the remaining human carbon budget. While from the year 2000, this figure is still widely cited, and represents a large increase from prior surveys which had typically counted carbon only in the first meter of soil. It follows that even small relative changes in this store are globally significant. There is much that is unknown about the dynamics of soil organic carbon (SOC), and how both human activities and natural events, such as fire, influence this carbon pool.

All SOC is ultimately derived from the breakdown of plant matter. As plant matter decomposes, the carbon is oxidized to CO_2 to generate energy by microbes and myriad other detrivores, such as worms. While most carbon is consumed relatively quickly for energy, some persists for many hundreds to thousands of years. On the most basic level, the quantity of SOC is determined by balance between the input from plant matter production and output due to decomposition. Environmental factors influence both these rates. On the regional scale, precipitation and temperature are the primary arbiters of this balance. As temperature goes up, plant productivity and decomposition both increase, but based on global surveys, the balance tends to favor decomposition, and SOC falls. Thus, large amounts of SOC are stored in the great northern boreal and arctic ecosystems [45].

The concentration of SOC is greatest closest to the soil surface, and steadily declines with depth. However, more carbon overall is stored in the deep soil than in the topsoil. As one goes down in depth, the rate of decomposition decreases, and thus the age of the carbon pool increases. Carbon in the subsoil is ancient, and may persist for many thousands of years.

Agriculture is the major human activity that perturbs soil carbon stores, and it is very consistently found that conversion of native ecosystems, such as prairie or forest, to crop or pasture results in marked soil carbon losses. As elaborated further in Section 20.3.1, the conversion of US prairies to arable crops (e.g. grains) may result in >50% soil organic carbon losses in the upper soil layers [379, 381].

Overgrazing is the leading worldwide cause of desertification, and, like conversion to cropping, can lead to marked soil carbon losses [405, 389]. It is possible that well-managed grazing systems under *light* grazing pressure, in parts of the Great Plains region of the US, can modestly increase SOC stores, but, as discussed extensively in Section 21.2.2, the weight of the evidence suggests little overall effect except when overgrazing severely degrades the land, and on balance grazing has almost certainly been harmful to both US and global soils.

Forest soils are also vulnerable to agriculture: Conversion of tropical forest to tree plantations of cash-crops (such as oil palm, rubber, and cacao) results in massive carbon losses in *above*-ground carbon stocks, but also leads to very appreciable soil organic carbon loss. Tropical forest soils store an estimated 692 billion Mg of carbon in the top three meters (roughly 10 feet) [44], and even small perturbations in this carbon pool from land-use change can translate into large atmospheric fluxes. Multiple mechanisms contribute to soil carbon loss. Land conversion dramatically increases soil erosion due to the loss of protective ground cover. With the loss of forest litter, a constant carbon input to the soil is lost. Without replenishment, the labile soil carbon pool is lost, and the deeper carbon pools may suffer ongoing decomposition without new inputs [48]. The loss of vegetative cover also increases soil temperature, which in turn increases carbon loss, and the trampling of soil alters its mechanical properties [49].

3.5.2   Forests

Worldwide, forests are a major carbon sink, and it has been found that even undisturbed old-growth forests have been absorbing large amounts of carbon over the last 50 years [46]. This is driven at least in part by global increases in atmospheric carbon dioxide and reactive nitrogen from fossil fuel burning and fertilizer manufacture, but also likely by historical land-use changes. While the worldwide transition to fossil fuels as civilization's principal energy sources over the last century and a half has largely fueled climate change, this transition did dramatically reduce the use of forest biomass for energy (and for other agricultural uses), allowing the forests of America and Europe, degraded from vast over-harvesting, to begin recovering, and in this ongoing recovery, to sequester large amounts of carbon [47].

The finding that old-growth forests are rapidly absorbing carbon undercuts a major rationale for biofuel use, i.e. that it is carbon-neutral, as this implies that not only must a harvested forest regrow to its prior state before the carbon-debt can be considered paid, but we must account for the additional growth that would have occurred in the absence of a harvest, and such dynamics are discussed in far more detail in Section 3.8.

While forest growth (and regrowth) is a carbon sink, deforestation, primarily in the tropics, is a major carbon source [46]. There is significant uncertainty concerning the sizes of both the global and tropical carbon sink from regrowth and source from deforestation. What is clear is that carbon losses, on a per-area basis, are truly massive when tropical forest is cleared for crops or pasture [70] (most often to support beef production).

Finally, the US EPA inventory provides detailed estimates on US forest area and carbon stocks, divided into above- and below-ground biomass, dead wood, litter, and soil organic carbon. This inventory is in rough agreement with a similar accounting by Smith et al. [617], and, overall, US forests store about 100 MgC per hectare (Ha, equal to 10,000 m^2, or 2.47 acres) in above-ground biomass.

3.5.3   The Seas

I note here that the ocean is the most significant sink of anthropogenic carbon, having absorbed 41% of all humanity's carbon emissions from fossil fuels and cement [71], also resulting in gradual acidification of the ocean, with deleterious effects on sea life. Furthermore, the ocean is, by far, the greatest sink of the Earth's excess heating due to warming emissions, with 93% of the excess heat from 1970 to 2011 being stored in the oceans [7].

3.6 An introduction to respiration, combustion, and heating values

- Respiration is the cellular conversion of glucose (and other organic hydrocarbon chains) and oxygen (O_2) to CO_2, H_2O, and heat.

- Combustion is a similar reaction involving O_2 and fossil fuel hydrocarbons with the general formula C_xH_y.

- Of fossil fuels, natural gas has the highest hydrogen content, highest energy content (heating value), and lowest CO_2 emitted per kWh of heat released by combustion. Coal has the lowest hydrogen content and highest CO_2 emissions per kWh; petroleum fuels are intermediate.

- Combustion byproducts result from nitrogen in air, sulfur contaminants in fuel, and incomplete combustion.

The cyclical conversion of energy, H_2O, and CO_2 to energy-storing carbohydrates and O_2, and visa versa, is the energetic basis for almost all life on earth, with photosynthesis representing the energy storage pathway, and respiration the energy-releasing pathway:

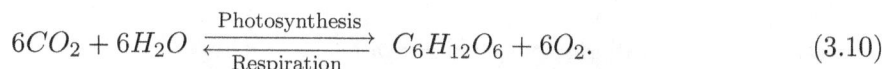

$$6CO_2 + 6H_2O \underset{\text{Respiration}}{\overset{\text{Photosynthesis}}{\rightleftharpoons}} C_6H_{12}O_6 + 6O_2. \tag{3.10}$$

Downstream of photosynthesis, glucose, $C_6H_{12}O_6$, can be converted to a variety of other biological molecules that act to store energy, such as lipids (long-chain hydrocarbons), or have a structural role, such as cellulose (these structural molecules still store useful energy that may be released upon digestion or burning). The carbon-based products of photosynthesis have provided almost all the energy powering human civilization, either as fresh biomass or ancient fossil fuels. Biological molecules and fossil fuels are extremely similar energy sources: both are based on carbon chains, and the latter result from anoxic breakdown of organic matter under high heat and pressure. Respiration is an oxidation reaction that is essentially a "slow burn," with products identical to those produced when glucose is combusted.

Combustion, an exothermic reaction between a fuel and an oxidant to yield heat and oxidized products, is similar to a fast version of respiration when a hydrocarbon fuel, with the general formula C_xH_y, interacts with O_2 to yield H_2O, CO_2, and heat:

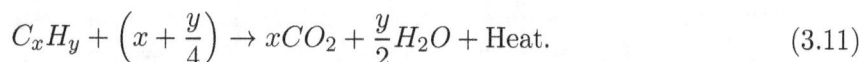

$$C_xH_y + \left(x + \frac{y}{4}\right) \rightarrow xCO_2 + \frac{y}{2}H_2O + \text{Heat}. \tag{3.11}$$

We can calculate the energy released in any particular reaction using the *enthalpy of formation*, H_f, defined as the enthalpy (or heat) change resulting when a compound is produced from its elements in their standard state. Typically, H_f is negative, indicating that heat is released, and the overall enthalpy of a reaction is the difference between H_f for the reactants and products. For example, for the combustion of methane, CH_4, we have

$$CH_4 + 2O_2 \rightarrow CO_2 + 2H_2O, \tag{3.12}$$

and our enthalpies of formation, under standard conditions (1 atm pressure, 298.15 K), are $H_{f(CH_4)} = $ -74.9 kJ/mole, $H_{f(CO_2)} = $ -393.5 kJ/mole, $H_{f(H_2O(l))} = $ -285.8 kJ/mole, and $H_{f(O_2)}$

= 0 kJ/mole [54]. Thus, the (product - reactant) difference is -890.309 kJ/mole, equal to -15.42 kWh/kg (this is the higher heating value, explained below). From the stoichiometry of Equation (3.12), we have that each kg of CH_4 yields (44.01/16.04) kg of CO_2, and, therefore, we can derive an emissions factor of 0.178 kgCO$_2$/kWh. That is, for every kWh of (gross) heat energy released by CH_4 combustion, we also get 0.178 kgCO$_2$.

At this point we must distinguish between the *higher heating value* (HHV) and the *lower heating value* (LHV). Water is a major combustion product, and at high temperature is formed as vapor, not liquid. Since water has a very high heat of evaporation, a significant fraction of the reaction energy is locked into water vapor as latent heat. This heat can only be usefully recovered if the water condenses to a liquid, but it is usually lost as part of the flue gas. For example, water vapor generated in internal combustion engines is invariably lost with the tailpipe exhaust, while advanced condensing furnaces can recover this latent heat. The HHV (or "gross energy") is defined as the heat released by a reaction occurring at 298 K (24.85 °C), with the reaction products returned to 298 K, and thus includes the latent heat in water vapor, and we use the enthalpy of formation for liquid water, as above. The LHV (or "net energy"), on the other hand, is the heat released when reactants return to 150 °C, and thus we lose the latent heat in water vapor, and it is appropriate to use the enthalpy of formation for water vapor [54]. We can repeat the calculation above to get the LHV by using the enthalpy of formation for water vapor, $H_{f(H_2O(v))}$ = -241.8 kJ/mole, giving us net energy of -802.3 kJ/mole, or -13.89 kWh/kg.

While HHV, or gross energy, may be more commonly used, LHV is also in widespread use, especially for transportation fuels, and there can be significant confusion when sources do not clarify. Using the LHV can inflate efficiency measures, and the HHV is generally the most appropriate for overall efficiency calculations.

Different fuels, obviously, have different energy (heat) contents, but more importantly, the amount of CO_2 per kWh of energy also varies significantly between fossil fuels, with coal the worst performer (at the point of combustion, woody biomass actually releases even more CO_2 per kWh than coal, as discussed in Section 4.10.1), and natural gas (methane) the best. Generally speaking, the more hydrogen by mass in a carbon-based fuel, the higher the energy-density, and the lower the CO_2 per kWh [51, 54], explaining why methane, with its four hydrogen atoms to a single carbon, is the cleanest-burning hydrocarbon. For an *n*-alkane, our general combustion reaction is

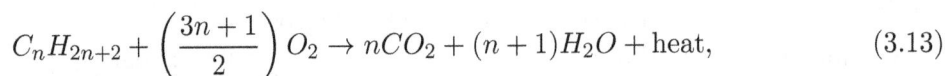

$$C_nH_{2n+2} + \left(\frac{3n+1}{2}\right)O_2 \rightarrow nCO_2 + (n+1)H_2O + \text{heat}, \qquad (3.13)$$

and the enthalpy of formation for the *n*-alkane increases in absolute value by almost exactly 5 kcal/mole (20.92 kJ/mole) with every additional CH_2 group, from -20 kcal/mole for ethane, C_2H_6 [53]. From this relation, and the enthalpies of formation above, we can compare the combustion characteristics of liquid/gaseous fossil fuels as in Figure 3.5. As seen, the lighter short-chain hydrocarbons are higher in energy (on a mass-basis), and release less CO_2 per kWh of energy generated. Since long-chain hydrocarbon mixtures are denser, on a volumetric-basis these fuels (e.g. diesel and fuel oil) are more energy dense, a factor salient, for example, when comparing miles per gallon achieved by diesel and gasoline vehicles.

Coal, unlike natural gas and petroleum-based fuels, is highly heterogeneous, and is composed of a mixture of non-combustible minerals (ash), water, and complex carbon chains with hydrogen and oxygen [54]. Coal also tends to contain 1–2% sulfur. Coals containing more hydrogen by mass emit somewhat less CO_2 per kWh [51], but all have a high CO_2/kWh ratio, as seen in Figure 3.5.

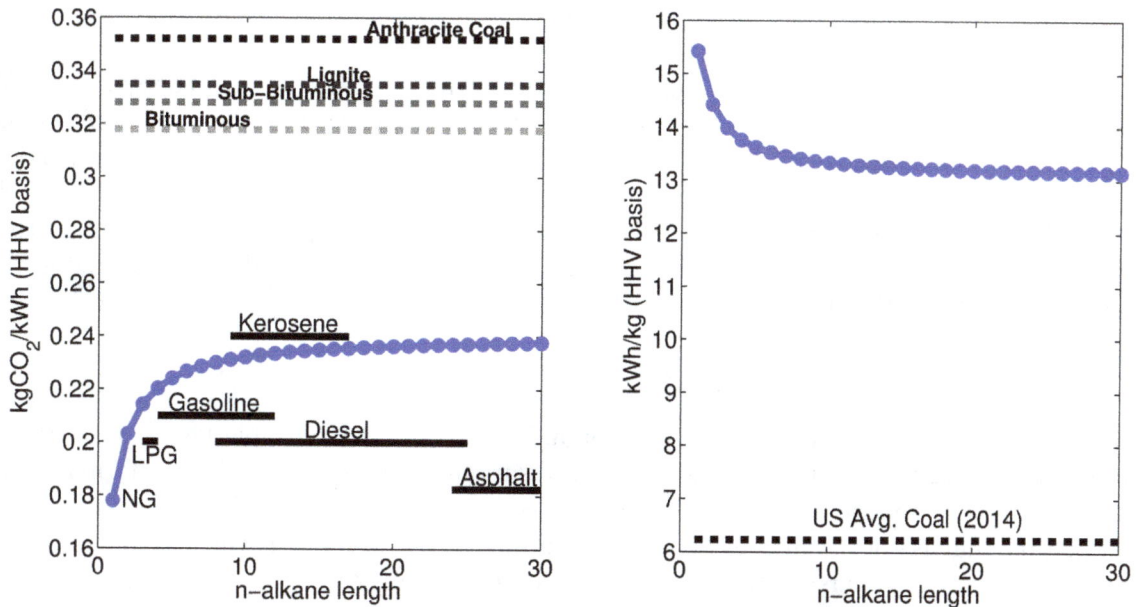

Figure 3.5: Theoretical emissions factors and energy-densities for alkanes of various lengths, on a HHV-basis. The left panel gives kgCO$_2$/kWh at the point of combustion for n-alkanes, along with this metric for different ranks of coal (per [51]). Approximate chain-lengths for common petroleum fuels are demonstrated. The right gives theoretical energy-density on a mass-basis, along with US average coal heat content in 2014 [114].

3.6.1   By-products of combustion

Fossil fuel combustion has several by-products, mainly sulfur oxides and nitrogen oxides, with incomplete combustion also yielding carbon monoxide and soot (black carbon). Coals are generally 0.5–5% sulfur by weight [54], crude oils average about 1.5% sulfur [207], while natural gas usually contains only scant amounts [55]). Oxidation of fuel sulfur leads to sulfur dioxide (SO$_2$), which ages to SO$_4$ and causes regional air pollution and cooling effects (see Section 3.4.5).

Nitrogen oxides (NO$_x$) are not related to any contamination, but a product of oxidized nitrogen (N$_2$) that naturally makes up 79% of the air. Hotter combustion conditions, which lead in general to more efficient engine operation and more complete combustion, are also more likely to yield NO$_x$. NO$_x$ aerosols have a mix of warming and cooling effects, as detailed in Section 3.4.5. Unoxidized carbon gives soot (a.k.a black carbon, which is strongly warming), while partially oxidized carbon gives carbon monoxide (CO) (also warming). Thus, we have ideal combustion as

$$O_2 + N_2 + C_xH_y + S \rightarrow CO_2 + H_2O + N_2 + S, \tag{3.14}$$

with no oxidation of fuel sulfur or air nitrogen, but actual combustion as

$$O_2 + N_2 + C_xH_y + S \rightarrow CO_2 + H_2O + NO_x + SO_x + CO + C_{\text{soot}}. \tag{3.15}$$

A variety of other particulates, oxidation products, etc., are also released depending upon the fuel and combustion conditions, e.g. mercury in coal, and various organic carbons released from wood burning.

3.7 An introduction to fossil fuels

Three major forms of fossil energy together provide almost all the world's energy: coal, oil (and the derived petroleum fuels), and natural gas. Furthermore, burning these fuels is the major driver of climate change. Here, I provide a brief introduction into the origins, properties, depletion dynamics (e.g. "peak oil"), and general emissions factors for these fuels. These fuels receive further dedicated attention elsewhere: coal and natural gas are discussed in further detail in the context of electricity generation, in Chapter 4, while tar sands oil is covered in the context of transportation fuels, in Chapter 6.

3.7.1 Origins and basic properties

All fossil fuels are formed via the burial of organic matter under multiple layers of inorganic sediments, leading to compression under high pressure and temperature, and the formation of carbon- and hydrogen-rich hydrocarbons. This deep burial and a lack of oxygen prevent the degradation that is the usual fate of dead organisms. Coal is formed when forest matter is buried in low oxygen marshes to form peat, which then transforms into successive grades of denser and denser sedimentary coal rock, as detailed further in Section 4.5; much coal dates from the carboniferous period, when the earth's land was largely covered in marshy tropical forests. Coal has the highest carbon:hydrogen ratio of the fossil fuels, the lowest heating value, and the highest CO_2/kWh emissions (see Figure 3.5).

Oil and gas, on the other hand, are generally formed when microscopic marine life, such as algae and zooplankton, are buried in deep anoxic ocean sediments, often in a deep *basin* [55]. *Conventional* oil and gas reserves form as follows. Organic matter in a *source* rock, often a shale (sedimentary rock made up of fine clay and some organic matter), is transformed by heat/pressure into oil. Between temperatures of about 65 and 150 °C, crude oil is formed (along with small amounts of gas), the so-called oil window, while at greater temperatures natural gas is formed, as shown in Figure 3.6. Now, oil and gas produced in shale expand, and with sufficient pressure the source rock is temporarily fractured, allowing these hydrocarbons to migrate through the subsurface, where they may encounter layers of more porous sedimentary rock.

Sedimentary rocks that are highly porous and permeability allow oil from source rocks to travel relatively freely, and these are referred to as *reservoir* rocks. Now, oft-times on its travels oil will encounter a *trap*, essentially a convexity in the reservoir rock overlain by an impermeable *caprock*, where these hydrocarbons can accumulate. Reservoirs may consist of either solely oil (with some dissolved natural gas, a so-called unsaturated pool); an oil reservoir overlain by natural gas–the free gas cap (a saturated pool); or a pure gas reservoir. Water, usually highly saline brine, is present at depth and occurs at the base of all reservoirs (water is much denser than oil or gas). This process is demonstrated in Figure 3.7.

Crude oils consist of a mix of straight hydrocarbon chains (alkanes), closed hydrocarbon rings (naphthenes), and aromatic hydrocarbon rings. Oils consisting primarily of shorter hydrocarbon chains are light, while longer, dense hydrocarbon mixes are heavy. Depending on its sulfur content, oil is referred to as sweet (low sulfur) or sour (high sulfur), with light, sweet oils the most valuable. All crude oil begins light, but when it seeps to the surface (or near surface), the lighter fractions may be lost to evaporation and water flows, and degraded by bacteria, yielding lower quality heavy oils. The Alberta tar sands are an example of a highly degraded, heavy bituminous oil.

At the refinery, crude oil is separated into fractions via distillation, such that the lightest fractions (composed of the smallest molecules) are gaseous, and include liquid petroleum gas (LPG) and propane, followed by gasoline. Kerosene, which is the basis for jet fuel, is of medium

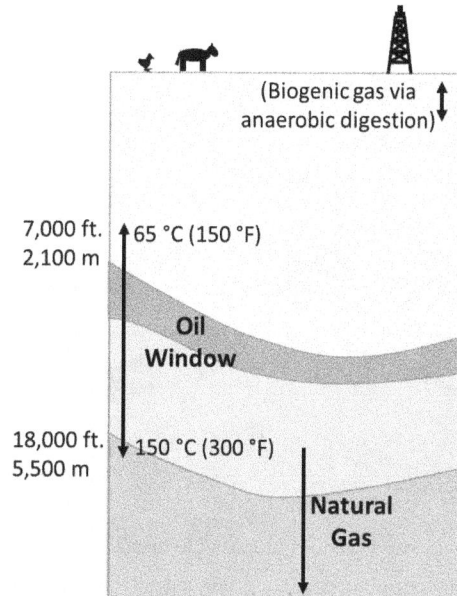

Figure 3.6: The oil window occurs between 65 and 150 °C, or about 2,100 to 5,500 m in depth on land (7,000 to 18,000 feet). Deeper and hotter, crude oil permanently degrades to graphite and natural gas. Note that natural gas can also be formed near the surface when bacteria anaerobically break down organic matter.

density, while heavier fractions make up diesel and heavy fuel oils, with bitumen/asphalt the densest and heaviest of all, as seen in Figure 3.8. Gasoline is generally the most valuable fraction, and therefore some heavier hydrocarbons are processed to gasoline via *cracking* [55].

Natural gas from the field varies in its exact composition. It is always mostly methane, but small amounts of ethane and nitrogen gas can be present, as may trace amounts of butane, propane, and hydrogen sulfide (a gas with significant sulfur content is a sour gas). Natural gas as delivered to customers is nearly pure methane.

3.7.2 Unconventional oil and gas

- Unconventional oil and gas resources are present in low permeability *tight* formations, either clay source shales or low permeability sandstone reservoirs.

- Hydraulic fracturing and horizontal drilling have markedly increased unconventional oil and gas production in the US since about 2000.

Conventional gas and oil reservoirs, as just discussed, are formed when hydrocarbons from low permeability source rocks, typically clay shales with high organic content, seep into higher permeability reservoir—sandstone or carbonate—rocks. Unconventional oil and gas come from *tight* formations, either sandstones or carbonate formations of low permeability, or the low permeability source shales themselves. By 2015, almost 70% of US natural gas and 50% of US oil came from unconventional sources.

Extraction from tight formations, especially shale, did not become economical until the advent of horizontal drilling techniques and hydraulic fracturing. With horizontal drilling, a single well-bore is drilled horizontally through a formation. Then, fracking fluid, a mixture of

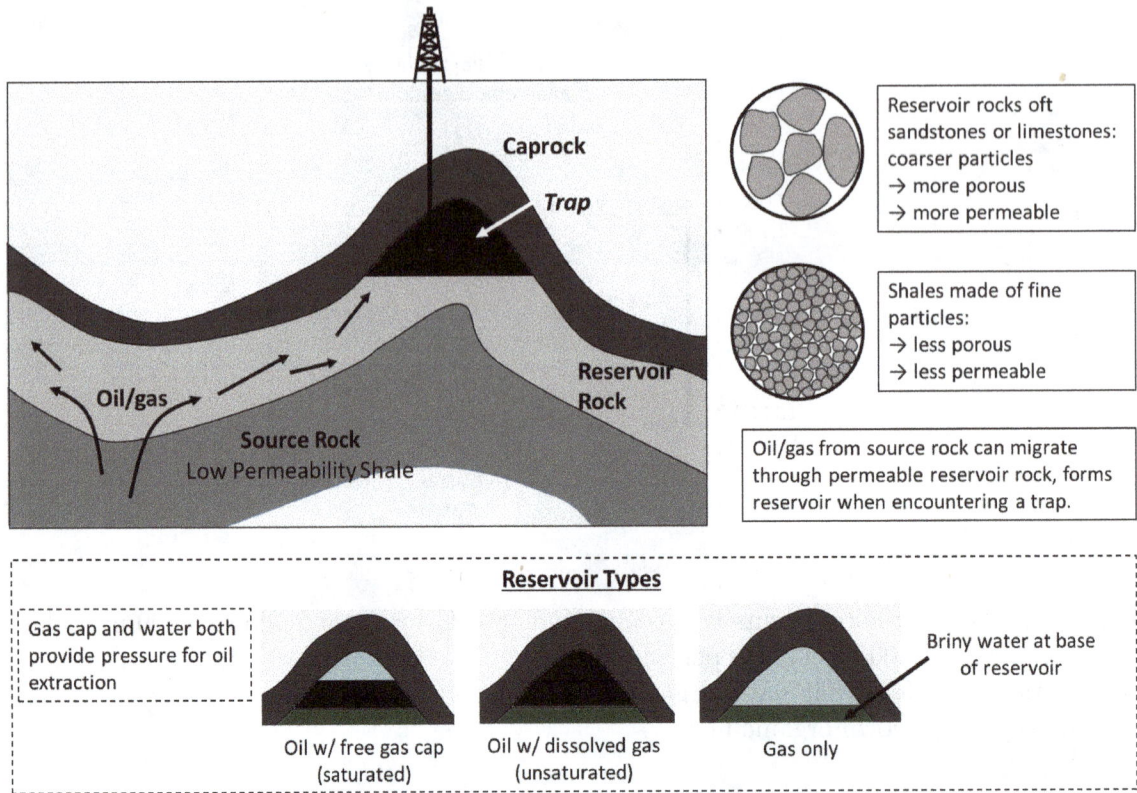

Figure 3.7: Schematic for the formation of conventional oil and gas reservoirs, whereby oil/gas from low permeability source rocks enters high permeability reservoir rocks, where it may migrate into a trap overlain by an impermeable caprock. Reservoirs so formed may consist of gas only, a gas layer above an oil layer (saturated), or gas mixed in with oil (unsaturated).

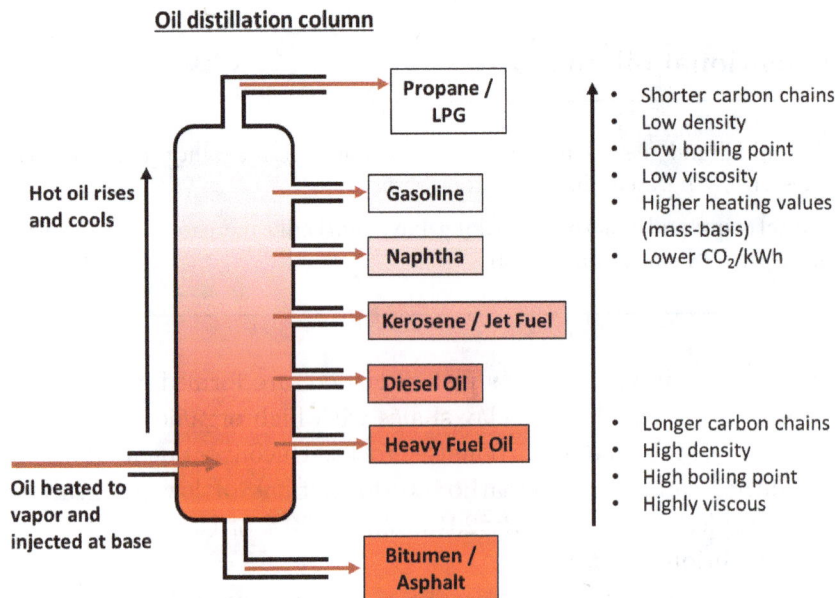

Figure 3.8: Distillation of crude oil into various fractions by density/boiling point.

mainly water and *proppants* (e.g. sand) is injected under high pressure, creating fractures in the shale and/or opening existing fractures. When the injection pressure is relieved, proppants hold the fractures open (see Figure 4.10 in Section 4.6.3).

US oil and gas production have increased dramatically in just the last few years, fueling, especially, increased use of natural gas for electricity generation at the (partial) expense of coal, but whether this is boon or bane is uncertain, for multiple reasons.

Natural gas leakage from fracking, and the larger natural gas system in general, is a point of controversy and uncertainty, and is a major factor undermining the notion of the shale gas revolution being a "green" energy revolution: methane combustion yields only about 55% the CO_2 as coal per kWh of energy produced, and combined cycle natural gas plants are also much more efficient than coal steam plants, yet methane is such a potent greenhouse gas that likely leak rates significantly offset these climate benefits (and at the extreme end of plausibility, overwhelm them entirely). Methane leak from gas exploration is discussed in much further detail in Section 4.6.

The other, and likely even more fundamental, challenge to the green nature of unconventional gas comes from basic economics. Cheap and abundant gas supplies can increase absolute fuel use in all sectors, which can in turn increase total emissions even if the CO_2 emissions per unit of energy decrease. Further, while low cost confers a competitive advantage over coal, gas also directly competes with wind and solar renewables, and building out gas-based infrastructure can "lock-in" gas use for decades to come; this issue is explored in Section 4.6.2. In addition to climatic effects, locoregional water quality may be affected by oil and gas exploration (both conventional and unconventional), and the effect of fracking on water supplies has been extremely controversial (see Section 4.6.3). Thus, there are a variety of negative externalities from both conventional and unconventional fossil fuel extraction. Compared to coal, unconventional gas is almost certainly better for both the climate and for regional water, wildlife, etc.: it is hard to imagine how gas wells could be worse than literal mountain-top removal for coal mining.

3.7.3   Hubbert and Peak Oil

- Fossil fuel reserve discovery and production tends to follows a bell-shaped curve, leading to the concept of "peak oil," as originally formulated by Hubbert.

- Multiple major peaks have already occurred: US conventional crude oil and natural gas production peaked in the early 1970s, and global oil discoveries peaked in the 1960s.

- Nonconventional oil and gas reserves are larger than conventional reserves and have yet to peak, but fossil fuel reserves are so vast that social or political factors must impose extraction peaks for dangerous climate change to be avoided.

While over the vastness of geologic time fossil fuels are to some extent renewable, on the scale of human civilization fossil fuel reserves are finite, and sufficient only for a few more centuries at best, regardless of the environmental consequences of burning them. M. King Hubbert, a geophysicist and geologist for Shell Oil and later the USGS, first warned in 1949 that the finite nature of fossil fuel reserves would shortly necessitate a transition to an alternative energy economy [12], and is most famous for his formulations on oil depletion and peak oil production. In 1956 [56], he correctly projected US conventional crude oil and natural gas production to peak by about 1970, and the idea of "Peak Oil" has since become quite popular in certain quarters, with some anticipating social and political chaos as production of fossil energy begins to fall, reminiscent of the "oil shocks" of the 1970s. While some seem to view Peak Oil as some inchoate apocalyptic vision of the future, Hubbert's original ideas were straightforward applications of mathematical reasoning to finite resource extraction.

The mathematical basis for Hubbert's work is actually quite simple. He observed that, for a finite resource of size Q, the rate of extraction (or production), P (equal to the rate of change in Q, i.e. $P = dQ/dt$), may initially increase, but at the point that the resource is entirely depleted, $Q = 0$, and P must then also fall to zero. Since P goes to zero, it must at some point hit a maximum and then decline, with this maximum necessarily occurring *before* resource depletion. Hence, "peak oil," in reference to the peak in production rate, or, alternatively, the discovery rate, which had *already* peaked for large US oil fields by the time Hubbert published his work [57]. Indeed, global oil discoveries peaked in the early 1960s, and have almost continuously declined since, with the overall discovery curve roughly bell-shaped [58]. A symmetric curve implies that roughly half of all reserves are depleted at the point of peak production.

Hubbart used a symmetric bell-shaped logistic curve to describe both the rates of oil discovery (i.e. discovery of fields) and production, as introduced in a second 1959 paper [57], which described well the discovery of large oil fields (his earlier 1956 work simply posited a general bell-shaped extraction curve, but did not give any particular mathematical form to it). Under this formulation, to estimate the time to peak oil, one only needs an estimate of total oil reserves and data for cumulative oil production from time zero.

We may conceptually consider our resource, Q, as discovered oil, Q_D, produced or extracted oil, Q_P, and known reserves, Q_R, all having units billion barrels (bbl). The *rates* of discovery and production are given as dQ_D/dt and dQ_P/dt, respectively; under Hubbert's model, discovered or produced oil can be described by a Logistic function,

$$Q_i(t) = \frac{Q_{max}}{1 + ae^{-k(t-t_0)}}.$$ (3.16)

Proved reserves are the difference between discovered and produced oil, $Q_R = Q_D - Q_P$. The general forms of the Hubbert curves are shown in Figure 3.9. There is always a lag between discovery and production, and the area under the production curve can never exceed the area under the discovery curve.

While there is debate over when global oil production will peak, it is clear that it must, and Chapman provides a good review of the recent controversies [59]. With respect to global warming, it is also clear that existing fossil reserves are sufficiently vast to cause extremely dangerous climate change well before they are exhausted [60]. Therefore, it seems we are faced with a choice: human civilization must hit fossil energy limits, and very soon (no more than a century or two), but we can hit this limit hard or soft. That is, we can push the climate system into a dangerous and unprecedented (on the scale of agricultural civilization) operating regime by maximally exploiting fuels only to face their exhaustion anyway, or we can preemptively make the transition to a non-fossil energy system. That is, it is essential that the timing of peak oil (and coal and gas) be determined socially and politically, and not by fundamental physical constraints.

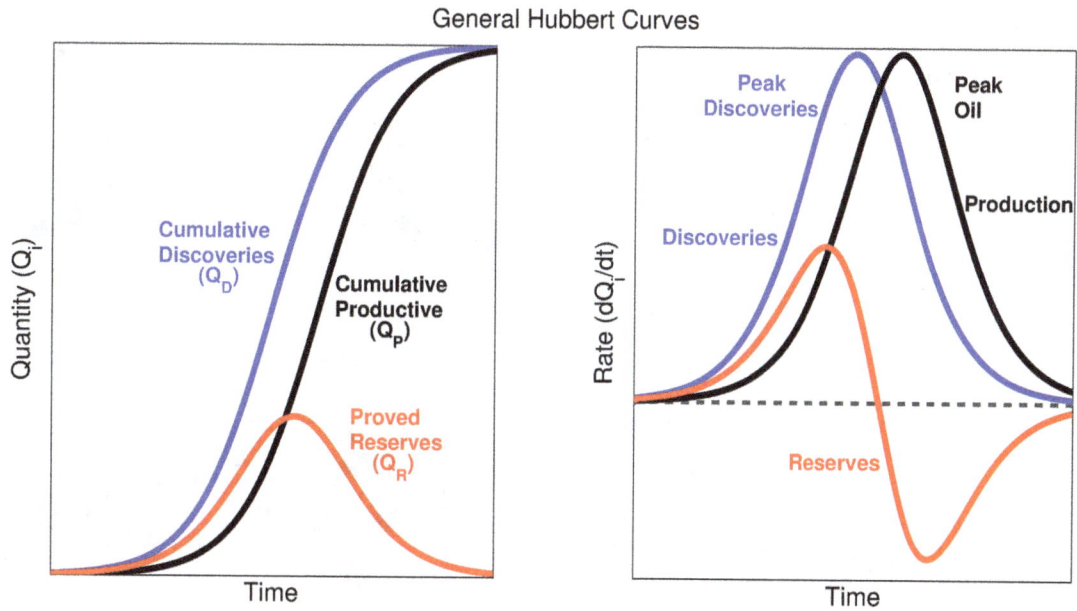

Figure 3.9: General form of Hubbert's curves describing cumulative oil discovery and production (left panel), and the *rates* of discovery/production (right panel). Peak Oil occurs (under the symmetric logistic law) when half of all oil has been produced.

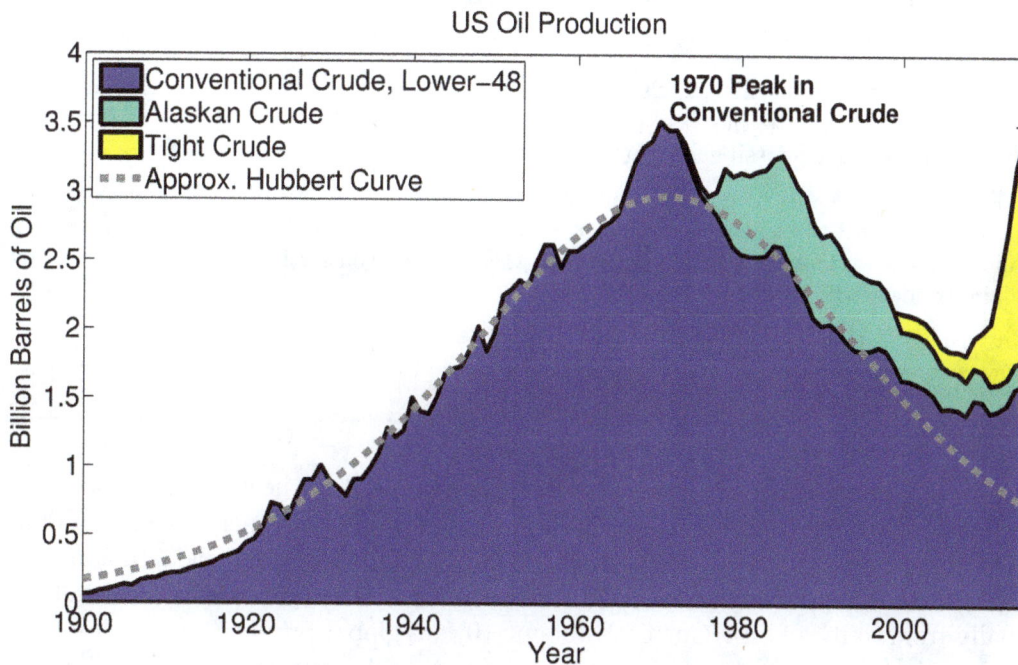

Figure 3.10: US crude oil production from 1900 to 2015. Conventional oil production peaked in 1970, as predicted by Hubbert, and has since fallen, with the recent uptick in production due to exploitation of unconventional tight oil plays. An approximation of Hubbert's curve is superimposed. Alaskan oil is treated separately, as the vast oil fields of the Alaskan North Slope were not discovered until 1968. Alaskan crude production peaked in 1988, and has since dwindled to only about 20% of the peak. Source: EIA.

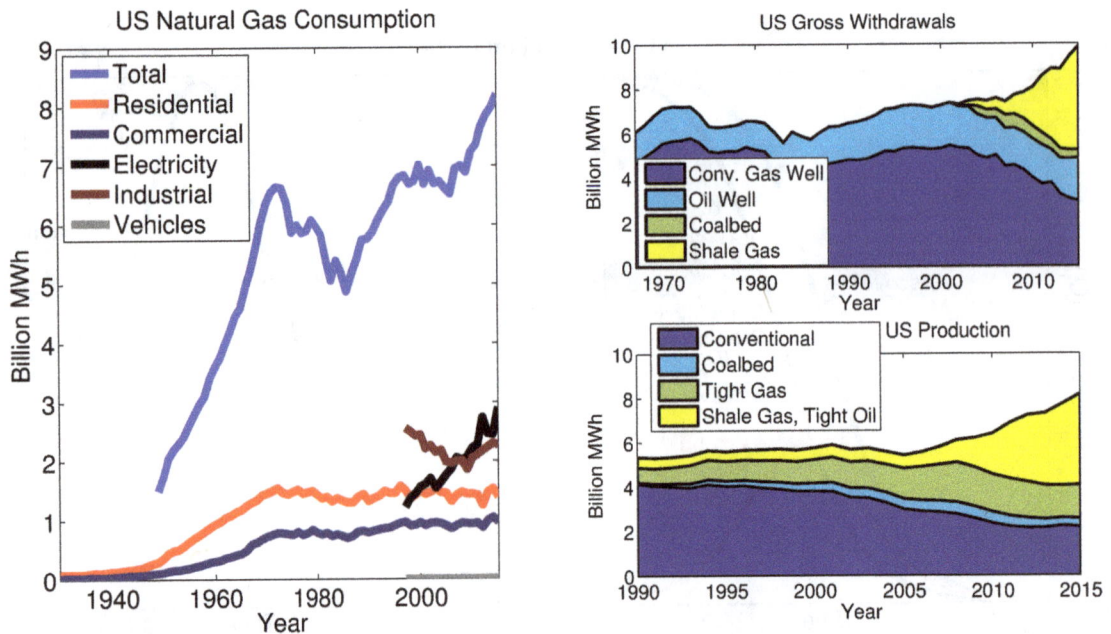

Figure 3.11: US natural gas production and consumption trends under several disaggregations. The left gives total consumption, and consumption by end-use. Note that, in concordance with Hubbert's predictions, conventional gas production peaked in the early 1970s. Source: EIA.

3.8   An introduction to biomass and biofuels

- Of the commonly used biofuels in the US, corn ethanol, soy biodiesel, and wood for heat/electricity, due to the vast land requirements for production, none is capable of significantly displacing US fossil energy use.

- All biofuels, especially corn ethanol and wood for electricity, deleteriously alter ecosystem carbon pools such that it takes decades to centuries before these carbon debts are paid off, if ever. Energy- and emissions-intensive agricultural inputs, such as nitrogen fertilizer, also undermine biofuels' claim to low-carbon status.

- In terms of carbon equivalent emissions, biofuels are probably closer to fossil fuels than to truly low-carbon energy sources, such as wind, solar, some hydro, and nuclear.

Biomass for energy generally consists of woody mass used for heat (either space heating or cooking) and electricity generation (currently a minor but growing energy source, and a major component of short-term renewable energy targets in Europe [47]), and modern transportation biofuels, namely ethanol (almost all of which is derived from corn in the US) and biodiesel (principally from soilbean oil). Currently, about 10% of global primary energy demand is met from biomass in general [62] (wood is the most important component at 6% of primary energy [61]), with the greater part of this use in highly inefficient cook-stoves and open fires throughout the developing world [63]. Such biomass is often harvested unsustainably and degrades forests, impairs indoor air quality, and is a major source of carbon dioxide, carbon monoxide, black carbon, and harmful particulate emissions [63]. In the US, wood is used as a heating fuel for several million households, and waste wood, agricultural byproducts, and forest products are used in wood-fired power plants, or they are co-fired in coal generators. About half the US corn crop is now devoted to ethanol for transportation.

While often promoted as a green and/or zero-carbon alternative to fossil fuels, biomass energy sources often fail to meet either criterion, due to high land and energy requirements for production, and the altered carbon dynamics that result from harvesting and burning. In this section, I review the basic energy conversion efficiencies of different bioenergy sources, and examine their scalability as fossil energy replacements. I also cover some of the alterations in carbon fluxes that result from large-scale biomass for energy projects. Corn ethanol is discussed in further detail in Section 22.3, wood as a residential heat source is treated separately in Section 12.5.6, and woody mass for electricity generation also gets a designated discussion, in Section 4.10.

3.8.1   Scalability of bioenergy sources

It is fairly simple to calculate the potential yields of current energy crops, and hence their feasibility as large-scale replacements for fossil energy. So let us make some of those calculations now.

Soybean biodiesel/fuel oil. Soybean yields in the US on are the order of 3 Mg/Ha, and with 20% of the dry mass lipids [334], we have (assuming 14% moisture content and 37.6 MJ kg^{-1} for lipid [65]) no more than about 6,000 kWh/Ha of oil energy, equivalent to just under 150 gallons of diesel fuel (HHV basis). Per the EIA, the US consumed 60.8 billion gallons of fuel oil/diesel in 2015, as well as 23.7 billion gallons of kerosene-based jet fuel (kerosene and diesel have nearly identical energy content); replacing this with soy oil (and generously assuming that no energy is lost from processing to biodiesel) would require about 575 million Ha, about 75% of the continental US landmass, and 3.5 times the area of *all* US cropland. Using all US land currently devoted to soy production (about 33 million Ha) for biodiesel would displace only 8% of US fuel oil use. Finally, if we wished to replace all US primary energy consumption (29.3 trillion kWa) with soy oil, we would need over six times the US continental land area.

Clearly, soy biodiesel can displace only a few percentage points of current fuel oil use, at best. Adding to the problem, the energy-return on energy-invested (EROI) for US soy biodiesel is probably no better than 5.5 [66], decreasing the effective energy yield of soy biodiesel by 20–25%, and increasing land requirements commensurately.

Soy cultivation for biofuels and animal feed is the major driver of deforestation of the rain forests in Brazil. In addition to being an ecological disaster, clearing of tropical rainforests releases vast quantities of carbon into the atmosphere, incurring a carbon-debt that takes, for soy biodiesel production, *several centuries* to repay even in the best case [70].

Corn ethanol. Corn ethanol suffers from an extremely low EROI, as discussed in Section 22.3, and likely yields no more energy than it takes to produce. This alone should condemn the technology, but for the sake of argument, let us assume that 100% of the energy is recovered. Using calculations detailed in Section 22.3, we see that a 10 Mg/Ha maize yield (typical for the US) gives us just under 21,000 kWh/Ha (HHV basis), or 576 gallons of gasoline equivalent (HHV basis). Therefore, the entire US corn crop could displace about 15% of gasoline used by the light-duty fleet (assumed to be 123.8 billion gallons), while replacing all gasoline would require almost 30% of continental land area and nearly 1.5 times the current US cropland area (for all crops). Again, these calculations are ridiculously biased in favor of ethanol because of its abysmally low EROI. If we assumed a *best-case* EROI of 1.3 [72], then it would effectively take 120% of the US continental land area to replace all gasoline.

Woody biomass. While soy-based fuel oil and ethanol are posited as replacements for liquid petroleum fuels, it would fall mainly upon woody biomass to replace natural gas and coal for heating and electricity generation. Let us assume that wood-fired electricity generation has a 25% thermal efficiency and that non-electrical heating is 70% efficient, and use data from the EIA on natural gas and coal consumption. Further supposing that forest woody net primary

61

productivity is on the order of 100–300 gC m^{-2} [67], with 200 gC m^{-2} the maximum harvestable (about one-third of mass is lost in the harvest [47]), and using a 11.12 kWh/kgC gross calorific value for wood, to displace all US coal and natural gas use, we would require, in all, 842 million Ha of forest land, while there exist only 302 million Ha of forest in all the US (including the vast forests of Alaska) [639].

Furthermore, appropriating all woody productivity from forests is clearly unsustainable and would decimate forest carbon stocks: an extraction rate of 200 gC m^{-2} corresponds roughly to frequent clear-cutting, and similarly intense logging severely degraded forests in the Pacific Northwest from the 1960s through 1980s [625]. Thus, while burning woody forest products for heat and electrical power is more scalable than modern biofuels, it is still unlikely that they could ever provide more than about 5–15% of such energy at current use rates.

It follows that, in some sense, it is almost irrelevant whether current biofuels are carbon-neutral or not (they are not), as they cannot hope to displace more than a tiny fraction of US (and global) energy use (with the very partial exception of woody biomass). On the other hand, even if they were truly carbon neutral, decreasing carbon emissions by a few percent could *never* (in my view) justify the truly vast tracts of land needed to provide this minuscule benefit. Even "next-generation" biofuels, such as algae, are hopelessly inefficient for meeting current US energy use (see calculations in [65]).

Note that even skeptical analyses, such as that of Schulze et al. [47], who argue that forest biomass cannot sustainably replace even 20% of the world's primary energy demand, typically fail to account for the generally lower thermal efficiency of wood power plants compared to either natural gas or coal. That is, the displacement energy between wood and fossil fuels is not one-to-one, as for example, to generate electricity it takes about 70% more primary wood energy than gas energy for the same electricity yield, under typical power plant efficiencies.

Figure 3.12 graphically illustrates the land bases that would be required to replace several different fossil energy sources with different biofuels (factoring in the EROI for soy biodiesel and corn ethanol), namely replacing diesel, kerosene jet fuel, and fuel oil with soy biodiesel, replacing light-duty vehicle gasoline with ethanol, and displacing natural gas and coal with wood. For comparison, the land bases required to replace all US primary energy or electricity with either solar photovoltaics or wind energy are given[2]. As can be seen, while it would take, in sum, the entire land area of the continental US almost four times over to replace the major fossil fuels with biofuel analogs, solar and/or wind could do the job with just a small fraction of US continental land area (and less than all urban land area, in the case of solar). Note also that wind harvesting directly uses only a small fraction of the land which is harvested, and turbines are commonly placed in agricultural fields.

3.8.2 Carbon dynamics

The burning of biomass to generate power (either motile or electric) has been widely considered a zero emission process, as it is assumed that all carbon liberated from burning plant matter will eventually be reincorporated into plant matter with regrowth. The zero-emissions assumption is erroneous for multiple reasons. In the case of annual crops with short-term carbon cycles, it may reasonably be assumed that all carbon liberated year to year is re-incorporated year to year. However, a seminal paper by Fargione et al. [70] pointed out that, while biofuels such as ethanol or soy biodiesel may be low carbon in this sense, a massive release of carbon occurs when natural ecosystems such as rainforest or grasslands are converted to producing food crops. This conversion creates a carbon debt that was estimated to take anywhere from 17 to 423 years to

---

[2]Assuming an average US solar irradiance of 1,800 kWh/m^2 and solar PV efficiency of 15% with a performance factor of 75%, and assuming that 1 W m^{-2} can be harvested by wind turbines.

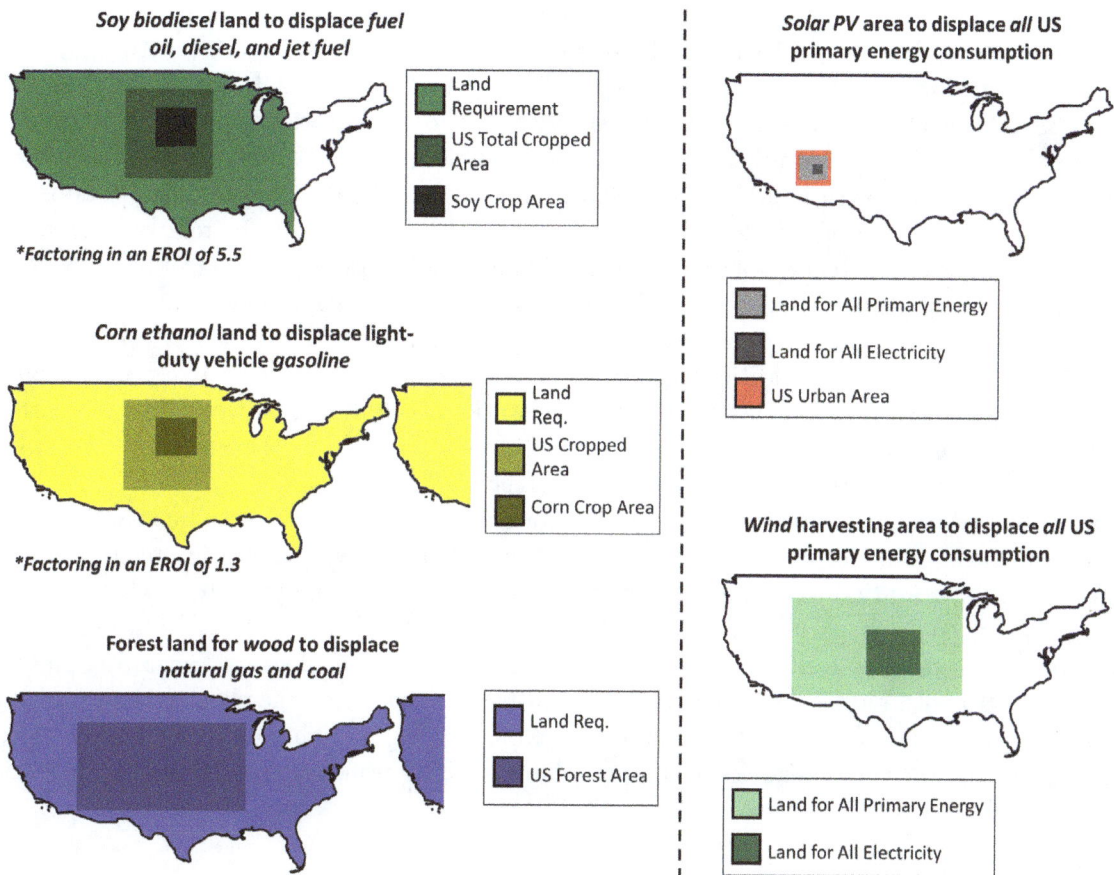

Soy biodiesel land to displace *fuel oil, diesel, and jet fuel*

Land Requirement

US Total Cropped Area

Soy Crop Area

Factoring in an EROI of 5.5

Corn ethanol land to displace light-duty vehicle *gasoline*

Land Req.

US Cropped Area

Corn Crop Area

Factoring in an EROI of 1.3

Forest land for *wood* to displace natural gas and coal

Land Req.

US Forest Area

Solar PV area to displace *all* US primary energy consumption

Land for All Primary Energy

Land for All Electricity

US Urban Area

Wind harvesting area to displace *all* US primary energy consumption

Land for All Primary Energy

Land for All Electricity

Figure 3.12: Land requirements for current biofuels/biomass options being pursued as replacements for different current fossil energy resources (left panels), and comparisons to land requirements to replace *all* US energy (or just all electricity) using solar or wind (right panels). For soy diesel and corn ethanol, the current cropped areas of soy and corn, as well as the total US cropped area, are also given for comparison. For wood, total US forest area is also provided for comparison, while solar is measured against current US urban area.

repay by offsetting carbon emissions from fossil fuel use, depending upon the habitat converted and the fuel produced. Even if no new land is directly cleared for production biofuels from food crops, this places pressure on the food system that can induce indirect habitat conversion.

Failing to return agricultural residues can also decrease soil carbon stores. Additionally, and as discussed further in Section 22.3.3, in the case of corn ethanol simply abandoning fields likely results in more carbon storage (via soil carbon build-up, and potentially increased above-ground biomass) than could be offset even under optimistic assessments for ethanol, and would have myriad ecology co-benefits.

Vegetation harvested from managed forests, on the other hand, does not necessarily have the problem of associated habitat conversion, but note that converting natural forests to tree plantations *does* decrease soil carbon, biodiversity, and total carbon stores, with one review finding forest plantations to have 28% less total ecosystem carbon [69]. Since this translates into a 291.5 $MgCO_2e$/Ha loss, converting natural forest to a plantation may release even more carbon than clearing grassland for crops [69, 70]. This issue aside, the assumption of zero emissions due to regrowth is highly problematic in the case of forest harvesting: while it may plausibly apply at a *very long-term* steady state, it is clear that, due to relatively slow tree growth, burning forest biomass creates an immediate carbon debt that may take decades or even centuries to repay via regrowth. Furthermore, when forests are continually harvested, there is permanent decrease in the carbon stock of the forest, and thus a permanent carbon shift to the atmosphere. Additionally, harvesting trees from forest also likely deleteriously alters multiple carbon pools other than living vegetation, including deadwood and soil carbon. Under harvesting, mineral nutrients are also continuously lost from the forest system, which will eventually degrade the productivity of the forest [47].

Finally, and perhaps most importantly, the world's forests have proved to be an efficient and continuous carbon sink, especially in Europe and North America, where they are recovering from centuries of deforestation that preceded the transition to a fossil energy system. Harvesting for bioenergy will at least partly undermine this sink. When fully accounting for the fact that old-growth forests likely continue to sequester carbon, and for the effects of harvesting on other forest carbon pools, biomass for electricity may in fact be *worse* for the climate than coal-fired electricity over a time-frame of several hundred years, but this is highly dependent on the forest growth rate and the characteristics of the wood harvest. This issue is explored further in Section 4.10.

3.8.3   Other inputs and emissions

Maintaining high productivity in cropping systems requires significant inputs in the form of fossil energy, fertilizer, and sometimes irrigation. In particular, embodied emissions in nitrogen fertilizer manufacture, and N_2O emissions from its application, can largely offset any carbon benefits from biofuel production. Fossil energy used for farming operations is also significant, while energy for ethanol processing largely cancels out any benefit to this particular fuel. These upstream factors are discussed extensively in Section 22.3.2.

3.9   US primary energy consumption

Having introduced the basics of fossil and biofuels, it seems apropos to close by showing just how dominant fossil fuels have been over the course of the last century and a half, and how dramatically they have increased the total energy available to civilization. Primary energy sources from 1775 through 2011, for the US, are shown in Figures 3.13 and 3.14, based on EIA data [73].

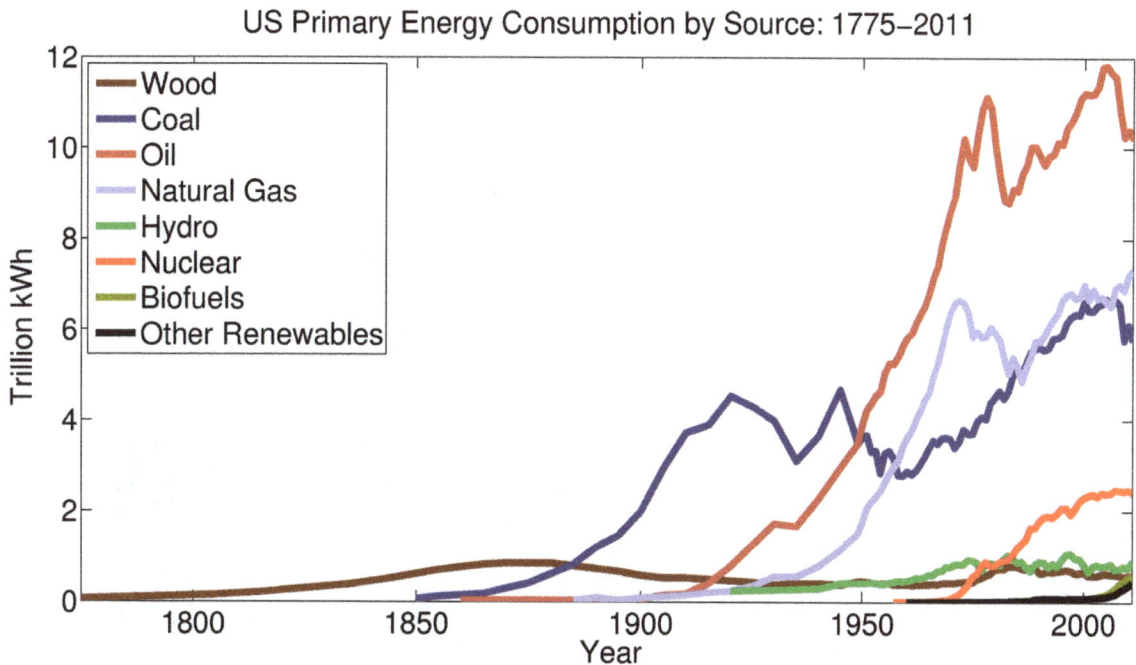

Figure 3.13: Historical primary energy consumption in the US, from 1775 to 2011, based on [73]. Wood was displaced by coal as the top energy source by 1885, while petroleum would overtake coal in 1950. Note the transient peaks in both oil and natural gas in the early 1970s, corresponding to peaking domestic production as well as the "oil shocks" of that era.

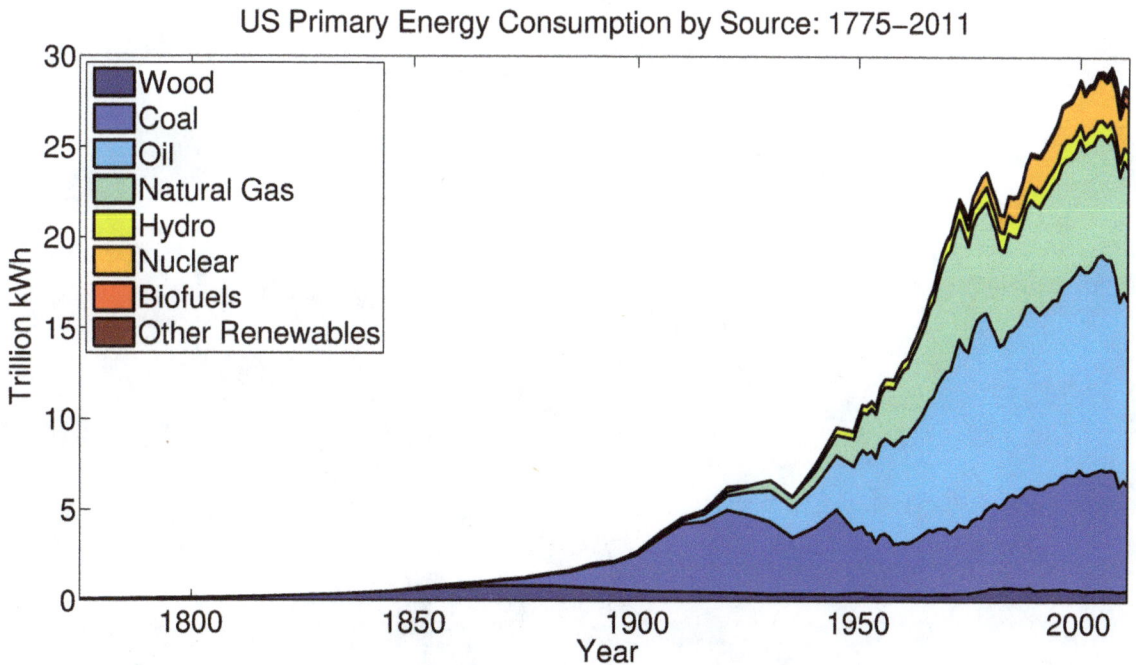

Figure 3.14: Historical primary energy consumption in the US, from 1775 to 2011, mirroring Figure 3.13 but now highlighting the dramatic increase in total energy consumption over the course of the twentieth century.

Chapter 4

Emissions from electricity generation

For all subsequent sections, it is necessary for us to determine, for various fuels and electricity sources, the *emissions factor* (EF), the mass of CO_2-equivalents produced per unit of fuel combusted or per unit electricity generated. For example, the *direct* EF for coal-fired electricity is roughly 1 kgCO$_2$e/kWh, meaning for every kWh of electricity generated, greenhouse gases equivalent to 1 kg of CO_2 are produced directly from coal combustion. In addition to combustion emissions, *lifecycle* EFs attempt to incorporate all upstream (e.g. production) and downstream (e.g. waste disposal) processes that also generate emissions, collectively referred to as *indirect* emissions. We must also distinguish between the EF for electricity generated and for electricity delivered, as transmission and distribution losses are not negligible; the latter is relevant for all household operations, except when power is provided by a local generator, e.g. rooftop solar, in which case such losses are near zero.

Another issue to consider is that, while we usually consider the *grid-average* emissions factor for carbon footprinting purposes, small changes in electrical loads, i.e. those acting on the margin, affect *marginal* electricity generation. That is, additional fossil-based generators must be dispatched to meet fluctuating demands, even under electrical grids with very high renewable or nuclear penetrance, and so the marginal EF is usually somewhat greater than the grid-average EF.

The EPA's eGRID database is a comprehensive characterization of nearly all grid-electricity generation in the US, and estimates state, regional, and national electricity-associated emissions using plant-level data. This database is widely used for GHG estimation by both governmental and non-governmental actors. The database considers only emissions at the point of electricity generation, ignoring any upstream or downstream emissions as well as electricity transmission and distribution losses (although these losses are reported). Furthermore, emissions from hydropower, nuclear, wind, solar, geothermal, and biomass are all assumed to be zero. However, there are emissions associated with renewable energy lifecycles, and although these are a tiny fraction of fossil fuel emissions, they are not quite zero. A notable exception to this rule is electricity derived from burning biomass, which likely has an overall carbon footprint comparable to fossil fuels, as I discuss, indignantly, at length.

For a full and honest accounting of electricity-associated emissions, we must account for all lifecycle emissions. In the following sections, I review the literature on lifecycle associated emissions for various electricity sources, and I derive general EFs for all power sources. I also determine crude emissions correction factors (to account for upstream emissions) which I apply to the eGRID database to yield corrected electricity EFs for all regions. I have also attempted to provide a fairly comprehensive understanding of each major power source, controversies surrounding them, and major externalities beyond carbon emissions. I close this chapter with an examination of how scalable renewable technologies are as a replacement for fossil fuels, primarily in terms of electricity generation, but also as replacements for all global primary energy consumption.

4.1 Recent trends in electricity generation

Coal has been the dominant fuel source for electricity generation since the inception of the grid, and the share of power generated by coal remained above 50% throughout the 1980s and 1990s, until cheap natural gas, largely from unconventional shale sources extracted via hydrofracking ("fracking"), began to significantly displace King Coal, and by 2012, coal had fallen to a 37% share, with NG up to 30% (eGRID), a near doubling over a decade previously; this decline may have stalled however, with coal back to 39% in 2014 (and NG at 28%). The overall effect of this fuel switch on global warming is controversial, due to upstream methane leaks and the effects of cheap fossil fuel on infrastructure and energy use patterns, and the reader is no doubt aware of the controversy surrounding fracking; I explore these issues further in Section 4.6. Despite its deep problems, the displacement of coal by NG has likely been of net climate benefit, although NG remains a carbon intensive fossil fuel source with multiple externalities.

We must understand that the relative demise of coal has little to do with any rise in renewables, and while coal is often referred to almost as a "dead fuel walking," in the media, it remains the top fuel source for electricity in the US and, especially, globally. Some regions of the US remain almost entirely reliant on coal for electricity. It is also important to understand just how little of the US generating mix comes from non-hydro renewables, i.e. wind (4.4% in 2014), solar (0.4% in 2014), biomass (1.6% in 2014), and geothermal (0.4% in 2014). However, wind and solar have rapidly expanded in just the last few years, and wind power is beginning to become a significant player nationally and especially in certain regions. Of these sources, geothermal is likely only to be developed in a few geologically active regions, and biomass, while renewable, is neither carbon neutral nor scalable (see Section 3.8) and the widespread use of biomass would likely amount to destroying the forest to save it. This leaves wind and solar, both of which truly are low-carbon, scalable technologies that can be deployed on a global scale. The barriers to their universal implementation are primarily social and political, in my view, not technological or environmental (see Section 4.14 for further discussion).

Hydropower (6.1%) and nuclear (19%) are currently by far the most important near-zero carbon electricity sources in the US, but whether they will expand significantly in the future is unclear. American river systems are already extremely fragmented and highly exploited, and thus a major expansion seems unlikely. Nuclear generation is stagnating, and since it relies on limited uranium supplies, it is not actually a renewable energy source. However, under fourth-generation nuclear technologies and breeder reactors, uranium reserves could last millennia, and despite significant emissions associated with plant construction and fuel production, nuclear is a genuinely low-carbon power source (see Section 4.8). Figure 4.1 summarizes recent trends in US electricity generation.

4.2 Lifecycle emissions by source

Figure 4.2 shows the approximate emissions factor for all major power sources, as determined in the following sections, divided into direct and indirect emissions. Error bars indicate the range of what I consider to be reasonably likely maximum and minimums, based on my review of the literature as presented in this chapter. Note that Figure 4.2 shows the EF in terms of $kgCO_2e/kWh$ *generated*, while to determine the impact of consuming a single kWh, we need the EF in $kgCO_2e/kWh$ *delivered*, which factors in transmission and distribution losses. These

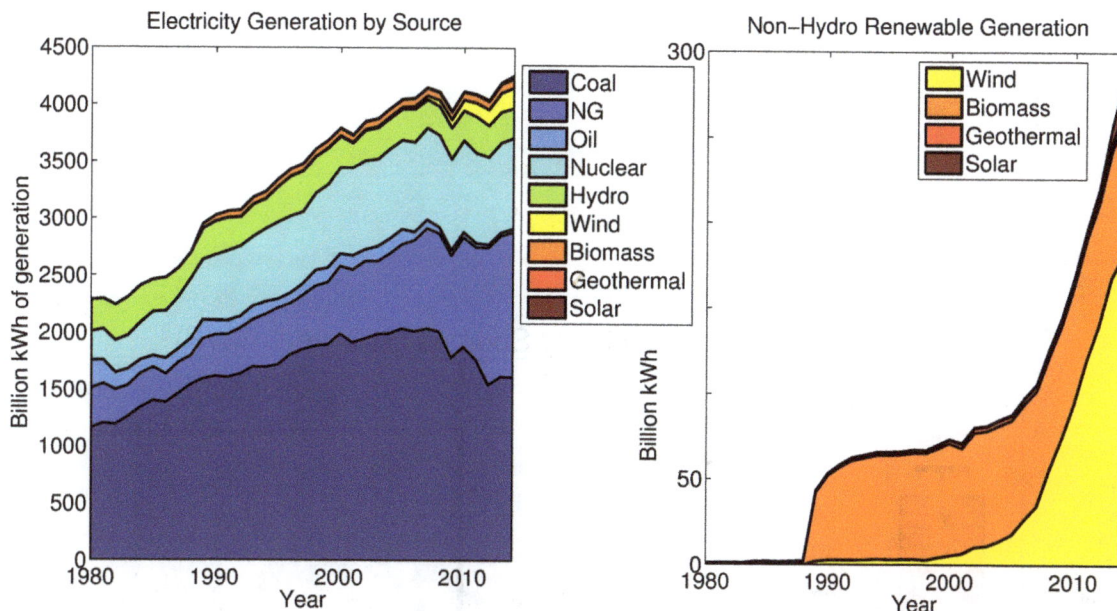

Figure 4.1: Trends in source for US electricity generation from 1980 to 2014 (source: World Bank). The left panel shows that, while coal continues to be the largest single electricity source, it has been partially displaced since about 2000 by cheap unconventional natural gas, while nuclear and hydropower have remained fairly steady. Non-hydro renewables are still little more than a rounding error, but solar, and especially wind, have increased dramatically (in a relative sense) in just the last few years, and wind will likely match or exceed hydro within a few years.

losses are typically around 4–8%, and I take 6.5% as a rough average (the mean T&D loss over the last three eGRID editions). That is, we simply increase the generation EF by 6.5%.

4.3   Overall power generation and emissions in eGRID

About 4.080 billion MWh of electricity were generated (in the US) in 2014, directly generating 2.0914 billion tonnes of CO_2. Coal was responsible for about 74% of this CO_2, with nearly all the balance from natural gas (23%). However, once we account for upstream lifecycle emissions, the US grid probably was responsible for around 2.60 billion $MgCO_2e$, almost 20% higher than direct emissions alone. In this case, coal and natural gas still account for nearly all emissions, but their respective shares shift to 66% and 27%, with upstream methane leakage accounting for NG's increased share, and the lifecycle emissions of other electricity sources diluting coal's share somewhat. Absolute generation and emissions by source are detailed graphically in Figure 4.3.

In terms of kWh generated, I have estimated the overall US grid-average EFs to be 0.5218 $kgCO_2/kWh$ in direct emissions, and 0.6378 $kgCO_2e/kWh$ including all lifecycle emissions. Assuming 6.5% transmission and distribution losses, this translates into direct and lifecycle EFs of 0.558 $kgCO_2/kWh$ and 0.682 $kgCO_2e/kWh$, in terms of kWh *delivered* (see Figure 4.4). The final number, 0.682 $kgCO_2e/kWh$, unless otherwise specified, is used throughout this book to determine the carbon footprint of various electricity uses.

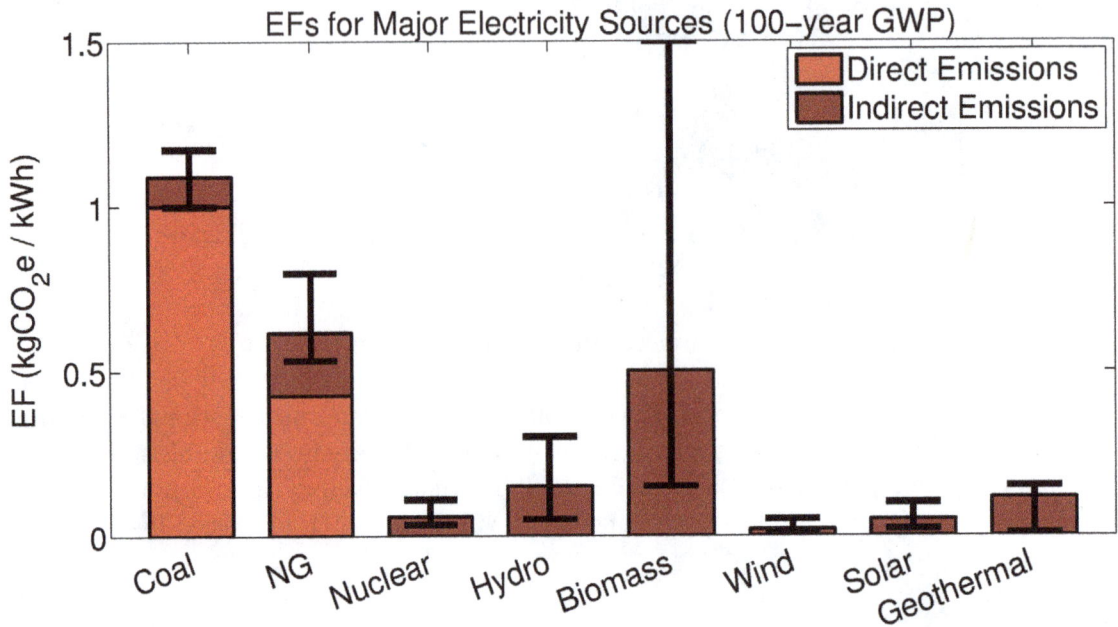

Figure 4.2: Approximate emissions factors for major sources of electricity in the US, over a 100 year time-horizon. Direct and indirect emissions are given, with indirect emissions accounting for 100% of the EF for non-fossil electricity sources; indirect emissions are the source of most uncertainty, and errors bars give my qualitative assessment of the likely maximums and minimums. These EFs are specific to the US grid, and may vary somewhat internationally. For example, tropical hydropower may in some cases be worse than coal, due to very high methane emissions from flooded forests. Out of fossil fuels, natural gas emissions are the most uncertain, varying significantly with the methane leak rate; I take a leak rate of 2.4% as my central estimate. Note that the EF for biomass is highlighted as being extremely uncertain, as most of the impact comes from uncertain effects on forest carbon storage dynamics. Numerically, and for typical power plant thermal efficiencies, my central estimates for upstream emissions are 89 gCO_2e/kWh for coal, 189 gCO_2e/kWh for NG, 58 gCO_2e/kWh for nuclear, 150 gCO_2e/kWh for hydropower, 500 gCO_2e/kWh for biomass, 19 gCO_2e/kWh for wind, 50 gCO_2e/kWh for solar, and 115 gCO_2e/kWh for geothermal power.

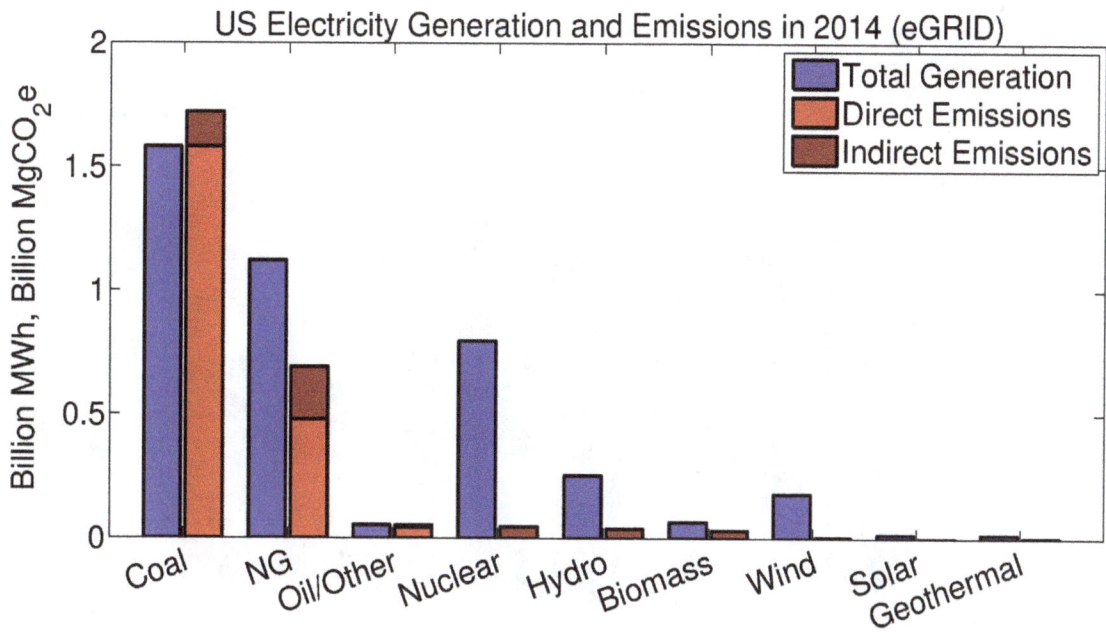

Figure 4.3: Total US electricity generation by source, and total associated (direct and indirect) CO_2e emissions, based on the 2014 eGRID database, and upon the electricity emissions factors derived in the chapter.

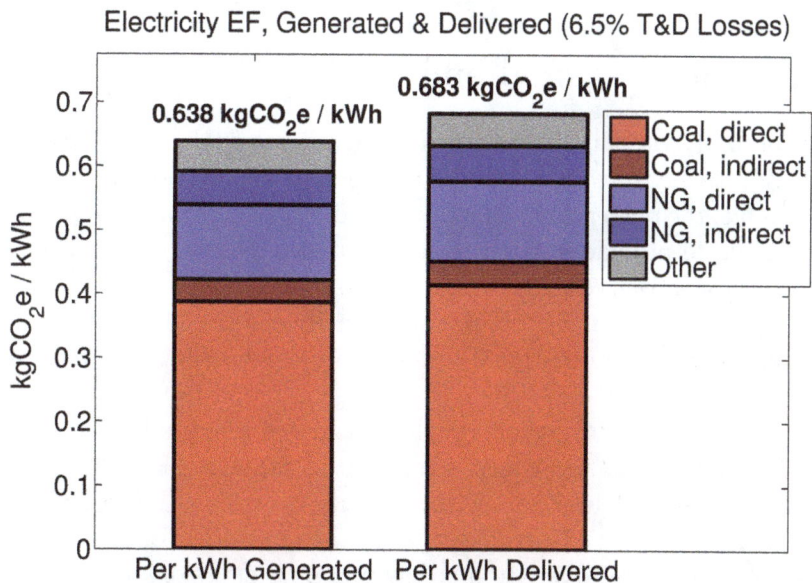

Figure 4.4: Final US grid-average lifecycle emissions factors for electricity generation, either on an as-generated or as-delivered (assuming 6.5% T&D losses) basis. The EFs are derived from the 2014 EPA eGRID database, and from assumed upstream emissions factors as determined in this chapter. The bars show the approximate breakdown of each EF into direct and indirect emissions from coal and natural gas, and all other emissions sources (other minor fossil fuels, and the indirect emissions from all other electricity sources). While other sources account for 33.8% of all generation, they are responsible for only 7.4% of emissions.

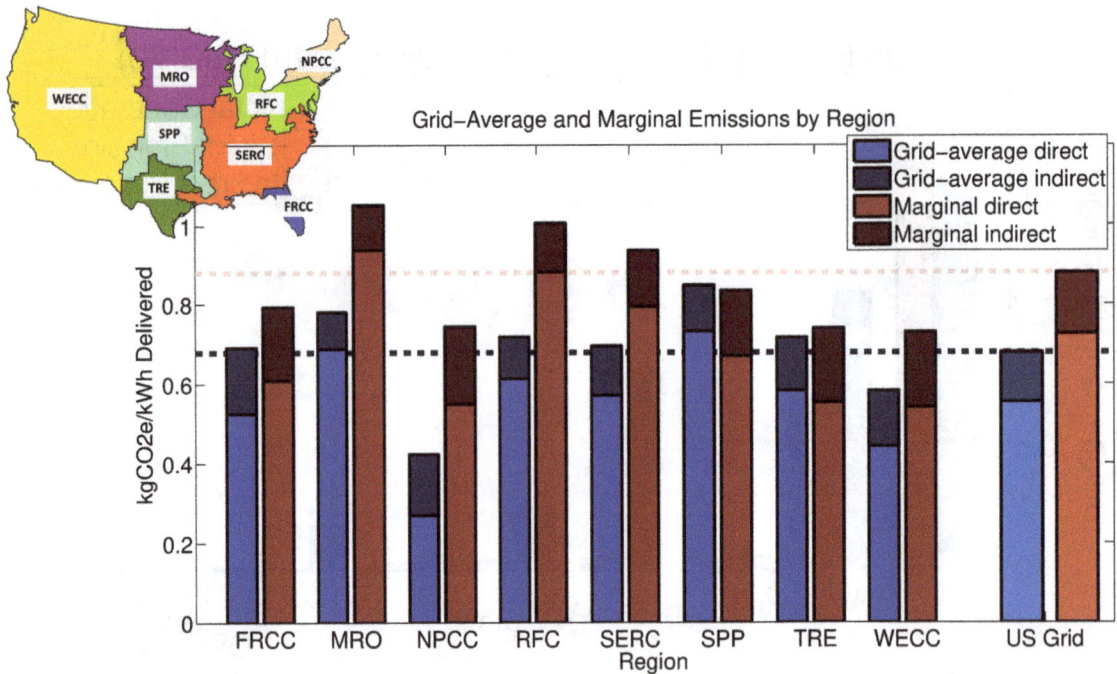

Figure 4.5: Approximate grid-average and marginal emissions rates, in kgCO$_2$e per kWh *delivered*, i.e. emissions factors include a 6.5% transmission and distribution loss factor, for the eight NERC regions of the continental US, as well as the approximate US averages (0.682 kgCO$_2$e/kWh and 0.880 kgCO$_2$e/kWh for grid-average and marginal electricity, respectively). Inset map showing NERC regions is adapted from the EPA's NERC region map.

4.4 Marginal and regional emissions

Throughout this book, I generally use (my own) derived national grid-average emissions factor (0.682 kgCO$_2$e/kWh) to measure the impact of electricity consumption. Since regional emissions factors vary quite a bit, the reader must be left to wonder how relevant this single estimate is, but it is not really feasible to provide region-specific estimates for every calculation. Furthermore, since most changes in electricity use represent a *marginal* change, it is worthwhile comparing region-specific marginal emissions factors to the US grid average. As we shall see, the US grid-average EF is a good surrogate for marginal emissions in many regions, but it is an underestimate in regions that are more heavily coal-reliant, especially the midwest.

Siler-Evans and colleagues [74] systematically calculated marginal power generation sources for the eight NERC regions in the continental United States. Marginal power is generated almost entirely by coal and natural gas, with the only exceptions being small amounts of oil-fired generation in Florida and the Northeast. Given the percentage of marginal generation from coal and gas, I calculate an approximate lifecycle marginal emissions factor for each region, as shown in Figure 4.5. I use 1.086 kgCO$_2$e/kWh for coal, 0.6162 kgCO$_2$e/kWh for NG (at 42.4% TE, 2.4% upstream leak), and 0.9939 kgCO$_2$e/kWh for oil-fired generation[1].

These calculations suggest marginal emissions on the order of 0.8798 kgCO$_2$e/kWh of power delivered (on a lifecycle-basis), almost 30% higher than US grid-average emissions.

---

[1] For oil I use 0.2508 kgCO$_2$e/kWh(t) combustion emissions (on HHV-basis) and add 0.0573 gCO$_2$e/kWh(t) of upstream emissions to sum to 0.3081 kgCO$_2$e/kWh(t). Assuming a 31% TE based, on EIA 2012 data, yields a total EF of 0.9939 kgCO$_2$e/kWh(e).

4.5 Coal-fired electricity

4.5.1 History and current use

> This age has been called the Iron Age...But coal alone can command in sufficient abundance either the iron or the steam; and coal, therefore, commands this age–the Age of Coal. It is the material energy of the country—the universal aid—the factor in everything we do. With coal almost any feat is possible or easy; without it we are thrown back into the laborious poverty of early times.
>
> *William Stanley Jevons, The Coal Question (1865)*

Coal was known to the ancient Britons and Romans, and it was used intermittently in Europe and Asia since at least the thirteenth and fourteenth centuries for heating, forges, and primitive steam engines [75], but the beginnings of an energy revolution began around 1600, when Britain began to mine and use coal on a larger scale [75, 19]. The first steam engine was patented in 1698 by Thomas Savory, and English coal mining accelerated in the latter 1700s, accompanied by the invention of the efficient Watt Engine. By 1800, 10% of the globe's commercial energy was provided by coal, as this fossil fuel began to displace wood [75]. Britain led the way, followed by continental Europe, and coal ushered in an age of iron and steam, powering the factories, rail networks, steam-powered fleets, and militaries of the age, in addition to providing heat for domestic use. Coal overcame wood as the world's chief source of energy by 1880 [75].

It is noteworthy that, from as early as several thousand years B.C.E. until about 1850, the continent of Europe underwent almost continual deforestation, driven by agricultural expansion and the demand for fuel- and construction-wood. With the industrial revolution and the shift to fossil fuels, the forests of Europe stabilized, and have in fact recovered appreciably since their nineteenth century nadir [76, 77].

Coal would itself be displaced by petroleum as the world's primary energy source by 1950 [75]. With electrification and the rise of oil as the globe's premier energy source, by the 1970s coal was rarely used residentially and electricity generation had displaced industry as its primary end-use, at least in the US. By 2011, 93% of coal use in the US was for electricity generation, with nearly all the balance industrial (EIA).

From the beginning of widespread electrification post-World War II, coal provided 45–55% of America's electricity, and indeed, as late as 2010, 45% of all US electricity was coal-generated (eGRID). In the last few years, driven largely by hydraulic fracturing of unconventional sources, natural gas supplies have expanded and displaced some coal generation, yet coal remains the single most important electricity feedstock, holding at 39% of generation in 2014 (World Bank).

There is also marked geographic variation in coal as a generating fuel. In a few states, no coal is used at all: in California < 1% of electrical generation comes from coal, but in much of the mid-west, south, and parts of the south-west, coal clearly remains king. In West Virginia, nearly 96% of all generation is derived from coal, per eGRID. Of course, electricity is widely sold and transmitted across state lines, so state-level generation does not accurately reflect the generation mix for customers in various interconnects (see Section 4.5 for regional emissions factors).

Evolution of Coal Grades

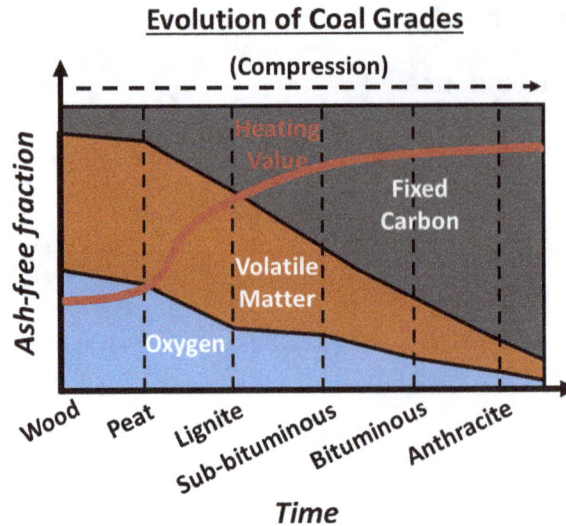

Figure 4.6: Qualitative schematic for the progressive transformation of woody vegetation to peat and, via burial and compression, through various coal grades. With time and pressure, oxygen and volatile matter content falls, leaving highly carbonaceous matter with a relatively high heat content (inset curve). Adapted partly from Figure 3.1 of [54].

Global coal resources are vast, with proven reserves estimated at 1,122 billion tonnes by the EIA (in 2014), while the total resource base may be over fifteen-fold higher [60]. In the US, the EIA reports a demonstrated resource base (DRB) of 433 billon tonnes, with technically recoverable reserves of 231 billion tonnes. While coal extraction and consumption has stagnated in the US, coal use in Asia expanded dramatically in just the last few years. In light of these recent trends, reports of coal's death may have been greatly exaggerated.

4.5.2   Coal properties

Coal derives, in the main, from ancient marshy forests, where decaying organic matter was buried in boggy, low-oxygen conditions to form peat. As the seas shifted, this peat would be buried under successive waves of ocean sediment. With time, pressure, and heat, water and oxygen are lost, as the peat breaks down into denser and denser carbon-rich material. Thus, we have a progression from wood/vegetation to peat, and then through successive coal grades: lignite, subbituminous, bituminous, and anthracite, as demonstrated in Figure 4.6. The geologically oldest anthracite is highest in carbon, and has the highest heating value of coal grades [54]; by virtue of its high carbon and low hydrogen content it is also the most carbon-intense coal grade (on a CO_2 per kWh basis) [51]. While coal can be found in every geologic strata, the greatest source is the carboniferous (named, in fact, in reference to its carboniferous coal deposits), an era of vast tropical and semi-tropical swampy forests spreading across the globe's landmass.

Coal is highly heterogeneous, and a large portion is composed of non-combustible minerals, or "ash." It is also contaminated by sulfur and mercury, and coal combustion causes serious locoregional air pollution, in addition to its high CO_2 emissions [54].

4.5.3 Lifecycle GHG emissions

- Coal is the foulest burning fossil fuel, and coal plants are relatively inefficient, emitting roughly 1 $kgCO_2$/kWh of electricity generated.

- Over 90% of coal lifecycle emissions are generated at the point of combustion, with coal mine methane the most significant upstream emissions source. Non-combustion lifecycle emissions sum to about 85 gCO_2e/kWh.

At the point of combustion, coal emits more CO_2 per kWh than any other fossil fuel. The EIA gives combustion emission factors (CEFs) of 0.3348, 0.3280, 0.3177, and 0.3520 $kgCO_2$/kWh(t) for lignite, sub-bituminous, bituminous, and anthracite coal, respectively, on the basis of thermal energy released, which gives us, assuming the average power plant thermal efficiency (TE) of 32.56% in 2012, emissions factors of 1.0283, 1.0074, 0.9757, and 1.0811 $kgCO_2$/kWh(e) on the basis of electrical energy generated. Thus, 1 $kgCO_2$/kWh(e) is an excellent rule of thumb for combustion emissions from coal generation. However, emissions also occur at other stages in the coal lifecycle, and many life-cycle EF estimates for coal-fired electricity have been published for multiple coal plant technologies.

While numerous upstream and downstream processes, including coal transportation, power plant construction, waste management, etc. contribute to coal's carbon footprint, nearly all emissions occur at four major points: (1) coal mine operations, (2) coal mine methane emissions, (3) coal transportation, and (4) direct combustion, with the last accounting for over 90% of the total. Of the processes upstream to combustion, coalbed methane from mines is the most important. A 2012 systematic review and harmonization of coal LCAs by Whitaker and colleagues [79] yielded a median EF of 1,030 gCO_2/kWh (IQR 1,000–1,090 g/kWh), using existing power-plant characteristics. This review estimated that 6% of life-cycle emissions are attributable to methane released during coal mining (point estimate of 63 gCO_2/kWh) and perhaps 2–3% from coal transportation. This work also determined the average CEF for 281 coal-fired plants in 2007 (using eGRID data) to be 970 gCO_2/kWh.

A widely cited study by Jaramillo and colleagues [78] estimated the coal CEF to be 975.0 gCO_2/kWh, and they estimated upstream emissions from coal to range from 39.0 to 78.0 gCO_2/kWh (mean 58.0 gCO_2/kWh), including CO_2 directly emitted from mining operations, methane emissions from coal mines, and the CO_2 emissions arising from coal transport[2]. Using 1997 data as reported by this group, 9.081 million barrels of fuel oil, 1.2 billion cubic feet of gas, 34 million gallons of gasoline, and 49.597 billion kWh of electricity were consumed in coal mining operations, which corresponds to emissions of roughly 36.43 million $MgCO_2$[3]. Just shy of 1 billion metric tons of coal were mined in 1997 (9.888×10^8 Mg), and using the reported heat content of 23.193 btu/kg coal, this corresponds to 2,188 billion kWh of electricity, yielding 16.64 gCO_2/kWh for mining operations-related emissions.

Data from the same work [78] also suggests coal mine emissions amounting to 31.10 gCO_2e/kWh, based on the 1997 EPA estimate of 75 million tons CO_2e; correcting from the EPA GWP for methane to IPCC values (i.e. using a GWP of 36 instead of 21) gives 53.31 gCO_2e/kWh. Coal transportation emissions sum to 15.33 gCO_2e/kWh, yielding total upstream emissions of 85.28 gCO_2e/kWh. The grand total of these point estimates is 1,060.28 gCO_2e/kWh. However, using the uncorrected mine methane number, we have 1,038.07 gCO_2e/kWh, which corresponds well to the median EF of 1,030 gCO_2/kWh reported by Whitaker et al. [79].

---

[2]These EFs are converted from Table 10S into gCO_2/kWh using a thermal efficiency of 32.56%

[3]Using conversion factors of 0.43 $MgCO_2$/bbl, 11.146 $kgCO_2e$/gallon gasoline, 5.26 $kgCO_2$/lb NG and 0.05 lb/ft^3, and 635 gCO_2/kWh for electricity (Jaramillo et al. value [78]).

A later work by the same group [80], using primarily 2010 data, suggests coal production, transportation, and coal mine methane emissions of 6.63 gCO_2e/kWh (95% CI 4.42–7.74), 14.37 gCO_2e/kWh (95% CI 2.21–35.38), and 72.03 gCO_2e/kWh (95% CI 7.58–227.47; corrected from 42.02 gCO_2e/kWh, using methane GWP of 36 vs. 21), respectively, and a CEF of 1,006 gCO_2/kWh. These numbers are derived from the original work assuming a coal plant thermal efficiency of 32.56% (EIA), and give an upstream total of 93.03 gCO_2e/kWh. Yet another study [81] examined lifecycle processes at the level of 364 coal plants in the US, in 2009, and estimated median emissions of 1,060 gCO_2e/kWh generated.

Briefly, we can also obtain a top-down estimate for coal mine methane emissions: the 2014 EPA GHG inventory estimated that 2,631 Gg of methane were emitted by coal mining operations in 2012, while 1,514.04 billion kWh of electricity were generated by coal-fired plants this year. Dividing, we get 62.56 gCO_2e/kWh over a 100-year time-frame (151.18 gCO_2e/kWh using 20-yr GWP), concordant with the studies reviewed above.

In sum, we may reasonably estimate about 7 gCO_2e/kWh, 15 gCO_2e/kWh, and 63 gCO_2e/kWh (151 gCO_2e/kWh using the 20-year GWP) for mining operation, coal transportation, and coal mine methane emissions, respectively, for a total upstream correction of 85 gCO_2e/kWh for coal-fired electricity, over the 100-year horizon (or 173 gCO_2e/kWh on a 20-year basis).

4.5.4 Surface mining, water, and land

As discussed in Section 4.6.3, unconventional gas and oil exploration have uncertain (but almost certainly at least marginally negative) effects on freshwater resources, and can also negatively affect local ecosystems. However, this must be compared to the dramatic, and unquestionably devastating effects that surface coal mining has on streams and ecosystems. Since 1975, coal production in central Appalachia has been increasingly via surface mining, with about two-thirds of coal extracted from surface mines in recent years. A common method is *mountaintop mining with valley fill* (MTM/VF), whereby whole mountaintops are cleared of forest, blasted away, and mined, while nearby valleys and streams are buried with the waste rock [82]. Valley fills can stretch for over a mile, and may be hundreds of meters deep [83]. Such mining is the major driver of land-use change in central Appalachia, and 5% of southern West Virginia had been converted to mine in 2005 [84].

Stream burial and drainage of mine waste clearly degrades regional water quality and deleteriously affects aquatic ecosystems and species. Large tracts of deciduous forest are lost directly to mining, while the forests that remain are highly fragmented. Furthermore, "reclaimed" mining sites support little woody vegetation or biodiversity, and sequester little carbon, even after many decades [83, 82]. The headwater streams buried by valley fills are ecologically sensitive and hydrologically important (as the primary water sources for most rivers), and their burial and contamination affects downstream river networks [84]. Such large-scale altering of the landscape topology also fundamentally alters the regional hydrology [83], and exposure to contaminated water and airborne dust in mined areas appears to negatively affect human health as well [82]. In sum, coal mining probably has regional environmental consequences far worse than any alternative fuel, except possibly tar sands oil (reviewed in Section 6.3.1), including unconventional natural gas extracted via hydraulic fracturing.

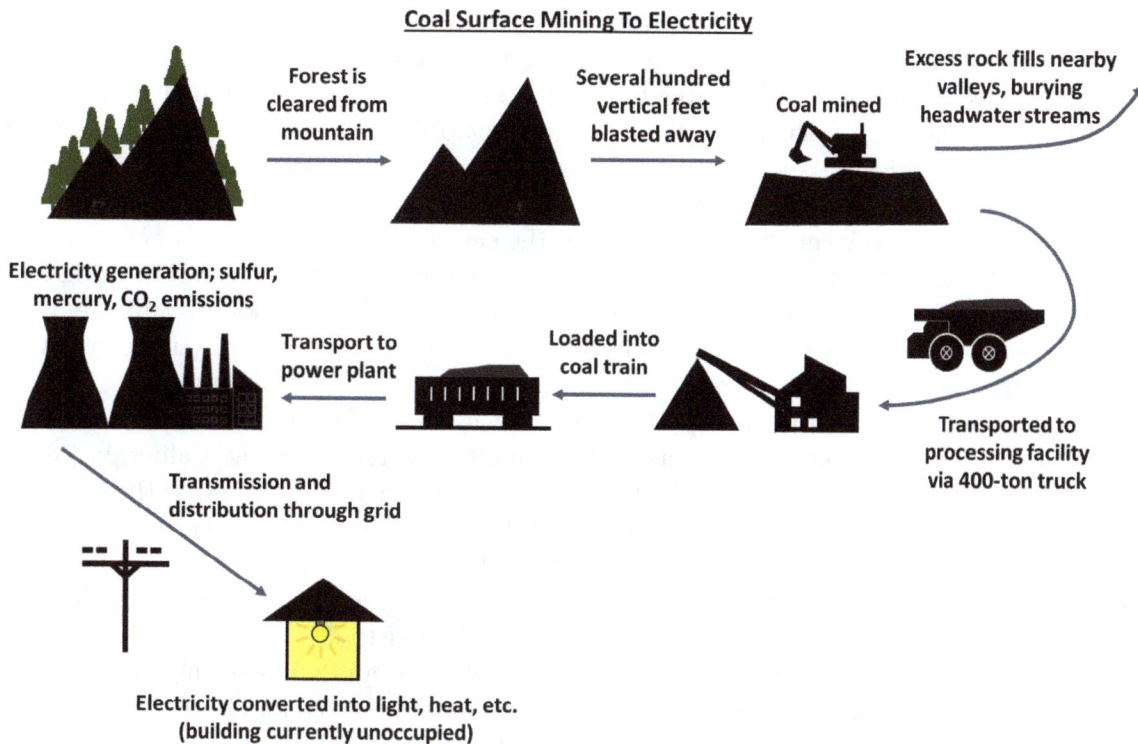

Coal Surface Mining To Electricity

Forest is cleared from mountain

Several hundred vertical feet blasted away

Coal mined

Excess rock fills nearby valleys, burying headwater streams

Transported to processing facility via 400-ton truck

Electricity generation; sulfur, mercury, CO_2 emissions

Transport to power plant

Loaded into coal train

Transmission and distribution through grid

Electricity converted into light, heat, etc. (building currently unoccupied)

Figure 4.7: Process for extracting coal via mountaintop mining and valley fill, and the downstream generation and delivery of electrical power.

4.6 Natural gas

4.6.1 Overview, the methane leak controversy, and overall GHG emissions

- Natural gas power plants generate only about 40–45% of the emissions of coal plants at the point of combustion, per kWh generated (427 gCO_2/kWh at 42.4% plant efficiency), but high upstream methane leaks, both at the extraction phase (for both conventional and unconventional gas) and distribution phase at least partially undermine this advantage.

- Methane is far more warming than CO_2, but also has a much shorter residence time in the atmosphere. If the leak rate is less than about 3–3.5%, natural gas for electricity is superior to coal over all timeframes, while it is superior at 100 years even for very high leakage (10%).

- The 2015 EPA GHG inventory suggests a leak rate of only about 1.2%, but a variety of observations and modeling studies suggest that methane inventories dramatically underestimate emissions, and a more likely leak range may be 2–4%; I take 2.4% as my central estimate. Including other upstream energy, this suggests upstream emissions on the order of 190 gCO_2e/kWh for typical gas generators.

As reviewed in Section 3.7, natural gas (NG) is formed at great depth, following the burial of organic matter (usually algae, phytoplankton, and other marine organisms) under multiple layers of ocean sediment. Under high heat and pressure, organic matter in low permeability *source* rocks, e.g. black shales, turns to gas and/or oil, from which it migrates to more permeable *reservoir* rocks, forming conventional reservoirs upon encountering a geologic *trap*. Conventional gas supplies, which continue to decline in the US, come from such reservoirs, while unconventional gas is extracted mainly via horizontal drilling and hydrofracking of the low permeability source rocks themselves (see Figure 4.10 for an illustration of the basic process).

By 2015, according to EIA data, about 70% of US natural gas (NG) production came from unconventional sources, i.e. shale gas, tight gas, and coalbed gas, with shale gas the dominant source. The EIA projects that unconventional gas supplies will continue to increase over the next few decades, while conventional supplies will decline even further. Natural gas has an intrinsically higher energy content than coal, and the carbon content per unit of energy is only 55% of coal's (0.3259 kgCO_2/kWh(t) for coal vs. 0.1809 kgCO_2/kWh(t) for NG, using EPA 2013 numbers). Moreover, modern combined cycle gas turbine (CCGT) power plants are much more efficient at converting heat to electricity than older steam generators or (single-cycle) gas turbines, which are comparable to coal-fired steam plants, in terms of efficiency.

EIA data indicates that the average thermal efficiency (TE) of a CCGT in 2012 was 44.8%, and the overall TE of gas generators was 42.4%. Similarly, a report by the California Energy Commission found a 2013 TE of 47.4% for CCGTs deployed in California, with an overall average TE equal to 40.0% for all gas generators. While somewhat inflated thermal efficiencies (TE) of over 50% are often cited for CCGTs in lifecycle analyses favoring gas as an electricity feedstock over coal, it is clear that, as deployed, gas generators are still far more efficient than coal plants, which had an average TE of only 32.5% in 2012 (EIA).

Expanding unconventional production, high energy density, and relatively high thermal efficiencies all have made NG-fired plants an attractive alternative to coal for electricity generation in recent years, and together these factors form the basis for the claim than NG can act as a relatively green "bridge fuel" to a low-carbon future. Indeed, a typical CCGT emits only about 404 gCO_2/kWh at the point of combustion, or almost 60% less than the 1,001 gCO_2/kWh emitted by a typical coal-fired plant, and converting all coal-fired generation in 2014 (1.58 billion MWh) to CCGT generation could thus save almost 950 million MgCO_2 at the point of generation annually, equivalent to removing about 200 million cars from the road (combustion emissions only), and avoid more emissions than we would by tripling all current wind and solar generation.

Yet, in 2011, a now rather famous analysis by Howarth and colleagues [85] concluded that using NG for electricity, derived from shale gas fracking, is actually *worse* for global warming than burning coal. This analysis had several important flaws, mainly that it compared coal and natural gas on the basis of primary energy, rather than delivered electricity (the main use of coal), and that it failed to account for coal-mine methane emissions. These problems aside, let us examine the fundamental claim, and the subject of much controversy, namely, that the fugitive emissions, or leakage, from natural gas systems increase the global warming potential of NG so much that it is as bad as or worse than coal.

As previously discussed, methane (NG is almost entirely methane) is an extremely potent GHG, especially over the short-term (100- and 20-year GWPs for fossil methane are, including climate-carbon feedbacks, 36 and 87, respectively [7]). Thus, natural gas leakage, which we typically express as a percentage of all gas produced, significantly reduces the relative (and absolute) advantage of NG vs. coal, or if it is great enough, can eliminate it altogether. Estimates for overall leakage, including leaks at both the production and distribution phases, range widely, from around 1–8%. As related in Section 4.6.1, natural gas is superior to coal over all time-frames if the total system leak is less than about 3–3.5%, and it is superior over the 100-year timeframe so long as the leak is less than about 10%, although such a high leak rate still yields extremely potent near-term warming. I now turn my attention to the long-running controversy in establishing this figure.

EPA estimates for the natural gas leak rate have changed dramatically over the years [86], and although EPA reports consistently give very small confidence intervals around the point estimate, this clearly cannot be valid, given these frequent and dramatic revisions. The still widely cited 2009 EPA estimate of natural gas leak corresponds to about 2.4% of US gas production [86]. The 2015 EPA inventory [10] gives 6.295 million Mg of CH_4 leaking from natural gas systems in 2013, which corresponds to a roughly 1.15% overall leak rate (based on EIA estimated 24,205,523 million cubic feet of NG production in 2013; using density of 0.05 lbs/ft^3 this is 598.97 million Mg CH_4).

The EPA estimate of NG leakage from natural gas systems is highly questionable, and it has been heavily criticized. It is largely based on a 1996 study conducted in collaboration with industry, with the 2014 inventory [640] reporting that adjustments were made on the basis of reductions in emissions reported to GasSTAR, a *voluntary* emissions reductions program, and from data on regulations. How these adjustments are made is not transparent. The EPA also relied on a 2012 industry study [641], which claimed much lower methane emissions than previous EPA estimates.

Another paper by O'Sullivan and Paltsev [97] argues that the *actual* methane emissions from bringing a shale gas well online are far less than the *potential* methane emissions. They assume that 70% of such emissions are captured, another 15% flared, and only 15% vented into the atmosphere, citing a 2012 EPA meeting as their source, but there does not seem to anything publicly available to support these numbers. This work also relies on the API/ANGA paper's [641] assertion that 93% of methane emissions from liquid unloading are captured or flared as its other primary source to support the assertion of high capture/flare rates. A recent study [91] that directly measured onsite well emissions at 190 shale gas sites, in collaboration with well operators, concluded that only 0.42% of gas leaked at the production phase (although additional leakage occurs at downstream transmission and distribution stages, perhaps 0.67% [88]). While these inventories and papers suggest overall leakage on the order of only 1%, almost all other lines of evidence indicate much higher leakage, as discussed now.

Several field studies have directly measured methane fluxes from NG fields, and have consistently measured leak rates far higher than those suggested by the EPA inventory. For example, airborne measurements over the Uintah Basin in Utah [90] suggested leakage of 6.2–11.7% of gas production, although this field is known to be a relatively high emitter. Caulton et al. [87] recently used aircraft to sample a 2,844 km^2 region in southwestern PA over the Marcellus Shale (by far the most productive play in the US) containing 3,438 wells (57.3% gas, 1.8% oil, 40.8% unknown). In this study, researchers were capable of sampling methane plumes corresponding to individual wells, and they estimated leakage at 2.8–17% of total gas production; a separate bottom-up inventory by the authors estimated 2.09–3.95% leakage. Even more concerning, a method based on remote satellite measurements by Schneising et al. [93] estimated NG leakage of 10.1±7.3% and 9.1±6.2% over the Bakken and Eagle Ford formations, respectively, both major shale plays. More modest (but still appreciable) leaks were inferred from 1-day aircraft-based measurements by Peischl et al. [94], who estimated production leakage of 1.0-2.1% over the Haynesville region, 1.0–2.8% from the Fayetteville region, but just 0.18–0.41% over a portion of the northeastern Marcellus Shale.

A comprehensive study by Miller et al. [25] of all anthropogenic methane emissions, using atmospheric methane measurements and weather modeling, estimated that US methane emissions are grossly underestimated by both the EPA and EDGAR, a global methane inventory. Furthermore, they found emission underestimation to be greatest over the south-central US, where there is a great deal of NG and other fossil fuel extraction. They concluded that methane from oil and gas operations is a factor of 4.9 ± 2.6 times too low in the EDGAR inventory and were heavily critical of the EPA's recent decisions to downgrade its estimates of oil and gas associated emissions.

Conforming with this conclusion, an extensive review by Brandt and colleagues [92] concluded that national inventories consistently underestimate methane emissions at all scales of measurement, a fact that may be partially explained by un-inventoried sources (such as abandoned oil and gas wells), using leak measurements taken on outdated equipment only from cooperating operators, and the strong likelihood that a small number of "superemitters," that would *necessarily* be underrepresented in inventories, account for a large fraction of total emissions. Various observations also show a 30% increase in US methane emissions from 2002 to 2014 [95], an era corresponding with dramatic expansion in shale gas (and tight oil) extraction.

It is finally important to note that unconventional gas sources do not necessarily suffer from higher fugitive emissions than do conventional wells. For example, a lifecycle analysis by Burnham et al. [88] estimated total methane leakage at 2.75% for conventional gas and 2.01% for unconventional, and Weber and Clavin [89] similarly estimated slightly higher lifecycle emissions for conventional versus shale gas. Still, a focus on shale gas is warranted, since such supplies are dramatically expanding, while conventional gas is nearing its twilight.

A direct comparison to coal

A key paper in the literature is that of Alvarez et al. [86], who compared the integrated warming over all times when using coal versus natural gas for electricity, and when replacing either light-duty or heavy-duty petroleum vehicles with NG-powered alternatives. They famously concluded that NG is better than coal at all time points if the leak rate is under 3.2%.

I perform a similar analysis, incorporating combustion and upstream emissions for both fuels, including coal mine methane emissions. The results are shown in Figure 4.8, and we see that, indeed, at a leak rate of around 3.5%, the short-term warming effect of methane is sufficiently powerful that NG has a stronger integrated warming effect for at least the first few years following combustion. However, only at a leak rate of about 10% or more does warming from NG exceed that of coal out to 100 years, although in this case the 20-year GWP for NG is about twice that of coal. At relatively low, but perhaps the most probable, leak rates, e.g. 2–4%, the 20-year GWP is about 65–95% that of coal, while the 100-year GWP is around 50–65% of coal's.

Bottom line and eGRID correction

Fugitive emissions from natural gas are uncertain, but likely on the order of several percent (say, plausibly between 1.5 and 5%). I conservatively use the 2.4% figure (based on the EPA's 2009 GHG inventory) used by Alvarez et al. [86] to derive adjusted emissions factors. I consider this a conservative estimate, despite the fact that the EPA has subsequently reduced their estimate of leakage rates, as the method is not transparent, and the downgrade is at odds with many recent studies that consistently suggest higher methane emissions than those tallied in national inventories. Natural gas for electricity generation almost certainly has a lower climate impact than does coal over the 100-year horizon, and most likely at the 20-year horizon as well, but is still a powerful CO_2e source overall.

Finally, using 1 gC/MJ (i.e. 13.2 gCO_2/kWh) for gas extraction operations [85], and at the 2012 EIA average TE of 42.4%, I calculate 31.13 gCO_2e/kWh for indirect CO_2e emissions from industrial activity associated with well operation. Therefore, for eGRID emissions, I add a crude correction of 158 gCO_2e/kWh + 31 gCO_2e/kWh = 189 gCO_2e/kWh to all gas-fired generating capacity.

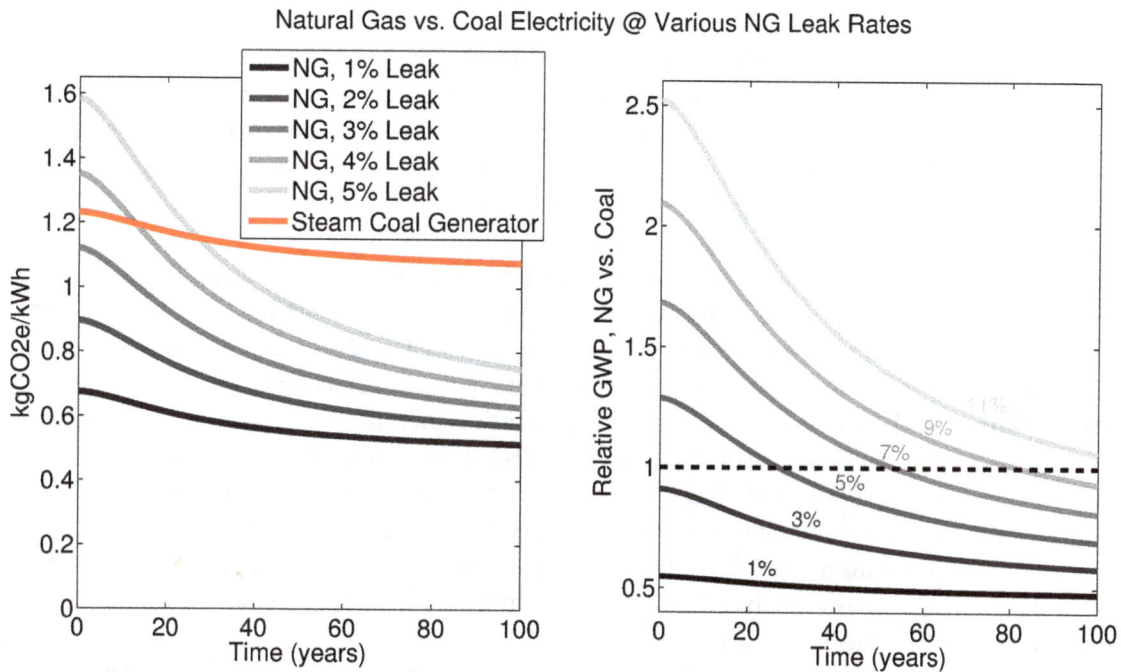

Figure 4.8: Lifecycle CO_2e emissions factors for natural gas and coal-fired electricity, under different NG leak rates (and including likely coal mine methane emissions). The left panel gives the absolute time-dependent $kgCO_2e/kWh$ emissions factors for coal and natural gas at several relatively modest leak rates, while the right panel shows the relative NG to coal GWP over a broader range of NG leak rates.

4.6.2 A bridge to nowhere? What is the wisdom of a fracking ban?

- Compared to restricted supplies, abundant, inexpensive natural gas from fracking is generally projected to displace some coal for electricity generation, but also increase total energy use and compete with renewable and nuclear energy, and have a very minor effect on total greenhouse gas emissions to the year 2050, *in the absence of major energy policy changes.*

- Natural gas from fracking will not reduce carbon emissions, and could increase them, unless it is used as one component of a rational, guided transition to low-carbon energy. However, restricting fracking is also likely to have only a very small benefit at best, if done in a larger policy vacuum.

- Combined with renewable portfolio standards (RPS) (mandated renewable energy targets) abundant natural gas can reduce overall carbon emissions; this is noteworthy as RPSs are widespread at the state level in the US.

- Because of its relatively low CO_2e/kWh (lifecycle) emissions factor, individual use of natural gas is still reasonably advisable for certain domestic applications (such as space heating).

There are two basic competing narratives on the role of natural gas in general, and natural gas derived via fracking from unconventional sources in particular, with respect to the world's energy future, and both largely focus on the role of NG in electricity generation. The first posits that NG is a bridge fuel to a renewable future that will act to displace coal for electricity generation in the short-term, while powering more flexible, load-following power plants that both complement renewables (that is, NG plants can cycle up or down more quickly than coal plants to meet the more variable loads associated with intermittent renewables [98]) and produce less than half the CO_2 of coal.

The counterargument, essentially that natural gas is a bridge to nowhere, has until recently largely turned on methane leakage from natural gas systems (as reviewed above), with this leakage undermining NG's relatively low combustion emissions. It is also argued that cheap natural gas, rather than (just) displacing coal, will also outcompete renewables, and furthermore, expanding NG generating capacity and other infrastructure represents infrastructure "lock-in" that will commit energy systems to fossil fuels for decades to come.

As examined at length above, the leakage is highly uncertain, but overall CO_2e emissions, per unit electricity produced, are most likely lower than coal. Further, NG has apparently reduced coal use in the US: the reduction in coal's share of US electricity generation from 52% in 2000 to 38% in 2014 can be attributed almost entirely to displacement by natural gas. Over the same time period, however, total electrical generation increased 12%, and combined lifecycle emissions from coal and NG electricity generation also ticked *up* about 5%. Thus, the specter of supply-induced demand is raised. It is also important to remember that gas-fired space, water heating, and cooking technologies will also be affected by the availability of NG, and are all generally less carbon-intensive (on a life-cycle basis, and even for relatively high methane leak rates) than the electric alternatives (except heat-pumps for heating). Industrial heating, too, is a major NG end-use affected by availability.

The issue, it would seem, runs much deeper than simply the CO_2e/kWh emissions factor, and we must look at the effect of cheap gas on the whole system-scale carbon emissions, as abundant natural gas is likely to have pervasive effects on world energy markets that could counterproductively increase total energy use (for heating, industrial use, and electricity generation), as well as encourage investment in new NG-based energy infrastructures instead of low-carbon alternatives, undermining (or at least delaying) a transition to low-carbon electricity. Overall, projections [99, 100, 101, 103] of economy-wide energy use under different levels of NG availability have generally concluded that increasing NG supplies *will indeed* slightly increase overall energy use, promote NG over other fuels, primarily coal but, to a lesser extent, nuclear power and renewables, and on balance could either slightly increase [100, 101] or decrease [104, 103] total carbon equivalent emissions, with an increase seeming the more likely outcome.

In one of the more masterly efforts at an integrated prediction, McJeon and colleagues [100] (see also the commentary in [102]) recently evaluated the effect of an "abundant" gas scenario (NG extraction at the likely upper bound of unconventional resources) vs. "conventional" gas scenario (NG extraction limited to conventional supplies) to the year 2050 using five global-scale models used to project energy use, and determined that, in the absence of new political policies, abundant gas would increase total energy use significantly across sectors, but have little effect on total CO_2e emissions. In fact, under most scenarios and models, abundant gas slightly increases global radiative forcing, and while it primarily displaces coal for electricity production, it also decreases renewable generation as well. The major caveat is that these are projections that assume no new energy policies, but importantly, demonstrate that in the unfettered market, fracking and increased exploitation of natural gas will not avert, and may slightly exacerbate, climate change due to market dynamics beyond the CO_2e/kWh metric.

Another very recent modeling effort by Lenox and Kaplan [101], using the EPA MARKAL database (specific to the US energy system), suggested that high availability of NG would slightly increase total electricity production by 2050 and incur higher total carbon emissions compared to a lower NG availability scenario, with this true *across a range* of upstream methane leak rates. Further, the low availability scenario would see greater deployment of renewables.

A general conclusion that can be drawn is that if the market is left to its own devices, increased availability of gas will accelerate the phase-out of coal-fired power, but delay the adoption of renewables [104], with little overall benefit to the climate. Given this, one policy measure that could synergize to decrease emissions with fracking would be a federal renewable portfolio standard (RPS), i.e. mandated implementation of renewable energy, forcing gas to compete mainly with coal instead of renewables. Even then, however, Shearer et al. [104] projected an RPS would yield only a small difference between high and low gas scenarios (in favor of the high gas scenario), using the EPA MARKAL model. Rather, these authors found that the existence of any carbon reduction policy was vastly more important than the relative availability of NG, with an RPS, carbon tax, or strong emissions reductions requirements all markedly reducing emissions regardless of NG abundance (the RPS scenario was the only one to meaningfully favor high NG).

Note that the conclusions of these analyses are actually quite insensitive to the methane leak rate, with the general conclusion that any climate benefit is equivocal even under very low leak rates [104, 101, 100], while higher leak rates would modestly exacerbate the warming potential of fracking [101, 100].

We can conclude that unconventional natural gas, if it is a bridge to a low-carbon future, is a narrow and treacherous one, but the analyses discussed here suggest this has more to do with the fact that, as an inexpensive fossil fuel, NG will be used on a large scale if available, than it does the controversy over the methane leak rate: the conclusion would be essentially unaltered even without upstream methane leakage. In the absence of a larger policy framework, or widespread voluntary rational use of this resource, shale gas fracking will probably not help the climate and may hurt. It is important to understand, however, that in the absence of either of the former, *with or without* fracking we face climate disaster. It is my conclusion that there would be, from a climate perspective, only a small benefit from a fracking ban, unless it is part of some broader carbon control policy. Under strong renewable mandates, a fracking ban could be slightly counterproductive.

Since the most rational use of NG as a bridge fuel would be as a short-term replacement for coal followed by a rapid phase-out, I offer my speculation that this might best be accomplished by mandated renewable energy targets to force NG to compete primarily with coal (and not renewables), combined with either a carbon tax or carbon cap.

From the individual's perspective, natural gas made abundant by fracking may be rationally used in the short-term for appropriate applications, such as space heating, *so long as one does not take its low price as a license to use it in excess.*

4.6.3 Fracking and water

Perhaps the most contentious public debate surrounding fracking is the one concerning water and this technology: concerns include water used for well development in arid regions and, especially, possible contamination of fresh groundwater supplies, as anyone who watched the 2010 film *Gasland* can attest. However, due to a lack of transparency from industry, the fact that groundwater may suffer methane contamination from both natural and anthropogenic processes, and wide variability in the geologic features of particular shales, it is difficult to reach broad conclusions concerning fracking and water supplies [106]. It is clear that fracking does pose more than a theoretical threat to water supplies, but the overall impact may be small relative to the current scale of oil and gas exploration via fracking; the greatest concern is probably not gas migration in the subsurface, but rather the fate of fracking wastewaters at the surface [108]. It must also be noted that conventional gas and oil exploration can cause similar problems, and conventional gas wells actually generate more wastewater per unit of gas produced than unconventional wells [108].

Fracking may affect freshwater resources via several mechanisms [105]: (1) contamination of shallow freshwater aquifers via migration of stray gas that is mobilized by high pressure

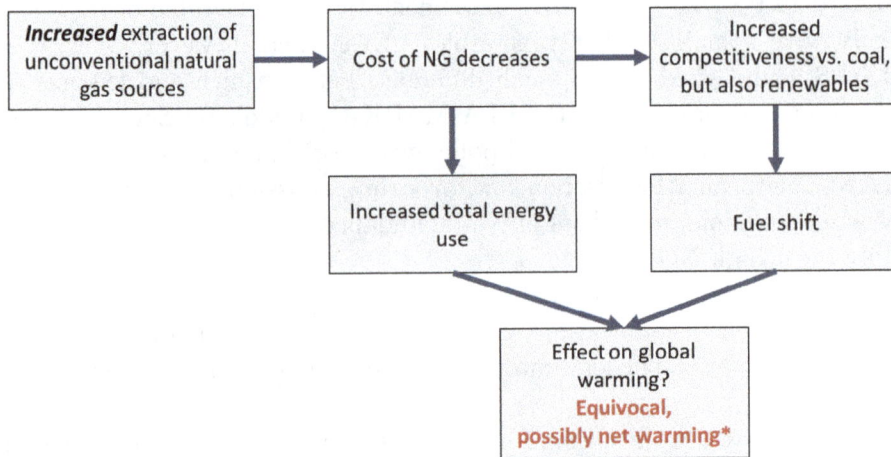

Figure 4.9: Outline for the market-driven outcome of increased extraction of unconventional gas resources, in the absence of any broader policy changes.

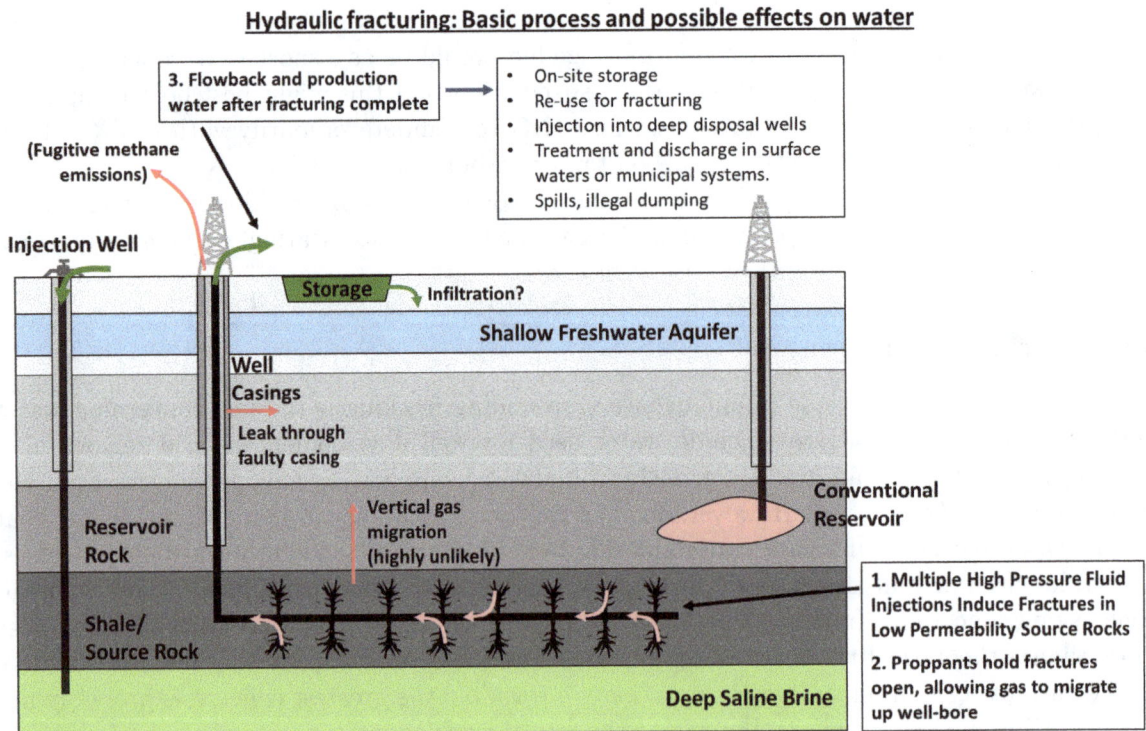

Figure 4.10: Schematic for the basic fracking process, along with some of the possible effects on ground and surface waters.

hydrofracking, or from leaking through faulty well casings, (2) contamination of surface and groundwaters via spills, leaks, or improper disposal of fracking wastewaters, and (3) accumulation of toxic elements near fracking sites. Stress on local water supplies is a fourth mechanism, and while the raw numbers involved in bringing a well online are impressive, on the order of 2–13 millions gallons per unconventional well [105], in the aggregate water used for fracking is but a minuscule fraction of freshwater consumption, accounting for only 0.2% of all water use in Pennsylvania (overlying much of the heavily exploited Marcellus Shale) [108], and less than 0.5% of water use in Texas and several other southwestern states [107]. Still, (sometimes illegal) acute water withdrawals can and have damaged local ecosystems, such as ecologically sensitive creeks [106].

There is evidence that, in the Marcellus Shale underlying northeastern PA, faulty well casings and/or faults induced by fracking have caused some groundwater contamination by methane and possibly by deep saline water as well [105]. Other shales seem less affected, and naturally occurring gas contamination is also common in northeastern Appalachia, so attribution remains controversial [105]. What is less uncertain is that flowback and production fluids (a mixture of water, proppants, other additives, and deep saline groundwater) that return to the surface once a new well is depressurized can contaminate local waters, either when leaching from local storage pools to groundwaters or via accidental, illegal, or even *authorized* discharge into local surface waters or municipal wastewater streams. Dealing with this highly saline wastewater is a major long-term challenge [108]. For now, it is either recycled for new fracturing operations (although this cannot be sustained indefinitely), disposed of in deep injection wells, treated and discharged, or even re-used on roads as a dust suppressant or for de-icing [105]. Some of these issues, and the general fracking process, are illustrated in Figure 4.10.

4.6.4 Induced earthquakes

Several mid-continental areas, most notably Oklahoma, have seen a sharp uptick in seismic activity (defined as frequency of magnitude 3 or greater earthquakes) in the last few years, and this increase is linked to increased oil and gas extraction in these areas. Hydraulic fracturing, *per se*, routinely (and by design) induces microearthquakes, but felt earthquakes from fracking are vanishingly rare in the US [109]. Highly saline wastewater is produced by all oil and gas wells, whether they were fracked or not, and this is routinely disposed of via deep well injection. Injection of water for *enhanced oil recovery* (EOR) can sweep oil towards production wells, and increases well pressure [109]. These latter two activities, wastewater injection and EOR, are more likely to induce felt earthquakes (although this is still extremely rare), and several earthquakes that caused damage have been linked to such activity, such as a 2011 magnitude 5.6 earthquake in OK [110].

Overall, induced seismicity related to well injections, while rather dramatic, has rarely caused felt earthquakes and even more rarely caused damage, and may be mitigated, for example, by well selection and coordinating high-volume injections [110]. Compared to the other externalities associated with the fossil fuel industry, this one seems relatively minor.

4.6.5 A note on natural gas as transportation fuel

While natural gas has been widely promoted as a clean transportation fuel, especially for buses, this notion is, at best, highly exaggerated. For heavy-duty vehicles, the lower efficiency of spark ignition gas engines compared to diesel engines, uncombusted methane escaping the tailpipe, and of course upstream methane leaks, together imply that these vehicles are probably marginally *worse* than the diesel-powered alternatives [86]; for further details see the discussion of natural gas buses in Section 9.1.2.

The lifecycle emissions of NG fueled passenger cars are probably very similar to those of comparable gasoline vehicles. At best, they are perhaps 3–10% less warming over a 100-year time frame, at the cost of increased warming early in time [86]. Gasoline-powered hybrid-electric or electric drivetrains, on the other hand, may generate 40–60% fewer emissions than comparable conventional vehicles.

4.7 Oil

Oil, being far more valuable as a transportation fuel, is increasingly being phased-out as a fuel for electricity generation, and was responsible for only about 0.7% of US electricity generation in 2014. There is little literature directly addressing the upstream emissions from fuel oil for electricity generation, but a 2004 study [111] estimated indirect emissions of 97 gCO_2e/kWh for an oil-fired power plant in Singapore, and a later review by Weisser [112] cites a range of 40–110 gCO_2e/kWh for upstream emissions. Wheel-to-pump emissions for diesel are estimated at 57.3 gCO_2e/kWh(t) by the GREET 1 model (on a HHV basis, see Section 6.3), which is essentially identical to fuel oil. Given an average TE of 31% for oil-fired plants in 2012, this suggests a correction of about 185 $kgCO_2e$/kWh.

4.8 Nuclear energy

Nuclear power plants generated about 19% of US electricity in 2014, making nuclear, *by far*, the largest source of near-zero carbon energy in the US: hydropower, wind, and solar together accounted for only 11% of 2014 generation. Globally, nuclear power was responsible for 11% of total generation (compared to 17% for hydropower), in 2012 [114]. Nuclear power has, as the reader is no doubt aware, been highly controversial among environmentalists, and it is claimed with some frequency that nuclear is not truly low-carbon, given the emissions associated with uranium mining, enrichment, etc. While nuclear lifecycle carbon emissions are non-zero, a large scientific literature (reviewed in a moment) shows that they are comparable to those of solar power, at around 25–50 gCO_2e/kWh, and calls to phase-out nuclear energy must be viewed in light of these facts. For example, replacing all 2012 nuclear generating capacity with natural gas (at a TE of 42.4%, and 2.4% upstream methane leak) would increase annual US electricity-associated emissions by about 16.9%, or 429.4 million $MgCO_2e$ (and equivalent to 73.3 million cars).

4.8.1 Nuclear plant operation

Let us first briefly review the basic principles of nuclear reactor operation, and the nuclear fuel cycle; for the interested reader, far more detail is available in the technical but readable introduction to nuclear energy by Murray and Holbert [113]. In essence, a nuclear reactor consists of a core, where enriched uranium-235 (U-235) contained in fuel rods undergoes a controlled fission chain reaction, generating heat. A coolant circulates through the core, taking up heat, and then in turn either (1) directly drives a turbine, or (2) passes through a heat exchanger, generating steam in a secondary fluid to drive a turbine and hence generate electricity. Light water reactors (LWRs), the only type currently in commercial operation in the US, use "light" water (i.e. normal water, as opposed to "heavy" water containing deuterium instead of hydrogen) as coolant. When the heated water directly drives the turbine, this is a boiling water reactor (BWR) (about one-third of US nuclear plants), while a system with a heat exchanger is a pressurized water reactor (PWR) (about two-thirds of US nuclear plants).

Note that unused waste heat must be rejected from the coolant water, often with the massive cooling towers frequently mistaken for reactors in the popular imagination. Thus, a nuclear plant is very similar in its broad operation to a coal plant, just with a core burning uranium via fission replacing the coal boiler.

Looking closer at the nuclear reaction, the process is initiated when a free neutron strikes a U-235 atom and is absorbed by the nucleus; the resulting U-236 atom is in an excited state, containing more energy than U-236 in ground-state. This extra energy is released via fission, yielding two fission atoms (which vary widely, but are generally heavy elements), several high-energy ("fast") free neutrons, and heat, as shown in Figure 4.11. So long as at least one neutron is released per fission event, a chain fission reaction may be sustained.

At this point, we must distinguish between "fast" (high-energy) and "thermal" (slow, low-energy) neutrons, and fissile vs. fissionable material. *Fissile* materials readily undergo fission in response to thermal neutrons, while any nucleotide that undergoes fission is *fissionable*. There is but one naturally occurring fissile isotope, U-235 (there are several "artificial" fissile isotopes, mainly Pu-239 and U-233); all other fission requires fast neutrons, with U-238 and thorium-232 (Th-232) common fissionable fuels. Now, enriched uranium fuel is 3–5% U-235, with the remainder U-238. Fast neutrons released by U-235 fission can be absorbed by U-238 nuclei, and then, via a series of radioactive decay events, give rise to plutonium-239 (Pu-239), which is itself fissile (see Figure 4.11).

Thus, a nuclear reactor can "breed" more fissile material (Pu-239) from the "fertile" U-238. We now have the requisite knowledge to distinguish between *thermal* and *breeder* (or *fast*) reactors. U-235 interacts more readily with thermal neutrons, and a "moderator" can be used to slow the fast neutrons released by U-235 fission, yielding mainly thermal neutrons that go on to interact with U-235 until all is consumed, but leaving most U-238 intact. All LWRs in the US are of the thermal class, use water both as coolant and moderator, and ultimately burn only a small fraction of the fuel uranium (most of the U-235, plus a small amount of converted U-238). Breeder reactors, on the other hand, lack a moderator, and thus rely more on fast neutrons. Since each U-235 fission event releases over two neutrons, on average, there are sufficient neutrons to both breed U-238 to fissile Pu-239, and to sustain the fission chain reaction. The Pu-239 then serves as the source of ongoing fission, until the entire uranium fuel base is consumed. Note that we only refer to a reactor as a breeder if more fissile material is produced than consumed. Thorium, which can be converted to fissile U-233 via absorbtion of a neutron (similar to the process producing Pu-239 from U-238), represents an alternative fuel base for breeder reactors.

Breeder reactors utilize fuel over 50 times as efficiently as thermal reactors, using nearly all fuel and dramatically cutting waste. Breeder reactors can also burn high-level nuclear waste from thermal reactors as fuel. They have not been widely deployed largely because of higher capital and operating costs compared to thermal reactors, and because of nuclear proliferation concerns (uranium fuel represents only 5% of the cost of nuclear power, and thus there is little financial incentive at present for fuel-efficiency).

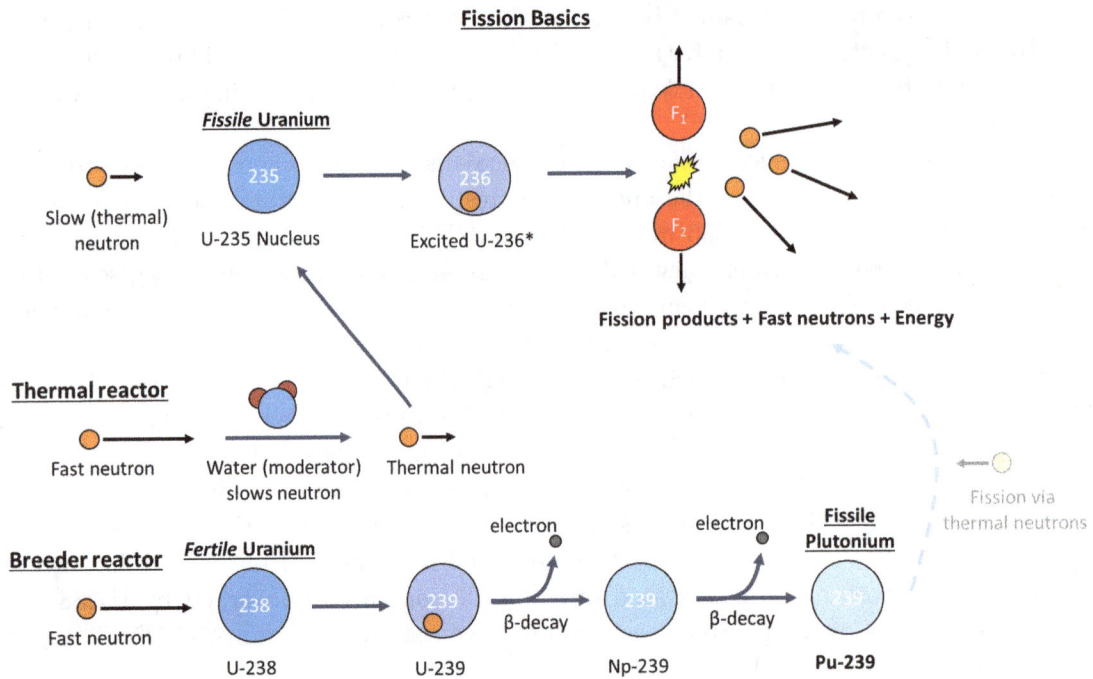

Fission Basics

Figure 4.11: Schematic for the basic process of nuclear fission induced in a *fissile* U-235 atom by a thermal (slow) neutron, to yield fission products, fast neutrons, and energy. Fast neutrons can either be slowed by a moderator (water, in most reactors) to induce further U-235 fission (thermal reactors), or may interact with *fertile* U-238 atoms to yield fissile plutonium (breeder reactors).

4.8.2 The fuel cycle

- Uranium from low grade natural ores is milled to U_3O_8, enriched such that the U-235 isotope concentration increases from 0.7% to 3–5%, and burned in power plants to yield spent fuel consisting of unburned uranium, plutonium, and fission byproducts.

- Reprocessing of spent fuel (as is done in France and several other countries) decreases fresh fuel use by 25% and greatly decreases the amount and radioactivity of the waste.

- The once-through fuel cycle employed in the US stores spent fuel without any recycling of plutonium, yielding waste that remains radioactive for hundreds of millennia.

Uranium, primarily as triuranium octaoxide (U_3O_8), is mined from naturally occurring deposits that vary widely in ore grade, from as little as 0.01% to 15% U_3O_8, with average ores around 0.1–0.15% U_3O_8 [115, 117]. At processing plants, the raw ore is milled to *yellowcake*: concentrated U_3O_8. Since only 0.7% of naturally occurring uranium is the U-235 isotope (99.3% U-238), it must be enriched for use in most reactors, to 3–5% U-235, a very energy-intensive process involving conversion of U_3O_8 to gaseous UF_6, and the major source of greenhouse gas emissions in the fuel cycle. This also leaves large stocks of depleted uranium (U-238) as a byproduct. Fuel fabrication follows, with enriched UF_6 converted to UO_2, which is assembled into fuel rods. After some months of fission, most of the U-235 has been burned, leaving radioactive fission by-products, large amounts of U-238, and a variety of radioactive plutonium isotopes (and other minor actinides) in the *spent fuel*. The plutonium and minor actinides are

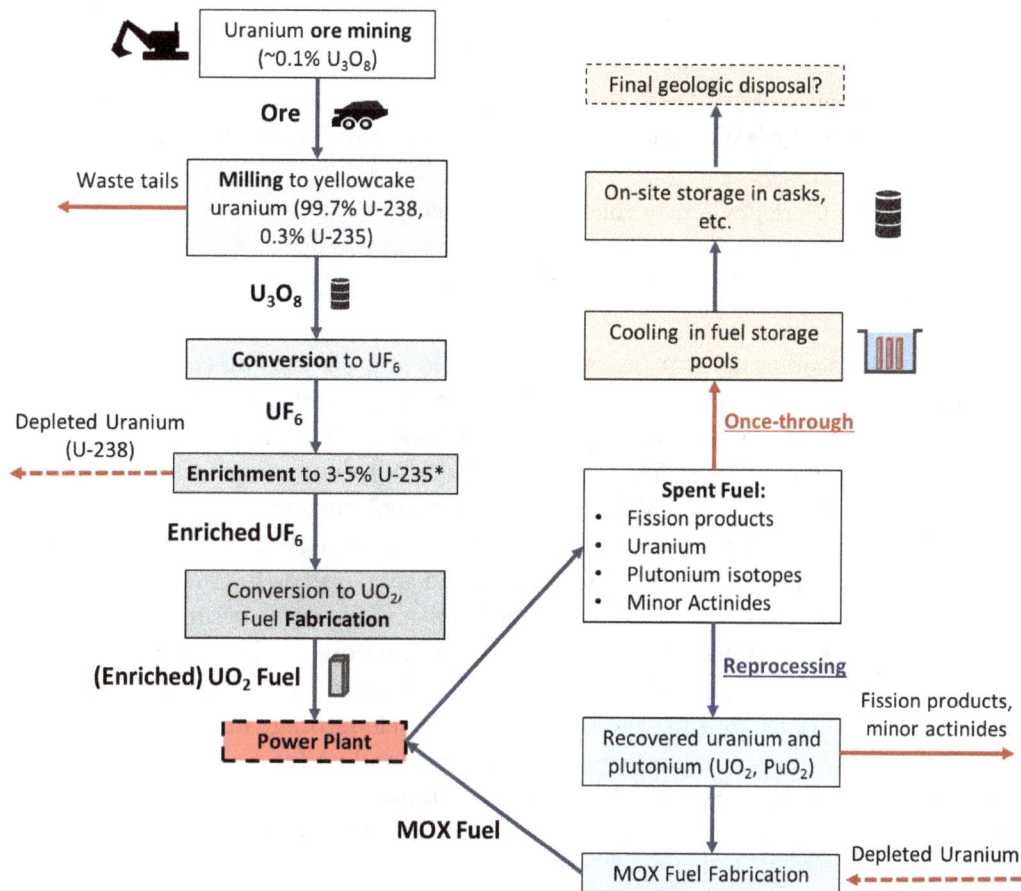

Figure 4.12: Basic uranium fuel cycle for thermal reactors. Spent fuel may either undergo reprocessing and be recycled into the fuel cycle (France, Russia), or disposed in the once-through cycle (USA).

by far the dominant contributors to long-term radiation hazard [113].

The fate of spent fuel varies by country. In the US, the fuel cycle is "once-through," and, after cooling on-site in water tanks for several years, spent fuel is packaged, stored on-site, and currently awaits final long-term geologic disposal. In many countries, most notably France, spent fuel undergoes reprocessing, where uranium and plutonium are separated from the useless fission products, combined with depleted or slightly enriched uranium to produce mixed-oxide (MOX) fuel (UO_2-PuO_2), and then mixed with fresh fuel. This reduces fresh fuel requirements by about 25%, reduces the volume of high level waste by about 85%, and, because long-lived plutonium isotopes are removed from the waste, waste radioactivity decreases by several orders of magnitude and largely dissipates within 100 years, as opposed to over 100,000 for once-through waste [116].

Potential fourth generation fast breeder reactors would recycle both plutonium and minor actinides, use fuel vastly more efficiently than thermal plants using either a once-through or plutonium-recycling fuel cycle, and would only generate small amounts of short-lived radioactive waste. Note that this is not a hypothetical technology, but has in fact existed for decades, and multiple operating fast reactors currently exist.

4.8.3 Greenhouse gas emissions

- There is significant variation in the literature, but most estimates for nuclear lifecycle emissions are less than 50 gCO_2e/kWh (amortized over a 30–40 year reactor lifetime), comparable to solar energy.

- Declining uranium ore grade may roughly double lifecycle emissions within 50 years, although this will still represent low-carbon energy.

There are myriad studies attempting to quantify the lifecycle carbon emissions from nuclear power, as reviewed by various authors [120, 117, 118, 119], and, for LWRs, estimates range from practically zero gCO_2e/kWh to over 200 gCO_2e/kWh; relatively recent reviews give mean estimates of 60 [120]; 66 [117]; 8, 58, or 110 gCO_2e/kWh depending on assessment method [118]; and a median of 12 gCO_2e/kWh in the most recent review [119]. The distribution of estimates is highly skewed to the right, with most studies estimating lifecycle emissions under 50 gCO_2e/kWh [117, 121]. Therefore, the overall mean may overstate true emissions.

Broadly speaking, emissions sources are (1) the front-end fuel cycle (uranium mining, milling, enrichment, and fuel fabrication), (2) power plant construction, (3) plant operation and maintenance, (4) eventual plant decommissioning after 30–40 years (sometimes including mine site clean-up and rehabilitation), and (5) the back-end fuel cycle (storage and disposal for the once-through fuel cycle). All phases require significant energy, and power plant construction requires on the order of several hundred thousands tonnes of carbon-intensive materials (e.g. over 150,000 tonnes of concrete) [117], but the comparative impact of each category varies significantly by source [117, 118].

One aspect worthy of special consideration is the influence of uranium ore grade on emissions, as average ore grade is likely to decline in the future [121]. Warner and Heath [119] estimated that lifecycle emissions could increase by 55 to 220%, while a more recent study by Norgate et al. [121] projected average emissions to increase from 34 gCO_2e/kWh to 60 gCO_2e/kWh over the next 50 years, assuming a concomitant decline in average ore grade from 0.15% to 0.01% U_3O_8. For my corrected eGRID emissions estimates, I have somewhat conservatively chosen 58 gCO_2e/kWh (middle estimate from [118]) as my correction for nuclear power.

4.8.4 Safety and health effects contra other power generation

While major nuclear disasters and waste toxicity are rightfully of concern, out of all major electricity sources, nuclear is associated with the *fewest* excess deaths per kWh of electricity generated [122], whereas coal, primarily via increased cardiac and respiratory mortality related to particulate emissions, causes several tens of thousands of deaths yearly in the US [124] and several hundred thousand deaths worldwide. On a per kWh basis, coal causes perhaps 500–1,500 times more deaths than nuclear [122]. Furthermore, nuclear power has likely acted to significantly displace coal as a power source historically, and on this basis Kharecha and Hansen [123] estimated that nuclear power has saved 1.84 million lives since its inception.

4.8.5 Is nuclear power renewable?

Uranium is a finite resource, and therefore nuclear power, while low-carbon, is not renewable. Under current rates of uranium use (almost universally using the once-through fuel-cycle) known and inferred uranium resources will last around 200 years. Once we take into account projected growth in demand, see e.g. EIA projections as discussed by Mudd [115], or assuming annualized growth of 1.9% [121], without improvements in efficiency reserves may last less than 100 years, and uranium supplies would necessarily peak before exhaustion of the fuel base. However, reprocessing of nuclear waste can reduce overall fuel requirements by 25% [116]. Breeder reactors, which have not been widely deployed largely because uranium proved to be far more abundant than initially expected, use but 1% the fuel consumed by current thermal reactors. Seawater represents a nearly infinite uranium source, although it is not currently economical to extract it, and thorium can also serve as a nuclear fuel, with the thorium cycle offering multiple potential advantages over uranium.

Since breeder and thorium reactors both currently exist, and it is economics (and proliferation concerns, in the case of breeder reactors) rather than technological barriers that have prevented their widespread adoption, nuclear power *could* act as a quasi-renewable technology, supplying power for many thousands of years (and possibly millions, if uranium could be extracted from seawater or other unconventional sources at large-scales).

4.9 Hydropower

- Hydropower is a low-carbon energy source overall, but has higher lifecyle emissions than other low-carbon energy sources, including solar, wind, and nuclear.

- The decay of flooded vegetation in reservoirs leads to significant carbon dioxide and methane emissions, and hydropower in tropical regions may actually be much worse than coal in terms of lifecycle CO_2e emissions.

- Large-scale damming of the world's river systems has other serious consequences for biodiversity, water quality, and the global hydrologic cycle.

Large-scale hydropower generated about 6.1% of all US electricity in 2014, making it the second-most important source of low-carbon energy, after nuclear. Globally, however, hydropower is more significant, at 17% of generation [114], and while there was little new installed capacity in the last several decades, there has been a recent rush to expand hydropower in emerging economies, especially in Amazonian regions, China, and southeast Asia, with a focus on mega-dams, and thus global hydropower capacity is projected to increase 73% in the next 10–20 years [130]. This is extremely controversial, as large-scale damming of over half the world's major river basins already profoundly affects the global hydrologic cycle with severe adverse effects on riverine and terrestrial ecosystems, biodiversity, and water quality. Dam mega-projects also displace large human populations, and catastrophic dam failures have resulted in thousands of deaths [125]. Finally, since tropical hydropower generates significant CO_2e emissions, it is unclear what benefit, if any, this hydropower boom will have for the climate [126, 127].

While there are no direct GHG emissions[4] associated with hydropower, flooding of vast areas and subsequent anaerobic decay of vegetable matter is a significant source of methane and CO_2, especially in tropical regions (where much new hydropower development is slated to occur), and it is likely that the global warming effect of at least some tropical hydropower is worse, and perhaps far worse, than coal [126], although hydropower in the temperate and boreal regions of North America is more benign. Other more minor lifecycle sources of carbon include dam construction, which requires large amounts of cement and other materials, and dam decommissioning at the end of life, which may release carbon from reservoir sediments.

---

[4]Although we could arguably consider the degassing of CO_2 and CH_4 from water passing through dam turbines as a direct emission, this is itself a consequence of anaerobic decay in the dam reservoir.

4.9.1 Greenhouse gas emissions from hydropower reservoirs

Dams and the resulting reservoirs alter carbon dynamics in several ways. Most importantly, dams flood wide expanses of land, submerging large amounts of vegetation (especially in former forests), and for the first 10–15 years after flooding, great quantities of CH_4 and CO_2 are released as this labile carbon pool degrades [129]. Detailed measurements of several individual reservoirs have shown these early emissions to decrease until a steady-state is reached, with further CO_2/CH_4 emissions coming from new biomass either growing in the reservoir or entering from outside (e.g. via rivers or runoff). The deep water of reservoirs is often anoxic, promoting anaerobic degradation of organic matter that favors methane production.

There are several pathways for greenhouse gas (CH_4 and CO_2) release from reservoirs. Most simply, these gases diffuse from the reservoir surface at the water-air interface. Second, bubbling is a major source of methane (methane has very low solubility in water, hence the formation of bubbles), where CH_4 produced in deep anaerobic sediments bubbles to the surface. Third, deep water passes through turbines near the base of dams. Since this deep water is cold and under high pressure, it can be supersaturated with respect to CH_4 and CO_2, and thus we have degassing from the turbulent flow through turbines and further on downstream. The latter two sources are significant, but have not been included in many studies [129], and thus GHG flux from reservoirs has probably been systematically underestimated. Finally, a generally neglected source of carbon emissions is the decomposition of above-water deadwood in flooded reservoirs, which Abril et al. [134] found to contribute about one-third of all carbon emissions in two tropical reservoirs. Figure 4.13 summarizes these GHG sources.

Reservoirs also alter global carbon flows, as significant amounts of organic matter enter freshwater from land, perhaps 1.9 PgC, and of this, Cole et al. [128] estimate that 0.23 PgC is buried in sediment (mainly in man-made reservoirs), 0.75 PgC is oxidized to CO_2, while 0.9 PgC finds its way to the ocean, where 25% may enter the deep ocean and be lost to the short-term carbon cycle [129]. Burial in reservoir sediment represents a carbon sink attributable to dams, although this may in turn reduce downstream sedimentation and carbon storage in the deep ocean [129], and the anoxic carbon sediments can also evolve to methane. Furthermore, sediment build-up in reservoirs eventually necessitates dredging or dam removal, at which point carbon contained in sediment may be released as CO_2 and CH_4 in amounts significant even when amortized over a 100-year dam lifetime [137].

We must further distinguish between gross and net emissions from reservoirs. *Gross* emissions are those discussed above, the CO_2 and CH_4 released from various points in the reservoir. *Net* emissions, on the other hand, are the net emissions that result from commissioning a hydroelectric dam after accounting for the previous carbon source/sink activity of the flooded land and, ideally, all downstream effects on the carbon cycle. For example, forests are generally considered a carbon sink, and so flooding of this sink increases net emissions independently of gross reservoir emissions.

4.9.2 Reservoir emissions by climate region

Several recent reviews [131, 129] have concluded that equatorial and, especially, tropical, reservoirs have far higher GHG emissions than temperate or boreal reservoirs. This is likely due to warmer water temperatures, increased biomass, and a tendency for tropical reservoirs to have a deep anoxic zone [131, 129]. From 150 literature estimates, Hertwich [129] determined globally averaged emissions factors of 85 gCO_2/kWh and 3 gCH_4/kWh, equating to 187 $kgCO_2e$/kWh. Assuming that power density (i.e. kWh generated per m^2 of reservoir) is similar across climate regions, from Table 2 of this work I calculate emissions of 233, 67, and 309 gCO_2e/kWh for boreal, temperate, and tropical regions, respectively. However, because many works did not

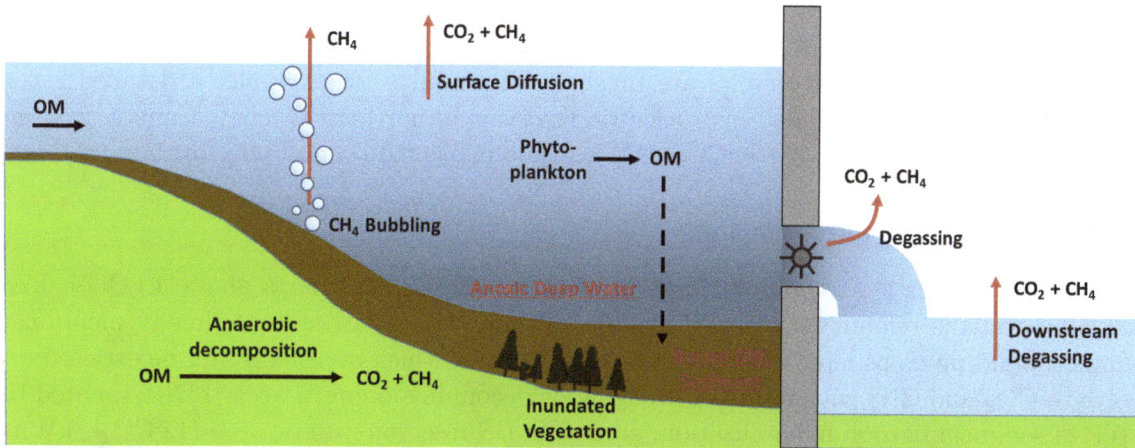

1. Organic matter (OM) from inundated vegetation, upstream river, and local phytoplankton is buried in anoxic sediments, leading to anaerobic decomposition and CO_2 and CH_4.
2. GHGs released via CH_4 bubbling, surface diffusion, turbine degassing, and further downstream degassing.
3. GHG emissions in first decade of reservoir creation dominated by decomposition of inundated vegetation.
4. Subsequent steady-state emissions from ongoing OM transport and reservoir production.

Figure 4.13: Schematic for major pathways by which CO_2 and CH_4 are emitted from hydropower reservoirs. Organic matter buried in deep, anoxic waters decomposes anaerobically, yielding CO_2 and CH_4, which may diffuse into the atmosphere across the water-air interface. Methane has extremely low solubility in water, and so much is released by bubbling. Further, these gases may dissolve into water at depth (under cold, high pressure conditions) but then de-gas after passing through turbines, both at the turbine and further on downstream.

measure methane bubbling and downstream bubbling/diffusion—a very significant methane source in tropical reservoirs [133]—is commonly neglected, these are assuredly underestimates and could easily be twofold higher. Further, above-water biomass decay has only been considered (to my knowledge) for two tropical dams (Petit Saut and Balbina [134]). It is therefore worthwhile to examine some particular studies more closely.

While overall, measurements are available for only a tiny fraction of equatorial and tropical dams, and these vary in how comprehensive they are, based on what data there is, many tropical hydroelectric reservoirs appear to be comparable or worse than coal (see, for example, Figure 4 of [132] and the discussion in [129]), with the few tropical dams for which detailed measurements are available showing very high CO_2e emissions indeed. The Petit Saut dam, located in French Guiana, was commissioned in 1994, and comprehensive field measurements [133, 134] over the subsequent decade suggest CO_2e emissions of, at a minimum, 0.8 kgCO_2e/kWh amortized over 100 years [134] (this estimate assumes CO_2e emissions fall ten-fold and remain constant after 10 years, which is almost certainly too conservative). Fearnside [126] estimated emissions of 6.683 kgCO_2e/kWh over the first twenty years of operation, for Petit Saut, implying an absolute minimum of 1.337 kgCO_2e/kWh over 100 years. The Balbina reservoir in Amazonian Brazil, commissioned in 1987, is even worse, emitting at least 2.5 kgCO_2e/kWh amortized over 100 years [134], and much more over the near term, with Hertwich giving 11.9 kgCO_2e/kWh [129].

It should be noted that the Petit Saut and Balbina reservoirs have a low energy density, in terms of kWh generated per m^2 reservoir area. Since GHG emissions are roughly proportional to the reservoir area, is follows mechanistically that higher energy density reservoirs should have lower emissions per power output; this is strongly supported by existing data [129]. A comprehensive simulation study of possible GHG emissions for 18 new and planned Brazilian hydroelectric reservoirs [136] suggested great uncertainty and an extremely wide range in emissions factors, with some high energy density projects showing low emissions, but some others possibly comparable to natural gas or coal plants.

Temperate and boreal reservoirs are much less harmful. For example, a detailed seven-year study [135] of a new boreal hydroelectric reservoir in northern Quebec, Canada gave *net* reservoir emissions of about 158 kgCO$_2$e/kWh, when projected over a 100-year dam lifetime.

4.9.3 Other lifecycle emissions

Dam construction is a very minor source of lifecycle emissions, perhaps about 4 gCO$_2$e/kWh (see [126] and the references reviewed therein). More significant, and as already mentioned, sediment build-up traps large amounts of organic matter, which eventually can be exposed and evolve to CO$_2$ and CH$_4$ with reservoir dredging or decommissioning. Pacca [137] estimated the GWP of sediment carbon mineralization, across six US reservoirs, to be 35–104 gCO$_2$e/kWh if 3% of sediment carbon mineralizes, and 128–380 gCO$_2$e/kWh if 11% mineralizes.

4.9.4 eGRID correction

The review by Hertwich [129] suggests reservoir emissions of 67 and 233 gCO$_2$e/kWh for temperate and boreal hydroelectric reserves, respectively, but based on many studies that likely systematically underestimated emissions. Teodoru et al. [135] suggested net emissions of 158 gCO$_2$e/kWh for a new boreal reserve, and dam construction and decommissioning likely add at least 40 gCO$_2$e/kWh. Therefore, I adjust eGRID hydroelectric emissions upward by 150 gCO$_2$e/kWh, which I consider conservative.

4.9.5 Other environmental consequences of dams

> - Extensive damming of a majority of the world's major river systems is a major component of global-scale alteration of the hydrologic cycle, with adverse consequences for many ecosystems.
> - Damming supports water withdrawals for agriculture and other human needs, and so adverse consequences are only partially attributable to hydroelectricity.

Dams are ubiquitous throughout the world's river systems: there are over 45,000 dams higher than 15 meters (49.2 ft) [142] and over 800,000 smaller dams [141], which collectively act to obstruct about two-thirds of the freshwater flowing to the ocean [141] and fragment half to two-thirds of the world's major river systems [142]. Essentially all major river systems in the continental US are heavily fragmented by damming [142]. This represents a fundamental alteration of the hydrologic cycle, with myriad consequences for both aquatic and terrestrial ecosystems [140]. Dams are constructed not just to provide electricity, but to control flooding, river flow, and divert large amounts of water for irrigation and other uses, and so the ecological consequences of dams are not attributable to hydroelectricity alone. In particular, humanity appropriates about half of all available freshwater via its various waterworks, overwhelmingly for agricultural irrigation [140]. Nevertheless, the focus of many new giant dams is on electricity generation.

The most obvious consequence of damming is the inundation of large areas of upstream land and wholesale destruction of the associated ecosystems. River fragmentation by dams blocks migration and dispersal of many species, and has caused the extinction of many fish populations and even species [142]. Water withdrawal adds further stress, with withdrawals

94

from many rivers so extreme that they no longer even reach the ocean. The reservoirs that result from damming also lose far more water to evaporation than do free running rivers [141].

Siltation of reservoirs reduces silt transport downstream, which in turn can increase erosion and alter coastlines that previously depended on silt influx. Dams chiefly act to reduce variability in river flow and eliminate flooding, whereas the natural cycle is typically one of annual flooding of a downstream floodplain, bringing nutrient rich sediment and water to a wide region: ancient Egyptian agricultural was famously dependent upon the annual flooding of the Nile for fertilization, a cycle that has now ceased. Floodplain ecosystems are usually highly adapted to this annual cycle, and its disruption has broadly negative effects [141].

4.10 Biomass

- Harvesting woody forest mass for electricity is not carbon neutral over a timescale of decades to centuries, and moreover, integrating the warming effect of harvesting on forest carbon dynamics over 100 years shows that emissions factors for live tree harvesting may range over an order of magnitude, from perhaps 0.2 $kgCO_2e$/kWh to 2.0 $kgCO_2e$/kWh (in terms of electricity generated), with the latter almost twice as bad as coal.

- Even if one only burns forest residues left after logging for other purposes, emissions factors could still be similar to natural gas over a 100-year time-horizon, as these residues would otherwise naturally decay only over decades (and also add to soil carbon stores). If their fate was simply to be burned on-site, then residues are indeed a relatively low carbon energy source.

- Woody biomass taken from forests for electricity generation is, overall, probably similar or only marginally better than alternative fossil fuels in its 100-year equivalent emissions profile (and some biomass sources will be significantly worse than coal); proposed biomass projects should be evaluated on an individual basis.

Woody biomass burned for electricity is an increasing component of renewable power portfolios in the US (but still accounted for just 1.6% of total electricity generation in 2014), and especially in Europe (which is aggressively pursuing bioenergy and has begun to import wood pellets produced in the Southern US), but the ecological and climate soundness of this practice is unclear. Woody biomass is often regarded as an immediately net zero-emissions energy source, because, it is assumed, the carbon liberated from wood burning will *eventually* be reincorporated into new biomass. This assumption is fundamentally flawed for multiple reasons, the simplest being that there is a long lag between the burning of forest matter and regrowth, implying that, at the very least, bioenergy is not immediately carbon neutral and has a short-term warming effect that is compensated for later. More fundamentally, this view fails to consider the fate of the forest in the absence of harvesting, which is generally to continue to absorb large amounts of carbon [46], with this true even for old-growth forests. Furthermore, continuous harvesting permanently reduces the amount of carbon stored in forests [150], can deplete forest soil carbon [145] and total carbon stock beyond the amount harvested [148], and may decrease forest productivity with corresponding reductions in carbon storage [143, 144].

Given all these factors, the ultimate effect of wood harvesting on the climate for bioenergy is extremely controversial, and the issue is surrounded by a great deal of genuine uncertainty. One may consult [151] or [152] for overviews of some aspects of the controversy. US forests are a major carbon sink, and offset 10–20% of the US's annual GHG emissions [46, 522]. The net effect of bioenergy harvesting on this sink is unclear. In some more arid regions, light thinning of forests could reduce wildfire risk and drought stress for net emissions reductions, but heavier harvesting is likely to undermine forest productivity and increase net carbon emissions

[143, 144]. Hudiburg et al. [143] concluded that, in the US Northwest, forest management for either fire prevention or bioenergy increased overall emissions compared to business-as-usual, with bioenergy the worst scenario.

Several recent studies that have attempted to compare the integrated warming of biomass harvested from re-growing forest have used the GWP_{bio} metric, introduced by Cherubini and colleagues [146], which purports to measure the relative warming effect of biomass versus an equivalent amount of fossil energy, and several studies have given GWP_{bio} in the 0.34–0.62 range (as reviewed in [150]), suggesting that the emissions from forest biomass are around half those of fossil fuels, although Holtsmark has derived larger values, about 1.5, for boreal forests [150]. The most obvious problem here, of course, is that fossil fuels are not a homogenous block, with combustion emissions for electricity generation from natural gas around 60% lower than from coal. Directly comparing wood-fired electricity generators, coal, and gas plants shows that the biomass emissions factor could plausibly range from as little as 0.2 $kgCO_2e$/kWh to almost 2.0 $kgCO_2e$/kWh, depending upon stand characteristics (see Section 4.10.3). Given the combustion characteristics of wood and wood-fired plants, a GWP_{bio} of 0.5 probably implies emissions similar to natural gas, while $GWP_{bio} = 1$ would be similar to coal. A recent global mapping effort by Cherubini et al. [147] yielded a global averaged 100-year GWP emission factor of 0.49 $kgCO_2e$ per $kgCO_2$ released via bioenergy, again suggesting forest bioenergy may be similar to natural gas (although the global temperature potential metric was much closer to zero, reflecting the fact that most warming from wood energy occurs early in time).

My own view is that, while woody bioenergy from fast-growing stands could be better than coal or natural gas over a 100-year timeframe, this is uncertain, depending strongly upon the growth characteristics of the particular forest being harvested and the fate of the forest under alternative management, and in many cases wood is still a relatively high-carbon energy source. Intense wood harvesting would probably have broadly negative ecological effects and could markedly undermine ongoing large-scale carbon sequestration in forest ecosystems, while *light* thinning may be beneficial to *some* forests. Furthermore, unlike wind, solar, and perhaps nuclear, wood is not a scalable replacement for fossil fuels [47]. For these reasons, individual woody bioenergy projects could be beneficial but should be carefully considered with respect to local conditions, and woody biomass in general should not be viewed as a valid long-term (or even short-term) fossil fuel replacement or as a major climate change mitigant. Managing forests for carbon sequestration (which, again, may involve some degree of lighter thinning with the products used for bioenergy) rather than fuel substitution strikes me as the generally superior strategy.

Fuel mixes for wood-fired plants are variable, but generally consist of a mix of urban tree trimmings, residues from forest clearing and thinning, low-quality lumber, and pulpwood or whole live trees, harvested either from natural forest or possibly tree plantations. Note that in the past a majority of wood-fired electricity was generated in industrial wood pulp plants in combined heat and power plants that used the heat on-site, and the electricity generated was thus, at least arguably, an additional benefit that one could credit as zero-carbon, but now utilities are increasingly generating electricity from designated biomass generators and via co-firing of wood with coal.

I now review more deeply the combustion emissions from wood, and the effects of forest harvesting on forest carbon pools, atmospheric carbon dioxide, and the integrated warming effect of biomass versus fossil fuel electricity.

4.10.1 Combustion emissions

> • At the point of combustion, typical wood-fired generators emit around 1.33 $kgCO_2$/kWh of electricity, while coal emits about 1 $kgCO_2$/kWh, but this fact means little in isolation of forest carbon dynamics.

As reviewed in my extensive discussion on firewood (Section 12.5.6), dry wood has a high heating value of about 5.56 kWh/kg (slightly more for softwood, slightly less for hardwood), corresponding to 11.02 kWh/kgC, assuming dry wood is 50% carbon by mass, and giving thermal combustion emissions of 0.33 $kgCO_2$/kWh(t). Now, wood-fired generators operate at lower temperatures than coal plants, and so thermal efficiency tends to be significantly lower (see Section 12.3 for a discussion of the Carnot Heat Engine and the importance of boiler temperature in determining engine efficiency), ranging from about 17–29%, with 24 or 25% typical [153]. This translates into combustion emissions anywhere from 1.147–1.957 $kgCO_2$/kWh(e), with perhaps 1.33 $kgCO_2$/kWh typical, and about 33% higher than coal. This by itself, of course, does not imply that biomass is a high-carbon fuel overall.

4.10.2 Overview of regrowth, carbon dynamics, and carbon payoff following harvesting

Harvesting forest biomass affects multiple carbon pools, and such changes must be considered in reference to a baseline counterfactual scenario where no harvest occurred. One method of carbon accounting, a "carbon-balance" approach, invokes the concepts of (1) carbon debt repayment, and (2) carbon sequestration parity, as shown schematically in Figure 4.15 [151]. We see that, because forests are likely to continue growing and absorbing carbon in the absence of harvesting, even once one has re-paid the carbon debt of harvesting compared to the pre-harvest forest mass, there is still a large debt relative to the counterfactual of no harvesting. True "carbon neutrality" is not reached until the carbon sequestration parity point (see Figure 4.15). However, as we shall see, even at this point, the cumulative warming effect of the harvesting scenario can be greater than the no-harvest scenario, and a "warming-debt" may still exist long after carbon parity.

While the carbon-balance approach is valid, it does little to tell us the absolute warming effect, either before or after reaching the carbon sequestration parity point, of biomass: what we would really like is to have a GWP-like metric for a more direct comparison to fossil fuels. Towards that end, several authors (e.g. [146, 150]) have used an alternative metric, the GWP_{bio}, as a means to quantify the warming effect of forest biomass compared to an equivalent amount of fossil energy. In the following section, I derive an expression for GWP_{bio} from a relatively simple mathematical model of forest growth and harvesting, and I offer my critique and modification of this metric.

4.10.3 A simple mathematical model for harvesting and biomass GWP

To gain some more direct insight, I follow several works by Holtsmark [149, 150] to explicitly describe the dynamics of live forest biomass, harvest residues, deadwood, and soil carbon, with and without harvesting. This model considers four carbon pools: (1) total live biomass, $B(t)$, (2) tree stem biomass, $G(t)$, which is some fraction, θ, of total biomass, assumed to be 48%, (3) natural deadwood, $D_N(t)$, and (4) deadwood as a harvest residue, $D_R(t)$. I take all variables to have units MgC (i.e. the absolute amount of carbon in each pool). We can model the growth

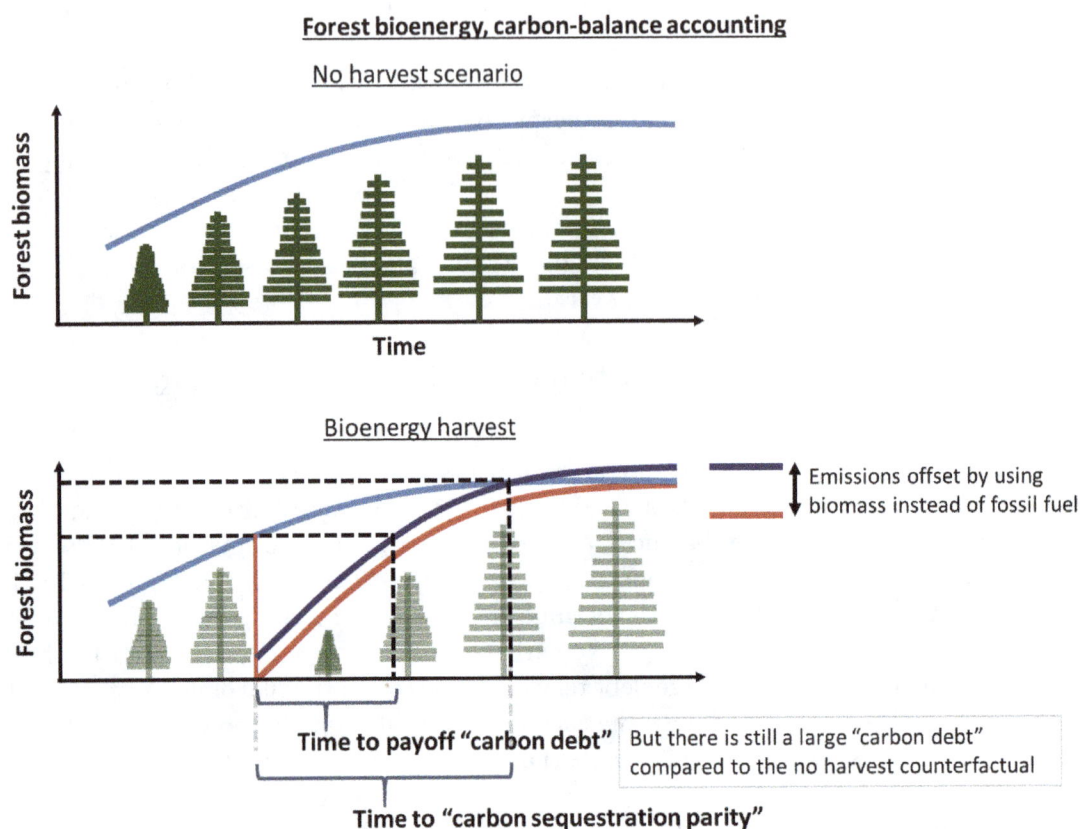

Forest bioenergy, carbon-balance accounting

No harvest scenario

Forest biomass

Time

Bioenergy harvest

Forest biomass

↕ Emissions offset by using
biomass instead of fossil fuel

Time to payoff "carbon debt"

But there is still a large "carbon debt"
compared to the no harvest counterfactual

Time to "carbon sequestration parity"

Figure 4.14: Schematic for the carbon-balance accounting method of assessing the carbon impact of woody bioenergy derived from forests. See also Figure 2 of [151]. At the point of harvest, forest biomass goes to zero (at least in schema, as shown by the red line), while some amount of fossil carbon is offset via the use of forest products for energy generation, and so net carbon storage is given by the dark blue line. With forest regrowth, equivalent carbon stores eventually match the pre-harvest forest (payoff time for the carbon debt), but it is not until equivalent stores equal to forest carbon stored under the counterfactual of continued growth that carbon sequestration parity is reached.

of tree stems with a Logistic function

$$G(t) = \nu_1 \left(1 - e^{-\nu_2 t}\right)^{\nu_3}, \tag{4.1}$$

where ν_1, ν_2, and ν_3 are parameters particular to any tree stand. Since θ is the fraction of live biomass that is stem, $B(t)$ follows simply as

$$B(t) = \frac{1}{\theta}G(t). \tag{4.2}$$

Natural deadwood is generated by live biomass, and we have litterfall given as βB, while we suppose first-order decay of deadwood at rate ω, yielding

$$\frac{dD_N}{dt} = \beta B - \omega D_N. \tag{4.3}$$

Now, if we harvest tree stems at time t_h, and σ is the fraction of non-stem residue (e.g. tree tops, stumps) taken with the harvest, then we extract (in MgC)

$$E(t_h, \sigma) = G(t_h) + \sigma(B(t_h) - G(t_h)). \tag{4.4}$$

That fraction of residue not taken, $1 - \sigma$, remains in the forest as residue deadwood, $D_R(t)$ and decays at first-order rate ω, giving

$$D_r(t, t_h, \sigma) = e^{-\omega t}(1 - \sigma)(B(t_h) - G(t_h)). \tag{4.5}$$

Now, we can calculate the net atmospheric carbon over time that results either without a harvest (given current forest conditions), or with a harvest of prescribed magnitude (e.g. clear-cut, light harvest, etc.) with stand regrowth according to the equations above, using the standard equations describing CO_2 concentration following a CO_2 bolus, and with $y(t)$ representing the concentration time-course for a single kg CO_2 bolus (see Section 3.3.1, or complete details in [149, 150]). If $C_0(t)$ is the atmospheric carbon (not CO_2) time-course without harvesting, and $C_H(t)$ is the time-course with harvest, then from the concentration difference,

$$C(t) = C_H(t) - C_0(t), \tag{4.6}$$

we can determine the additional integrated radiative forcing, $\text{AGWP}_{\text{bio}}(t)$, as

$$\text{AGWP}_{\text{bio}}(t) = \int_0^t A_{CO_2} \frac{44}{12} C(s) ds, \tag{4.7}$$

where A_{CO_2} is the radiative efficiency of CO_2. Now, the GWP_{bio} metric, as used by Holtsmark [150], Cherubini [146], and others has been taken as

$$\text{GWP}_{\text{bio}}(t) = \frac{\text{AGWP}_{\text{bio}}(t)}{\text{AGWP}_{CO_2}(t)}, \tag{4.8}$$

where $\text{AGWP}_{CO_2}(t)$ is the integrated radiative forcing that results directly from the combustion of the forest harvest

$$\text{AGWP}_{CO_2}(t) = \int_0^t A_{CO_2} \frac{44}{12} E(0, \sigma) ds, \tag{4.9}$$

assuming harvesting at time 0. Typically, $\text{AGWP}_{CO_2}(t)$ has been interpreted as the warming that would have, without combustion of wood, resulted from an equivalent degree of fossil fuel combustion, and therefore $\text{GWP}_{\text{bio}}(t)$ gives us the relative amount of warming, at any particular time, resulting from using forest wood over fossil fuel [146, 150]. If $\text{GWP}_{\text{bio}} < 1$, then wood is less warming, while if $GWP_{bio} > 1$ then fossil energy is better, and, in keeping with the IPCC GWP framework, it is probably most appropriate to evaluate GWP_{bio} at 100 years post-harvest.

However, this interpretation of GWP_{bio} is fundamentally flawed, in my view, as it implicitly assumes that a carbon atom from biomass corresponds to exactly the same amount of electricity (or heat, or automotive power, etc.) as a carbon atom from fossil fuel. This is obviously false, as the intrinsic heat content of wood is lower than that of fossil fuels (i.e. 1 kg of wood carbon yields less energy than 1 kg of fossil carbon), the efficiency of wood-fired power plants is significantly lower than that of typical fossil plants, and fossil fuels themselves vary markedly in their emissions profiles (e.g. coal vs. natural gas). Fortunately, it is a fairly simple matter to account for these factors, and we can simply derive $kgCO_2e/kWh$ emissions factors for bioenergy that are directly comparable to similar EFs for any fossil fuel, as follows.

Given E, the mass of carbon harvested, we determine the amount of electricity generated, assuming a HHV of 11.02 kWh/kgC and a power plant thermal efficiency of 25%. Given the kWh produced, which we denote E_{kWh}, we then use the formula

$$ \text{EF}(t) = \left(\frac{1}{E_{kWh}} \right) \left(\frac{\text{GWP}_{bio}(t)}{\text{AGWP}_{1kgCO_2}(t)} \right), \tag{4.10} $$

to yield a time-dependent emissions factor with units $kgCO_2e/kWh$, with $\text{AGWP}_{1kgCO_2}(t)$ the integrated RF from a single kg of CO_2. Since we have already constructed time-dependent emissions factors for coal and natural gas (see Figure 4.8 in Section 4.6.1), we can directly compare electricity generation from forest products to these two fossil fuels, which I do in a moment, finding that forest bioenergy is generally just as bad as fossil energy. Note that in the following results, I have omitted any negative perturbation of the forest soil carbon pool by harvesting (this was considered by Holtsmark [150]), and so they are actually biased in favor of bioenergy.

Model results and emissions factors

Figure 4.15 shows the evolution of different forest carbon pools following clear-cutting of the trees, with 75% of harvest residues left to decompose in the forest, along with the total change in atmospheric CO_2 resulting from this clear-cut, compared to the counterfactual of leaving the forest be. I use parameter values from [150], including ω (decay rate for deadwood) at 0.04 year^{-1}.

Figure 4.16 compares forest growth dynamics, and equivalent emissions factors, when harvesting either a slow-growing boreal forest with either clear-cutting, heavy harvesting, or light harvesting, or a faster-growing forest under the same set of harvests. Note that the emissions factor actually increases early in time: this is due to carbon released via decay of residual deadwood. We see that harvesting from slow-growing forests is often worse than coal, while fast growing forest biomass is comparable to or better than natural gas by the 100-year mark.

Finally, let us model a common scenario, and one that has often been viewed as a net zero-emissions activity, namely the use of unmarketable residues (branches, tree-tops, etc.) from a lumber harvest for energy. As pointed out by Ter-Mikaelian et al. [151], such residues are often left to decay naturally, and even when burning is a common practice, often less than 50% of slash piles are actually set afire. In any case, I find that, compared to the counterfactual of simply leaving residues to rot, the residue 100-year EF is on the order of 0.3 $kgCO_2e/kWh$ for a first-order decay rate of 0.04 day^{-1}, as used by Holtsmark [150], and closer to 0.6 $kgCO_2e$ when $\omega = 0.02$ day^{-1}, which is a more likely overall average for coarse woody debris [559]. If the counterfactual scenario is burning about half of residues on-site, then these EFs are in the 0.1–0.3 $kgCO_2e/kWh$ range, which is indeed fairly low carbon. However, not modeled here is

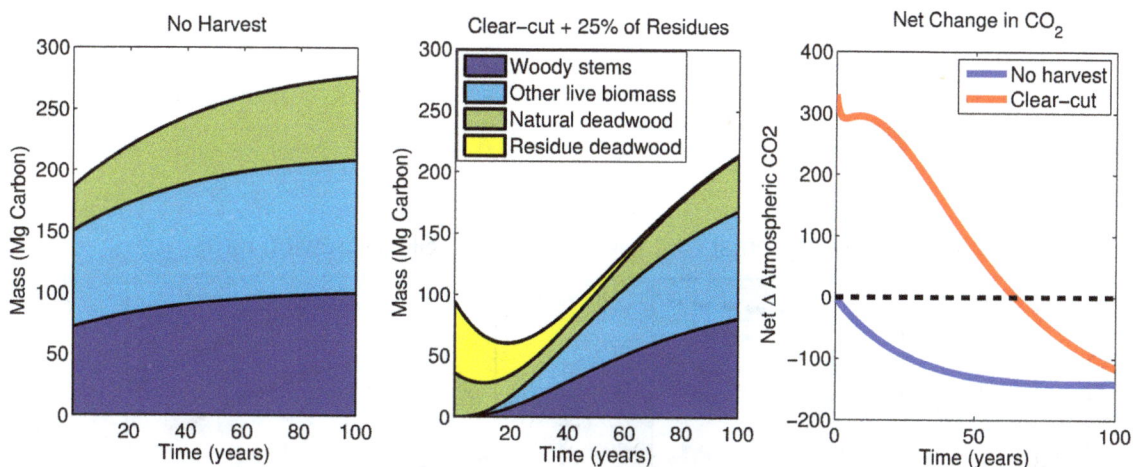

Figure 4.15: Approximate carbon pools for a slow-growing boreal forest either under no harvesting (left panel) or under clear-cutting of the woody stems, leaving 75% of harvest residue (branches, stumps, etc.) to decay in place (middle panel). The right panel shows the net change in atmospheric CO_2 under either scenario: without harvesting the forest continues to slowly take up carbon, while harvesting results in large early CO_2 release that is eventually mostly compensated for.

the effect residue harvesting has on depleting forest soil carbon pools, which Repo et al. [145] found to be quite significant.

Now, the above results do not account for forest soil carbon dynamics, the inclusion of which makes bioenergy even more unfavorable [150, 145]. Additionally, the logistic growth model assumes that old-growth forests stop taking up carbon, an assumption that is probably false. Harvesting also can decrease forest productivity, and continuous harvesting leads to a permanent forest carbon deficit. Accounting for any of these behaviors would lead to higher emissions factors for biomass. Furthermore, I have not considered any of the energy/emissions required for harvesting, power plant construction, or the forest land that must be cleared for access roads, etc. Therefore, I feel fairly confident in the conclusion that biomass electricity is closer to fossil fuels in its emissions profile than it is to other major renewable energy sources.

4.10.4 eGRID correction

I use 500 gCO_2e/kWh as an extremely crude eGRID correction, given the great uncertainty as to the true impact of biomass on climate.

4.11 Geothermal

Geothermal power plants generated only 0.37% of all US electricity in 2010 (eGRID), but geothermal plants are a significant source of energy in California and Nevada, where they generated 6.17% and 5.89% of all electricity, respectively. Non-trivial geothermal capacity exists only in three other US states, namely Hawaii, Utah, and Idaho, where it accounted for 1.85%, 0.66%, and 0.60% of electrical generation, respectively. Geothermal plants convert the heat of geofluids from great depths to electricity, and conventional plants are all installed over geologically young and active areas containing unusual reservoirs of hot geofluids located relatively close to the surface. In general, wells are drilled to a depth of 1–3 km, and superheated fluid or steam is pumped to the surface, where it drives turbines that in turn generate electricity.

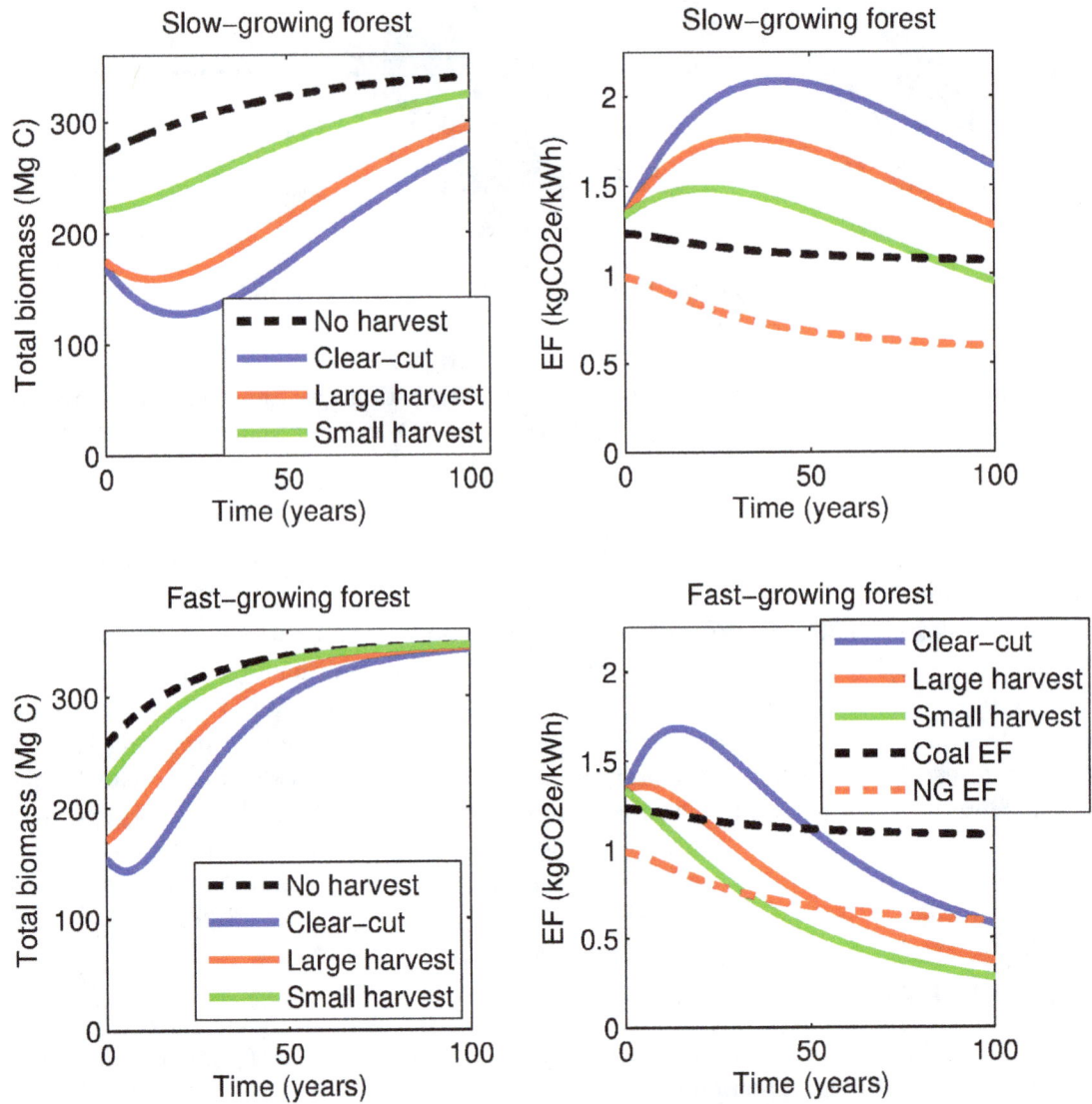

Figure 4.16: Changes in forest biomass and equivalent $kgCO_2e/kWh$ emissions factors, when forest biomass is harvested for electricity generation. The top gives results for a slow-growing forest, showing that this scenario may be worse than coal even out to 100 years, while the bottom panels show results for a fast-growing forest, where lighter harvests are clearly superior to natural gas at the 100-year horizon. Even in the fast-growing forest case, it is still generally at least 50 years before any benefit to bioenergy over fossil fuels becomes manifest.

Cooled fluid is then pumped back via an injection well to replenish the geothermal reservoir.

There are three major classes of conventional geothermal plants—dry steam, flash, and binary—each designed for use with steam/fluid of different temperatures. Dry steam plants are employed in the hottest areas, where superheated steam is directly pumped from the well to drive a turbine. Flash plants use superheated fluid that is composed of a mixture of liquid and vapor; the mixture is "flashed" via centrifugal action separating the two phases, and the resulting vapor drives a turbine. The liquid may be flashed a second time ("double-flash" plant) to increase efficiency. Binary plants use the coolest geofluids (< 150 °C), which are pumped through a heat exchanger to transfer heat to a secondary fluid with a lower boiling point; this secondary fluid vaporizes and drives a steam turbine.

All dry steam plants in the US are located in the Geysers, the largest geothermal field in the world, located just north of San Francisco. This field alone accounts for about 55% of CA geothermal capacity and 47% of overall US capacity. Most flash plants in the US are also located in CA, while outside CA, most plants are of the binary type, and nearly all growth in geothermal power over the last 20 years has been in binary generation. California accounts for 83% of US geothermal power, and 93% of steam and flash power [156].

Geofluids contain significant quantities of non-condensible greenhouse gases. In binary plants, the secondary fluid cycles in a closed loop through the vapor and liquid phase, while the geofluid is extracted and re-injected without ever making contact with the atmosphere. Thus, binary fluids are considered to have operating emissions that are effectively zero [156]. In contrast, the geofluid in steam and flash plants comes into direct contact with the atmosphere, allowing significant fugitive CO_2 and CH_4 emissions to occur.

A 2003 survey of US geothermal plants by Bloomfield et al. [157] reported a weighted average of 90.7 gCO_2/kWh and 0.753 gCH_4/kWh (25.6 gCO_2e/kWh, using GWP 34), for an overall average of 116.3 gCO_2e/kWh. Assuming that the emissions from binary plants (14% of generation in 2003) are nil, we have an EF of 135.3 gCO_2e/kWh (including 29.8 gCO_2e/kWh due to methane) for steam and flash plants. A widely cited work by Bertani and Thain [155] reportedly sampled 85% of 2001 world generating capacity, and reported a range of 4-740 gCO_2/kWh with a weighted average of 122 gCO_2/kWh (see discussion in either [156] or [154], I have been unable to locate the original work); this survey considered CO_2 only, omitting any methane emissions. Other estimates are thoroughly reviewed in a recent overview by Bayer et al. [154].

Recently, Sullivan and colleagues extended the GREET model to incorporate geothermal energy [156]. They concluded that plant-cycle (i.e. emissions from plant construction, well-drilling, etc.) are insignificant, at perhaps 4 gCO_2e/kWh for steam/flash plants, and 6 gCO_2e/kWH for binary plants. Using data reported to CARB, they further concluded mean emissions of 97.9 gCO_2e/kWh for operating plants in California; it is unclear if this data includes methane emissions or represents CO_2 emissions alone. Given this uncertainty, and the fact that this estimate is lower than other major estimates cited above, I suggest an EF of 125 gCO_2e/kWh for steam/flash plants, and an EF of 6 gCO_2e/kWh for binary plants. Binary generation by state is 6%, 25%, 63%, 100%, and 100% for CA, UT, NV, HI, and ID, respectively [156]. Since eGRID subregions do not partition strictly by state, however, and since most geothermal energy is generated in California where binary plants are rare, I use an eGRID correction of 115 gCO_2/kWh across the board.

4.12 Wind

- Wind power has the lowest embodied carbon emissions of any major power source: the likely range for utility-scale wind is 10–30 gCO_2e/kWh.

- Other negative environmental impacts, mainly bird and bat deaths, are of concern but are probably low compared to fossil-fuel power plants (and other human activities and structures), although harms to bat populations in particular remain poorly understood.

Wind power probably has the lowest lifecycle emissions of any electricity source, likely in the 10–30 gCO_2e/kWh range for utility-scale wind. This source has rapidly scaled up in recent years, increasing from less than 0.5% of US generation in 2005 to 4.3% in 2014 (World Bank), and wind is becoming a major power source in the Midwest: South Dakota and Iowa both derived about 25% of their electricity from wind in 2012, while in six other states wind was at least 10% of the generation mix (eGRID).

Wind has been generally viewed, accurately, as a truly low carbon energy source with relatively low environmental harms. It remains true that wind farms can negatively affect wildlife, and increased bird and bat mortality due to collisions with turbines has been fairly widely publicized. As discussed below, wind turbines may actually be safer overall for birds (per kWh generated) than fossil fuel plants, but this hazard may be greater for bats and should be considered when siting future wind farms. By harvesting significant amounts of wind energy and inducing turbulence, very large-scale wind farming also has the potential to alter regional climate and ground surface temperatures, but these effects are very minor [158].

I briefly introduce the physics of wind turbine, review lifecycle greenhouse emissions, turn to a brief discussion of the scalability of this intermittent power source, and close with a discussion of other negative environmental impacts.

4.12.1 Wind harvesting fundamentals

Wind resources

On a planetary scale, wind is generated by sunlight, which disproportionately heats air near the equator, inducing it to rise and move toward the poles, and by the Coriolis force from Earth's rotation. The latter induces clockwise rotation in the Northern Hemisphere, and counter-clockwise rotation in the Southern Hemisphere. At the local scale, the landscape strongly affects winds, with rougher surfaces (e.g. cities) significantly slowing air flow; because wind speed must equal zero at ground level, we also have a boundary layer effect such that wind speed increases with vertical height for several hundred meters. At low altitude, wind speed, V, increases logarithmically with height,

$$V \propto \log\left(\frac{z}{z_0}\right) \tag{4.11}$$

where z is height, and z_0 is the *roughness* (measured in m) of the terrain, and may be 10–100 over mountains, 0.5-2 over cities and forests, 0.01 over plains, and 0.0001–0.001 m over the sea. This helps explain why wind resources in the US are greatest off-shore and over the Great Plains region, as depicted in Figure 4.17, and why turbines towers are typically 35–125 meters in height (not including the rotors), with newer installations and off-shore turbines trending higher and higher.

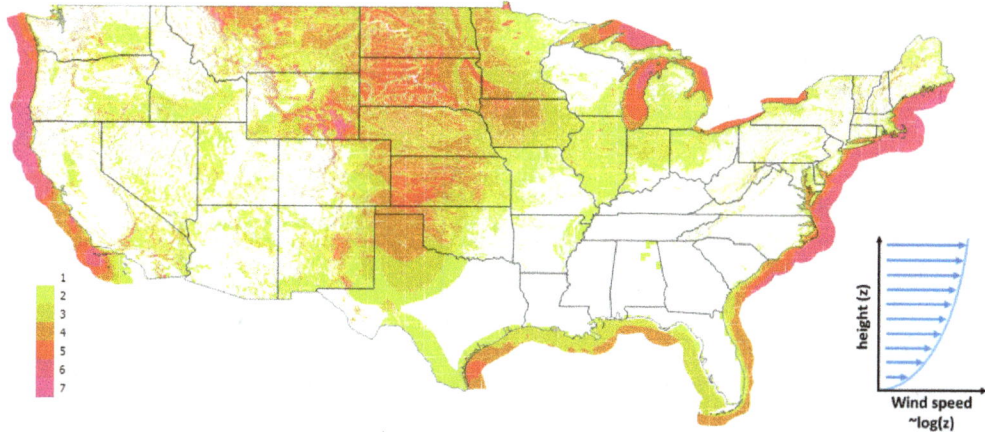

Figure 4.17: Map of wind resources in the US, which are concentrated in offshore and the Great Plains regions. Also inset is the logarithmic relationship between height and wind speed. Map data is from publicly available NREL databases.

Maximum harvestable wind: from individual turbines to the regional scale

The amount of power that a turbine generates is proportional to the rotor area, air density, and wind speed cubed, and it can be shown that it is theoretically impossible for a turbine to extract more than 59% of the power in wind (the *Betz Limit*) [162]. To see this, first note that the kinetic energy contained in a mass of air of volume V moving at velocity u is given as

$$E = \frac{1}{2}\rho V u^2,\tag{4.12}$$

where ρ is the density of air. If this air is moving through a circular cross-section such as, say, a turbine rotor, of area A, then it moves at volumetric rate $dV/dt = Au$, and the power of the moving air is

$$P = \frac{dE}{dt} = \frac{1}{2}\rho A u^3.\tag{4.13}$$

Now, when a parcel of air passes through a rotor that harvests energy, we must have that both the velocity of exiting air decreases, while the cross-sectional area increases (a consequence of the conservation of mass). Since the air cannot exit at velocity 0 (the passage of air would cease), nor can it maintain the same velocity (no energy should be harvested in this case), power will be optimally extracted when the exit velocity, u_2, is some non-zero fraction of the wind stream velocity, u_1. By arguments outlined in Figure 4.18, this optimum occurs when $u_2 = u_1/3$, at which point the power coefficient, C_p, i.e. the fraction of power extracted, is exactly 19/27, or about 0.59. This maximum is known as the "Betz Limit." There are additional limits on power extraction related to rotational inertia, and one may consult Grogg [162] for a far more detailed treatment.

Since wind is markedly slowed and as much as 50% of its energy extracted upon its encounter with a turbine, to place turbines in close succession would be fey, and so these giants must array themselves across many leagues. Significant extraction of kinetic energy from wind, combined with limited replenishment from higher altitude winds, acts to limit large-scale wind electricity generation to about 1 W m^{-2} in windy regions of the US [159]. For perspective, this harvesting rate over the entire area of Kansas (213.1 billion m^2, or 2.8% of the Continental US) would yield 1.87 billion MWh/year, or a little under half annual US electricity generation; it would similarly

105

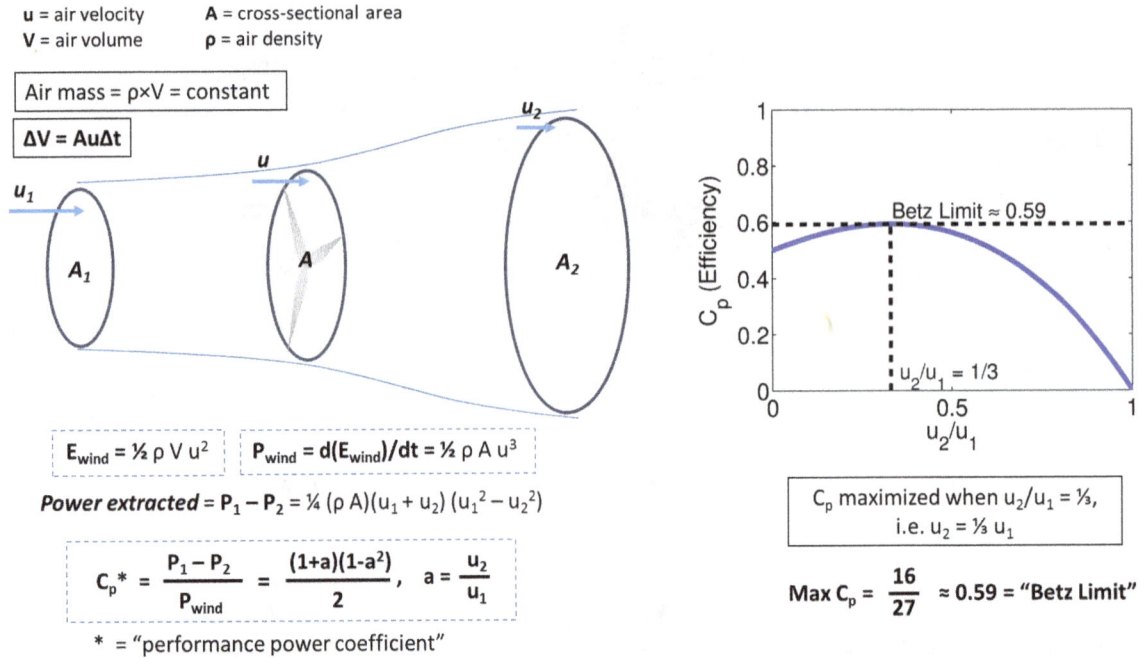

u = air velocity   A = cross-sectional area
V = air volume   ρ = air density

Air mass = ρ×V = constant

ΔV = AuΔt

$E_{wind} = \frac{1}{2} \rho V u^2$ $P_{wind} = d(E_{wind})/dt = \frac{1}{2} \rho A u^3$

Power extracted = $P_1 - P_2 = \frac{1}{4}(\rho A)(u_1 + u_2)(u_1^2 - u_2^2)$

$$C_p{}^* = \frac{P_1 - P_2}{P_{wind}} = \frac{(1+a)(1-a^2)}{2}, \quad a = \frac{u_2}{u_1}$$

* = "performance power coefficient"

Betz Limit ≈ 0.59

$u_2/u_1 = 1/3$

C_p maximized when $u_2/u_1 = \frac{1}{3}$, i.e. $u_2 = \frac{1}{3} u_1$

Max $C_p = \frac{16}{27} \approx 0.59$ = "Betz Limit"

Figure 4.18: Outline of the physical argument underlying the Betz Limit, the maximum fraction of power that may be harvested from a wind stream by a rotor. This maximum is extracted when the exit air stream's velocity is one-third that of the entering stream, and is equal to exactly 19/27, or about 0.59.

take about one-third of Texas to provide half the US's electricity[5]. Since turbines themselves take up only a very small fraction of the land area harvested for wind, and may be placed on land co-used for agriculture, this land footprint is not in itself prohibitive.

Capacity factor

The *nameplate capacity* gives the maximum power output of a turbine, while the *capacity factor* is the ratio of actual power generated to nameplate capacity. For example a 2 MW turbine with a capacity factor of 0.30 will, on average, generate 0.6 MW, yielding 5,256 MWh of electricity annually. The average capacity factor for US wind power was 0.325 in 2015, per the EIA. The capacity factor obviously varies with location, but it is generally larger for larger turbines, as they reach the higher and faster winds [160].

---

[5]Per the American Wind Energy Association, installed wind capacity in 2014 was 61,327 MW. Assuming a capacity factor of 34% (EIA) and a maximum of 1 W per m^{-2}, this implies about 20.9 billion m^2 of land were devoted to wind-farming in 2014, or just under 10% of the area of Kansas (of course, <1% of this land is actually turbine).

4.12.2 Lifecycle GHG emissions

- Utility-scale turbines generate on the order of 20 gCO_2e/kWh or less over a 20-year lifecycle, and the initial carbon investment is usually paid off within six months.

- Small residential-scale wind installations are slightly more carbon-intensive, but still low-carbon overall (perhaps 50 gCO_2e/kWh for a grid-tie system, and >100 gCO_2e/kWh for an off-grid system with batteries).

The harvesting of wind energy generates no emissions directly, but turbines are massive structures whose fabrication requires up to several thousand tonnes of carbon-intensive materials. Nevertheless, the literature is uniform in finding that the overall impact is ultimately nearly trivial for most turbines: a review by Arvesen et al. [160] gave 19±13 gCO_2e/kWh, Dolan et al. [163] gave a range of 3.0 to 45 gO_2e/kWh with a median of 11 gCO_2e across 126 estimates (after harmonization), and Nugent and Sovacool give an average of 34.1 gCO_2e/kWh across 41 studies [165].

Carbon emissions come from (1) material production, (2) materials transport and turbine construction/assembly, (3) operation and maintenance, and (4) decommissioning after a roughly 20-year service life. Materials, especially steel and concrete, are the major source of embodied carbon emissions. Decommissioning, while requiring some energy, acts overall to offset some lifecycle emissions if steel, iron, and other metals are recycled. The relative GHG impact of these categories is summarized in Figure 4.19, and Figure 4.20 shows the material makeup of one large 2 MW geared turbine, based on [166].

A general trend seen in LCAs is one toward lower lifecycle emissions for larger turbines. As noted above, larger/taller turbines tend to have higher capacity factors [160], and it is generally more efficient on a materials basis to build a single turbine with a high capacity than multiple smaller turbines [161]. However, this trend is slight and disappears once turbines reach the MW scale (see Figure 4.19), and even very small-scale residential installations are likely to emit only around 50 gCO_2e/kWh, comparable to rooftop solar. Note that *off-grid* residential wind-systems will have significantly higher lifecycle emissions, due to the need for batteries. For example, one study [164] of an off-grid installation suggested 116.3 gCO_2e/kWh, 61% of which was attributable to the battery system (although even this is still very low compared to fossil energy).

Finally, the carbon payoff time for large-scale wind turbines is less than one year, and likely only a few months. Suppose a 2 MW turbine has associated lifecycle emissions of 2,000 $MgCO_2e$ and operates with a capacity factor of 0.30, generating 5.256 million kWh/year. Assuming it displaces grid-average electricity, then the entire initial carbon investment is paid off in 6.6 months (displacing marginal electricity, and we have payoff at 4.9 months). Note that this example turbine has 19 gCO_2e/kWh amortized over a 20 year lifetime.

For corrected eGRID emissions, I use an EF of 19.0 gCO_2e/kWh for wind power.

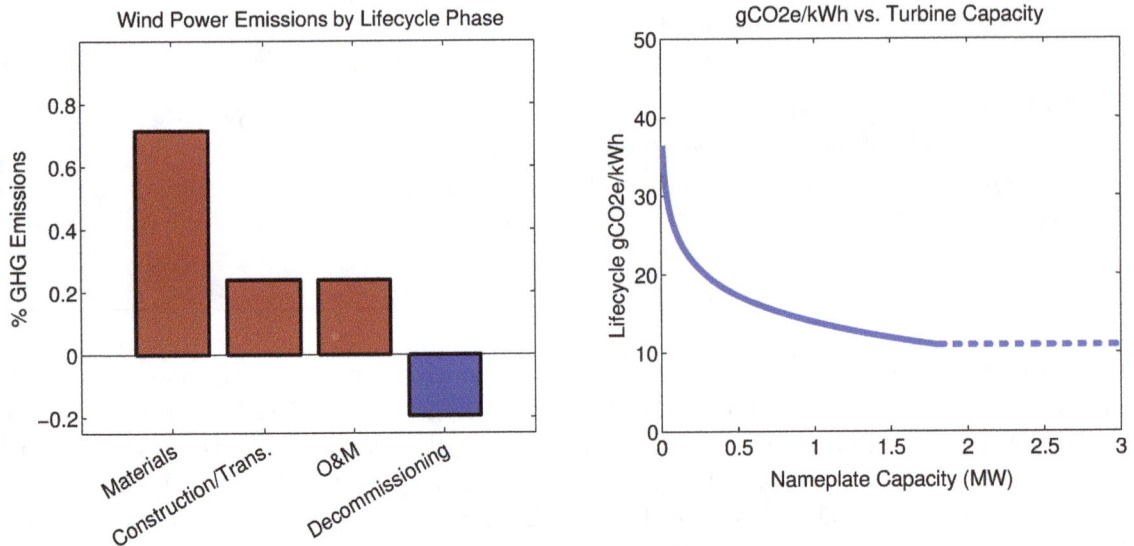

Figure 4.19: The left panel shows the relative contribution to GHG emissions from each major phase of the wind turbine lifecycle (based on [165]). The right panel shows that turbines with lower nameplate capacities tend to have higher associated emissions (based on the relation given in [160]).

4.12.3 Other externalities

Bird and bat mortality

- Wind power does kill birds via collisions, but probably causes fewer overall deaths than fossil power (coal, in particular), for a net benefit to birds. Bat mortality due to turbines is more concerning and less well-understood.

- Wind turbines are responsible for only about 0.015–0.035% of about one billion annual bird deaths attributable to (American) Man.

Bird deaths from wind turbines are probably the best-publicized downside to this power source, and there are likely several hundred thousand such deaths yearly [171]. However, analyses by Sovacool [168, 169] suggested that, both in absolute terms and per kWh of electricity, fossil fuel plants kill far more birds (including direct collisions with cooling towers, power lines, etc., and other ecological effects from fossil fuel extraction and combustion, e.g. mountain-top removal for coal mining), implying wind power is actually an overall boon to birds in terms of direct mortality (although one may wish to read Waldien and Reichard [170] for a clear-minded critique of this work). Furthermore, climate change, driven largely by fossil power plants and countered by wind power, represents a long-term, fundamental threat to the survival of myriad bird species [169].

All this being said, turbine-related avian mortality varies widely with season, location, and species, and wind farms can be very destructive to local bird populations. In particular, California demonstrates much higher mortality than the Greats Plains region, and raptors, which rely on thermals, are especially vulnerable to turbines [171]. Therefore, appropriate siting is an important consideration for future installations [167]. Regional aviation mortality rates for wind power, along with comparison values for nuclear and fossil power plants are given in Figure 4.21.

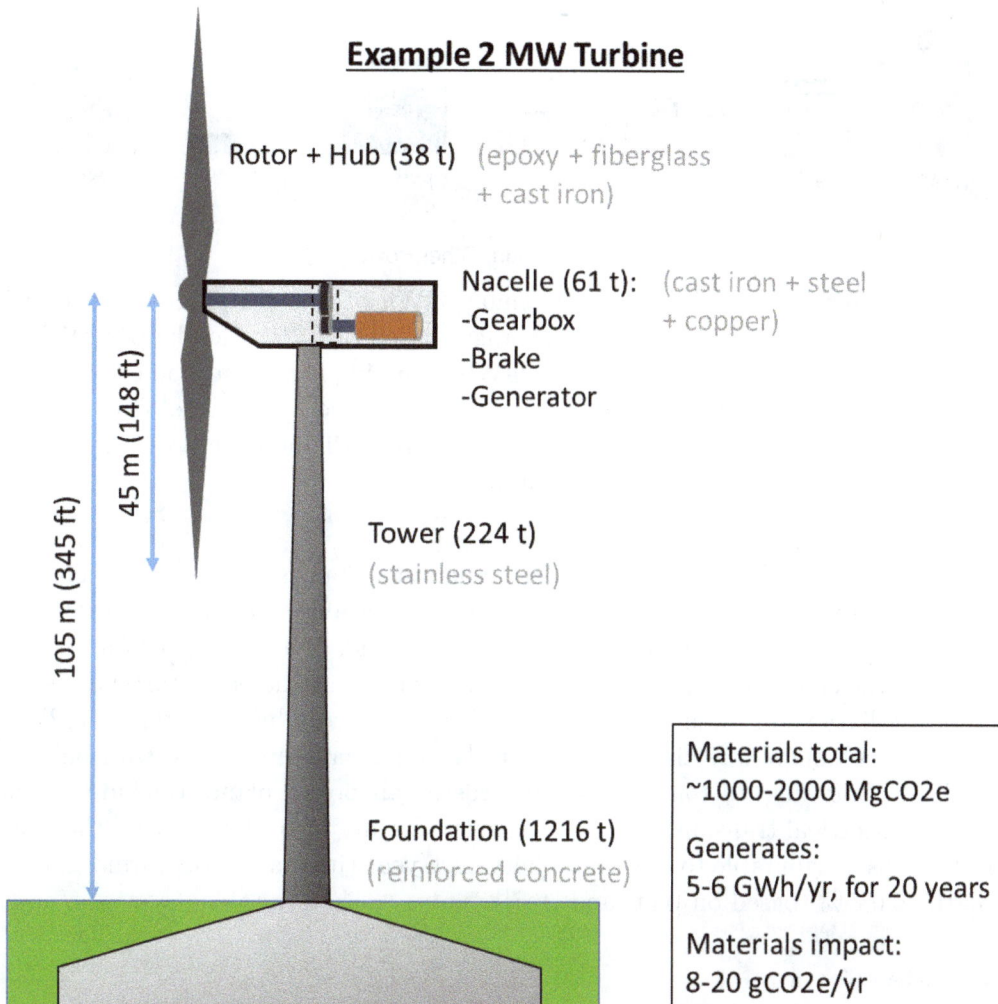

Example 2 MW Turbine

Rotor + Hub (38 t) (epoxy + fiberglass + cast iron)

Nacelle (61 t): (cast iron + steel
-Gearbox   + copper)
-Brake
-Generator

45 m (148 ft)

105 m (345 ft)

Tower (224 t)
(stainless steel)

Foundation (1216 t)
(reinforced concrete)

Materials total:
~1000-2000 MgCO2e

Generates:
5-6 GWh/yr, for 20 years

Materials impact:
8-20 gCO2e/yr

Figure 4.20: Basic schematic, materials requirements, and approximate materials carbon footprint for an example large 2 MW turbine, based on data presented in [166].

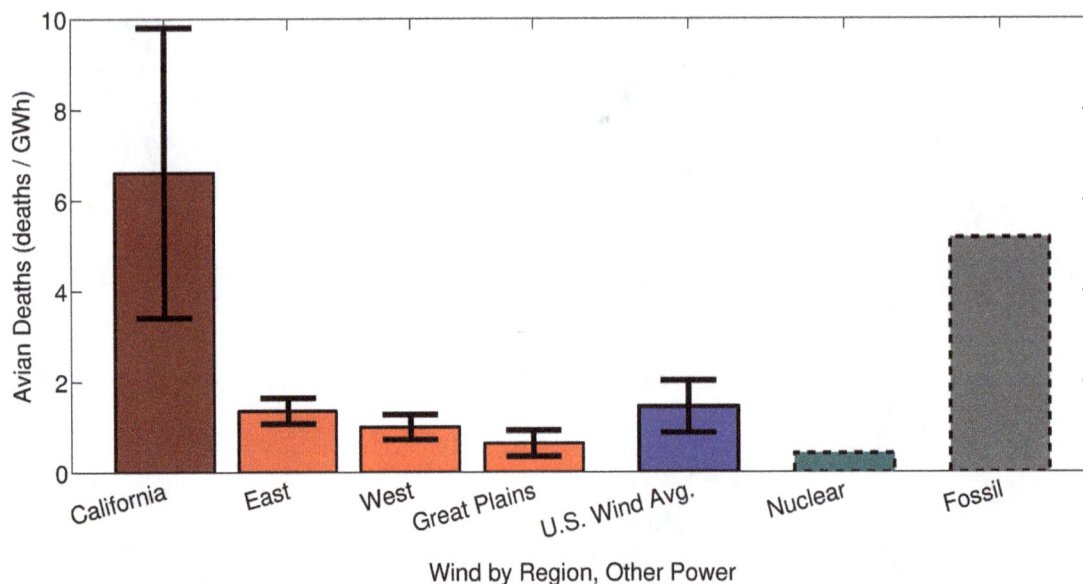

Figure 4.21: Approximate avian deaths per annual GWh of electricity generated by wind for four different regions and as an overall US average, based on [171] (deaths per MW capacity given by [171] have been converted to annual GWh assuming a capacity factor of 32.4%). For comparison, avian deaths from nuclear and fossil power derived by Sovacool [169] are included, but are very uncertain. Note that, fortuitously, the Great Plains have both the lowest avian death rate and the greatest wind power potential.

Bat mortality from turbine collisions is less understood, but is potentially much more concerning, since even though overall death rates may be comparable to those from birds [172], only three major bat species are affected (compared to thousands, for birds), bats do not interact with turbines in the same manner as birds (e.g. bats may be attracted to turbines as potential roosts), and installations in some landscapes could result in very high mortality [172].

Finally, power plants of any kind are a relatively minor hazard to birds, with buildings and windows killing about *twenty times* as many birds as all power plants (including wind) put together [167]. Electrical transmission lines, however, are major bird killers. Of the nearly *one billion* bird deaths attributable to human activity each year (just in the US), wind power causes perhaps 0.015–0.035% (based on [171] and [167]).

Direct climate effects

The turbulent wake of turbines mixes near-surface air to cause daytime cooling and nighttime warming, and wind harvesting may affect local to regional weather and precipitation patterns [174]. The net effect of harvesting wind for 100% of civilization's energy needs could be to very slightly increase global mean temperature (about 0.1 °C) and slightly reduce precipitation (1%) [173]. Recent climate models have confirmed that temperature and precipitation effects are extremely minor on both a regional [174] and global scale [158], under any remotely plausible wind power deployment scenario.

4.13   Solar

While solar panels generate no greenhouse gases during their operation, they are not completely carbon free, as manufacture, materials mining, and maintenance have associated carbon emissions. Most emissions occur from electricity use at the manufacturing stage, where the magnitude of emissions is also dependent on the energy mix of grid electricity. At both the global and US scales, most panels are constructed from crystalline silicon (c-Si), and this is the only solar technology scalable to a significant fraction of global energy use. On average, within the US, c-Si panels likely have associated carbon emissions of about 40–50 gCO_2e/kWh amortized over a 30-year system lifetime, or <5% of the carbon impact of coal. Furthermore, this figure is likely falling with improving technology, and will be lower in sunnier areas of the country, especially the Southwest, where the same panels may generate around 33% more energy over their lifetime as compared to the US average. In this section, I first introduce the physical fundamentals of the solar cell and fundamental limits to its efficiency, and then review the lifecycle carbon emissions associated with solar power.

4.13.1 Basics of solar PV

Several parameters describe the characteristics of a solar system. The *solar irradiance* (or insolation), is the amount of energy that falls upon a unit of area in a given amount of time, and is typically expressed in units of $kWh/m^2/year$. Irradiance varies from as little as 1,000 $kWh/m^2/year$ in northern European countries to 2,400 $kWh/m^2/year$ in the American southwest; 1,800 $kWh/m^2/year$ is the US average, and 1,700 $kWh/m^2/year$ is the average for southern Europe [177, 179]. The *conversion efficiency* gives the fraction of incident energy that is converted into electricity, and is in the range of 14–21% for commercial PV panels. For example, one square meter of panels with a conversion efficiency of 15%, and located in the southwest (solar irradiance of 2,400 $kWh/m^2/year$), would yield 360 kWh in a single year. Conversion efficiency varies with panel type, and is likely to continue to increase in the coming years.

Operational factors such as varying weather conditions and shade can decrease actual solar PV output, and the ratio between actual and maximum power output is referred to as the *performance ratio*. Standard values are 0.75 and 0.8 for rooftop-mounted and ground-mounted systems, respectively [177]. Thus, applying a performance ratio of 0.75 to the previous example, the actual energy generated by one square meter of panels would be 270 kWh instead of the theoretical max of 360 kWh.

4.13.2 Physical fundamentals

Modern solar photovoltaics are based on p-n semiconductor junctions, which spontaneously convert sunlight to electrical current, with silicon the fundamental semiconductor material for almost all PV technology. The essential idea of a silicon solar cell is that (1) light striking silicon generates free electrons (a consequence of the "photoelectric effect," the discovery of which earned Albert Einstein the 1921 Nobel Prize in physics), (2) there is a permanent difference in electrical voltage across the p-n semiconductor junction, and therefore (3) free electrons flow across this voltage difference, generating an electrical current which can be attached to an external load.

To understand this process more fully, first recall from basic chemistry that electrons surrounding an atomic nucleus are restricted to certain "orbitals" which can contain only a fixed number of electrons, proceeding outward from the nucleus in "shells." Silicon (Si), like carbon, has four "valence," i.e. outermost, electrons, but to completely fill its outermost orbitals requires eight electrons. Thus, these four electrons can interact to form bonds and a tetrahedral crystal lattice, as illustrated in Figure 4.22. Now, when a photon strikes Si, if it is of sufficient energy (about 1.1 eV or more), then it may excite an electron, freeing it from the valence band (the electron is then said to enter the conduction band) and allowing it to travel freely (and randomly) through the lattice. This leaves behind a positively charged "hole," i.e. an unpaired valence electron. Like the free electron, as electrons un-pair and re-pair, the hole can also migrate randomly through the lattice. This is an example of the photoelectric effect, but with a pure Si lattice, light does nothing more than create electron-hole pairs that diffuse randomly until recombining, generating no useful current, as also shown in Figure 4.22.

To utilize the photoelectric effect, we must force excited electrons to flow in a preferential direction (and hence generate current), a task accomplished through the introduction of impurities, or "dopants," into our crystal lattice. There are two basic dopant classes: (1) n-type, which adds un-paired electrons to the lattice, and (2) p-type, which creates holes in the lattice. If we stack n-doped silicon on p-doped silicon, un-paired electrons from the n-side fill holes on the p-side, and thus the n-side develops a permanent net positive charge (as it has lost electrons), while the p-side becomes negative. It follows that shining light on such a p-n junction will generate excited electrons that flow unidirectionally to the positively charged n-type silicon. We then place wires on the silicon surface to collect up electrons, connect to an external load, and back to the p-type silicon, and thus electrons are driven in a closed loop, generating electrical current. This process, using phosphorus and boron as dopants, is outlined in Figure 4.23.

Limits to solar cell efficiency

Under standard test conditions, the conversion efficiency for silicon-based solar cells faces a theoretical upper limit of about 32% (the limit is 34% for any p-n solar cell), known as the Shockley-Queisser limit, first described in detail by those authors in 1961 [175] (who initially gave 30% as an upper limit) [176], with more recent refinements by Rühle [176]. Given this, it is rather amazing that actual monocrystalline silicon solar panels have achieved up to 25% efficiency. Thus, commercial efficiencies on the order of 15–20% are actually quite good, and further increases in efficiency are really just icing on the cake. It bears repeating: currently available photovoltaics are remarkably efficient at converting sunlight to electricity, and *it is completely unnecessary to wait for further gains before widespread commercial deployment.*

Fundamental limits. Now, a quick look at the theoretical underpinning of this 32% limit. For p-n solar cells, light energy is lost via three fundamental mechanisms: spectrum losses, radiative recombination, and current impedance. Spectrum losses are the dominant factor. To excite a valence electron to the conduction band requires a specific amount of energy, known as the *band gap energy*, E_g, which is about 1.1 eV for silicon [183]. Now, individual photons have a fixed amount of energy determined solely by the wavelength (or frequency) of light, given as

$$E_{photon} = \frac{hc}{\lambda}, \tag{4.14}$$

where h is the Planck constant, c is the speed of light, and λ is the wavelength of light, and photons of energy less than 1.1 eV are unable to induce photoexcitation. Thus, longwave infrared solar radiation is lost. Furthermore, since it requires 1.1 eV and *only* 1.1 eV to generate an electron-hole pair, any photon energy above the band gap is wasted as heat. Given the spectrum of light hitting the Earth's surface (at solar noon), we can determine that only about 50% of the Sun's energy can be converted into electron-hole pairs, as demonstrated schematically in Figure 4.24; silicon's E_g of 1.1 eV is actually quite close to the optimum.

Radiative recombination is the spontaneous filling of holes by electrons (this releases energy as a photon, essentially the inverse of photoexcitation and the basis for LEDs), and some fraction of our electron-hole pairs will inevitably be lost to this process before they can generate a useful current. For silicon, about 25% of light energy is lost this way. Finally, we lose about 14% of energy to circuit resistance [175, 176]. In total, then, the fraction of light energy converted to electrical current is 50% × 75% × 86% = 32%.

Silicon (Si) Lattice and Photoelectric Effect

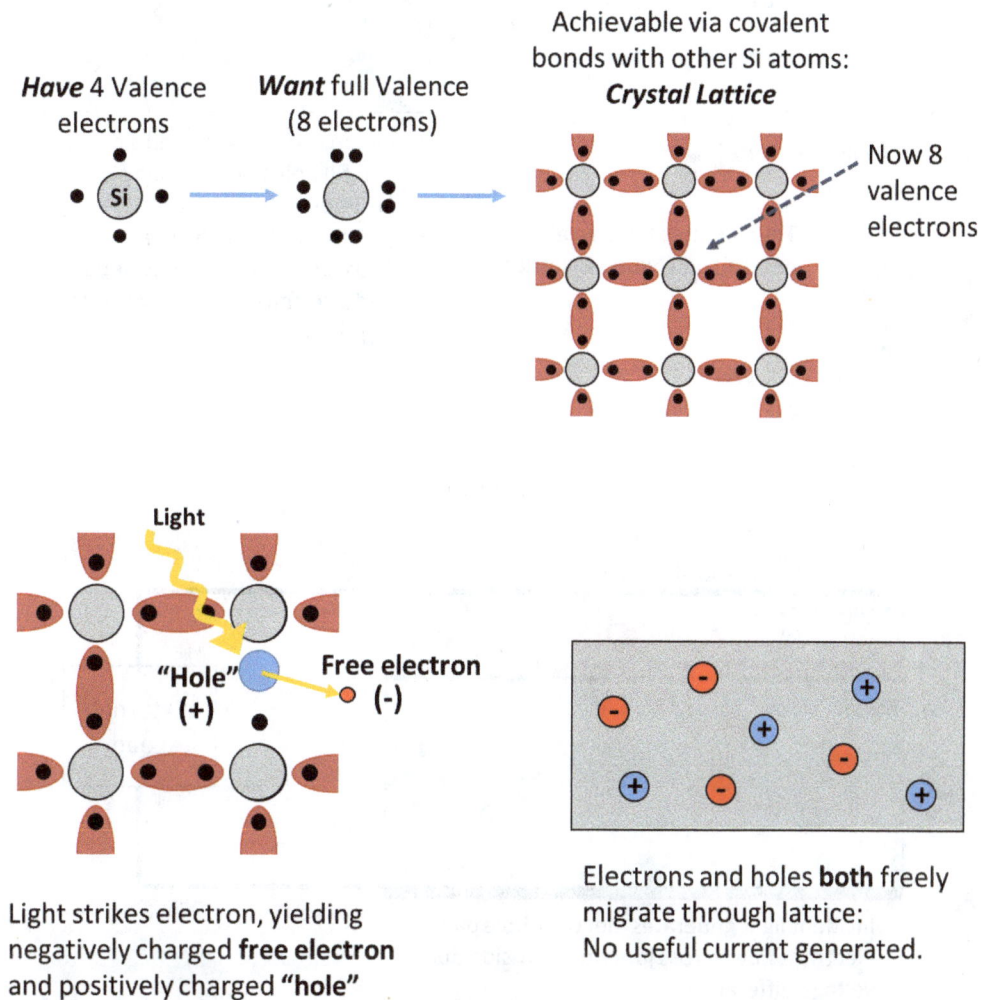

Have 4 Valence electrons

Want full Valence (8 electrons)

Achievable via covalent bonds with other Si atoms: **Crystal Lattice**

Now 8 valence electrons

Light

"Hole" (+)

Free electron (-)

Light strikes electron, yielding negatively charged **free electron** and positively charged **"hole"**

Electrons and holes **both** freely migrate through lattice: No useful current generated.

Figure 4.22: Silicon (Si) has four valence electrons, while a full valence consists of eight. Thus, electrons from four Si atoms can join pairs (covalent bonds), yielding a tetrahedral lattice (a 3-D structure but shown schematically in 2-D). Sufficiently energetic light excites electrons to yield a free electron and hole, which migrate randomly until recombining.

Dopants and the p-n Junction

n-type Dopant

Phosphorus (P) has 5 valence electrons

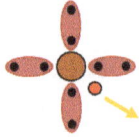

Thus, unpaired electron when in lattice → freed easily
Leaves:
(+) nucleus, (-) free electron

p-type Dopant

Boron (B) has 3 valence electrons

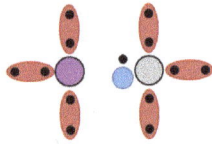

Thus, creates (electrically neutral) hole when in lattice

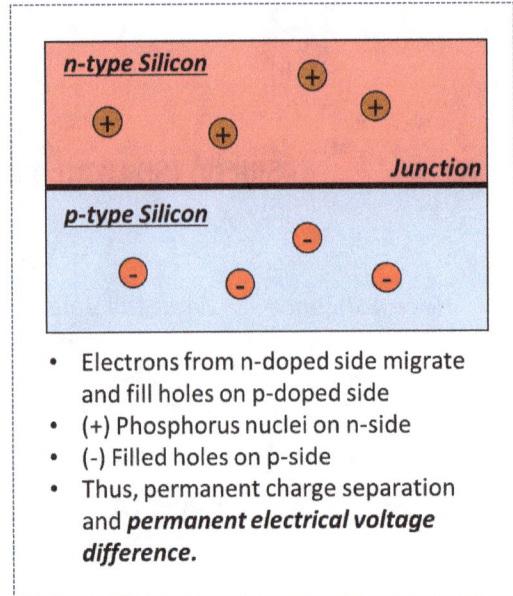

n-type Silicon

Junction

p-type Silicon

- Electrons from n-doped side migrate and fill holes on p-doped side
- (+) Phosphorus nuclei on n-side
- (-) Filled holes on p-side
- Thus, permanent charge separation and *permanent electrical voltage difference.*

Silicon Solar Cell

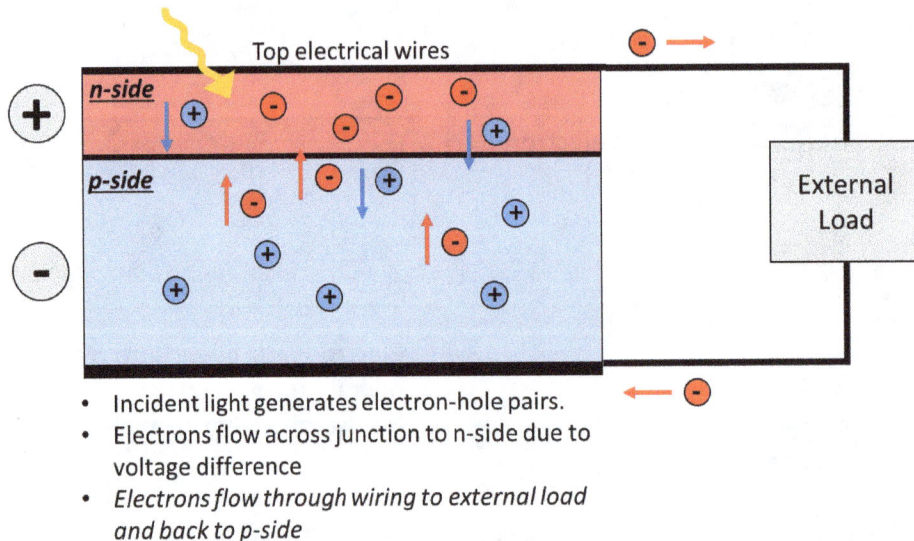

Top electrical wires

n-side

p-side

External Load

- Incident light generates electron-hole pairs.
- Electrons flow across junction to n-side due to voltage difference
- *Electrons flow through wiring to external load and back to p-side*

Figure 4.23: As shown in the top, n-doping of silicon entails adding phosphorus (P), which has five valence electrons. Within the Si lattice, P impurities yield unpaired electrons which readily dissociate from their nucleus. Conversely, p-doping with the addition of boron (B), which has only three valence electrons, leaves holes in the Si lattice. Upon sandwiching a layer of n-type and p-type silicon, free electrons from the n-layer fill the the p-layer holes, leaving excess positive charges in the n-doped silicon, and thus we set up a permanent charge separation and voltage drop across the p-n junction. When light shines upon a p-n junction solar cell, excited electrons are driven across the p-n junction to the n-side by the voltage difference, and then may pass through wiring across an external load and back to the p-side. Thus, an electrical current is generated.

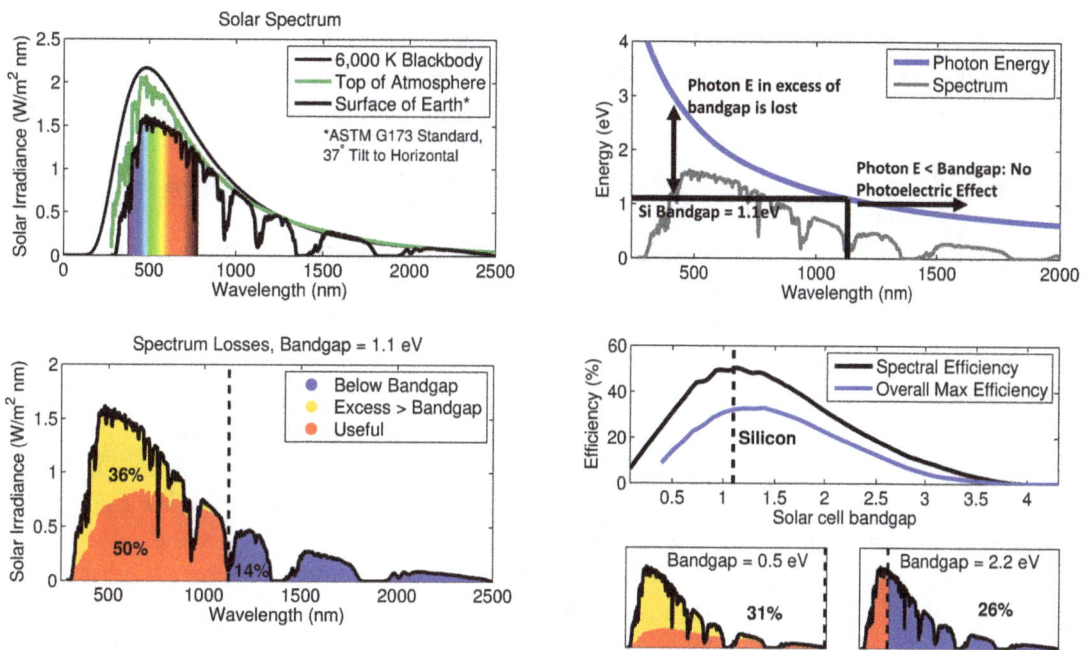

Figure 4.24: Schematic for spectral losses in a solar cell based on a single p-n junction. Photons with a long wavelength and photon energy below the bandgap, which is 1.1 eV for Si, are unable to induce photoexcitation and thus their energy is lost, while any photon energy in excess of the bandgap is also lost, leaving only about 50% useful energy across the solar spectrum when the bandgap is 1.1 eV (spectral efficiency). On the right we see how maximum spectral and actual efficiency vary with the bandgap, and spectral efficiency under two example alternative bandgaps, 0.5 and 2.2 eV, is also demonstrated (under which only 31% and 26% of light energy is useful, respectively).

Other efficiency losses. In practice, several other mechanisms also limit efficiency. Untreated silicon reflects 36% of incident light, but coating and texturing can reduce reflection losses to nearly zero. Electrical wires overlying the solar cell (necessary to collect electrons) "self-shade," blocking a small fraction of sunlight. High temperatures also degrade solar cell performance by increasing lattice energy and interfering with the movement of charge carriers [183].

4.13.3 Emissions factors and energy pay-back time

The fabrication of crystalline silicon(c-Si) PV modules is an energy-intensive process, requiring on the order of 1,000–1,500 kWh/m^2 of panel area of *primary* energy. About 75% of energy used is electrical, where each kWh(e) corresponds to about 3 kWh of primary energy, and monocrystalline-silicon (mo-Si) modules require perhaps 15–35% more energy than polycrystalline-silicon (Si) to fabricate [177]. Energy payback times for silicon panels are reported to be on the order of 1.5 to 3 years [177]. For example, an array exposed to 1800 kWh/m^2/yr with 15% efficiency and a 75% performance factor generates 202.5 kWh(e)/yr, which is roughly equivalent to 550 kWh of primary fossil energy (assuming 37% generating efficiency), and yielding energy payback after about two years.

The upstream emissions from solar module fabrication translate into roughly 40–50 kgCO$_2$e/ kWh amortized over a 30-year system lifetime for c-Si-based technologies [177, 178], while thin-film technologies are probably slightly cleaner, at around 20–30 kgCO$_2$e/kWh [179, 177]. While some earlier (1990s) analyses suggested much higher GHG emissions associated with solar PV, marked reductions in silicon requirements, lower energy silicon processing, recycling technology, and increasing panel efficiency since then have significantly improved solar PV's environmental profile (as reviewed by Fthenakis and Kim [177]), and lifecycle emissions are likely to continue to fall by as much as 50%. Lifecycle emissions, being total emissions divided by lifetime energy generation, are also strongly affected by solar irradiance and hence can vary more than twofold by region, with the US, and especially the southwestern US, a relatively favorable region (this regional dependence also holds for energy payback time). I now briefly review the production process and embodied energy for c-Si panels, since these account for over 90% of solar capacity globally, omitting a detailed treatment of thin-film panels (see, e.g. [179] for an in-depth discussion), as well as novel PV technologies not currently in widespread use.

Silicon PV lifecycle emissions are nearly all generated at the fabrication phase, with purification of crystalline silicon dominant. As schematized in Figure 4.25, and discussed in detail in a moment, the fabrication process entails (1) silica (SiO$_2$) mining, (2) purification to metallurgical grade silicon (99% pure), (3) purification to electronic- or solar-grade silicon (99.9999% pure), (4) silicon ingots cut to thin wafers, (5) wafer doping, (6) wafer etching and texturing, (7) electrode screen printing and anti-reflective coating, and finally (8) assembly into modules. Subsequently, modules are deployed to form the basis of a full PV system, which requires additional mounting and electrical components ("balance of system" components), and is illustrated in Figure 4.26.

Silicon is the second-most abundant element in the earth's crust, and it is commonly found as silicon dioxide (SiO$_2$, "silica") in sand and gravel ("silica sand" or "quartz sand"). Mined quartz sand is purified to metallurgical grade silicon (MG-Si, about 98–99% pure silicon) via reaction with carbon in an electric arc furnace at 1,500–2,000 °C [180], consuming about 12 kWh/kg MG-Si [182]:

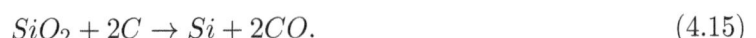

$$SiO_2 + 2C \rightarrow Si + 2CO. \tag{4.15}$$

Now, for a solar cell p-n semiconductor we require a silicon crystal lattice pure to one part

116

in one million (solar grade silicon, SG-Si), while for electronics the lattice must be pure to one part in a billion (electronics grade silicon, EG-Si). Either is achieved by processing MG-Si in a Siemens reactor, whereby silicon metal is volatilized via reaction with hydrochloric acid (HCl) into trichlorosilane ($SiHCl_3$, TCS) gas, which is then purified by distillation. In the traditional Siemens process, TCS is converted to extremely pure Si metal via chemical vapor deposition onto a U-shaped silicon filament electrically heated to about 1,100 °C; this yields electronics-grade silicon [182].

A modified Siemens process can be used instead to produce SG-Si of lower purity. Trichlorosilane is converted to silane (SiH_4) via a series of reactions, and silane then deposits upon the seed silicon filament at 850 °C, representing a significant energy savings. This purification step, either via the Siemens or modified Siemens process, yields a large polycrystalline rod, and represents the largest use of energy in the solar production, with the Seimens process consuming on the order of 100 kWh/kg EG-Si [182]. An alternative technology, the fluidized bed reactor (FBR), uses only 10–30% the energy, but currently has only about a 10% market share.

Next, the polycrystalline rod is melted down and faces one of two fates. For polycrystalline solar cells, it is cast into a mold and then cut into 15.6 × 15.6 cm wafers. Alternatively, monocrystalline silicon can be produced by the *Czochralski process*, in which a seed crystal is dipped into our molten silicon and slowly withdrawn and rotated, giving a large mono-crystalline ingot, which is then cut into wafers. Wafers must then be "etched," via a chemical bath to remove saw damage, doped with impurities (e.g. phosphorus), and the surface is randomly textured and given an anti-reflective coating to reduce reflectance from 35% to < 5% [183]. A silver electrode screen is applied to the cell's surface, and production is complete. Each cell produces about 0.5 V, so typically 60 or 72 cells are wired together to produce a single module.

Balance of system (BOS) components account for around one-third of the overall energy and carbon footprint of solar PV, with Fthenakis and Kim giving 15 gCO_2e/kWh attributable to BOS [177]. These components include aluminum frames, mounting systems, electrical wiring, and inverters for converting the DC power produced by solar cells to AC.

I add 50 gCO_2e/kWh to all solar capacity for eGRID corrections, although given how little power is solar-generated on the national scale, this has no appreciable effect on electricity EFs.

4.14 Limits to renewable energy and energy transition

The public conversation around renewable energy is rife with controversy, and very frequent assertions to the effect that (1) renewable energy is not scalable, (2) it is too intermittent, and (3) it is too expensive. Solar energy, in particular, as a highly (but predictably) intermittent source that lends itself to distributed generation is subject to competing claims. Let us examine these objections in some detail, with the goal, as always, not to prove one side right or wrong, but to understand the issues with a clear mind. Note that in this discussion, I generally exclude biomass, the scalability of which is addressed in Section 3.8. Furthermore, it is important to remember that while most public conversation in the US centers around electricity generation, this accounts for only about 40% of primary energy use, and decarbonizing heating and transportation fuels is a further challenge. In general, I conclude that renewable resources, especially wind and solar, *are* sufficient to provide all electricity and all energy in general, from a basic resource and materials standpoint, from a cultural (i.e. grid integration) standpoint, and from a financial perspective. Nuclear is an "honorary" renewable technology that can, at least in principle, also play a role in a low carbon energy future.

Silicon purification

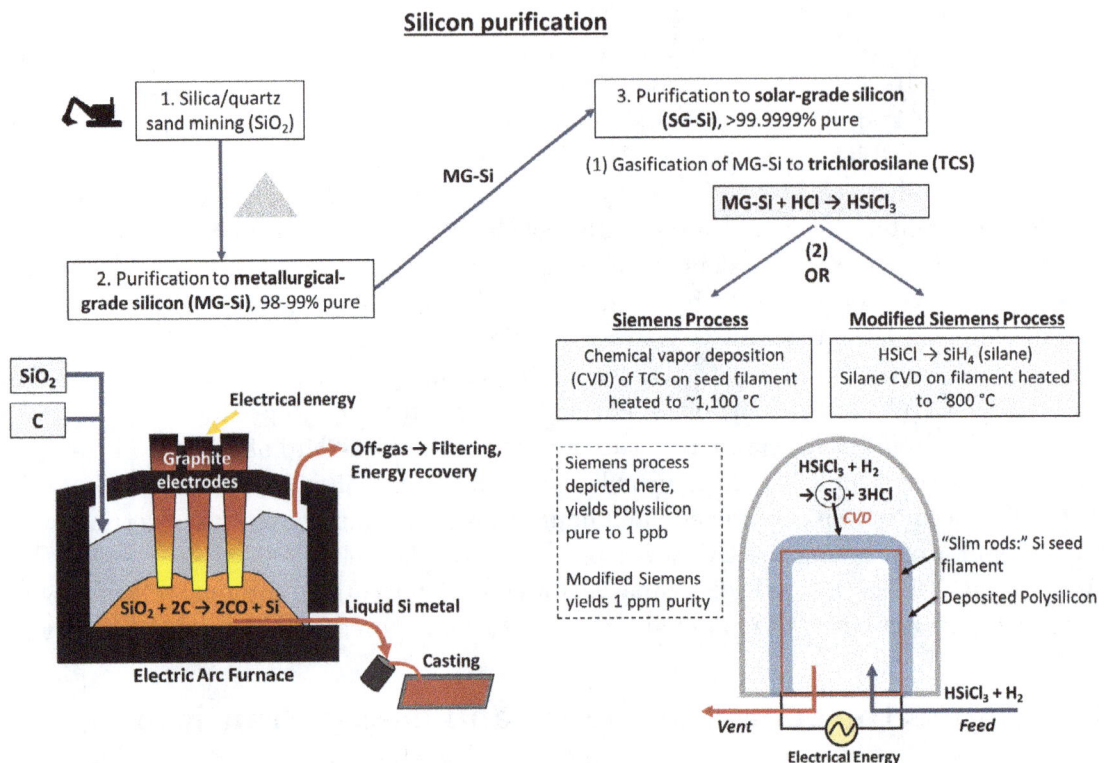

Figure 4.25: Schematic for the process of purifying quartz sand to solar- or electronics grade silicon. First, raw mined silica feedstock is fed, along with carbon, to an electric arc furnace, to yield 98–99% pure metallurgical-grade silicon (MG-Si). Much higher purity is then achieved at the major energy-consuming stage, by which either the Siemens or modified Siemens process, volatilized Si is heated and deposits upon a seed silicon filament.

From Purified Polysilicon to a Complete Solar System

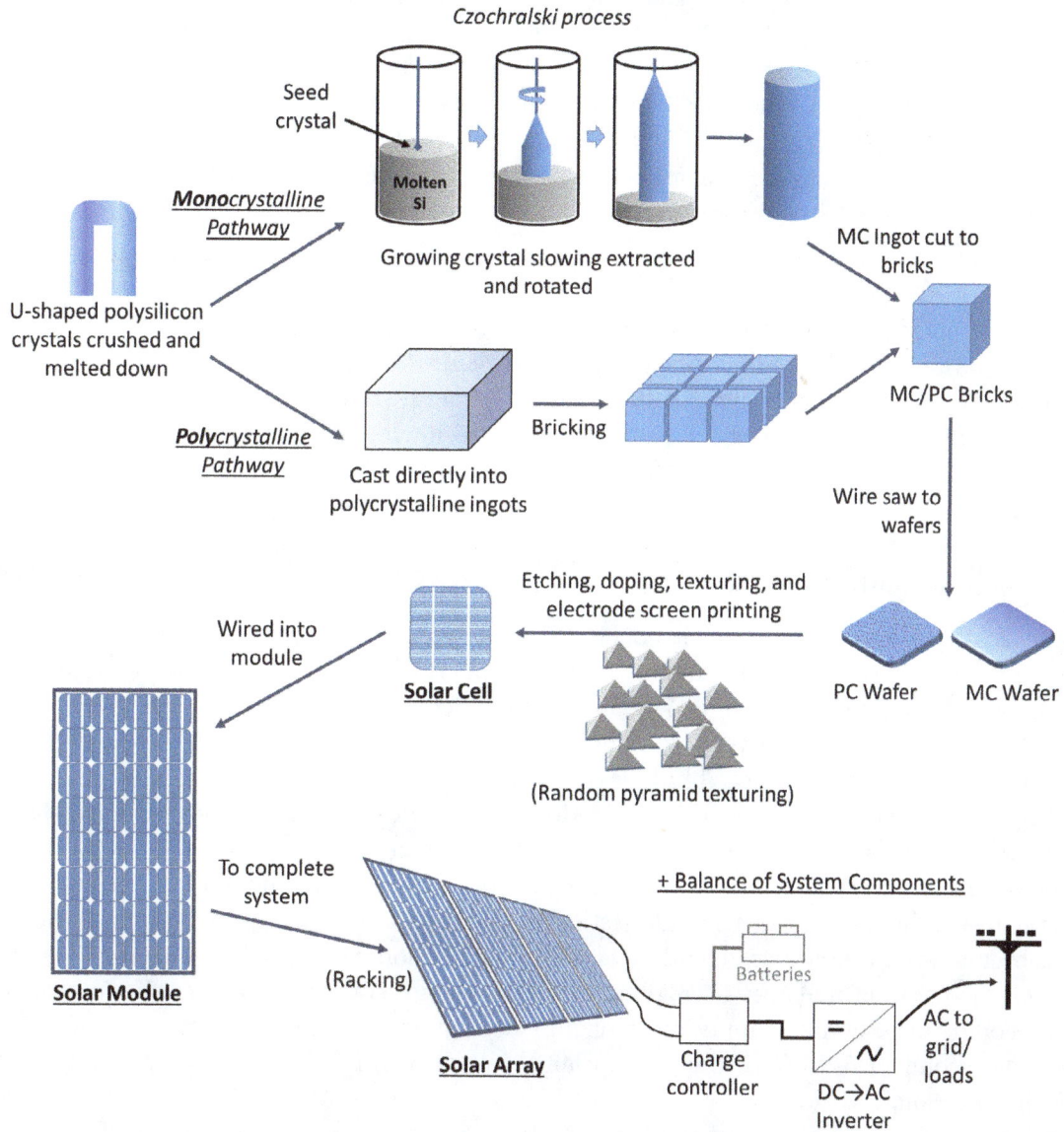

Czochralski process

Seed crystal

Molten Si

Monocrystalline Pathway

Growing crystal slowing extracted and rotated

MC Ingot cut to bricks

U-shaped polysilicon crystals crushed and melted down

Polycrystalline Pathway

Cast directly into polycrystalline ingots

Bricking

MC/PC Bricks

Wire saw to wafers

Etching, doping, texturing, and electrode screen printing

Wired into module

Solar Cell

PC Wafer   MC Wafer

(Random pyramid texturing)

To complete system

+ Balance of System Components

Batteries

Solar Module

(Racking)

Solar Array

Charge controller

DC→AC Inverter

AC to grid/ loads

Figure 4.26: Schematic for solar cell fabrication from a polycrystalline ingot, and then incorporation into a deployed solar system.

4.14.1 Fundamental resource, land, and material limits to scaling up of renewables

---

- Global geophysical resources, land, and materials are all sufficient to power the world entirely with wind and solar. This also holds at the US national scale.

- The global land base is sufficient to provide all of earth's *electricity*, but possibly not all *energy*, from wind alone at plausible levels of deployment (1.5–2.0% of global land).

- Material demand, mainly steel and concrete, is not a significant constraint to large-scale wind power. Rare earth element supplies may limit the expansion of turbines that use permanent magnet generators, but these are a minority of all turbines.

- About 0.5% of global land could supply all earth's energy (including electricity) from solar PV. Solarizing about one-third of US rooftop area could provide all the nation's electricity.

- Metal availability limits thin-film PV technologies, but is unlikely to constrain crystalline-silicon PV (the dominant technology).

Intermittency, demand-supply balancing, incorporation into existing energy transportation and distribution networks, etc. are at least partly cultural, rather than technological, limits to renewable energy. The physical availability of energy resources, and the land-base and materials required to extract them (under existing technologies) represent more fundamental limits. Fortunately, as we shall see, all of these are sufficient for a world powered entirely with wind and solar.

Fundamental limits to wind

Geophysically, Marvel et al. [173] concluded that harvestable near-surface wind resources amount to over 400 TW, equal to about 3.75 trillion MWh/yr, or around 24 times global primary energy consumption and 174 times global electricity generation. However, this calculation assumes uniform harvesting across the entire globe, both land and sea, and the real question is, how much could feasibly be extracted?

Only 14.894 billion Ha out of 51.007 billion Ha of global area is land (CIA World Fact Book), and perhaps 13% of land experiences average wind speeds $\geq$ 6.9 m/s at 80 m [184], where wind harvesting is relatively economical. Turbines extract kinetic energy from the air, with the result that large-scale wind harvesting is clearly limited to around 1 W(e)/m^2 [186, 159], and therefore such a land-base could yield about 170 billion MWh, roughly eight times global electricity generation and just over global primary energy consumption. It follows that wind energy could, in principle, supply most or all electricity using only a fraction of the suitable land area, around 1.5–2.0% of the globe's land, but providing *all* energy from wind is a less likely proposition.

Note that several published estimates, e.g. [184, 185], by ignoring the limiting effect of large-scale turbine deployment on wind energy extraction (turbines sap the air of kinetic energy), have likely significantly overstated feasible wind resources severalfold, as elaborated by Adams and Keith [186]. For example, an estimate by Archer and Jacobson [184] of 72 TW extractable from the 13% of land with economically sufficient wind speeds corresponded to a power density of 4.3 W/m^2: too high by a factor of four.

Wind turbines require large amounts of steel, concrete, fiberglass and epoxy for the tower, foundation, and blades, respectively, but supply of these materials is not likely to be a major barrier to large-scale wind harvesting [188]. The rare earth elements (REEs) neodymium (Nd) and dysprosium (Dy) are components of permanent magnet electric generators used in gearless direct-drive wind turbines, which were introduced in 2003, and reached a 14% market share by 2007 [190] (induction generators, used in the majority of turbines, do not require REEs [191]). Because exploitable reserves of these elements are rather scarce, and nearly all REEs are mined and produced in China, which recently instituted export quotas, they represent a potential material limitation to upscaling of *direct-drive* wind power and, far more importantly, motors for electric vehicles [190]; several authors have examined REEs as limiting factors for renewable energy technologies, e.g. [189, 190].

In 2010, wind turbines accounted for only 1% of Nd and Dy use, and were projected by Habib and Wenzel [190] to remain a minor end-use in 2050 under multiple energy scenarios, including a 100% renewable energy scenario (electric vehicles were the dominant user in this case). Therefore, high REE demand due to background uses (e.g. consumer electronics) and electric vehicles could limit the penetration of direct-drive turbines in the future, but REEs are *not* a material constraint on wind power in general.

Limits to solar

- Geophysical resources and the land base do not limit solar; existing urban rooftops are more than sufficient to provide all global electricity.

- Metal availability strongly limits some thin-film PV technology (CdTe, CIGS), but silicon reserves are vast.

- Silver (used for electrodes) is the only significant material limitation to very large-scale (terawatt) crystalline silicon PV, but PV silver content is falling rapidly, and copper is a viable replacement metal.

Geophysical solar resources vastly exceed civilization's energy consumption. Jacobson [187] reports 1,700 TW of solar PV potential over the globe's land, corresponding to nearly 100 times global primary energy demand. The continental US alone (766 billion Ha in area, about 5% of earth's land), assuming an average solar insolation of 1,800 kWh/yr, energy conversion efficiency of 15%, and a performance factor of 75%, could in principle supply about ten times the global primary energy demand. It follows that, using reasonably well-sited solar PV, about 0.5% of earth's land could supply all global energy (supplying all electricity would require a miniscule 0.07% of global land); even in energy-hungry America, just 0.27% of the continental US could supply all the nation's electricity. Note that urban rooftop area is likely sufficiently large to meet all solar land requirements, and thus, if distributed solar is seriously pursued, few new lands need be devoted to solar farming (see calculations in Section 4.14.2 below).

Existing metal reserves and extraction rates *strongly* limit the scalability of CdTe and CIGS thin-film PV panels (currently <10% of PV), and these technologies can never provide more than a few percentage points of the world's electricity [192]. Cystalline silicon-based PV, on the other hand, is not limited by silicon availability: silicon reserves are vast, and 100% of global electricity could by supplied by c-Si PV in 2030 without Si production growth exceeding historical growth rates [192].

Silver (Ag), which is screen-printed onto panels to form the top electrodes in c-Si cells, does potentially limit the deployment of silicon PV at terawatt scales [193]: it would take on the order of 50–100% of existing silver reserves (about 570 kilotonnes [195]) to provide all global electricity with c-Si PV at current use rates, which were reported to be about 4 g Ag/m^2 in 2015 by [196], a marked reduction from even a few years earlier (some slightly earlier estimates are 57 mg Ag/W in [192] or 8.2 g Ag/m^2 [194]). Silver is also used at lower rates in concentrating solar power (CSP) as a highly reflective mirror coating, at about 1 g Ag/m^2 (CSP achieves electrical conversion efficiencies in the 15–20% range, similar to PV) [194].

In general, silver is a major limiting material for industrial civilization, as it is widely used in electronics (silver has the highest electrical conductivity of any metal), photography (silver halides are light-sensitive, but are being replaced by digital photography), and other industrial applications; jewelery is the other major use of silver, recently demoted to second place after electronics. Rather remarkably, about 75% of all silver reserves have already been mined (with ongoing mining since at least 4,000 B.C.E.), and at current mining rates (27,300 tonnes/year in 2015 [195]), all reserves will be exhausted in only two decades; nearly all silver must come from recycled sources in the very near future. One may consult Grandell and Thorenz [194] for further discussion of the silver cycle and implications for solar technology.

Despite the rapidly approaching "peak silver," this metal is unlikely to ultimately limit solar PV (or CSP) [192], as its use in c-Si PV is projected to fall by over half from 2015 to 2026 [196], and copper will probably eventually displace silver, with silver PV content falling by an order of magnitude. Copper is a viable replacement metal that has been used in several commercial (but sadly now defunct) panels: "Saturn" panels by BP-solar, "Pluto" panels by Suntech, and, most recently, highly efficient TetraSun panels employing copper electrodes were briefly produced by First Solar until discontinuation in the summer of 2016. Concentrated solar power uses silver at a low rate compared to PV, and aluminum is an alternative mirror coating if needed.

4.14.2   Is rooftop solar scalable? Some back of the envelope calculations

- Solarizing roughly one-third of the US's urban rooftop area (0.27% of total continental US land area) could, in principle, supply all US electricity. Concentrating generation in the South and Southwest would reduce this fraction.

Total US electricity generation in 2010 was 4,125.847 billion kWh (eGRID), and therefore, using the US average irradiance of 1,800 kWh/m2/yr, module efficiency of 15%, and a performance ratio of 0.75, it would take about 7,900 square miles of PV panels to replace all 2010 generation. The fraction of urban area that is rooftop is 20–25% [197]; assuming 20% roof area, we would then require 39,333 square miles of urban area, with every single rooftop completely covered by PV panels.

Total urban area in the US is, as of the 2010 Census, 106,000 square miles. Thus, solarizing 37% of US urban rooftop area (conservatively assuming 20% of urban area is rooftop) could theoretically meet the entire US electricity demand (only 27% of rooftop area would be needed if average module efficiency increased to 20%). Consider as a local example the case of California, which generated 204 billion kWh in 2010. Under the higher southwest irradiance of 2,400 kWh/m2/yr, the entire state's demand could be met if 84% of rooftop area in the Los Angeles/Long Beach/Anaheim urban area (1,736 square miles total) was solarized.

4.14.3   Some notes on renewable variability and grid integration

The variability of wind and solar energy resources is a barrier to adoption, but one that does appear to be insurmountable (and even if it is, there is no alternative to an eventual fossil-free future). Several studies have concluded that using existing technologies, wind and solar, along with nuclear and large hydropower, can provide essentially 100% of electricity (and potentially all heat and transportation energy as well) at US regional, US national, various European country, continental, and global scales on a decadal timescale, at modest cost or even net savings compared to fossil-based generation [198, 199]. That does not mean that this will happen, but it does mean that the challenge facing an energy transition is primarily one of political will, and not one of technological (or even economic) constraints.

Briefly, to understand the challenge wind and solar represent, regional electricity grids classically operate on under a dispatch model, where to match electricity supply with demand, we have a hierarchy of generating plants—baseload, load-following, and peaking—that are *dispatched* as the load on the grid increases. At the bottom, baseload generating plants, classically coal-fired or nuclear steam plants, operate near full capacity at nearly all times (excepting scheduled maintenance). Combined cycle natural gas (CCNG) plants also generally operate as baseload generators, but have some ability to load-follow, as do some nuclear plants. Load-following plants are intermediate and may ramp up or down to meet demand. Other natural gas turbines may operate either as load-following plants or peaker plants. Economically, there is a clear relationship between resource type and generating costs, with baseload generation cheapest, and peak resources by far most expensive, and hence the economic incentive to avoid energy use at peak hours.

To cope with regularly fluctuating demand that does not coincide with fluctuating and sometimes erratic generation, the so-called *supply-demand balancing* problem, a *local* renewable electricity system (disregarding flexible hydropower generation) will generally require either supplemental fossil-fuel generation and backup, or significant energy storage capacity (or both). Even under such a model of localized renewable generation, intermittent renewables can significantly displace fossil generation, with several European countries already generating around 20% of their power portfolios with wind and solar. While wind and solar may vary unpredictably at local scales, at the large regional scale variability smoothes out, and with regional grid interconnections most US electricity demand could, in principle, be met purely by wind, water and solar, *without* energy storage or significant fossil backup using the *current* electrical grid [198]. In a future system, complete energy system electrification in combination with thermal energy and hydrogen storage could also meet 100% of global demand without any fixed battery storage [200]. A well-designed near-term regional system could still likely provide 33–50% of power with primarily wind and solar renewables with little over-generation [201], and one study concluded, for example, that Germany could generate up to 50% of its electricity with wind and solar without any energy storage [202]. In sum, renewable integration is a genuine but not intractable technical challenge, and this should highlight the value of energy conservation, which can both reduce the total amount of fossil energy that must be replaced, and synergize with renewable generation to decrease carbon emissions.

Part II

Transportation

Chapter 5

Total per capita transportation emissions

- On average, each American generates 5.87 $MgCO_2e$ per year from personal transportation, with 73% of this due to the production and combustion of fuel in personal vehicles and another 8% attributable to vehicle manufacture; including infrastructure, this total increases to 6.58 $MgCO_2e$ per year. Aviation is the only other significant emissions source, at just under 1 $MgCO_2e$ per year per person (including private aviation).

- Average transportation household emissions are on the order of 14.9 and 16.4 $MgCO_2e$ without and with infrastructure, respectively.

- Given that only about two-thirds of Americans are licensed to drive, the average *driver* is responsible for 40% more transportation CO_2e than the average American (about 8.25 $MgCO_2$ per driver, from all transportation excluding infrastructure).

Before addressing various transportation modes in greater depth, I take this opportunity to summarize the per capita emissions sums derived in much more detail in the following chapters. As an overall average, gasoline combustion in personal vehicles generated (both directly and indirectly) about 4.36 $MgCO_2e$/capita in 2013, with about 80% of this total due to the tailpipe emissions, and the balance due to upstream gasoline extraction, refining, etc. Carbon equivalents attributable to the manufacture and maintenance of the light-duty fleet was on the order of 0.49 $MgCO_2e$/capita[1], and thus light-duty vehicle emissions sum to just under 5 $MgCO_2e$ per person. This is when averaging across the US populace; if we consider emissions for typical *drivers*, then we get about 6.51 $MgCO_2e$ from fuel combustion and 0.72 $MgCO_2e$ from vehicle manufacture, or over 7 $MgCO_2e$ per driver.

Including non-CO_2 forcing from high-altitude jet fuel combustion, US per capita emissions attributable to commercial aviation were on the order of 0.87 $MgCO_2e$/capita in 2015, and including private general aviation bumps this total to 0.95 $MgCO_2e$/capita. However, aviation CO_2e attributable to individuals will vary widely, given the high variation in flights per person, and there is also uncertainty in the magnitude of non-CO_2 aviation forcing.

On an overall per-capita basis, public and collective transportation, including Amtrak, intracity rail, bus, and long-distance bus travel add a trivial amount of CO_2e: Heavy bus transit systems may generate a total of 2.16 million $MgCO_2e$ (fuel-cycle only), with heavy, commuter,

---

[1]Derived using 123.77 billion gallons of gasoline, a fuel-cycle gasoline EF of 11.146 kgCO_2e/gallon, 236 million registered light-duty vehicles, and assuming 9.2 $MgCO_2e$/vehicle amortized over 14.1 years; these factors are developed in Chapter 6.

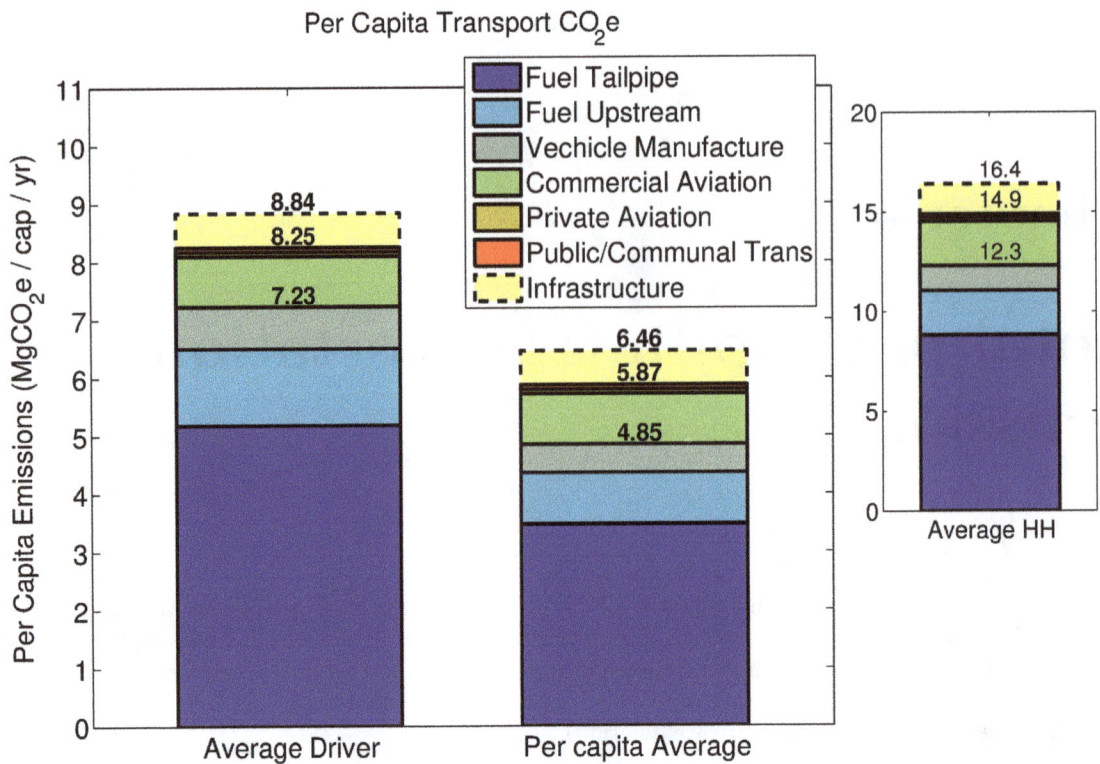

Figure 5.1: Annual US per capita emissions due to transportation, either for the average person or driver, and for the average 2.54 person US household. Numbers inset into bars give the sums for personal vehicles, everything but infrastructure, and for everything.

and light rail transit systems together adding about 5.69 million $MgCO_2e$. Amtrak emissions sum to 1.47 million $MgCO_2e$, while the school bus system and intercity bus (motorcoach) travel generate 10.28 and 3.80 million $MgCO_2e$, respectively (fuel-cycle only). Taken together, these all sum to just 23.4 million $MgCO_2e$, equivalent to about 0.073 $MgCO_2e$/capita (in 2014).

Finally, the construction and maintenance of the transportation infrastructure also carries a non-trivial carbon cost, perhaps on the order of 0.55–0.60 $MgCO_2e$/capita for roadways, parking, and street lighting combined (around 190–200 million $MgCO_2e$ in all), while airport and rail infrastructure add almost nothing. Although individual-level choices are unlikely to directly affect these infrastructure emissions except in the aggregate and over the longer-term, it is important to at least recognize their existence. Figure 5.1 summarizes all these numbers, giving emissions for the average American, average driver (assuming all other modes are equal to the average), and the average 2.54 person household.

Chapter 6

Fundamentals and the personal automobile

- The combustion of primarily gasoline, and upstream fuel production, in light passenger vehicles generates about 1,375 million $MgCO_2e$, or about *one-fifth* of all US territorial emissions. The personal automobile is the single greatest source of greenhouse gas (GHG) emissions for a typical adult American.

- Each gallon of gasoline burned generates a little over 11 kg of CO_2e, with 80% from direct tailpipe emissions, while the other 20% is from upstream "well-to-pump" (WTP) processes, i.e. oil extraction and refining. Tar sands oil at the center of the Keystone XL controversy are associated with about twice the WTP emissions of typical oils.

- The average driver generates about 6.5 $MgCO_2e$ yearly from fuel consumption (the average vehicle generates about 5.9 $MgCO_2e$ yearly, but there are more vehicles than drivers), and the average passenger vehicle on the road gets 21.6 MPG. Additionally, about 0.7 $MgCO_2e$ per year are attributable to vehicle manufacture (for an average car), implying that a typical driver ultimately generates a full 7.2 $MgCO_2e$/year.

- There is a *hyperbolic*, not linear, relationship between MPG and GHG emissions. It follows that upgrading from a low-MPG to mid-MPG vehicle is much more beneficial than upgrading from a mid- to high-MPG vehicle.

- Fuel consumption scales essentially linearly with vehicle weight, implying that, for conventional gas-powered (non-hybrid) vehicles, size is fuel.

- Vehicle manufacturing is responsible for about 10% of lifetime vehicle emissions, with manufacturing emissions, like fuel use, scaling linearly with vehicle weight.

- Emissions from gasoline hybrid-electric vehicles are 30–45% lower than those of comparably-sized conventional vehicles, and the best hybrids are 50–60% less emitting than the average passenger vehicle; this includes the entire vehicles lifecycle.

- Electric vehicles (EVs) are *not* zero-emissions, due to the power-plant emissions from generating electricity. Emissions-intensity varies geographically, but on average most EVs are comparable to gasoline-only hybrids. The upstream emissions associated with EV battery manufacture are also non-trivial, especially for larger long-range EVs.

- "Eco-driving" techniques can reasonably improve fuel efficiency by about 10% (and at least 5%). Aggressive driving can decrease fuel efficiency by as much as 30%.

- Reducing miles driven by 10%, adopting reasonable eco-driving techniques that increase fuel efficiency by 5%, and upgrading from a typical 21.6 MPG vehicle to a 40 MPG-equivalent vehicle would together reduce an individual driver's footprint by over 50%, or about 3.7 $MgCO_2e$; if all drivers made these changes, it would be equivalent to removing 133 million of today's vehicles from the road.

Personal transportation represents the single largest component of the typical household carbon footprint, and most personal transportation-associated emissions (about 70%) come from combusting fuel in personal automobiles (including the upstream emissions associated with fuel production). Smaller portions are attributable to vehicle manufacture and air travel, while public transportation contributes virtually nothing. The energy and emissions related to the construction and maintenance of the roadway and parking infrastructure are also nontrivial, although it is difficult for an individual to directly affect such activities. Due to its dominance as a transportation mode, ubiquity, and profound influence on urban design and American culture and lifestyle, any discussion of transportation-associated emissions begins, and nearly ends, with the personal automobile.

Conventional, i.e. non-hybrid, non-electric, vehicles, account for the vast majority of vehicles already on the road as well as new-car sales, and in 2014 hybrid-electric vehicles accounted for less than 3% of overall market share[1]. While diesel is quite popular in Europe, in the US nearly all conventional vehicles are gasoline-powered, with the diesel market share similar to that of hybrids, at 2.8% in 2014[2]. Therefore, my primary focus, in terms of carbon-footprinting, is on gasoline-powered vehicles, and if not otherwise noted, all figures and results are derived using gasoline as the fuel. I do discuss extensively the potential of and lifecycle emissions associated with alternative (i.e. hybrid and electric) drive-trains, which are markedly lower than those of conventional vehicles. Indeed, I find that gasoline-only hybrid-electric vehicles have manufacturing emissions similar to conventional vehicles, while they consume 40–60% less fuel. Electric vehicles (EVs) are generally comparable to gasoline-electric hybrids in their lifecycle emissions profile, although in certain regions, especially the upper Midwest, the dirty electricity generating mix tilts the balance (in the near-term, at least) towards hybrids, and battery manufacturing for larger EVs is also a major source of emissions (hybrid batteries are one to two orders of magnitude smaller).

This chapter is organized as follows. I first examine how much and for what purposes Americans drive, followed by a brief overview of historical trends in automobile use and fuel consumption in America, and the likely prospects for the future. I then move to basic concepts in MPG, emissions factors, etc. Moving to more technical discussion, I examine the tailpipe emissions that result from burning fuel in cars, and then consider in some detail the upstream, or "well-to-pump" emissions that result from oil extraction and refining. I perform an extended analysis of the upstream emissions from Canadian tar sands, which are at the center of the ongoing Keystone XL controversy.

Emissions associated with vehicle manufacture are then explored, and the potential of alternative drive-trains, i.e. hybrid-electric, plug-in electric, and fully electric vehicles is discussed. The discussion of automobiles closes with a brief examination of how driving style affects fuel economy. I also offer a brief discussion on the relationship between personal wealth and automotive transportation in America.

I would also like to briefly note the use of brand name vehicles in this chapter, as when comparing electric and hybrid vehicles to conventional vehicles I focus largely on the top-sellers, without this focus meant as any kind of endorsement. Nearly 50% of all hybrids ever sold in the US belong to the Toyota Prius family, and while market share is declining, over 40% of new hybrid sales are still in this family; the Prius also has the highest EPA MPG rating of any new gasoline-only vehicle. The Chevrolet Volt is the best-selling plug-in hybrid, and the Nissan Leaf and Tesla Model S together have accounted for the vast majority of pure EV sales.

---

[1]http://www.usatoday.com/story/money/cars/2014/06/09/hybrid-cars-market-share-polk/10238155/

[2]http://www.foxbusiness.com/industries/2014/07/14/can-diesel-cars-make-inroads-in-america/

6.1 How much do Americans drive?

> - The average passenger vehicle travels 11,346 miles per year, while the average driver puts in 12,621 miles. At 21.6 MPG (mean for 2013), this translates into fuel-related emissions (including fuel production) of about 5.9 MgCO$_2$e/car, and 6.5 MgCO$_2$e/driver.

The primary source I rely in this section is the annual highway statistics series published by the U.S. Federal Highway Administration (FHWA), which uses data from the national Highway Performance Monitoring System, state-level data on motor vehicle registration, driver licenses, fuel use, etc., and modeling to arrive at detailed overall estimates on miles travelled on US roadways. The periodic National Household Transportation Survey (NHTS), also performed by the FHWA, is a very useful adjunct data source, as it gives a much more detailed breakdown on the daily transportation habits of Americans: where they go (work, store, pleasure), how many trips and by what mode (car, bicycle, etc.), distances for typical trips, etc. One drawback of the most recent 2009 survey [642] is that it is focused on daily travel, omitting long-distance trips. For that, we must turn to the 2001 NHTS survey [643] or other sources.

In 2013, the FHWA reported that there were 236,010,230 registered light duty vehicles in the US, each putting in 11,346 miles and burning through 524 gallons of fuel, with a total of 123,769,719,000 gallons of fuel consumed by light duty vehicles. Assuming 100% gasoline, this represents 1.1 million MgCO$_2$e of carbon emissions from direct combustion, and 1.3795 MgCO$_2$e when the upstream emissions that go into producing fuel are also accounted for (see Section 6.3).

I focus principally on light duty vehicles, as they mainly represent personal transportation, whereas medium and heavy duty vehicles are primarily used for commercial applications. Light vehicles accounted for 89.6% of all vehicle-miles travelled and about 75% of transportation fuel use, by volume. Note that motorcycles, officially excluded from the light vehicle category, make up about 3.44% of all vehicles but consumed only 0.28% of all fuel in 2013 (on a volumetric basis). Since the error introduced is so small, for simplicity I follow the FHWA convention and leave motorcycles out of most light vehicle calculations.

There are more (light duty) vehicles than licensed drivers, with 212,159,728 licensed US drivers in 2013, and as a consequence there is a discrepancy between miles per vehicle and miles per driver; simply dividing the number of licensed drivers by the total number of light-duty vehicle-miles suggests 12,621 miles per driver, and implies 22,088 vehicle-miles per household (using 1.75 licensed drivers per HH); dividing the total number of vehicle-miles in 2013 by 122.46 million households similarly gives an almost identical 21,867 vehicle-miles per household. It should also be noted that the 2009 NHTS suggests a slightly lower 19,850 vehicle-miles per household, but the NHTS restricted itself to daily travel, presumably omitting some long-distance trips. For simplicity, I usually assume an even 12,500 miles per driver per year for demonstrative calculations. Further, I assume a typical vehicle lifetime of 160,000 miles, equal to 12.8 years if driven 12,500 miles/yr (and 14.1 years when using 11,346 miles per year per car).

The average MPG of all light-duty vehicles in 2013 was 21.6 MPG, implying the typical passenger *vehicle* generates 5.86 MgCO$_2$/yr from fuel consumption per year, while the average *driver* likely generates about 6.51 MgCO$_2$/yr, using 12,621 miles-per-driver and assuming gasoline as the fuel. Broad historical trends in vehicle miles travelled, gasoline consumption, the number of vehicles in the US fleet, and household vehicle characteristics are summarized in Figures 6.1 and 6.2.

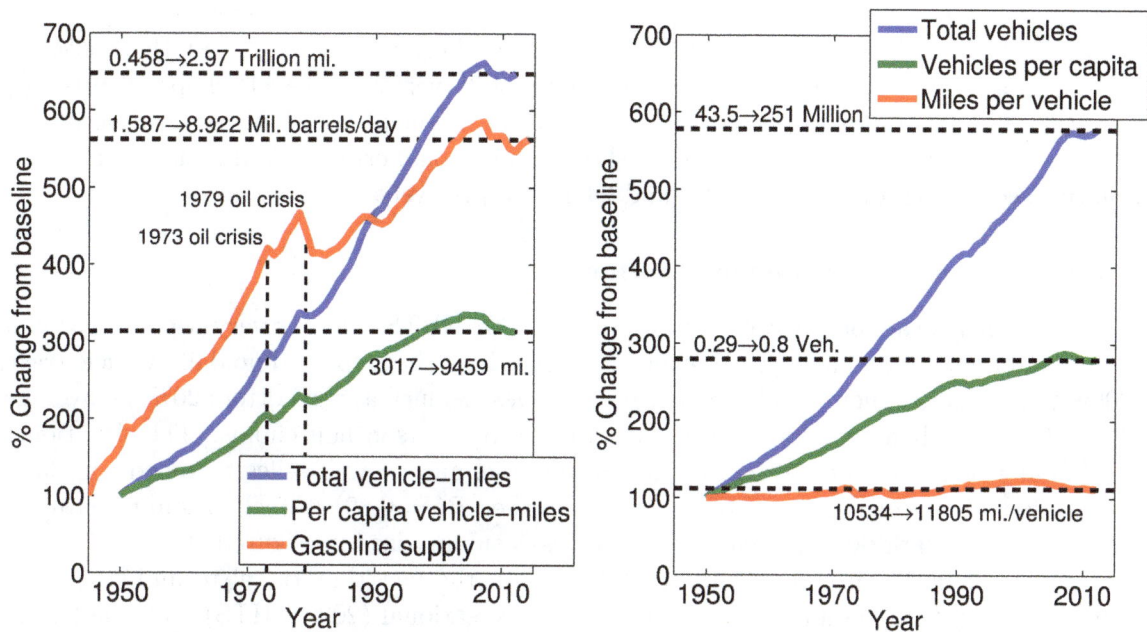

Figure 6.1: The left panel gives annual US vehicle-miles travelled, per-capita vehicle-miles travelled, and the gasoline supply, while the right panel shows total number of vehicles, vehicles per capita, and miles travelled per vehicle, from 1950 through 2012. Source: 2009 NHTS.

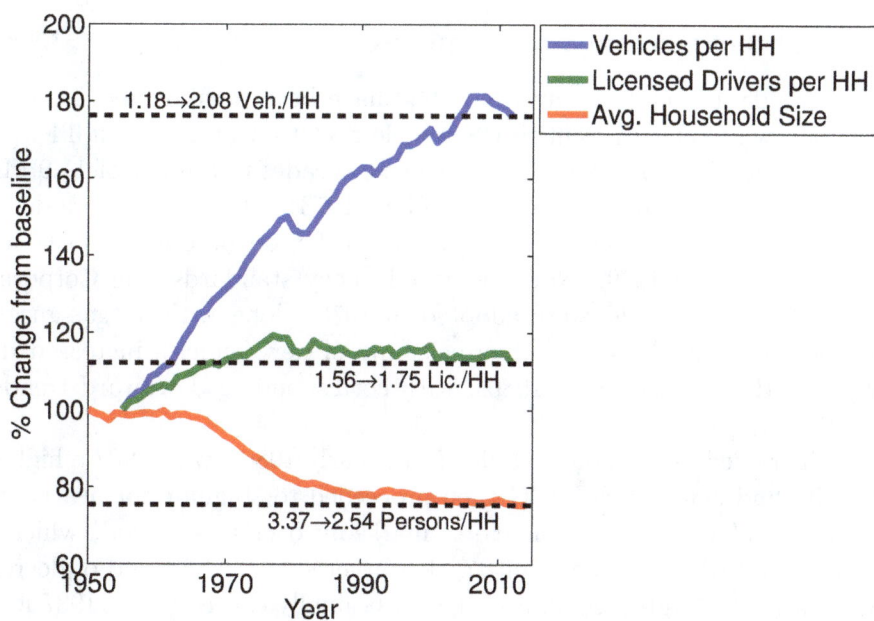

Figure 6.2: This figure shows how the number of vehicles per household has increased almost two-fold since 1955, while the number of licensed drivers per household has remained nearly flat and average household size has fallen 25%.

6.1.1 American driving patterns

Based on the 2009 NHTS [642], the typical driver takes just over three one-way trips a day, each of which averages 9.72 miles, for a total of 28.97 daily vehicle-miles. At the household level, each day saw 5.66 vehicle-trips, summing to 54.38 vehicle-miles. Perhaps surprisingly, commuting to and from work, while the single largest driver of vehicle-miles, still accounted for only 27.8% of all vehicle miles, with the shopping, personal errands, and social/recreational categories accounting for 15.0%, 17.8%, and 24.4%, respectively.

Variation in annual vehicle-miles and fuel use

Based on publicly available survey results from the 2009 NHTS, there is very wide variation in the number of miles individuals drive: even *excluding* those that reported no daily vehicle travel in 2009, the top 20% of households drove over *ten times* as much as the bottom 20%, on average. This is also true when restricting our comparison to one-person households. The situation is very similar when examining estimated annual fuel use, and overall, miles travelled correlated well with fuel use, explaining 89% of the variation (i.e. $R^2 = 0.89$) in fuel consumption under linear regression. Variation in annual vehicle miles is summarized in Figure 6.3.

This huge variation in annual mileage, along with the fact that travel to and from work accounts for only 28% of daily travel, while 24% is recreational (2009 NHTS), and the further finding from the 2001 NHTS [643] that about 50% of long-distance trips are either for vacation or visitation, suggests that it is within the power of individuals to rather dramatically cut their annual driving and fuel consumption. Failing that, higher mileage conventional vehicles and alternative drive-trains, i.e. hybrids and electric vehicles, can reduce driving-associated emissions by 40–60% compared to typical vehicles (and by as much as 75% compared to the worst conventional vehicles). *Ideally, these two strategies (i.e. decreased miles and increased efficiency) should be pursued in tandem, rather than in opposition.*

6.1.2 Historical fuel economy, and emissions

Broad historical trends in fuel economy from 1923 have been documented in detail by Sivak and Schoettle [203], who found the entire US vehicle fleet to average about 14 MPG in 1923, with this number slowly decaying over the subsequent decades to a nadir of 11.9 MPG in 1973 (passenger cars hit their low point of 13.4 MPG in 1973). The 1970s heralded a new age of resource-limitation, with the decade seeing domestic US oil production peak in 1970, and global oil crises in 1973 and 1979. New energy efficiency standards, the Corporate Average Fuel Economy (CAFE) standards, were adopted in 1975, along with a "gas-guzzler" tax for especially inefficient cars [204], and the fuel economy of new vehicles increased dramatically from the late 1970s through the early 1980s, with corresponding downward trends in vehicle weight and horsepower [205].

This trend of increased MPG largely stalled in the early 1980s, with 1987 a high water mark for MPG (at 22.0), and new vehicle MPG then proceeded to stagnate for nearly two decades. Vehicle horsepower and weight both increased markedly over this period, which also saw a dramatic expansion in SUV and other light truck market share. The trend of decreasing MPG did not reverse until 2005, and in 2009, average MPG finally exceeded the 1987 level [205]. In 2014, the average new passenger vehicle averaged about 24.2 MPG [205], implying that, over the course of nearly three decades, fuel economy ultimately increased by a remarkably scant 10%.

It does appear that the widespread adoption of light trucks (mainly SUVs, but also pickups, vans, and minivans) for personal transportation contributed to the decline in fuel economy seen

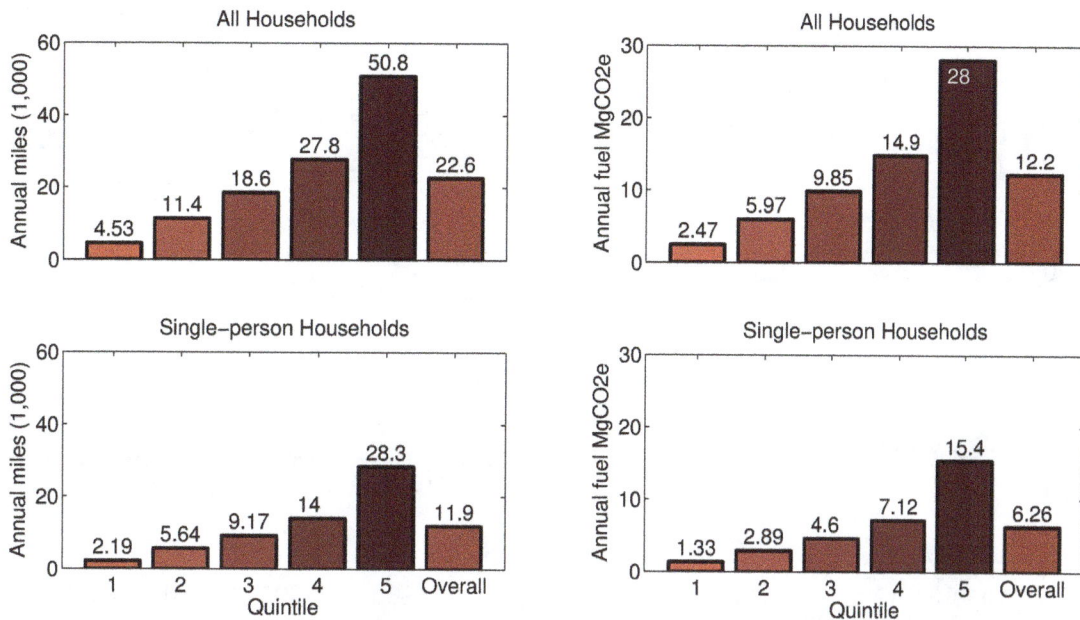

Figure 6.3: Variation in daily vehicle miles travelled among US households that reported any daily vehicle travel, based on the 2009 NHTS. The left panels give the mean miles (in 1,000s) travelled for each quintile, along with the overall mean. The right panels show the corresponding fuel-related emissions (including upstream fuel-production) in $MgCO_2e$. The top panels use data for all households, while the bottom are restricted to single-person households.

from the late 1980s through the early 2000s. This can be seen in Figure 6.5, which shows that the gap between the average new *car* and new *vehicle* (i.e. cars+trucks) MPG grew markedly from the 1980s to a peak in 2004. The year 2004 also represents a local nadir in MPG, being the lowest new vehicle MPG since 1980. However, average car MPG also decreased during this period, as these also vehicles grew larger and more powerful. This may be partly due to the increasing popularity of car-based SUVs (also known as "crossover" SUVs), which tend to have fuel economy better than truck-based SUVs, but still worse than most other cars.

Figure 6.5 also shows the share of new vehicles sales that are pickup trucks/SUVs, as well as the average household size. Strangely enough, even as Americans have continued to favor larger vehicles over the last 40 years, household size continues to fall, and jobs where a heavier-duty work vehicle may be necessary make up a smaller portion of employment (e.g. construction or farming). Furthermore, while in 1970, 26.4% of the US population lived in rural areas, this percentage fell to 19.3% by 2010 (per the US Census), so massive rural demand seems unlikely to explain the shift to SUVs either.

Vehicles with fuel economy comparable to today's hybrids have, in fact, been available for nearly 30 years. The 1986 GM Sprint has a revised EPA rating of 48 MPG city/highway combined, and the original 2000 Honda Insight achieved 53 miles per gallon (EPA revised estimate). This was not exceeded until 2016, when the Toyota Prius Eco posted a combined 56 EPA rating.

New fuel efficiency standards, through CAFE, were established by the Energy Independence and Security Act of 2007, while the EPA was compelled to regulate carbon dioxide emissions under the Clean Air Act by the *Massachusetts v. EPA* Supreme Court decision. This led to the establishment of a goal of a CAFE fleet-average 54.5 MPG for new cars and light trucks by 2025 [206]. However, since fuel consumption under CAFE standards is measured using highly outdated standards, MPG is grossly inflated, and this 54.5 figure translates into an actual fuel

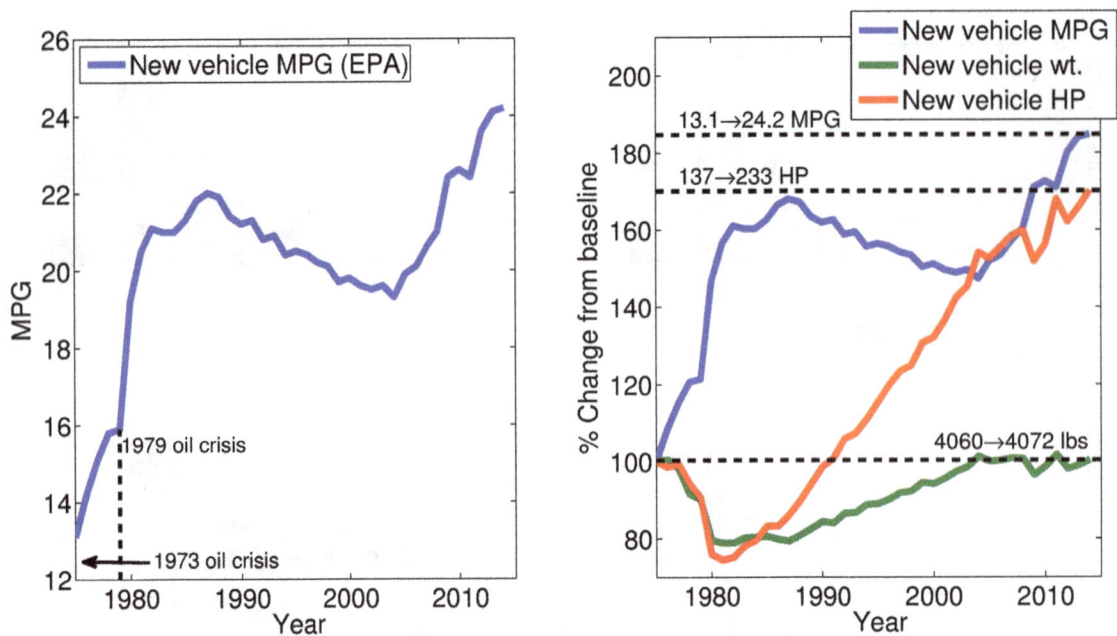

Figure 6.4: New passenger vehicle MPG from 1975 through 2014, per the EPA [205]. The right panel shows how weight and horsepower reductions accounted for most of the decrease in fuel consumption in the early 1980s, following the initial adoption of CAFE standards. Weight and horsepower trends reversed, however, correlating with the drop in MPG through the 1990s. The later plateau in weight (and a slow-down in the power trend) imply that improvements in car and engine technology explain the more recent fuel economy gains (since about 2004).

economy of 35–40 MPG. Given the very recent reductions (2016–2017) in gasoline prices and the continued strong performance of the light truck and SUV market, it remains to be seen whether these standards will actually be met.

6.2 Basic concepts

6.2.1 MPG: city, highway, and combined

Fuel consumption, which is jointly determined by MPG and miles driven, accounts for 90% of vehicle-related emissions, an obvious but deeply important fact. As I discuss later, the MPG rating of a diesel vehicle is equivalent (from a CO_2e standpoint) to an 11% lower MPG under gasoline, and the EPA's MPGe ratings for electric vehicles are are not comparable to gasoline engine MPGs in terms of emissions; the MPG_{GWP} metric, which is explained in Section 6.6.3, does compare electric vehicle emissions to gasoline vehicles, and is determined using upstream emissions from electricity generation.

The EPA gives all vehicles produced in the US with a gross vehicle weight under 8,500 lbs a city, highway, and combined MPG rating; the former two are determined by running a vehicle through a series of standardized duty cycles. The combined MPG rating is a (weighted) harmonic mean of city and highway MPG, with city and highway MPG weighted 55 and 45%, respectively. The *harmonic*, and not the arithmetic mean is the appropriate average, and the logic is explained in a moment.

In determining how many emissions one can expect to generate through driving, it is most appropriate to use the combined MPG rating (in the absence of a log of actual MPG achieved). Using the higher highway MPG rating will invariably underestimate one's personal emissions.

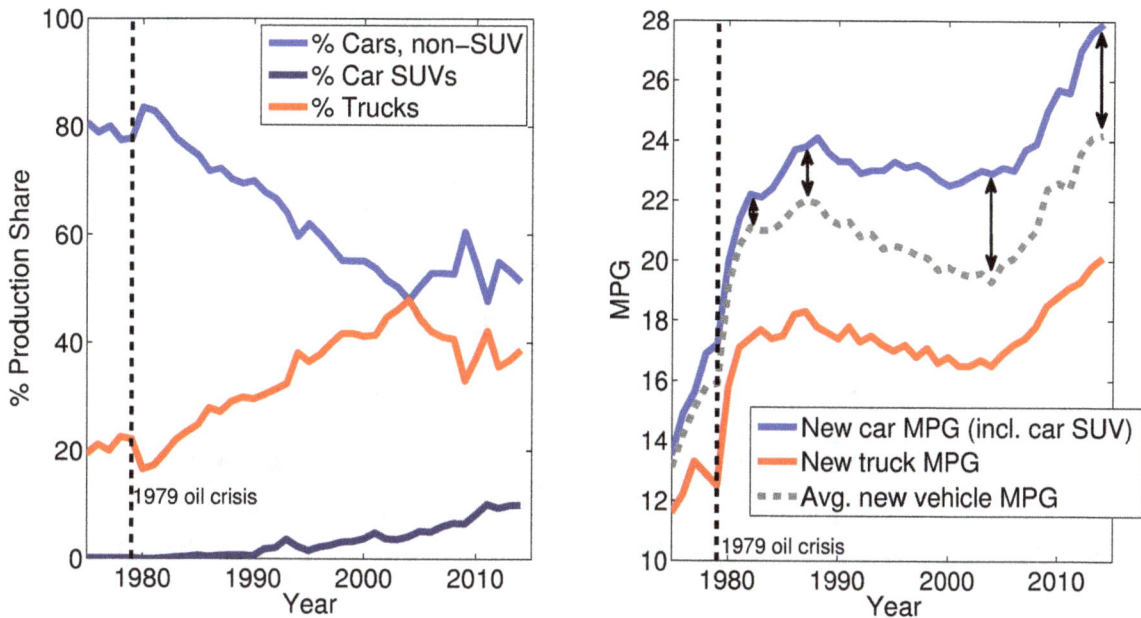

Figure 6.5: New car, car-based SUV, and light truck (pickup, van, SUV) market shares from 1975 through 2014 (passenger-vehicles only) [205]. The right panel shows MPG disaggregated into car- and truck-based vehicles, with the average (gray line) being "pulled" toward the truck curve (red line) as the truck market share increased through the early 2000s.

Throughout this chapter, I use the EPA's combined MPG ratings for all calculations that consider a specific vehicle model.

6.2.2 The hyperbolic relationship between MPG and emissions

The US FHWA estimated an average MPG of 21.6 for all light duty vehicles in the US, in 2013. New vehicles get slightly better MPG, with the EPA estimating an average 27.9 MPG for new cars, 20.1 for trucks, and 24.2 MPG for new cars and trucks combined, in 2014 [205]. It is probably common to assume that the effect of MPG on one's carbon footprint is linear. In other words, one might assume that the difference between 30 MPG and 20 MPG is the same as 20 MPG versus 10 MPG. However, the relationship is not *linear* but *hyperbolic*. That is, in evaluating fuel use, it is much better to think of the gallons of gas consumed per 100 miles travelled than the miles travelled per gallon of gas. For example, cars getting 10, 20, and 30 MPG will consume 10, 5, and 3.3 gallons of gas to travel 100 miles, respectively. This more clearly demonstrates that, at the lower MPG range, even small changes in MPG have a dramatic effect on fuel consumption and hence carbon emissions, while the effect is much diminished at higher ranges. For example, going from 40 to 50 MPG changes fuel consumption from 2.5 to 2 gallons per 100 miles, only a tenth of the fuel saving achieved by a 10 to 20 MPG improvement (10 to 5 gallons per 100 miles). This general hyperbolic pattern is illustrated in Figure 6.6. Figure 6.7 illustrates how small increases in MPG have a far more profound effect on fuel consumption at the lower MPG range than at higher levels.

This hyperbolic relationship means that, when calculating average MPG, we must use the harmonic mean and not the arithmetic mean, as noted previously. For example, consider a household that has a large truck averaging 10 MPG, and a small car that gets 40 MPG, each driven 12,000 miles per year. Intuitively, we might assume that the average MPG for the household is 25 MPG (the arithmetic mean), but in fact it is only 16 MPG (the harmonic

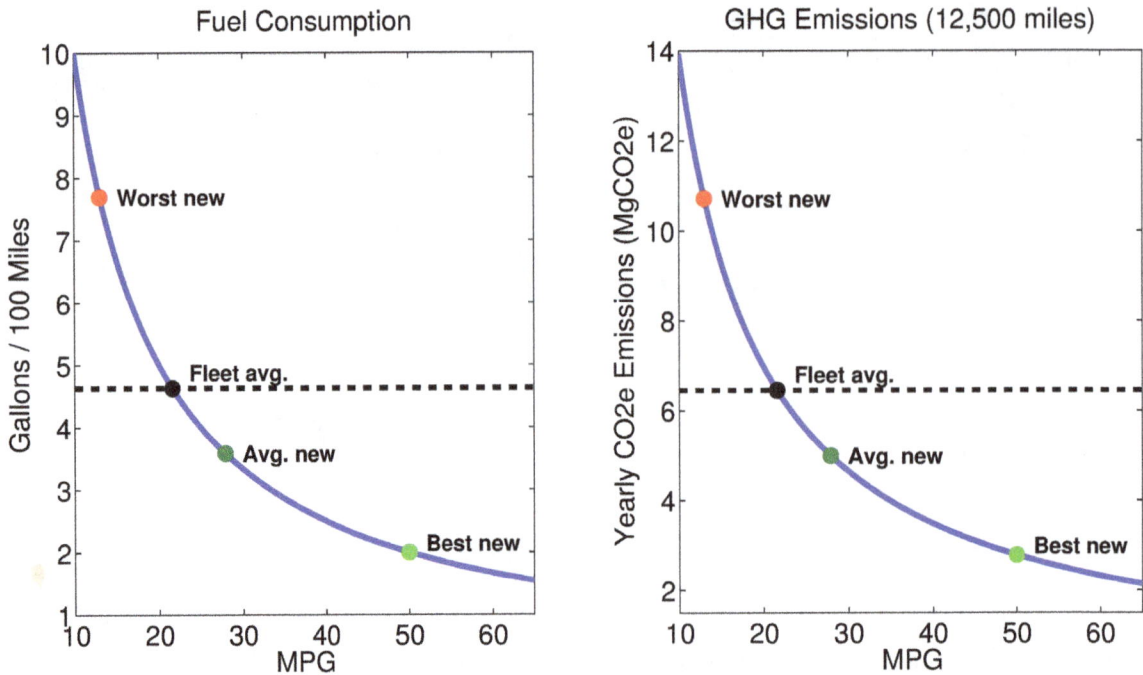

Figure 6.6: The hyperbolic relationship between MPG and fuel consumption (left panel), assuming 12,500 miles per year. Since the amount of fuel consumed is directly proportional to GHG emissions, there is an identical hyperbolic relationship between MPG and carbon emissions, as shown in the right panel.

mean). To see that this is true, it is easier to consider gallons of gas consumed: 1,200 gallons of gas are consumed by the truck (12,000 miles / 10 MPG), and 300 gallons of gas are consumed by the car (12,000 miles / 40 MPG), giving a total of 24,000 miles and 1,500 gallons of gas. Dividing, we see that the household fleet actually got only 16 MPG (and corresponding to annual emissions of 16.7 $MgCO_2e$). Suppose that, aghast at this calculation, our household shifted their driving habits to 6,000 truck miles, and 18,000 car miles. The average fleet MPG then becomes about 23 MPG, lowering emissions by 30% to 11.7 $MgCO_2e$, an improvement but still weighted towards the inefficient truck MPG; if all miles were travelled in the car, fleet emissions would fall 60% relative to baseline, to 6.7 $MgCO_2e$. These calculations are meant to further emphasize the fact that low MPG vehicles have a seemingly disproportionate effect on fleet MPG and fuel consumption, and our collective priority should be to increase fuel efficiency, and to decrease miles driven, at the *lower* end of the MPG scale.

6.2.3   Operational emissions as a function of MPG and miles driven

Figure 6.8 shows annual fuel-cycle (tailpipe plus well-to-pump) emissions as a function of miles driven and vehicle MPG, using gasoline as the fuel. In Section 6.5.2, I add estimated emissions from vehicle manufacture and disposal ("vehicle-cycle") to the table, which increases per annum emissions by 10–15% for any given cell. The latter table is probably better referenced for overall carbon footprinting purposes.

6.2.4   A note on "fuel consumption" versus "fuel economy"

It should be noted that the term "fuel economy," which corresponds to the familiar vehicle MPG ratings, is *not* synonymous with "fuel consumption," which refers to the number of gallons

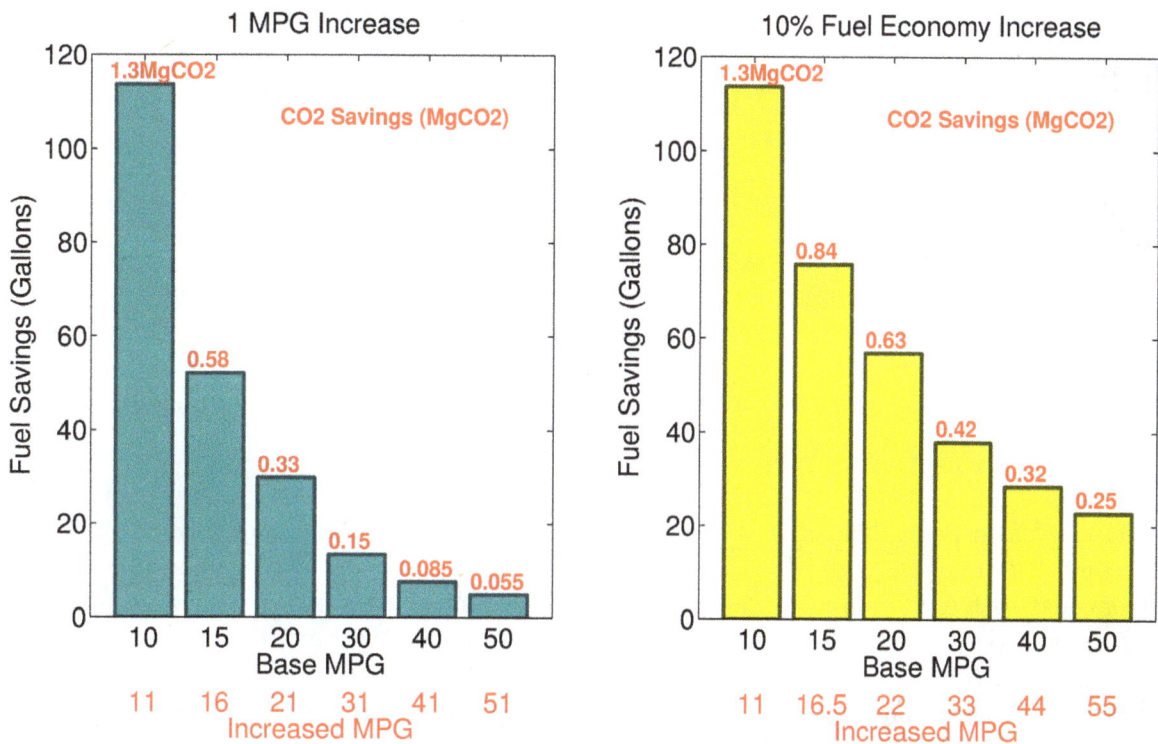

Figure 6.7: The left panel gives the yearly fuel (and CO_2e) savings for a 1 MPG increase for different base MPGs (assuming 12,500 miles per year), while the right panel gives fuel and CO_2e savings for 10% increases in fuel efficiency from the same base MPGs. In both cases the emissions savings are far more pronounced for improvements from low base MPGs, a consequence of the hyperbolic relationship between MPG and fuel consumption. The difference in GHG emissions between 10 and 11 MPG is 1.27 $MgCO_2$e. This difference alone is over 45% of the yearly emissions of 50 MPG vehicle.

Fuel–Cycle Only

MPG	Miles (Thousands)							
	5	7.5	10	12.5	15	17.5	20	22.5
12	4.6	7	9.3	12	14	16	19	21
15	3.7	5.6	7.4	9.3	11	13	15	17
18	3.1	4.6	6.2	7.7	9.3	11	12	14
21	2.7	4	5.3	6.6	8	9.3	11	12
24	2.3	3.5	4.6	5.8	7	8.1	9.3	10
27	2.1	3.1	4.1	5.2	6.2	7.2	8.3	9.3
30	1.9	2.8	3.7	4.6	5.6	6.5	7.4	8.4
33	1.7	2.5	3.4	4.2	5.1	5.9	6.8	7.6
36	1.5	2.3	3.1	3.9	4.6	5.4	6.2	7
39	1.4	2.1	2.9	3.6	4.3	5	5.7	6.4
42	1.3	2	2.7	3.3	4	4.6	5.3	6
45	1.2	1.9	2.5	3.1	3.7	4.3	5	5.6
48	1.2	1.7	2.3	2.9	3.5	4.1	4.6	5.2
51	1.1	1.6	2.2	2.7	3.3	3.8	4.4	4.9

Figure 6.8: Annual CO_2e emissions (MgCO$_2$e) for various vehicle MPGs and annual mileages. Red text signifies annual emissions worse than average—6.45 MgCO$_2$e, based on 12,500 miles at 21.6 MPG—while green signifies emissions totals better than average. The greener (redder) the text, the better (the worse). The 21 MPG row and 12,500 mile column are highlighted as being typical of American cars and drivers.

of fuel consumed per gallon, and is mathematically the inverse of fuel economy. This is an important distinction, as highlighted by the discussion above, and carbon emissions are directly proportional to fuel consumption but not to fuel economy. Furthermore, as I discuss later, fuel consumption (and hence carbon emissions) increases linearly with vehicle size, whereas the relationship is hyperbolic with respect to MPG.

Finally, consider that "eco-driving" can decrease fuel consumption by about 10% on average (see Section 6.8). The effect of eco-driving on a vehicle getting 18 MPG at baseline would then be to increase mileage to 20 MPG and save 0.56 gallons of fuel per 100 miles, while eco-driving would shift a 36 MPG vehicle to 40 MPG and save 0.28 gallons per 100 miles. The MPG increase is twice as great for the higher mileage vehicle, yet the *actual fuel savings* are twice as high for the low-mileage vehicle; Figure 6.7 demonstrates how such relative changes in fuel economy translate into fuel savings. All this should highlight the fact that fuel consumption is a much "cleaner" metric than economy. Nevertheless, since MPG is so familiar and ubiquitous, I still present most results in terms of MPG (and fuel consumption as well where appropriate).

6.3 Petroleum fuels and tailpipe, well-to-pump, and total emissions

- Gasoline emits 8.887 $kgCO_2$/gallon at the point of combustion, while upstream emissions total 2.259 $kgCO_2e$/gallon, giving a fuel-cycle emissions factor of 11.146 $kgCO_2e$/gallon. Diesel is more emitting, at 12.504 $kgCO_2e$/gallon over the fuel-cycle, and one should discount the sticker MPG of a diesel vehicle by 11% to get the gasoline-equivalent MPG.

- Upstream processes generating emissions include oil extraction (about 40%), crude transport (5%), refining (50%), and final distribution (5%). Tar sand bitumen generates significantly more emissions at the extraction phase than does conventional oil.

At the point of combustion, petroleum fuels emit CO_2 and H_2O (along with other products of imperfect combustion), and it is straightforward to calculate the resulting direct, or "tailpipe," emissions from automobile use. This can be done using knowledge of the basic chemical composition of fuels, as discussed in Section 3.6, but I adopt standard EPA emissions factors (EFs), given as 8.887 $kgCO_2$/gallon gasoline and 10.180 $kgCO_2$/gallon diesel, assuming 100% of the carbon in fuel is oxidized to CO_2 [208]. Note that diesel, as the more carbon- and energy-dense fuel (on a volume-basis, on a mass-basis gasoline is actually slightly more energy-dense), has the higher combustion EF, and therefore, while diesel-powered vehicles typically have higher MPG ratings than comparable gasoline-powered cars, comparisons of such vehicles should adjust for this fact by reducing the diesel MPG by 11% to obtain gasoline-equivalent MPG (this factor includes the upstream fuel-cycle emissions discussed below). Thus, for example, the 2015 Volkswagen Jetta, which has a combined sticker rating of 35 MPG, adjusts downward to 31.2 MPG, lowered but still 8–20% better than the 26–29 MPG obtained by various gasoline models.

Upstream processes in gasoline production, including extraction, refinement, and distribution, are not at all negligible, and these are referred to as "well-to-pump" (WTP) emissions; they equal about 24–27% of gasoline tailpipe emissions, with fuel refining and oil extraction the major sources of WTP emissions. The basic process of conventional crude oil extraction and refining via distillation, whereby hydrocarbons of different density are separated, with gasoline representing a light distillation fraction, and diesel a heavy fraction, is outlined in Section 3.7.

In terms of associated emissions, the Argonne National Laboratory has developed a comprehensive lifecycle model for various transportation fuels, termed the GREET model, and I draw directly from this work: the 2014 GREET 1 model estimates WTP emissions of 2.259 $kgCO_2e$/gallon of gasoline (25% of tailpipe emissions) and 2.324 $kgCO_2e$/gallon of diesel (23% of tailpipe emissions) (converted from 19.814 $kgCO_2$/mmBtu and 17.948 $kgCO_2$/mmBtu, assuming lower heating values of 0.114 mmBtu/gallon and 0.1295 mmBtu/gallon for gasoline and diesel, respectively[3]). Other estimates, many site-specific, of WTP emissions exist, but the GREET model is widely used and is representative of the national average.

Adding the tailpipe and WTP EFs gives the total well-to-wheels (WTW), or "fuel-cycle," EFs of 11.146 $kgCO_2e$/gallon gasoline and 12.504 $kgCO_2e$/gallon diesel. Emissions associated with the fuel-cycle are summarized in Figures 6.9 and 6.10. Oil extraction and refining are the two major processes that generate upstream WTP emissions, with extraction generating about 40% of emissions and refining about 50% (about 9–13% of oil energy is lost at the refining stage [209]); crude oil transportation and distribution are minor contributors, at around 5%

---

[3]I use GREET 1 estimates directly, but do note that using updated AR5 GWP values for CH_4 and N_2O suggest WTP emissions 5.0% and 4.4% higher for gasoline and diesel, respectively, but these respective differences translate into only 1% and 0.8% errors on an overall WTW basis.

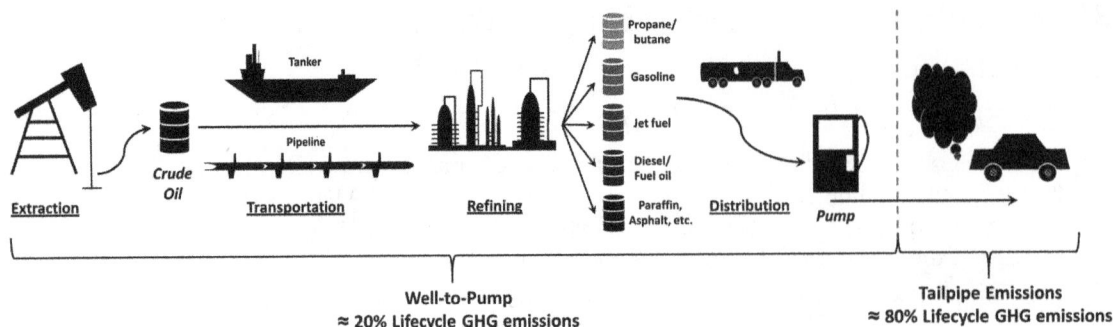

Figure 6.9: Schematic of the gasoline/diesel fuel cycle. The well-to-pump process proceeds from extraction at the well to yield crude oil, which is transported to refineries where it is refined into various fuels, which are it turn distributed to the customer. This whole process, along with the emissions embodied in developing the infrastructure for oil extraction, contributes about 20% of the fuel lifecycle emissions. Direct combustion of the fuel in automobiles generates tailpipe emissions, which account for roughly 80% of the fuel lifecycle emissions.

of emissions each [207]. The exact contribution of each of these processes varies significantly between oil sources. For example, extraction of conventional Kuwaiti crudes takes relatively little energy, while transport of this crude is more energy-intensive than average. Generally speaking, refining is consistently the most emissions-intensive process, with the major exception being tar sands bitumen, where extraction is dominant and highly emitting, as discussed next [207].

6.3.1 Tar sands vs. conventional crude and the Keystone XL

- Bitumen, found in the Alberta tar sands, is a form of extremely heavy oil or tar that may be found mixed with sand and clay. Hot water or steam is used to extract the bitumen, either from surface mined tar sands or via deep steam injection, and this extremely energy-intense process increases WTP emissions by 80–90%, and total lifecycle emissions by 15–25%, relative to typical crude oils.

- Surface mining of boreal forests is extremely locally destructive, and forest carbon losses may further increase tars sands WTP emissions by 15–25% and lifecycle emissions by 5–10%.

Overview

As the reader is likely aware, there has been significant controversy over the Keystone XL pipeline (unapproved at the time of this writing, but this may change with the Trump presidency), with the increased energy and emissions associated with extraction of bitumen ("extra heavy oil" or "tar") from the Alberta tar sands as one of the major arguments for blocking its construction. Briefly, the tar sands (or "oil sands") are geologic formations composed of a mix of sand, clay, and bitumen. Bitumen is a form of extremely viscous heavy oil, and it is essentially identical to asphalt. While oil extraction is in general an energy-intensive process, extracting crude oil from the viscous bitumen of the tar sands is even more energy-intensive, with WTP emissions likely 80–90% greater than other typical North American crudes, and overall lifecycle emissions roughly 15–25% greater [211, 210, 212]. If carbon emissions resulting from boreal forest degradation related to oil sands exploitation are included in our accounting,

142

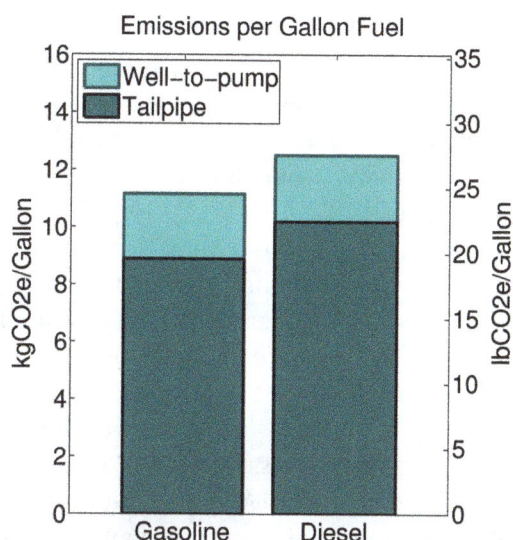

Figure 6.10: Upstream well-to-pump and direct tailpipe GHG emissions for gasoline and diesel, on a per-gallon basis.

lifecycle emissions may be 25–35% greater compared to conventional crude, and WTP emissions 2–2.5 times greater (see Section 6.3.1), but there is great uncertainty in this.

Bitumen is extracted from oil sands by either open-pit surface mining of shallow deposits, or via so-called *in situ* methods (Latin for "in place"), for deeper deposits. With surface mining, the oil sands are simply scooped up by heavy equipment, and the bitumen is extracted through a hot water process [210]. *In situ* methods involve injection of hot steam into the oil sands, which decreases the viscosity of bitumen, allowing it to be pumped out of the reservoir like conventional crude. That is, the basic difference between the methods is whether hot water is used to separate bitumen from sand that has already been mined in bulk (surface mining), or hot water is used to separate the bitumen from the sand while it is still in the geologic reservoir.

While historically most Canadian bitumen was extracted via surface mining, about 80% of oil sands reserves are only accessible via *in situ* methods, and in 2013, the *in situ* share reached 53% of total bitumen extraction in Alberta [4]. While surface mining is slightly less emissions-intensive than *in situ* extraction, it is far more locally destructive (and the destruction of forest could well tip the carbon balance in favor of *in situ* extraction). The two major *in situ* process are steam assisted gravity drainage (SAGD) and cyclic steam stimulation (CSS), with CSS, also known as the "huff-and-puff" method, the older technology.

Once extracted, bitumen is still much too viscous to flow in a pipeline, so it is diluted with a lighter material, such as synthetic crude oil (SCO), and then it is either sent to an upgrader facility, where the bitumen is processed into lighter SCO that is then sent on to a refinery, or delivered directly to a refinery that can accept raw bitumen; most bitumen is upgraded before refining.

Lifecycle analyses

The US State Department, in its final analysis, estimated that fuels derived from Canadian tar sands oil transported by the Keystone XL pipeline would ultimately have 17% higher WTW emissions than several reference crude oils [212], implying that the upstream WTP emissions

---

[4]Alberta Office of Statistics and Information, https://osi.alberta.ca/osi-content/Pages/OfficialStatistic.aspx?ipid=916, accessed 3/28/2015

143

are over 80% greater for this source.

A 2012 analysis by Bergerson and colleagues [210] used confidential data (and hence their results cannot be replicated) from mining companies to estimate WTP emissions ranges of 122.04–186.84 gCO_2e/kWh and 93.24–154.44 gCO_2e/kWh of reformulated gasoline (RFG), for SAGD and surface mining, respectively, when bitumen was upgraded to SCO upstream of refining. The authors cite a range of 21.24–136.44 gCO_2e/kWh RFG for conventional crude, drawing upon multiple works of literature. Note that these ranges all overlap, and omitting upgrading from the bitumen pathway (shipping directly to refinery) results in lower overall emissions. However, since most bitumen is upgraded, I only compare SCO bitumen pathways to conventional crude. Furthermore, CSS is a commonly used *in situ* technology and appears to generate about 36 gCO_2e/kWh more than SAGD, largely because of increased extraction emissions.

The above ranges suggest average WTP emissions of 4.138 $kgCO_2e$/gallon, 5.160 $kgCO_2e$/gallon, 6.363 $kgCO_23e$/gallon, and 2.6341 $kgCO_2e$/gallon for surface mined, SAGD, CSS, and conventional gasoline, respectively. Note that the latter figure is reasonably comparable to the GREET estimate of 2.259 $kgCO_2e$/gallon. Weighting *in situ* extraction 53% (in the absence of any concrete data, I make the crude assumption that SAGD and CSS each account for 50% of this total) and surface mining 47% suggests that WTP emissions and lifecycle emissions are 90% and 22% higher, respectively, for gasoline sourced from bitumen versus conventional crude.

An earlier review of studies by the same research group [211] suggested that total lifecycle emissions for surface mining and *in situ* techniques exceed, on average, those for conventional gasoline by 9% and 26%, respectively. Weighting the techniques by 47% and 53% as above suggests 18% higher lifecycle emissions overall. A 2009 Department of Energy/National Energy Technology Laboratory [207] study examined WTP emissions for diesel fuel for a variety of oil sources, and found Canadian oil sands to have the highest emissions of any source, at 4.403 $kgCO_2e$/gallon (34.0 $kgCO_2e$/mmBtu). This is 85% higher than the US average in the study (2.383 $kgCO_2e$/gallon), and over 2.5 times the *domestic* oil average (1.748 $kgCO_2e$/gallon, 13.5 $kgCO_2e$/mmBtu).

Two reports examining lifecycle GHG emissions were commissioned by the Alberta government [214, 215]. One report [215] concluded that lifecycle emissions for the dilbit/synbit oil sands pathway (this is the pathway that excludes the emissions-intensive bitumen upgrade process) are about 10% higher than typical conventional crudes; the other [214] gives a range of results, but reports a lower gap between conventional crudes and bitumen than other studies, at 6 to 18% over the lifecycle (omitting questionable co-generation credits). Given that these are non-peer-reviewed reports and commissioned by a government with an active interest in promoting oil sands exploitation, skepticism is warranted. Other reports are fairly consistent in finding bitument to be 15–20% more emitting (over the lifecycle) than typical US crudes [213, 212, 207, 210, 211].

Emissions from forest degradation

While the analyses reviewed above indicate that *in situ* methods generate more emissions, largely due to the truly massive amounts of natural gas that must be combusted to generate steam in sufficient quantities, open pit surface mining is locally far more devastating, and requires the wholesale destruction of the boreal forest that overlies the oil sands. Boreal forests are a major global carbon sink, have been estimated to store 239 Mg carbon per hectare (876 $MgCO_2e$/hectare) [46], and continue to take up large amounts of carbon. Emissions from forest destruction is typically not accounted for by the analyses reviewed.

Data from Global Forest Watch[5] suggests that 775,500 hectares were cleared or degraded in the Canadian tar sands region from 2000–2012. There are 475,000 hectares of surface-minable area, and in this zone, forest loss is at 20%. If we assume that all carbon from the surface-mined areas has been oxidized to atmospheric CO_2, and that 40% [216] of the carbon from the other cleared and degraded areas is lost, we have roughly 322 million $MgCO_2$ from forest degradation, which is about 1.261 kgCO2e/gallon bitumen[6]. Supposing that only half of forest degradation is attributable to tar sands exploration (the other half being due to logging and other industrial activity), we would have 0.794 kgCO2e/gallon bitumen. In sum, forest degradation may increase upstream WTP emissions by an additional 15–25% and lifecycle emissions by 5–10%, although these figures are quite uncertain. An earlier report by Global Forest Watch also explores this issue in depth [216].

6.3.2 WTP emissions vary widely by crude oil source

From the above analysis, it should be clear that Canadian oil sands crude is markedly more emissions-intensive than the national US average. However, it is also important to note that both domestic and international sources of crude vary markedly in WTP emissions intensity. A DOE/NETL report [207] concluded that domestically produced conventional oil has low WTP emissions, by global standards, while, as noted above, Canadian tars sands and Venezuelan "extra heavy oil" have the highest WTP emissions, as summarized in Figure 6.11.

6.4   Emissions embodied in vehicle manufacture

- Vehicle materials, manufacture, disposal, and maintenance (the "vehicle-cycle") scale directly with vehicle size, and are on the order of about 5 kgCO2e/kg vehicle. Emissions for the average new passenger vehicle in 2014, weighing in at 4,072 lbs, amount to about 9.2 $MgCO_2e$ under the GREET 2 model, which translates to 0.72 $MgCO_2e$ amortized over 12.8 years, or 0.49 $MgCO_2e$ amortized over 14.1 years.

- Conventional and hybrid drivetrains yield similar vehicle-cycle emissions on a per-weight basis, although electric-only drivetrains may have appreciably higher vehicle-cycle emissions due to battery production.

- Vehicle-cycle emissions are about 10% of lifecycle emissions for conventional vehicles, and perhaps 15–20% of lifecycle emissions for hybrids (due to lower use-phase emissions, not higher production emissions).

The embodied emissions of the vehicle manufacturing process (and delivery to customers) should be included in any assessment of the carbon footprint of vehicle ownership and operation. In this section, I mainly restrict discussion to conventional vehicles, i.e. non-electric drivetrains (emissions from battery manufacture for such vehicles are reviewed in Section 6.6.2). About two-thirds to three-fourths of conventional vehicle manufacturing emissions stem from material production, with the remainder coming from vehicle assembly [217]. By weight, standard vehicles are composed primarily of ferrous metals (iron and steel) (67%), aluminum (6%), plastics (7.5%), rubber (4%), and glass (2.7%) [217]. While there is only a tenth the aluminum as ferrous metal, virgin aluminum is 5–15 times as emissions-intensive to produce. Therefore, the carbon

---

[5]see http://www.wri.org/blog/2014/07/tar-sands-threaten-world%E2%80%99s-largest-boreal-forest, accessed 3/28/2015

[6]Determined from bitumen extraction statistics for 2003–2012 from the Alberta Office of Statistics and Information. Yearly extraction from 2000–2002 was assumed to be equal to that in 2003, a likely overestimate.

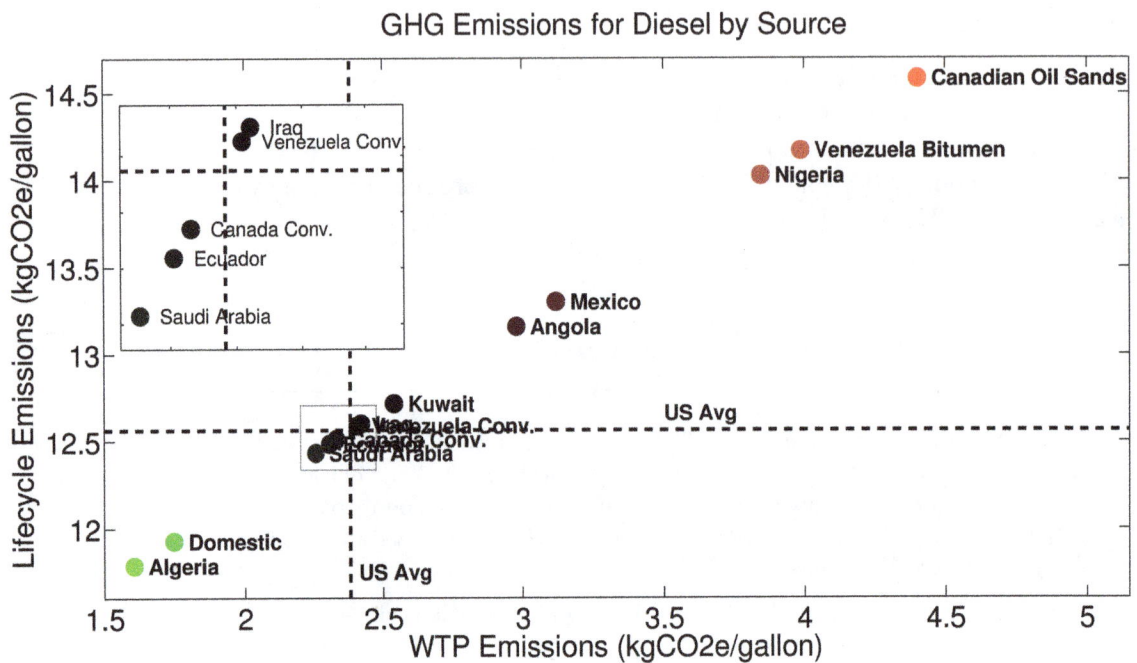

Figure 6.11: The WTP and lifecycle emissions for diesel fuel for various sources of crude oil, based on 2005 data. Pump-to-wheel emissions are identical, regardless of oil source, and so all variation is due to differences at the WTP level. Note that this data largely pre-dates the increase in shale/tight oil exploitation in the US (domestic), and so may be slightly out-of-date. However, the conclusion that bitumen from Venezuela and the Canadian oil sands has the highest associated emissions is expected to be robust.

footprint of parts manufacture is strongly dependent on the rate of aluminum recycling, as nearly all energy use (as well as other environmental harms) associated with aluminum occurs at the mining and refining stages. Moreover, aluminum is increasingly being used to reduce vehicle frame weight, and it is especially popular for electric vehicles (while more emissions-intensive at the production phase, vehicle lightweighting reduces use-phase emissions) [219]. The carbon emissions from material manufacturing scale linearly with total weight; energy and emissions at the vehicle assembly step likewise increase in linear proportion to weight.

Emissions from vehicle production have been estimated in multiple publications, with the GREET 2 model, developed by Argonne National Laboratory, probably the most widely used. Unlike most other works, it also considers the entire "vehicle-cycle," i.e. in addition to manufacturing-associated emissions, it includes the emissions from oil and other fluid changes, as well as those from vehicle disposal. Under the 2014 GREET 2 model, the vehicle-cycle emissions for a 4,072 lb (1,847 kg, 2014 average weight) conventional vehicle are 9.194 $MgCO_2e$, of which about 78, 13, and 9% is attributable to materials, assembly and disposal, and fluids (e.g. oil, etc.), respectively. Excluding fluids, manufacturing emissions are 8.367 $MgCO_2e$. Total emissions scale linearly with vehicle weight and are roughly 5 $kgCO_2e/kg$ (within a reasonable weight range) for the total vehicle-cycle, or about 4.5 $kgCO_2e/kg$ excluding fluids. The more exact empirical relation is shown in Figure 6.12; this figure also shows GREET 2 vehicle-cycle emissions for hybrid vehicles, which are nearly identical, as a function of weight.

Note that, since the average driver generates about 6.5 $MgCO_2e$ per year from the fuel-cycle, over a lifetime of 12.8 years (160,000 miles), vehicle manufacture and maintenance increases lifetime emissions by about 10%. Amortized over the vehicle life, this is an additional 0.72 $MgCO_2e/yr$, for total emissions of 7.2 $MgCO_2e/year$ attributable to a typical driver.

The results from GREET 2 are generally consistent with other studies. A well-done and widely cited report performed by the MIT energy laboratory ("On the road in 2020") [217] projected emissions of 5.39 $MgCO_2$ for a 1996 baseline vehicle, or per-weight emissions of 4.07 $kgCO_2/kg$ vehicle. However, this is probably an underestimate for several reasons: (1) the assumed weight of the 1996 vehicle (1,323 kg) is significantly less than the average 2014 passenger vehicle (1,847 kg, nearly 40% greater), (2) relatively high rates of materials recycling were assumed, and (3) projected 2020 US grid electricity emissions were used, rather than current emissions. The first factor is the most significant, and if we adjust to the average vehicle weight in 2014 of 1,847 kg [205], we have 7.52 $MgCO_2$ per vehicle.

As the report was largely concerned with projections for the year 2020, projected recycling rates of 95% for metals and 50% for plastics were assumed. These are reasonable, but perhaps a bit high, for the present day, as the EPA reports almost 90% of aluminum in vehicles is recycled, with nearly 80% of vehicle material recycled overall [218]. Several analyses by vehicle manufacturers have given similar emissions estimates. For example, a process-level LCA by Volkswagen [221] estimated that producing a 1999 Golf A4 took 85.6 GJ of primary energy (37.6 GJ for car fabrication, 48 GJ for materials) with associated emissions of 4.402 $MgCO_2e$ (1.89 and 25.12 $MgCO_2e$ for fabrication and materials, respectively), yielding 4.157 $kgCO_2e/kg$ vehicle (using a vehicle curb weight of 1,059 kg).

Results using EIO-LCAs are also in rough agreement with the above figures. Using an EIO-LCA, Samaras and Meisterling [220] estimated that 102 GJ of primary energy and 8.5 $MgCO_2e$ of emissions (using 0.3 kg CO_2e/kWh primary energy) are embodied in a 2006 Toyota Corolla (which has similar dimensions as a Toyota Prius), yielding per-kg emissions of 7.39 $kgCO_2e/kg$. This is reasonably similar to EIO analysis focusing on household carbon footprints by Jones and Kammen [2], who estimated 9 $MgCO_2e$ per household vehicle.

Finally, despite controversy over increased energy requirements in hybrid-electric vehicles, the GREET 2 model suggests that, on a per-kg basis, hybrid vehicle-cycle emissions are essen-

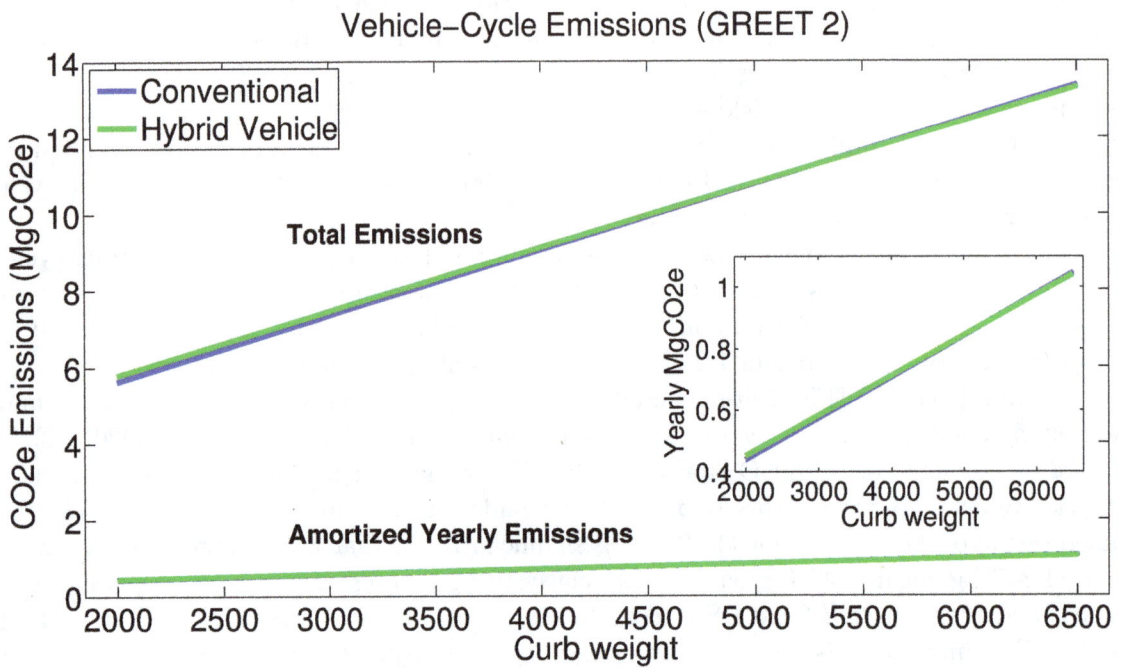

Figure 6.12: Vehicle-cycle (i.e. fabrication, maintenance, and disposal) emissions for conventional and hybrid vehicles as a function of weight (curves are nearly identical), under the 2014 GREET 2 Argonne National Laboratory model. Both lifetime emissions and emissions amortized over a 12.8 year vehicle lifetime are shown (with the inset magnifying the latter). The average new passenger vehicle in 2014 had a curb weight of 4,072 lbs, and associated emissions of 9.19 $MgCO_2$e lifetime, which amortizes to 0.72 $MgCO_2$e yearly.

tially identical to those of a conventional vehicle. The hybrid traction battery, while perhaps the single most emissions-intensive component of the vehicle, is so small as to make this additional burden negligible over the lifecycle of this vehicle. This does not necessarily hold for pure EVs, and their manufacturing emissions are discussed in Section 6.6.2.

6.5 Emissions and vehicle weight

6.5.1 Fuel consumption and weight, or "size is fuel"

From the above discussion, it is clear that embodied emissions from manufacture (and overall vehicle-cycle) scale linearly with vehicle mass. Furthermore, for conventional vehicles there is a linear strong relationship between fuel consumption, and therefore carbon emissions, and vehicle weight [205]. Figure 6.13 shows fuel consumption and economy (i.e. MPG) plotted against vehicle curb weights for the 30-bestselling car models in the US in 2014 (using 2015 model-year weight and MPG data, sales data are from http://www.goodcarbadcar.net/).

It is reasonably clear from this plot that, at least when it comes to conventional drive-trains, "low-emissions" is essentially synonymous with "small" and although MPG varies somewhat for a given weight, there are no glaring outliers[7]. Indeed, weight explains 84% of the variation in fuel consumption (i.e. $R^2 = 0.84$). The rule that "size is fuel" is underscored when we consider that the best mileage (40 MPG combined) of any gasoline-powered car is achieved by the diminutive Mitsubishi Mirage, which weighs in at just over 2,000 pounds and generates all of 74 horsepower from a 1.2 L, three-cylinder engine.

The relationship between weight and fuel consumption is weaker for hybrid vehicles, with weight explaining only a little over half (56%) of the variation in consumption (Figure 6.14). As can be seen by the curves in Figure 6.15, the hybrid weight-MPG curve is shifted upward compared to conventional vehicles. Note that while a significant number of individual hybrid vehicles have MPGs that are actually worse than the best conventional vehicles, these are generally SUV crossover-type vehicles, and the vast majority of hybrid vehicles actually sold achieve mileages over 40 MPG. Overall, the "size is fuel" rule is relaxed somewhat for hybrids, with fuel economy depending more on individual vehicle characteristics.

6.5.2 Overall emissions

I now examine the overall yearly emissions for typical *new* conventional vehicles as a function of weight. As always, I assume 160,000-mile/12.8-year lifetime, with vehicle-cycle emissions amortized over these years, and I use the previously derived empirical relationship between curb weight and MPG for the 30 most popular conventional vehicles to derive MPG for a given weight. Hence, given vehicle weight we can calculate both the vehicle-cycle emissions using the GREET 2 model, and infer likely fuel-cycle emissions. Note that our weight-MPG relation relates fuel economy and weight for *new* 2015 model-year vehicles, and the existing vehicle fleet will presumably have somewhat higher emissions for a given weight. Figure 6.16 shows the so-derived emissions estimates.

Using our weight-MPG relation, We can also infer vehicle weight when given fuel economy, and thus determine likely total lifecycle emissions for a vehicle achieving any particular fuel economy. Such calculations are tabulated in Figure 6.17, and are expected to hold fairly well for both conventional and hybrid-electric vehicles.

---

[7] with the possible exception of the Jeep Wrangler, which gets a mere 18 MPG combined for a minimum curb weight of 3,760 lbs; there are no apparent outliers in the other direction.

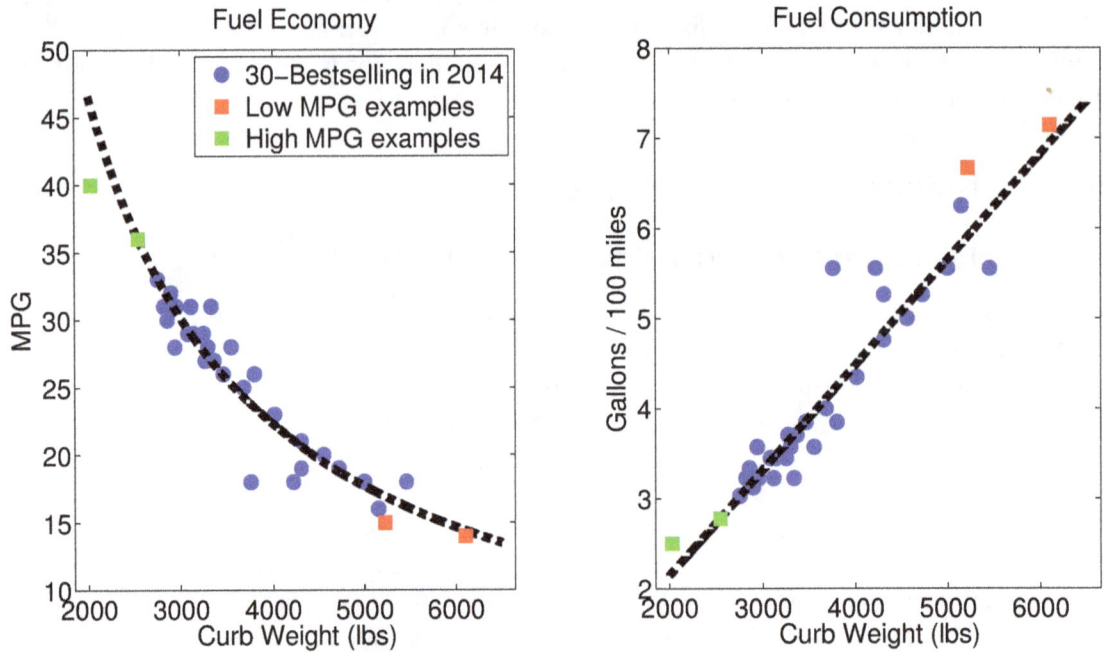

Figure 6.13: Relationship between curb weight and fuel consumption and economy (MPG) for the 30 bestselling vehicles in the US in 2014, all of which are conventional gasoline-powered vehicles. All data is for *new* 2015 MY vehicles.

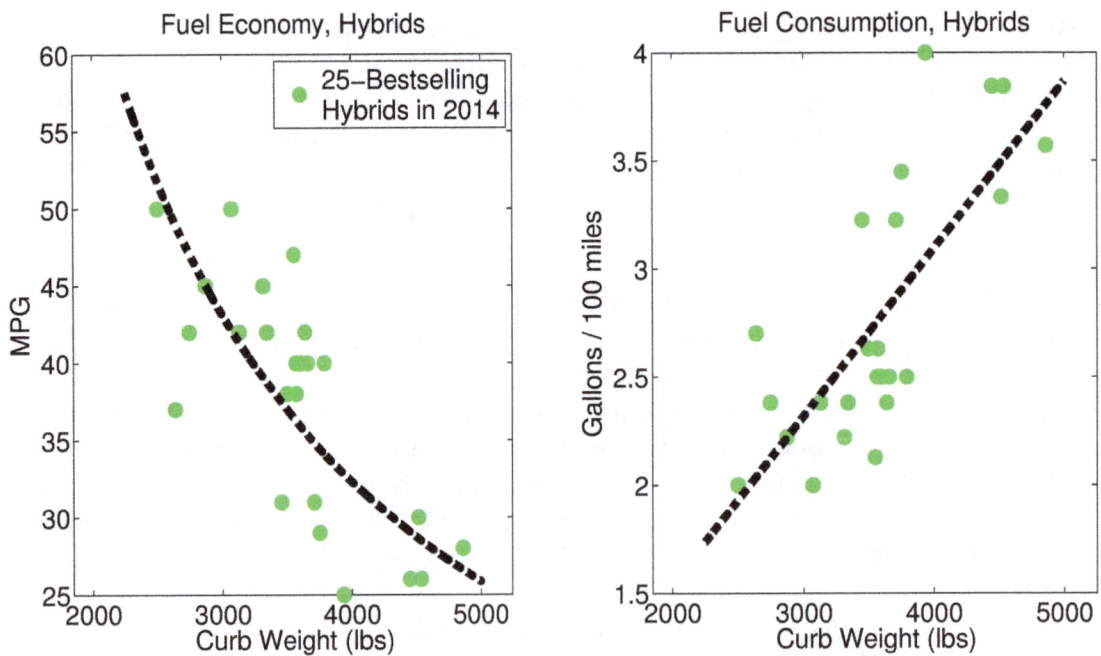

Figure 6.14: Relationship between curb weight and fuel consumption and economy (MPG) for the 25 bestselling hybrid vehicles in the US in December, 2014 (source: http://www.hybridcars.com/december-2014-dashboard/). Note that the relationship between weight and fuel consumption is much weaker relative to conventional vehicles, and that the figure scales are different compared to Figure 6.13. All data is for *new* 2015 MY vehicles.

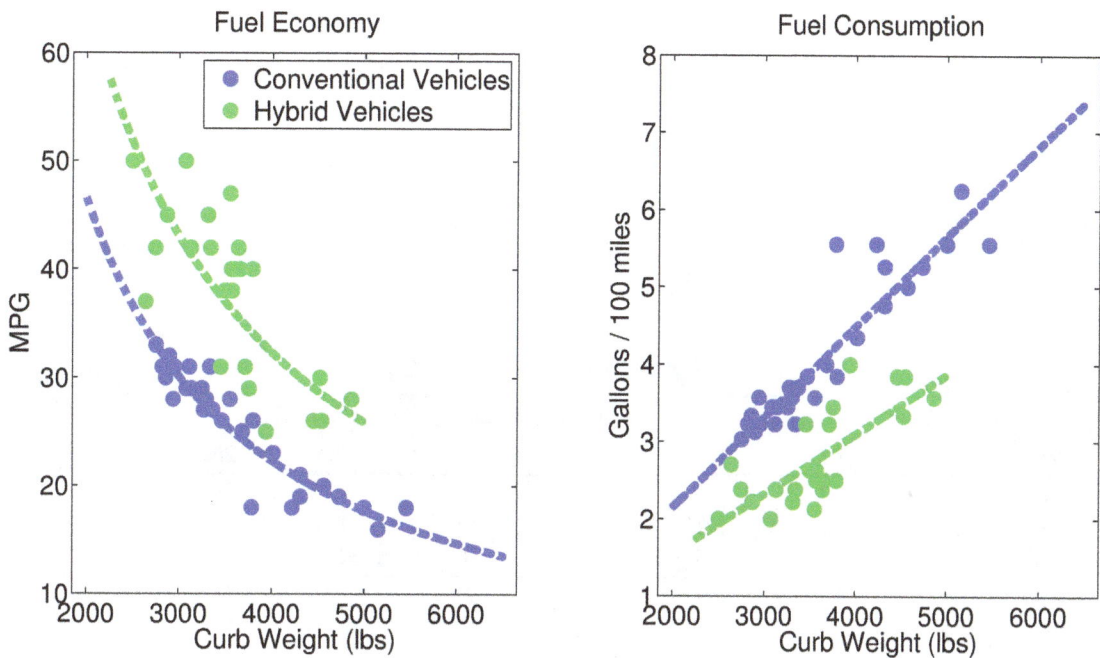

Figure 6.15: Fuel use as a function of weight for conventional and hybrid vehicles. All data is for *new* 2015 MY vehicles.

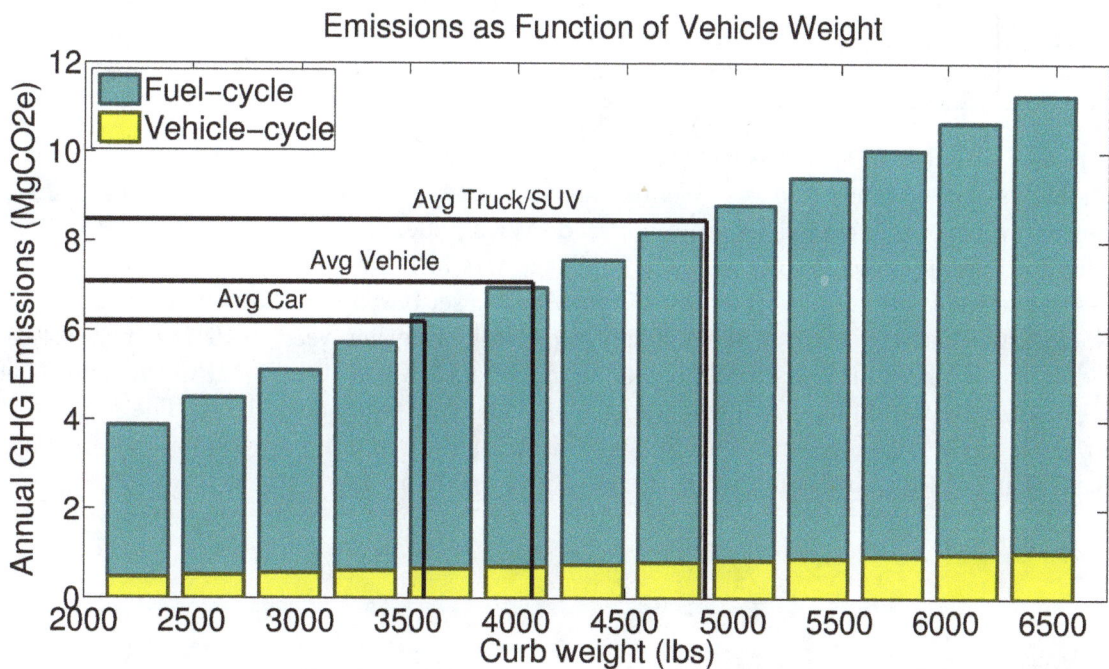

Figure 6.16: Expected yearly emissions as a function of weight for typical conventional vehicles sold in 2014/2015. Due to the close relationship between MPG and weight, fuel combustion-associated emissions increase linearly with weight (fuel-cycle), as do the amortized emissions from vehicle manufacture, disposal, and maintenance (vehicle-cycle).

Fuel–Cycle + Vehicle–Cycle

MPG	Miles (Thousands)							
	5	7.5	10	12.5	15	17.5	20	22.5
12	5.1	7.7	10	13	15	18	20	23
15	4.1	6.1	8.2	10	12	14	16	18
18	3.4	5.1	6.9	8.6	10	12	14	15
21	3	4.4	5.9	7.4	8.9	10	12	13
24	2.6	3.9	5.2	6.5	7.8	9.1	10	12
27	2.3	3.5	4.6	5.8	6.9	8.1	9.2	10
30	2.1	3.1	4.2	5.2	6.3	7.3	8.4	9.4
33	1.9	2.9	3.8	4.8	5.7	6.7	7.6	8.6
36	1.8	2.6	3.5	4.4	5.3	6.1	7	7.9
39	1.6	2.4	3.2	4.1	4.9	5.7	6.5	7.3
42	1.5	2.3	3	3.8	4.5	5.3	6.1	6.8
45	1.4	2.1	2.8	3.5	4.3	5	5.7	6.4
48	1.3	2	2.7	3.3	4	4.7	5.3	6
51	1.3	1.9	2.5	3.2	3.8	4.4	5	5.7

Figure 6.17: Annual CO_2e emissions (MgCO$_2$e) for various vehicle MPGs and annual mileages, including estimated amortized yearly emissions from the vehicle-cycle, i.e. manufacturing, etc. Note that vehicle-cycle emissions are amortized over a 160,000 mile lifetime, which equates to a different lifetime in years for each yearly mileage. As in Figure 6.8, red text signifies annual emissions worse than average—7.17 MgCO$_2$e in this case, based on 12,500 miles at 21.6 MPG (6.45 MgCO$_2$e) and a 4,072 lb curb weight (0.72 MgCO$_2$e per year, 9.19 MgCO$_2$e total)—while green signifies emissions totals better than average. Again, the 21 MPG row and 12,500 mile column are highlighted as being typical of American cars and habits. The inclusion of the vehicle-cycle increases emissions by 10–15%, with the relative increase greater for higher mileage vehicles (due to the sharper decrease in tailpipe emissions).

6.6 Hybrids, plug-in hybrids, and electric vehicles

6.6.1 Overview

While they still represent only a small fraction of market share, alternative drive-trains, including hybrid electric vehicles (HEVs), plug-in hybrid electric vehicles (PHEVs), and battery-electric vehicles (BEVs, a.k.a. electric-only vehicles) are becoming more popular. Indeed, if we are to transition from a fossil-based economy, then all vehicles will eventually need to be powered by electricity, or possibly other non-fossil-derived fuels, such as hydrogen. As discussed in Sections 3.8 and 22.3, current US biofuels (soy biodiesel and corn ethanol) cannot replace petroleum fuels at scale. In the shorter-term, however, it is essential to understand how these alternative drivetrains compare to each other and to traditional internal combustion vehicles (ICEVs) in terms of their lifecycle emissions profiles, in order to rationally move toward short-term emissions reductions.

In this section, I discuss emissions embodied in battery manufacture, discuss the MPGe and MPG$_{\text{GWP}}$ concepts for electric vehicles that allow a rational comparison of the in-use emissions of these drivetrains (electric vehicles are manifestly *not* zero-emissions even during the use-phase, as electricity generation is a highly carbon-intensive process), and also address the issue of electric vehicles acting as an additional, or marginal load, on the grid and the relatively high emissions-intensity of such electricity use. I find that, overall, HEVs, PHEVs, and BEVs are all reasonably similar in total emissions per mile, but this varies with particular model and with geographic region. In particular, larger long-range BEVs tend to be closer to conventional economy vehicles in their emissions profile, while most smaller PHEVs and BEVs are similar to HEVs. Geographically, the West, Texas, and Northeast are the best regions for EVs, while the upper Midwest is the worst.

The physical principles underlying hybrid drivetrains are discussed in Section 7.2.3, and these drivetrains are compared on the basis of their primary energy efficiency in Section 7.3.

6.6.2 Emissions embodied in battery manufacture

- Lithium-ion battery manufacture is emissions intensive, generating perhaps 100–200 kgCO$_2$e/kWh of battery capacity. This figure is uncertain and may fall as battery technology matures, but the only studies on commercial systems consistently suggest about 150 kgCO$_2$e/kWh.

- Additional emissions from battery manufacture are negligible for hybrid-electric vehicles, but can be significant for larger pure EVs and equivalent to several years of operational emissions: probable estimates for 30 kWh and 85 kWh battery packs are 4.5 and 12.75 MgCO$_2$e, respectively.

- Even in the most pessimistic case, EVs are still superior to *comparable* (but not necessarily *all*) conventional vehicles on an overall lifecycle basis, but battery-related emissions could tip the scales towards hybrids as the more favorable near-term technology.

Some readers will recall headlines from a few years ago claiming that the much-maligned Hummer is actually greener that the (also much-maligned) Toyota Prius, due to the environmental cost of manufacturing the latter. As I discuss in a moment, this is patent nonsense, and stems mainly from a non-peer-reviewed report from an advertising firm that makes laughably absurd conclusions (this "Dust to Dust" report is not worth refuting in detail). Sadly, the idea is "out there" now, and seems to refuse to die. Like most good lies, it contains at its heart a grain of truth: HEVs do require slightly more energy than comparable conventional vehicles to produce, perhaps 5–10% more. This manufacturing cost is recouped in gasoline savings after only a few months of typical driving (or less), and the lifetime emissions of a typical HEV are vastly lower than those of a typical conventional vehicle. Further note that, on the basis of total vehicle weight, HEVs and standard internal combustion engine vehicles, have nearly identical vehicle-cycle emissions, as shown in Figure 6.12. For pure EVs, however, the additional manufacturing emissions from the battery may be more appreciable.

The battery is the major component of hybrid and electric vehicles that contributes to its greater manufacturing energy and emissions costs. Numerous lifecycle analyses have addressed this issue, but there remains significant uncertainty. First, note that two major battery classes are used in EVs, namely nickel-metal hydride (NiMH) and lithium-ion (Li-ion), and emissions may either be reported on the basis of battery weight (kg) or energy capacity (kWh). Li-ion batteries store twice or more the energy per unit weight, with the GREET model suggesting about 65 Wh/kg for NiMH and 133 Wh/kg for Li-ion (although many newer Li-ion batteries achieve slightly higher energy densities). It follows that battery packs for pure EVs are quite heavy, with the Nissan Leaf's 30 kWh Li-ion battery pack weighing in at 218 kg (480 lbs, energy density of 140 Wh/kg) [223], while the Tesla Model S 85 kWh Li-ion battery pack is a massive 544 kg (1,200 lbs and 156 Wh/kg). The batteries for HEVs are much smaller, with a 1.3 kWh capacity typical.

A 2010 review of earlier studies by Sullivan and Gaines [222] gave average emissions factors of 13.6 and 12.5 $kgCO_2e$/kg (or about 200–250 and 100–125 $kgCO_2e$/kWh) for NiMH and Li-ion batteries, respectively. Ellingsen and colleagues [227] recently determined a minimum of 172.9 $kgCO_2e$/kWh directly using data from a commercial Norweigan manufacturer of Li-ion batteries; several other studies reviewed indicated production emissions roughly 150 to 250 $kgCO_2e$/kWh. Similarly, Zackrisson et al. [226] calculated Li-ion battery production to generate 166 $kgCO_2e$/kg, and a 2013 lifecycle analysis by the EPA [229] estimated an average of 112 $kgCO_2e$/kWh for Li-ion batteries (ranging from 63 to 151 $kgCO_2e$/kWh for different battery chemistries).

Most significantly, Kim and colleagues [228] very recently published an assessment for the commercially produced Ford Focus EV, estimating emissions of 140 $kgCO_2e$/kWh for the vehicle's 24 kWh Li-ion battery pack (totalling 3.4 $MgCO_2e$). This and the work by Ellingsen et al. [227] are the only studies on actual commercial systems.

Lower estimates are given by Dunn et al. [225] who, using a bottom-up method, calculated only 39.2 $kgCO_2e$/kWh for Li-ion batteries (converted from 5.1 $kgCO_2e$/kg, at 130 Wh/kg), and a lifecycle assessment by Notter et al. [224] suggested 53 $kgCO_2e$/kWh. Nevertheless, the balance of the literature seems to support 100–175 $kgCO_2e$/kWh as the more likely range for Li-ion batteries (NiMH batteries may be around 200 $kgCO_2e$/kWh or more), with the only two studies examining actual commercial systems consistent at roughly 150 $kgCO_2e$/kWh, which I consider as my central estimate.

For HEVs, the small battery size (about 1.3 kWh on average) indicates that the additional emissions from the battery are negligible, even under the most pessimistic assessment. The impact may, however, be fairly significant for pure EVs, especially those with longer ranges, and the range of estimates above (39 to 250 $kgCO_2e$/kWh) suggest an additional 1.2–7.5 $MgCO_2e$ for the Leaf's 30 kWh battery pack, and 3.3–21.25 $MgCO_2e$ for the 85 kWh Model S; 4.5 $MgCO_2e$ and 12.75 $MgCO_2e$ are reasonable mid-point estimates. Even amortized over a 12.8-year vehicle lifetime, towards the upper range such impacts are not trivial.

6.6.3   Electric vehicles: understanding MPGe and the MPG$_{\text{GWP}}$ concept

- Electric vehicle fuel economy is given by the EPA in miles per gallon equivalents (MPGe), which relates the miles the EV can travel on an amount of electrical energy equivalent to the thermal energy contained in one gallon of gasoline (33.7 kWh on LHV basis); EVs get on the order of 100 MGPe.

- To recast MPGe in terms of global warming potential, we define MPG$_{\text{GWP}}$ as the equivalent MPG a gasoline-powered vehicle would get if it had the same overall per-mile CO_2e emissions, and using the carbon-intensity of electricity generation, I determine that MPG$_{\text{GWP}} = \alpha$ MPGe, where α = 0.485 for US grid-average electricity (and α = 0.365 for marginal electricity). Regionally, α varies from 0.39–0.78 under grid-average electricity and from 0.31–0.43 for marginal electricity, with α lowest in the Midwest and South.

Electric vehicle fuel economy is rated by the EPA in *miles per gallon equivalents* (MPGe), and most EVs get on the order of 100 MPGe. However, equivalents here does not refer to global warming or CO_2 emissions equivalents, but energy equivalents. That is, it is a comparison based on the thermal energy content of gasoline, and the electrical energy required for an EV to go one mile. Gasoline contains 33.7 kWh(t) per gallon (LHV basis), and EVs travel about three miles per kWh of electrical energy. Multiplying these two numbers gives 101.1 miles / (energy in a gallon of gas) as the MPGe. To re-state the idea, if an EV gets 100 MPGe, you would need electrical energy equivalent to the total energy content of a single gallon of gas to go 100 miles. It does not matter whether the source of the electricity is solar or coal, the MPGe is constant and is a property of the vehicle only.

Thus, it is clear that the MPGe tells us little about the vehicle's equivalent MPG from the perspective of GHG emissions. On vehicle labels, the EPA only gives a semi-quantitative 1 ($\leq$ 12 MPG) to 10 (38+ MPG) scale measuring of the GHG impact, and this is based on tailpipe emissions only, which are, of course, zero for pure EVs. Since, for any EV, we know the kWh used per mile, if we know the carbon intensity of electricity generation and delivery (about 0.682 kgCO_2e/kWh for the US grid as a whole), we can easily determine CO_2e emissions per mile. From there, we can derive the equivalent MPG for a gasoline-powered vehicle; the Union of Concerned Scientists has denoted this metric the MPG$_{\text{GWP}}$.

So for example, assume an EV rated at 100 MPGe. We divide by 33.7 kWh to get 2.97 miles/kWh, and electrical energy consumption of 0.337 kWh/mile. Now, charging at grid-average emissions rate, 0.682 kgCO_2e/kWh, we have upstream emissions of 0.23 kgCO_2e/mile. Since each gallon of gasoline has associated lifecycle emissions of 11.146 kgCO_2e/gallon, a per-mile emissions-rate of 0.23 kgCO_2e is thus equivalent to a gasoline-powered car getting 48.5 MPG, i.e. MPG$_{\text{GWP}}$ = 48.5. Note that this calculation yields a constant conversion factor for MPGe to MPG$_{\text{GWP}}$, which I denote α, equal to 0.485 for grid-average US electricity.

Now, regional electricity grids vary in their emissions-intensities (see Section 4.4), and thus the MPGe to MPG$_{\text{GWP}}$ conversion ratio, α, necessarily varies geographically. Furthermore, calculated emissions from EV operation vary depending on whether we use the grid's average generating mix, or the marginal generating mix. That is, EVs represent an additional load upon the grid, and therefore the additional electricity that must be generated will be drawn from (generally) more emissions-intensive marginal generating sources. For example, the Nissan Leaf, the best-selling electric car in 2014 (followed closely by the Tesla Model S), gets 114 MPGe; using grid-average EFs, this is equivalent to 44.5–89.2 MPG$_{\text{GWP}}$ (55.3 on average), with $\alpha = 0.39 - 0.78$, while marginal EFs give 35.6–48.9 MPG$_{\text{GWP}}$ (41.6 on average), and $\alpha = 0.31 - 0.43$. Figure 6.18 plots MPG$_{\text{GWP}}$ and estimated yearly emissions under four different

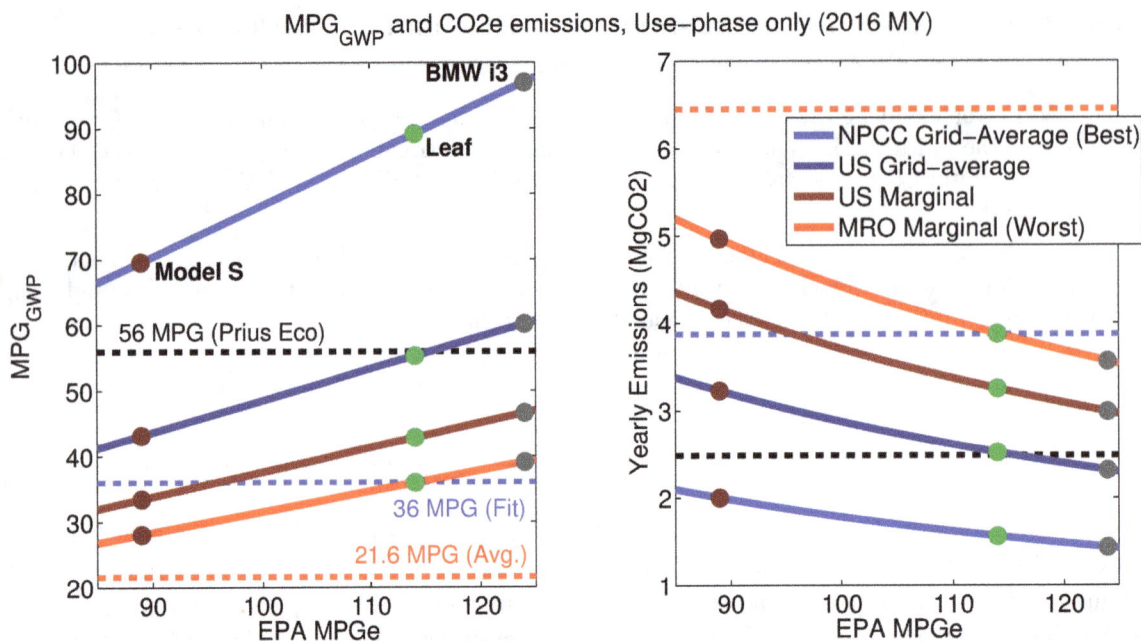

Figure 6.18: Electric vehicle gasoline-equivalent MPG (MPG_{GWP}) as a function of MPGe rating (left panel), and yearly emissions (right panel, assuming 12,500 miles), under four different US grid emissions factors. The best-case is the grid-average for the northeast NPCC region, while the worst-case is marginal emissions in the upper midwestern MRO region. The overall US grid-average and the average US marginal emissions are also shown. Included for reference, we plot $MPGe/MPG_{GWP}$ and emissions points for several EVs, along with MPG/emissions under the best gasoline-powered HEV (2016 Prius Eco), best conventional vehicle (36 MPG Honda Fit), and fleet-average vehicle (21.6 MPG).

grid EFs. Note that all calculations in this section relate electricity generation for EVs to the fuel-cycle for gasoline only, and do not take into account differing vehicle manufacturing emissions for EVs and conventional vehicles.

6.6.4 Marginal emissions and electric vehicles

- As an additional load on existing electrical grids, to determine their direct effect on carbon emissions, plug-in EVs are arguably best evaluated using marginal emissions factors for electricity generation/consumption, not grid-average EFs.

- Under marginal emissions scenarios, at the use-phase EVs are better or similar to high-MPG HEVs in the West, Texas, and Northeast, closer to economy conventional vehicles in much else of the country, and almost as bad as typical conventional vehicles in the upper midwest (MRO subregion). Colder northern temperatures also degrade EV performance.

- Charging late at night (as is typical) usually relies more on coal-fired generation and leads to higher emissions.

As already mentioned, when deployed, plug-in electric vehicles represent a large additional load on existing power grids, and thus, arguably the most appropriate way to compare different drivetrains, at least over the relatively short-term time horizon, is to use marginal emissions rates, rather than the grid-average, for electricity generation [230, 231]. Zivin et al. [230],

156

drawing from several databases, calculated marginal emissions rates on an hourly basis across the eight NERC geographic subregions, and used these to determine how operational emissions for a plug-in EV would vary both geographically and with charging time-of-day. They determined that, in the western (WECC) region, marginal emissions are quite low (just 0.364 kgCO$_2$e/kWh, which strikes me as rather implausible: a natural gas plant would have to operate at 54% thermal efficiency to achieve this, far better than what is generally achieved), and here an EV would generate fewer emissions than a Prius, regardless of charging time; in Texas, the EV was marginally superior to a Prius. However, in other subregions, the EV was either comparable to a Prius or in-between a Prius and an economy car getting 31 MPG, while in the MRO region (upper midwest), EVs approached typical gasoline vehicles. Furthermore, in the south and east, marginal emissions were significantly higher at night, when EVs are expected to be charged. This study considered emissions only at the point of generation (tailpipe or smokestack).

Tamayao et al. [231] performed a similar but more comprehensive analysis, considering marginal emissions estimates by both Zivin et al. [230] and Siler-Evans et al. [74], grid-average emissions, and additionally considered all lifecycle emissions in their comparison of two EVs, the Nissan Leaf and (plug-in hybrid) Chevy Volt, versus the Prius. Consistent with Zivin et al., they determined the Leaf to be superior to the Prius in the west, marginally better in Texas, and inferior in the rest of the country, with the Leaf almost as emitting as a typical conventional car in the MRO region. The Volt, on the other hand, was comparable to or worse than the Prius everywhere. Again, charging late at night was found to increase emissions, as the marginal mix relies strongly on coal during these hours [231]. For the interested reader, Tamayao et al. [231] also provide a useful review of several other recent studies comparing ICEVs, HEVs, and EVs.

Axsen and colleagues [233] studied the effect of adding 1 million PHEVs to the California grid, using an electricity-dispatch model to realistically model the resulting emissions. They concluded that emissions would be lower than those of typical conventional vehicles, but roughly equivalent to a fleet of gasoline-only HEVs getting 42.5 MPG. Thus, even in a relatively low-carbon grid, when implemented at scale, EVs are likely to be similar to HEVs (again, at least in the short-term).

Finally, I note here that low temperatures can strongly degrade the performance of EVs, both via decreasing battery efficiency and power, and by adding a large cabin heating load. While waste heat from gasoline combustion heats ICEVs and HEVs, BEVs must use significant amounts of electrical energy. Higher temperatures also add a cooling load, and temperature-dependence was studied in some depth by Yuksel and Michalek [232], whose results show colder northern temperatures to further increase EV emissions under the already relatively dirty mid-western and eastern grids.

Some further thoughts on grid-average vs. marginal carbon accounting

In many studies and publications, the average electricity emissions rates, rather than the marginal rates, are used to determine the carbon footprint of PHEVs and EVs. The basic problem with this, I think, is that any time we talk about MPG$_{GWP}$, we are always, or at least I am, implicitly comparing this MPG$_{GWP}$ value to that of a gasoline-fueled car. I am not considering, say, the MPG$_{GWP}$ of an electric vehicle in isolation, but am implicitly comparing two alternatives: the EV versus a conventional or hybrid-electric vehicle. Because of this fact, I take the view that we are essentially comparing two *additional* loads on the system, either electric or liquid fuel. Therefore, it seems most appropriate to use the marginal electricity generation emissions, and not the grid-average. After all, if I go with the gasoline-fired vehicle, I add emissions from fuel combustion but avoid additional electricity use, and visa versa.

Now, some may argue that grid-average emissions rates should be used when calculating the carbon emissions of EVs, as it is not fair to say that other societal demands for electricity should be privileged over vehicle charging. And indeed, we generally use grid-average EFs in our accounting of other electricity uses. The difference, arguably, from a policy and personal level here, is that EVs are not already part (to any great degree) of the electricity-consuming infrastructure. And disregarding what accounting scheme is "fair" for the moment, it is sensible to simply consider the actual expected effect on GHG emissions that occur when an individual purchases (and uses) a plug-in EV. Fair or not, the increased demand *will* effect marginal electricity generation, and so we must at least seriously consider the marginal emissions in our decision making. In any case, I present results for both grid-average and marginal emissions rates in subsequent sections. We must also always keep in mind the goal of a low-carbon transition, and what the appropriate role of electrical vehicles is with respect to this. It is certainly arguable that PHEVs are a more appropriate bridging technology than HEVs, as they may help to establish an electricity-based transportation infrastructure while avoiding the costs of very large battery packs. Additionally, a scenario where an EV or PHEV may be very desirable is in conjunction with rooftop solar (appropriately scaled up to cover the EV load), as utilities are often hostile to rooftop customers generating more electricity than they use, and in this case net carbon emissions for operating one's EV may truly be nearly zero.

6.6.5 Comparison of major PHEVs, BEVs, HEVs, and ICEVs

In this section, I compare total lifecycle emissions for several different top-selling vehicles, the HEV Prius, PHEV Chevy Volt, and the BEVs Nissan Leaf and Tesla Model S, to each other and to new conventional vehicles. This comparison is highly salient, as the Prius hybrid family is by far the best-selling HEV brand, while the Leaf, Volt, and Tesla account for the vast majority of EV/PHEV sales (at least historically). All calculations are made assuming a 160,000 mile lifetime, with 12,500 miles per year, for a 12.8 year vehicle lifespan. The 2014 GREET model is used to calculate vehicle-cycle emissions for ICEVs and HEVs, while I use the GREET model to obtain a regression equation for vehicle-cycle emissions without the battery for PHEVs and BEVs. To this base vehicle-cycle EF, I add 150 gCO_2e/kWh of battery capacity, based upon the discussion in Section 6.6.2.

An important factor in evaluating the emissions-attributable to a plug-in hybrid, and the utility of the technology in general, is the fraction of vehicle travel miles that can be taken in electric-only mode. As is frequently pointed out by proponents of the technology, 95% of daily (one-way) vehicle trips are under 30 miles [642], which would suggest that PHEVs can meet almost all demand using electricity only. However, since the average driver takes over three one-way trips a day, each driver averages almost 30 miles of daily travel. Moreover, daily travel over 60 miles accounts for over 50% of total vehicle miles [234]. It follows that many plug-in hybrids on the market, with electric-only ranges of about 20 miles or less, will operate in electric-only mode for less than half of all miles, assuming that they are charged once per day; the only widely available PHEV (as of this writing) that can go over 50% of total miles on electricity is the Chevy Volt (38-mile electric-only range for the 2015 model year, and now 53 miles for the 2016 model). Figure 6.19 shows the approximate percentage of vehicle-miles run on electricity as a function of electric-only range. A low EV-only range is not necessarily a negative, as such vehicles have the advantages of supporting an EV-based infrastructure while having lower materials, energy, and financial costs due to battery manufacture, do not necessitate a second vehicle for longer distance trips (as many pure EVs do in practice), and represent a lower marginal load on the electrical grid.

Having all salient information, we can calculate all use-phase (or fuel-cycle) emissions, as well as amortized manufacturing emissions, as demonstrated in Figure 6.20. From this figure, it is clear that under most electricity sources, the Leaf and Volt are actually more emitting overall than the Prius and likely comparable to a decent hybrid vehicle, but are still usually

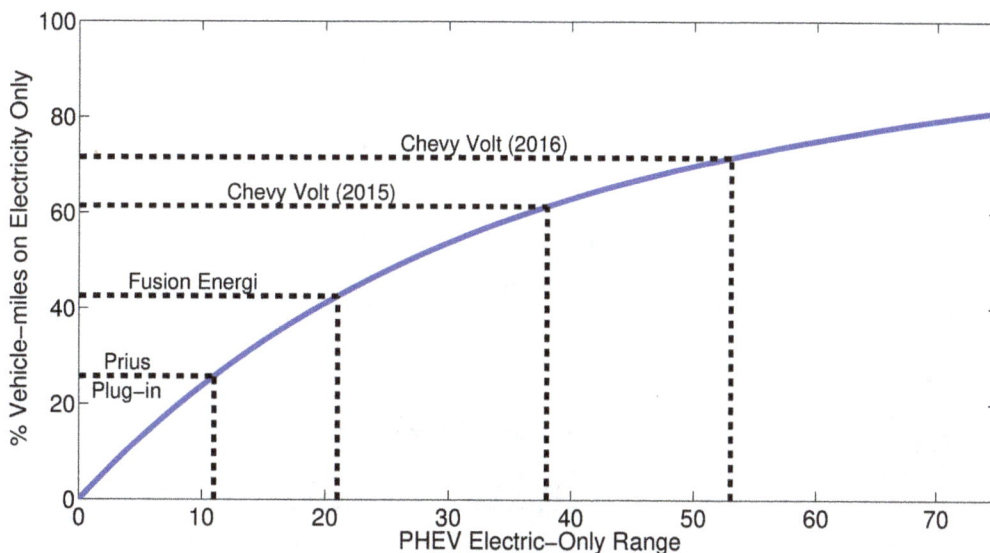

Figure 6.19: Percentage of total vehicle miles travelled in electric-only mode for a plug-in hybrid vehicle, as a function of vehicle electric-only range. Curve is based on Table 7 and Figure 8 of [234], inferred from vehicle-mile travel reported in the 2001 NHTS. Several example vehicles (2015/2016 model year) are included, demonstrating that most PHEVs will operate in electric-only mode less than 50% of the time, although, as noted in the text, this is not necessarily a negative.

better than a best-case conventional vehicle (36 MPG Honda Fit). The lifecycle emissions of the larger Tesla Model S are generally similar to an economy gasoline vehicle, but in the heavily-coal reliant grids of the upper midwest and east, it may actually be slightly worse than a typical *new* gasoline vehicle (although still slightly better than the current fleet-average).

Comparing to conventional vehicles, overall the Nissan Leaf and Chevy Volt are roughly equivalent to a 40–50 MPG vehicle (most other smaller BEVs/PHEVs are similar to these two vehicles), within the range of the most popular hybrids. The Tesla Model S is more like a 30–36 MPG vehicle, comparable to an economy ICEV. This is a bit disappointing, but still, if one compares the Model S to competing luxury sedans at roughly the same price point, these heavy vehicles often get only around 16 miles per gallon. Furthermore, other luxury brand PHEVs also perform relatively poorly. Figure 6.21 presents approximate lifecycle emissions for 16 top-selling BEVs and PHEVs.

The consumer's bottom line

The above analyses suggests that the most popular electric and plug-in hybrid-electric vehicles are roughly comparable to decent gasoline-powered HEVs, with the larger Tesla models more similar to economy conventional vehicles. This is not to say that EVs are *bad*: they are still generally low-emissions vehicles compared to conventional gasoline-powered cars, but their advantage is not as great as the popular imagination might have it. At least over the nearer-term (say 10 to 20 years), HEVs may actually be the best choice overall for reducing GHG emissions, especially because large-scale deployment of EVs will affect relatively carbon-intensive marginal electricity generation. This conclusion will vary geographically, and is almost certainly true that a hybrid is preferable to an EV throughout most of the midwest, east, and south. In parts of the west, Texas, and upper northeast, EVs are comparable to the best hybrids and be somewhat, but probably not dramatically, superior.

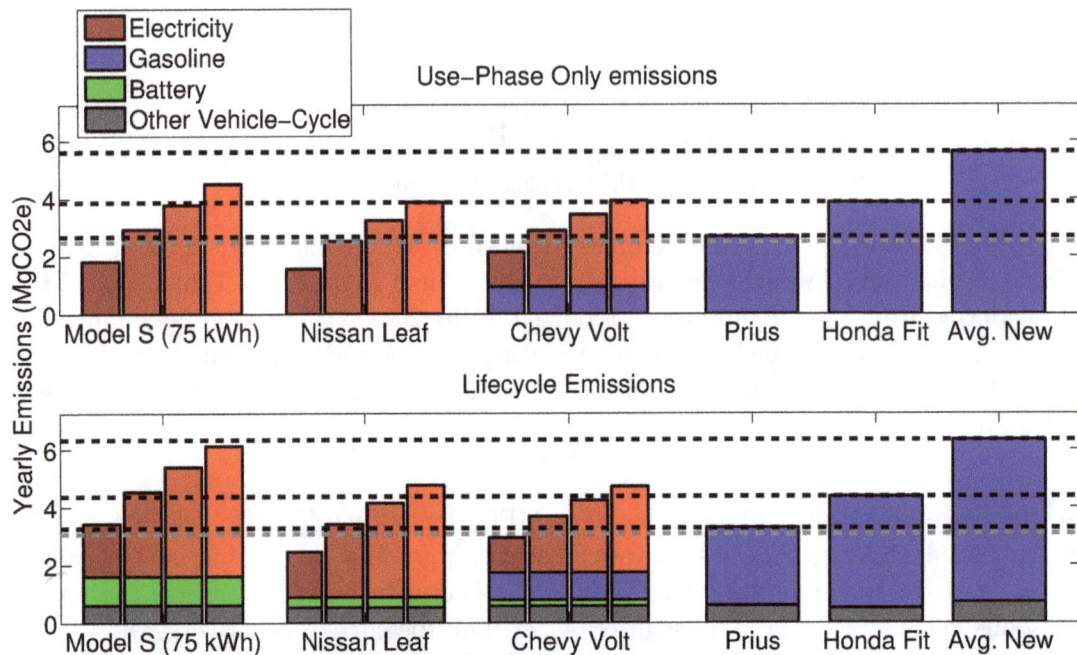

Figure 6.20: The top panel compares yearly emissions (assuming 12,500 miles/yr) from just the use-phase (i.e. fuel-cycle) of the 2016 model year Tesla Model S (75 kWh version, 98 MPGe), Nissan Leaf (114 MPGe), Chevy Volt (106 MPGe, 42 MPG gasoline-only), Toyota Prius (52 MPG, 56 MPG for Eco version), Honda Fit (36 MPG), and an average new vehicle getting 24.8 MPG and weighing 4,072 lbs (for 2015). Four cases for each EV give emissions under four representative electricity EFs, which are, from left to right, NPCC subregion grid-average (best-case), US grid-average, US marginal, and WRO subregion marginal (worst). The Prius and Fit represent best cases for HEVs and ICEVs, respectively. Note that the gray dashed line shows emissions for a 56 MPG Prius (Eco version), while the bar shows emissions for the standard 52 MPG version. The lower panel compares these vehicles over the entire vehicle-cycle, with battery and other vehicle-cycle emissions amortized over 12.8 years.

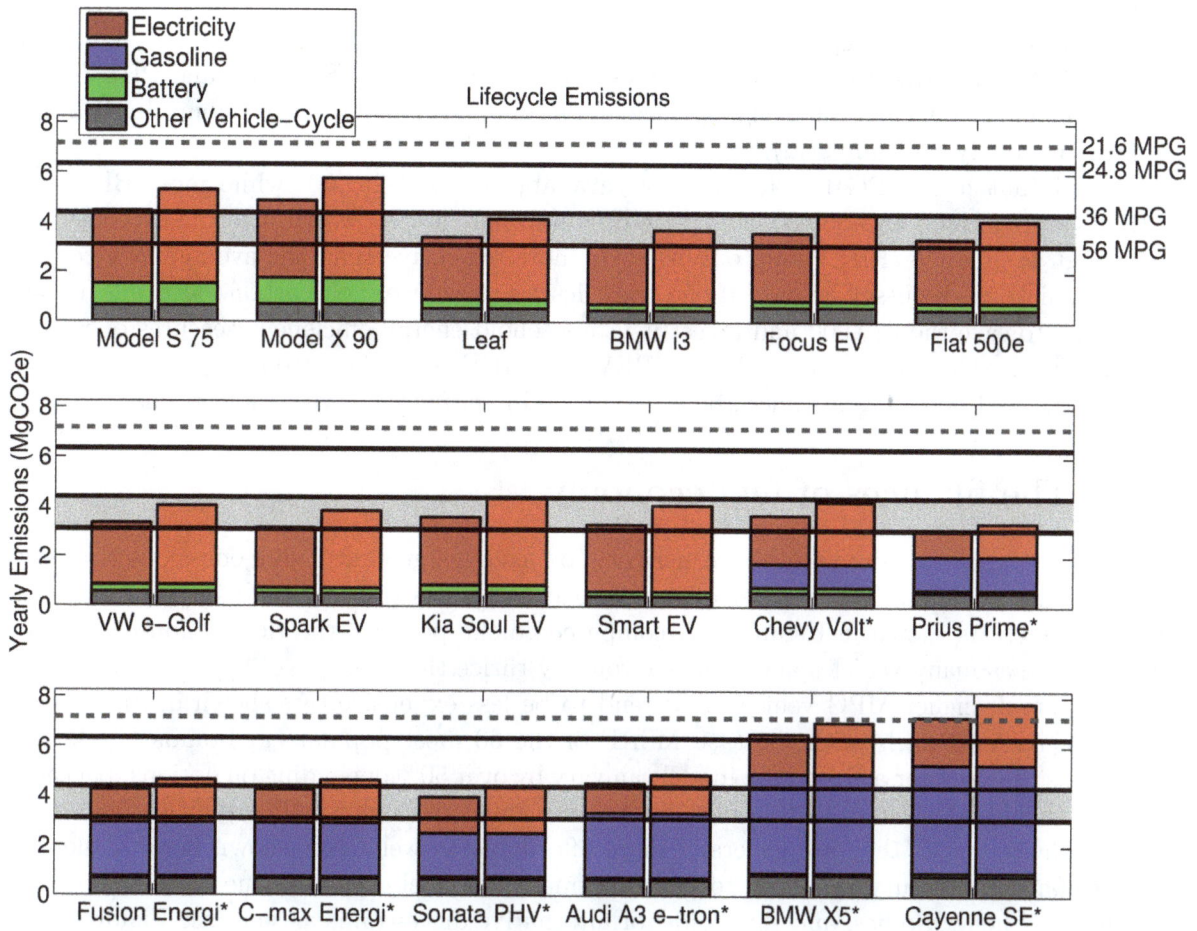

Figure 6.21: Lifecycle carbon emissions over a year (for 12,500 miles/yr, 12.8 mile vehicle lifetime), for 16 top-selling pure EV and plug-in hybrid EV vehicles (asterisk indicates PHEV), using either US grid-average electricity emissions, or US average marginal emissions (left and right bar for each vehicle, respectively). The inscribed horizontal lines, from bottom to top, give emissions for the best HEV (56 MPG Prius Eco), best conventional vehicle (36 MPG Honda Fit), average new vehicle in 2015 (24.8 MPG), and fleet-average vehicle (21.6 MPG, dotted line). The grayed region between the 56 and 36 MPG lines represents the majority of HEVs (the top-ten selling hybrids in 2014 had a weighted harmonic mean MPG of 45 MPG and an arithmetic mean weight of 3,356 lbs, placing a typical HEV right in the middle of the gray region). We see that the majority of EVs and PHEVs are similar to typical HEVs in their overall emissions profile, the heavier Tesla models are more similar to an economy car, while luxury PHEVs can be worse than the average new passenger vehicle.

Generally speaking, there is little immediate advantage, if any, to a plug-in hybrid over a straight hybrid, and the plug-in version may actually be more emitting under marginal emissions scenarios, although short-range PHEVs may help support infrastructure and can, like pure EVs, be deployed in conjunction with residential rooftop solar.

One should also note that the *absolute* emissions differences between EVs and HEVs under different electricity generating scenarios are usually much less than the differences between the worst, or even average, conventional vehicles and the best. Thus, the biggest short-term priority is a shift away from low-MPG ICEVs, with economy ICEVS, HEVs, and PHEVs/EVs all reasonable alternatives (although we are at the point where even an economy ICEV seems difficult to justify over a hybrid).

Given that high-MPG hybrids are now available for $20–25,000, while most BEVs and PHEVs start around $30–35,000, this price-premium may not be worth it for many consumers (although even BEVs/PHEVs are only slightly more expensive than the average new conventional vehicle, as discussed below). If one truly desires to reduce their carbon footprint, it could be better to save the $10,000 and invest it in household energy efficiency measures or rooftop solar. The major circumstance where a BEV or PHEV might be strongly preferable in the short-term, as I note several times above, would be in *conjunction* with rooftop solar.

6.7   The finances of fuel economy

As in every other domain, wealthier Americans, on average, generate more emissions related to personal automobiles. The number of cars owned and miles traveled both increase markedly with income. As shown in Figure 6.22, compared to the poorest, the richest households own about twice as many vehicles and consume roughly thrice the fuel.

In general, higher MPG vehicles also tend to be less expensive. As shown in Figure 6.22, MPG decreases linearly with the *base* MSRP of the 30 most popular car brands. I use base MSRP, as the price for a given base model can vary by over 50% depending on options, packages, etc. It should be clear from this picture that not only are higher MPG vehicles less taxing on the environment, they are generally more affordable as well. Not shown here, higher-end performance and luxury cars tend to have abysmally low fuel economy, due to their oversized engines and heavy bodies, and this holds for alternative drive-trains as well (see Figure 6.21).

While hybrid and electric vehicles are, in general, more expensive than comparable conventional vehicles, many are still appreciably less expensive than the average new car. Indeed, Kelley Blue Book reported the average new car price to be $33,666 in March, 2016, with $26,772 the average price of new hybrid vehicles; pure EVs exceeded the average by just $3,000 ($36,644), a surprisingly low margin, to my mind.

More expensive vehicles also cost more to insure, with price being a primary determinant of insurance rates. Larger, heavier vehicles also have higher maintenance and parts (e.g. tires) costs, and lifetime fuel costs vary by $15–20,000, even between conventional vehicles. For example, vehicles getting 16, 36, and 50 MPG will cost $25,200, $11,200, and $8,064, to fuel over a 160,000 mile lifetime, at the 2015 average retail price of $2.52/gallon (for gasoline, per the EIA). Longer-term, prices will inevitably increase, despite the recent dip in prices that is largely related to increased (but unsustainable) conventional oil extraction by OPEC nations. In all, then, it is rather remarkable the financial resources Americans have proven willing to devote towards high-priced vehicles that are, in general, actively worse for the larger world than perfectly functional alternatives.

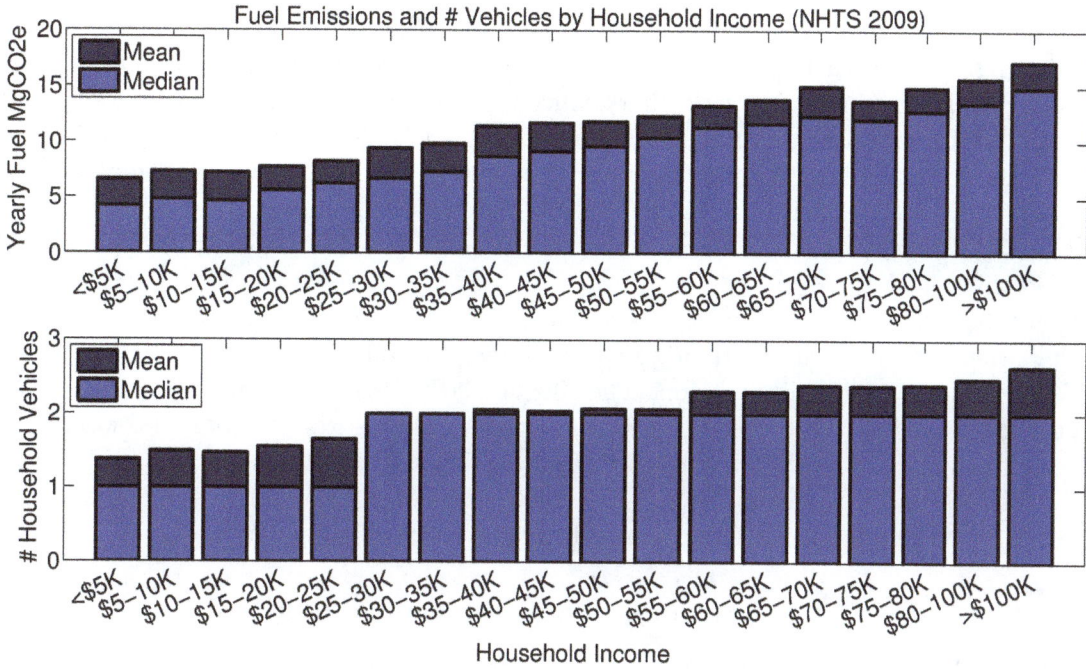

Figure 6.22: Annual fuel emissions (top panel), and number of vehicles owned (lower panel), by household income using data from the 2009 NHTS. The darker bars give the mean, while the lighter show the median. Richer households own on the order of twice as many vehicles and use around three times the fuel.

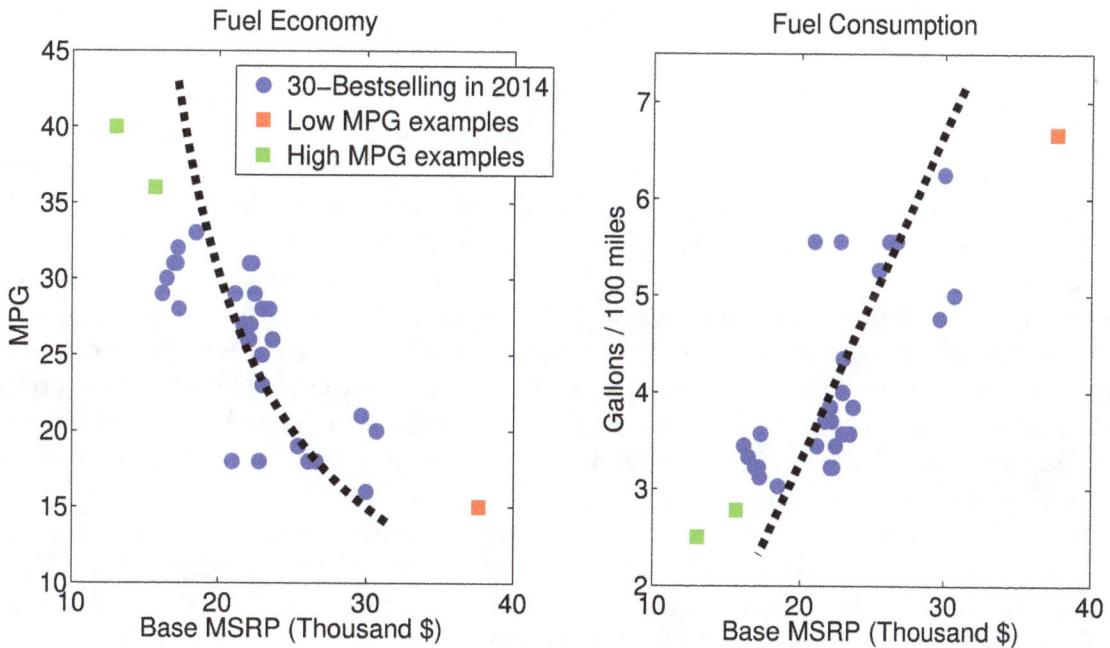

Figure 6.23: Fuel consumption and economy as a function of base MSRP, for the 30 bestselling models in 2014 (MY 2015), all of which are conventional internal combustion vehicles. MPG tends to decrease linearly with base vehicle MSRP, essentially because larger vehicles, which have lower MPGs, also have higher sticker prices. Note that the association between price and MPG is not as strong as that between weight and MPG, with $R^2 = 0.55$.

6.8 Driving style

"Eco-driving," a term that has come into more widespread use recently, refers mainly to relatively simple changes in driving style that reduce fuel consumption, and has been promoted as a potential adjuvant strategy for emissions mitigation, and indeed, widespread adoption of the practice could meaningfully reduce emissions immediately. Conversely, while there is less in the scientific literature, aggressive driving habits may reduce fuel efficiency up to 30%. Thus, a car rated at 25 MPG may reasonably achieve anywhere from 19 to 28 MPG in practice, solely as a result of driver habits.

Along those lines, one study [235] tracked fuel use for 117 drivers, each driving an identical vehicle (either a 2006 or 2007 V6 Honda Accord), under naturalistic conditions, for several weeks each in the Detroit metropolitan region. The 10% and 90% percentile in fuel consumption varied from the mean (23.6 MPG) by 13 and 16%, respectively. That is, the 10% best and worst drivers achieved 27.4 versus 20.5 MPG, a difference of about 33%, and equivalent to a gap of 154 gallons and 1.7 $MgCO_2e$ over a year of driving (at 12,500 miles/year). Part of this difference may be attributed to route differences, but much is likely due to driving style as well.

6.8.1 Overview and effectiveness of general eco-driving techniques

Broadly speaking, eco-driving entails simple driving techniques to decrease fuel consumption, optimal route selection, and proper vehicle maintenance (the EPA has stated that failure to properly maintain a vehicle may reduce fuel efficiency by up to 8%). The first is the primary focus of most literature and outreach efforts, and eco-driving techniques generally consist of moderate acceleration, anticipating traffic flow and signals to minimize sudden stops and starts, coasting as much as possible to a stop, using cruise control on the highway, reducing time spent idling, and driving at or below the speed limit [236]. Chapter 7 introduces the basic physics of internal combustion engines, and an understanding of these principles can lead to more deeply informed eco-driving strategies.

Reductions in fuel consumption from baseline in the range of 5–30% have been reported, but eco-driving is generally estimated to reduce fuel consumption by 5–10% under sustained, real-world driving conditions, for most drivers. In the late 1970s, the DOE instituted its Driver Energy Conservation Awareness Training (DECAT) training program, and Greene [237] estimated that such training could yield fuel savings in excess of 10% for average motorists (the DECAT program was discontinued during the Reagan years). Kurani et al. [238] found that simply providing instantaneous feedback on MPG to drivers increased fuel economy by a statistically significant 2.7% overall, but the increase was 9.3% in drivers who had expressed a prior desire to save gas (most hybrid, electric, and even many conventional vehicles now provide some kind of instantaneous fuel economy feedback or coaching to drivers). Curiously, sex was a strong determinant of fuel savings, with men averaging 1.9% but women saving 5% of fuel. Perhaps women are more willing to forgo aggressive driving habits?

While short-term increases in fuel economy do tend to exceed long-term increases following an eco-driving course, these increases can be durable. A recent study in Singapore found that fuel consumption decreased about 12% among 116 participants immediately after such a course [239]. A small study of 10 drivers [240], using onboard monitoring of fuel use, found that most drivers immediately decreased fuel consumption following a short eco-driving course, and that fuel consumption remained about 6% lower after four months of follow-up.

Note that the above studies have typically focused on reducing MPG via tactical changes in driving patterns. Strategic changes, or "eco-routing" by avoiding congested or hilly paths, and choosing the most economical route may also lead to significant, but highly variable, fuel savings. One may consult Alam and McNabola [241] for a recent review.

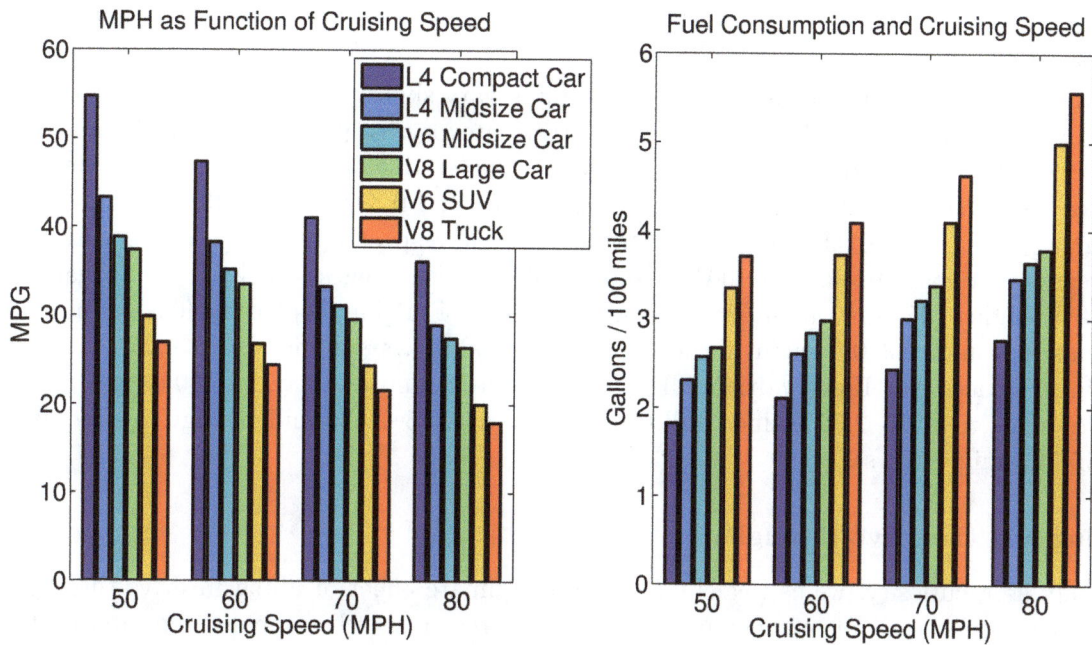

Figure 6.24: Empirical MPG (left panel) and fuel consumption (right panel) as a function of constant cruising speed for multiple vehicle classes, based on [242]. Across vehicle classes, increasing cruise speed from 50 to 80 MPG increases fuel consumption by 40–50%, while each 10 MPH increment is responsible for a 10–20% increase in fuel burn.

We may reasonably conclude that eco-driving techniques, as implemented by more "typical" drivers may reduce fuel consumption by 5–10%, while more highly motivated drivers may see savings of 10–15% and possibly more.

6.8.2   Highway speed and fuel economy

When traveling at sustained highway speeds, peak MPG occurs around 50 MPH for most vehicles. A recent study [242] of 74 vehicles by Oak Ridge National Laboratory researchers found fuel economy to decrease, on average, by 13.9% for each 10 MPH increment over 50 MPH. Fuel consumption very similarly increased 13.8% with each 10 MPH increment, with the difference greatest when increasing from 70 to 80 MPH, especially for larger SUVs and trucks. Average MPG and fuel consumption across cruising speeds is given graphically for six sample vehicles in Figure 6.24.

6.8.3   Idling

Idling is widely decried as unnecessary by both governmental agencies, including the EPA, DOT, and DOE, as well as major car manufacturers. However, idling is often believed to be necessary to "warm-up" a car, and it is widely believed that turning off and restarting a vehicle consumes a large amount of fuel. A series of simple experiments by Argonne National Laboratory have refuted both claims (see, e.g. [243]).

If one turns a vehicle off and then restarts it, there is indeed a small surge in fuel use relative to idling, but fuel use returns to baseline within 10–15 seconds [243]. The excess fuel consumed in a restart is roughly equivalent to 10 seconds of idling, and therefore it follows that one should ideally kill the engine and restart for any stop anticipated to be longer than 10 seconds. Contrary to conventional wisdom, under most conditions a vehicle warms up quicker if immediately driven rather than idled, and thus *there is no reason to idle an engine for the purpose of warm-up.*

Reported rates of fuel consumption while idling vary slightly between Argonne documents, and most documents fall within the gray literature category. A white paper [243] gave numbers of 0.2, 0.265, and 0.5 gal/hr for 2.0, 2.5, and 4.6 L engine vehicles, respectively, while the ANL's idling calculator cites number of 0.16/0.29, 0.17/0.39, and 0.39/0.59 for idling without/with a load (e.g. AC) for 2.0, 2.5, and 4.6 L engine vehicles, respectively. Thus, idling with a load increases fuel consumption 50–100%. Another figure also suggests fuel use of 0.5 gal/hr for a 3.0 L engine. In sum, about 0.1–0.15 gal/hr per L of engine displacement is a safe rule of thumb. I assume 0.3–0.5 gal/hr for idling for a typical car, assuming about 3 L as an average engine displacement. If every driver idles just five excess minutes a day, for 300 days a year, this corresponds to 1.575–2.625 billion gallons of fuel and 17.55–29.26 million $MgCO_2e$ (equivalent to 2.7–4.5 million cars).

Idling and the drive-through

Largely as a curiosity, we can calculate the approximate effect of idling in drive-throughs on American fuel consumption. The NPD group, a market research company, reported that 12.4 billion drive-through visits where made in 2011 (39.8 visits per-capita, or 59.05 visits per licensed driver). QRS magazine has performed an annual "Drive-thru Performance Study," since 1998. Its 2014 report surveyed 23 top brands, and found an average wait time of 203.29 seconds, although this appears to only cover the time from placing an order to food delivery, i.e. it does not include time waiting to order (and certainly seems quicker than my anecdotal experience would suggest). Guessing that at least 30% of the average drive-through visit must be spent waiting to order would yield about five minutes per visit. Using this higher figure suggests that each visit to the drive-through consumes 0.025–0.0417 gallons.

Further calculations suggest about 310–520 million gallons of fuel are consumed in drive-throughs annually, corresponding to 3.5–5.8 million $MgCO_2e$ of emissions. This is equivalent to roughly 600,000–1,000,000 vehicles, or 0.25–0.4% of all personal vehicle fuel-use. I suppose to be fair, we should include the 10 seconds worth of emissions from restarting your car if you park, but the calculations come out essentially the same. Therefore, eliminating the drive-through from one's life will, on average, save 1–1.5 gallons of gasoline per year. This is a relative pittance at the individual level, but it is not entirely insignificant in the aggregate. Although, since a single half-pound hamburger may generate upstream emissions equivalent to almost a gallon of gasoline (see Chapter 21), the meat-heavy items typically consumed in fast food restaurants are far more harmful to the climate than the idling of engines in anticipation of these meaty morsels.

The aggregate effect of eco-driving

On average, each light-duty vehicle consumed 524 gallons of fuel in 2013, which corresponds to a total of 1.375 billion $MgCO_2e$ of emissions (assuming all gasoline engines). If every driver reduced fuel consumption by 10%, which is achievable via eco-driving techniques, this would be equivalent to removing about 23 million vehicles from the road. Relative to a baseline of 6.45 $MgCO_2$/yr for a typical car, a driver could save around 0.645 $MgCO_2e$ via eco-driving, or alternatively add nearly 2 $MgCO_2$ (1.94 $MgCO_2e$, a 30% increase in fuel consumption) by driving extremely aggressively. Thus, the potential difference, based only on driving styles, is 2.58 $MgCO_2e$, equivalent to 5,000 miles of regular driving.

6.9 MPG$_{\text{GWP}}$ equivalent of walking, bicycling, and e-bikes

As a simple exercise, we can quickly derive MPG$_{\text{GWP}}$ values for walking and cycling when all energy consumption to power such locomotion is *additional*. Walking has a MPG$_{\text{GWP}}$ of about 45 under "diet-average" emissions, many 100s when wheat or other grains provide the additional energy, but perhaps <10 MPG$_{\text{GWP}}$ (i.e. much worse than driving) if additional beef is consumed. Similarly, biking yields 93 MPG$_{\text{GWP}}$ under the dietary average, and on the order of 1,000 MPG$_{\text{GWP}}$ if grains provide additional energy. These MPG$_{\text{GWP}}$ values will obviously be (much) greater when only a fraction of calories burned is additional.

It has occasionally been argued that, due to the high carbon footprint of growing food, that driving could actually be "greener" than walking. This is generally a gross exaggeration, but it is actually plausible under limited circumstances. Based on data in Chapter 18, we have roughly 2–4 gCO_2e/kcal of food consumed under a typical American diet (including the portion wasted), with about 3 gCO_2/kcal a best guess, and according to CDC data, the average American weighs 82.2 kg (180.9 lbs; 195.5 for men, 166.2 lbs for women). Using 2.59 J/kg/meter cited in [244] implies an energy cost to walking of 82 kcal/mile, finally yielding emissions of 246 gCO_2e/mile walked, equivalent to 45 MPG$_{\text{GWP}}$. Thus, under an average diet, walking could generate almost 50% of the emissions of a typical vehicle, and be no better than a decent hybrid, assuming all calories consumed by walking are additional to one's baseline diet, a highly dubious proposition.

In reality, the marginal emissions due to diet from walking relatively short distances (e.g. less than two or three miles) are likely zero, as most Americans already consume excess calories at baseline, and little additional caloric consumption can be expected after reasonable levels of exercise. It is still notable that the emissions from our food system are such that walking could plausibly generate emissions even on the order of those generated by gasoline-powered vehicles. Note that bicycling is about twice as energetically efficient as walking, and so bicycling costs roughly 40 kcal/mile, and 120 gCO_2e/mile (93 MPG$_{\text{GWP}}$) if all calories are additional and derived from an average American diet.

It is also true that if additional calories truly are consumed to fuel walking/cycling, the source will matter. For example, if we suppose 1–2 kgCO_2e/kg for typical grain and oil products with energy densities on the order of 3,000–4,500 kcal/kg, then the MPG$_{\text{GWP}}$ for walking will be in the 200–600 MPG range, whereas for beef, with an EF in the 40–50 kgCO_2e/kg range and 3,230 kcal/kg carcass, then walking achieves an MPG$_{\text{GWP}}$ on the order of 10 or less. Thus, in the highly unlikely scenario that after, say, walking to the store for a two mile round-trip, one consumes 50 grams of beef that they would not have without the walk, then driving would indeed have been more environmentally sound.

It is important to note an important second-order effect of walking (or bicycling) for daily transportation purposes: trip-length is necessarily limited to several miles. In the 2009 NHTS, the average one-way trip length for shopping was 6.4 miles, and the average weighted trip length for all purposes other than commuting to work (a distance I assume is inflexible) was 8.59 miles. Thus, due to trip-length limitation, walking or cycling forces one to patronize shops and other attractions closer than average.

6.9.1 E-bikes

It is worth noting that electric bikes (or e-bikes) may achieve the best MPG$_{\text{GWP}}$ of any powered vehicle in common use. Manufacturers advertise 12–24 Wh/mile; in my own experience of riding an e-bike occasionally, a 240 Wh battery could probably yield 5–10 miles without pedaling (24–48 Wh/mile), although it is advertised at 10–15 miles (16–24 Wh/mile). In any case, let us assume 24 Wh/mile (without any supplemental pedaling). Then, for grid-average electricity we should have 17.1 gCO_2e/mile, or an MPG$_{\text{GWP}}$ of 652 MPG; even at 48 Wh/mile we would have

34.2 gCOe/mile and 326 MPG_{GWP}.

6.10   The bottom line

The average US driver generates roughly as many greenhouse gas emissions from personal vehicle use alone as the average global citizen generates in a year. Indeed, the personal automobile is the single largest source of GHGs under consumption-based carbon accounting, and Americans simply must drive less, and in dramatically more efficient cars than they currently do. The world turned before the era of the SUV, and while smaller cars may be less fun to drive, it seems a poor argument that the world must be brought to its knees for the sake of a marginally more enjoyable driving experience. Similarly, aggressive driving can dramatically increase emissions, and so is a thrill purchased at a high price indeed.

A broad shift to small, high mileage conventional vehicles and HEVs has the potential to reduce GHG emissions associated with personal transportation by nearly 50%. Plug-in hybrid electric vehicles and pure EVs can yield GHG savings similar in magnitude, but because they are expected to affect electrical power generation on the margin, the effect of large-scale EV deployment would be, in the shorter-term, a shift in transportation-associated emissions from direct fuel combustion to fossil power plants. The average and marginal generating mixes vary significantly by geography, with the Midwest, Wouth, and most of the East relatively unfavorable regions for (short-term) EV deployment. Emissions related to battery production are also non-trivial for larger EVs. Electric vehicles remain a good consumer choice, and in the longer-term, all personal vehicles must be powered by electricity or other renewable fuels, but the short-term utility of investing in PHEVs or EVs over the most efficient HEVs is somewhat questionable, although a co-investment in rooftop solar (appropriately sized to meet the additional EV load) could dramatically reduce both residential and EV-associated emissions, and promoting electric infrastructure is also of value. Thus, plug-in hybrids may be the optimum technology in the nearer future.

At the individual level, trading a low MPG vehicle in for one that gets over 35 MPG combined is possible for essentially any consumer who can afford a new car, and in any case, household wealth is already strongly correlated with both overall vehicle-related emissions and miles driven. Of more basic import is reducing miles driven. Given that household miles driven varies by an order of magnitude between the bottom and top 20%, and that a majority of vehicle-miles are at least "semi-optional" (i.e. only 28% of daily travel is for commuting to work), one should be mindful in this arena. One aspect that gets little attention in both the popular media and scientific literature is the importance of driving style, and "eco-driving" strategies also have the potential for immediate and non-trivial emissions reductions. Walking and bicycling are essentially zero-emissions alternatives to driving under likely daily transportation and eating habits, although in rare cases when many additional calories are consumed, the heavy carbon footprint of food must be considered.

Increasing one's vehicle MPG from the fleet-average 21.6 to 35 MPG and reducing miles driven by about a third from 12,500 to 8,000 miles/year would reduce annual fuel-related emissions by 60%, from 6.45 $MgCO_2e$ to 2.55 $MgCO_2e$. The absolute savings are nearly 4 $MgCO_2e$, or almost 20% of the average American carbon footprint. Increasing MPG further to 50 would drop annual fuel-cycle emissions by 72%, to 1.78 $MgCO_2e$.

Chapter 7

Physical principles of vehicle travel and alternative drivetrains

In this chapter, I develop the basic physical and mathematical principles underlying vehicle fuel consumption. Such a discussion helps us understand why fuel efficiency varies with driving patterns (e.g. city versus highway, low- vs. high-speed), why fuel consumption also varies among different classes of conventional vehicle, e.g. compact, full-size, and SUV, and why alternative drivetrains, e.g. hybrid-electric drivetrains, achieve far higher fuel efficiencies across driving conditions. Furthermore, physically informed eco-driving tactics are likely to be more efficacious, and the more empirical discussion of eco-driving in the previous chapter can be better grounded in the topics presented here.

7.1 "Road-load:" vehicle parameters and driving cycle

Fuel consumption is determined by two general factors [245]: 'vehicle load," i.e. the absolute power demand ("load" is synonymous with the power demand), which depends on vehicle mass, velocity, acceleration, road grade, drag, accessory use (heater, AC, console, etc.), and of course the actual driving cycle, and by the efficiency of the powertrain, which is the engine plus the transmission (the engine generates power, and the transmission *transmits* the power to the wheels). In this section I discuss how the intrinsic vehicle factors that determine engine load affect theoretical energy consumption, and how the driving cycle affects energy use per mile. The section after explains the basics of engine operation, how drivetrain efficiency varies with the vehicle load and engine factors, and why alternative drivetrains yield better fuel economy.

The power required to propel a vehicle increases in direct proportion to mass, and as shown empirically in Section 6.5.1, variation in mass explains nearly all the variation in MPG for conventional vehicles sold today: "size is fuel." *Vehicle specific power* (VSP) is the power demand per unit of vehicle mass, and, this is key, VSP increases non-linearly with velocity and acceleration. Several forces must be overcome to propel a vehicle: (1) rolling resistance, or road friction, acting on the tires, which is constant regardless of velocity, (2) air drag, which increases with the square of velocity, (3) gravity via road grade, and (4) the rotational inertia of rotating and reciprocating parts. Acceleration is an additional forward force that must be generated.

The major *forces* on the vehicle are expressed formally as:

$$F = \underbrace{C_R M g}_{\text{Rolling resistance}} + \underbrace{0.5 \rho C_D A v^2}_{\text{Air drag}} + \underbrace{M g \sin \theta}_{\text{Road grade}} + \underbrace{M a (1 + \epsilon)}_{\text{Acceleration}}, \tag{7.1}$$

where M is the vehicle mass (kg), A is the frontal area of the vehicle (m^2), ρ is the density of air (1.2 kg m^{-3}), g is the acceleration due to gravity (9.8 m s^{-2}), v is the velocity (m s^{-1}), a

Forces upon a vehicle in motion

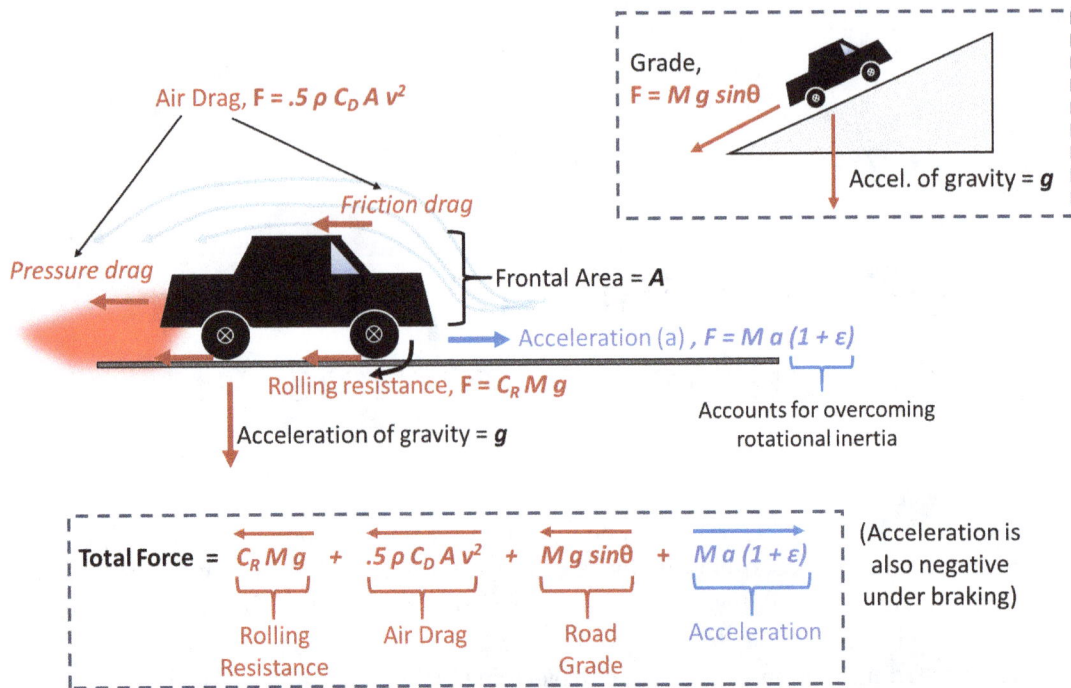

Figure 7.1: Schematic for the major forces acting upon a vehicle in motion. Aerodynamic drag increases in proportion to velocity squared (v^2), and has two basic components: friction drag and pressure drag, where the former results from the friction of air against the vehicle body, while the latter results from local changes in air flow, and is the dominant form of drag on vehicles [257]. The rolling resistance and force from the road grade are both proportional to the downward force created by the vehicle's mass ($M \times g$). The force of forward acceleration opposes these "road-loads."

is acceleration (m s^{-2}), C_R is the dimensionless coefficient of rolling resistance, and C_D is the dimensionless coefficient of air drag. Omitted in the above equation is the rotational inertia term, accounted for instead by the $(1 + \epsilon)$ multiplier. This force is generally proportional to velocity and can be either negative or positive. Instead of considering it explicitly, we can assume that rotational inertia acts to resist changes in velocity (much like a gyroscope), i.e. acceleration, and thus we can approximate its effect by adding the multiplier $(1 + \epsilon)$ in the acceleration term.

Now, we have in general that power is force times velocity:

$$P = F \times v \tag{7.2}$$

And so the equation for overall power demand, the so-called brake power P_b (this confusingly named term is *not* the power applied by the vehicle's brakes, but power output at the crankshaft), follows naturally as follows, with the only addition the power load from accessory use (e.g. lights, AC, etc.):

$$P_b = VSP \times M = \underbrace{MC_R vg}_{\text{Rolling resistance}} + \underbrace{Mgv \sin\theta}_{\text{Road grade}} + \underbrace{0.5\rho C_D Av^3}_{\text{Air drag}} + \underbrace{Mva(1 + \epsilon)}_{\text{Acceleration}} + P_{\text{accessory}}. \tag{7.3}$$

Note that most terms in Equation (7.3) depend directly on vehicle mass. The main exception is air drag, although the frontal area of the vehicle is, in reality, correlated with mass, and the experimentally determined drag coefficient is likely to be larger for larger vehicles. The accessory load is also formally independent of vehicle size and mass, although clearly larger vehicles will tend to use more energy for air conditioning, the major accessory power drain.

Armed with these equations, we can determine power and energy demand for different vehicles over different driving cycles, and we can determine which forces/power demands are dominant at different speeds and driving conditions. Moreover, we can estimate how changes in vehicle parameters, such as mass or drag coefficient, affect energy consumption. Note that actual fuel use is a function of both the power demand and the efficiency of the engine; the latter varies not only between vehicles, but with vehicle speed, engine speed, and engine load (power demand). Here I am only discussing the absolute power and energy demands, and I present results for a "perfect engine," i.e. one that converts 100% of fuel energy into energy at the wheel, without coasting or idling losses. Obviously this is quite impossible, even for all-electric drivetrains, but it gives an idea of how vehicle parameters and driving conditions affect the theoretical maximum MPG.

7.1.1   Overall energy for EPA driving cycles

The EPA currently uses five driving cycles to establish city and highway MPG estimates. I examine power and energy use primarily under three of these cycles: the urban drive cycle (UDC), the (gentle) highway drive cycle (HWY), and the aggressive driving cycle (US06). An additional short cycle, US03, is conducted with the vehicle AC on, and it is similar in its characteristics to the UDC. The speed over time, average speed, and time spent idling for these four cycles is illustrated in Figure 7.2. The UDC, for example, has an average speed of about 20 MPH, and about 20% of the total trip is spent idling.

The power required of the engine and the brakes over time for the UDC, HWY, and US06 cycles, using vehicle parameters representative of a midsize sedan, is shown in Figure 7.3. As can be seen, the power needed is usually less than 30 horsepower (22 kW). Even under aggressive driving, the peak power demand rarely exceeds 70 HP, with a brief peak at about 100 HP. Thus, most new vehicles are dramatically overpowered, at a 233 HP average in 2014 [205], and larger engines are less efficient (as explained in Section 7.2.2).

I calculate the engine and braking energy required to cover these driving cycles for several representative vehicles: compact car, midsize/large car, and SUV/Pickup, as shown in detail in Figure 7.4. Mass, rolling resistance, and air drag parameters are all selected to be representative for such vehicle classes, but obviously there will be individual variation within a class. This figure also details the fraction of engine work that is devoted to each of the four major loads: acceleration, rolling resistance, air drag, and accessory use (assuming a low 400 W accessory load, implying the AC is off). I use this figure to guide the following discussion.

Across drive-cycles, at the theoretical maximum MPG, typical compact cars consume only about 55–60% as much fuel as typical SUVs/pickups. This corresponds fairly well to an average of 15–22 MPG for truck-based SUVs, and perhaps 30–36 for compacts, but still underestimates the actual advantage of smaller cars, as these vehicles also have more efficient engines.

From Figure 7.4, it is also seen that a more aggressive driving pattern can require 50–75% more energy per mile than a gentler pattern. This difference is especially pronounced for the SUV/pickup class, with their higher weight and air drag. Moreover, the acceleration/braking cycle is dominant for urban and aggressive driving, but the rolling resistance and air drag become more important at highway speeds, with air drag dominant. This is sensible, as air drag increases nonlinearly with velocity.

The small accessory load of 400 W is only a minor component of energy use, but note that it is a much larger fraction of energy use in the urban drive cycle, where the average speed

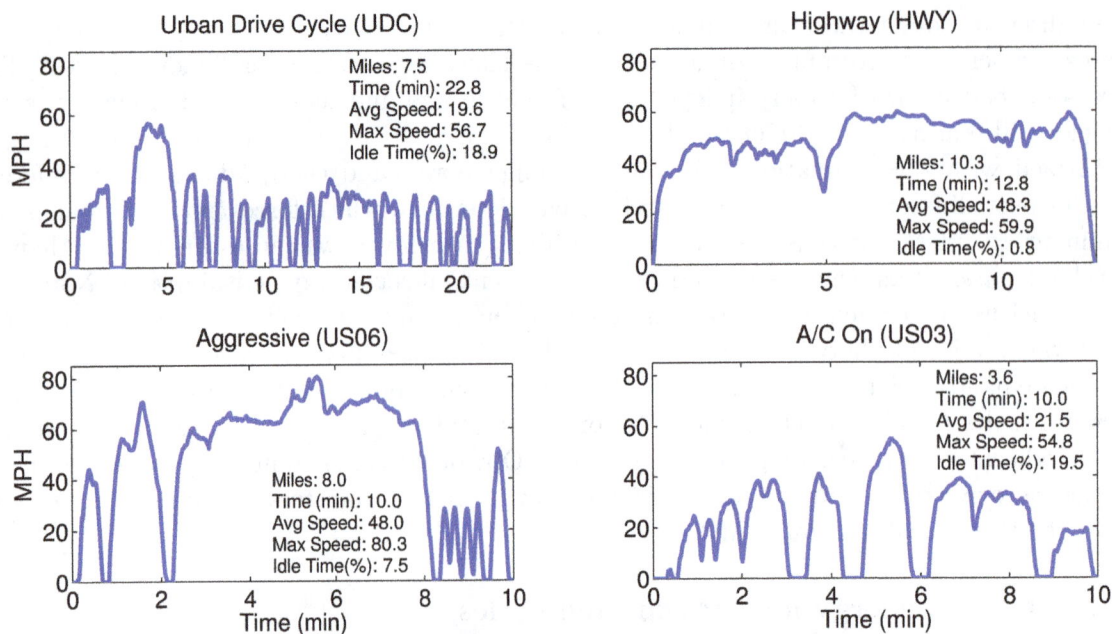

Figure 7.2: Four of five EPA testing drive-cycles, with MPH given as a function of time, along with total time, distance, average and maximum speeds, and time spent idling.

is lower. This is intuitive: suppose speed went to zero, then *all* energy would go towards the accessory load (disregarding energy lost to idling). We can also experiment with higher loads. Going from 400 W to 3,000 W, representative of the AC on at full blast, energy consumption increases 50–75% under urban driving, and 25–33% under the highway cycle. I explore the effect of AC in depth in the next section.

7.1.2 Air conditioning and accessory loads

Air conditioning is, by far, the most important accessory load contributing to fuel consumption. Heating, in gasoline-powered vehicles, is not a significant additional load on the engine, as waste heat from the engine is used to heat the air (heating can, however, consume a large amount of energy in all-electric vehicles, which rely on either resistive electrical heating or an electrical heat pump; this issue was examined in some depth by [232]). Air conditioning, on the other hand, uses a compressor motor that cycles on and off depending on the cooling demand, and draws around 3–6 kW (4–8 HP) of power when on [246]. Following [246], I assume 3 kW for AC operating in peak mode for a typical sedan. While this may sound small, it takes as little as 6 kW (9 HP) to maintain a 50 MPH cruising speed for a subcompact car, implying that full AC could increase fuel use by 50% in this setting. The relative burden may be lesser for larger vehicles: an SUV may take 15 kW (20 HP) to maintain 50 MPH, and so 3 kW of AC yields a 20% increase in fuel consumption. However, in SUVs the AC motor is more powerful and the cooling load greater, and so full-blast AC would generally draw more power. Across a range of vehicles and conditions, both theoretical calculations and actual tests [247] suggest that AC increases overall fuel use by 15–30%, with around 10% of *all* passenger vehicle fuel going towards the AC.

A 2000 study by the National Renewable Energy Laboratory [246] estimated that, on average, the AC is on 43–49% of the time a personal vehicle is in operation, amounting to 107 to 121 hours of AC per year (out of 249 hours of vehicle operation per year). This reports claims 62 gallons of fuel per vehicle per year are consumed to operate the AC. This is quite plausible:

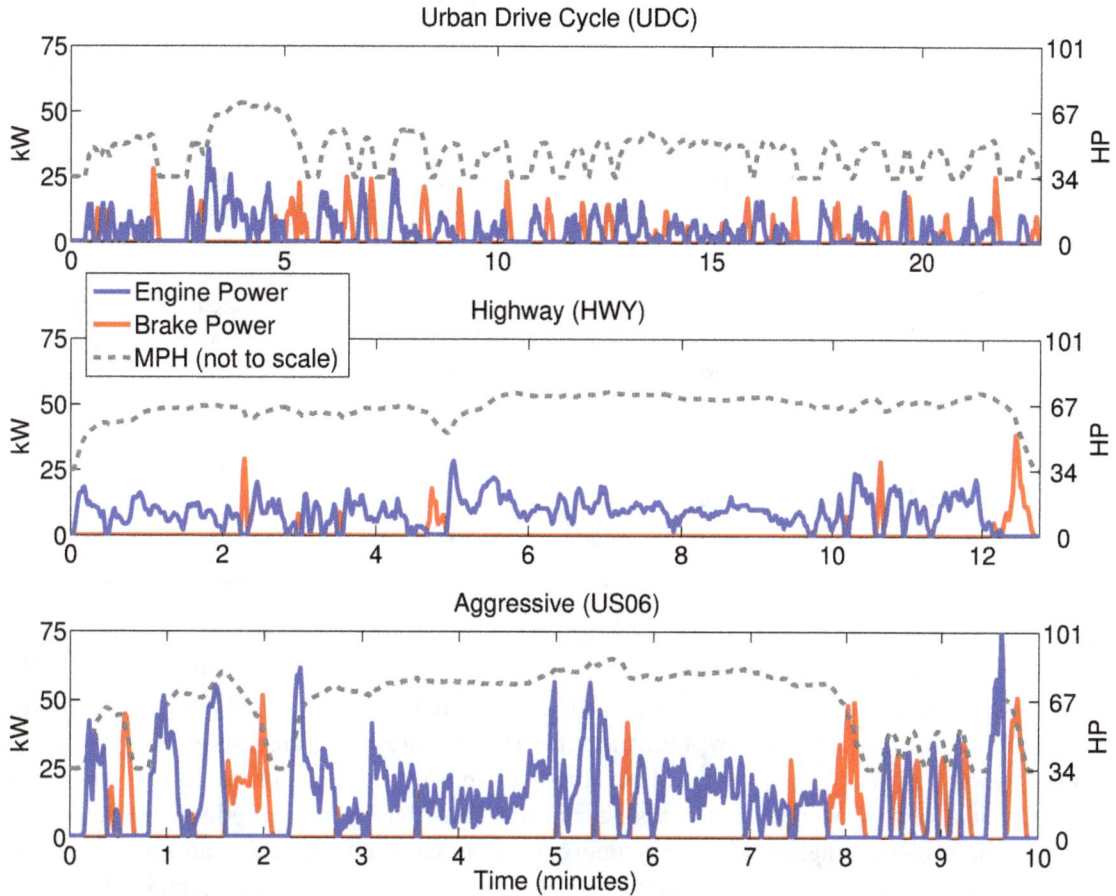

Figure 7.3: Vehicle engine and brake power requirements under the three major EPA drive cycles, using physical parameters representative of a midsize sedan (see caption of Figure 7.4 for values). Note that the brakes are not activated at all times of deceleration, but only when aerodynamic, grade, and tire drags do not sufficiently slow the vehicle as prescribed (that is, vehicles naturally slow when the accelerator is not actively engaged, and brakes need only be activated when this background drag is insufficient for desired deceleration). Aggressive driving can more than treble peak power demands, and far more energy is lost to the acceleration/braking cycle under urban and aggressive driving compared to (gentle) highway driving.

Vehicle	Engine Work (kWh/mile)	Brake Work (kWh/mile)	Theoretical Max MPG	Max MPG w/ 80% Regen Braking	Engine Work Components (% of Total)			
					Accel	Rolling	Air Drag	Accessory
Compact								
US06	0.25	0.074	135	177	47.5	14.4	38.1	3.37
UDC	0.16	0.077	206	334	65.1	19.8	15.1	12.6
HWY	0.15	0.017	223	245	28.6	26.1	45.2	5.54
MidSize								
US06	0.29	0.091	115	154	49.1	14.8	36.1	2.87
UDC	0.19	0.094	173	282	66	20	14	10.6
HWY	0.17	0.021	192	212	29.8	27	43.2	4.76
SUV/Pickup								
US06	0.45	0.12	74.7	94.6	43.7	13.5	42.8	1.87
UDC	0.28	0.13	120	188	62.9	19.4	17.8	7.34
HWY	0.27	0.026	122	132	25.9	24.1	50	3.02

Figure 7.4: Shown is the energy expenditure required to drive the vehicle forward (engine work) and to brake the vehicle (brake work) under three different EPA driving cycles: US06 (aggressive driving up to highway speeds), UDC (urban drive cycle), and HWY (gentle highway driving under 60 MPG). Energy expenditures for three generic vehicle classes are given. Note that engine work is simply the energy that must ultimately be transferred to the wheels, and the massive loss of fuel energy in this transfer is not considered: we are effectively assuming a perfect, 100% efficiency engine. Given the energetic requirements, the third and fourth columns show the theoretical maximum MPG for perfect engines with and without regenerative braking (assumed 80% efficient). The final four columns show that, under city driving conditions, energy requirements are dominated by acceleration (and braking), while at highway speeds the rolling and air drags become more important, with air drag dominant. Moreover, the constant accessory load (assumed to be a low 400 W here, implying AC off) is more important in the slower urban driving cycle. All vehicles have C_R fixed at 0.009; the compact car has $M = 1,250$ kg, $C_D = 0.28$, $A_R = 2.15$ m^2, midsize car has $M = 1,515$ kg, $C_D = 0.3$, $A_R = 2.25$ m^2, and the SUV/pickup has $M = 2,075$ kg, $= C_D$ 0.4, $A_R = 3.0$ m^2. These parameters are reasonably representative for *new* vehicles. Older SUVs and trucks, especially, are expected to be heavier with higher coefficients of drag, C_D.

operating an AC at 3 kW for 121 hours at 15–20% overall engine efficiency amounts to the energy equivalent of 54–72 gallons of gasoline. Compare this to 524 gallons of fuel per annum for the average passenger vehicle, and the rather remarkable conclusion is that on the order of 10% of gasoline consumption goes to operate on-board ACs. This is compatible with tests by the Clean Air Vehicle Technology Center [247] showing a roughly 20% average increase in fuel use across seven vehicles and five test conditions with AC on: A 20% increase in fuel use for about half of the operating time implies that about 10% of total fuel goes towards AC.

These tests [247] found fuel use to increase 29% in 95 °F weather with the sun out, but it is perhaps more noteworthy that fuel consumption also increased 16–22% in 75 °F conditions without a solar load. Thus, we can infer that the all too common habit of simply leaving the AC on regardless of ambient conditions is actually quite harmful. Indeed, automobile ACs do not necessarily shut down at cold ambient temperatures unless manually turned off [248], and a European study estimated that two-thirds of AC-related CO_2 emissions in Central Europe result from AC operation at ambient temperatures below 64.4 °F (18° C) [248].

Figure 7.5 shows how theoretical fuel consumption increases with accessory load under EPA drive cycles for a representative compact car, midsize car, and SUV. Also shown is how the approximate MPG changes, assuming overall drivetrain efficiency of 20%. The interesting conclusion is that the relative (but not absolute) effect of AC is much greater in smaller more efficient vehicles. Under the urban drive cycle, the AC on high (3 kW) increases theoretical energy demand by 90% in a compact car, versus 55% in an SUV. Since urban driving is characterized by a slower overall speed than gentle highway driving, the effect of AC is much greater. Aggressive driving, with its high speed and high power demands, is least affected.

Theoretically, the *absolute* increase in energy demand for a given AC level is equal regardless of vehicle size, although this effect would be much less apparent under MPG calculations at the pump for an SUV (recall the hyperbolic relation between fuel consumption and MPG). However, in reality, larger vehicles are likely to use more fuel for AC for two main reasons: (1) smaller engines are more efficient than larger engines and can therefore generate the energy to meet an equivalent auxiliary demand with less fuel, and (2) a larger surface area and interior volume imply a larger cooling load, as there is more vehicle surface area for heat exchange and more air to cool (for example, the interior volume of a Ford Explorer is about 1.5 times that of a Honda Civic). Indeed, the calculations below suggest that an SUV requires 30–40% more energy than a compact car to maintain the same cabin temperature.

Calculating the cooling load and strategies for reduction (recirculation is essential!)

We can do some quick back-of-the-envelope calculations to demonstrate what the approximate cooling load is for different vehicles. There are two major heat loads on the vehicle: solar irradiation and heat conductance through the car body (primarily through the windows). As outside air is exchanged for inside air, we have the additional effect that new air must be cooled as well.

Solar load. The solar load is calculated from the solar irradiance, about 1 kW/m^2 at noon (and obviously less at other times), multiplied by the effective window area. If we assume that at most the sun is shining through half the window area, and correct slightly for the 30 degree angle of light (because light rays are not perpendicular to the window, this reduces the effective window area somewhat), then *effective area = 0.44 × window area*; given a window area of 3.41 m^2 for a 1990 Pontiac Grand Prix [249], we arrive at perhaps 1–1.5 kW as the peak solar load for a compact vehicle.

Heat conductance. In the cold, heat is lost from the vehicle to the surroundings primarily by convective heat transfer, and this transfer increases roughly linearly with vehicle speed. That is, the faster the vehicle moves, the more heat is stripped from its skin, and this phenomenon

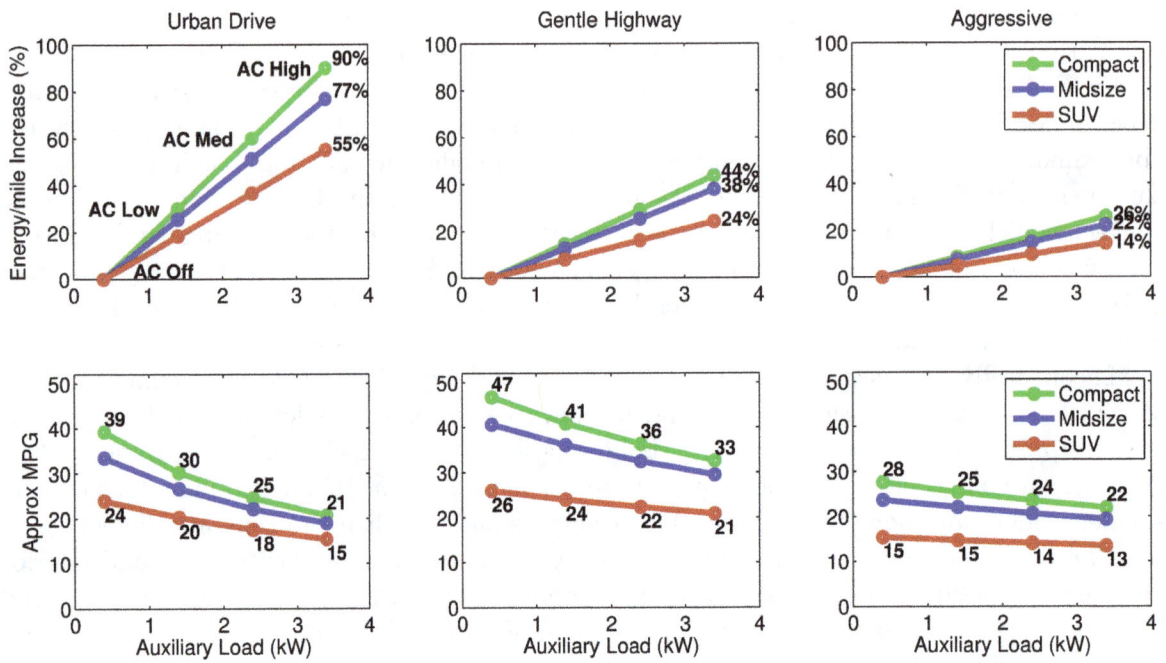

Figure 7.5: Theoretical increase in energy demand under different accessory loads, 400 W (baseline), 1,400 W, 2,400 W, and 3,400, corresponding to differing levels of air conditioning. The top panel gives relative energy/mile demand under three EPA drives cycles, for three representative vehicles (compact, midsize, SUV/truck, as defined by parameters in Figure 7.4). The bottom panel gives approximate equivalent MPGs achieved, crudely assuming an overall 20% drivetrain efficiency.

is likely familiar to anyone who has ridden a bike in chilly weather. The same principle applies when it is hot outside, and heat is being transferred into the car. Since most heat transfer occurs through the windows, we can infer, from numbers given in [249] (80 W/K at 55 MPH with 1.5 m^2 effective window area), a coefficient of heat transfer on the order of 1 W / (m^2 effective window area) / K / MPH. Thus, a midsize sedan (1.5 m^2 effective window area) moving 20 MPH and maintaining a 20 °F difference between inside and out gains heat at a 333 W rate; if it were traveling at 80 MPG, heat gain is 1,333 W.

Air exchange. On a medium fan setting, a typical vehicle experiences around 100 air exchanges (AEs) per hour (note that air exchange varies widely with vehicle model and speed); this decreases by about an order of magnitude to perhaps 10 AE/hour when air is recirculated [254] (it does not go all the way to zero because air infiltration into the vehicle still results in significant inside-outside air exchange, and this leak is also highly variable between vehicles).

The *approximate* cooling loads (and AC power requirement, assuming a COP of 2.35) for a small car and SUV maintaining differences of 10, 25 and 50 °F between the inside and outside are given in Figure 7.6; these are the power demands after a steady-state temperature has been reached in the cabin. The effects of using the air recirculation setting and a 30% tint, alone and together, are also given. It is clear that, at higher cooling loads, the recirculation setting is essential and can roughly halve power demand. Not only does this trivially simple setting change save fuel, it also meaningfully reduces exposure to airborne pollutants that adversely affect human health [255].

These numbers only represent broad likely ranges, as all parameters used in their calculation will vary widely with vehicle model and driving conditions. These calculations confirm that, from basic physical principles, using 1 kW, 2 kW, and 3 kW as low, medium, and high AC settings, respectively, is a reasonable choice and helps lend credence to the previous conclusions concerning AC.

These calculations are made assuming an interior volume of 3 m^3 for a compact car, and 4.5 m^3 for an SUV (based on publicly available manufacturer specifications). I assume that total surface and window area is 30% higher for the SUV, based on published vehicle dimensions, and hence heat conductance and solar load are also 30% higher.

AC or open windows?

Popular advice varies on this topic, but most sources suggest windows-down is better at lower speeds, while AC on and windows-up is preferable on the highway. The logic is that windows-down increases the aerodynamic drag coefficient, C_D, and since air drag increases with the square of velocity, the extra power to overcome air drag outweighs the power needed to run the AC only at higher speeds. We can do some basic calculations to test this notion, but our conclusion hinges crucially on how much windows-down affects C_D. In short, I find that open windows actually have a very minimal effect on drag, windows-open is greatly superior to AC at low speed, and there is little meaningful difference at highway speeds, although even then windows-open may be slightly better. Windows-open (as a uniform condition) is also superior under all simulated EPA drive cycles.

While one author [250] assumed a 20% increase in drag with open windows, there was no clear rationale. Tests by the Florida Solar Energy Center (FSEC) reported a fuel savings of 3% at highway speeds with windows closed versus open, but no further detail was provided [251]. If 50–75% of energy is used to overcome air resistance at highway speed, then this translates into a 4–6% increase in the drag coefficient. As reported in [250], tests on a 1986 VW GTI found fuel use to increase 2.5% with windows down (and AC off) at 67 MPH, similarly suggesting a 3–5% increase in drag. A 1982 paper on aerodynamic drag suggested a 5% increase in drag when windows are open [252], and a 1980 JPL/NASA publication [253], which performed tests

Temp. Difference:	10° F		25° F		50° F	
	Compact	SUV	Compact	SUV	Compact	SUV
Conductive	0.13–0.58	0.17–0.78	0.31–1.5	0.42–1.9	0.63–2.9	0.83–3.9
Solar	1.1–1.5	1.5–2	1.1–1.5	1.5–2	1.1–1.5	1.5–2
Air Exchange	0.42–0.69	0.63–1	1–1.7	1.6–2.6	2.1–3.5	3.1–5.2
Cooling Load (kW)	1.7–2.8	2.3–3.8	2.5–4.7	3.5–6.5	3.8–7.9	5.5–11
AC Load* (kW)	0.71–1.2	0.98–1.6	1.1–2	1.5–2.8	1.6–3.4	2.3–4.7
w/ Recirc.	0.55–0.92	0.74–1.2	0.66–1.3	0.88–1.8	0.83–2	1.1–2.7
w/ 30% Tint	0.57–0.99	0.78–1.4	0.91–1.8	1.3–2.5	1.5–3.2	2.1–4.5
w/ Both	0.41–0.72	0.54–0.97	0.51–1.1	0.69–1.5	0.69–1.8	0.93–2.5

*COP = 2.35

Figure 7.6: Table showing approximate cooling loads (in kW) resulting from conductive transfer, solar irradiance (at the solar peak), and air exchange, under inside-outside temperature gradients of 10, 25, or 50 °F. The cooling load is calculated for both typical compact car and SUV, and this is converted into a power demand upon the AC, assuming a 2.35 coefficient of performance (as heat pumps, ACs move more energy than they use, as discussed in Section 12.3). AC loads are then determined under air recirculation, 30% tint, or both, with results indicating that the greatest benefit is had simply via air recirculation.

on vehicles with the front windows open, also reported a 3–5% net increase in drag. Thus, the balance of the seems to suggest a 5% increase in C_D (at most) is reasonable, a much smaller difference than what appears to be commonly assumed (about 20%). I refer to the increase in C_D with windows open as ΔC_D, for brevity.

As shown in Figure 7.7, if $\Delta C_D = 5\%$, across fixed cruising speeds using the AC at any level uses more energy than having the windows down, even at high highway speeds. The disparity in energy is far greater at low speeds, where the AC represents a relatively larger load on the engine. At high speed, the three AC levels and windows-open all use almost the same amount of fuel. For completeness, I have simulated $\Delta C_D = 20\%$, and if this is the case, AC becomes marginally preferable to windows-down between 50 and 65 MPH, but again, the difference in fuel use is nearly trivial, and ΔC_D is highly unlikely to be so large.

Simulating our three base vehicles—compact, midsize, and SUV/pickup—on the UDC, HWY, and US06 EPA drive cycles, I have found the windows-down/AC-off setting with ΔC_D = 5% to consume less energy than low-level AC (1 kW) for all vehicles and driving cycles. If ΔC_D is 20% (again, very unlikely), then windows-down is approximately equal to windows-up with medium AC (2 kW) across vehicles and cycles.

There have been few reliable empirical tests directly addressing this issue (although what is available in the scholarly literature supports $\Delta C_D \approx 5\%$, as above). A 2004 conference paper [256] reported curves of fuel use from field test runs that clearly show that AC with closed windows consumed more fuel than windows-down up to 110 KPH (68.3 MPH), the maximum tested speed (see figures on slides 14 and 15 [256]), for both an SUV and sedan, at 86 °F ambient temperature. The document seems to indicate that AC was the more efficient option only when the ambient temperature dropped below 59–68 °F, hardly a relevant scenario.

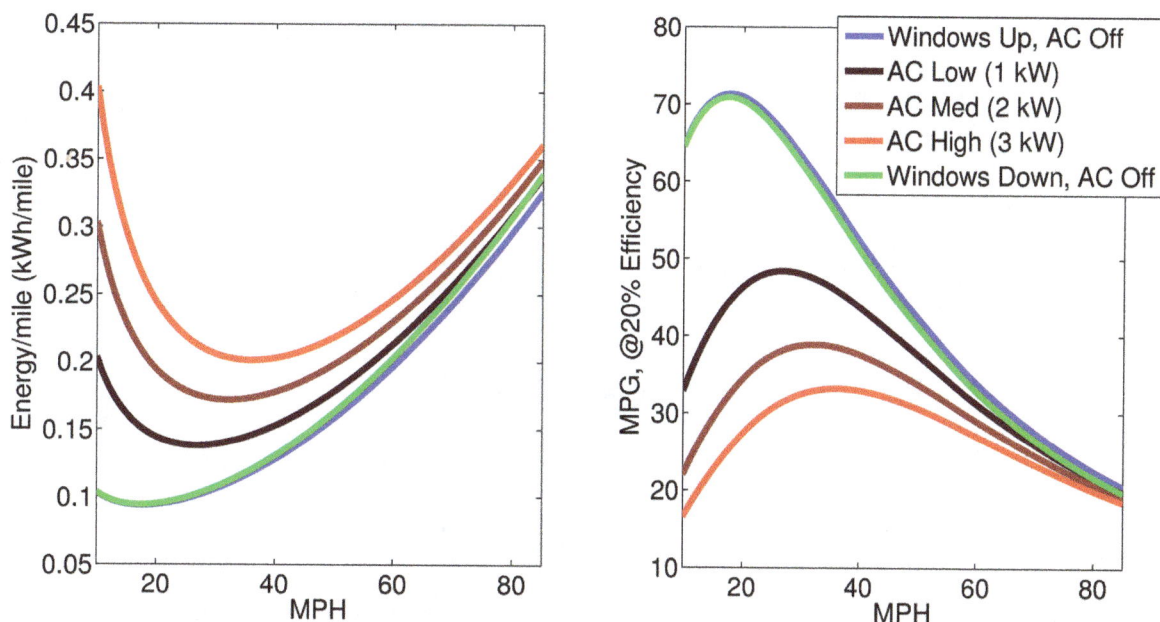

Figure 7.7: Theoretical vehicle energy requirements and approximate MPG (at 20% efficiency) under different AC setting/windows combinations, as a function of constant cruising speed (MPH) (for a midsize sedan). As seen, there is virtually no difference between the windows up and down with AC off until one hits 60 MPH, and even then the difference is minimal. The AC, on the other hand, dramatically affects fuel consumption when cruising below highway speeds.

In sum, opening the windows minimally affects the aerodynamic drag coefficient, and it is probably always more efficient to roll down the windows than to use the AC, even up to 85 MPH cruising speeds.

7.1.3   Relative loads: urban driving and cruising

At low constant speeds, the dominant loads upon the vehicle are the accessory load and the rolling resistance of the tires. Depending upon accessory (e.g. AC) use, the accessory load may be dominant up to as much as 40–50 MPH cruising speeds (under very high AC use). However, beyond 40–50 MPH, aerodynamic drag always becomes dominant. Figure 7.8 demonstrates how the loads upon a vehicle change dynamically under active driving, with acceleration and braking dominant under stop-and-go conditions, while rolling resistance and then aerodynamic drag increasing in important at higher speeds; the same trend is seen under constant cruising speed (no acceleration or braking).

7.1.4   Vehicle Mass

Empirically, for a given drivetrain class (e.g. conventional or hybrid drivetrain), vehicle mass is the dominant determinant of fuel consumption across popular production vehicles (see Section 6.5.1). This is especially true for conventional vehicles. With all else held constant, Figure 7.9 shows how fuel consumption varies by weight for a theoretical midsize sedan across EPA driving cycles, and we see that with all other vehicle parameters held constant, a doubling in weight increases fuel use by about 50%.

Yet the empirical relationship between weight and fuel consumption derived directly from MPG ratings for top-selling vehicles, also shown, suggests in contrast that doubling weight

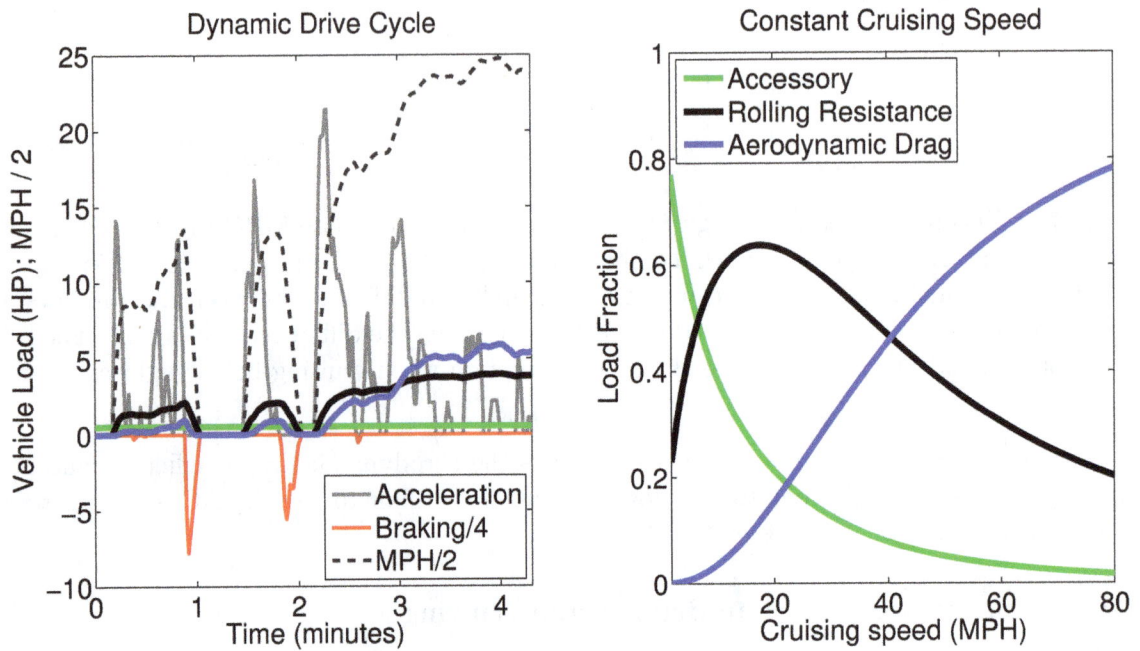

Figure 7.8: The left panel shows vehicle loads under active driving (note that, for display purposes, the braking load is quartered, and the inscribed MPH speed curve is halved in magnitude). The right panel gives relative load at constant speed, for a midsize sedan using a low (400 W) accessory load. We see that the accessory and acceleration demands are dominant at low speed, rolling resistance is the major load at intermediate speeds, while aerodynamic drag takes over at high speed.

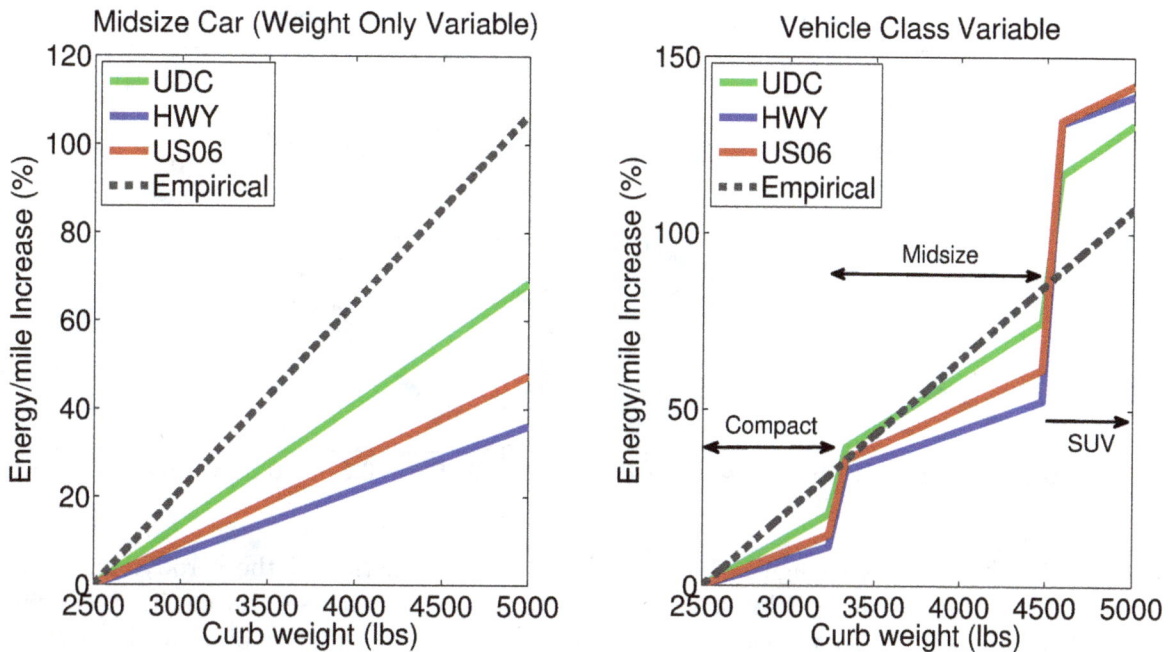

Figure 7.9: The left panel shows the theoretical increase in fuel consumption for a midsize sedan as weight increases, with all other vehicle parameters held constant, under three three EPA drive cycles. Also shown is the empirical relation between weight and fuel consumption across 30 new 2015 MY vehicles. We see that, when other vehicle parameters are constant, weight as an *independent* variable increases fuel consumption by about 50% with each doubling. The right panel demonstrates that, if we plot the increase in fuel consumption from a baseline 2,500 lb compact car, but shift other vehicle class parameters to a midsize vehicle at 3,250 lbs, and to an SUV at 4,500 lbs (I also have drivetrain efficiency decrease by 10% as we move through each class, consistent with results presented in Section 7.2.2), then we match the empirical curve reasonably well, indicating that weight is a confounding factor for other vehicle characteristics that also decrease fuel economy.

increases fuel use by over 100%. This is likely because weight is not an independent parameter, with other parameters, such as frontal area and drag coefficient, also increasing with weight and larger vehicle class (larger engines are also less efficient). Confirming this notion, if we vary vehicle class with weight, we can actually theoretically reproduce the empirically observed weight-fuel consumption curve fairly well, as shown in the second part of Figure 7.9.

Avoiding hauling unnecessary mass can marginally reduce fuel consumption, as every additional 100 lbs of payload likely increases fuel consumption by about 1–2% (the relative increase is larger for smaller cars and under urban driving), translating into perhaps 5–10 extra gallons of gasoline on a yearly basis. Note that this implies that the extra fuel burned per each additional passenger in a car is roughly equivalent to the fuel burned by a 1,000 MPG car. In other words, the weight of extra passengers in no way undermines car-pooling.

7.1.5 Aerodynamics

Unlike other vehicle loads, aerodynamic drag increases with velocity squared, and is thus the major determinant of fuel consumption at higher cruising speeds. Drag is, in turn determined by the product of the drag coefficient and the vehicle frontal area, $C_D \times A$. The earliest vehicles had C_Ds approaching 0.8, this fell to about 0.5 by the 1950s, while drag coefficients for cars

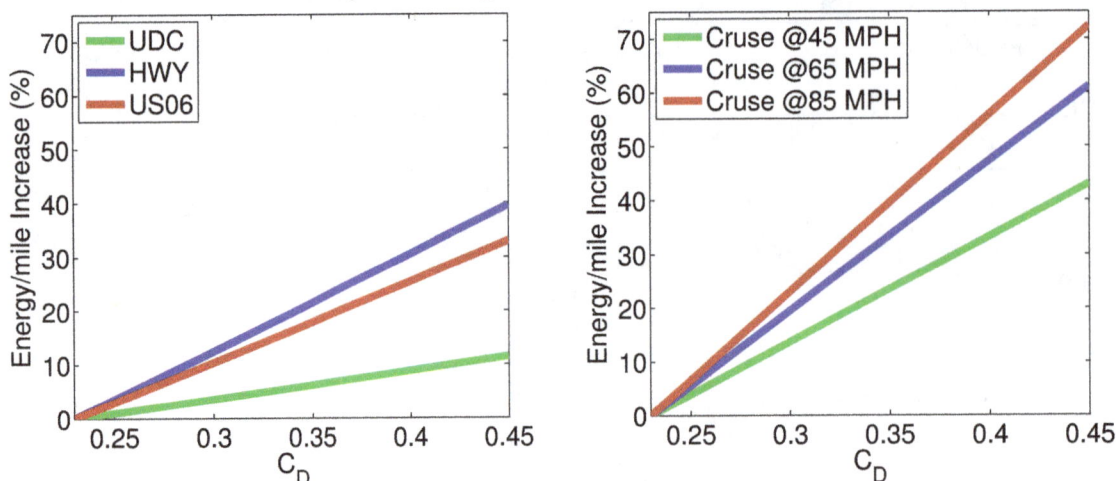

Figure 7.10: Theoretical change in fuel consumption as a function of the aerodynamic drag coefficient, C_D, under three EPA drive cycles (left panel), and for several constant cruising speeds (right panel). Other vehicle parameters are fixed as representative of a midsize car.

have been on the order of 0.3 since the 1980s [257]. The best modern cars have a C_D in the 0.24–0.28 range; C_D tends to increase with vehicle size and may be on the order of 0.35–0.45 for larger SUVs and pickups. Furthermore, frontal area increases with vehicle size, perhaps by 30–40% from the compact to SUV class, and thus the $C_D \times A$ product may plausibly range almost threefold across existing passenger vehicles.

Figure 7.10 shows how theoretical energy consumption changes as a function of C_D for our three major EPA drive cycles, with all other vehicle parameters held constant. This figure also demonstrates changes in energy consumption with C_D at several constant cruising speeds.

7.1.6 Rolling resistance

The coefficient of rolling resistance, C_R, is within the 0.008–0.013 range for most tires on the market (one can assume about 0.009 as a baseline value); a 2003 Green Seal report measured a minimum and maximum of 0.0062 and 0.0152, respectively, in a test of 51 tire models. Furthermore, C_R varied by 20–30% between tires of similar size, type, and performance. A smaller C_R value implies that less power is lost to rolling resistance, and several reports and theoretical calculations suggest low C_R tires would reduce fuel use by 1.5–4.5%. While C_R values are not readily available for many tires, high MPG tires with low C_R values are now fairly widely marketed; the C_R may be as low as 0.006 for such models.

Using our theoretical model, under the three EPA driving cycles considered above, about 10–25% of power is used to overcome rolling resistance, with rolling resistance a more important component at highway speeds; for the worst values of C_R, over 30% of energy may be lost to the tires. Thus, switching from, say, tires with C_R of .009 to .006, i.e. a 33% reduction, would result in about 6% fuel savings, and as a rule of thumb, every 0.001 decrement in C_R corresponds to a 1.5–2% fuel savings. For a 22 MPG vehicle, 4.5% less fuel translates into a new MPG value of 23.04, a change that could be hard to detect by the casual observer, but would nevertheless translate into 25.6 gallons of fuel per year (at 12,500 miles per year).

Going from very poor tires to the best could yield 8–10% fuel savings (a quick web search shows that 8% savings are advertised by mainstream high MPG tire manufacturers). Since tires with deeper treads designed for increased grip also lose more energy to friction, one would do well to lose any off-roading tires on a street vehicle. Insufficient tire inflation also increases the effective C_R, perhaps by 10–20% (although the penalty could be almost arbitrarily high for severely deflated tires). It follows that poorly inflated tires might increase fuel use by a few percent, roughly equal to full tank of gas each year.

7.2 Understanding engines and drivetrains

I take the view that a basic understanding of the principles of internal combustion engine (ICE) operation and efficiency is valuable in understanding the potential for efficiency gains, and especially allows an informed understanding of just why hybrid and other alternative drivetrains are advantageous.

At the most basic theoretical level, an ICE is a heat engine. The chemical energy stored in fuel is turned into heat in the internal combustion chamber, and this heat flows to the ambient environment. Some fraction of this heat flow can be siphoned off to do useful work; the maximum fraction is subject to thermodynamic limits and is around 50% for a gasoline-powered ICE operating under the Otto cycle (see below). Thus, heat engines are intrinsically inefficient: turning fuel into heat always leads to significant energy loss to the environment. The idea behind a fuel cell vehicle is to convert stored chemical energy directly into electricity or work without combusting it to heat as an intermediate step, thus yielding greater efficiency than an ICE [245].

As opposed to an external combustion engine, e.g. a steam turbine in which a working fluid is heated separately from fuel to drive a turbine, in an internal combustion engine, air acts as a working fluid that interacts directly with the fuel. Therefore, ICEs are also sometimes described as air pumps: to generate power, we require more fuel, and to combust more fuel one needs more air; the efficiency and power output of an ICE is largely determined by its ability to pump air.

An excellent technical discussion of the basic physics of efficiency in automobiles for the non-specialist can be found in Ross [245], and the following discussion is indebted to this work. The EPA's Physical Emission Rate Estimator (PERE) model, as extensively documented in [258, 259], also serves as a valuable resource. Freely available online [54], Chapter 6 of the textbook Energy Conversion, is also a useful reference for the interested reader.

7.2.1 Basic classes of engines: spark ignition (Otto cycle) and compression ignition (Diesel cycle)

Engines are divided into two basic classes by the method of fuel ignition. Spark ignition (SI) engines, including nearly all gasoline and natural gas engines, inject a fuel-air mixture into the cylinder where it is ignited by a park plug. Compression ignition (CI), which is the basis for diesel engines, entails spraying droplets of fuel directly into the cylinder which then spontaneously combust under high pressure as the cylinder compresses the air/fuel mixture.

In practice, CI engines have several efficiency advantages over SI engines. CI engines operate at higher compression ratios (explained in a moment), and the fuel-air mixture is *lean*, i.e. it has a low fuel to air ratio (SI engines generally operate at a stoichiometric ratio of fuel to air); both measures increase the thermal efficiency of the engine. Second, diesel engines lack a throttle, decreasing engine friction losses by perhaps 25% compared to SI engines [245]. The primary drawbacks of diesel engines are that they are heavier and have higher nitrous and particulate emissions (an unavoidable result of droplet combustion chemistry), and filtering these byproducts reduces efficiency somewhat. Since the vast majority of passenger vehicles sold in the US are gasoline-powered, as are all hybrid-electric passenger vehicles available in the US, I will restrict discussion mainly to SI gasoline engines.

The Otto Cycle

The four-stroke Otto cycle is the theoretical basis for SI gasoline engines. In essence, an engine consists of one or more reciprocating cylinders, whose sliding motion is translated into rotational motion at the crankshaft (see Figure 7.11); this rotational motion turns the wheels to drive the car forward (it also provides the energy for various belt-driven accessories). Every slide of the reciprocating cylinder head represents a single stroke, two strokes translate into a full revolution at the crankshaft, and therefore four strokes translate into two full revolutions at the crankshaft.

To begin the Otto cycle, the cylinder volume expands at *constant pressure*, drawing in the fuel-air mixture (stroke 1). The piston then slides up to compress the fuel air mix (stroke 2), and ignition occurs. The ignition occurs at constant volume, increasing the pressure, and driving the subsequent *power* stroke (stroke 3). We then have idealized heat rejection (pressure loss at constant volume), the final piston stroke (stroke 4) expels the exhaust gas (volume loss at constant pressure), and the cycle repeats. This whole process can be visualized as a pressure-volume loop, the construction of which is demonstrated in Figure 7.11; the final result is reiterated in Figure 7.12.

A key point to notice is that as the fuel-air mix is compressed (compression stroke), the internal volume in the cylinder goes from V_1 to V_2, and the volume change during the subsequent power stroke is identical, as $V_1 = V_4$, and $V_2 = V_3$ (see labels on Figure 7.12). It can be shown by thermodynamic arguments that the maximum efficiency of the cycle is determined by the compression ratio, $r_C = V_1/V_2 = V_4/V_3$, as follows:

$$\eta_{t_I} = 1 - \frac{1}{r_C^{\gamma-1}}, \tag{7.4}$$

where r_C is the compression ratio, and γ is about 1.40 for air [245]. For actual fuel-air engines, γ depends on the fuel/air mixture, and thus maximum efficiency for an Otto Cycle is is determined by how lean the fuel-air mixture is, according to

$$\eta_{t_{FA}} = (1 - .25\phi)\eta_{t_I} = 0.75\eta_{t_I} \text{ for } \phi = 1, \tag{7.5}$$

where the ϕ parameter is the fuel-air ratio relative to stoichiometric; $\phi = 1$ for most actual (gasoline) engines, and $\phi < 1$ implies less fuel than air and more efficient operation. Therefore, with a compression ratio of 12, we would have an ideal Otto cycle efficiency of $\eta_{t_I} = 0.63$, and a physical maximum $\eta_t = 0.47$.

As the compression ratio, r_C, becomes arbitrarily large, efficiency η_{t_I} approaches one, thus begging the question, why not simply increase the compression ratio? The answer is that in actual gasoline engines, the compression ratio is limited by *knock*. This occurs when fuel autoignites in the cylinder prematurely, which damages the engine and reduces power and efficiency. As one may know, the octane number is a measure of resistance to knock[1]. The other obvious efficiency measure would be to change ϕ, the fuel-air ratio to a *lean* mix, i.e. one with more air than fuel (on a stoichiometric basis) and $\phi < 1$. All diesel engines are lean-burning, and a handful of lean-burn gasoline engine have been produced, but no current production vehicles in the US use this technology [245].

---

[1]It is important to note that there is no advantage to using fuel with a higher octane number than the engine is rated for, as the compression ratio will be tuned to that particular number, and gasoline mixtures with different octane numbers have extremely similar heat contents.

Returning to our pressure-volume loop which, once stared at for a bit, actually helps to clarify the operation of the Otto cycle markedly, and it is especially useful once we realize that energy (work, W) is the integral of pressure over the change in volume,

$$W = \int P dV. \tag{7.6}$$

Put in simpler terms, the energy delivered by the cycle is the area enclosed by the pressure-volume loop, as illustrated in Figure 7.12. Now, with this we can understand how different engine modifications can yield more energy and/or more efficient engine operation. First, we see that changing the compression ratio, as illustrated in Figure 7.12, increases the area of the loop, thus yielding more energy. This visual analysis is far more intuitive than the thermodynamic argument. Second, we see how the Atkinson cycle, a modification of the Otto cycle in which the expansion stroke is longer than the compression stroke, yields more energy per cycle, as explained in Figure 7.14, and many modern hybrids use a modified Atkinson cycle.

With the pressure-volume diagram, we can also visualize pumping energy losses, which are mainly due to the throttle. That is, an actual engine must do work to expel exhaust gases, and it must also expend energy to suck air in through the intake manifold. When the throttle, which controls the amount of air entering, is partially closed, it takes more energy to suck the air in. This exhaust/induction "pumping loop" is visualized in Figure 7.13, and the *volumetric efficiency*, η_V, of an engine is a lumped measure of how easily the engine "breathes." Pumping energy losses can be reduced by adding additional inlet and outlet valves to the cylinder, allowing easier air flow, and also explaining why marketing materials sometimes advertise, say, 16 valves on a four-cylinder engine. This implies two inlets and two outlets, contra the minimum eight-valve total. Throttling losses can also be avoided through variable valve timing, where the opening and closing of valves is timed to modulate air intake.

Turbocharging (or "turbosupercharging") is a form of *forced air induction*, which is physically realized when hot exhaust gas, carrying energy that would normally be wasted, drives a turbine (via the thermodynamic cycle known as the *Brayton cycle*, also the basis for jet engines and one half of the combined Brayton-Rankine cycle forming the basis of high efficiency natural gas turbines). This turbine then drives a supercharger, which provides compressed air at high pressure at the cylinder inlet [54]. Thus, our pumping loop essentially inverts: instead of air being sucked in at low pressure, it is forced in at high pressure, and now the pumping loop *provides* energy rather than siphoning it off, increasing both the power and efficiency of the engine. This is an example of how combining two thermodynamic cycles can effectively increase overall engine efficiency.

7.2.2   Drivetrain efficiency

The overall efficiency of a drivetrain, η, is determined as the product of the engine thermal efficiency, η_t, engine mechanical efficiency, η_m, and the transmission efficiency, ϵ. The first two factors together determine the overall engine efficiency. Thermal efficiency is the efficiency with which fuel energy is converted into kinetic energy *within the cylinder*. Some of this energy is then lost via friction and pumping losses, so only some fraction actually manifests as power at the crankshaft. The fraction of energy that makes it from the cylinder to the crankshaft is the mechanical efficiency. Since engine efficiency is measured at the crankshaft, overall engine efficiency is the product thermal and mechanical efficiency. Now, there is an additional loss as energy is transmitted from the crankshaft to the wheels, quantified as the transmission efficiency. In mathematical terms

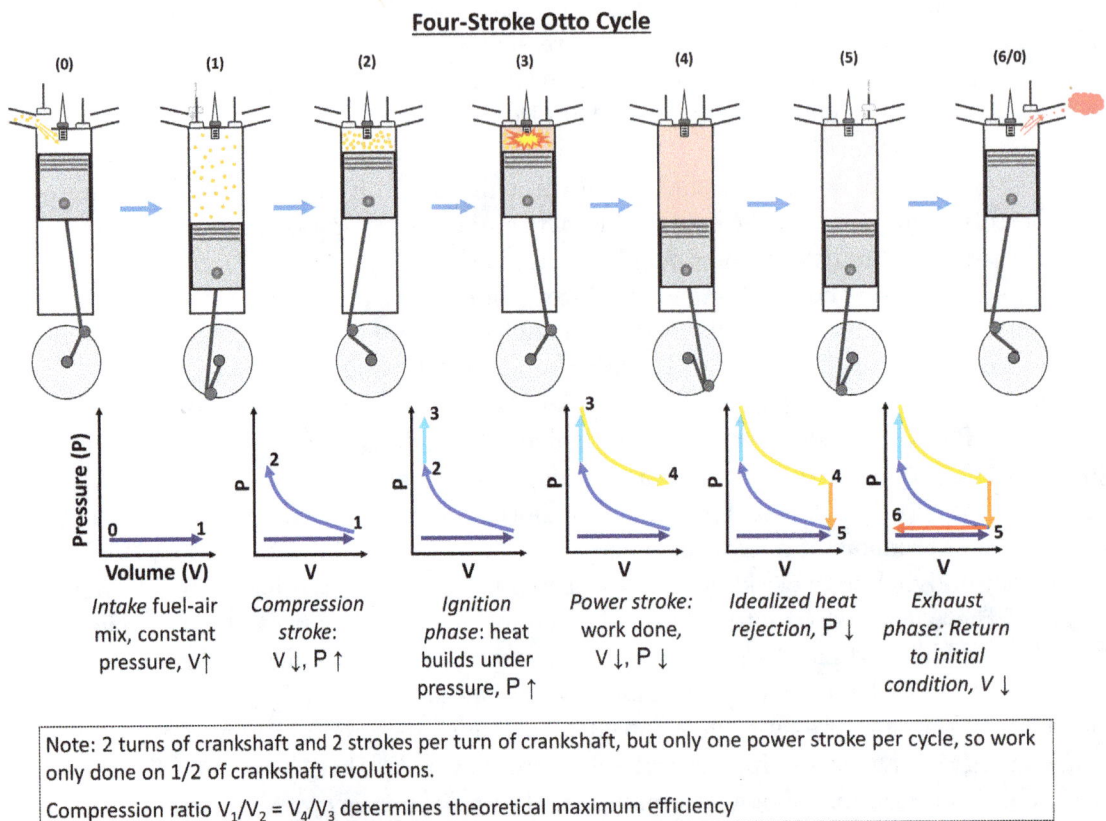

Four-Stroke Otto Cycle

| (0) | (1) | (2) | (3) | (4) | (5) | (6/0) |

Pressure (P)

Volume (V)

Intake fuel-air mix, constant pressure, V↑

Compression stroke: V ↓, P ↑

Ignition phase: heat builds under pressure, P ↑

Power stroke: work done, V ↓, P ↓

Idealized heat rejection, P ↓

Exhaust phase: Return to initial condition, V ↓

Note: 2 turns of crankshaft and 2 strokes per turn of crankshaft, but only one power stroke per cycle, so work only done on 1/2 of crankshaft revolutions.

Compression ratio $V_1/V_2 = V_4/V_3$ determines theoretical maximum efficiency

Figure 7.11: Schematic for the four-stroke Otto thermodynamic cycle, as realized (ideally) in an actual piston. The top sequence shows the piston positions and states, whereas the bottom sequence gives the step-wise construction of the pressure-volume curve describing the cycle.

186

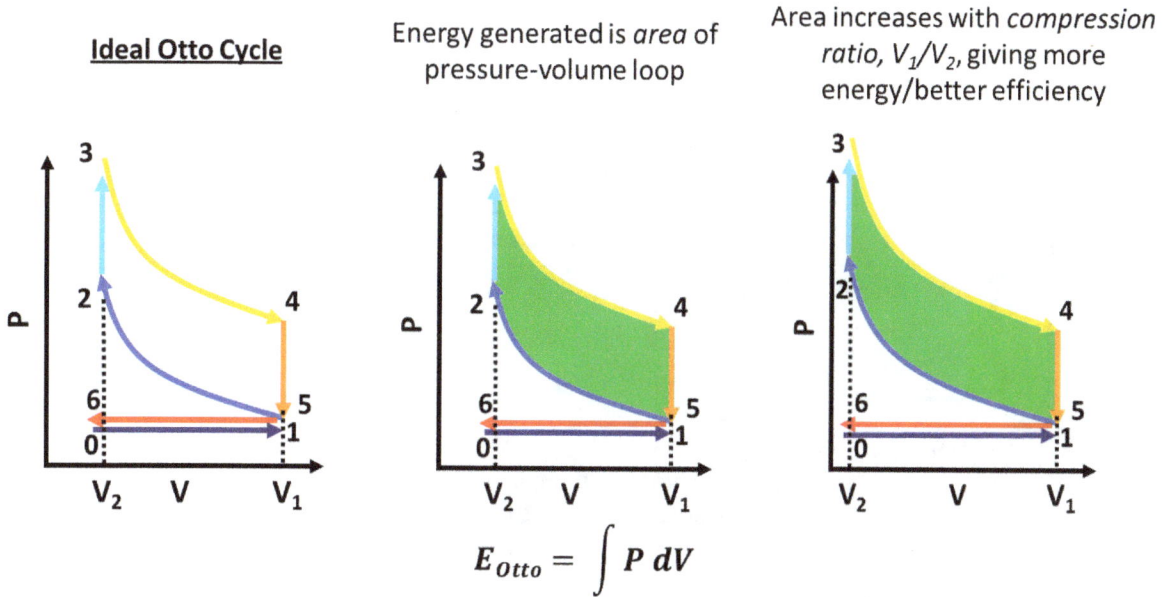

Figure 7.12: The ideal Otto cycle, with key points in the cycle labelled numerically, as constructed in Figure 7.11. We see that the area enclosed by the pressure-volume curve is the total amount of work done by the engine, and that increasing the compression ratio, V_1/V_2, increases the distance between these two points and hence increases the area enclosed by the curve, implying more energy is harvested.

$$\eta = \underbrace{\eta_t \times \eta_m}_{\text{engine efficiency}} \times \epsilon. \tag{7.7}$$

As an overall average, we have $\eta_t \approx 0.38$, $\eta_m \approx 0.45 - 0.55$, and $\epsilon \approx 0.85 - 0.95$, giving our overall $\eta \approx 0.15 - 0.20$ [245]. That is, no more than 20% of the energy contained in the fuel is converted into motion. As I discuss, all these factors vary with the instantaneous driving conditions, and our maximum η is around 0.3–0.35.

Thermal efficiency

Let us consider each of these factors in sequence, beginning with thermal efficiency, which we have already mostly covered in our discussion of the Otto cycle. The power that develops within the cylinder is referred to as *indicated power*. As already discussed, maximum thermal efficiency is determined by the compression ratio, r_C, as visualized by the pressure-volume loop, and by the leanness of the air-fuel mixture. Recall the theoretical maximum thermal efficiency:

$$\eta_{t_I} = 1 - \frac{1}{r_C^{\gamma - 1}} \tag{7.8}$$

with $r_C = V_1/V_2 = V_4/V_3$, and $\gamma = 1.40$ for air. The maximum efficiency for a fuel-air mixture is:

$$\eta_{t_{FA}} = (1 - .25\phi)\eta_{t_I} = 0.75\eta_{t_I} \text{ for } \phi = 1, \tag{7.9}$$

where ϕ is the fuel-air ratio relative to stoichiometric, and is one for all production gasoline engines. For a typical compression ratio of 10.5, we thus have a maximum thermal efficiency

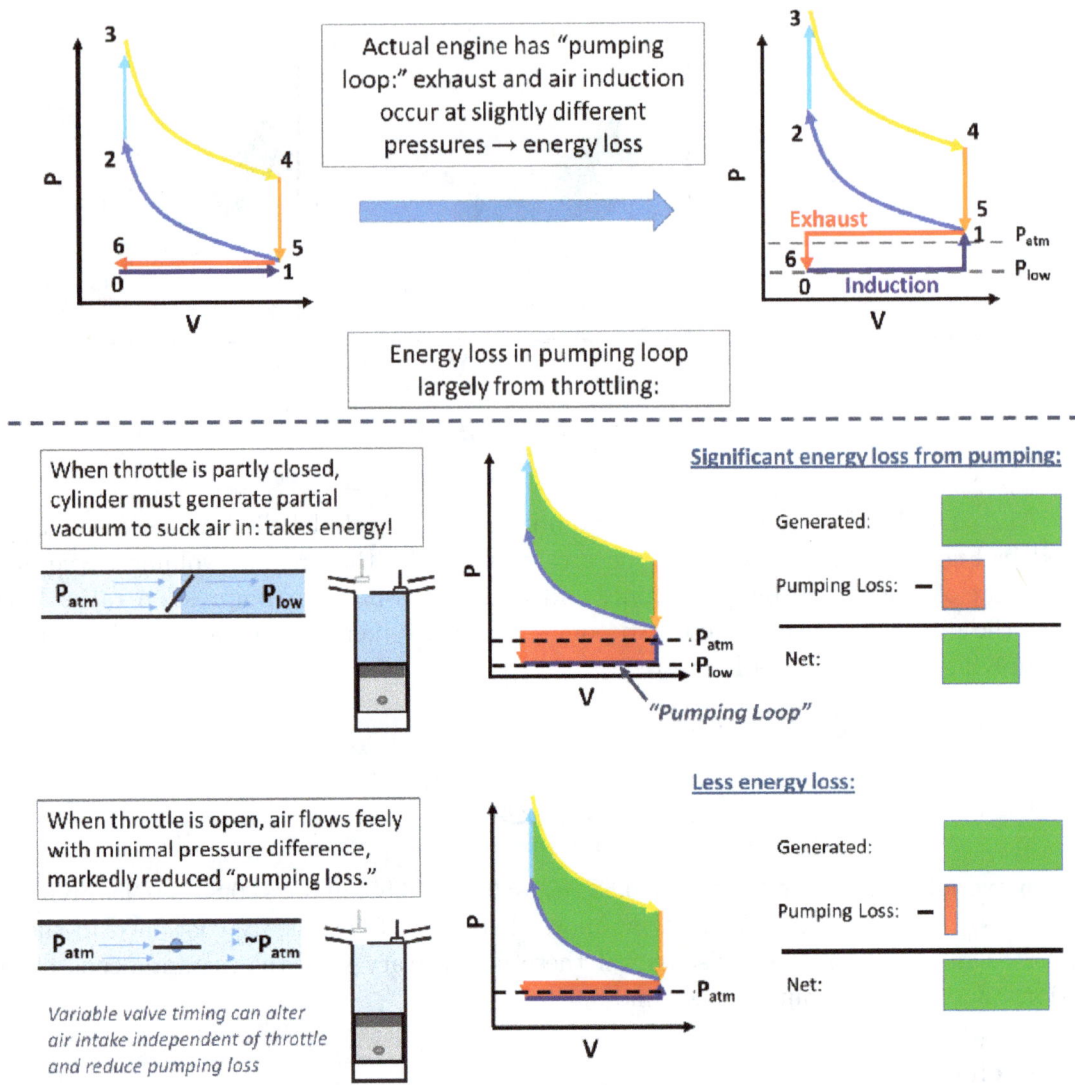

Figure 7.13: The ideal Otto cycle has the exhaust and air induction processes take place at constant pressure. In reality, the exhaust is under slightly higher pressure than atmospheric (i.e. ambient), while air must be sucked in under a partial vacuum, i.e. the pressure is less than atmospheric. This gives rise to the "pumping loop," and the area of the pumping loop is work that must be performed by the engine to move air in and out of the cylinders, lowering overall efficiency. Air intake into the engine must be regulated, and in most engines the throttle controls air flow. When it is partially blocking flow, a greater partial vacuum must be generated by the cylinder to suck in the same amount of air, reducing efficiency. Note that a turbocharger harvests waste exhaust heat to force air into the cylinder under high pressure, "inverting" the pumping loop such that it is traversed clockwise (vs. counterclockwise) to actually provide useful energy, rather than siphon it off.

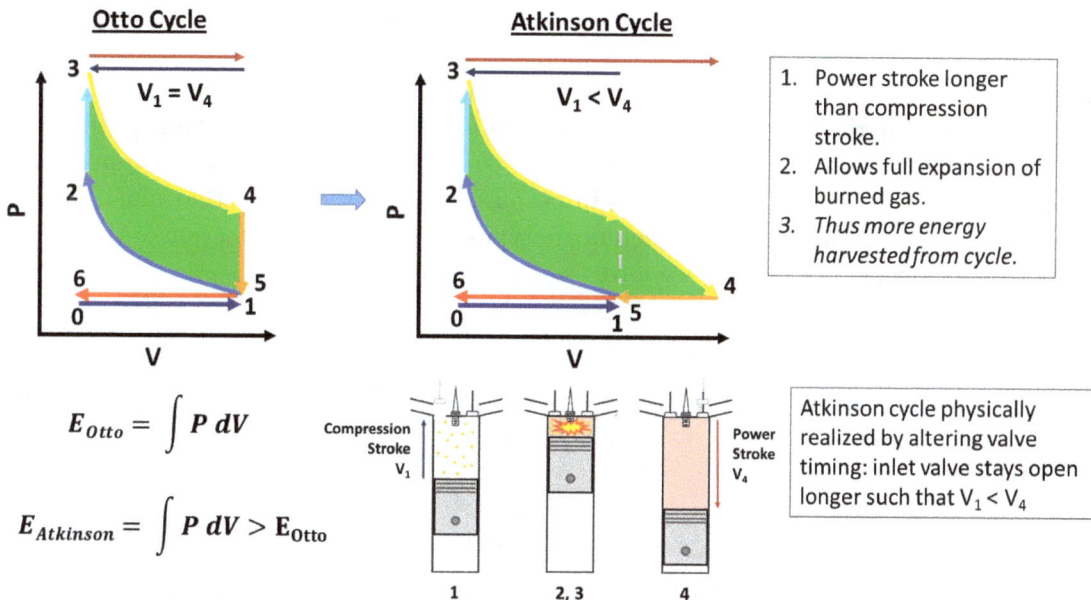

Figure 7.14: The Atkinson cycle is a more efficient modification of the thermodynamic Otto cycle, wherein the power stroke is longer than the compression stroke, allowing fuller expansion of the hot gas and thus more energy output. A modified Atkinson cycle is implemented in a number of modern hybrid engines, with a shorter effective compression stroke realized by having the inlet valve stay open partly into the (physical) compression stroke.

of 46%. Now, there are several other sources of energy loss in an actual cylinder, the most important being heat loss through the cylinder wall, with perhaps 15% of the energy lost this way. Larger engines, with their lower surface area to volume ratio, have relatively less heat loss (note that smaller engines are still more efficient overall because of friction losses, discussed in a moment). Higher RPMs imply less time for heat to diffuse out, and thus heat loss decreases with engine speed. A leaner fuel-air mix also decreases heat loss [245]. Several other minor factors decrease thermal efficiency, including about 1–2% of fuel remaining unburned. Following Ross [245], we have $Q \approx 0.15$ as the heat loss, and other losses subsumed into a constant efficiency, $\eta_c \approx 0.95$, to get a final thermal efficiency

$$\eta_t = \eta_{t_I}(1 - .25\phi)\eta_c(1 - Q) = 0.75\eta_{t_I}\eta_c(1 - Q) = \eta_{t_{FA}}\eta_c(1 - Q). \qquad (7.10)$$

Plugging in numbers, with a compression ratio of 12, we have $\eta_t \approx 0.38$. Thus, a little less than 40% of the energy in the fuel is converted into usable energy within the cylinder, and a highly optimized gasoline engine could conceivably achieve η_t in the 0.40–0.45 range.

Mechanical efficiency

Now, let us discuss *mechanical efficiency*, η_m, which is defined as the fraction of energy generated in the cylinder that is converted into work at the crankshaft. Mechanical efficiency is about 0.45–0.55 overall, so although about 40% of the fuel energy is converted into work within the cylinder, only about 20% becomes useful work at the crankshaft.

While thermal efficiency does not vary much with driving conditions, mechanical efficiency varies dramatically, ranging from perhaps 0.85 at optimum, to 0.0 when idling and coasting. Note that mechanical efficiency must, by definition, be zero under the latter conditions, because while the engine is still converting fuel to work within the cylinder, none is manifesting itself at the crankshaft. Hybrids are so efficient largely because they maintain engine operation in a narrow range where mechanical efficiency is high.

Before continuing, let us clarify a few terms. The term *indicated power*, $P_{\text{indicated}}$, is the power that is generated within the cylinder, and this is determined by the thermal efficiency of the engine. The somewhat confusingly named *brake power*, P_{brake}, is the power output measured at the crankshaft. The "brake" term comes in because P_{brake} is the power that must be applied to brake the engine, and many dynamometers are types of brakes. Some fraction of the indicated power is lost to friction, P_{friction}, so we have

$$P_{\text{indicated}} = P_{\text{brake}} + P_{\text{friction}}. \tag{7.11}$$

It also follows that mechanical efficiency is the ratio of brake power to indicated power (as stated above in slightly different terms), or in equations

$$\eta_m = \frac{P_{\text{brake}}}{P_{\text{indicated}}}. \tag{7.12}$$

Essentially, we have several sources of friction within the engine that reduce mechanical efficiency. For a given engine load (load, not speed[2]), the overall friction power loss is nearly constant. Since friction loss is (nearly) constant, as the load on the engine increases (and power generation within the cylinder increases), the *fraction* of power that is lost to friction decreases, and mechanical efficiency, by definition, increases. Thus, we see why engines achieve optimal efficiency under relatively high loads. This is demonstrated schematically in Figure 7.15.

Furthermore, friction losses increase in direct proportion to engine speed, as shown in Figure 7.16. This is quite intuitive: a cylinder head at 1,000 RPMs "rubs" on the wall 1,000 times, and 500 pumping loops draw power. At 2,000 RPMs, the cylinders rubs twice as much, and 1,000 pumping loops draw power. Thus, maximum mechanical efficiency occurs at high load and low engine speed (RPM). An example of a moderately high-load, low-RPM setting is cruising at highway speed in overdrive. A low-load (zero-load, actually), high-RPM setting might be revving the car in neutral.

Note that under wide-open-throttle (WTO) conditions, efficiency falls, so very high loads reduce efficiency. Furthermore, a high RPM is necessary to draw more power under high demand conditions. In other words, don't floor it, we want a moderately high load at a mid-low RPM. I say mid-low instead of low because there is a small benefit to thermal efficiency with increased RPM, and empirical engine performance maps support this.

Finally, friction losses increase in direct proportion to the engine displacement, i.e. total (effective) internal cylinder volume, a quantity that is ubiquitously advertised on vehicle bumpers and side-panels, e.g. 3.0 L. Halving the engine displacement roughly halves friction losses, and this is the major reason smaller engines are so much more fuel-efficient (for a given power demand). Smaller engines also operate at higher relative load, all being else equal, thus further increasing efficiency (to a point). The general effect of engine displacement and the interaction between displacement and engine load on mechanical efficiency are demonstrated in Figure 7.17. Thorough discussions of the mathematics of friction loss motivating Figures 7.15, 7.16, and 7.17 can be found in [258] and [259], but for space I omit the mathematical details here.

Transmission efficiency

Finally, let us briefly discuss transmission efficiency, ϵ. Not all power from the crankshaft is ultimately transmitted to the wheels, and $\epsilon \approx 0.85 - 0.95$. Because an ICE can only operate

---

[2]It is important not to confuse engine *load*, which is the power demand, with engine speed, which is the revolutions per unit time, e.g. RPM or RPS.

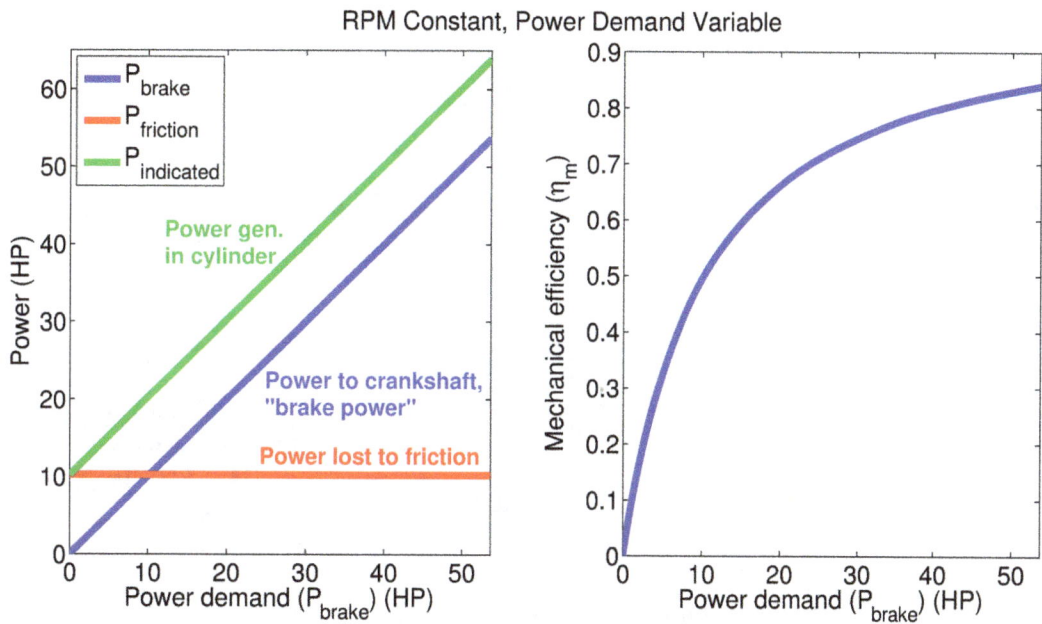

Figure 7.15: Effect of increasing load on engine mechanical efficiency, when engine speed (RPM) is held constant. Since engine friction losses are essentially constant (for a given engine RPM), efficiency increases monotonically with load. In reality, efficiency falls at very high load, so this relationship only only holds up to moderately high loads.

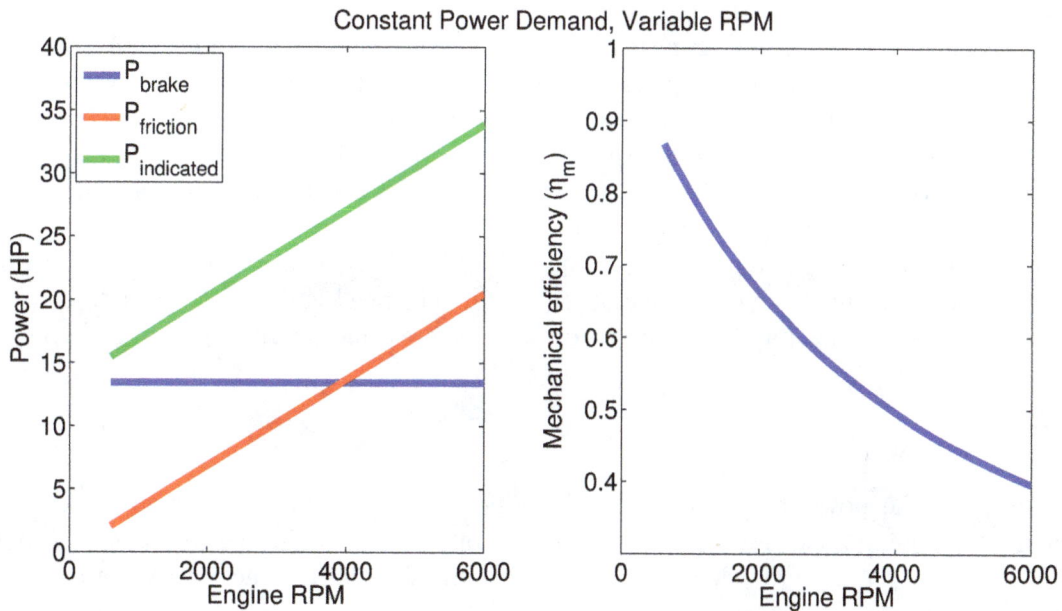

Figure 7.16: Effect of increasing engine speed (RPM) on mechanical efficiency, when the power demand is held constant. Friction loss increases in direct proportion to RPM, and so efficiency falls the faster the engine is turning over. Note that at very low engine speeds, relatively more heat can escape through the cylinder walls, thus reducing thermal efficiency (not seen here).

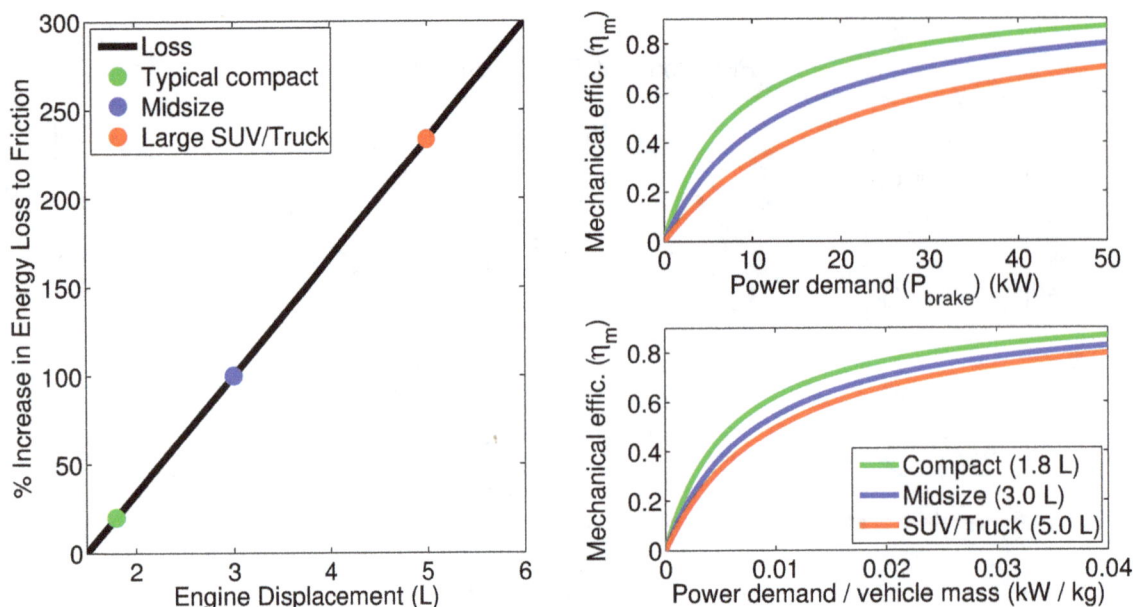

Figure 7.17: The left panel shows how relative friction losses increase with engine displacement, under a constant engine speed (RPM). The right panels show how mechanical efficiency, μ_m, varies with load for three representative engines displacements (1.8 L, 3.0 L, 5.0 L). The top shows μ_m as a function of absolute power demand, while the bottom gives μ_m in terms of power demand per unit vehicle mass, confirming that even after adjusting for the larger mass and thus relatively higher power demands of larger vehicles, they still see lower mechanical efficiency.

over a limited range of speeds, we have multiple gears, where each gear is characterized by a different ratio of engine revolutions to axle revolutions. Manual transmissions are generally highly efficient across all gears, and this explains why manuals have traditionally been more fuel efficient. Traditional automatic transmissions have tended to have a lower overall efficiency, and efficiency is lower in lower gears [258]. However, automatic transmissions have improved, and the difference between the two transmissions for new cars is equivocal.

Finally, continuously variable transmissions (CVTs) are a new technology that tends to yield appreciably greater efficiency, for two reasons. First, transmission efficiency ϵ is relatively high, but second and more importantly, the continuously variable gearing ratio allows the engine to operate in mechanically more efficient modes, i.e. engine speed is optimized to reduce friction loss for a given engine load and vehicle speed. Hybrids are almost uniformly are equipped with CVTs.

Torque and power

Understanding the power-torque relationship is necessary to modeling engine power output under actual driving conditions, and is central to optimizing engine performance. *Torque* is developed at the crankshaft, while the *power* is essentially the product of torque and engine speed.

Torque, τ, also known as the moment of force, has SI units of Newton-meters (N-m). Loosely speaking, it is a rotational force applied to an object, and for a force applied at a right angle to a lever arm, torque is just the product of force and lever distance. Mathematically torque is defined as the cross product of the lever arm position, $\mathbf{r}$, and force, $\mathbf{F}$,

$$\tau = \mathbf{r} \times \mathbf{F}, \tag{7.13}$$

with the magnitude given by

$$|\tau| = |r||F|\sin\theta, \tag{7.14}$$

where θ is the angle of force relative to the lever arm. Now, we have torque in N-m, and if we multiply by the rate at which the object turns, or angular velocity ω, in s^{-1}, we get power, in units N-m s^{-1}. This is nearly analogous to power as the product of force and velocity for road loads, $P = Fv$, and so,

$$P = \tau\omega. \tag{7.15}$$

By analogy, we may think of tightening a bolt with a wrench. The harder you push, the greater the torque, and torque only relates to how hard you push. The rate at which you're doing work, i.e. the power, is determined both by how hard you push and by how fast you turn the bolt. The same is true for an engine. Now, in reality a given engine will have a torque-RPM curve that can be determined experimentally. That is, torque generated by a physical engine varies somewhat with RPM, but it is important to understand that there is no fundamental mathematical/physical reason for this to be so: it is a function of the physical characteristics of the engine.

Gasoline engines have a slightly convex but fairly flat RPM-torque curve with a peak at mid-high RPM, while diesel engines have the highest torque at low engine speed with a dramatic drop-off at higher RPMs [258]. The Power-RPM curve is then determined simply as the product of the engine speed and torque (at every point). Note that this is why vehicle labels report peak torque and peak horsepower at a specified RPM.

Finally, given vehicle speed and the current transmission gearing ratio, we can solve for engine speed, in RPM. If we have the engine's RPM-torque curve, we can then directly solve for the power output under any particular driving condition.

An integrated understanding

In sum, the internal combustion engine is a rather finicky beast. We see that over half the energy embodied in fuel is lost to the intrinsic inefficiencies of thermodynamic cycles based on heat engines, while several other heat loss mechanisms collaborate to reduce maximum thermal efficiency to roughly 0.40. The mechanical efficiency of the drivetrain then varies vastly depending on driving conditions, and the transmission adds further losses, so that in the end only around 15–20% of fuel energy is converted to motion. There is a relatively narrow torque and RPM range where engine efficiency is maximal, and certain driving conditions yield zero efficiency, e.g. coasting, braking, and idling.

Engine friction losses are minimized under moderately high loads at lower RPMs, and this can be confirmed by experimental engine performance maps, a hypothetical example of which is given in Figure 7.18. Friction losses increase in direct proportion to engine size and, for a given vehicle load, the load on the engine is lighter the larger the engine (implying lower efficiency). One can conclude that the basic eco-driving task is to maintain the drivetrain in a high-efficiency operating mode, and this task is greatly abetted by the hybrid-electric drivetrain, as discussed next.

7.2.3 Hybrid drivetrains

A hybrid drivetrain typically includes an internal combustion engine that powers both the wheels and an electrical motor/generator connected to a battery. This configuration can increase the

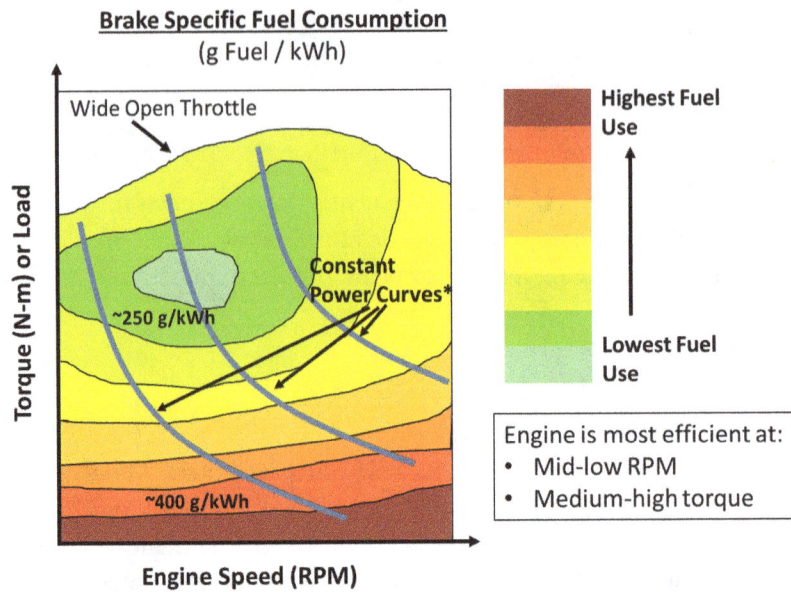

Brake Specific Fuel Consumption
(g Fuel / kWh)

Wide Open Throttle

Torque (N-m) or Load

~250 g/kWh

Constant
Power Curves*

~400 g/kWh

Engine Speed (RPM)

Highest Fuel
Use

Lowest Fuel
Use

Engine is most efficient at:
- Mid-low RPM
- Medium-high torque

*Since $P \approx torque \times RPM$, the same power output can be realized either at low
torque and high RPM, or visa versa.

Figure 7.18: Hypothetical engine performance map, showing how fuel consumption changes with engine speed (RPM) and torque, based on [245]. As can be seen, fuel consumption is minimized at mid-low RPM and mid-high engine load.

overall drivetrain efficiency from perhaps 17% for a conventional ICE to around 30%. All energy driving a hybrid ultimately comes from the liquid gasoline fuel, and hybrids come in two basic architectures, serial and parallel [258]. With the serial architecture, the gasoline engine acts only as an onboard electric generator for the battery-motor system, with only the latter directly driving the wheels. Under the parallel architecture, the engine and battery/motor systems operate in tandem, with both providing power to the wheels and the battery charged either by regenerative braking or the engine. Finally a series/parallel architecture, the design used in the Toyota Prius and many other production hybrids, is essentially a parallel architecture but with the advantage that the battery/motor system can power the wheels alone (as in the serial architecture), allowing the gasoline engine to turn off during much of the drive cycle. For a typical series/parallel architecture, efficiency gains occur at several major points in the system:

1. The engine[3] operates more in higher efficiency modes (mid-high load, mid-low engine speed), as the adjunct battery-motor system takes over (turning off the engine completely) during low efficiency modes (e.g. low-load cruising or light acceleration, coasting, braking, and idling).

2. The engine can be downsized, as the electric motor provides supplemental power at times of high demand (as extensively discussed above, smaller engines are intrinsically more efficient).

3. Some energy normally lost to the acceleration-braking cycle is recovered via regenerative braking (the electric motor siphons power from the wheels to both slow the vehicle and charge the battery).

---

[3]Note that "engine" refers to the gasoline-powered (or diesel, etc.) internal combustion engine, while "motor" refers to the electric motor, which also acts as an electric generator.

To reiterate, the electric motor/generator-battery system acts as a buffer that allows the engine to operate in a more efficient operating range. When less power is needed and the battery state of charge (SOC) is high, the electric motor runs alone and the engine need not operate at the less efficient low-RPM/low-load range. If the battery charge is low the engine can still operate in a higher, more efficient mode (higher power, mid-RPM) but shunt some power to the battery.

Furthermore, because the electric motor can assist the engine when the power demand is high, the engine need not be sized for peak power demand and can be significantly down-sized compared to a comparable ICE vehicle. This engine down-sizing is a major reason for improved fuel efficiency, and many hybrid engines also use the more efficient Atkinson thermodynamic cycle (see Figure 7.14).

7.3 An energy-based comparison of internal combustion, hybrid, and battery-electric vehicles

We may summarize the differences in energy consumption and emissions between different vehicle classes via an energy-based analysis. There are multiple papers that have compared the primary energy efficiency of different drive-trains, e.g. [260, 262]. That is, what fraction of the energy embodied in the primary fuel, either crude oil for ICEVs and HEVs, or natural gas, coal, etc. for battery electric vehicles, ultimately is converted to power at the wheel? One problem with many such analyses, in my view, is that they are overly focused on efficiency/emissions when the power source is a renewable: marginal generation will probably be supplied by fossil sources for decades to come, and thus should be the focus of any near- to medium-term analysis. Another problem is overly optimistic estimates of power-plant efficiency, which is also misleading.

I think a more informative analysis is to examine "fuel-to-wheel" efficiency, i.e. the fraction of energy embodied in the fuel that is directly burned to yield energy for the vehicle (gasoline or diesel for ICEVs/HEVs vs. a power-plant fuel for BEVs) that makes it to the wheel. I subsume further upstream energy losses into emissions estimates.

Now then, essentially, a conventional internal combustion vehicle is a platform for a gasoline power generator, which converts fuel energy into motion at an overall efficiency of 15–20%, at an average of about 17% [245]. Since, on an embodied energy basis, well-to-wheel emissions for gasoline are about 0.334 $kgCO_2e$/kWh, we have overall emissions per kWh of energy delivered to the wheels as 1.963 (1.669–2.225) $kgCO_2e$/kWh delivered.

A hybrid-electric vehicle is a platform for a modified gasoline generator coupled to an on-board battery-motor system, allowing the generator to operate more efficiently, and further, allowing some energy normally dispersed through braking to be recovered. This combined system is reported to achieve about 29–30% efficiency [260]. We can see how this might be true: suppose 50% of the energy generated by the engine is ultimately routed through the adjuvant battery-motor system in city driving, with a subsequent conversion efficiency to motion of 75%, denoted η_B (see BEV discussion below), while 50% goes directly to motion (also assume 80–100% will go directly to motion under highway driving). Assuming a reasonable range for ICE thermal efficiency of $\eta_t = .38 - 0.40$, mechanical efficiency $\eta_m = 0.85$, and transmission efficiency $\epsilon = 0.88 - 0.95$, we would have a plausible efficiency range of 25–30%, with 28–29% perhaps most likely. Newer hybrids systems with lower mechanical losses and improved transmissions may achieve 30% efficiency.

There is an additional trick here, however: some energy is recovered from braking that would be lost with an ICE. If 25% of all forward energy is lost to braking in city driving, about 75% is available to be re-captured and re-converted to forward motion at 70% efficiency (regenerative braking has a slight efficiency cost compared to other charging [259]), and 55% of drive-time is spent in city vs. highway driving, then about 7.22% of the fuel requirement is displaced, increasing *effective* drivetrain efficiency from 29% to 31.26%, and effective emissions per useful kWh are 1.0675 kgCO$_2$e/kWh, around half of the ICE emissions.

Finally, consider the battery electric vehicle. Energy is stored on-board in batteries, which is converted into motion by the electric motor drivetrain (also equipped with regenerative braking). The electric drivetrain converts 85–90% of the energy delivered by the battery into motion [261], with 87.5% a reasonable average estimate [262]. However, the efficiency of charging a battery from the grid varies from 89–96% [261], with 90% a more likely value [262], and battery discharge is perhaps 96% efficient [262], implying an overall charge-discharge cycle efficiency of 86%, and thus the complete drivetrain is about 75% efficient (0.875 × 0.86). Including regenerative braking, again at a 70% overall conversion of braking energy to motion, we would have an adjusted efficiency of 80.84%.

If we assume electricity is generated by a combined cycle natural gas turbine, the most efficient plausible marginal generating source, at a thermal efficiency of 44.8% (EIA 2012 average), and we also include a 6.5% transmission and generation energy loss, we have an overall efficiency, in terms of primary energy, of 31.4%, and this increases to 33.9% when including the effect of regenerative braking; thus, the primary energy efficiency is essentially equal to a HEV. Now, since the emissions of natural gas are (assuming 2.4% upstream leakage) 0.2613 kgCO$_2$e/kWh, our battery electric vehicle, in this case, gives 0.7717 kgCO$_2$e/kWh, 27.7% lower than the HEV analysed above, and just shy of 60% less than the conventional vehicle.

If, on the other hand, we assume a coal-fired plant at 32.5% efficiency, we have overall energy efficiency of 22.8% that improves to an effective 24.6% when accounting for regenerative braking, which is at about the midpoint between the ICE and HEV efficiencies, and we have emissions of 1.4408 kgCO$_2$e/kWh, 35% higher than an HEV, but still over 25% lower than the conventional engine. Using US grid-average emissions, the EV emissions are 21% lower than those of the HEV, but under the approximate US marginal emissions average, they are nearly equal, coming in at 2% *higher*. Note that these calculations include the energy offset provided by regenerative braking, and are all summarized in Table 7.19.

So, we see that HEVs and EVs are nearly comparable in terms of emissions per kWh delivered to the wheel. There will still, of course, be significant variation in the energy demand for any given vehicle of either class, based on vehicle weight, profile, aerodynamic drag, etc. A lighter more aerodynamic HEV could easily be superior, even markedly superior, to a large EV in terms of emissions per mile (or equivalently, MPG$_{GWP}$). On the other hand, since they generate almost twice the emissions per kWh compared to either the HEV or EV, the only time an ICE vehicle with be superior to an HEV/EV is when comparing a compact ICE to a large HEV or EV, or under a particularly dirty generating mix. Not captured in this energetic analysis, the emissions associated with large battery pack production can also shift our assessment to favor HEVs or even economy ICEs over larger EVs (see Section 6.6.2).

Anytime we compare two comparable vehicles, e.g. compact vs. compact, SUV vs. SUV, hybrid and electric vehicles will generate significantly fewer emissions at the use-phase than the conventional version, even when, in the case of an electric vehicle, all electricity is coal-fired. Thus, HEVs and EVs should be thought of as in the same basic low-emissions category, while ICEs are the general high-emissions category. Note that fuel consumption still varies nearly threefold among conventional passenger vehicles, and absolute emissions savings are much greater going from worst to best in the ICE class versus best ICE to typical, or even best, HEV/EV.

Vehicle Class	Tank −> Wheel Effic (%)		Upstream Efficiency (%)	Fuel −> Wheel Effic (%)		Emissions per kWh −> Wheel (kgCO2e/kWh)	
	w/o Regen	w/ Regen		w/o Regen	w/ Regen	Absolute	vs. ICE (%)
ICE	17	17	100	17	17	1.9629	0
HEV	29	31.3	100	29	31.3	1.0675	−45.6
EV							
CCGT	75	80.8	41.9	31.4	33.9	0.77166	−60.7
Coal	75	80.8	30.4	22.8	24.6	1.4408	−26.6
US Marginal	75	80.8	35.4	26.5	28.6	1.0883	−44.6
US Grid−Avg	75	80.8	–	–	–	0.84364	−57

Figure 7.19: Table showing the efficiency of a given drive-train (gasoline ICE vs. gasoline HEV vs. EV) in converting fuel to energy at the wheel, where this efficiency varies with the fuel type used to generate the electricity that powers the electric vehicle (EV) (not calculated for the average US generating mix). Vehicle efficiency is given with and without regenerative braking, although emissions per kWh delivered to the wheel are only provided with regenerative braking. As highlighted, the most important comparison between the HEV and EV is that using the US marginal generating mix, in which case the emissions per kWh are actually slightly lower for the HEV. Note that emissions estimates, as usual, are based on the full fuel lifecycle.

Chapter 8

Air travel

8.1 Overview

- Fuel use and production in commercial aviation directly generated about 200 million $MgCO_2e$ in 2015. However, this figure increases to about 240–360 million $MgCO_2e$ when additional non-CO_2 forcing is accounted for (with 280 million $MgCO_2e$ a best estimate). Raw per capita numbers are 0.62 $MgCO_2e$ per capita, or 0.75–1.12 $MgCO_2e$/capita including non-CO_2 forcing, with 0.87 $MgCO_2e$/capita an intermediate value.

- Considering CO_2 emissions alone, commercial air travel can yield 50–95 passenger-miles per gallon (PMPG), normalized to standard gasoline (jet fuel is somewhat more energy dense than gasoline, on a volume basis), with domestic travel averaging 55 $PMPG_{GWP}$.

- However, jet engines emit nitrogen oxides, sulfur, water vapor, and soot that affect high altitude ozone and methane chemistry, create contrails, and affect high altitude cirrus clouds, yielding a mix of cooling and warming effects; the overall (including CO_2) warming effect is, very crudely speaking, about 1.25–2.0 times that of CO_2 alone; assuming a 1.5 multiplier, domestic air travel adjusts to 39 $PMPG_{GWP}$, international travel adjusts to 30 $PMPG_{GWP}$, and US commercial air travel overall achieves an adjusted 36 $PMPG_{GWP}$.

Since its inception in about 1940, commercial aviation has grown rapidly worldwide, with growth in aviation-related CO_2 emissions faster than that of overall anthropogenic CO_2 [263]. While growth in this sector has been relatively flat for the last 15 years in the US, an era that has seen consolidation and increasing system efficiency, on the global scale aviation is projected to grow at about 5% per annum until 2030 (or nearly a trebling in global air traffic) [278]. On a per-capita basis, Americans average just under 2,800 air miles per year, and air travel generates about one-fifth the CO_2e of passenger cars, making it a fairly distant, but still significant, second when it comes to transportation emissions. Frequent flyers, however, may dwarf their driving footprint, and a single international round-trip to Europe can easily generate 2–4 $MgCO_2e$ (or about half the yearly emissions of a typical driver).

Indeed, while as a crude average Americans fly around 2,800 air miles yearly, few Americans are average and there is marked variability in air travel among American adults. According to surveys by Airlines for America [279], just under 50% of US adults travel by air in any given year, and although American adults averaged 2.2 trips by air in 2016, those who actually took any flights at all averaged 4.5 trips. Moreover, a majority of flights are now for personal *leisure*, not business or other personal matters. Thus, the typical flyer is responsible for around 2 $MgCO_2e$ of aviation-related emissions, while this sum is zero for a slight majority of all Americans.

Like all petroleum fuels, jet fuel is converted to CO_2 upon combustion, and on the basis of direct CO_2 emissions alone, air travel is quite efficient, achieving passenger-miles per gallon (PMPG) better than most (single-occupancy) hybrid gasoline vehicles, and even comparable to passenger rail in the very best cases (45–55 passenger-miles per gallon typically, but up to 99 PMPG). Also similar to other fuels, the upstream fuel-cycle, i.e. oil extraction and refining, increases jet fuel emissions by about 25%, to 11.95 $kgCO_2e$/gallon, compared to emissions at the point of combustion alone of 9.59 $kgCO_2$/gallon (see below).

While efficient on the basis of fuel consumption, multiple other chemicals released by jet engines affect high atmospheric chemistry with both warming and cooling effects that vary with altitude and geographic location, and remain poorly understood. Particularly uncertain is the effect of jet contrails and induced cloudiness (warming) and sulfur emissions (cooling) on global cloudiness, which may be the dominant factors. On balance, non-CO_2 radiative forcers are likely to have a warming effect significantly greater than than of the CO_2 emissions alone. This has motivated efforts to define an "uplift factor," a simple multiplicative factor that one can apply to aviation CO_2 for carbon footprinting and/or emissions trading schemes [268]. This factor is on the order of 2.0 or more when considering the integrated effect of aviation from 1940 to the present, but it is probably smaller, around 1.25 to 1.75, when assessing the *future* warming effect of any given plane trip.

In sum, arriving at a reliable emissions factor for air travel is a challenge beset by significant uncertainty, but I suggest a mid-range uplift factor for planning and carbon-footprinting purposes of 1.5, applied to jet fuel combustion CO_2 only (and not to upstream fuel-cycle emissions), indicating that most domestic air travel achieves 35–45 MPG-equivalent, while international travel gets more like 30–35 MPG-equivalent. Using such an uplift factor, jet fuel emissions are equivalent to 16.75 $kgCO_2e$/gallon, and US per-capita aviation emissions increase from 0.62 to 0.87 $MgCO_2e$ per year. When comparing an air trip to a road trip, one should therefore view domestic flying as roughly equivalent to driving a high-MPG conventional vehicle or average hybrid-electric vehicle (with a single occupant).

This chapter is organized as follows. Raw and gasoline-equivalent passenger- and seat-mile per gallon figures for US aviation are discussed in the following section. Overall fuel use, flights, and passenger miles for the US fleet are also presented. The issue of the "marginal passenger" and proper carbon accounting for an individual's flight is then addressed. Non-CO_2 factors are discussed in detail in Section 8.3, as is their integration into alternative uplift factors. The basic physical principles underlying flight and the jet engine are introduced in Section 8.4, and a basic physical model for commercial flight is used to inform a discussion of how different flight parameters (e.g. flight length, payload, aircraft class) affect fuel efficiency.

8.1.1 Air travel MPG, fuel-cycle CO_2e emissions, and the "uplift" factor

- Domestic air travel achieves 69.1 and 58.7 seat- and passenger- miles per gallon, respectively, in terms of kerosine jet fuel.

- These figures adjust to 64.4 SMPG and 54.8 PMPG on a gasoline-equivalent basis (kerosine is denser than gasoline).

- After including a (combustion) uplift factor of 1.5 to account for non-CO_2 forcing, our final gasoline-equivalent MPGs are 46.0 and 39.1, on seat- and passenger-mile bases respectively, for domestic air. International travel achieves 36.8 and 29.8 seat- and passenger-MPG equivalents, respectively.

On the basis of raw fuel consumption, air travel actually achieves very good MPG: in 2015, in terms of raw jet fuel, the domestic US fleet achieved 69.1 seat-miles per gallon (SMPG), yielding, at an 85% load factor, 58.7 passenger-miles per gallon (PMPG). International travel is not as efficient, getting 55.3 SMPG and 44.8 PMPG (81% load factor), in terms of raw jet fuel. On a gasoline-equivalent basis, these MPG numbers are 64.4 SMPG and 54.8 PMPG for domestic travel, and 51.6 SMPG and 41.8 PMPG for international travel.

These are overall averages, but fuel burn and equivalent MPG figures will vary significantly with individual flights, depending on aircraft, seating arrangement, flight distance, payload, etc. For example, very short trips (on the order of a few hundred miles) may consume 30–50% more fuel per mile than longer (1,500+ miles) trips. All-economy seating arrangements are more efficient than those that include first-class seats (these take up about twice the space of an economy seat). As a nearly best-in-class example, the narrow-body Boeing 737-800, with a manufacturer reported range of up to 3,386 miles and a useable fuel capacity of 6,875 gallons could achieve 99 seat-miles per gallon (92 gasoline-equivalent SMPG) in an all-economy 184 seating-arrangement [1].

The above gasoline-equivalent adjustments are necessary because aviation jet fuels are formulated from kerosene, and are slightly more energy-dense than gasoline (on a volume-basis), implying that, as with diesel, MPG figures are not directly comparable. From Table 4 of [264], I calculate combustion emission factors of 9.59 $kgCO_2$/gallon (3.16 $kgCO_2$/kg) for jet fuel, and from Figure 15, I estimate an additional 18 gCO_2e/MJ of well-to-pump emissions (for jet fuel from mixed crude oil), or 2.361 $kgCO_2e$/gallon WTP (using 43.2 MJ/kg and a density of 0.802 kg/L), for a grand total of 11.95 $kgCO_2e$/gallon for the jet fuel fuel-cycle. On a volume-basis, this is 7% higher than the fuel-cycle emissions for gasoline.

Now, as discussed elsewhere, we must adjust for non-CO_2 forcers for final estimates of the emissions per mile, and using an uplift of 1.5 (applied to combustion CO_2 only) as a central estimate, we finally get 16.745 $kgCO_2e$/gallon. This uplift factor may plausibly vary from 0.75 (i.e. non-CO_2 emissions are slightly cooling) to over 2.0, and Figure 8.1 shows final adjusted emissions factors and gasoline-equivalent MPG values for commercial air travel.

8.2   Recent trends in fuel use and passenger patterns

Commercial aviation has existed since about 1940, while the first jet-powered aircraft were introduced in the 1950s. The subsequent decades saw major expansion in the US, but with plateaus and consolidation in the last decade or so, including a transient drop in air traffic during the Great Recession of 2008. On the global scale, however, aviation continues to growth rapidly.

General trends in US and global aviation fuel use are given in Figure 8.2, and Figure 8.3 gives a detailed (monthly) breakdown of recent US commercial aviation capacity and utilization, in terms of available seat-miles, actual passenger-miles, and the load-factor.

8.2.1   The marginal cost of commercial air travel

This section is intended as a brief discussion and guide for those contemplating the emissions of, say, a road trip versus a plane trip, and how best to account for the marginal emissions cost they incur via air travel. In brief, available air seats track demand closely, and so one should, in my view, "take credit" for most or all associated emissions. Load factors are lowest in winter, and

---

[1]This calculation assumes a minimum fuel reserve of 1,800 kg (593 gallons), based on cruising fuel consumption of about 1 kg fuel/second and 30 minutes of cruising as a minimum safety factor. This back-of-the envelope calculation corresponds extremely well to a maximum of 97 seat-miles per gallon calculated in Section 8.4.4 from the basic physics of flight for an all-economy Boeing 737-800.

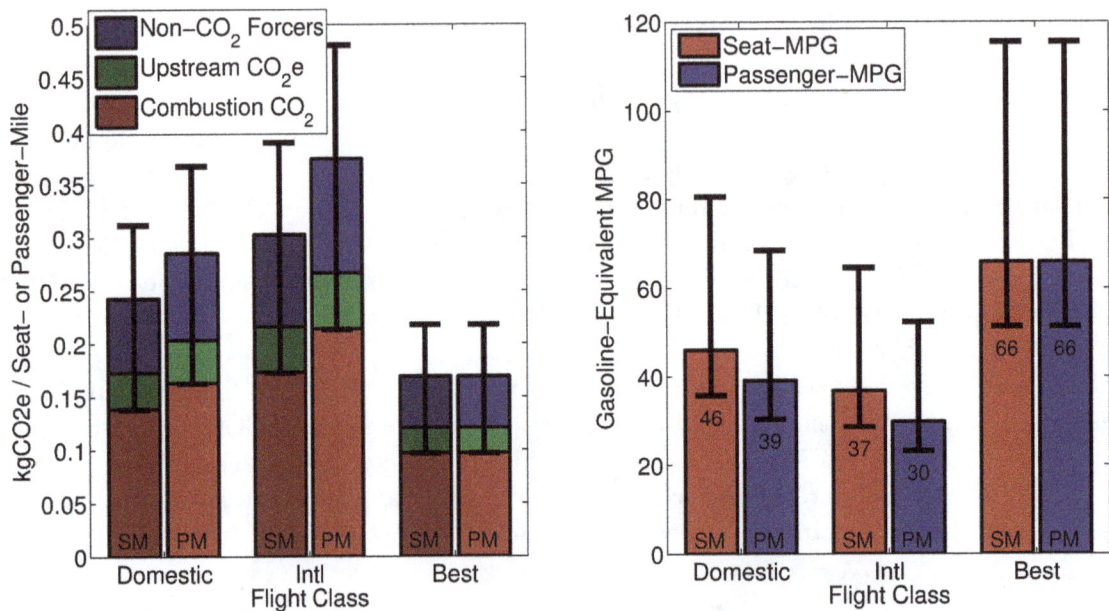

Figure 8.1: The left panel gives fuel-cycle (divided into upstream fuel-cycle and combustion) and non-CO$_2$ emissions, on a seat-mile and passenger-mile basis, for average US domestic and international flights in 2015. Also included is a best-case scenario of a new narrow-body jet (e.g. Boeing 737-800 with winglets) achieving 99 MPG (jet fuel) with a 100% load factor (implying seat-MPG = passenger-MPG). The central non-CO$_2$ estimate is based on 1.5 uplift factor for combustion CO$_2$, with error bars giving equivalent emissions for an uplift factor from 0.75 to 2.0. The right panel shows total emissions in terms of the gasoline-equivalent MPG; it is appropriate to use these MPG numbers when comparing the carbon impact of flying to driving.

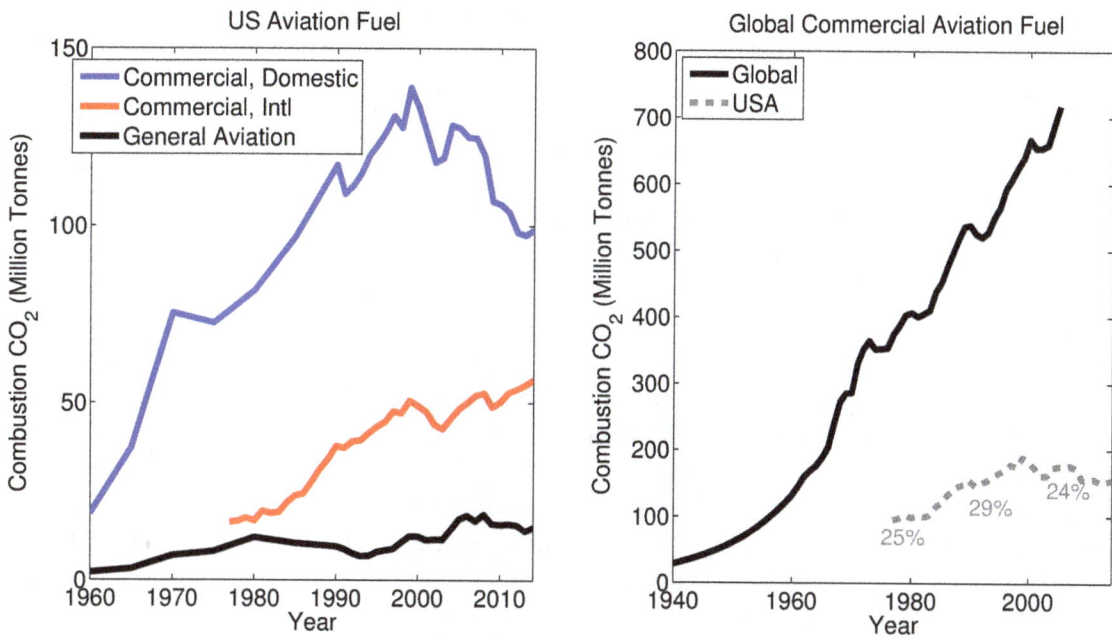

Figure 8.2: Trends in US and global aviation fuel use, based on [269, 280] and BTS figures. The left panel gives US fuel use, divided into domestic and international commercial aviation, and domestic general aviation (e.g. small private and corporate aircraft). The general aviation category includes both jet fuel and aviation gasoline, and all fuel use is reported in terms of combustion CO_2. The right panel shows global commercial aviation fuel from 1940 to 2005, along with US commercial fuel from 1977 to 2014; the US alone accounted for between 24 and 29% of the world's aviation fuel during the years shown.

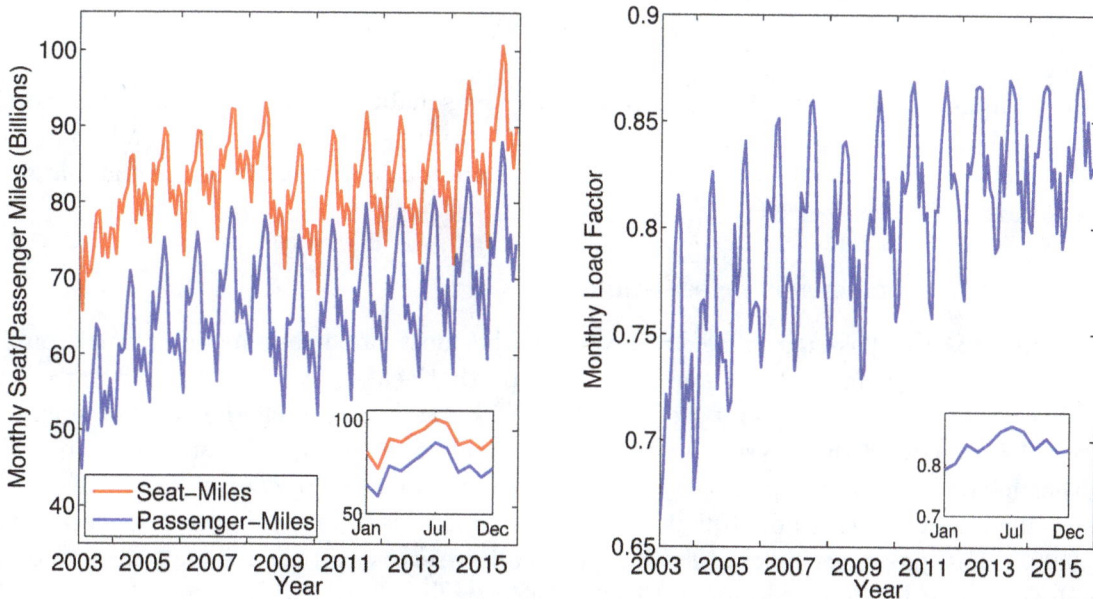

Figure 8.3: Monthly passenger- and seat-miles from 2003 through 2015 (source: BTS), for US carriers, including both domestic and international flights. The right panel gives the load factor (fraction of seats occupied), and the insets in either panel give data for the year 2015.

thus winter trips are least likely to directly increase aviation emissions. The weight of a single passenger and their luggage increases aircraft emissions by about 0.03 $kgCO_2e$/mile (equivalent to 2.7 gallons of gasoline over 1,000 miles of travel), representing a lower bound for marginal air travel emissions.

The "marginal passenger argument"

One must perennially confront the assertion that traveling by plane (from an individual perspective) carries no cost whatsoever: that plane was scheduled to fly, whether or not one person buys a ticket will not affect this, nor could a single ticket plausibly add further air routes. In fact, since that seat could go unclaimed, one is practically obligated to travel by air. I will refer to this notion, that there is no marginal cost to individual air travel, as the "marginal passenger argument." Scale this argument up, of course, and it is nonsensical. The entire industry exists because of consumer demand, and it continues to grow rapidly at the global scale because of increasing demand.

As seen in Figure 8.3, US flights and seat-miles closely track, and it is obvious from visual inspection alone that passenger-miles explain almost all the variation in seat-miles. This is confirmed by a series of regression analyses: for the last five years, passenger-miles explain 96–99% of the variation in seat-miles, with each additional passenger-mile corresponding to an additional 0.78–0.88 seat-mile, while the total number of passengers explains 85–96% of the variation in flight number, and 100 passengers additional passengers corresponds to 0.79–0.89 more flights. Therefore, we might conclude that a "marginal passenger," rather than having no effect, typically adds around 0.85 seats.

We may break our analysis down further, and examine the marginal cost by month of travel. We see from Figure 8.3 that not only are demand (passenger-miles) and supply (seat-miles) lower in the winter, but the load-factor is smaller as well, implying more empty seats. Regressing on monthly data, I find that demand best explains supply during the peak summer months (roughly May through August), with each passenger adding up to 0.89 seats, while in January and February each passenger adds only 0.49 to 0.67 seats. Thus, the marginal passenger argument is most applicable in winter. As a caveat, winter trips also cause more warming contrails than summer trips [266].

For planning purposes, given the above one may reasonably apply a discount factor to air travel emissions of, say 0.85 for general travel, and perhaps 0.50 for low-demand routes or seasons. However, one should clearly avoid anything approaching a full embrace the marginal passenger argument.

Accounting on a passenger- or seat-mile basis

While emissions per passenger-mile are probably the most important metric from a public policy perspective, there is an argument to be made that seat-miles are more relevant on an individual scale. After all, it does seem reasonable to "assign" oneself the carbon emissions associated with moving their particular seat independent of the overall aircraft load factor. On the other hand, one is participating in a system that, as a matter of course, must leave some seats unfilled, and it seems doubtful that system-wide load factors can improve much beyond 0.85–0.90. Thus, passenger-miles are still a personally relevant metric. I would encourage the individual to consider both, and weigh them as they see fit.

Minimum marginal cost

As shown in Section 8.4.4, each 100 kg (the approximate weight of a passenger with luggage), increases fuel burn by about 0.0017 kg fuel/mile (for a Boeing 737) and emissions by about .02–0.03 kgCO_2e/mile (including non-CO_2 forcers), and thus this is the absolute minimum marginal cost of air travel. This emissions rate is equivalent to 372–557 gasoline-equivalent MPG, which is trivial for an individual. On the other hand, with five passengers, we should have obtain a combined "five-passenger" MPG of 74–111, and since five passengers might feasibly achieve 50 MPG in a hybrid vehicle, traveling by air could result in *direct* marginal emissions only about 33–50% lower.

8.2.2 Some notes on general aviation

- Single-engine (gasoline-powered) piston aircraft get no better than 10–16 MPG. Business and private jets get only 0.67–1.67 MPG-equivalent, and usually carry only a handful of passengers. General aviation is therefore extremely carbon intensive.

My focus, unless otherwise stated, is on commercial aviation in this chapter. General aviation, a catch-all for private aviation, including business travel and private single-engine planes, is a very minor source of emissions at the national scale. However, for individuals who partake, it is an appreciable source of CO_2. Small single-engine planes use gasoline-powered piston engines, generally fly at comparatively low altitudes (12,000–25,000 feet) and speeds (<150–200 MPH), and achieve MPGs comparable to or worse than a very large SUV, about 10–16 MPG at best, and can carry 4–6 occupants[2], for about 40–60 seat-MPG (although one may assume that, as with cars, all seats are rarely filled). Given the lower flight altitudes, non-CO_2 forcers are less significant and can probably be ignored.

Business/private jets, such as the Gulfstream series, tend to get about 1–2.5 MPG overall (raw jet fuel basis), at best, and can carry between 8 and 19 passengers. Together, this amounts to 20–25 seat-miles per gallon in terms of raw jet fuel, but after adjusting to gasoline and accounting for non-CO_2 forcers (business jets travel at the same altitudes as commercial jets), we arrive at an adjusted 0.67–1.67 gasoline-equivalent aircraft-MPG, or 13–17 seat-MPG. Since, according to a private general aviation company [267], on average 3.4 out of 8 seats were filled in business jets (load factor of 42.5%), this class of aviation likely achieves only about 5–7 passenger-MPG equivalent.

It follows that general aviation, especially with business or private jets, is extremely carbon-intensive, and moreover, there is no "marginal passenger" argument to be made: all trips are clearly additional. If you happen to number among the few for whom this is actually an option, remember this, and as a final example, a round-trip from New York to Paris in a Gulfstream G450 generates carbon-equivalent emissions of almost 100 MgCO_2e, or about 35–45 times the emissions that the same trip via commercial aviation would yield (for a single passenger).

8.3 Understanding non-CO_2 emissions

Aviation is fairly unique among anthropogenic activities in that it affects climate via high-altitude fossil-fuel combustion, a unique setting where *short-term* effects on climate are dominated by very short-lived combustion by-products, rather than CO_2. Since CO_2 is a long-lived

---

[2]Based on review of piston-powered Cessna models, using maximum range, fuel capacity, and assuming a minimum 10% fuel reserve. MPG figures for private jets are similarly derived.

species, quantifying the comparative contribution of powerfully warming but short-lived by-products and CO_2 has been an challenge. In 1999, the IPCC [263] published a special report on aviation and global warming that attempted to address this issue by calculating the radiative forcing from the beginning of commercial aviation up to 1992 attributable to different aviation-related emissions, and found the aviation-related forcing, in 1992, to be 2–4 times that attributable to CO_2 emissions alone (with 2.7 as a best guess). This multiplicative factor is referred to as the "radiative forcing index" (RFI). Several more recent estimates of the RFI are slightly lower, on the order of 2.0 [269], when excluding the highly uncertain but likely warming effect of aviation-induced cirrus (AIC); including AIC gives an RFI of 3.0 or more. Thus, in a crude sense, aviation as a system can be thought to have at least twice the global warming impact we would expect from the CO_2 emitted by jet fuel combustion alone.

We must understand, however, that the RFI reflects the *current* warming effect of a *past* emission stream, and it is *not* a multiplier we can apply to determine the future warming impact of a single plane trip (or even a single year of aviation), as is frequently done. Furthermore, the RFI is a single-year "snapshot" of warming from an emission stream, not an integration of that warming. The global warming potential (GWP), which is integrating and forward looking, is the more appropriate metric (global temperature potential is also an option). Looking at the *future* impact of a plane trip, the ratio of the GWP for all aircraft emissions to the GWP for CO_2 alone, which I refer to as the GWP_{ratio}, can be estimated as roughly 1.3–1.4 excluding AIC, or 1.8–2.0 including AIC [270], values appreciably lower than the RFI. The past-looking RFI and forward-looking GWP_{ratio} for air travel are compared in detail in Section 8.3.4. Either can be used to inform an uplift factor for aviation CO_2, but I generally view GWP_{ratio} as the more relevant, and have used a central estimate of 1.5 for all calculations concerning the overall impact of flying, unless otherwise stated.

The primary non-CO_2 forcers that give us an RFI > 1, as discussed below, are nitrogen oxides (NO_x) which increase tropospheric ozone (O_3) and decrease methane (CH_4), for a net warming effect; linear contrails resulting from jet exhaust that have a warming effect; and, finally, induced cirrus clouds that also have a warming effect, although the magnitude of this effect remains highly uncertain. More minor forcers include the water vapor and soot (black carbon) that result from jet fuel combustion; sulfate aerosols have been considered to have a minor direct cooling effect [269, 270, 39], but several recent works have suggested that sulfate-cloud interactions could induce major cooling [275, 276, 278], perhaps sufficient to outweigh the positive warming from contrails and cirrus.

Best estimates of radiative forcing (in mW m^{-2}) for each major forcer and 90% likelihood ranges, as integrated from 1940 to 2005, are shown in Figure 8.4, based primarily on Lee et al. [269], and supplemented by [39]. I discuss the particulars of the major forcers in the following sections.

8.3.1 Forcing from products of combustion

Jet fuel is formulated from kerosene, a relatively heavy distillate of petroleum, and is a mix of hydrocarbon chains around 10 to 16 atoms long, along with small amounts of sulfur. Ideally, combustion oxidizes hydrocarbons to yield only H_2O and CO_2, both of which are potent greenhouse gases. Within the troposphere, H_2O generated by combustion is overwhelmed by the hydrologic cycle, and although small amounts of aviation H_2O enter the lower stratosphere, the direct warming effect of aviation H_2O is nearly trivial. Since, like all petroleum fuels, jet fuel is contaminated with sulfur (about 0.05% sulfur [263]), this too is oxidized to sulfur dioxide (SO_2), which ages to form aerosols that reflect incoming solar radiation and exert a cooling effect. Incomplete combustion at high temperature yields soot, or black carbon (BC), which absorbs sunlight to warm the air. These forcers all have very short atmospheric lifetimes (days), and

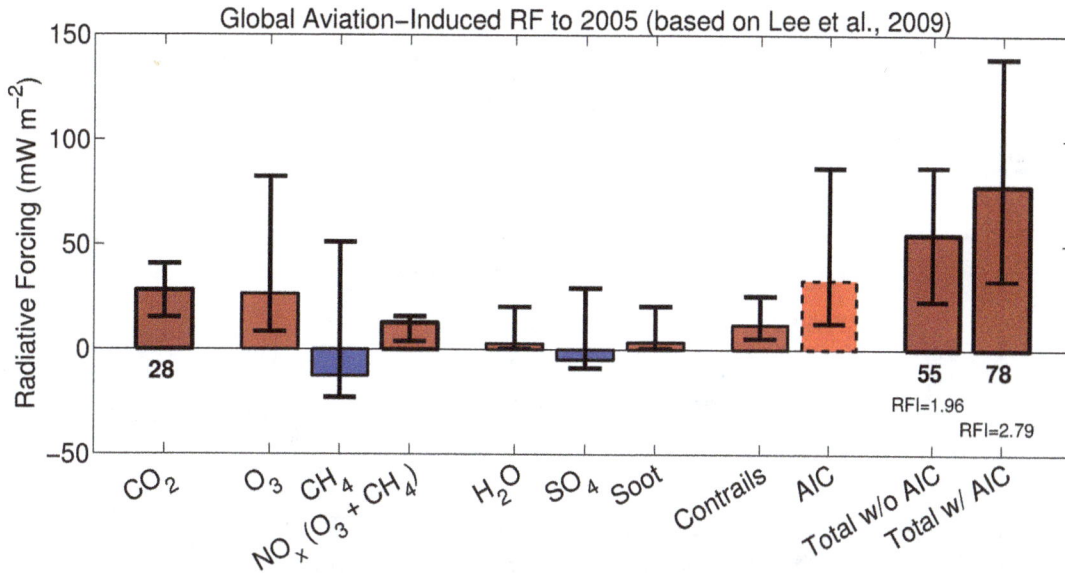

Figure 8.4: Radiative forcing attributable to major forcers resulting from commercial aviation, as integrated from the beginning of commercial aviation to 2005 [269]. Because aviation-induced cirrus (AIC) is poorly understood, it is often excluded from estimates of the RFI. Also note that the effect of NO_x is the sum of its downstream influences on O_3 and CH_4, and not an independent term. All forcers except CO_2 and CH_4 are extremely short-lived, exerting their warming or cooling effects over only days to months. Methane is also relatively short-lived, acting over a few decades versus centuries to millennia for CO_2.

so their direct warming/cooling effects are extremely short-lived. The direct warming/cooling effects of water, sulfates, and soot are all relatively small, but these molecules can interact to form or enhance contrails, cirrus clouds, and low-altitude liquid water clouds, as discussed in the following section. The effect on liquid water clouds has only been recently studied, but may be one of dramatic cooling.

Depending on combustion conditions, a small fraction of the nitrogen (N_2) in the ambient air is oxidized to NO_x. At the altitudes where jets fly, at the upper tropopause (60–80%) and lower stratosphere (20–40%) [263], conditions are such that NO_x emissions have a stronger warming effect than anywhere else in the atmosphere. NO_x has two major downstream effects that influence radiative forcing: increased ozone (warming) and decreased methane (cooling). The major mechanism for ozone creation is the catalytic oxidation of carbon monoxide (CO) to O_3 by NO_x. Secondarily, NO_x also catalyzes the oxidation of CH_4 to O_3 and CH_3, thus acting as both an ozone source and methane sink. Note that the background NO_x level strongly determines the effect of added NO_x on ozone production and destruction: at high background levels new NO_x actually increase ozone destruction, while at low levels, such as are encountered in the upper troposphere, NO_x emissions increase ozone concentrations almost linearly [263]. The radiative efficiency of ozone increases with altitude, and therefore not only does aviation NO_x very effectively produce ozone, but the ozone at aviation altitudes is also more warming than that nearer the land. The lifetime of ozone in the upper troposphere is only about one month, so the ozone-mediated warming attributable to NO_x is very strong but also very brief.

NO_x increases the background level of the hydroxyl radical (HO), which in turn decreases the lifetime of methane, lowering its concentration in the atmosphere. This lowered concentration in turn leads to higher HO levels, reducing methane further via a feedback loop [263]. This is the major NO_x cooling mechanism, and acts over the course of several decades, contra ozone.

207

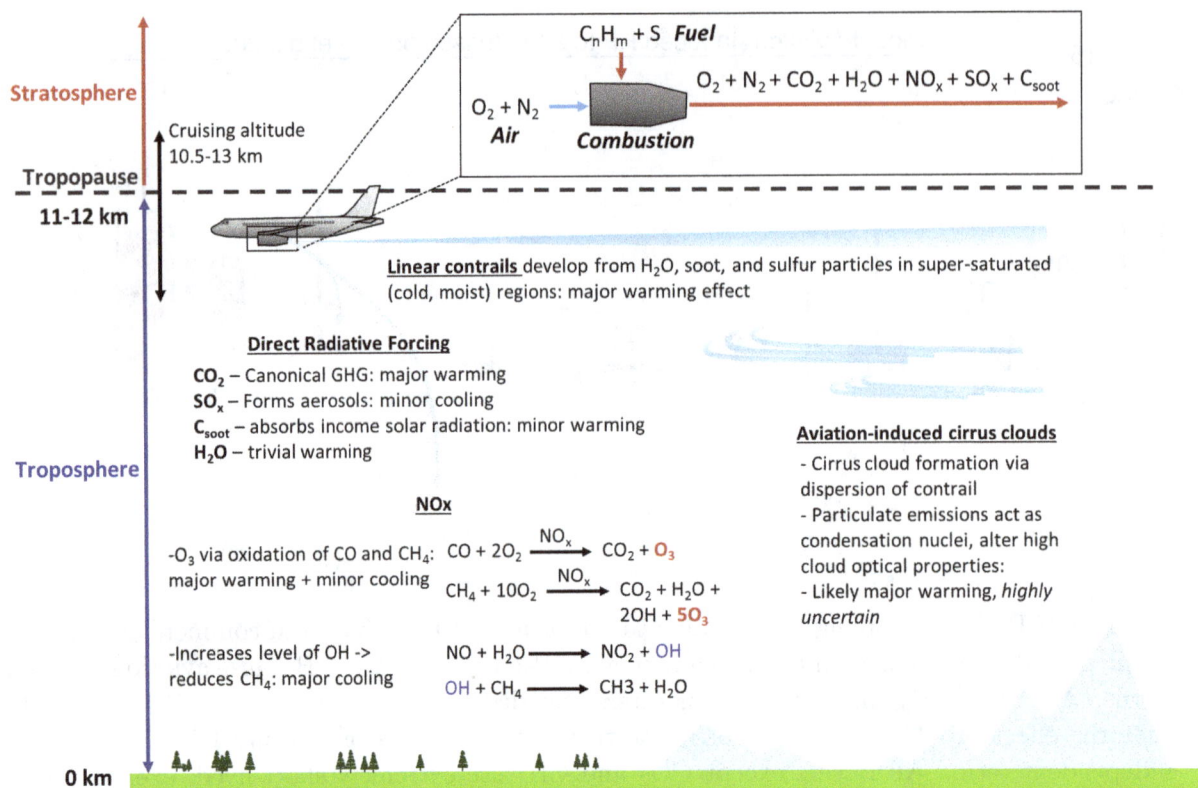

Figure 8.5: Schematic for the products of jet fuel combustion, and their direct and indirect downstream effects that lead to a mix of warming and cooling.

The products of imperfect jet fuel combustion are summarized graphically in Figure 8.5, as are their downstream warming and cooling effects.

8.3.2 Forcing from contrails and clouds

When the air is particularly cold (less than about -40 °C), as it is at the top of the troposphere, and the air is supersaturated with respect to water, the water contained in jet exhaust can condense into ice crystals to form **linear contrails**. If the air is supersaturated with respect to ice, these contrails may *persist* [268]. Contrails are optically similar to high cirrus clouds, also composed of ice crystals. Both reflect incoming solar radiation, but also reflect and absorb outgoing infrared radiation, for a net warming effect, although there is significant uncertainty as to the exact balance of these two factors [7, 263]. Persistent contrails can spread and promote cirrus cloud formation ("contrail cirrus," a major component of "aviation-induced cirrus"), but because such clouds lack a characteristic linear shape, there has been marked uncertainty in quantifying warming from AIC, although it is generally estimated to have a stronger effect than linear contrails. Aviation cirrus also decreases natural cloud formation. Nearly all water in contrails/AIC comes from the ambient air [271], implying less water available to other clouds, and this dehydrating effect may offset about one-fifth of contrail/AIC warming [272].

Other jet engine particulate emissions, principally soot and sulfur aerosols also act as nucleation niduses, enhancing ice crystal formation, as well as affecting their size distribution and optical properties. Thus, they not only promote contrails, but may induce cirrus clouds in the absence of contrails [272]. Note that the cirrus-enhancing effect of sulfur aerosols at high altitudes is opposite the strongly cooling effects of sulfur aerosol-cloud interactions at low altitudes.

The IPPC [263] estimated aviation-induced contrails to cover approximately 0.1% of the earth's surface in 1992, with a net radiative forcing of 20 mW m^{-2}; estimates for more recent years have been significantly downgraded, however, with Lee et al. [269] giving 11.8 mW m-2 for 2005, and the most recent IPCC assessment report uses 10 mW m^{-2} as its central estimate of contrail RF in 2011 [7].

Warming from AIC is highly uncertain; Lee et al. [269] give a range of 12.5 to 86.7 mW m^{-2} (best estimate 33 mW m^{-2}) for 2005. Several more recent estimates include a 2011 work by Burkhardt and Krcher [272], who estimated that contrails and contrail cirrus together had a net warming effect of 31 mW m^{-2}, and a 2015 estimate by Schumann et al., who give a combined 60 mW m2 [271]. Other recent estimates are not as large, with Chen and Gettelman [273], for example, giving only 13 mW m^{-2} for contrail cirrus (and just 3.1 mW m^{-2} from linear contrails).

8.3.3 Indirect aerosol effects

The indirect effects of sulfate and black carbon aerosols have not been extensively studied, but several recent studies have suggested that, by serving as nuclei for lower altitude liquid water clouds, sulfate (and secondarily, black carbon) aerosols could exert a strong cooling effect that may match the warming effect of (ice-crystal) contrails and AIC. Aviation aerosols can affect low-altitude water clouds either via emissions during climb and approach, or may be transported downward from cruise altitudes [277]. Contra cirrus clouds, liquid water clouds reflect more incoming solar radiation than outgoing longwave radiation, to yield a net cooling effect.

A 2013 climate model by Gettelman and Chen [275] gave RF due to sulfate, black carbon, and contrails/AIC as -46, +8, and +16 mW m^{-2}, respectively, with the net effect a cooling - 21±11 mW m^{-2}, and a later work by these authors [274] projected that, in 2050, sulfate aerosol cooling would more than outweigh warming from aviation contrail cirrus and black carbon. Righi et al. [276], also in 2013, estimated aviation aerosols (soot and sulfates) in the year 2000 to have an RF between -69.5 and 2.4 mW m^{-2}, across four scenarios, with -15.4 mW m^{-2} the forcing in the reference case. Kapadia et al. [278] estimated -23.6 mW m^{-2} via increased cloud albedo from aviation aerosols. It follows that this highly uncertain aerosol effect may be similar in magnitude, but opposite in sign, to the also highly uncertain AIC effect. These results also highlight the concern that the adoption of lower-sulfur jet fuels for health reasons could undermine aerosol cooling [278], and represent environmental problem-shifting (similar to the adoption of lower sulfur bunker fuel for international shipping, as discussed in Section 9.4.5).

While the qualitative conclusion of this chapter remains that the net climate effect of aviation emissions is likely warming greater than that of the CO_2 alone, these recent studies suggest that the extra warming may not be as great as initially thought, and the overall effect of non-CO_2 emissions could even be one of net cooling.

8.3.4 RFI vs. GWP

It is important to understand that the RFI is a *backward-looking* metric, as opposed to a forward-looking metric such as GWP. That is, while the GWP gives the integrated radiative forcing over some future time-span, the RFI tells us how much radiative forcing, *at the present moment*, can be attributed to a past emission stream. This is a very important distinction, and it weighes short-lived forcers more strongly than longer lived ones, especially if the emission stream is increasing in magnitude, as with aviation.

With the aviation emission stream, since about 1940 CO_2 has been building up in the atmosphere, continuously trapping heat, since, as a long-lived species, most of the CO_2 emitted in 1940 remains in the atmosphere. This is in contrast to most other aviation emissions, which are short-lived. Thus, while the RF for aviation-related CO_2 depends on the entire 75+ year history of aviation, the RF from short-lived forcers depends only on the current year (or decade, in the case of methane perturbations); these recent years are ones of historically high air travel, and are thus weighted higher.

The implication of the above is that a *forward-looking* radiative forcing index, which measures the net warming attributable to a single year of aviation, out to, say, 100 years, in relation to the radiative forcing from CO_2 alone, will be smaller than the backward-looking RFI. Looking forward in time is precisely the GWP concept, and we can derive two numbers, GWP_{total}, measuring the GWP of all aviation climate forcers out to 100 years, and GWP_{CO_2}, measuring the GWP of CO_2 alone, for a single year of flight (or alternatively, a single kg of fuel, a single flight, etc.). The ratio, $GWP_{total}/GWP_{CO_2} = GWP_{ratio}$, is loosely analogous to a forward-looking RFI, but will be smaller in magnitude, because CO_2 becomes relatively dominant later in time. Thus, it is not appropriate to use the RFI as a stand-in for GWP, as is sometimes done, especially when assessing the relative impact of a single air trip versus some other transportation mode.

Now, let us examine some actual numbers and emissions-streams. Figure 8.2 gives us global CO_2 emissions from 1940 to 2005 due to jet fuel consumption. From this time-series, we can then determine the magnitude of non-CO_2 forcers. Emissions of non-CO_2 forcers are determined from Fuglestvedt et al. [39] and Lee et al. [269], and from these we can derive radiative forcing perturbations for O_3, methane, contrails, and aviation-induced cirrus, as well as the more minor forcers (all in terms of jet $kgCO_2$ emitted). As seen in Figure 8.6, yearly radiative forcing tracks CO_2 closely, with about half attributable to non-CO_2 factors when AIC is excluded, and over two-thirds due to non-CO_2 factors when AIC is included, giving us an RFI of about 2.0–4.0.

Now, to get a quantitative feel for the future effect of individual flights, I run the following two computational experiments. In the first, suppose all aviation ceased after 2005, perhaps as a consequence of some zombie-related cataclysm. This is shown in Figures 8.6 and 8.7, and we see that radiative forcing from contrails and other positive but very short lived forcers instantly goes to zero, while the negative methane perturbation persists for a decade or two, causing RFI to transiently drop below 1, before exponentially relaxing to 1 with longer time.

Thus, RFI is seen to be fairly meaningless as a forward-looking metric, and we must sum the total warming effect of our experimental emissions streams: we integrate radiative forcing from the very beginning to 100-years post-aviation, and we see that our GWP_{ratio} is about 1.25 without AIC, and 1.75 with AIC, and excluding indirect aerosol-cloud cooling (the exact GWP_{ratio} varies with our estimates of contrail and AIC warming, as well as the exact effect of NO_x on O_3 and CH_4). Performing the more obvious, and relevant, experiment of a single year of constant aviation emissions, one gets GWP_{ratio}s of around 1.25–1.30 and 1.6–1.9, without and with AIC, respectively. Including, in turn, indirect aerosol-cloud cooling and maximum AIC radiative forcing, GWP_{ratio} may plausibly range from 0.75 to 2.5.

These estimates for GWP_{ratio} correspond well to those of Lee et al. [270], who derived GWP_{ratio} factors (using my parlance) of 1.3–1.4 without AIC, and 1.9–2.0 with AIC. While AIC has typically been excluded from final RFI estimates, it is probably more correct to account for at least some degree of AIC. Furthermore, indirect aerosol cooling is only recently being elaborated on, but is likely a nontrivial factor. Given these considerations, I have chosen to use 1.5 as my central estimate for GWP_{ratio} and the combustion CO_2 uplift factor.

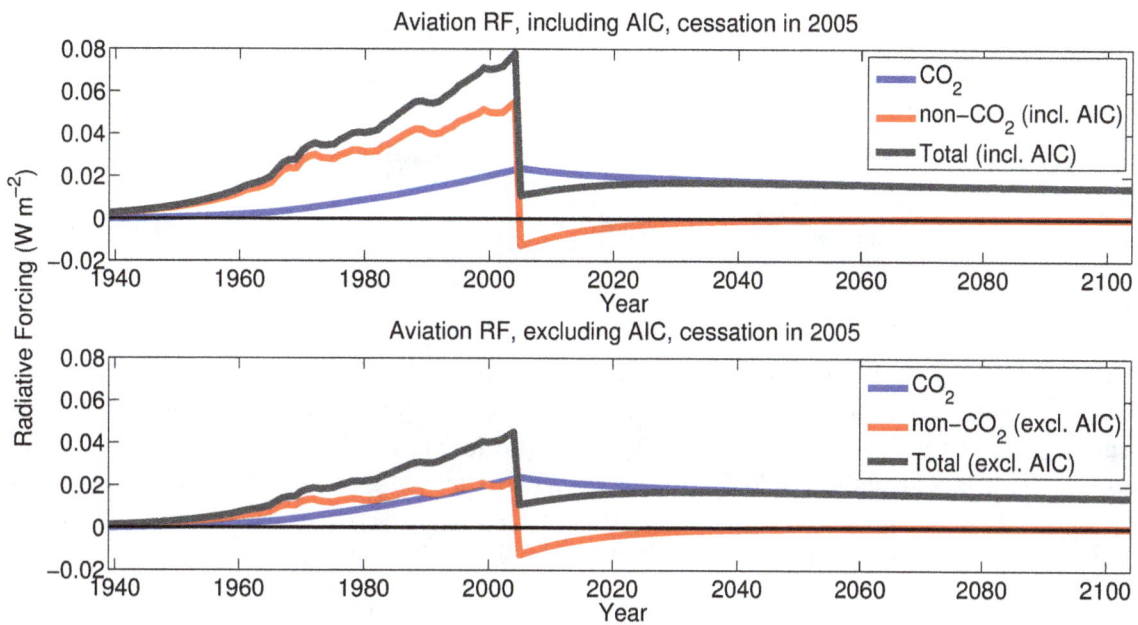

Figure 8.6: Total radiative forcing attributable to CO_2 and non-CO_2 forcers (as a lumped sum) due to commercial aviation, from 1940 to 2005, with a simulated cessation in emissions in 2005 onward. The top panel includes the uncertain effect of AIC, while the bottom excludes it.

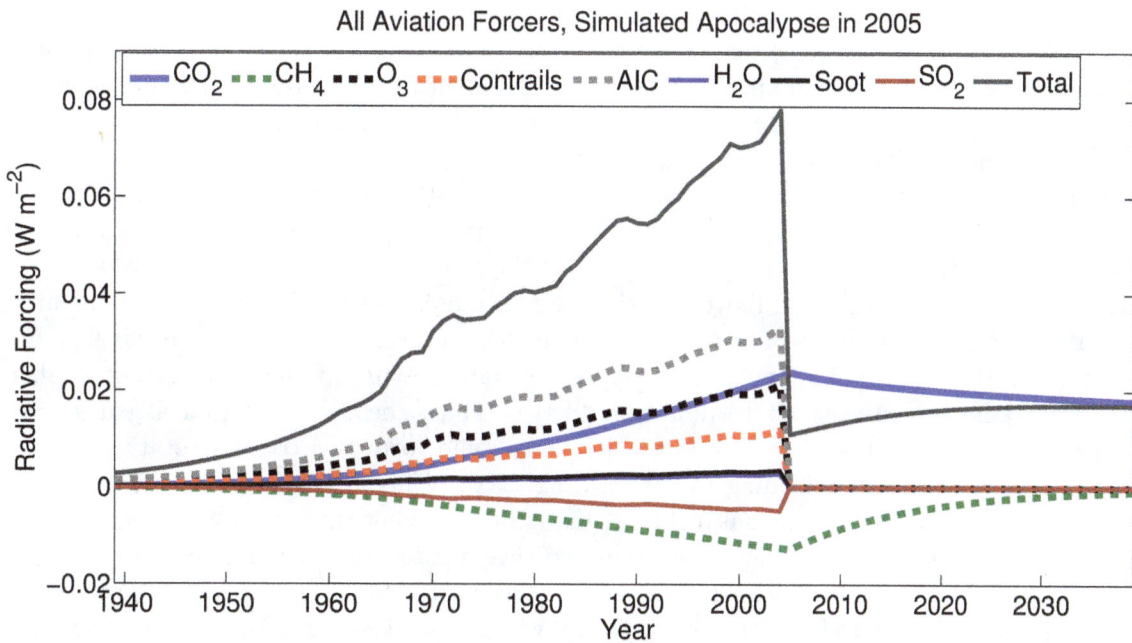

Figure 8.7: Approximate radiative forcing attributable to all major CO_2 and non-CO_2 forcers from commercial aviation, from 1940 to 2005, with a simulated cessation in emissions in 2005 onward.

Forces acting on aircraft at cruise

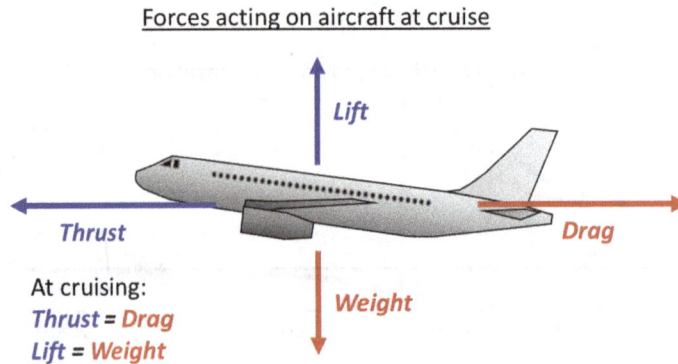

Lift

Thrust

Drag

At cruising:
Thrust = Drag
Lift = Weight

Weight

Figure 8.8: The four basic forces acting on an aircraft at cruise. *Drag* is directed along the x-axis, opposite the direction of flight, and is a combination of the parasitic drag (friction and form drag) and the lift-induced drag. *Thrust* is the forward force generated by the engines, and equals drag at constant velocity. *Weight* (recall that weight is properly understood as a downward force, equal to $M \times g$, where M is mass and g is the acceleration of gravity), is balanced by *lift*, the upward force generated by the flow of air over the wings.

8.4 Physical principles of aviation

8.4.1 Lift and drag

For an aircraft cruising at a constant speed and altitude, there are four basic forces acting in balance on the craft: (1) forward *thrust* (F) generated by the engine, (2) *drag* (D) from the air, (3) the *weight* (W) of the craft, and (4) *lift* (L) generated across the airfoil. Since velocity is constant, the forces must be in balance such that lift equals weight, and thrust equals drag. Let us focus on drag, the aerodynamic resistance to movement through air. *Parasitic drag*, which acts on all moving bodies (cars, ships, runners, etc.) is the sum of the *friction drag* resulting from the friction between the air and body surface and *form* or *pressure* drag that is created by changes in local air pressure induced by the moving craft (e.g. the low pressure wake); this is discussed in the context of automobiles in Section 7.1.5.

For an airplane, the generation of lift by the wings creates an additional drag force, referred to as *induced drag*. Very briefly, lift is generated because the pressure above an airfoil is lower than that beneath, yielding a net upward force. At the termination of the wing tip, however, the pressures must be equal. Thus, above the wing there is a pressure gradient of low pressure at the wing base increasing to atmospheric pressure at the tip, and this gradient induces air flow from tip to base. Below the wing, we have an opposite: high pressure at the base and atmospheric pressure at the tip, causing air to flow base to tip. Thus, these opposing air streams create vortex sheets (one from each wing) which destabilize and form two counter-rotating vortices, strongest at the wing tips, leading to vortex drag [281]. Winglets fitted to the wing tip, now fairly commonplace on commercial airliners, dissipate the wing tip vortices and hence reduce induced drag. Also, spatially separating these vortices via longer wings reduces induced drag, and in general the level of induced drag is inversely proportional to the wing aspect ratio (AR), the ratio of wing length to breadth (i.e., long thin wings induce less drag than short squat ones); one may consult Asselin [281] or other texts for a much more in-depth discussion.

Turning to the mathematics, our basic equation for drag, D, is

$$D = \frac{1}{2} C_D \rho A V^2,$$
(8.1)

where ρ is the density of air, A is the wing area, V is velocity, and C_D is our dimensionless drag coefficient. This drag coefficient, in turn, is determined as the sum of the so-called *zero-lift* drag coefficient, C_{D0}, and the induced drag coefficient (drag due to lift), C_{Di}:

$$C_D = C_{D0} + C_{Di}. \tag{8.2}$$

The induced drag coefficient is given by the expression

$$C_{Di} = KC_L^2, \tag{8.3}$$

where

$$K = \frac{1}{\pi \text{AR} e}, \tag{8.4}$$

and AR is the wing aspect ratio and $e \leq 1$ is the "span efficiency factor," on the order of 0.80 for commercial jets. Finally, the lift coefficient, C_L, is given as a function of lift (L):

$$C_L = \frac{2L}{\rho A V^2}. \tag{8.5}$$

Putting everything together, we can expand Equation 8.1 to our final equation for total drag force, as

$$D = \underbrace{\frac{1}{2}C_{D0}\rho A V^2}_{\text{parasitic drag}} + \underbrace{\frac{1}{2}K\frac{L^2}{\rho A V^2}}_{\text{induced drag}}. \tag{8.6}$$

Thus, drag decomposes into two-terms, the first of which (parasitic drag) increases with the square of velocity, while the second (induced drag) decreases with velocity squared. These two parabolic curves sum, as shown in Figure 8.10, to give an optimum velocity where drag is minimized. This is in sharp contradistinction to ships and automobiles, for which drag increases nonlinearly with velocity regardless.

8.4.2 Cruising altitude

Let us examine the effect of cruising altitude on drag and cruising fuel consumption. The density of air decreases by about 70% from sea level to 35,000 feet, a typical cruising altitude, as shown in Figure 8.11. It is intuitively obvious that this thinner air is easier to "punch" through, and so flight at altitude should be more efficient. Note from Equation 8.6, however, that while lower air density implies lower parasitic drag, it also implies higher induced drag. For typical aircraft parameters, the optimum cruising altitude corresponds to actual cruising altitudes, around 35–43,000 feet, as is also demonstrated in Figure 8.11.

8.4.3 The jet engine

Basics. The modern era of jet-powered commercial aviation began in the late 1950s, as new turbojet-powered aircraft replaced older propeller-driven aircraft. These new aircraft flew much faster and higher, and indeed, modern cruise speeds, about 500–575 MPH, and altitudes, 10.5–11.5 km, have not changed appreciably since. In essence, a jet engine acts to draw air in from the surroundings, accelerate it via the addition and combustion of fuel, and then use this accelerated air stream to drive the aircraft forward. This is physically realized via a gas turbine, whereby air is drawn in, compressed, mixed with fuel and combusted, with the resulting combustion gas passing through a turbine. Now, some energy is extracted from the gas stream by this turbine to drive the air compressor, while the rest has a variable fate. In the case of a *turbojet*, the gas stream passes through a nozzle, where thermal energy is converted to kinetic energy, producing

Lift-Induced Drag

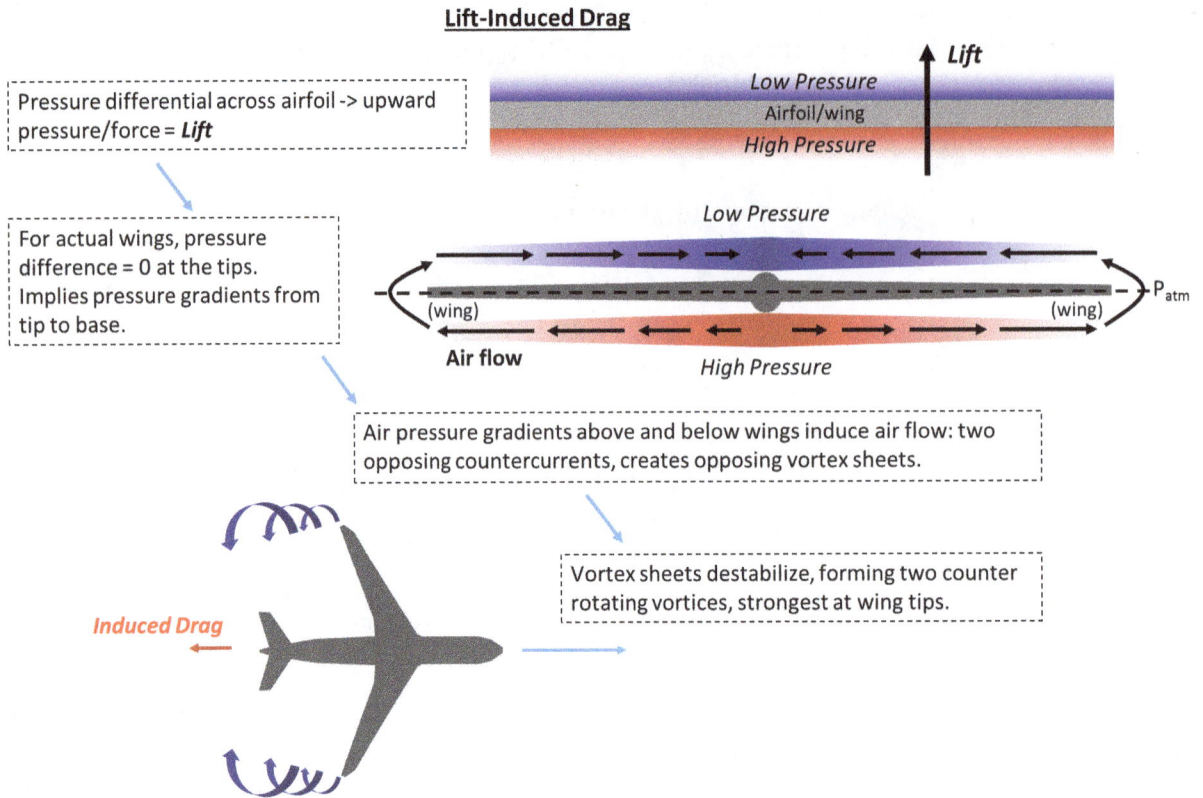

Pressure differential across airfoil -> upward pressure/force = **Lift**

For actual wings, pressure difference = 0 at the tips. Implies pressure gradients from tip to base.

Low Pressure
Airfoil/wing
High Pressure
Lift

Low Pressure
(wing)
Air flow
High Pressure
P_{atm}
(wing)

Air pressure gradients above and below wings induce air flow: two opposing countercurrents, creates opposing vortex sheets.

Vortex sheets destabilize, forming two counter rotating vortices, strongest at wing tips.

Induced Drag

Figure 8.9: Schematic for the physical basis of "induced-drag," a drag that is unique to aircraft and related to the "wing-tip vortices" induced by pressure gradients across the physical wings, as seen in the figure and discussed in the text.

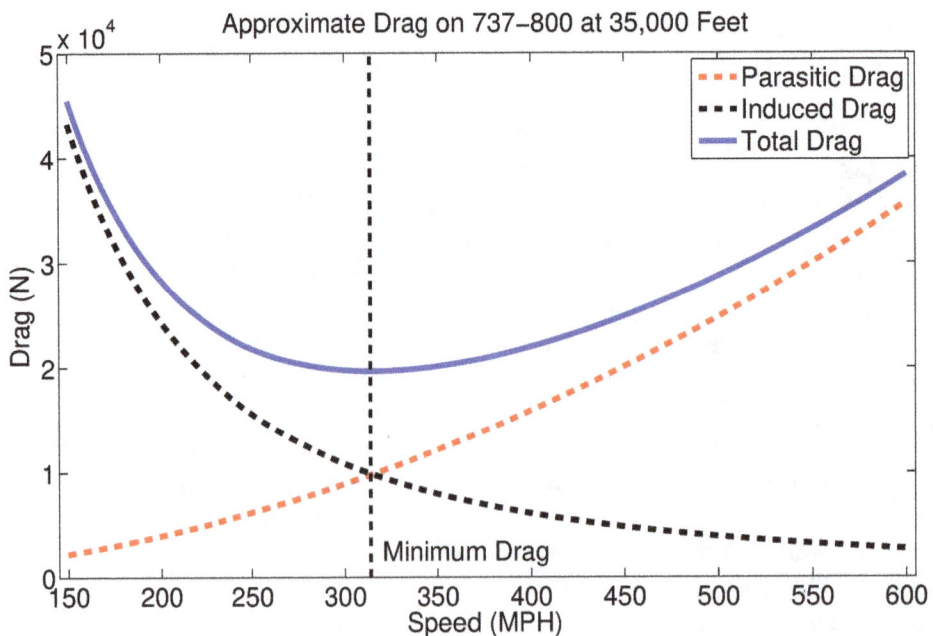

Figure 8.10: Approximate magnitude of parasitic and induced drag forces, as a function of cruising speed, for a Boeing 737-800 at 35,000 feet. Parasitic drag increases nonlinearly with speed, while induced drag falls nonlinearly with speed. An optimum that minimizes total drag is achieved at slightly over 300 MPH.

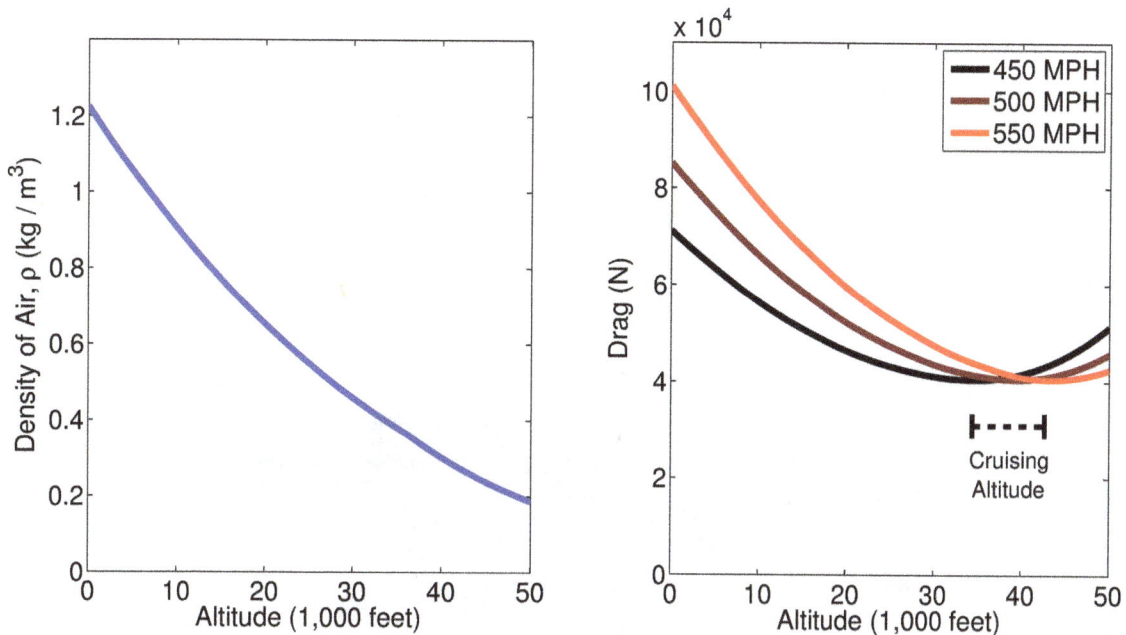

Figure 8.11: The left panel shows how the density of air decreases with altitude, while the right panel shows overall drag on a Boeing 737-800 as a function of altitude, when cruising at either 450, 500, or 550 MPH. Across typical cruising speeds, drag is minimized within the actual cruising altitude range of 35–43,000 range (10.5–13 km), with higher cruising speeds favoring a slightly higher altitude.

a high speed exhaust jet. With a *turboprop*, on the other hand, nearly all the energy in the gas stream is used to drive the turbine, which then drives a shaft connected to a propeller.

Finally, a *turbofan*, a kind of hybrid of the former two, is used on most modern commercial airliners. With this design, we have a fan in front of the core compressor and turbine. Some air passes into the core engine, while a much larger fraction *bypasses* the core engine through the fan. The core engines acts like a turbojet but with a second turbine that extracts some mechanical energy from the gas stream to drive the fan. Thus, the air stream exiting the core turbojet is slowed, while a much larger volume of surrounding air is also accelerated. The advantage to this hybrid design is that propulsion efficiency is highest when the velocity of the jet air stream exiting the engine is nearest the airplane air speed. Pure jet engines produce a very fast jet stream, and so are inefficient at typical cruising speeds. The turbofan design slows the stream, thus increasing efficiency. Turbofans are characterized by the *bypass ratio*, the ratio of air that bypasses the core engine to that entering, with higher bypass ratios implying greater efficiency. Turboprops are analogous to extremely high bypass ratio turbofans, but are generally only used for short trips, because these engines are maximally efficient at 200–400 MPH.

Basic mathematics. For the interested reader, a more in-depth discussion of the thermo-dynamic cycles and efficiencies involved in automobile engines is presented in Chapter 7, but my discussion on gas turbine efficiency here is more limited. In the most general terms, a jet aircraft moving at airspeed V_a, moving a stream of air through its engine at a mass rate $\dot{m}$ (in kg of air per second), with jet exit stream velocity V_{jet}, creates a change in air momentum, i.e. a *thrust* (F), given as

$$F = \dot{m}(V_{jet} - V_a). \tag{8.7}$$

Note that we necessarily have $V_{jet} \geq V_a$. Since power is force times velocity, propulsive power

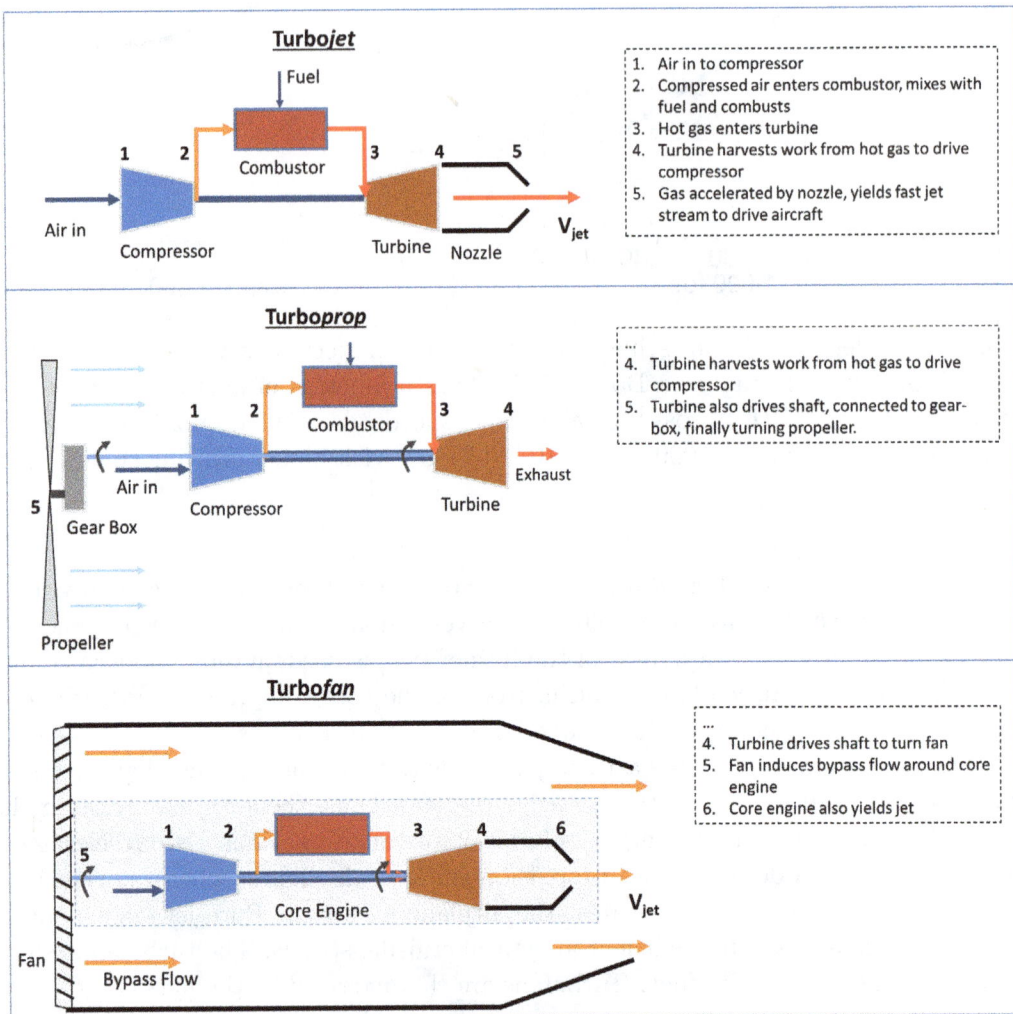

Turbojet

Fuel

1 2 Combustor 3 4 5

Air in

Compressor Turbine Nozzle V_{jet}

1. Air in to compressor
2. Compressed air enters combustor, mixes with fuel and combusts
3. Hot gas enters turbine
4. Turbine harvests work from hot gas to drive compressor
5. Gas accelerated by nozzle, yields fast jet stream to drive aircraft

Turboprop

1 2 Combustor 3 4

Air in

5 Gear Box Compressor Turbine Exhaust

Propeller

...
4. Turbine harvests work from hot gas to drive compressor
5. Turbine also drives shaft, connected to gearbox, finally turning propeller.

Turbofan

5 1 2 3 4 6

Core Engine V_{jet}

Fan

Bypass Flow

...
4. Turbine drives shaft to turn fan
5. Fan induces bypass flow around core engine
6. Core engine also yields jet

Figure 8.12: Schematics for the basic physical operation of the three classes of gas turbine engines.

is given simply as

$$P = F \times V_a = \dot{m}(V_{jet} - V_a)V_a. \tag{8.8}$$

Now, let us briefly examine the efficiency of jet engines in converting chemical fuel energy into aircraft motion. The overall efficiency, η, is given as

$$\eta = \eta_t \times \eta_p, \tag{8.9}$$

where η_t and η_p and are the thermal and propulsive efficiencies. First, the thermal efficiency is the fraction of potential energy contained in the jet fuel that is converted to kinetic energy in the gas stream exiting the engine, and it is on the order of 0.40–0.50 [263]. The fundamental thermodynamic cycle at work in a gas turbine is the *Brayton* cycle, and one may consult either Chapter 7 of [263] or [54] for further details. Putting that aside, if $\dot{m}$ is the mass of air passing through the engine (kg/s), then the difference in kinetic energy between inlet and and outlet is

$$\frac{1}{2}\dot{m}(V_{jet}^2 - V_a^2) \tag{8.10}$$

Since the fuel potential energy is simply $\dot{m}_f \times h$, where h is the heat content of fuel (42.8 MJ/kg) and $\dot{m}_f$ is the fuel consumption rate (kg/s), it follows that

$$\eta_t = \frac{\frac{1}{2}\dot{m}(V_{jet}^2 - V_a^2)}{\dot{m}_f h}. \tag{8.11}$$

Note that V_{jet} and $\dot{m}$ are not independently varying parameters: supposing η_t is roughly constant, then the larger V_{jet} is, the smaller $\dot{m}$. That is, we can either dramatically accelerate a small quantity of air (e.g. in the case of turbo*jets*) , or modestly accelerate a large parcel of air (as with turbo*fans*).

Turning now to propulsive efficiency, η_p, this is the fraction of kinetic energy in the gas stream that is converted into forward propulsive power. That is,

$$\eta_p = \frac{FV_a}{\frac{1}{2}\dot{m}(V_{jet}^2 - V_a^2)} = \frac{2V_a}{V_{jet} + V_a}. \tag{8.12}$$

Thus, we see that the closer the jet stream velocity is to the aircraft's air speed, the higher our propulsive efficiency. Thus, a turbofan, which moves a large amount of air at low velocity ($\dot{m}$ large, V_{jet} small) is much more efficient than a turbojet, for subsonic flight. Of note, the supersonic Concorde, which was in service until 2003 and capable of Mach 2.04 (1,354 MPH), used turbojets, a sensible choice in light of this discussion. Figures 8.13 and 8.14 summarize these mathematics in a more graphical format.

To sum up, for a modern high bypass turbofan, thermal efficiency is about 0.40–0.50, propulsive efficiency is around 0.80, and, including efficiency losses from the fan and turbine, overall efficiency finally evaluates to 0.30–0.37 [263]. For comparison, this overall efficiency is about twice that of a conventional internal combustion automobile, and comparable or modestly superior to a hybrid-electric vehicle.

8.4.4 From physical principles to relevant results

The flight cycle and trip length

Fuel burn and non-CO_2 radiative forcing both vary with trip length, cruising altitude, and aircraft payload. Fuel burn can be determined from the basic physical parameters of any given flight, and is affected by aircraft characteristics, cruising speed and altitude, payload, and trip

Jet Propulsion

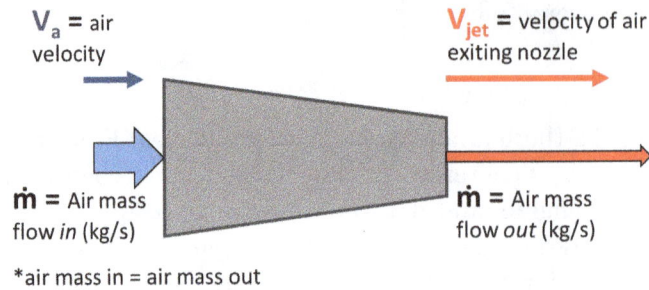

Change in air momentum = **Thrust** = F = $\dot{m}$ (V$_{jet}$- V$_a$)

Propulsive power = F × V$_a$ = $\dot{m}$ (V$_{jet}$- V$_a$) V$_a$

Figure 8.13: Basic mathematics for jet propulsion.

Basic Mathematics for the Efficiency of Jet Propulsion

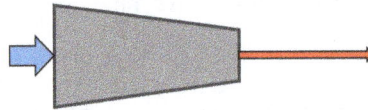

Kinetic energy in < Kinetic energy out

Change in kinetic energy of air passing through engine = ½ × $\dot{m}$ (V$_{jet}$²- V$_a$²)

Thermal efficiency (η_t) = fraction of potential energy contained in fuel that is converted into kinetic energy.

$$\eta_t = \frac{½ \times \dot{m}\,(V_{jet}^2 - V_a^2)}{\dot{m}_f \times h}$$

$\dot{m}_f$ = fuel mass consumption

h = heat content of fuel

Propulsive efficiency (η_p) = fraction of propulsive kinetic energy produced that is transformed into forward propulsion, *i.e. the fraction of power produced by the engine imparted to the aircraft, rather than the exhaust gas.*

$$\eta_p = \frac{F \times V_a}{½ \times \dot{m}\,(V_{jet}^2 - V_a^2)} = \frac{2\,V_a}{V_{jet} + V_a}$$

Where Propulsive power = F × V$_a$ = $\dot{m}$ (V$_{jet}$- V$_a$) V$_a$

Overall efficiency (η) = fraction of potential energy contained in fuel transformed into propulsive power
= **thermal efficiency × propulsive efficiency**

$$\eta = \frac{F \times V_a}{\dot{m}_f \times h} = \eta_t \times \eta_p$$

Figure 8.14: Basic mathematics for the efficiency of jet propulsion. Note that η_t is roughly constant, and so any increase in V_{jet} implies a corresponding decrease in $\dot{m}$.

length. Let us first consider the effect of trip length, which is the primary parameter the individual passenger has direct control over (that, and their own weight). Broadly speaking, journeys are divided into the landing/takeoff (LTO) cycle, and the cruise. The reference LTO is defined as takeoff (engines at 100% power for 0.7 minutes), climb (85% power, 2.2 minutes, up to 3,000 feet), the approach (30% power, 4 minutes), and ground taxi/idle (7% power, 26 minutes) [263]. For any given trip, LTO cycle fuel consumption will be similar, while cruising fuel burn obviously varies with overall journey distance. It follows that the fuel consumed per mile is highest for very short journeys. However, longer flights require more fuel, and hence more mass to transport, which eventually outweighs the advantage of a longer cruise and implies that optimum flight distances are in the medium-haul range.

We can calculate the fuel burned by any prescribed flight by calculating the fuel used during several distinct segments:

1. Takeoff and acceleration to cruising speed.

2. Climb to cruising altitude, during which air density drops.

3. Constant cruise at altitude.

4. Descent and deceleration, during which air density increases.

5. Final powered approach.

6. Ground idle/taxing (calculated from the specified thrust for a given aircraft engine, the engines thrust-specific fuel consumption, and the assumption of 7% engine power for 26 minutes of idle/taxing).

This algorithm, executed using publicly available parameters for two aircraft, the Boeing 737-800, a two-engine narrow body jet widely used for short- to medium- range domestic air travel that can seat up to 184 passengers, and the Boeing 747-400, a wide-body, four-engine jet capable of seating up to 400 and used for long-haul flights, is used to inform the following discussion. Both these aircraft are relatively new and in widespread use, and I have chosen them because they represent good, but not state-of-the-art, fuel economy. All calculations are only approximate, but the purpose here is to derive general principles combined with reasonable, not perfect, emissions estimates.

The gasoline-equivalent seat-MPG, and the associated fuel-cycle CO_2e emissions, for a Boeing 737-800 and Boeing 747-400 as a function of trip distance are given in Figure 8.15. Note that I assume every seat is filled, and that each passenger and their luggage averages 100 kg in weight. From the figure it is clear that, for a particular aircraft, fuel consumption per mile will decrease with trip length, but only to the point where the increased fuel load required begins to outweigh the efficiency advantage of a longer cruise time. Indeed, medium-range (roughly 750 to 2,500 mile) trips are more efficient than either short hops or long-haul international flights. However, there is very little difference in fuel consumption between, say, a 2,000 and 3,500 mile trip taken in a Boeing 737, and the lower fuel economy of long-haul flights that is realized in practice appears to have more to with the characteristics and seat configurations of the larger wide-body jets used for this purpose.

Note from Figure 8.15 that seat-miles per gallon, as is obvious, depends on how many seats one packs into the plane, with a 737-800 all-economy seating arrangement (with 184 seats, the maximum), using about 11% less fuel than a two-class arrangement (e.g. 16 first-class and 146 economy seats, totaling 162). Similarly, the actual *passenger-miles* per gallon achieved depends both on the load factor, and on the seating arrangement.

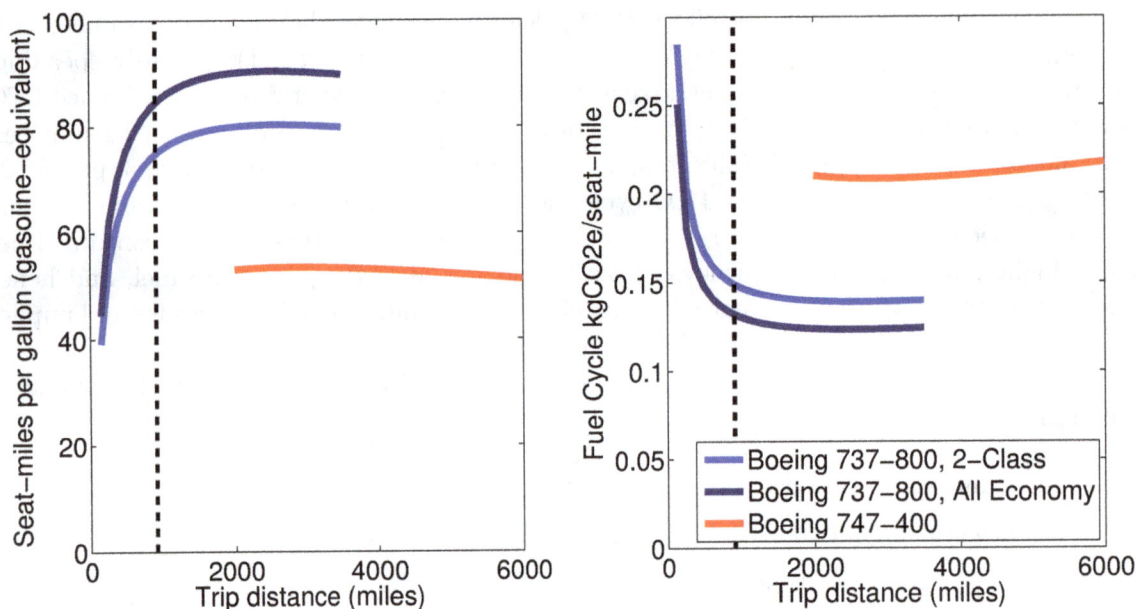

Figure 8.15: Seat-miles per gallon and fuel-cycle emissions as a function of trip distance for three different aircraft: (1) Boeing 737-800 with 162 seats (two-class seating), (2) Boeing 737-800 with 184 seats (economy-only seating), and (3) a Boeing 747-400 with 400 seats. The dotted line gives the average US flight distance in 2014, 912.4 miles [265]. For a particular aircraft, very short trips are highly inefficient, while the longest trips see a slight efficiency drop.

Weight and fuel consumption

- Over half the weight of a loaded aircraft comes from the empty weight of the craft itself, while the fuel load is anywhere from 5 to 30% of total weight.

- It follows that additional passenger and/or cargo weight has a very small effect on total fuel use, and a Boeing 737 carrying its maximum payload uses only about 15% more fuel than an empty Boeing 737.

- The marginal fuel-cycle emissions cost of 100 kg of passenger and luggage is about 20 gCO_2e/mile (or about 30 gCO_2e/mile including non-CO_2 forcers), for a Boeing 737.

Depending on fuel load, over half to two-thirds the weight of a fully loaded aircraft is that of the aircraft itself, and perhaps another quarter is fuel. For example, the empty weight of a Boeing 737-800 is 41.4 tonnes, while the maximum payload (passengers + cargo) and maximum fuel load are 20.3 and 20.9 tonnes, respectively. Thus, passengers and cargo add relatively little to overall weight, and it follows intuitively that additional passengers affect fuel consumption but little. Figure 8.16 demonstrates the fractional breakdown of empty weight, fuel, passengers, and extra cargo for several fully loaded aircraft and trip lengths.

Now, for an aircraft any additional weight, in the form of passengers or cargo, will increase fuel burn both as a direct result of the added cargo load, and from the weight of the extra fuel needed to move this cargo (and the extra fuel needed to move this extra fuel weight, and so on *ad infinitum*). One can solve for this increased fuel burn (and required initial fuel load) iteratively, from zero to maximum payload; Figure 8.17 shows how fuel burn increases from zero to maximum payload for two example trips. Across plane types and trip distances, a fully

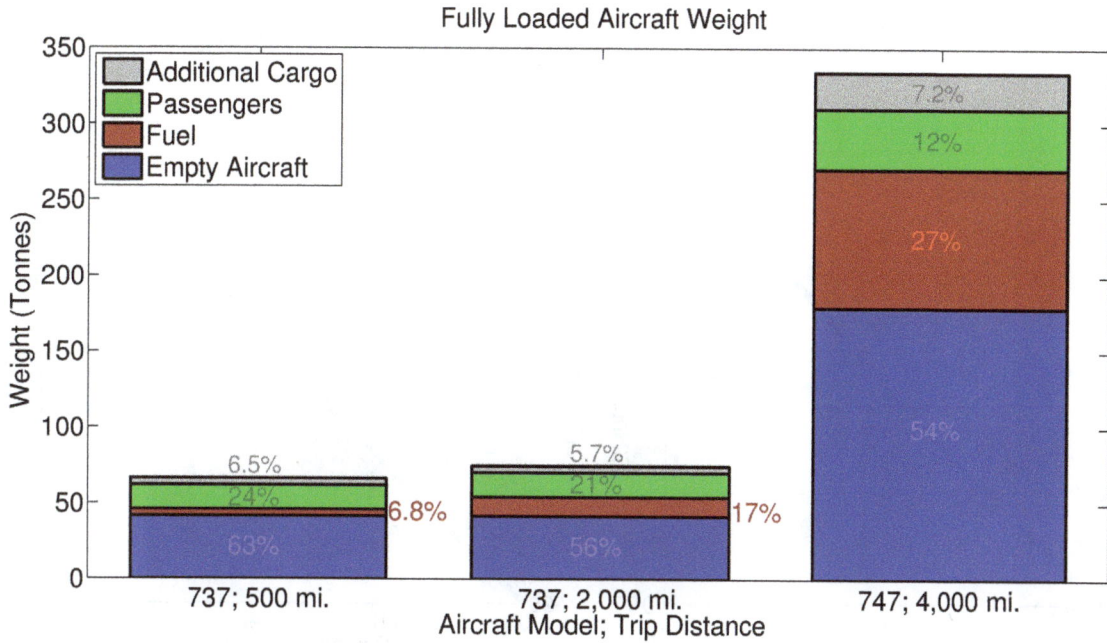

Figure 8.16: Weight of several aircraft loaded to maximum payload, divided into empty aircraft weight, fuel, passengers, and additional cargo. The Boeing 737 and 747 are assumed to carry full loads of 180 and 400 passengers, respectively, with each passenger and luggage weighing 100 kg; belly cargo makes up the difference between passenger and maximum payloads. Fuel may make up as little as 5% of total weight for short-distance trips, or over 30% for long-haul flights; empty operating weight dominates regardless.

loaded airplane uses only 15–20% more fuel than one empty.

Now let us examine how overall fuel burn increases with the passenger load, and how this affects both the seat-miles per gallon and passenger-miles per gallon. I again assume a typical passenger with luggage weighs 100 kg, and I also assume no additional belly cargo for these simulations. Figure 8.18 shows how these metrics change as passenger count increases from 90 to 180 in a Boeing 737-800 (with 180 seats total). As can be seen, the emissions per seat-mile increase only by about 7% with a doubling in the passenger load, while emissions per passenger-mile decrease by 46%.

It is obvious by now that more passengers per plane implies overall fuel savings. We might, however, still ask ourselves how much additional fuel is consumed as a *direct* consequence of one's being on board, as opposed to an empty seat in our place. Well, using the same simulations above, I calculate that the marginal increase in emissions when adding a single 100 kg passenger is generally in the range of 18 to 27 gCO_2e per airplane-mile. In other words, as a rule of thumb one could assume that their own weight increases aircraft fuel-cycle emissions by about 0.02 $kgCO_2e$ per mile travelled, equivalent to 0.03 $kgCO_2e$ once non-CO_2 forcers are accounted for. Using the latter, the marginal cost of a 2,000 mile journey would be 60 $kgCO_2e$, equivalent to 5.4 gallons of gasoline. For comparison, a 2,000 mile journey in a hybrid vehicle getting 50 MPG would use 40 gallons of gasoline, about a 7.5-fold difference. This, however, is not really the correct comparison to make, given the high load factor and demand-driven supply in commercial aviation, as I discuss in Section 8.2.1.

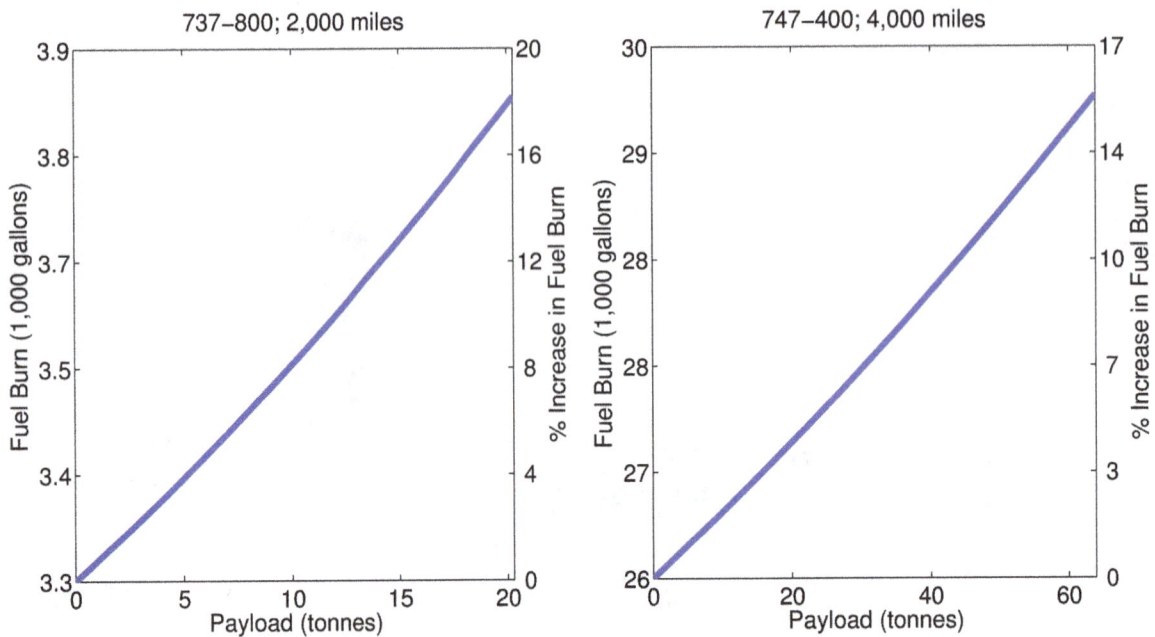

Figure 8.17: Fuel burn as a function of payload (up to the maximum payload), for 2,000 and 4,000 miles trips in a 737 and 747, respectively. The left y-axis gives the absolute fuel burn (in 1,000 gallons), while the right y-axis shows the percentage increase from an empty craft. As can be seen, fuel burn increases only by about 15% moving from an empty to full aircraft.

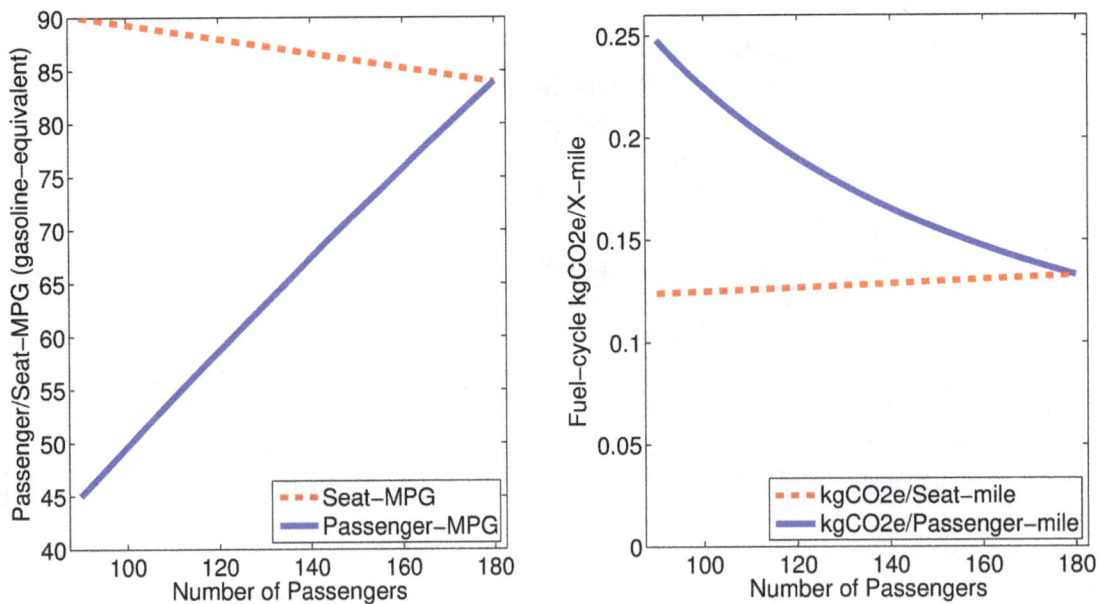

Figure 8.18: Seat-miles per gallon and passenger-miles per gallon, given in terms of gasoline-equivalents, along with the associated fuel-cycle emissions, for a 737-800 carrying from 90 to 180 passengers (each, with their luggage, weighing 100 kg). Doubling the number of passengers increases total fuel burn by 7%, hence a 7% increase in emissions per seat-mile. However, this slight increase in fuel burn is essentially trivial, and passenger-miles per gallon nearly double with a doubling in passengers.

Impact of first vs. business vs. economy class

Finally, I simply note that, based upon publicly available seating maps, each first class seat takes up as much space as about two economy-class seats; business-class seats equal about 1.5 economy seats. It follows that first-class also has roughly double the climate impact. Therefore, I would advise against indulging in the relatively minor (it seems to me) comforts of first class, and furthermore, patronize carriers that employ all-economy seating arrangements.

Chapter 9

Public transportation, infrastructure, and freight

9.1 Public and mass transportation

Public and non-public mass transportation (e.g. private bus travel) account for a trivial portion of household transportation emissions, barely amounting to a rounding error [2], and only a tiny fraction of Americans use these transport modes with any regularity [642]. The 2009 NHTS reported that only 1.9% of all personal travel was via public transit, and although commuting to/from work was the most significant subcategory, transit still only accounted for 5.1% of these trips. Federal highway statistics suggest that about 2.8% of all passenger miles are via transit, Amtrak, and long-distance motorcoach combined. The National Transit Database reported 10.366 billion public transit trips in 2015, a large number, but nearly trivial next to the 233.8 billion household vehicle trips reported in the 2009 NHTS.

Despite these diminutive numbers, it is important to understand the energy and emissions associated with various transit modes for at least two major reasons: (1) public transit can be a locally important transportation option, especially in dense cities, and (2) public transit is widely promoted as a low-carbon, sustainable alternative to the personal vehicle-based transport system, and so we must be sure this underlying hypothesis is correct if we are to rationally plan at a policy level.

While in Chapter 6, I mainly focus on emissions at the single driver level, in comparing mass transit to personal vehicles it is important to consider actual vehicle occupancy. This is because transit emissions, on a passenger-mile basis, are extremely sensitive to the load factor, or occupancy rate, and a proper comparison between personal vehicles and transit takes the occupancy of either mode into account. Federal highway statistics suggest about 1.38 person-miles per vehicle-mile across light-duty vehicles, while the 2009 NHTS reports vehicle occupancy (in person-miles/vehicle-miles) of 1.13 for commuting trips, and an average of 1.67 across all daily travel. At a fleet-average of 21.6 MPG, and occupancy rate of 1.38, we have an overall average of 0.3739 $kgCO_2e$/passenger-mile in fuel-cycle emissions (or 0.2981 $kgCO_2$/passenger-mile in tailpipe emissions only). Thus, the average passenger-miles per gallon achieved by the light duty fleet is about 29.8 PMPG. As seen below, this almost identical to the per-passenger emissions achieved by metropolitan transit bus systems, while rail-based transit systems generally get better than 50 PMPG.

In this section, I mainly review the *use-phase* emissions (i.e. fuel-cycle emissions for vehicles powered by liquid fuels, and lifecycle electricity emissions for electrified rail) of the two major mass transit modes in use, buses and rail, looking separately at long-distance (i.e. motorcoach and Amtrak) and local area (e.g. transit and school buses, and light rail) transit modes.

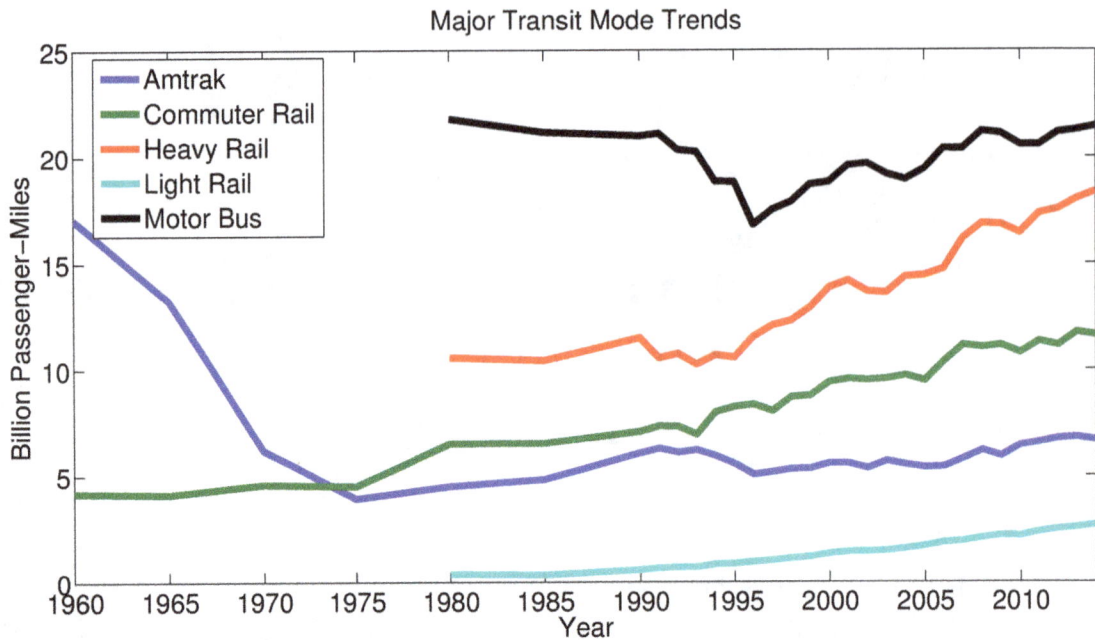

Figure 9.1: Trends in passenger-miles travelled via different transit modes over the last few decades. Source: BTS.

The emissions attributable to transit infrastructure, and transportation infrastructure more generally, are examined further in Section 9.2, a grand summary of emissions related to personal transportation modes is given in Section 9.3, and the chapter is rounded out with a brief discussion of freight transport.

9.1.1 Overview of transit modes

Restricting our discussion to public transit systems, nearly all passenger miles are travelled via bus and commuter or heavy rail, with light rail still a relatively minor but growing component. Figure 9.1 summarizes recent trends in transit ridership, and shows total trips and passenger miles for the major modes. Figure 9.2 compares approximate carbon emissions for metropolitan bus, heavy rail, commuter rail, and light rail, alongside emissions for typical and best-case passenger-vehicle configurations, on an achieved passenger-mile basis, and on a seat-mile basis (i.e. emissions under full ridership). As seen, under typical ridership levels, buses are only slightly better than cars, while rail is markedly superior. As for long-distance transportation options, motorcoach and Amtrak are both significantly superior to air travel and typical vehicle configurations on both a passenger- and seat-mile basis, as discussed later.

9.1.2 Buses

Buses are perhaps the canonical form of communal transit, with the stereotypical image of bus transportation as one of an inconvenient and unreliable form of transportation, utilized only by those without financial (or perhaps geographic) access to alternatives, or by those rare persons both noble and foolish enough to sacrifice the convenience of the automobile for some greater good. Furthermore, because public transit systems are often under-utilized and must serve routes with low ridership, the overall ridership for bus transit can be quite low, leading some to question whether this is even a "green" mode of transportation at all, given the large amounts of fuel that must be used to drive often mostly empty buses. Here I examine both the potential

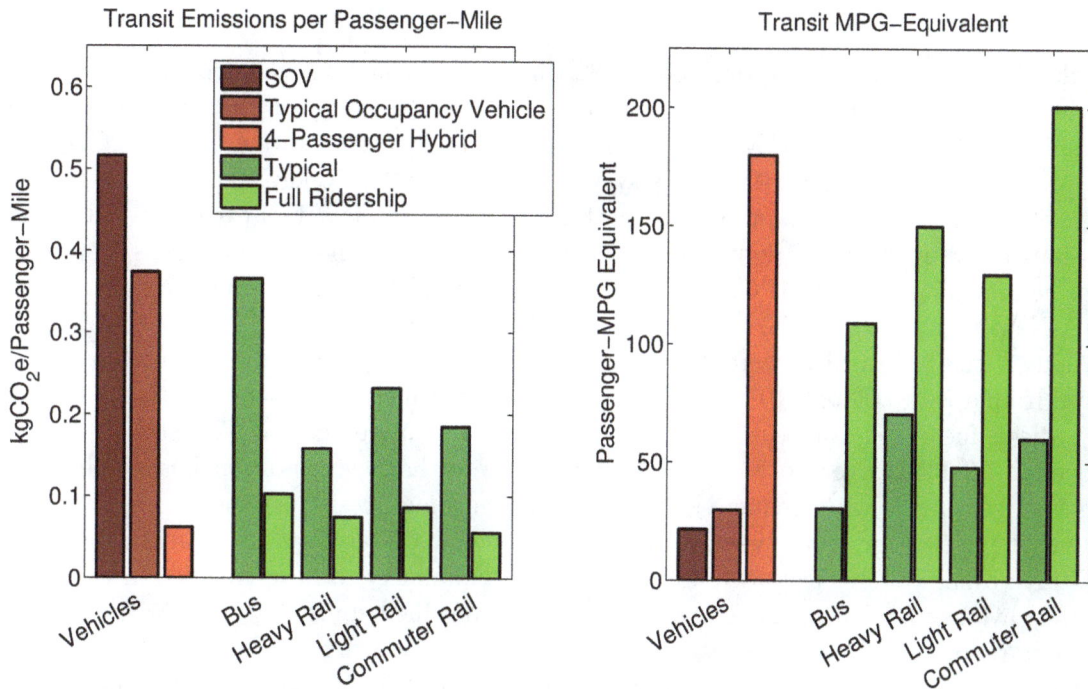

Figure 9.2: The left panel summarizes approximate fuel-cycle emissions on a passenger-mile basis for common transit options, under either typical ridership or full ridership. These are based on national averages, and individual systems may be significantly better or worse. For comparisons, emissions are shown for typical single-occupancy vehicles (getting the fleet-average 21.6 MPG), typical occupancy vehicles (1.38 passengers), and a nearly best case scenario of a 4-passenger hybrid electric vehicle getting 45 MPG. The right shows the same data but on a $PMPG_{GWP}$ basis. Rail transit is markedly superior to passenger vehicles under typical conditions.

and actual emissions of the major bus systems in existence in the US.

Buses are the workhorses of three major classes of transportation: metropolitan transit, longer-distance motorcoaches ("intercity" bus), and the oft-ignored school bus system, which is actually the largest of all three. Metropolitan transit buses are characterized by low fuel efficiency (mainly attributable to driving conditions) and very low ridership, and as a result, the CO_2 emissions per passenger-mile are indeed only marginally better than those of the light duty vehicle fleet. Long-distance motorcoaches and school buses, on the other hand, are among the most efficient transport systems there are, achieving passenger-MPGs in the hundreds. School buses probably reduce light duty vehicle miles and fuel consumption more than any other transit system.

Metropolitan bus

A Federal Transit Administration report [282] thoroughly characterized the (direct) emissions from a number of metropolitan transit systems. In 2008, across 412 bus systems in the US, ridership (i.e. seat occupancy) was 28%, and CO_2 tailpipe emissions were 0.0803 $kgCO_2$e/seat-mile (or 3.2 MPG-diesel for a 40-seat bus), translating into 0.2916 $kgCO_2$/passenger-mile, and equivalent to a single occupant vehicle getting 30.5 MPG (on a fuel-cycle, or WTW basis, emissions are roughly 0.3657 $kgCO_2$e/passenger-mile). Out of the 50 largest bus systems, the worst achieved just 16.2 $PMPG_{GWP}$, the best got 154.3 $PMPG_{GWP}$, and the largest system (MTA New York City Transit) checked in at 34.7 $PMPG_{GWP}$.

Aside from a few outliers, across almost all bus systems tailpipe CO_2 emissions per seat-mile ranged from about 0.05 to 0.10 $kgCO_2$, suggesting most transit buses achieve just 2.5–5.5 MPG (on a diesel-fuel basis, and assuming 40 seats per bus). Note that FHWA statistics suggest an overall 7.2 MPG average for buses, but it is likely that metropolitan buses operating in congested conditions with frequent starts and stops get much less than the average.

Out of the 50 largest bus systems, the highest occupancy rate was 58%, a clear outlier, and only three other systems exceeded 40%. Sadly, this is unlikely to have changed in the last few years, as average bus occupancy rates have actually declined somewhat: the National Transit Database reported that the bus load factor, expressed as passenger-miles travelled per vehicle revenue mile, had declined to 10.1 in 2015, down from 12.1 in 2006, and equivalent to 25% occupancy for a 40-seat bus.

These numbers seem to undermine somewhat the green credentials of urban bus systems, which currently yields per-passenger emissions roughly comparable to the light-duty passenger fleet. Given the very low bus load factor, there is much room for improvement. At 100% occupancy (an unachievable goal), the 2008 bus fleet would achieve 111 $PMPG_{GWP}$. A plausible best-case scenario for the nationwide bus system might be tailpipe emissions of 0.05 $kgCO_2$/seat-mile (corresponding to a bus getting about 5 MPG-diesel) and a 45% occupancy rate, yielding 80.0 $PMPG_{GWP}$.

As an aside, buses are quite expensive, with new diesel buses having a base cost around $300,000 [289]. Given that on average only about 10 people are riding a bus at any given time, a depressing thought experiment suggests that it could actually be more cost-effective, and lead to lower fuel consumption, for transit authorities to simply purchase high-MPG, $15,000 economy vehicles for 20 regular bus riders in lieu of new equipment.

In sum, transit buses are only slightly better than personal vehicles, as currently used, but systems with moderately higher utilization and fuel economy than the national average could reasonably achieve 40–60 $PMPG_{GWP}$. It should also be kept in mind, of course, that public transit also serves purposes other than emissions reductions, such as providing mobility for those who cannot drive, and transit systems also can relieve congestion, providing a second-order fuel consumption benefit. Buses using alternative fuels and drivetrains, e.g. compressed natural gas

(CNG), hybrid-electric, and battery-electric buses are increasingly being deployed, with natural gas dominant and widely advertised as a "clean-burning" and green alternative. As I review in Section 9.1.2, however, CNG buses likely have similar or even greater lifecycle emissions than standard diesel buses, while hybrid-electric buses may reduce carbon emissions significantly, perhaps by 20–25%.

Long-distance bus, or "motorcoach" travel

- Overall, long distance bus travel is the most climate-friendly transportation option available in the US, with the system currently achieving about 200 MPG_{GWP} on a passenger-mile basis, better than US intercity rail even at 100% occupancy. Given an overall load factor in the range 0.6–0.7, a completely occupied motorcoach may achieve in the 300–400 PMPG_{GWP} range, which is superior even to European electrified rail systems.

- Motorcoaches travelled 63.17 billion passenger-miles in 2013, or almost exactly 200 miles per capita, and generated emissions of about 3.80 million MgCO_2e via fuel consumption (assuming 100% diesel fuel). If all these trips had been taken in single-passenger automobiles, the carbon emissions would have been 8.5 times greater.

Historical perspective. Buses were once a major mode of intercity transportation, and while they suffered a precipitous decline in both service and reputation since their heyday in the post-WWII years, this trend has actually modestly reversed in the past few years. The first scheduled intercity bus route was inaugurated in 1913 [286], and the sector rapidly expanded over the next five decades. By 1960, there were over 300,000 route-miles in service, and scheduled carriers serviced over 15,000 communities [283]. During its height, intercity bus travel was also regulated on both the federal and state levels as an essential public service. However, intercity bus would follow the same general pattern of demise that befell intercity rail, albeit delayed one or two decades.

Following 1960, multiple factors, including the expansion of the new Federal Highway System, rising household incomes coupled with the rise of the personal automobile, the later expansion of low-cost commercial air-travel, and the hollowing out of city cores that was driven at least partly by redlining and white-flight, drove a rapid decline in intercity bus ridership, and bus travel was increasingly seen as an unsafe and unreliable option used only by those without the financial means for alternatives [283].

Between the 1960s and 1990, ridership decreased from 140 million to 40 million, while many routes went out of service; ridership and service continued its rapid decline into the early 2000s, until over two-thirds of intercity bus service had disappeared compared to 1960. However, since 2006 there has been a modest revival, driven especially by non-traditional curbside bus services (versus traditional services such as Greyhound, which operate out of established terminals) [283].

Recently, several authors have documented the nascent rebound in intercity bus travel with many arguing that this frequently ignored modality is the most sustainable and environmentally friendly mode of long-distance travel available in the US today. Woldeamanuel [286] in particular has argued that buses meet several criteria for sustainability beyond their low carbon emissions, namely, supporting both economic development and equitable access to long-distance transport. In terms of equity, the intercity bus network connects a significant number of smaller and rural communities that lack access to rail or air, and covers in total about 89% of the US rural population. Moreover, intercity buses generally serve the lowest-income customers of any long-distance transport mode [643]. Those traveling by commercial air are on average wealthiest, with rail and personal automobile travelers in the middle.

MPG and emissions overview. Generally speaking, intercity bus is a highly efficient mode of transportation. A 2015 industry report by the American Bus Association [284] reported that, in 2013, industry motorcoaches travelled 1.68 billion miles in service (i.e. under trips with passengers) and 1.86 billion miles overall, yielded 63.17 billion passenger-miles, and consumed 303.6 million gallons of fuel for an overall fuel efficiency of 6.13 MPG; assuming all fuel was diesel, this is equivalent to 5.39 MPG under gasoline. This report give an average ridership of 37.6 passengers, and thus we have an average of 202.9 MPG_{GWP} for on-road passenger-mile MPG. However, we should include the 9.9% of motorcoach miles travelled without passengers, giving the system an overall passenger-mile 182.8 MPG_{GWP}. This does not, however, include the upstream emissions from vehicle construction or emissions attributable to bus terminal operation and other associated infrastructure.

These numbers compare favorably with the US's intercity rail (Amtrak) system, which achieves about 65 MPG_{GWP} and 112 MPG_{GWP} in terms of passenger- and seat-miles, respectively. Thus, the motorcoach industry, as it exists, generates only a third of the emissions Amtrak does, on a passenger-mile basis, and 30–50% less than even European rail networks. Furthermore, we should consider that most motorcoaches have a 56-seat configuration, implying that maximum operating efficiency is on the order of 300 MPG_{GWP}.

The motorcoach industry's 63.17 billion passenger-miles (in 2013) is about an order of magnitude greater than the (2013) 6.81 billion Amtrak passenger-miles, but still an order of magnitude lower than the 595.34 billion passenger-miles provided by the domestic commercial air industry. Nevertheless, this suggests that motorcoach service could be far more easily expanded to the scale of the airline industry. Furthermore, since expanding the industry does not require new large-scale infrastructure projects, it is a good short- to medium-term option for rapid emissions reductions.

School bus system

- The school bus system is larger in scale than any other mass transit system, achieves the system-level equivalent of almost 150 passenger-miles per gallon (gasoline-equivalent basis), and may displace as much vehicle travel as all other mass transit modes combined.

There are reportedly over 480,000 school buses in the US, which collectively transport 26 million children daily, and annually travel 5.76 billion miles (this is over twice the 2 billion vehicle-miles reported for metropolitan transit buses, and almost 50% more than the 3.9 billion transit vehicle-miles of any kind [285]), consuming 822 million gallons of diesel fuel (equivalent to 922 gallons of gasoline) [287]. These figures suggest school buses achieve an average of 7 MPG-diesel, and that each bus transports 54 children.

If we conservatively assume an average one-way distance of 6.3 miles from home to school (average one-way distance for school/church trips in [642]), then if each child that currently rides the bus were transported in a private vehicle, and assuming 1.5 children per vehicle, then we should add 87.4 billion miles of vehicle travel (assuming 200 school days per year), and consume roughly 4 billion gallons of gasoline. Thus, school buses may annually save on the order of three billion gallons of gasoline, or about 35 million $MgCO_2e$.

Note that these calculations also suggest that the school bus system provides the equivalent of 131.1 billion passenger-miles, and thus achieves the equivalent of 142 $PMPG_{GWP}$, lower than the 337 $PMPG_{GWP}$ we would get from naively averaging 54 children per bus and 5.76 billion bus miles over 922 million gallons of gasoline-equivalent, but still excellent.

Overall, the school bus system plausibly reduces total passenger vehicle fuel consumption by as much as 2.5% compared to what it would be (this is reasonably consistent with 6.2% of daily person-miles being to school or church [642] and about half of children taking the bus), and may offset nearly as much fuel as all other public and mass transportation modalities combined: the NTS reports 57 billion passenger-miles via transit and 6.675 billion passenger-miles on Amtrak in 2014, while long-distance motorcoach provided 63.17 billion passenger-miles in 2013 [284]. To replace all this with passenger vehicle transport at an occupancy rate of 1.38 would require 92 billion vehicle-miles, roughly as many miles as are potentially avoided via the school bus system. This calculation does not, however, take into account the possible transit multiplier posited for public transit, whereby development and behavioral patterns associated with transit, especially rail-based transit, reduce vehicle miles by multiple indirect means beyond direct mode-switching [294], as discussed in Section 9.1.4.

Alternative fuel and powertrains: natural gas, hybrids, and electric buses

There are three principle alternatives to diesel-powered buses coming into use: natural gas (typically compressed natural gas, but occasionally liquid natural gas), diesel hybrid-electric, and battery-electric (i.e. all-electric) drivetrains. Several recent studies have analyzed both lifecycle and actual operational emissions of these alternatives, generally finding that hybrid diesel buses consume about 25% less fuel than conventional diesel buses. CNG buses consume about 36% to 50% more energy than diesel buses, as spark-ignition CNG engines are intrinsically less efficient than diesel engines. Since natural gas is less carbon intensive on a per unit energy basis than diesel, CO_2 tailpipe emissions for CNG buses may still be slightly lower, but because natural gas vehicles emit significant quantities of unburned methane (as much as 3% of all fuel methane is released in the exhaust across London buses [290]), total CO_2e emissions are probably higher even at the tailpipe level [290]. This is further exacerbated by upstream methane leaks associated CNG.

Zang and colleagues [288] performed a very detailed inventory of on-road energy consumption and emissions for 75 Chinese urban transit buses with diesel, diesel-electric hybrid, and compressed natural gas drivetrains under typical Beijing conditions, i.e. slow and congested traffic with a great deal of idling, and certainly comparable to the worst driving conditions in US cities. Of note, diesel and CNG buses generated similar tailpipe emissions, while diesel hybrids had 25% lower emissions than conventional diesel. Driving conditions strongly affected fuel consumption; of greatest import, lower average speeds dramatically increased fuel use, while AC on full (relative to no AC) increased fuel use by about 23% for diesel buses and 48% for hybrid buses. Of all drivetrains, hybrid fuel consumption was most strongly affected by AC load and suboptimal driving conditions, i.e. lower average speed. Numbers from Zhang et al. suggest, in gasoline equivalents, 6.33 MPG_{GWP} for diesel buses in Beijing without AC, and 5.15 MPG_{GWP} with AC. Hybrid diesel buses achieved 8.40 MPG_{GWP} at baseline, but only 5.67 MPG_{GWP} using AC. Thus, hybrids used in the range of 10–30% less fuel, but the advantage was at the low end when using AC.

A similar study performed by the NREL in NYC [289] comparing diesel, hybrid, and CNG transit buses examined real-world fuel consumption over a 12 month period, and found *energy* use (not emissions) in CNG buses to be about 37% higher than diesel (1.70 MPG compared to 2.33 MPG in diesel equivalents). Translating this into emissions, this suggests, over a 100-year timeframe, 7.5% lower *tailpipe* emissions for the CNG buses, but *lifecycle* emissions differing by -5.5 to +8.7% for 1.2 to 2.4% upstream methane leak rates, a *conservative* rage. Therefore, the CNG buses were marginally better to marginally worse relative to diesel even before accounting for methane emitted in the CNG vehicle exhaust. On the other hand, hybrid diesel buses had 22.3 to 27.6% lower emissions than conventional diesel (and thus were also clearly superior to CNG buses). Of note, fuel economy of hybrids followed a clear periodic trend, with fuel consumption higher in the summer months, when AC could be expected to be in use.

Finally, a comprehensive simulation analysis of the London bus transit system by Chong et al. [290] with a thorough lifecycle analysis of CNG upstream emissions and, especially, rogue methane emissions from CNG vehicles, concluded that converting to a CNG-based bus fleet would *increase* net CO_2e emissions.

Marginal emissions

Given that the average passenger load is extremely low for most bus lines, the marginal emissions cost of riding the bus is zero, under almost any reasonable scenario. Large increases in ridership during peak hours could lead to more buses operating, in which case it may be reasonable to attribute an emissions factor under peak ridership to the individual transit user. The basic point is, unlike the case of air travel, public transit seats are massively under-utilized, and so the "marginal passenger argument" is generally applicable.

To reiterate, while every time you drive, even in a very fuel efficient vehicle, you are directly adding carbon to the atmosphere, using public transit rarely if ever leads to increased carbon, due to the low load factor of the system as she currently exists. But by the same token, it is also entirely valid to assess the carbon footprint of transit systems as they currently exist to guide public policy.

9.1.3 Passenger Rail

Unlike buses, passenger rail systems are clearly superior to typical personal vehicles for both intracity and intercity travel, and generally achieve the equivalent of 50 passenger-miles-per-gallon or better. Overall, the Amtrak system, the only existing intercity rail system in the US, achieves on the order of 50–65 PMPG$_{GWP}$; commuter and metropolitan rail similarly get 45–65 PMPG$_{GWP}$. High-performing metropolitan rail, such as the NYC subway system, may yield well over 100 PMPG$_{GWP}$. I now consider various rail systems in turn.

Transit rail: light, commuter, and heavy

Transit heavy rail (an electric railway operating on an exclusive electrified track) and commuter rail (an electric or diesel-powered railway operating on general railway lines) provide the vast majority of rail transit, although light rail (electric rail, usually powered by overhead power lines, that intersects with vehicular traffic crossings) is increasing in importance. All forms of transit rail are clearly superior to transit bus, being more energy-efficient and enjoying much higher ridership. The equivalent passenger-MPGs achieved are on the order of 45–65 across most systems, with the huge NYC subway (almost 60% of all heavy rail) over twice as efficient as the average rail system. Ridership and energy-efficiency numbers below are drawn from [282].

Heavy rail. In 2008, electricity consumption to drive heavy rail averaged 0.109 kWh/seat-mile, and ridership averaged 47%, suggesting 0.1582 kgCO$_2$e/passenger-mile (under US grid-average electricity), equal to 70.5 PMPG$_{GWP}$. The NYC Subway system accounted for 59.3% of all heavy rail passenger-miles in the US, in 2008, implying that this system alone is responsible for about 19% of *all* US transit (passenger-mile basis). The NYC system is extremely efficient, with an energy requirement of 0.107 kWh/seat-mile, 59% occupancy, and at a regional grid-average electricity emissions factor of just 0.423 kgCO$_2$e/kWh, this yields 0.0767 kgCO$_2$e/passenger-mile, equivalent to 145 PMPG$_{GWP}$; even under the marginal generating mix we get 83 PMPG$_{GWP}$.

Note that the low emissions factor of the NYC subway necessarily implies that, on average, most other heavy rail systems are more carbon intensive, achieving on the order of 45–55 $PMPG_{GWP}$ (0.2–0.25 $kgCO_2e$/passenger-mile).

Light rail. Light rail is slightly more energy intensive than heavy rail, requiring on average 0.126 kWh/seat-mile, and with somewhat lower ridership (37%), we get 0.239 $kgCO_2e$/passenger-mile at US grid-average emissions, and 47 $PMPG_{GWP}$. Note that a few very low ridership systems have emissions profiles much worse than passenger vehicles, while the largest system, in Los Angeles (shocking, to be sure), had 46% ridership and emissions roughly equivalent to 64 $PMPG_{GWP}$.

Commuter rail. Commuter rail is similar to other systems, generating perhaps 0.15 $kgCO_2e$/passenger-mile of direct emissions, which adjust to about 0.185 $kgCO_2e$/passenger-mile on a lifecycle basis, and about 60 $PMPG_{GWP}$. While the fewest fraction of seats is occupied on commuter rail (30%), the larger cars carry the most passengers on an absolute basis.

Amtrak and US intercity rail

It is instructive to perform a simple top-down estimate of the per-passenger and per-seat MPG equivalent of the much-maligned interstate rail system, Amtrak. From Amtrak's 2013 Sustainability Report, we can calculate that the Amtrak system in all, including train operation, station operations and maintenance, etc., generates about 1.47 million $MgCO_2e$ annually, with over 95% from diesel and electricity consumption (the remainder is from gasoline for maintenance vehicles and other heating fuels). Now, direct locomotive operation accounts for 80% of this total, at about 1.17 million $MgCO_2e$ (including all fuel-cycle emissions). For a comparison to air and vehicle travel, it seems fair to primarily consider the emissions from train operation, as maintenance and station operation are part of the supporting infrastructure, but I give both calculations. Necessary to these calculations, we have in 2013, Amtrak trains moving 11.80 billion seat-miles and 6.81 billion passenger-miles, for an overall load factor of 0.577.

For Amtrak train operation, we get 64.7 MPG_{GWP} (0.172 $kgCO_2e$/mile) on a passenger-mile basis, and 112.1 MPG_{GWP} (0.099 $kgCO_2e$/mile) on a seat-mile basis. Including all Amtrak emissions, we get the somewhat worse per passenger-mile 51.6 MPG_{GWP} (0.216 $kgCO_2e$/mile) and per seat-mile 89.4 MPG_{GWP} (0.125 $kgCO_2e$/mile).

We might ask, what is the marginal emissions cost of riding an Amtrak train? As for all modes of mass transportation, it is obviously in the most proximate sense zero, but if we don't consider this notion valid for air travel, we should question it for rail as well. Data on monthly ridership and available seats are available for the decade spanning 2002 to 2012, and as shown in Figure 9.3, both passenger miles and seat miles increased over that time, but with a disproportionate increase in passenger-miles and an uptrend in the load factor.

The data suggests that, at present, an Amtrak trip likely counts as a near carbon-free ride, given the excess capacity of the system, the fact that seasonal peaks in ridership correspond with a minimal increase in train trips, and the rather poor correlation between passenger and seat-miles. At most, we might assess a 50% marginal emissions fee, giving about 200 MPG_{GWP} for intercity rail travel. Note also, if we assume that seat-miles will be relatively decoupled from passenger-miles up to a system-wide load factor of 0.85, the Amtrak system can absorb 47% more passengers (3.22 billion passenger-miles) until seat-miles and passenger-miles strongly couple.

Now, even if all riders are "marginal passengers," as would be the case under dramatically expanding rail demand, one would achieve close to 100 MPG_{GWP}, or about two to three times the (adjusted) fuel efficiency of domestic air travel, and thus Amtrak is always the preferable option regardless. However, one wonders whether Amtrak represents a viable near-term replacement for air travel? A quick review of the numbers is disheartening: domestic air travel

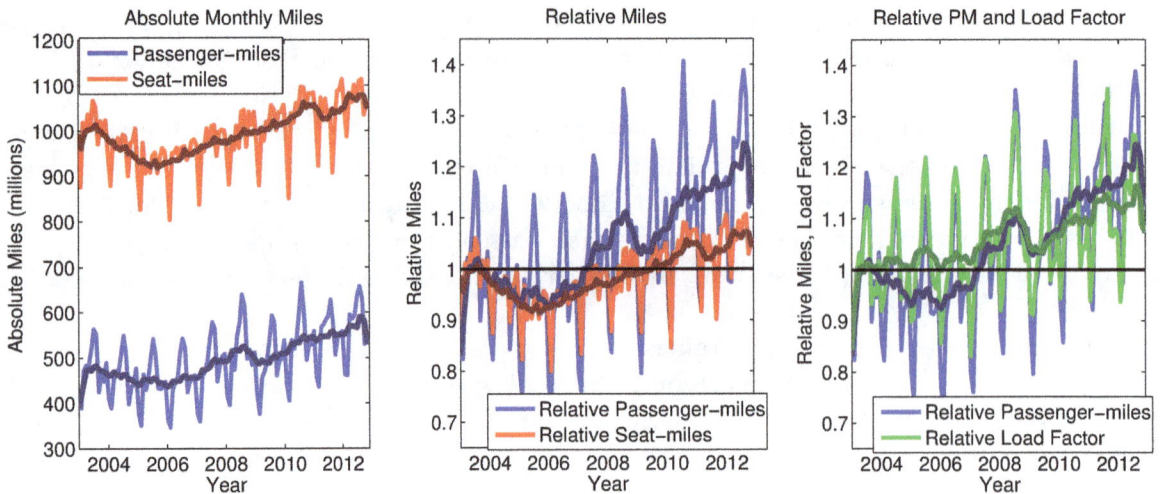

Figure 9.3: The leftmost panel gives monthly Amtrak seat- and passenger-miles (along with the 12-month moving average) from 2002 to 2013; the center panel gives these metric relative to 2002 ridership, while the rightmost panel gives the relative load factor and passenger-miles. We see that passenger-miles increased out of proportion to the increase in seat-miles, with a corresponding relative load factor increase. Source: BTS.

accounted for 595.3 billion passenger-miles in 2014—87 times the 6.81 billion Amtrak passenger-miles. Even at 100% ridership, this is still 50 times the current Amtrak capacity, and 35 times the historically high ridership in 1960 (17.064 billion passenger-miles). Furthermore, throughout much of the middle of America, Amtrak lacks designated right-of-ways and must share freight rail lines, so scaling these lines up could be difficult.

A brief comparison to European rail

European electrified rail systems probably generate only half to a third of the emissions that typical US rail does, thus showing that mass transit in Western countries can achieve very low emissions profiles. For example, a report by Network Rail [291], the state-owned company that manages most of the rail network in the UK, estimated total emissions associated with conventional and high-speed rail (HSR) at around 45 gCO_2e/seat-mile and 50 gCO_2e/seat-mile, respectively (for European electric rail systems, using about 0.55 $kgCO_2e$/kWh for UK electricity, and including upstream infrastructure-associated CO_2). The report gives load factors of about 40% for typical UK conventional rail, and over 60% for European HSR systems.

Thus, we would have conventional rail getting 247.7 MPG_{GWP} on a seat-mile, and 99.1 MPG_{GWP} on a passenger-mile basis; and HSR getting 222.9 and 137.8 MPG_{GWP} on seat- and passenger-mile bases, respectively. This report also estimates that rail infrastructure accounted for about 20% of lifetime emissions (these are included in the above numbers), with concrete and steel responsible for most infrastructure emissions, consistent with the larger literature.

9.1.4   Second-order effects on vehicle-miles, development, and congestion

Transit is not merely a one-to-one stand-in for passenger vehicles, as transit- versus vehicle-oriented development patterns can vary markedly, affecting the pattern of life and personal transportation in a broad way. This is mainly true for rail-based transit systems (especially newer light rail lines), as unlike the bus system, they entail permanent infrastructure development generally perceived as aesthetically and economically appealing, and that can promote other pedestrian-friendly and dense mixed-use development [292]. Cities with rail transit tend to have stronger overall transit systems and higher transit ridership, and rail is often built to attract additional riders that are not economically restricted to public transportation [292]. Even under auto-focused development, however, any transit mode may help to relieve congestion and in turn reduce overall fuel consumption [282], with several studies concluding that transit can modestly decrease vehicle traffic on nearby roadways (see, e.g., Bhattacharjee and Goetz [293] and references therein), although the overall effect may be small.

Transit may reduce overall vehicle-miles travelled (VMT) beyond simply those avoided by direct mode-shifting from auto to transit via several mechanisms, and several authors have attempted to quantify a so-called *transit multiplier*, defined as the ratio of the total VMT reduction to the direct VMT reduction attributable to transit. Mechanisms for the multiplier effect include (1) "trip chaining," whereby multiple errands are combined into a single transit trip, (2) concentration of employment and populations reducing vehicle trip length, (3) compact development also increasing walking and bike trips, and (4) decreased vehicle ownership among households near transit lines [294]. As recently reviewed by Ewing and Hamidi [294], transit multiplier estimates have ranged from 1.29 to 9, with different metropolitan areas giving disparate results, and even the same study's conclusions can change markedly with assumptions, suggesting that these results are neither robust nor generalizable. It seems sensible, of course, that transit multipliers will vary between particular transit systems, will strongly depend upon the overall level of transit development in a city, and will evolve with time.

Ewing and Hamidi [294] performed a natural experiment in Porland, OR, by comparing VMTs, walk, bike, and transit trips in a light rail transit corridor and a similar highway corridor. Travel surveys were performed in both regions, both before rail construction in 1994, and at long-term follow-up in 2011. The presence of rail appeared to increase household transit trips by 0.60 per day with a direct reduction of 0.47 VMT per day, while increased walking and density associated with rail led to an overall 1.43 VMT per day reduction, giving a transit multiplier of 3.04. It seems safe to assume, then, that rail-based transit can reduce vehicle miles several-fold beyond those avoided via direct mode-shifting, but the exact multiplier will be particular to each city and system.

9.2 Infrastructure

Fully understanding the climate impacts of transportation requires assessing the emissions that result from infrastructure construction and maintenance. Indeed, I roughly estimate that roadway and parking infrastructure-related emissions are on the order of 10–15% of the fuel-cycle emissions of the entire vehicle fleet, and if we include these in our emissions accounting, then the carbon footprint of personal transportation increases a further 5–10%.

While infrastructure is only very indirectly alterable via individual-level behavior, it is important to complete our understanding of the topic, and it is directly relevant from a policy perspective, especially when it comes to assessing the wisdom of new infrastructure projects. For example, high speed rail in California has been extremely contentious, and it is legitimate to question whether the emissions resulting from rail construction (primarily from cement production) outweigh the benefits of lower operational emissions. Of course, such an analysis cannot be performed in isolation, and increased maintenance or expansion of highway and/or airport infrastructure that would likely result in the absence of HSR should be considered as the counterfactual. Despite such controversies, most infrastucture emissions by far are related to vehicle transport and the truly vast infrastructure needed to support it, as discussed next.

9.2.1 Roadways, other pavements, and the vehicle infrastucture

---

- Roadway and parking infrastructure is emissions-intensive, with the vast majority of emissions attributable to hot-mixed asphalt (HMA) and Portland cement concrete (PCC) materials production; asphalt is a petroleum product that must be mixed with aggregate at high temperatures, while Portland cement derives from burning lime at extraordinarily high temperatures, a process that uses large amounts of fossil energy and also directly emits CO_2 from burning lime.

- In sum, annualized roadway/parking infrastructure emissions may amount to about 190–200 million MgCO$_2$e per year, with 90 million, 70 million, and at least 30 million MgCO$_2$e due to roadways, parking, and roadway lighting, respectively.

Up to 45% of urban land area may be paved [295], and the paved highway and interstate systems that cross America are truly vast in scale. These pavements are composed principally of either cement-containing concrete or bituminous asphalt, and both materials are extremely energy and emissions-intensive. The overall global warming impact of roadways and other pavements is dominated by the extraction and production of construction materials, at both the initial construction and maintenance phases. Material transportation and equipment operation are further, but less significant, sources of emissions. Additionally, a full life-cycle analysis may be extended to include, among other factors [298, 299], the effects of construction/maintenance on traffic congestion, energy use from roadway lighting, the effect of pavement surface type and roughness on vehicle fuel consumption, and albedo (i.e. roadway reflectivity) and urban heat island effect. Furthermore, paving land over for roads and parking infrastructure is a nontrivial land-use change, with its attendant costs.

Roadways are composed of multiple layers, the logic being that force is transmitted from the hard top layer to be dispersed in the deeper, thicker, and coarser sub-surface layers. These lower layers are composed of various sands, gravels, or other aggregates, and although it does take energy to produce, transport, and compress these layers, most energy and emissions go into the top layer, which may either be composed of hot-mixed asphalt or concrete (or occasionally a mix).

There is a significant scientific literature devoted to lifecycle analysis of energy and global warming impact of pavements, with the focus frequently on comparing asphalt and concrete surfaces. However, conclusions consistently vary, and even within the same work whether concrete or asphalt is preferable varies with the exact roadway specifications. It follows that neither can be concluded to be generally superior, and the best surface will be very much project-dependent. It is not my purpose or interest to exhaustively review this literature, but to inform the reader of the general principles of roadway construction, maintenance, and give a gross estimate as to the climate impact of the vast liquid stone-scapes that man hath built for his vehicles. Santero and Horvath [298] provide an excellent literature review and discussion of the aforementioned factors that may be included in a full LCA, and the following discussion is indebted to their analysis.

Overall emissions estimates

The Federal Highway Administration estimated that, in 2013, there were a total of 8,656,070 lane-miles of public road and 4,115,462 road-miles, for an average of 2.1 lanes per road. However, only about 66% of public road (2.678 million miles) is paved; since paved lane-miles is not given by the FHA, we might reasonably assume that paved roads average a slightly higher 2.25 lanes, giving just over six million paved lane-miles (6.0255 million). If we conservatively assume 12 foot wide lanes with an overall average 2 feet of paved shoulder per lane, we get around 16,000 square miles of paved road in the US. From the discussion below, the construction and maintenance of this network may plausibly embody between one and six billion $MgCO_2e$, with a middle estimate of perhaps 4.5 billion $MgCO_2e$. Amortized over 50 years, this amounts to about 90 million $MgCO_2e$ per year.

Estimates for global warming impact of materials extraction and production vary. For example, Santero and Horvath [298] suggest that emissions may range from an bare minimum of 101 $MgCO_2e$/lane-mile to a plausible extreme of 1,200 $MgCO_2e$/lane-mile, and they cite five previous studies that together give a potential range of 129 to 801 $MgCO_2e$/lane-mile for roads of various construction and 40 to 50 year planned lifespans. This wide range is attributable to the wide variation in materials requirements: more intensely used pavements are typically much thicker and may require more frequent maintenance/re-surfacing. Also, note that these values include materials for pavement maintenance and periodic partial and complete re-surfacings. A review of the maintenance schedules (for asphalt-surfaced Canadian freeways) in an Athena Institute report for the Cement Association of Canada [300] suggests that maintenance consumes 1–1.5 times as much asphalt as initial construction.

Furthermore, a range of works summarized by Zapata et al. [297] suggest asphalt pavements could last up to 35 years before a major reconstruction, but may have a lifespan as low as 10–15 years; concrete pavements may last somewhat longer, around 25–35 years. My own review of popular sources, maintenance schedules given in the Athena study [300], and the assumed pavement lifespans in White et al. [295] also suggest that asphalt pavements may last on the order of 10–25 years before a complete reconstruction is required. If we assume a 15-year lifespan, then at least twice as much asphalt is consumed over 50 years as is invested in the initial construction.

Using 15 $MgCO_2e$/lane/mile/year, a midpoint estimate based on [295], and 6.0255 million lane-miles, this would suggest annualized construction and maintenance emissions on the order of 90 million $MgCO_2e$, or about 6.5% of the 1.3795 billion $MgCO_2e$ attributable to passenger vehicle fuel use (including upstream fuel-cycle). Since production of hot mixed asphalt and portland cement concrete is the major source of these emissions, I review these production processes next.

Materials Production

The majority of roadway emissions are attributable to material production (80%), with material transportation of secondary important (15–20% of emissions) [295]. The two major materials involved in roadway construction are hot-mixed asphalt (HMA), and Portland cement concrete (PCC). Both are composed chiefly of sand and gravel aggregate that is cemented together into a smooth, hard surface by one of two binders, or "cements:" asphalt, a sticky heavy oil, or Portland cement (what one would commonly think of as "cement"). Both materials are produced in two basic steps: (1) the binder is produced, either via petroleum extraction and refinement for asphalt, or by heating lime to extreme temperatures (1,500 °C/2,732 °F) to produce Portland cement, and (2) the binder is mixed with an aggregate of sand and gravel. In the case of HMA, asphalt production is energy- and emissions-intensive to be sure, but mixing the asphalt with the aggregate actually uses several times as much energy, as this must be done at high temperature and large amounts of energy are required to heat and dry the sand and gravel that is added to the mix. For PCC, nearly all energy and emissions occur at the cement production stage, with mixing a relatively trivial process [295].

Aggregates. Gravel or stone must be quarried and crushed to produce coarse aggregate, and sand must be quarried as well. While not trivial, aggregate extraction and processing represents only a minor component of HMA and PCC emissions. White et al. [295] give

Principles of the Pavement

Surface course, the "pavement proper:"
- Hot-mixed asphalt (HMA) >90% of cases
- <10% Portland cement concrete (PCC), mix of HMA and PCC layers, or other
- Vast majority of roadway emissions attributable to production of asphalt and cement for this layer

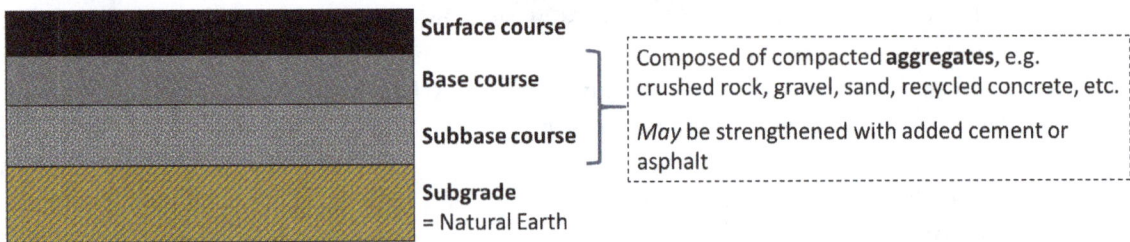

Surface course

Base course

Subbase course

Subgrade
= Natural Earth

Composed of compacted **aggregates**, e.g. crushed rock, gravel, sand, recycled concrete, etc.

May be strengthened with added cement or asphalt

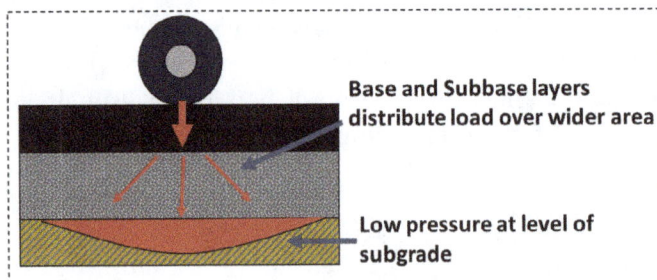

Base and Subbase layers distribute load over wider area

Low pressure at level of subgrade

Figure 9.4: General schematic for pavement construction and materials. Production of hot-mixed asphalt and portland cement concrete for roadways and parking infrastructure is a significant source of US carbon emissions, and yet another upstream process adding to the overall burden of personal transportation.

emissions factors of 2.8 $kgCO_2e$/tonne gravel and 2.5 $kgCO_2e$/tonne sand, which are reasonably close to the reported Canadian average of 4.0 $kgCO_2e$/tonne for aggregate (coarse and sand combined) [300]. The fine/coarse aggregate mix is roughly 50/50 for both HMA and PCC, and thus 3 $kgCO_2e$/tonne aggregate is a reasonable mid-estimate for production. Transporting aggregate is more significant, with White et al. [295] giving 0.454 $kgCO_2e$/tonne-mile, or almost 9 $kgCO_2e$/tonne if aggregate is transported 30 miles on average. Overall, this sums to 0.012 $kgCO_2e$/kg of aggregate.

Hot-mixed asphalt. Over 90% of pavements worldwide are surfaced with HMA, sometimes referred to as "flexible roadway," as HMA is more elastic and deformable than PCC; this material consists of about 5% bitumen ("asphalt" or "extra-heavy oil") and 95% gravel and sand aggregate. The viscous bitumen is a thick, sticky heavy oil that may be obtained in naturally occurring formations (for example, the Canadian tar sands) or as a heavy distillation fraction of petroleum refinement. The terminology here can be a bit confusing: "bitumen," "asphalt," or "asphalt cement" all refer to the petroleum product that binds the gravel together to form a hard road surface, while in common parlance "asphalt" is often, as one well knows, used in reference to the HMA gravel-asphalt mixture.

Extracting and refining bitumen is an energy- and emissions-intensive process. White et al. [295] give an emissions factor of 0.426 $kgCO_2e$/kg of refined bitumen[1], and the Athena report [300] similarly suggests 0.391–0.423 $kgCO_2e$/kg bitumen.

Drying, heating, and mixing the sand and gravel aggregate with bitumen is actually about *twice* as emissions-intensive as obtaining the bitumen binder. The aggregate is heated to 150–170 °C and must be thoroughly dried before mixing with the bitumen, also kept at 150 °C. Upon mixing, the product is transferred to hot storage bins to await transfer. This process can be performed in batches or via continuous processing in a rotating hopper [296], with continuous processing slightly more efficient. An older analysis of two HMA plants [296] found that by far the greatest use of energy is drying the aggregate (for continuous processing, diesel fuel use broke down as 47% for drying, 12% for heating, and 36% lost as waste heat), as sand and other small aggregates with a very large collective surface area can store a great deal of moisture, and water has a very high heat of evaporation (this fact may be exploited for space-cooling via the evaporative cooler, as discussed in Section 12.4).

Adding up bitumen production, HMA mixing, and transportation to job sites, HMA has an overall emissions factor of about 0.1 $kgCO_2e$/kg HMA [295]. As an example, a highway with 12-foot lanes, a seven inch HMA layer, and a 10-inch aggregate base layer would then generate (assuming HMA and aggregate have densities of 2,275 and 1,700 kg/m^3, respectively) on the order of 270 $MgCO_2e$ per lane-mile (about 90% related to the HMA), or 18 $MgCO_2e$/lane-mile/year over a 15-year road lifetime.

Portland cement concrete. Portland cement concrete (PCC) is a mixture of Portland cement (the binder) and sand and gravel aggregate. Producing Portland cement is actually a major source of CO_2: globally, Portland cement production accounts for 5–7% of anthropogenic CO_2 emissions [21]. Its share is smaller in the US, but still significant, at perhaps 2–3% of emissions. Multiple estimates of the production emissions factor exist, with most between 0.8 and 0.9 $kgCO_2e$/kg cement; a recent review gave a mean value of 0.842 $kgCO_2e$/kg cement over eight studies [21], and 0.9 $kgCO_2e$/kg is commonly used rule-of-thumb. Note that typical portland cement *concrete* is only about 13% cement by weight (the rest is mainly sand and gravel aggregate), and so per kg of concrete, we have a cement emissions factor of about 0.1 $kgCO_2e$/kg.

---

[1]This equates to 1.6 $kgCO_2e$/gallon, assuming a density of 1 kg/L for bitumen. This value is quite reasonable, given that upstream emissions for gasoline are on the order of 2.26 $kgCO_2e$/gallon, the EPA estimates that refining bitumen is about twice as energy-efficient as refining gasoline, and roughly half of gasoline emissions are attributable to extraction and half to refining.

Including aggregate, transport, and mixing emissions, this increases to perhaps 0.13 kgCO$_2$e/kg PCC; higher strength concretes with higher cement fractions will have proportionately higher emissions factors.

In producing Portland cement, carbon dioxide is released principally from the conversion of the calcium carbonate (CaCO$_3$) in limestone to calcium oxide, or "quicklime" (CaO), via the reaction:

$$CaCO_3 + \text{Heat} \rightarrow CO_2 + CaO \tag{9.1}$$

This reaction is called limestone "calcination" (from the Latin *calcinare*, "to burn lime"), it is the key step in cement production, and the process *directly* releases about 50–60% of cement production CO$_2$ [21]; most of the remainder is attributable to the large amount of fossil energy required to provide heat. In more detail, the production process requires first quarrying limestone; the quarried rock is crushed and then mixed with smaller amounts of silicon- and aluminum-containing clay and sand, iron, and several other minor additives. This mixture is fed into massive kilns which are heated to 1500 °C (2700 °F), converting lime (CaCO$_3$) to quicklime (CaO) and finally yielding a product known as "clinker." The clinker, along with a bit of limestone additive, is ground to an extremely fine consistency, yielding Portland cement. The cement is mixed with water and aggregate on-site, creating Portland cement concrete.

Cement carbonation. As PCC ages, water reacts with CaO to yield Ca(OH)$_2$ (calcium hydroxide, or "slaked lime"), which in turn reacts with ambient CO$_2$ to re-form CaCO$_3$, sequestering the CO$_2$:

$$CaO + H_2O \rightarrow Ca(OH)_2, \tag{9.2}$$
$$Ca(OH)_2 + CO_2 \rightarrow CaCO_3 + H_2O. \tag{9.3}$$

This process is known as *carbonation*. Over a 50-year lifetime, exposed PCC pavements can re-sequester from a few percent, to perhaps 25% at most, of the CO$_2$ liberated in the original calcination process. However, if concrete is crushed and exposed to air at the end of life, up to 75% may be rapidly carbonated [298].

"Green" cements. Several industrial by-products, most notably coal fly ash and blast furnace slag, are "cementitious" materials that can partially substitute for Portland cement in blended cement mixtures. Since these materials are by-products that would otherwise simply be discarded, their emissions impact is essentially nil, and they can displace up to about 30% of portland cement to reduce overall cement manufacture emissions by a comparable amount.

Coal fly ash is generated by coal combustion in power plants, and it contains a significant quantity of quicklime (CaO, 1.4–22.4%) along with large amounts of silicon, aluminum, and iron oxides (37.8-58.5% SiO$_3$, 19.1-28.6% Al$_2$CO$_3$, and 6.8-25.5% Fe$_2$O$_3$ by mass in the US [301]). Traditionally, fly ash is disposed of in landfills or stored in large lagoons, which can have adverse effects on the local environment [301], and thus its incorporation into cement represents a beneficial re-use at multiple levels.

9.2.2   The parking infrastructure

It is clear that roadway construction, maintenance, and lighting generate a large amount of emissions, all towards the end of allowing vehicles to *travel* at will throughout urban spaces and across the country. But it turns out that the parking infrastructure to *store* these cars, both at residences and almost every school, place of business, etc., also embodies significant emissions and social costs.

A near universal feature of modern American life is the expectation of free parking, and minimum parking requirements for new business and residential developments are ubiquitous

[302]. This is a high, added cost to development, and it very arguably has had a corrosive effect on the urban and suburban built environment—and civic life—since World War II: parking requirements have prevented re-development in denser cities cores and pushed development into suburban sprawl, where land is more easily acquired [303]. Indeed, the second-order effects of minimum parking requirements on driving patterns, etc. no doubt have a high cost, but it is beyond me to attempt to directly quantify this. Parking policies may be an effective way to alter driving patterns and residential environments, but there are multiple trade-offs and pressures, and Marsden [304] provides a good review for those interested.

While it is unknown how many parking spots there are in the US, a decent guess is that each car requires at least one residential spot, plus two to four spots elsewhere [302, 305]; Chester et al. [305] give a range of estimates based on multiple observations and studies, with 3.4 spots per vehicle as a reasonable point-estimate. This translates into 840 million parking spots covering roughly 8,800 square miles, and the annualized lifecycle emissions for the associated asphalt and concrete requirements was estimated at 70 million $MgCO_2e$, or about 5% of the fuel- and vehicle-cycle emissions for the US passenger fleet.

Finally, a comparison between parking and solar occurs to me. Shoup [302], in 1999, gave an estimate of \$40/square foot to build above-ground parking, or \$431/m^2. Adjusted for inflation, this is \$613/m^2 in 2015 dollars, and a reasonable estimate of total land area devoted to parking (in the US) is 8,880 square miles [305] (scenario 4 in the paper). Now, let's conservatively assume a solar panel with 15% efficiency, giving 150 W/m^2 (at 5 kWh/m^2/day), and use the average installed cost for residential solar of \$3.46/W (in Q1 of 2015), giving \$519/m^2 of installed solar panels (commercial-scale solar was appreciably cheaper, at \$2.19/W). I have also conservatively estimated that it would take about 7,900 square miles of solar PV to provide all US electricity generation in Section 4.14.2. So, in sum, we as a society have built perhaps 8,800 square miles of parking that costs \$613/m^2, whereas as a society we would need 7,900 square miles to generate all the US's electricity, at a cost of \$519/m^2. I leave the implications of this calculation to the reader.

Roadway and parking lighting

I cover roadway and parking lighting in Sections 14.4.1 and 14.4.1, concluding that the associated emissions are uncertain, but likely on the order of 20–70 million $MgCO_2e$/year, with 30 million $MgCO_2e$/year a conservative estimate.

9.2.3 Rail infrastructure

Rail infrastructure (i.e. track, stations, etc.) is probably about as emissions-intensive to construct as roadway (mainly due to concrete and steel production), and when amortized on a per-passenger mile basis, the *absolute* upstream infrastructure emissions for rail are generally less than for passenger car, and may be on the order of 0.03–0.05 $kgCO_2e$/passenger-mile for a typical rail system (based on [306, 307]), increasing lifecycle emissions by perhaps 20–50%, depending on the rail system. Infrastructure operation could further increase emissions by around 15–25% (based on [306] and Amtrak sustainability report). Summing over about 39.4 billion rail passenger-miles (sum of Amtrak and all transit rail passenger miles), this amounts to only 1.2–2.0 million $MgCO_2e$ in aggregate, and <1% of roadway and parking infrastructure emissions.

Therefore, while infrastructure can potentially increase *relative* emissions associated with rail systems by 25–100%, the absolute emission levels are small and generally comparable to or better than those related to passenger car. One should still note that particular routes may be much more emissions-intensive than average, such as the proposed California high-speed rail

241

route, which would require a great deal of tunnelling and elevated track [307, 308], but even so, so long as ridership is reasonable, these emissions would be offset within a few years of operation (by shifting ridership for more emissions-intensive modes, namely auto and air) [308].

9.2.4 Air infrastructure

I simply note here that, compared to fuel burn by aircraft, the emissions associated with airport infrastructure construction and operation are likely trivial and add almost nothing to the lifecycle emissions of this transport mode [306].

9.3 Final summary of personal transportation emissions

Transportation as a system embodies far more carbon than generated just via the direct combustion of fuel, about 55–60% more for vehicle-based transport, 25–30% more for air, and perhaps 50–150% for rail [306], when all upstream fuel production, vehicle production, and other infrastructure development is fully accounted for. This also implies that while sectorial inventories give transportation as responsible for 26% of US CO_2e emissions (EPA), this share is likely closer to 40% under consumption-based accounting (and when including freight transport); a non-trivial amount of emissions are also imported from other countries, embodied in foreign-manufactured vehicles and imported oil.

In sum, and as already elaborated in Chapter 5, annual consumption-based emissions for personal transportation in the US, excluding all freight transport (which generally is a component of the consumption footprint of various goods and services), amount to, approximately

1. 1,100 million $MgCO_2e$ due to tailpipe fuel combustion.

2. 275 million $MgCO_2e$ from upstream fuel extraction and refining.

3. 154 million $MgCO_2e$ from vehicle production and maintenance.

4. 280 million $MgCO_2e$ from commercial aviation.

5. 25 million $MgCO_2e$ from private general aviation.

6. 23.4 million $MgCO_2e$ for all mass (public and private) transit modes: 2.2 million $MgCO_2e$ from heavy bus transit, 3.8 million $MgCO_2e$ from intercity bus, 10.3 million $MgCO_2e$ from school buses, 5.7 million $MgCO_2e$ from transit rail, and 1.5 million $MgCO_2e$ from Amtrak.

7. 190 million $MgCO_2e$ from roadway and parking construction maintenance, roadway lighting, and a pittance from rail and air infrastructure.

Excluding infrastructure, this sums to about 1,860 million $MgCO_2e$, or just under 6 $MgCO_2e$ per capita, while if we include infrastructure, our total is 2,050 million $MgCO_2e$, closer to 6.5 $MgCO_2e$ per capita.

9.4 Freight

While my focus remains individual-level consumption, it is important to review the emissions associated with freight transport, and especially to derive per-tonne-mile emissions factors, so that we can quantify the contribution of upstream transportation to the impact of goods consumption, and especially the impact of food transportation, given the prominent billing of "food miles" in today's discourse.

Several excellent data sources for freight exist, including the FHWA's annual Freight Facts and Figures publications and the ORNL Transportation Energy Book (see Chapter 5 on heavy vehicles); unless otherwise noted, all figures on fuel use and miles in this section are drawn from the former. Truck is the dominant mode of freight transport, with medium- and heavy-duty trucks responsible for 44.8 % of all freight tonne-miles in 2011, followed by rail (29.2%), pipeline (17.3%) and water transportation (8.5%). It is far more efficient to move goods by rail or river, and so trucks dominate freight emissions, accounting for about 85–90% of domestic freight emissions. Furthermore, while truck tonne-miles have more than doubled since 1980, per-mile fuel consumption has decreased only 20%. Note that air freight, while extremely emissions-intensive, accounts for only 0.2% of all tonne-miles (with a significant fraction transported as belly cargo in passenger jets). Even so, this tiny fraction still likely accounts for around 5% of US freight emissions overall. I review emissions from each major domestic freight mode in the following section, and close with brief review of international shipping, given the vast quantities of goods that are now imported.

9.4.1   Truck

> - Medium- and heavy-duty trucks consumed about 25% of all vehicle fuel to move 2.3982 trillion tonne-miles of freight in 2011, for a lifecycle emissions factor of 0.2209 $kgCO_2e$/tonne-mile (assuming all fuel was diesel).

Taking a closer look at trucks, they are divided into eight vehicle classes, based on gross vehicle weight rating (GVWR), the maximum weight when *loaded*, with light-duty trucks (essentially pickups, vans, and SUVs) defined as those under 10,000 lbs (classes 1 and 2), medium-duty trucks 10,000 to 26,000 lbs (classes 3–6), and heavy-duty trucks over 26,000 lbs (classes 7 and 8). The commercial freighters we are interested in are divided into single-unit trucks (SUT) with three axles and a maximum GVW around 51,000 lbs, and combination trucks (usually five-axle), the familiar tractor-trailers of America's interstates, which have a maximum GVW of 80,000. The former get about 7.3 MPG, while the latter get 5.8 MPG (6.3 MPG is the combined average), and together these vehicles consumed 42.38 billion gallons of fuel in 2011, or 25% of all vehicle fuel, by volume. Divided among 2.3982 trillion tonne-miles, 267.21 billion truck miles, and assuming all diesel fuel, we have a truck freight emissions factor of 0.1799 $kgCO_2e$/tonne-mile at the tailpipe (similar to EPA emissions factor of 0.1609 $kgCO_2$/tonne-mile), or 0.2209 $kgCO_2e$/tonne-mile considering the full fuel-cycle. Overall, medium and heavy trucks generated roughly 530.13 million $MgCO_2e$ from fuel-cycle emissions, in 2011.

Based on FHWA numbers, a freight truck is carrying just shy of 9 tonnes (20,000 lbs) at any time as a crude average, including the 20% of the time that a truck is empty of cargo. A tractor-trailer carrying a maximum payload of 54,000 lbs, and getting 4.64 MPG (diesel), calculated based on a presumed 20% increase in fuel consumption from the average of 5.8 (see below), would emit 0.1018 $kgOC2e$/tonne-mile, or less than half the typical emissions. Clearly, heavier payloads are more efficient.

To understand why larger loads use less fuel per unit freight, recall from Section 7.1, that fuel energy is used to overcome two basic resistive forces: rolling resistance, which is proportional to vehicle weight, and aerodynamic drag, which is proportional to velocity *squared*, but *not to weight*. Since air drag increases with the square of velocity, it becomes the dominant driver of fuel consumption at highway speeds, and trucks, with their large profiles and "boxy" shapes, have unfavorable aerodynamics, with drag accounting for perhaps 40–65% of fuel use over long-distance drive cycles [309]. Since drag is completely independent of weight, it follows that a fully-loaded truck may use only marginally more fuel than one carrying a partial load, and indeed, simulations based on the road-load model presented in Section 7.1, using parame-

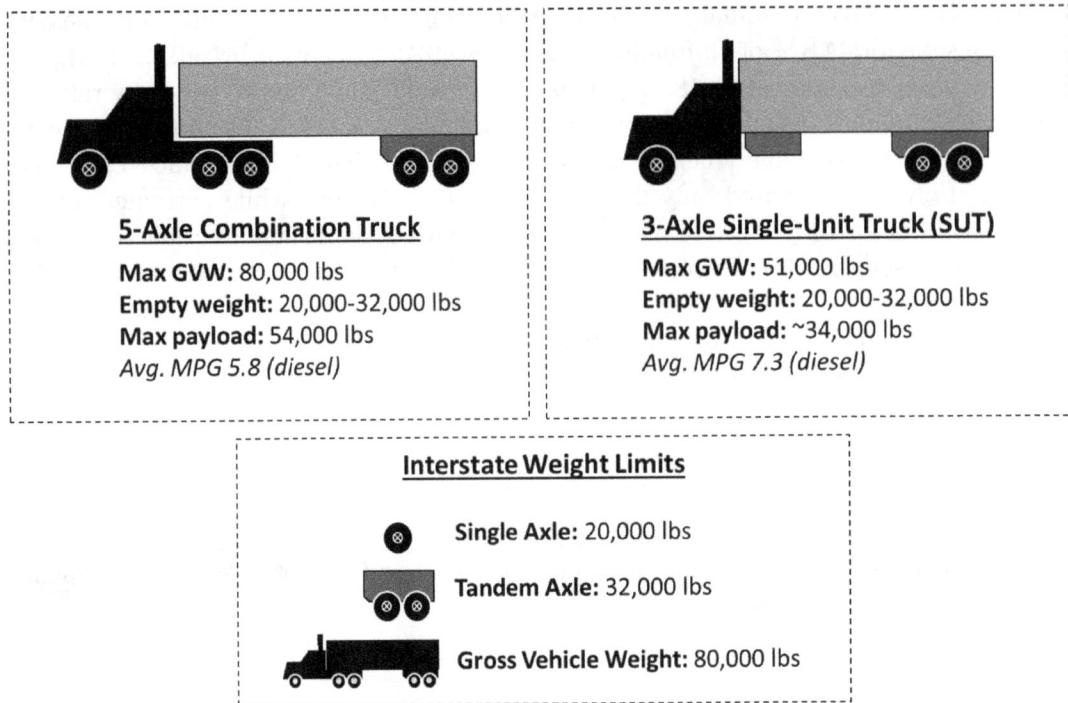

Figure 9.5: Basics of class 7/8 heavy duty trucks, which haul most freight in this country and are broadly divided into combination trucks, usually a five-axle tractor-trailer combination, and single-unit trucks (SUTs), which are a bit less common. The former are generally limited to 80,000 lbs gross weight, due to federal interstate weight limits, and are the most efficient way to move loads by truck. One may consult the ORNL Transportation Energy Book for many more details and statistics on heavy trucks.

ters representative of a typical truck (see [309, 310]), suggest that a doubling of gross vehicle weight increases fuel consumption by only about 25% when cruising at 65 MPH. A more detailed analysis by Mohamed-Kassim and Filiponne [309] similarly found a doubling in weight to increase fuel consumption by less than 25% over long-distance drive cycles.

As an aside, the reader has no doubt noticed various trailer side skirts and other modifications on long-haul trucks. These, and other drag-reducing devices are one strategy for increasing fuel economy, and are actually rather effective at modifying air flow around the vehicle to reduce pressure drag, which is responsible for >80% of the total drag [309]. A baseline tractor-trailer combination has a drag coefficient, C_D, of about 0.8 [310, 311], and the addition of a front deflector, side fairings, and trailer side skirts can decrease C_D by about 25% [311, 309]; throwing in a boat tail could decrease C_D by another 5% [309], as summarized in Figure 9.5, and thus, an aerodynamically optimized tractor-trailer could easily yield a very significant 10–15% fuel reduction on long hauls compared to the baseline.

It is worth noting that heavy duty trucks, while inefficient compared to rail, vastly outperform passenger vehicles and light trucks: A light truck hauling one tonne might get 15–20 MPG at best, while a combination tractor-trailer can transport 25 times as much cargo for only about four times the fuel.

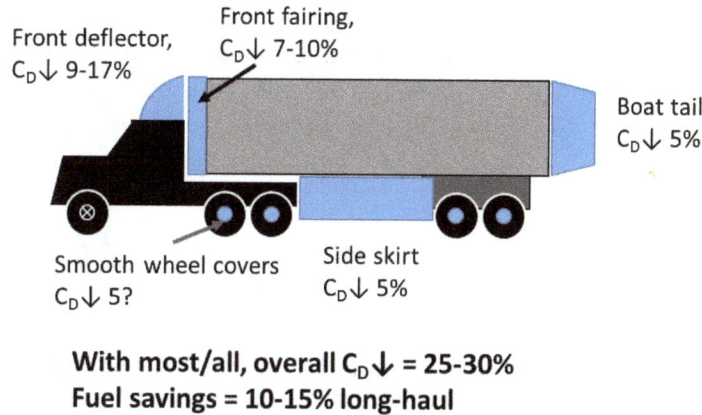

Figure 9.6: Common aerodynamic modifications to tractor-trailers, and expected reductions in the aerodynamic drag coefficient, C_D, based on [311, 309].

9.4.2 Rail

Rail remains a major, growing means for freight transport, and it is especially popular for long-distance transportation. Freight rail consumed 3.710 billion gallons of diesel in 2011, and dividing between 1.5655 trillion tonne-miles gives 0.0296 kgCO$_2$e/tonne-mile on a lifecycle basis, almost 7.5 times lower than truck transport. It should be noted that rail is disproportionately used to transport lower value commodities, such as coal and grain, while trucks more often move higher value goods. And that's about all there is to say about that.

9.4.3 Domestic water

Domestic waterways are still used to transport goods to the tune of 453.4 billion tonne-miles, but the rivers are plied less every year. It is unclear what the actual emissions-intensity of this transport mode is, as figures on watercraft energy consumption in the FHWA Facts and Figures publication do not agree with explicitly reported energy intensity (see Tables 5-7, 5-8, and 5-11), but this publication does give 0.0701 kWh/tonne-mile for water transport, translating into 0.0231 kgCO$_2$e/tonne-mile, assuming diesel is the fuel. This is similar to the emissions intensity of international shipping (see below), and while much smaller than the EPA emission factor (0.065 kgCO$_2$e/tonne-mile), I use it for all calculations.

9.4.4 Air freight

Calculating the carbon footprint of air freight is complicated for two reasons: (1) aviation has multiple non-CO$_2$ effects on the atmosphere that act to amplify the warming effect beyond the classical greenhouse gases, and (2) a large portion of air freight is transported in the bellies ("belly freight") of passenger aircraft, and thus we must decide how to apportion these emissions between the passenger and freight components of the business. Based on Section 8.4.4, the minimum CO$_2$e cost for an additional tonne being hauled by a 737-800 is about 0.3 kgCO$_2$e/mile, already higher than likely truck freight emissions. Supposing a fully loaded Boeing 737-800 freighter, carrying a maximum payload of 20.28 tonnes and burning 3,875 gallons of jet fuel over a 2,000 mile trip, we arrive at 1.6 kgCO$_2$e/tonne-mile (including non-CO$_2$ forcing). Repeating the same calculations for fully loaded Boeing 747-400 freighter carrying 63.92 tonnes and burning 29,770 gallons over 2,000 miles gives 3.9 kgCO$_2$e/tonne-mile. If the freight load factor is a more likely 50%, then emissions factors are 2.9 and 7.2 kgCO$_2$e/tonne-mile, for the 737 and 747 respectively.

In conclusion, cargo carried by air freighters is at least an order of magnitude worse than truck freight, and even belly freight in passenger aircraft adds enough weight to be worse than truck, as an overall average. Since over 50% of air cargo is reportedly carried on designated freighters [312], the overall average air freight emissions factor may be on the order of 2–4 $kgCO_2e$/tonne-mile.

9.4.5   International shipping

> - Seaborne shipping generates about 3% of global CO_2 emissions, and about 0.0125 to 0.025 $kgCO_2e$/tonne-mile from combustion and upstream fuel cycle greenhouse gas emissions, similar in magnitude to rail freight.
> - Extremely high sulfur dioxide and nitrogen oxide emissions from marine bunker fuel have an atmospheric cooling effect that largely cancels out the warming CO_2 effect over 100 years, but new lower sulfur fuel standards will sharply reduce this offset.

Ocean-going vessels dominate international transport (moving 90% of world trade), for the sea has, since antiquity, been the dominant means of trade among human societies, and the whale road remains among the most efficient means of moving the many products of Man's industry. While efficient, the sheer volume of shipping is such that carbon emissions from international shipping are not insignificant, and have typically been estimated at around 3% of global CO_2. Most shipping emissions come from three sources: (1) bulk transport ships, (2) oil tankers, and (3) container ships [313]. It is a testament to the modern world's fundamental dependence on fossil fuels that almost half of all seaborne tonnage is either oil (32% in 2010) or coal (about 12% in 2010, based on EIA numbers).

Marine bunker fuel, a low quality fuel oil used for propulsion, is by far the dominant source of shipping emissions, with a recent International Maritime Organization (IMO) study estimating yearly shipping emissions of 1,036 million tonnes CO_2 (2007–2012 average) [313]. These emissions are only from direct fuel combustion, and do not include upstream fuel cycle emissions, or emissions embedded in ship production, etc. Of note, over the 2007 to 2012 period, CO_2 emissions fell slightly (about 14%) despite an increase in demand, with this drop almost entirely attributable to the widespread adoption of "slow steaming," i.e. going slower: just as air drag increases with the square of velocity, so does drag in the water (air and water can similarly be modeled as fluids), and thus there is a non-linear increase in ship fuel consumption with speed.

Now, in general, the larger the ship, the more efficient the transport. For container ships, Psaraftis and Kontovas [314] estimated emissions factors between 0.0174 $kgCO_2$/tonne-mile and 0.0322 $kgCO_2e$/tonne-mile for all but the smallest ships (emissions are 2–3 times lower for oil tankers and bulk carriers). Supposing that the upstream fuel cycle increases emissions by 23% (similar to diesel) and an average combustion EF of 0.020 $kgCO_2$/tonne-mile, we might reasonably assume 0.025 $kgCO_2e$/tonne-mile for products moved by container ship, and 0.0125 $kgCO_2e$/tonne-mile for those moved by bulk carrier. However, as discussed next, sulfur dioxide and nitrogen oxide co-emissions from bunker fuel have a cooling effect that may more than outweigh CO_2 warming over the next century, although new sulfur controls will eliminate most of this offset in the future.

Bunker fuel was about 2.51% sulfur by mass in 2012 [313] (compare this to 0.0015% for ultra-low sulfur truck diesel), and thus shipping has been a major global source of sulfur dioxide, which has a strong cooling effect (see Section 3.4.5), and a 100-yr GWP on the order of -40 to -79 (the latter number includes indirect aerosol-cloud interaction). The IMO estimated annual

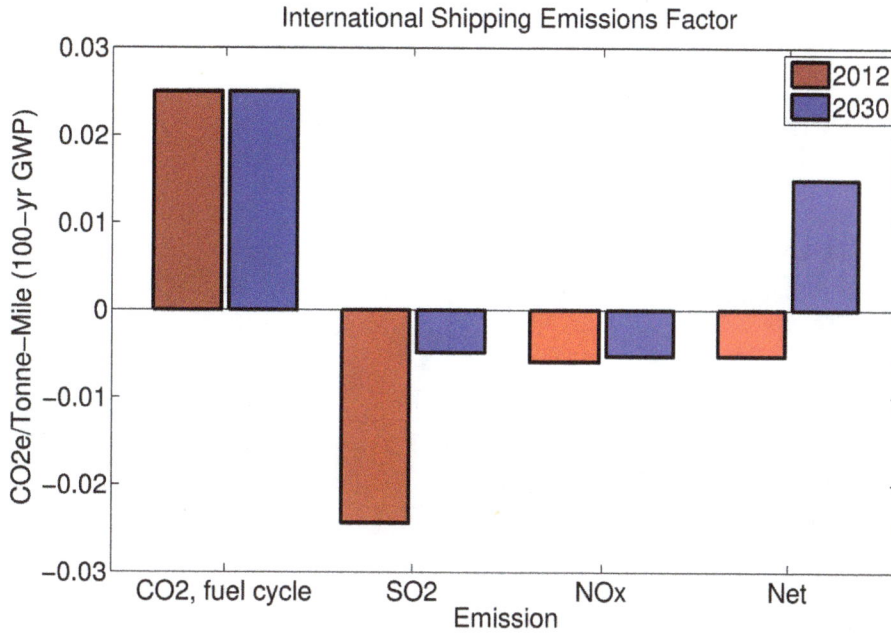

Figure 9.7: Approximate emissions factor for international shipping (container ship) on a tonne-mile basis for 2012 and 2030, where the lifecycle CO_2 emissions are derived from [314], as in the text, and using fuel sulfur and nitrogen contents and anticipated changes for marine bunker fuel given in [313]; 100-yr GWPs used for SO_2 and NO_x are -76 and -32, respectively. As seen, under the 2012 fuel mix, international shipping could actually have a net cooling effect (the harmful effects of sulfur and nitrogen emissions notwithstanding), while the cleaner 2030 fuel mix will be net warming.

sulfur emissions of 11.3 million tonnes SO_2 (2007–2012 average), as well as 6.3 million tonnes (as elemental N) of NO_x, which also has a strong cooling effect when emitted at sea (100-yr GWP of -32). Based on IMO emissions, the above GWPs, and similar emissions estimates by Eyring et al. [315], the cooling effect of nitrogen and sulfur co-emissions outweigh the warming effect of CO_2 over the short-term (20 years) and offset perhaps 55% to 100% of CO_2 over the 100-year time-frame. Figure 9.7 shows how the effective emissions factor for shipping changes when sulfur and nitrogen emissions are included, for both 2012 and 2030 under anticipated fuel changes.

Since 70% of shipping emissions occur within 250 miles of coastline, sulfur emissions cause significant coastal air quality problems, and beginning in 2015, ships operating in designated Emission Control Zones (ECAs), must burn fuel with 0.1% sulfur content or less. As a result of ECAs, and new limits on sulfur emissions outside ECAs (MARPOL Annex VI), the IMO projects overall sulfur content to decrease from 2.51% in 2012 to 0.5% by 2030. At the same time, IMO projections give a 30–50% increase in shipping CO_2 from 2012 to 2030 [313]. Thus, addressing sulfur emissions in isolation of a larger carbon reduction strategy is arguably problematic environmental problem-shifting.

Part III

The residence: household energy use, water, construction, and lawns

Chapter 10

Total per capita emissions from the residence

- Direct consumption of electricity and fossil fuels (primarily natural gas) accounts for most ($\geq$ 90%) residential shelter emissions, at about 12.0 MgCO$_2$e/household, or 4.7 MgCO$_2$e/person; overall, >60% of CO$_2$e is due to electricity (and most of the remainder results from natural gas combustion). Space heating and general electric loads are the two largest sources of emissions.

- Emissions attributable to housing construction and the building cycle are about 5–10% of those from residential energy when amortized over a 100-year building lifetime. Note however, that construction of an average new single-family still likely generates well over 100 MgCO$_2$e in very short order, an impact that is relatively small only when dividing over many decades of building life.

- Municipal water provision and (non-food waste) waste management are additional minor emissions sources attributable to the residence, although water provision is more carbon-intensive in parts of the Southwest than the rest of the nation.

- Residential (and institutional) lawns represent a non-trivial use of urban land, fertilizer, fuel, and irrigation water. Due to soil organic carbon accumulation, low-input lawns may act as weak net carbon sinks in some areas, but may be net carbon sources in others, especially in arid areas where large amounts of irrigation water are required, or when grass clippings are landfilled (due to resultant methane emissions).

- Including direct energy use, construction, water provision, and waste management, annual carbon-equivalent emissions attributable to residential shelter sum to about 13.125 MgCO$_2$e per household, or 5.2 MgCO$_2$e per person. Single-person households generate a higher 9.2 MgCO$_2$e annually.

The carbon footprint of residential shelters is dominated by electricity and direct fossil fuel (and wood) combustion for various household operations, mainly space heating and cooling, water heating, refrigeration, cooking, lighting, and various general electrical loads, which together sum to almost exactly 12.0 MgCO$_2$e/HH/year, or 4.72 MgCO$_2$e/person/year; of this total 64% is attributable to electricity, with the rest to direct fuel use.

The construction footprint is of secondary importance, with my best guess about 63 to 84 MgCO$_2$e for 1,500 to 2,000 ft^2 dwellings; using 75 MgCO$_2$e as a midpoint, this amortizes to 0.75 MgCO$_2$e/HH/year over a 100-year building lifetime, or 0.295 MgCO$_2$e/person/year. Emissions embodied in water conveyance and wastewater treatment sum to about 0.25 MgCO$_2$e/HH/year, or just under 0.1 MgCO$_2$e/person/year. Waste and recycling is treated separately, in Chapter 24, but the direct footprint of waste management is nearly trivial, with the exception of methane

Per Household and Capita Residential CO_2e

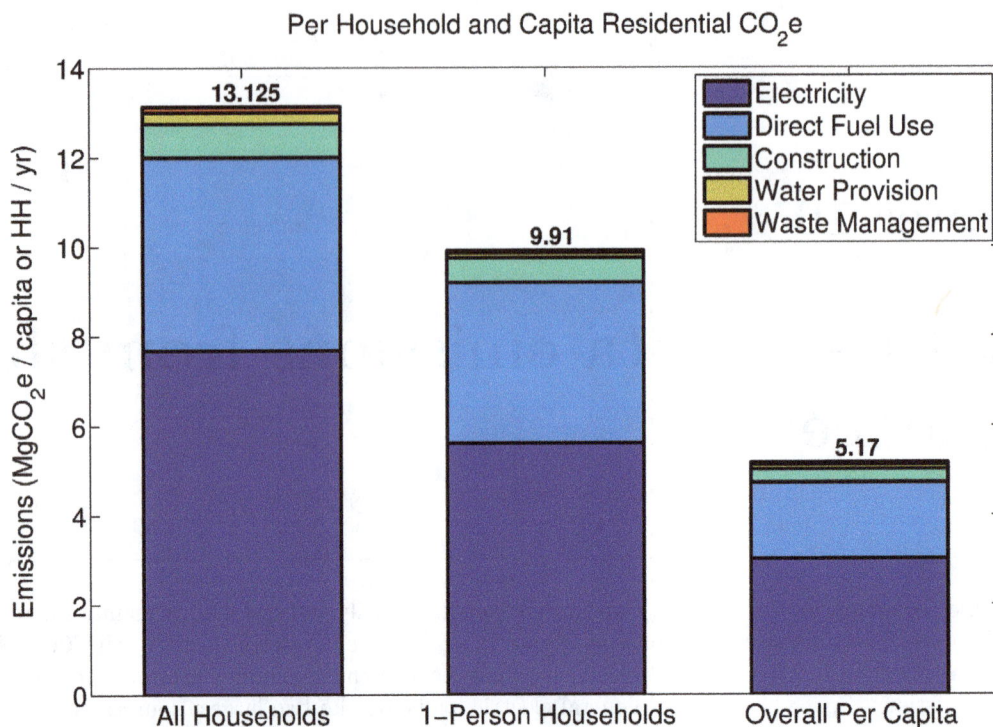

Figure 10.1: Overall average residential emissions attributable to electricity, direct fuel use (principally natural gas, but also fuel oil, propane, and wood), the construction footprint, upstream water provision, and downstream waste management (excluding landfill methane from food waste, which is attributed to diet), either in terms of overall household average (leftmost), single-person households (center), or overall per capita average (rightmost).

due to landfilling organic matter, and almost all of which is likely due to food wastes (0.1–0.2 $MgCO_2e$/capita), but this portion is attributed to diet in my accounting; waste collection and methane from paper waste might add about 0.05 $MgCO_2e$/year/person, or 0.125 $MgCO_2e$/HH. Together, then, we have about 13.125 $MgCO_2e$/HH/year, and almost 5.2 $MgCO_2e$/person/year attributable to residential shelter.

If we examine single-person households, we find that per-capita residential emissions for this subset are appreciably higher than the overall per capita average, with, based on the 2009 RECS, about 8.51 $MgCO_2e$/year for household energy use (with 61% due to electricity). Furthermore, in the 2009 RECS, dwelling square footage for single-person households was about 73% that of the mean household, suggesting about 0.55 $MgCO_2e$/person/year for the construction footprint. Assuming the above figures for per capita water withdrawals and waste management, we have 9.21 $MgCO_2e$/person/year, 75% higher than the overall per capita average. Overall household mean, single-person household, and overall per-capita residential carbon emissions are summarized in Figure 10.1.

Chapter 11

Overall energy/emissions from household energy use

In 2013, according to the 2015 EIA Energy Outlook [529], American households directly consumed 1,392 billion kWh of electricity; 1,480 billion kWh of natural gas; and 445.4 billion kWh of fuel oil, propane, and other fuels combined[1], directly and indirectly generating emissions of about 1.464 billion $MgCO_2e$[2]. This is about 22% of the EPA's estimated 6.673 billion $MgCO_2e$ US territorial emissions (in 2013). Note that 21.5% (0.315 billion $MgCO_2e$) of household energy-related emissions are *indirect*, and their omission would underestimate household energy emissions at 17% of the EPA territorial emissions.

Now, at the household level, the 2009 EIA Residential Energy Consumption Survey (RECS) [530] indicated that, on average, households consume 27,568 kWh of energy (including 1,300 kWh of wood energy), of which 11,320 kWh is electrical, 12,109 kWh is natural gas, and the remainder is other heating fuels (propane, fuel oil, kerosene, and wood); this corresponds to almost exactly 12.0 $MgCO_2e$, of which 7.72 $MgCO_2e$ is attributable to electricity, and 4.29 is attributable to fuel (including wood) use. Note that electricity accounts for only 43% of energy, but 64% of emissions, as it is more emissions intensive to convert fuels such as natural gas to electricity, transmit the energy to a residence, and then use this electricity, than it is to simply burn the fuel on-site.

These number represent crude averages. There is marked heterogeneity in fuel and energy consumption among households, and energy and emissions vary markedly with income, household size, square footage, climate region, and how individuals *choose* to use energy. The bottom 20% of households generate, on average, only about 40% of the mean (4.86 versus 12.0 $MgCO_2e$/HH, based on the RECS), and so it is clear *from the data* that very significant reductions in energy use are possible for most individuals. If everyone in the US used household energy like the bottom 20%, this would translate into a roughly 13% reduction in US territorial emissions (based on EPA and RECS numbers). Note that even if Americans collectively shifted to using household energy as the bottom 50% of households do, this would still reduce household emissions by 39% (average of 7.30 vs. 12.0 $MgCO_2e$/HH), and US emissions by 8.6%.

In the following section I review the general breakdown in energy/emissions by end-use (e.g. space heating, lighting, etc.), and in the subsequent section I review how overall energy varies with some of the aforementioned factors. Far more detailed discussions of each major end-use

---

[1]To be precise, 146.5 billion kWh of fuel oil, 126.0 billion kWh of propane, and 172.9 billion kWh of other fuels

[2]Using emissions factors of 0.6820 $kgCO_2e$/kWh electricity, 0.2613 $kgCO_2e$/kWh NG, 0.3081 $kgCO_2e$/kWh fuel oil, 0.2486 $kgCO_2e$/kWh propane, and 0.3 $kgCO_2e$/kWh other fuel. The last EF is chosen as an approximate average across fuel oil, propane, and kerosene.

are then given in the following chapters.

11.1 Overall energy/emissions by end-use

About 72% of the energy used by American households is devoted to either *generating* heat (i.e. space heating and water heating), or *moving* heat (i.e. air conditioning, refrigeration, and sometimes space heating). The remaining 28% of household energy use goes toward lighting and a variety of mostly electricity-powered appliances, such as televisions, computers, cooking appliances, etc. Reflecting the dominant role of heat, the EIA RECS divides energy use into these five broad categories: space heating, water heating, air conditioning, refrigeration, and *other*; four of five, and the named four at that, are exclusively heat generating or moving applications.

Note that, if we include clothes drying and most cooking appliances, an even greater share of household energy is devoted to heat generation or movement than as tabulated above, probably approaching 80%. Now, while 70% of energy is devoted to the four primary heat-related categories, only about 60% of household emissions are attributable to these categories. This is because heaters often directly use fuel combustion, while appliances (except some ovens, stoves and clothes dryers) rely more exclusively on electricity, which is associated with higher emissions per unit of energy than is fuel combustion. It is important throughout this chapter to be mindful of the difference between *energy* and *emissions*.

Figure 11.1 demonstrates the average energy devoted to the five EIA RECS energy use categories, along with the associated emissions.

11.2 Energy/emissions and major household characteristics

11.2.1 Square footage

All else being equal, larger houses consume more energy, both from the larger construction footprint and, much more importantly, from increased operational requirements. Energy use scales roughly linearly with square footage, as demonstrated in Figure 11.2; this figure also shows how the association with shelter area and emissions is robust when looking at households of different sizes (in terms of persons) and across climate zones. Figures 11.3 and 11.4 demonstrate how, since 1950, new US home sizes have ballooned dramatically.

11.2.2 Household size

Household size, i.e. the number of persons within the household unit, is obviously a meaningful determinant of residential emissions and energy, with both trending up strongly to about a three or four person household, will smaller increases thereafter. On average, single-person households have the highest per-capita residential emissions. These general trends are unsurprising, as heating and cooling loads vary very little with a dwelling's occupancy, a single refrigerator may serve many, and lighting also scales more with floor area than population. Figure 11.5 gives absolute and per-capita emissions under the 2009 RECS as a function of household head count.

11.2.3 Climate region

Heating and cooling (as a lumped category) is the dominant driver of emissions across most residences; since climate largely determines heating/cooling loads, it is apropos to examine how climate zone affects overall emissions. Perhaps surprisingly, heating+cooling emissions are

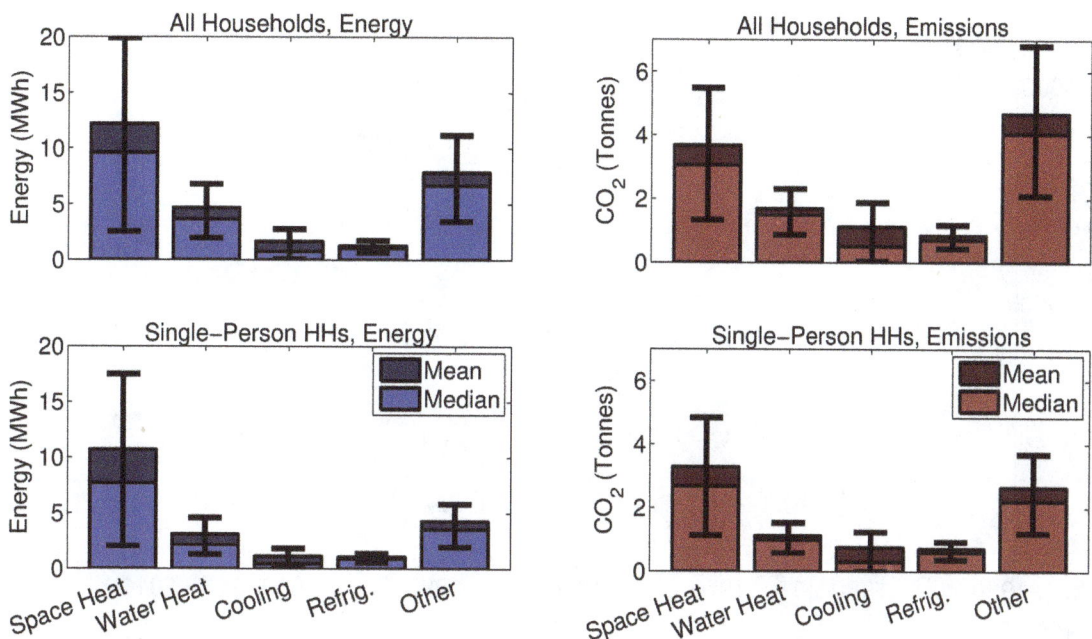

Figure 11.1: Average energy (left panels) and associated CO_2e lifecycle emissions (right panels) by major energy end-use category, as averaged across all households (top panels) and for single-person households (bottom panels), based on the 2009 EIA RECS. The mean and median, along with the upper and lower quintiles (error bars), are given. Space heating consumes the most energy, but since much of this is energy comes from natural gas or other heating fuels, the lumped category *other* (primarily lighting and other electrical loads that are more emissions-intensive on a CO_2e per kWh basis) is the largest single source of emissions overall. However, for single-person households, space heating is also the top emitter.

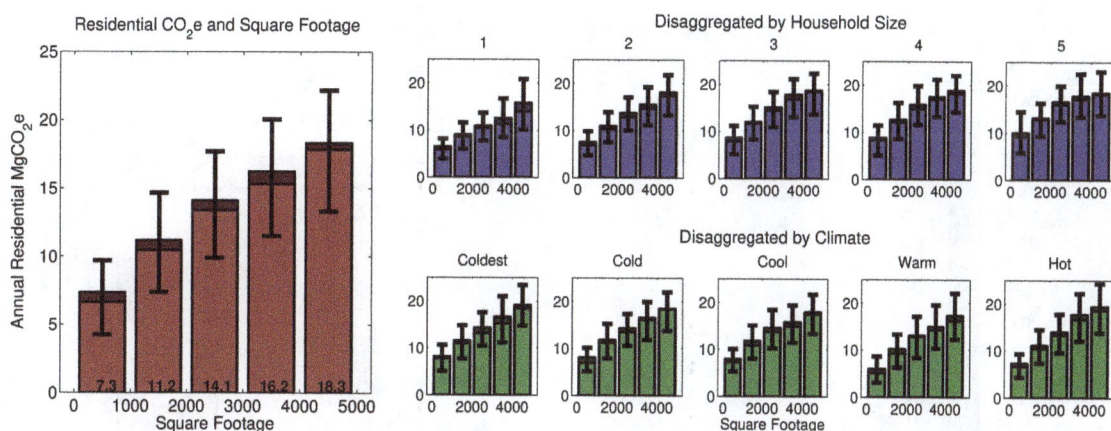

Figure 11.2: Carbon-equivalent emissions attributable to residential energy use as a function of dwelling square footage, under the 2009 RECS, with the overall trend on the left, while trends are given for as functions of household size and climate zone on the right. Clearly, dwelling area is a powerful determinate of energy use regardless of the number of persons in the house or climate. The lighter bars give the median, the darker the mean, and the error bars are the 20% and 80% percentiles.

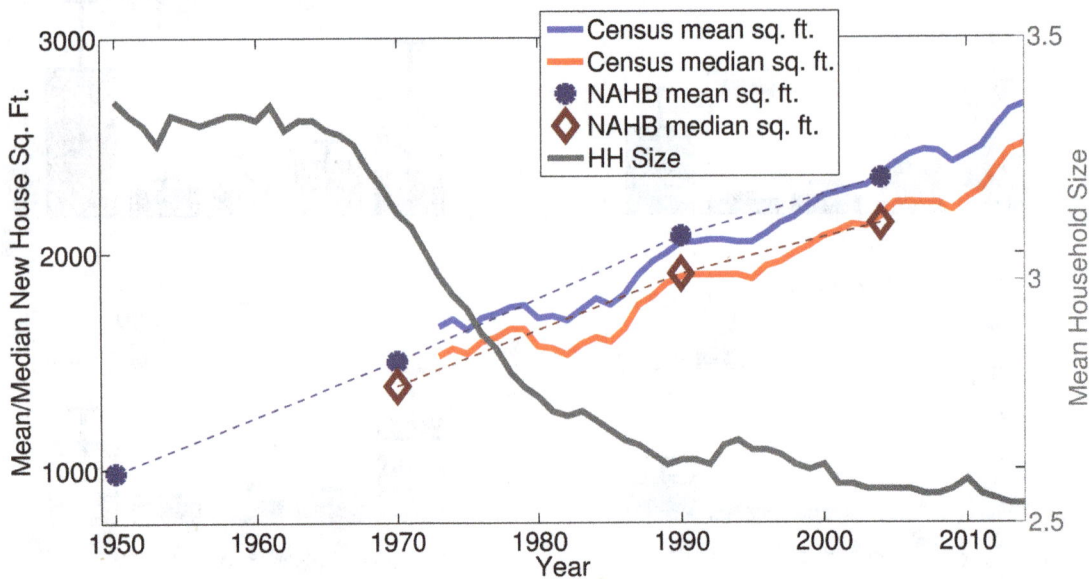

Figure 11.3: Figure demonstrating how new single-family home square footage has grown over threefold since 1950, even as mean household size has declined significantly. Data-points from 1950, 1970, 1990, and 2004 come from a widely cited NAHB report [531], while the time-series for 1973 through 2014 are from the US Census. Note that the median home size is somewhat less than the mean, as very large houses skew the distribution. All data is for new, single-family residences. The data suggests about 292 sq. ft. per occupant in 1950 rising to 1,059 sq. ft. in 2014, a 363% increase.

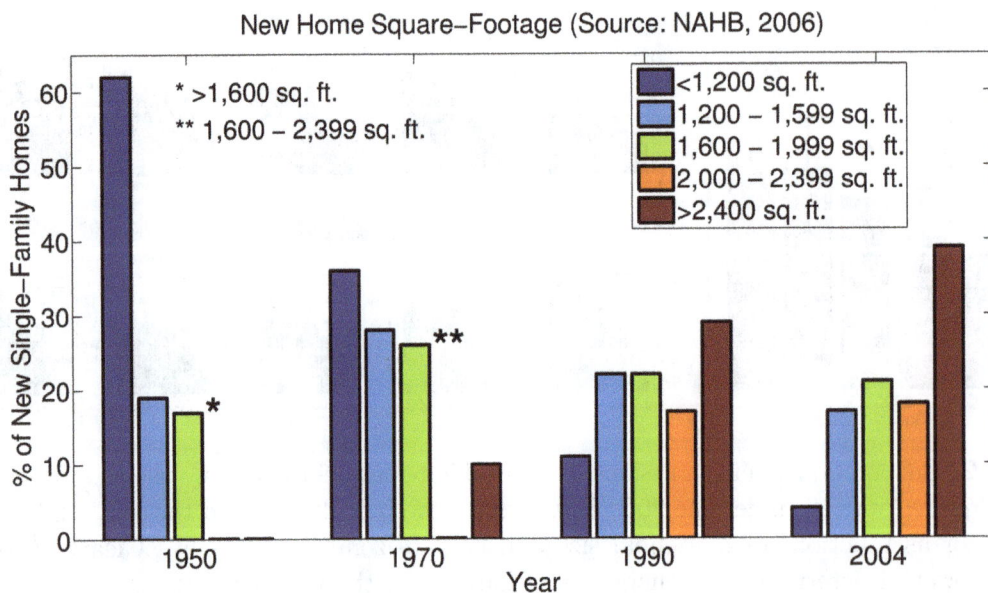

Figure 11.4: Distributions of new single-family house sizes in 1950, 1970, 1990, and 2004.

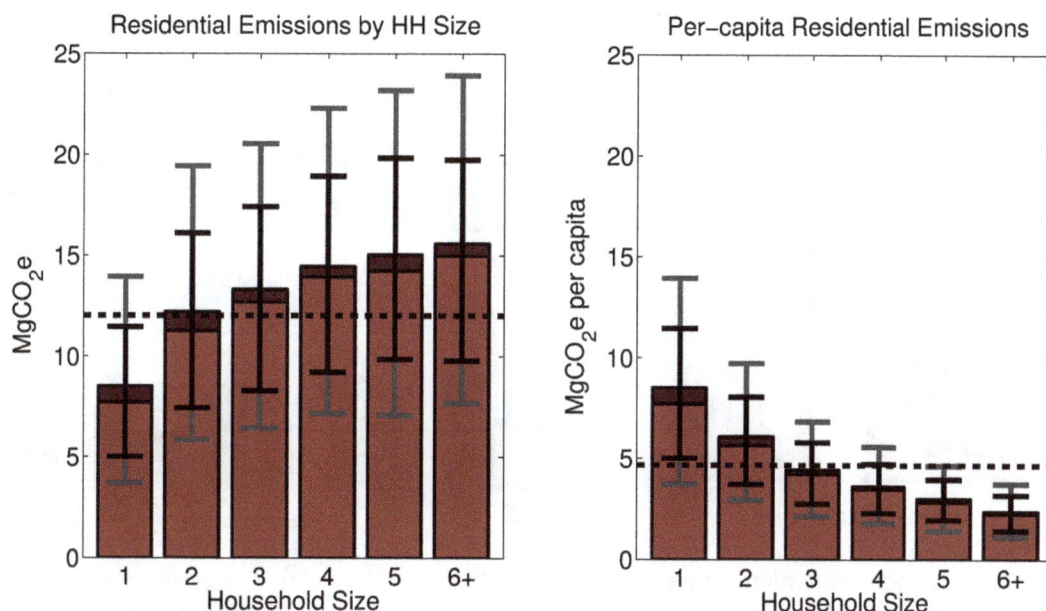

Figure 11.5: Absolute and per-capita residential energy-use emissions as a function of household size, based on the 2009 RECS. Dark and light error bars indicate quintiles and deciles, respectively, the lighter bar gives the median and the darker the mean, while the black dotted horizontal line signifies the overall average across all households. Clearly, larger households have higher CO_2e emissions, but are also lower-emitting on a per-capita basis.

highest in the coldest regions. That is, winter heating emissions in cold areas outweigh summer cooling emissions in hot areas, and this is presented in far more detail in the subsequent chapter.

11.2.4 Income

As expected, higher incomes predict higher residential emissions. In the 2009 RECS, residential emissions increase roughly linearly with household income, such that emissions for the top quintile (roughly >$100,000 annual household income) are about 64% higher than the bottom quintile (<$20,000). Income does, however, correlate with household size, with 2+ person households appreciably richer than singletons, and this partially explains the income-carbon association. Stratification by household size shows that the richest (by quintile) single-person households generate about 30% more carbon than the poorest, while the difference is around 50% for all other households.

Interestingly, climate also seems to interact with income, such that in the mildest, warm climate zone, there is less difference between rich and poor, while being wealthy in the hottest climate predicts far higher emissions compared to one's impoverished counterparts. Figure 11.6 shows the emissions gap between the top and bottom quintiles as a function of household size and climate zone. Note that these calculations are restricted to residential energy use, and do not include the inevitably larger (on average) construction footprints associated with wealth.

11.3 Emissions reductions strategy

In brief, it is likely a relatively simple matter for an average household to reduce its residential emissions by 25–50%, equivalent to a highly nontrivial 3–6 $MgCO_2e$/year per household, or 1.2–2.5 $MgCO_2e$/year per person, and some general strategies for this follow.

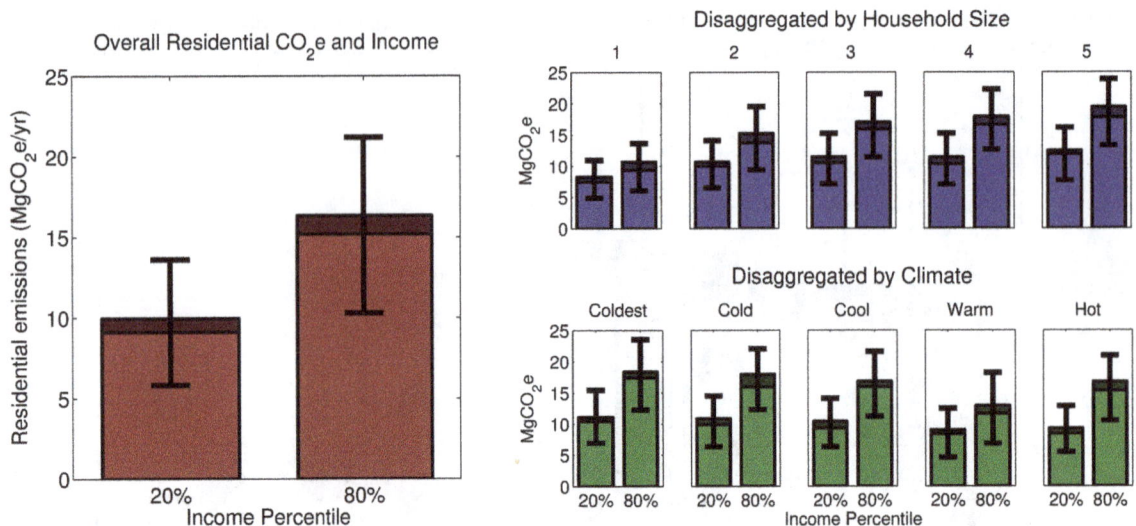

Figure 11.6: Carbon-equivalent emissions attributable to residential energy use for the bottom and top quintiles of the income distribution (error bars give the top and bottom quintile within each group), based on the 2009 RECS. The left panel gives results for the lumped population, with the top quintile generating 64% more CO_2e. The series on the left show how this disaggregates with household size and with climate zone, with the gap larger for bigger households and in more severe (colder/hotter) climes.

In the 2009 RECS, a variety of "other uses," mainly general electric loads, including lighting, washers and dryers, and a variety of plugged-in devices such as TVs, electronics, etc., actually make up the largest single lumped category of energy use other than space heating, and the number one CO_2e source (owing to the high carbon-intensity of electricity relative to heating fuels), with about 4.3 $MgCO_2$e/HH (6,300 kWh) due to general electricity use. Even subtracting off likely lighting, cooking, and washing/drying loads, we still have over 3,000 kWh/HH (and over 2 $MgCO_2$e/HH) per year attributable to other electric loads. Of these, phantom (or standby) loads from various devices and appliances turned "off" but in standby mode, may account for 500–600 kWh/year on average (see Section 15.4). Therefore, prudence in eliminating unnecessary appliances, keeping them off when not in use, and removing the wall-plug to avoid phantom energy loads may be an underappreciated way to avoid excess energy use, and one might easily reduce these miscellaneous electric loads by well over 50% (and even approach a 100% offset, depending upon how Spartan one's lifestyle is) via such a strategy, and save at least 1 $MgCO_2$e/HH/yr.

Heating and cooling together account for 40% of CO_2e from household energy use (4.78 $MgCO_2$e/HH), and so simply manipulating the thermostat may yield impressive carbon savings, at 20–50% of heating/cooling emissions. First, even in the coldest climates, every degree (in Fahrenheit) one sets the thermostat lower reduces heating emissions by at least 3–5%; setting the thermostat lower (e.g. 5–10 °F) at night and while gone may save an additional 5–15% in cooler climes (and more in warmer areas), and so sacrificing several degrees of warmth and some conscientiousness may reasonably save 20–30% on heating energy. Relative cooling savings are even greater, on the order of 7–15% per degree (°F), and daytime setbacks can save an impressive 15–30% even in Phoenix, AZ (with relative savings much larger in colder cities). Thus, shifting the base thermostat combined with night (winter) and daytime (summer) setbacks in a typical region may save 25–30% on heating, and a rather remarkable 50–80% on cooling, for net savings of 35–40% on the combined categories.

Additionally, multiple energy-efficiency improvements can independently reduce heating/cooling loads. The best heaters, either condensing gas furnaces with efficiencies approaching 100% or electrical heat pumps, yield about 20% fewer CO_2e emissions compared to typical units; similarly, the best new ACs are generally 20% less emitting than the average existing unit. Reducing air infiltration, sealing and insulating ductwork, and ensuring the house is optimally ventilated may all independently reduce heating/cooling costs by at least 10–20% compared to an average baseline, and so, taken together, several permutations on household improvements and behavioral changes can cut heating and cooling emissions by at least 50–75%, with absolute savings on the order of 2–3 MgCO$_2$e/HH.

Using ultra-low flow shower heads, washing dishes in tepid water, and avoiding unnecessary general tap hot water use may easily save 33–50% of hot water, or about 0.55–0.85 MgCO$_2$e/HH (from a base average of 1.69 MgCO$_2$e/HH). Highly efficient heat pump and solar water heaters can reduce water heating energy/emissions at least two or threefold, while ensuring adequate tank insulation or using a gas heater in lieu of an electric resistance heater can also reduce water heating CO_2e. A combination of conservation and a high-efficiency upgrade (heat pump or solar heater) would together eliminate 80% or more of water heating CO_2e.

A complete transition from traditional incandescent (or halogen incandescent) lighting to a mix of CFLs and LEDs may save over 75% of lighting energy relative to a 2008/2009 baseline, or about 0.885 MgCO$_2$e/HH/year. Line-drying clothes can save over 0.5 MgCO$_2$e/HH/year, and upgrading any old refrigerators (and not using the old inefficient model as a "beer fridge" or some such nonsense) could save something like 0.3–0.5 MgCO$_2$e/year.

11.3.1 Rooftop solar

Investing in rooftop solar, in much of the country but especially in the Southwest, can offset nearly 100% of one's residential electricity, and synergize with the efficiency/behavioral changes outlined above. When one's house also uses direct heating fuels, electricity accounts for a little over half of CO_2e emissions; when energy is all-electric, a near 100% carbon offset is possible with solar. As discussed in Section 4.13, the lifecycle emissions for solar are about 0.05 kgCO$_2$e/kWh, and so displacing 15,000 kWh per year (the mean for all-electric households in the 2009 RECS) with solar would reduce CO_2e emissions from 10.23 MgCO$_2$e to 0.75 MgCO$_2$e, a 93% savings. Supposing an installed cost of 3\$/W, a yearly average of 5 peak sun-hours per day, and an 80% performance factor, a \$30,000, 10 kW solar system could meet the needs of a typical household. Behavioral change and efficiency measures that reduced energy demand by 40% would then necessitate only an \$18,000, 6 kW system to meet all remaining need; combining these efficiency measures with a relatively inexpensive \$9,000, 3 kW system would still reduce overall residential energy emissions by about 67% from baseline.

Chapter 12

Heating and cooling

12.1 Overall energy and emissions: magnitude and variability

Based on the 2009 RECS, as an overall average (and including all wood use as a heating fuel), a full 44.3% of household energy was devoted to space *heating* (12,201 kWh/HH, or 4,744 kWh/person), equating to 30.5% of household CO_2e emissions (3.666 $MgCO_2$e/HH, 1.425 $MgCO_2$e/person), while space *cooling* was a smaller 5.9% of energy (1,638, kWh/HH, 637 kWh/person) and 9.3% of emissions (1.117 $MgCO_2$e/HH, 0.434 $MgCO_2$e/person). The crude carbon-intensities of these two activities are 0.3005 $kgCO_2$e/kWh and 0.682 $kgCO_2$e/kWH for heating and cooling, respectively. Note that these are the carbon intensities for energy used, and not the intensities in terms of heat added or removed from the house (a distinction that is elaborated on in our discussion of heat pumps and engines in Section 12.3).

These overall averages mask marked variation with climate, as expected. Figure 12.1 shows how emissions scale across five defined climate zones due to heating, cooling, and the sum of these two. Perhaps surprisingly, warm and hot regions actually have the lowest emissions from the lumped category, *heating plus cooling*, despite their increased dependence on AC.

What is also notable from Figure 12.1, is just how much variation in emissions there is *within* a given climate region. Across climate regions, the bottom and top 20% of households used 19–37% and 186–214% of the overall mean, respectively, for heating, 0–19% and 222–360% for cooling, and 18–38% and 184–217% for combined heating+cooling. Now, residence area (i.e. square footage) is, for obvious reasons, a primary determinant of the energy used for space heating/cooling, as demonstrated in Figure 12.4, and so it is useful to examine space heating/cooling emissions *per unit area* of space. Surprisingly, the variation of heating/cooling emissions among households on a CO_2e per area basis is *just as great* as on an absolute basis, and these variations are summarized in Figure 12.1.

The above analysis looks at absolute emissions and per-area heating emissions as independent variables, showing huge variation, and we also have that emissions scale strongly with residential square footage. We further have the simple mathematical identity

$$\text{Total } CO_2e = \text{Area} \times CO_2e/\text{Unit Area}, \tag{12.1}$$

begging the question: are lower emitting households low-emitters because the residence is small, or because they use very little energy per unit area? The answer is both, as a closer look at the data shows. Consider the bottom 20% of heating emitters in the coldest climate zone: they have both smaller residences (27.2% below the mean) and have per area heating emissions 48.8% lower than the mean, for an overall savings of 62.8% relative to the mean. The least users of AC in the hottest climate zone have residences 41.1% smaller than average and per area cooling emissions 61.4% lower, for emissions savings of 76.9% versus the mean. The rest of the numbers are summarized further in Figures 12.3 and 12.4.

Note that looking at per-area emissions in isolation could possibly lead one to the deeply misleading conclusion that larger houses are actually preferable, as they have lower per-area emissions on average, as seen in Figure 12.4. This is likely due to a combination of both physical and social factors, including the decreasing surface area to volume ratio and increasing thermal mass of larger houses, but also the facts that newer houses are both larger and more energy-efficient, larger houses are more likely to be better insulated (based on RECS data), and wealthier households, which own larger residences, may have newer, more efficient furnaces, etc. But even without controlling for these factors, it is clear that any benefit in emissions per area in *no way* overwhelms the advantage of a smaller heating/cooling volume offered by a smaller house.

Finally, while many emissions sources, such as hot water and lighting, scale with household size (i.e. number of people in the household, not square footage of the dwelling), space heating and cooling do not: we are principally heating the house and not the people within. It follows that per-capita heating/cooling emissions fall markedly for larger households: it may be madness to heat and cool a 5,000 square foot mansion for one, but not, perhaps, for ten.

12.1.1 The outsized influence of heavy consumers

It is a general theme that even relatively modest reductions by the heaviest consumers across consumption categories have a disproportionate effect on overall carbon emissions, and this clearly holds true here too. By way of demonstration, consider the following two thought experiments: (1) those 50% of households below the median reduce their heating emissions to *zero*, or (2) those 50% of household above the median pare back their heating costs to precisely the median, i.e. they make the profound sacrifice of tolerating only a typical level of consumption in the most energy-rich and wasteful society to ever exist.

Restricting ourselves to the coldest climate zone for fairness, the first scenario would reduce residential heating emissions by 29.0%, while the latter yields a similar 25.4% savings. If we perform the same experiment for cooling, in the hottest climate zone, the first scenario saves 22% of cooling emissions, but the second reduces emissions by a full 35.5%. Thus we see that curtailing excess among heavy users is just as (and sometimes more) effective than even complete abstinence among light users.

12.1.2 Conclusion

I believe we have reached the practical end of what the RECS can tell us about this topic and the sources of variation among heating/cooling emissions. It is enough to see that yes, indeed, even in very cold and very hot areas, many actual individuals use much less than their typical fellow. This variation is affected by myriad factors, and to understand how one can best join their ranks, I believe we are best served by approaching the topic from the basic physics and first principles involved, as I proceed to do in the remainder of the chapter.

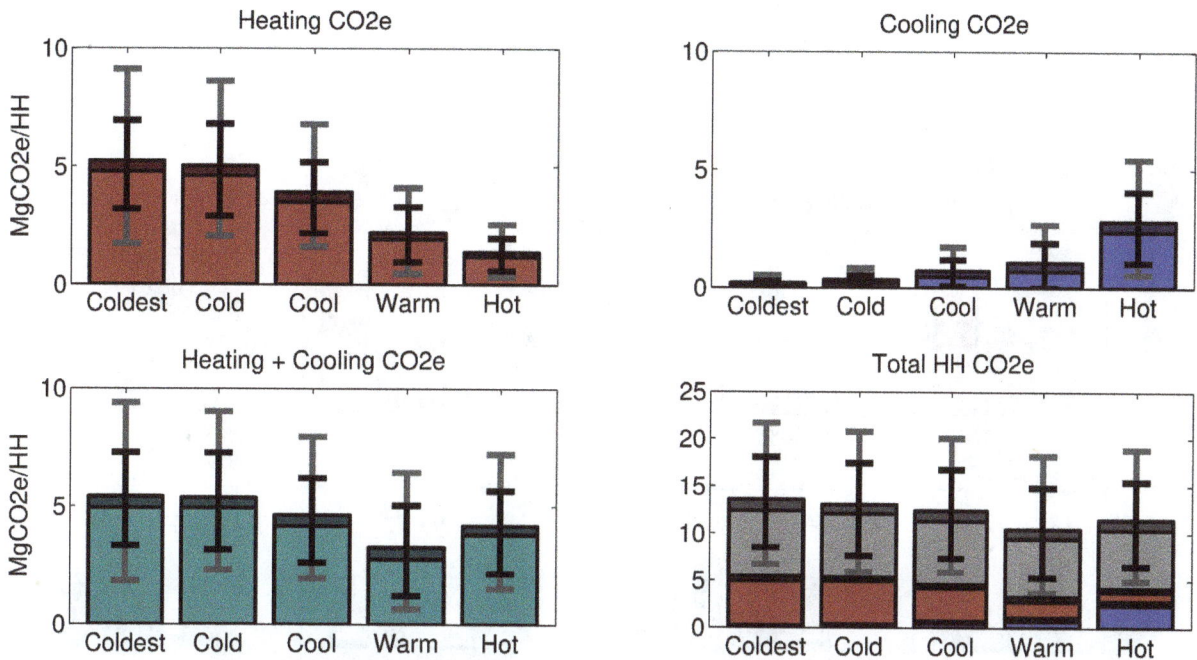

Figure 12.1: Carbon emissions attributable to household heating and cooling (top two panels) across five climate regions, defined in terms of heating and cooling degree days relative to 68 °F (abbreviated HDD and CDD, respectively) as (1) "coldest," <2,000 CDD and >7,000 HDD; (2) "cold," <2,000 CDD and 5,500–7,000 HDD; (3) "cool," <2,000 CDD and 4,000–5,499 HDD; (4) "warm," <2,000 CDD and <4,000 HDD; and (5) "hot," >2,000 CDD and <4,000 HDD. The light bar gives the median, the dark shows the mean (always slightly higher than the median, given that the distribution of emissions is skewed right by high consumers), while the dark and light error bars are, respectively, the top and bottom quintiles and deciles. Now, heating and cooling emissions scale with climate in an utterly unsurprising manner. What is a bit surprising is that the *sum* of these two (bottom left) is highest in the coldest regions. The hottest regions actually use appreciably less energy for air conditioning than the coldest regions use for heating, and this data suggests that settling warm and hot climes may actually be preferable, in terms of carbon emissions, despite great angst over the AC-dependence of Southwestern cities. The bottom right panel shows that this trend in heating+cooling emissions is reflected in total household emissions as well.

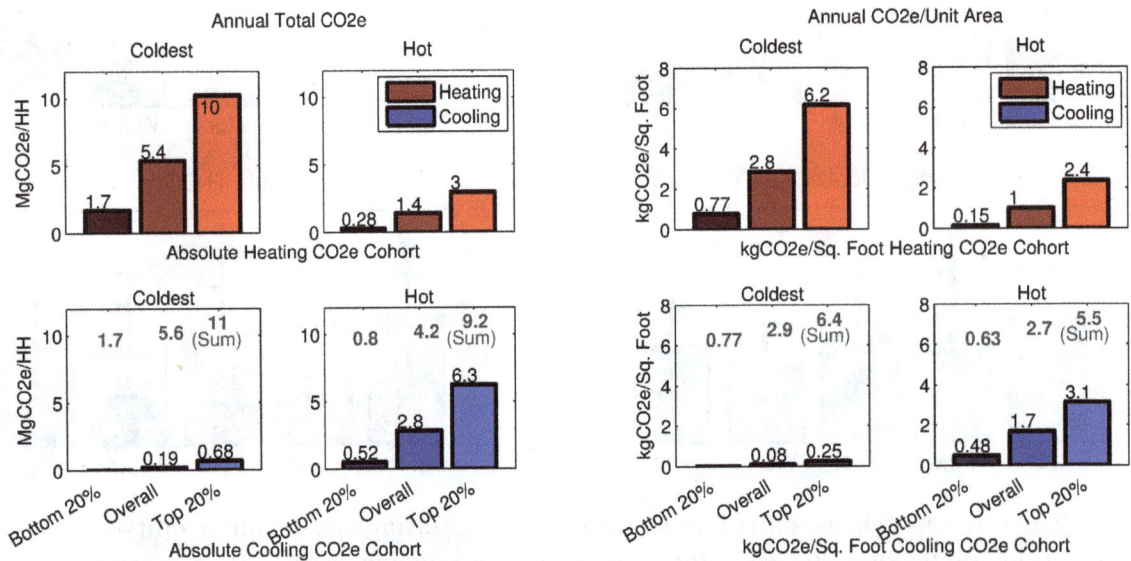

Figure 12.2: Carbon emissions attributable to household heating and cooling, stratified into the bottom 20%, all, and the top 20%, by either absolute emissions (left panels), or emissions per square foot (right panels). The left shows the absolute heating (top) and cooling (bottom) emissions for these cohorts, in the coldest and hottest climate zones (the sum of heating and cooling is also given in gray text in the bottom panels). We see that the bottom 20% of households generate only 20–30% as much CO_2e as the average, while the top 20% generate about twice the average. The right panels similarly show the CO_2e per square foot of residential space stratified into similar cohorts and we again see that the bottom 20% of households use only about 20–30% as much carbon to heat/cool each square foot as the average.

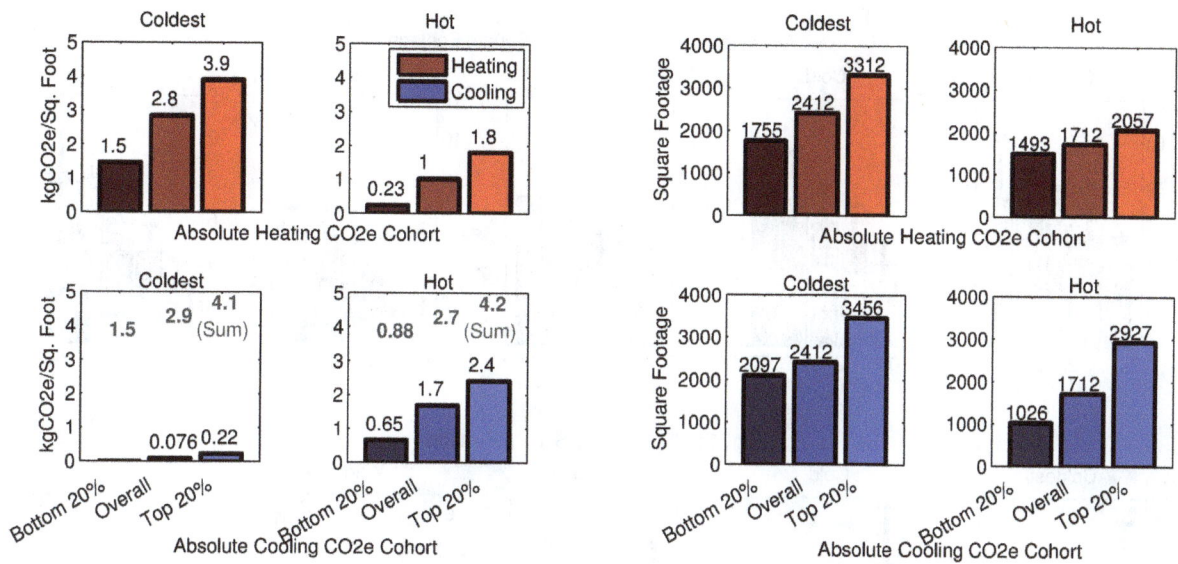

Figure 12.3: A closer look at the size (square footage) and per square foot heating/cooling emissions of those households in the bottom 20% of total heating/cooling CO_2e emissions (results shown for coldest and hottest climate zones). As we see on the left, $kgCO_2e/ft^2$ are about 50–75% lower than the average household, while the right shows that total housing square footage is 30–40% smaller than average. Note that there is less variability in housing size when it comes to cooling houses in the coldest zone or heating houses in the hot zone, and more variability when heating houses in the cold or cooling them in the heat, which makes physical sense. Finally, from this figure we also incidentally see that houses in the coldest climate zone are over 40% larger than those in the hottest (not shown, but there is also a consistent trend from larger to smaller house size as we go from the coldest to hottest climate zone), contributing to the large emissions impact of heating in cold areas.

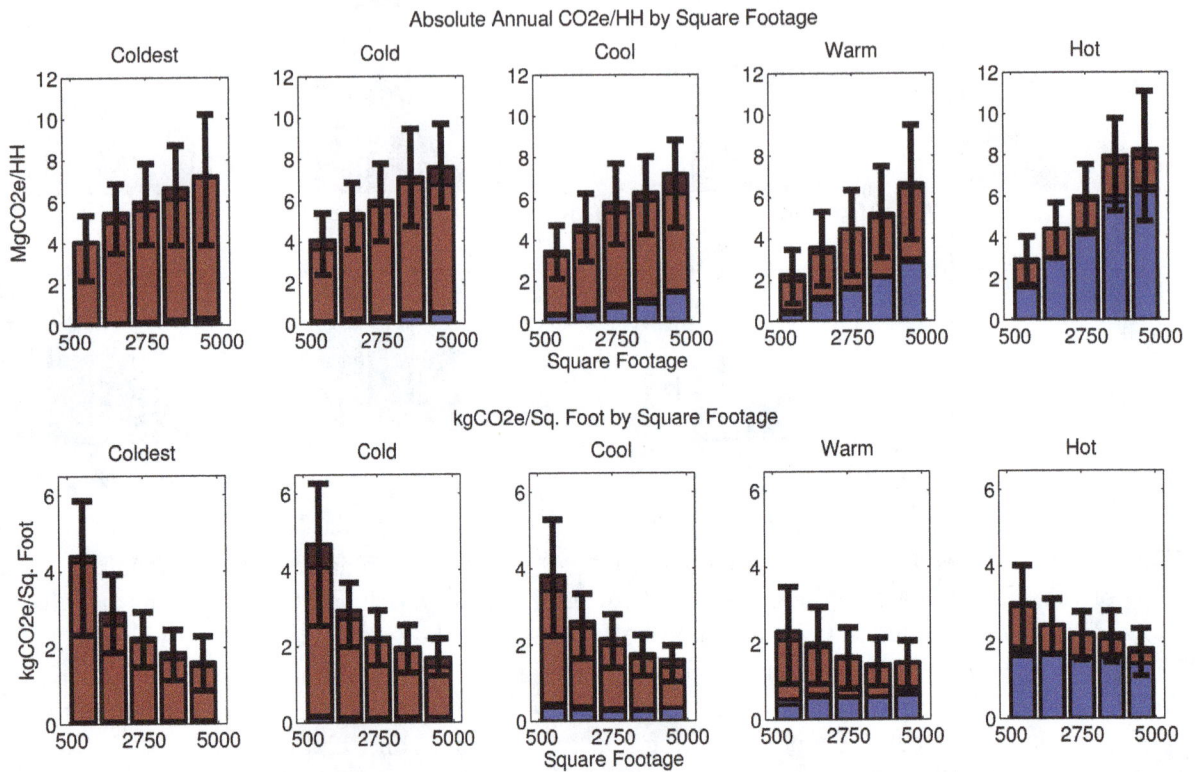

Figure 12.4: The top panels show how absolute heating (red) and cooling (blue) emissions scale with residential area across our five climate zones, while the bottom series shows how the per square foot emissions change with floor area. Error bars give the top and bottom quintiles for the sum of heating+cooling.

12.2 Seasonal temperature, heating and cooling degree days

- The cooling degree day (CDD) metric, a summation of the difference in outside temperature and thermostat set point over the year, approximates the total cooling load, and changing the thermostat by a single degree Fahrenheit can reduce yearly CDDs by about 7–15%.

- Relative cooling savings are lower in hotter climates, but absolute savings are greater. Increasing the thermostat from 73 to 78 °F in Phoenix, AZ, can lower CDDs by 30%; increasing to 83 °F saves over 50% of CDDs.

- The heating degree day (HDD) metric mirrors the CDD; because heating loads are much higher than cooling loads most places, every 1 °F change in the thermostat reduces HDDs by 3–9%; relative savings are least but absolute savings greatest in cold climes, with, e.g., a 5 °F drop eliminating about 15% of HDD in Minneapolis, MN.

Heating and cooling operate on very similar principles, and I begin with a simple examination of the thermostat and approximate energy consumed for *cooling*, reflecting, I suppose, my personal bias towards cooling, probably as a result of growing up in the American Southwest. Now, raising the thermostat is the easiest immediate way to save energy from AC, and while some popular sources cite energy savings of just about 1% per °F, a more detailed analysis shows that actual savings are far greater. In the subsequent sections I use detailed simulations to demonstrate that for each degree, in Fahrenheit, one adjusts their thermostat up by, there is, roughly, a corresponding 10% energy savings (over the entire year). But this can also be seen, and more intuitively understood, from a simple inspection of seasonal temperature curves.

Annual temperature varies roughly as a sine wave. Daily temperature, too, varies as a sine wave over 24 hours, and solar irradiance also is described by a sine wave during the daylight hours. This is demonstrated, for temperature, in Figure 12.5, which shows temperature over a typical meteorological year (TMY) for Phoenix, AZ (based on publicly available NREL data [532]).

Now suppose our thermostat set point (T_{set}) is 78 °F, so whenever the outside temperature (T_{out}) is above 78 °F, the AC must work to maintain this inside temperature; otherwise the AC is off. Furthermore, the greater the difference between the set point and outside temperature, the more cooling must be done. So, for a higher set point there will be fewer days that the AC need be on at all, and on those days when it is on, it works less than it would have. Thus there is, in a sense, a dual advantage to raising the thermostat. The overall amount of work that the AC must do is approximated by the metric of *cooling degree days* (CDD):

$$CDD = \int T_{out} - T_{set} dt \tag{12.2}$$

where the integral is calculated only when the outside temperature is greater than the inside temperature. We can approximate CDD by simply using the difference between the thermostat and average daily temperature:

$$CDD \approx \text{Sum over each day(Average outside temperature} - \text{Inside temperature).} \tag{12.3}$$

The two methods are shown graphically in Figure 12.5; using the latter introduces about a 10% underestimation, and while figures typically show average daily temperatures, all calculations are made using the more accurate hourly data.

As an example of the utility of the CDD concept, consider a single day where the average temperature was 90 °F, and the set point was 78 °F. Then the CDD total is approximately 90

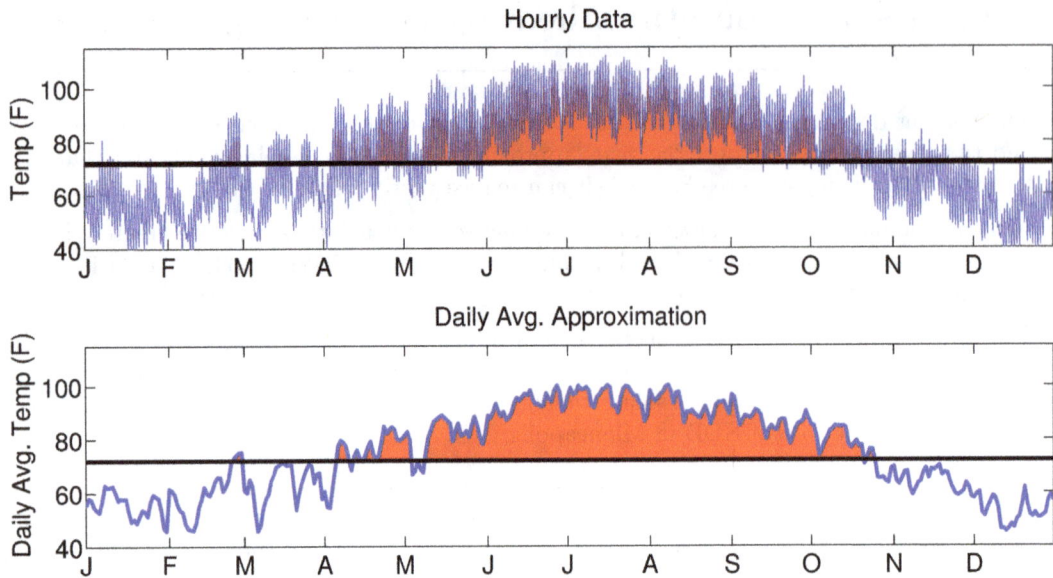

Figure 12.5: Integrated CDD (with a thermometer set-point of 78 °F) using either hourly temperature data (top panel), or average daily temperature (lower panel), under Phoenix, AZ TMY data, based on [532].

- 78 = 12. Over a month of such days we would have CDD = 30 × (95 - 78) = 360, and total energy needed for cooling proportional to 360 CDD. If we instead set the thermostat to 79 °F, then we would have CDD = 30 × (90 - 79) = 330, about 9.3% lower, and thus we would expect about 9.3% energy savings.

Figures 12.6 and 12.7 show CDD for different set points using TMY data for Phoenix, AZ. It is easily seen from these that rather dramatic energy savings are achievable with only a few degrees difference. Compared to a baseline of 78 °F, increasing to 83 °F yields 29% fewer CDD, while dropping to 73 °F increases CDD by 34%. An indoor temperature of 70 °F results in almost twice the CDDs as 80 °F (3,705 CDD vs. 1,877 CDD), and therefore roughly twice the energy consumption.

Overall then, in Phoenix, AZ, the hottest city in the US, each degree change in the thermometer corresponds to 7–8% fewer CDDs, and similar energy and emissions savings. As we see in the next section, the *absolute* CDD (or energy) savings per degree are greatest in the hottest climates, although the *relative* CDD/energy savings are greater in colder climes. *It follows that, the hotter the area you live, the more important it is (in the absolute sense) to accept a higher indoor temperature during the summer.*

Cooling degree days for different climes

In the prior section we saw that, in Phoenix, AZ, a 1 °F increase in the thermostat set-point results in a 7–8% reduction in CDD (using 78 °F as a baseline). However, in Chicago, IL, moving from 78 to 79 °F results in an 18% decrease in CDD. Intuitively, it is clear why this should be. Suppose the average outside temperature (when the AC is on), is 92 °F, as might be the case in Phoenix. Then changing from an indoor temperature of 78 to 79 °F results in an outside-inside temperature change of 7%. If the outside temperature was only 83 °F, however, we would see a 20% reduction in the indoor-outdoor temperature change going from 78 to 79 °F inside. Thus, the higher absolute temperatures in hotter cities imply smaller relative differences in CDD per thermostat degree change.

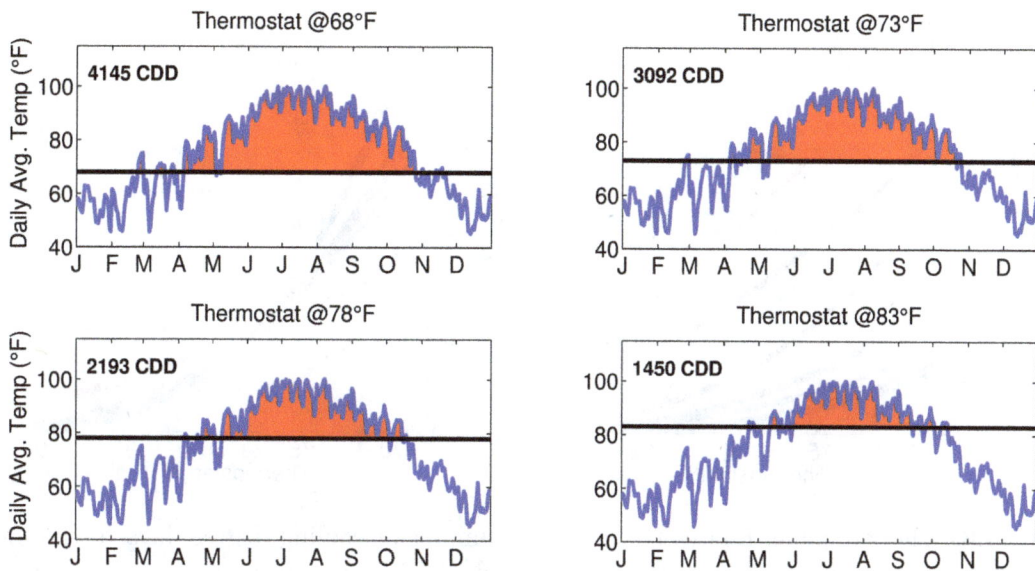

Figure 12.6: Cooling degree days for Phoenix, AZ, under different indoor temperature set-points, spanning 68 to 83 °F. The CDD value is given by the red area under the curve, and correlates to the total amount of heat that must be removed from inside the house, and thus energy use (and emissions). As can be seen, each five degree (in °F) increase results in an incremental increase of 35–50% more CDDs (the relative difference is greatest at higher thermostat set points). Increasing the thermostat from 68 to 83 °F reduces CDDs by 65%; both set-points are plausible in the Phoenix area.

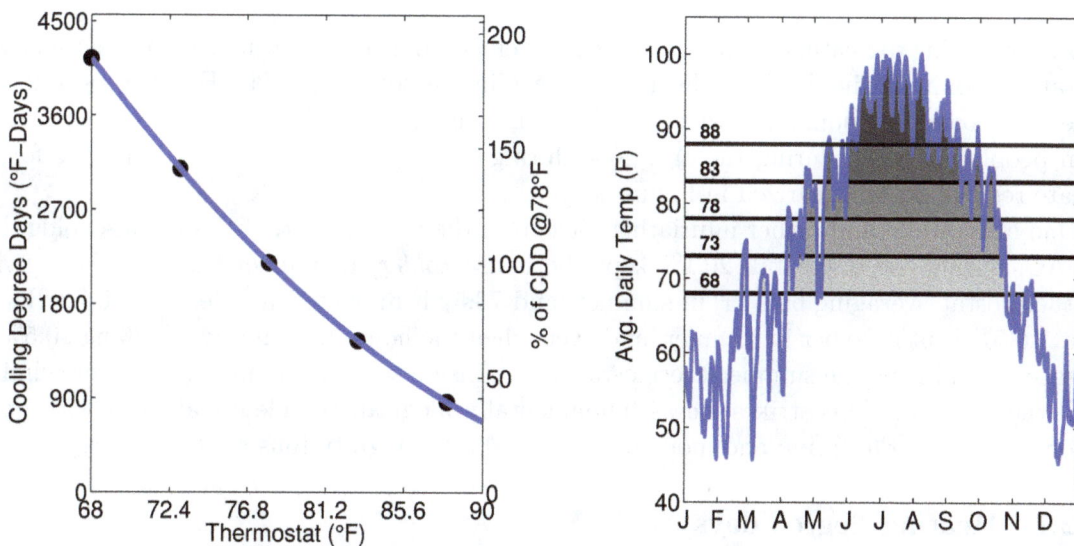

Figure 12.7: Cooling degree days for Phoenix, AZ, under different indoor temperature set-points, as a continuous curve on the left, and with representative points shown on the right.

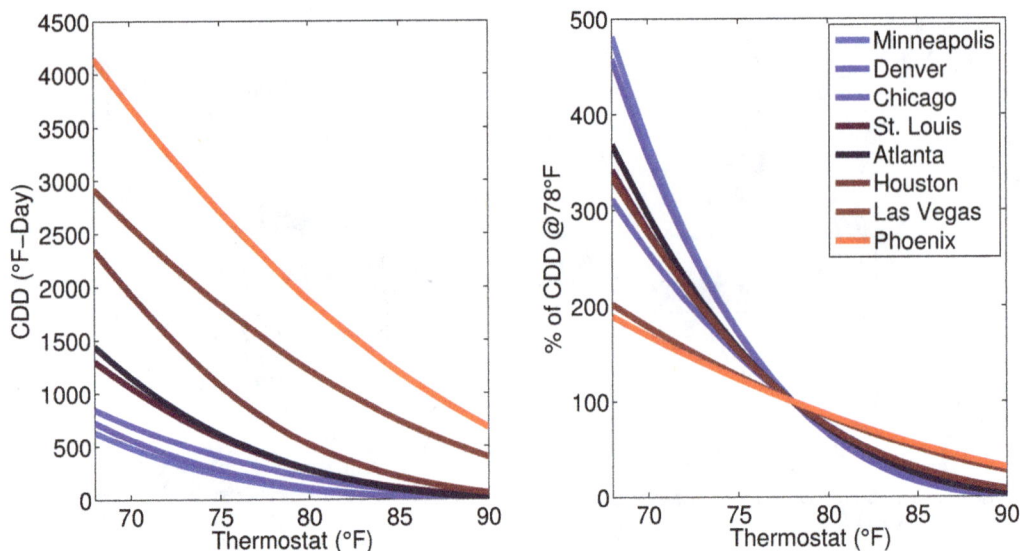

Figure 12.8: The left panel relates the absolute number of cooling degree days to thermostat set-point for eight cities across climate zones (using NREL typical meteorological year data), while the right panel shows the relative change in CDDs compared to a baseline of 78 °F.

This would seem to imply that altering the thermostat set-point (in summer) is more efficacious in colder climes. However, as is clearly demonstrated in Figures 12.8 and 12.9 (using TMY for representative cities), increasing the thermostat results in far greater *absolute* CDD savings in hot climates. Again, this should make intuitive sense. In Phoenix, AZ, there are 2,193 CDD for a thermostat at 78 °F, and 2,032 CDD at 79 °F, implying an absolute savings of 161 CDD (relative savings 7.3%). In Chicago, the corresponding numbers are a mere 158 CDD at 78 °F, 129 CDD at 79 °F, and an absolute savings of 29 CDD (relative savings 18.3%).

Typical set-points

Before continuing to heating degree days, let us briefly detour to tabulating common actual thermostat set-points during both the heating and cooling seasons. The 2009 EIA RECS collected three thermostat set points for winter and summer (assuming AC is used): the temperature when people are home during the day, gone during the day, and at night. The results for all climate regions are summarized in Figure 12.10.

One interesting, and rather infuriating, fact from the data is that, of those households that had AC, during the day a full 20.1% kept the house *colder* in summer than in winter, with the thermostat averaging 69.3 °F in summer, and 73.5 °F in winter, a difference of 4.2 °F. At night, 16.5% kept it colder in summer, and even when the house was empty, a baffling 10.5% of households still kept the summer thermostat lower (inside set-point temperatures were similar in all these cases). This strikes me as fundamentally illogical, completely at odds with one's seasonal thermal adaptation and mode of dress, and a truly gratuitous waste of energy.

12.2.1 Heating degree days

The heating degree day (HDD) concept mirrors that of cooling degree days, being an integration of the approximate heating load for a given thermostat setting. Generally speaking, the number of HDDs in very cold cities is far greater than the CDD burden in even the hottest cities. Using Minneapolis, MN, as one of the coldest cities in the US, we have 8,691 HDD at a thermostat

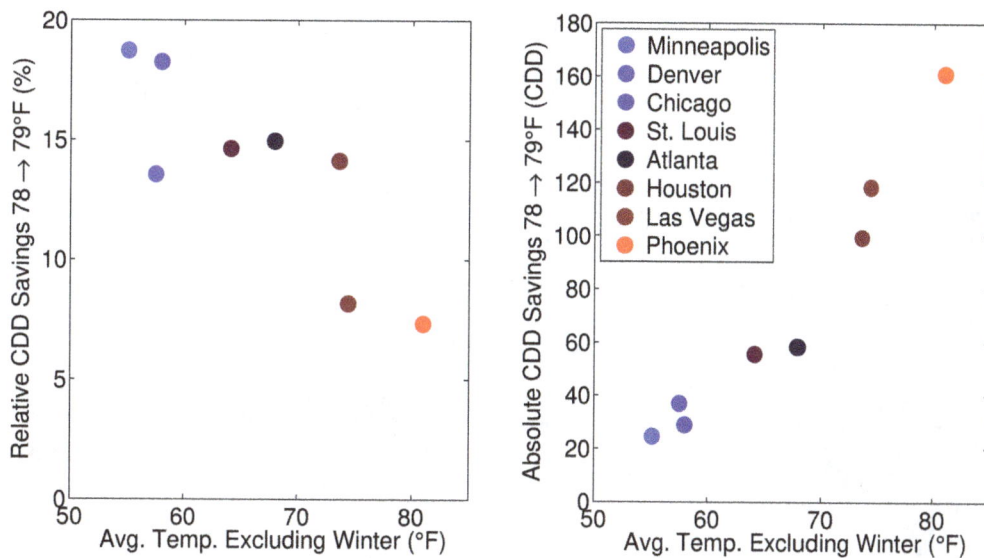

Figure 12.9: The relative and absolute change in CDDs resulting from a 1 degree thermostat change, from 78 to 79 °F, in eight different cities, plotted as a function of the city's average temperature (excluding winter). As we see on the left, the relative difference is greatest in colder cities, while the absolute difference is maximized in hot climes.

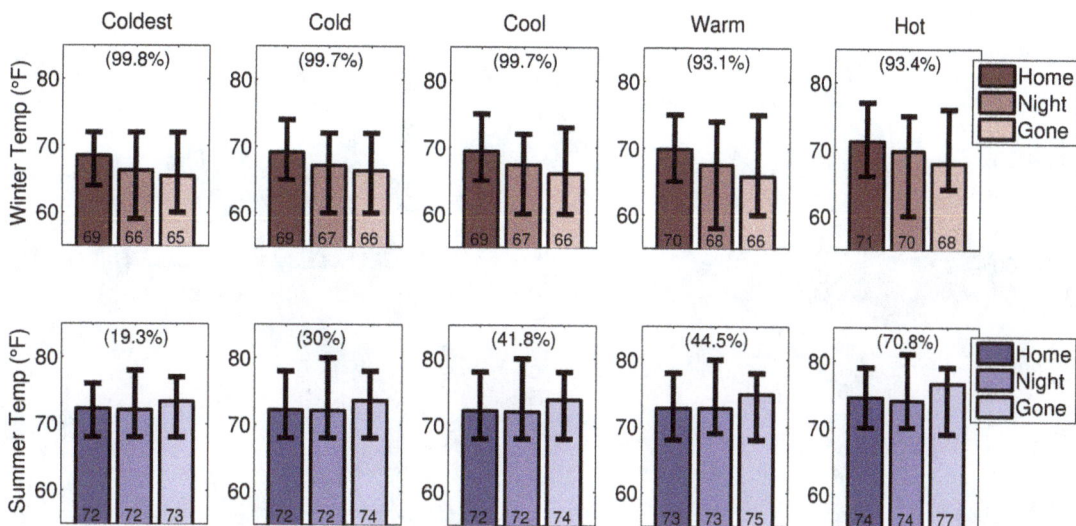

Figure 12.10: Average thermostat set-points for five climate regions under three different conditions: (1) "Home" (people home during the day), (2) "Night", and (3) "Gone" (nobody home during the day), as determined from the 2009 RECS. The percentage in each plot indicates the number of households that used either heating or cooling (i.e. AC) technology, and set-points are derived only from that subset of households. The bars show the mean set-point, and error bars give the upper and lower deciles.

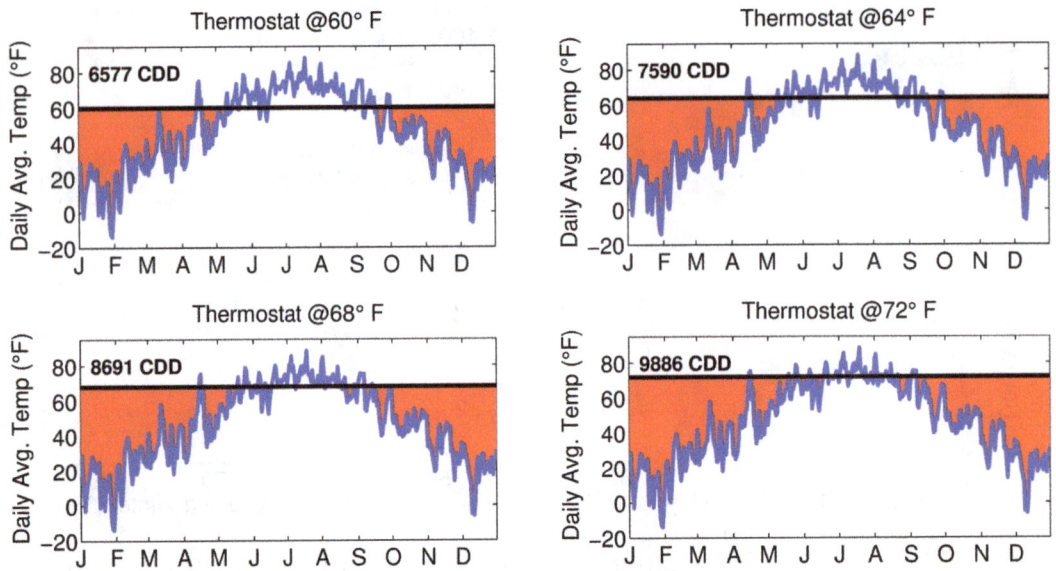

Figure 12.11: Heating degree days for Minneapolis, MN, under different indoor temperature set-points, spanning 60 to 72 °F. Mirroring Figure 12.6, HDDs are given by the red area under the curve, correlating with the heating load.

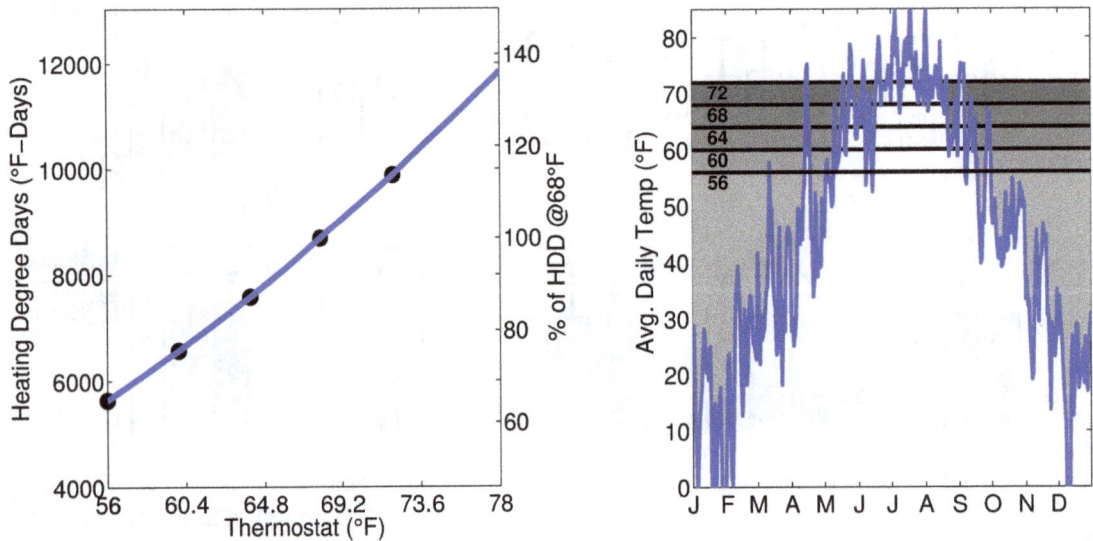

Figure 12.12: Heating degree days for Minneapolis, MN, under different indoor temperature set-points, as a continuous curve on the left, and with representative points shown on the right.

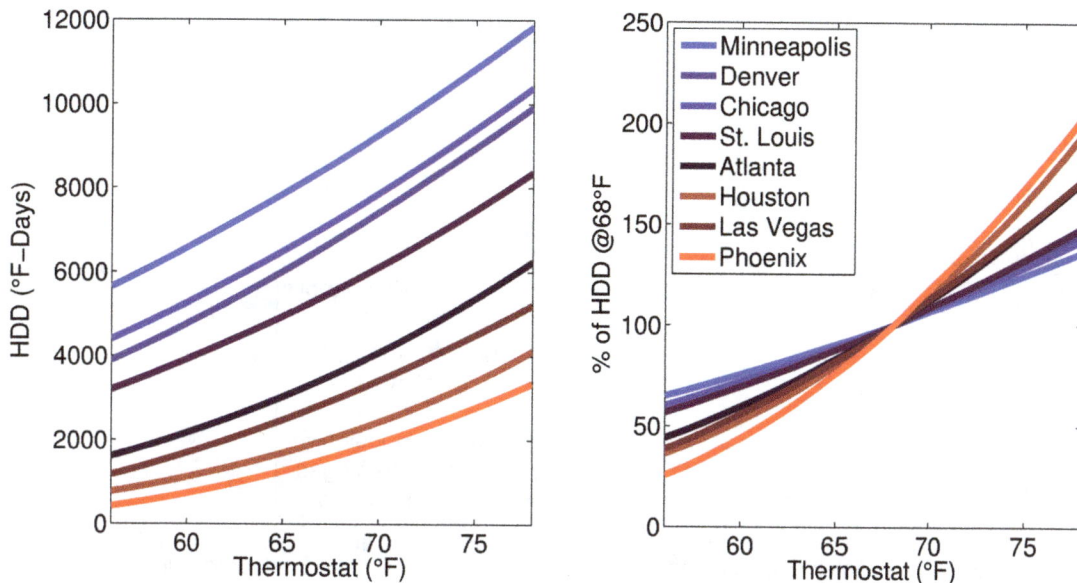

Figure 12.13: The left panel relates the absolute number of heating degree days to thermostat set-point for eight cities across climate zones (using NREL typical meteorological year data), while the right panel shows the relative change in HDDs compared to a baseline of 68 °F.

set-point of 68 °F, which is thrice the 2,899 CDD we would have in Phoenix, AZ with the thermostat at 74 °F. Total HDDs for Minneapolis under different thermostat set-points are demonstrated in Figures 12.11 and 12.12.

Thus, in the coldest cities, there may be three to four times as many HDD in winter as there are summer CDD in the hottest cities. Furthermore, for all but the warmest cities, there are far more heating than cooling degree days; in many cities 90% or more of all degree days are heating rather than cooling. This corresponds well to RECS data, in which 88% of energy devoted to either heating or cooling is for heating (i.e. 7.4 times as much energy goes towards heating).

Because more energy, in absolute terms, is typically devoted to heating compared to cooling, we see that the relative energy savings of altering the thermostat set-point are not as great as they are when cooling, but the absolute savings are greater. In Minneapolis, MN, moving from 68 to 67 °F gives a 3.3% HDD drop. Four degree deviations from 68 °F (i.e. to 64 or 72 °F), result in 13% increases or decreases in HDD. Per degree savings are around 4–6% for other reasonably cold cities, and Phoenix sees an 8.3% saving per degree. Thus, a four degree winter thermostat difference translates to a 15–20% energy difference in most colder areas, and a 25–30% difference in warmer climes.

While degree days are a useful metric, there is not a one-to-one relationship between heating or cooling emissions and degree days, due to the different technologies and physics involved in actually altering and maintaining a temperature difference between the environment and a building interior. Let us now turn to understanding one such fundamental technology: the heat pump, or air conditioner.

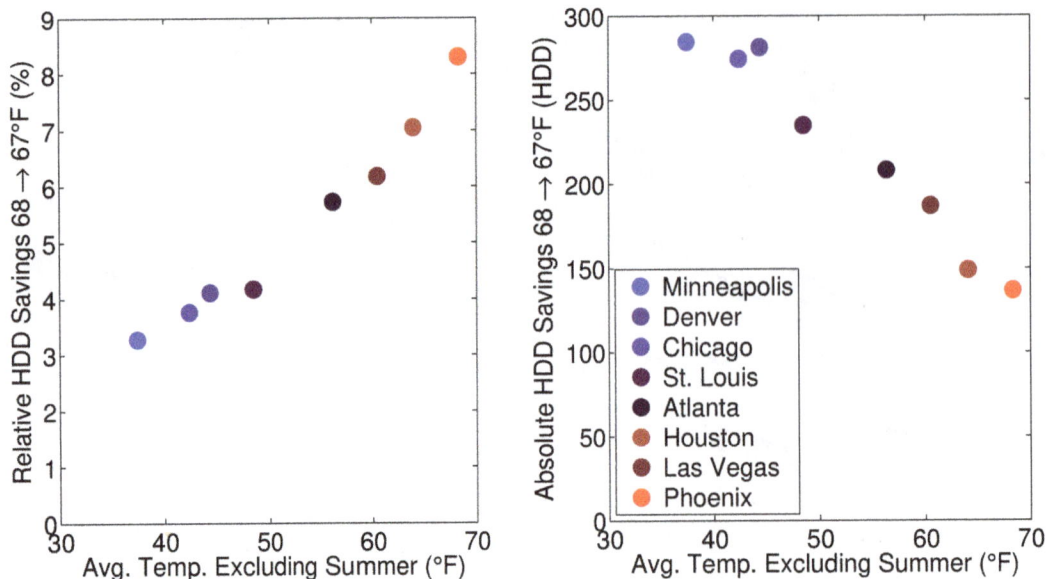

Figure 12.14: The relative and absolute change in HDDs resulting from a 1 degree thermostat change, from 68 to 69 °F, in eight different cities, plotted as a function of the city's average temperature, excluding summer. The relative benefit is greatest in hot cities, while the absolute benefit is maximized in the cold.

12.3   Understanding heat pumps and air conditioning

- The Carnot heat engine is the theoretical basis for heat pumps/air conditioners: heat pumps do work on a system to *move* heat from a cold body to a hot body. For an AC, the cold body is the inside of the house, while for commercial "heat pumps" the opposite is true.

- More heat energy can be moved than is invested, with the ratio called the coefficient of performance (COP); commercial ACs/heat pumps typically achieve a COP in the 2.5–5 range. The greater the temperature difference between the inside and outside, the lower the COP, and so there is an additional efficiency advantage to higher summer thermostat settings.

An air conditioner is a *heat pump*, that is, it moves heat from one pool of air to another. This is distinct from the process of generating heat, e.g. by burning fuel. Moving heat requires energy, but the energy requirement is much less than the energy of the heat moved, an almost magical result[1]. To understand heat pumps (and air conditioners) requires an introduction to the notion of the *Carnot cycle* and the *Carnot heat engine*, a fascinating theoretical construct that underlies the theory of thermodynamics. As shown in Figure 12.15, suppose we have two large bodies (or "baths"), each at a constant temperature, one hot and one cold[2]. It is rather

---

[1]An imperfect analogy is that of pumping water. This water obviously contains a great deal of heat energy, but we clearly will not need as much energy as is contained in the fluid to move the fluid around.

[2]We must take care here to distinguish between heat and temperature. Heat is a measure of the thermal energy contained in a body, while temperature is a measure of the average kinetic energy of particles in a body, i.e. the average particle speed. Consider as an example a 100-ton boulder at, say 10 °C, and a 1-pound rock at 20 °C. There is vastly more thermal energy, i.e. heat, contained in the boulder than the rock despite its lower temperature. To raise the temperature of the boulder to 20 °C would require a massive infusion of heat energy, while the rock cooling to 10 °C would represent only a tiny release of heat. Similarly, heating or cooling the mass of air in a mansion by 10 °C takes far more energy than heating/cooling the air in a studio apartment.

intuitive that one could move heat from the hot body to the cold body, using a "working fluid" and in the process siphon some off to do useful work. A steam engine, for example, does just this. Using thermodynamic principles, the theoretical maximum efficiency with which work can be extracted from this process can be derived as

$$\text{Maximum efficiency} = \eta = \frac{T_H - T_C}{T_H}, \tag{12.4}$$

where T_H and T_C are the temperatures of the hot and cold bodies, respectively. Note that temperature here is measured on the absolute Kelvin (K) scale, where the minimum value of 0 K corresponds to absolute zero, and degrees Celsius are related to Kelvins by $^\circ C = K - 273.15$.

Now, as an example, suppose we moved 10 kWh of energy from a hot boiler, held by coal or gas combustion at $T_H = 830$ K, to a condenser held at $T_C = 300$ K by cooling fluid. Then our maximum efficiency is $(T_H - T_C)/T_H = 530/830 = 0.64$, so about 6.4 kWh of energy could be extracted from the 10 kWh energy investment. This process, when using water as a working fluid that is converted to steam and then back to liquid in a continuous cycle to extract usable energy from the boiler is known as the *Rankine Cycle*, forms the basis of steam turbines, and is illustrated in Figure 12.15.

Thus, we extract work energy from this half of the thermodynamic cycle. This cycle can also be reversed: by performing work on the system, it is possible to move heat energy from the cold body (again using a working fluid) to the hot body. The magic of this process is that the amount of heat energy that can be moved is greater than the energy invested as work. So for the example above, we extracted 6.4 kWh of work from moving 10 kWh of heat, and a perfect reversal would imply investing 6.4 kWh to move 10 kWh.

The ratio of energy moved to energy invested is referred to as the *coefficient of performance* (COP). That is, COP units of heat energy can be moved by a single unit of work energy. If our goal is to warm the hot body e.g. a commercial "heat pump," and a conceptual reversal of the Rankine Cycle), then this is expressed more formally as

$$\text{COP} = \frac{Q_H}{W}, \tag{12.5}$$

where Q_H is the amount of heat energy moved to the hot body, and W is the amount of work energy input required to move this heat. It can, from a thermodynamic argument similar to the one used to derived the maximum efficiency of a heat engine, be shown that the theoretical maximum COP is:

$$\text{COP} = \frac{T_H}{T_H - T_C} = \frac{1}{\eta}. \tag{12.6}$$

On the other hand, if we are interested in cooling the cold body (e.g. an air conditioner), our COP is:

$$\text{COP} = \frac{Q_C}{W}, \tag{12.7}$$

where Q_C is similarly the amount of heat removed from the cold bath, W is the work invested, and the theoretical maximum cooling COP is:

$$\text{COP} = \frac{T_C}{T_H - T_C} = \frac{1}{\eta} - 1. \tag{12.8}$$

Performing both halves of the thermodynamic cycle of a heat engine coupled to a heat pump, with perfect efficiency, is the Carnot cycle. Note that, the higher the temperature difference between the two bodies, the more energy we can extract as work in the first half (heat engine mode), but the more energy must be invested in the second half (heat pump mode). Under

275

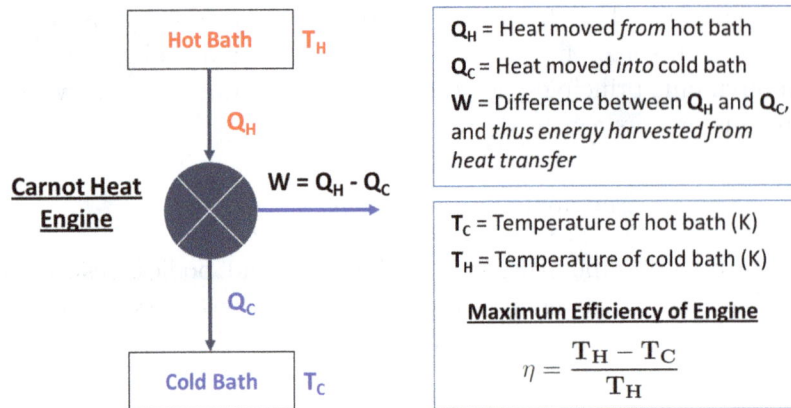

Figure 12.15: Schematic for theoretical operation of a Carnot heat engine.

perfect efficiency no energy is lost or gained by running the Carnot cycle. In reality, of course, imperfect efficiency means running such a cycle will always drain energy from the system.

So to summarize, *heat engines* are devices that harvest usable work from heat transfer from hot to cold, working with the heat gradient, and actual examples include internal combustion engines (e.g. diesel or gasoline) that convert heat flow to kinetic energy (e.g. to move a car's wheel) or steam turbines that generate electricity. *Heat pumps* are devices that perform work to transfer heat from cold to hot bodies, i.e. they work against the heat gradient. Thus, an AC moves heat from the cold interior of a house to the hot outside air, while a "heat-pump" moves heat from the cold outside air to the warm interior.

In physical terms, an AC is composed of a compression and evaporation coil which contain the working fluid (e.g. the halocarbons, HCFC-22 or HFC-134a). When fluid is compressed in the compression coil, it gets extremely hot and thus heat radiates into the hot environment. The now cooled fluid then evaporates in the evaporation coil, air is passed over the coil with a fan, and heat from the air is deposited into the coil fluid. Thus, heat is *transferred* via the working fluid. A "heat-pump" simply reverses the process, and we can think of it as "air conditioning" for the outdoors, dumping the waste heat indoors.

12.3.1   Efficiency as a function of ambient temperature

The greater the temperature difference between the interior and exterior, the less efficient a heat pump/AC is, both in terms of theory maximum efficiency and under experimental conditions. Recall that, for air conditioners, the maximum theoretical COP is $T_C/(T_H - T_C)$, and thus the larger the denominator, the lower the COP. Figure 12.20 shows how maximum COP varies with outdoor temperature for a fixed indoor temperature (T_C) of 80 °F.

Physically realized COPs are, of course, much smaller than the theoretical maximum, but they do follow the same basic temperature-dependent pattern. Based on the data presented in [533] and [534] (and references therein), the COP for *physical* ACs scales linearly with the indoor-outdoor temperature gradient in approximate accordance with the relation,

$$COP = COP_{8.33}(1.25 - .0293\Delta T), \tag{12.9}$$

where ΔT is the inside-outside temperate difference in °C, and $COP_{8.33}$ is the COP when the temperature difference is 8.33 °C (15 °F), as this corresponds to the EER (defined below), which is widely available to characterize individual AC units. In general, for every 10 °F increase in the indoor-outdoor temperature difference, the COP decreases by about 15–20%. Note then,

276

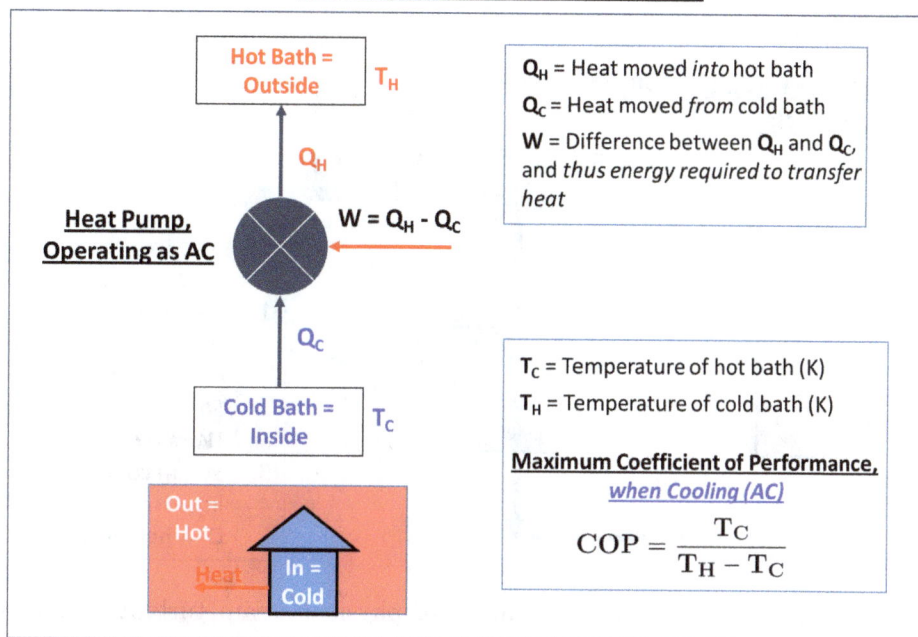

Heat Pump as Cooler (Air Conditioner)

Hot Bath = Outside, T_H

Q_H

Heat Pump, Operating as AC

$W = Q_H - Q_C$

Q_C

Cold Bath = Inside, T_C

Out = Hot
Heat
In = Cold

Q_H = Heat moved *into* hot bath

Q_C = Heat moved *from* cold bath

W = Difference between Q_H and Q_C, and *thus energy required to transfer heat*

T_C = Temperature of hot bath (K)

T_H = Temperature of cold bath (K)

Maximum Coefficient of Performance, *when Cooling (AC)*

$$COP = \frac{T_C}{T_H - T_C}$$

Figure 12.16: Schematic of the operation of a heat pump moving heat from a cold body to a hot body, where the goal is to cool the already cold body. That is, the heat pump is an air conditioner and the cold body is the dwelling interior. Figure 12.17 similarly shows a heat pump moving heat from a cold body to a hot one, but the goal is to warm the hot body.

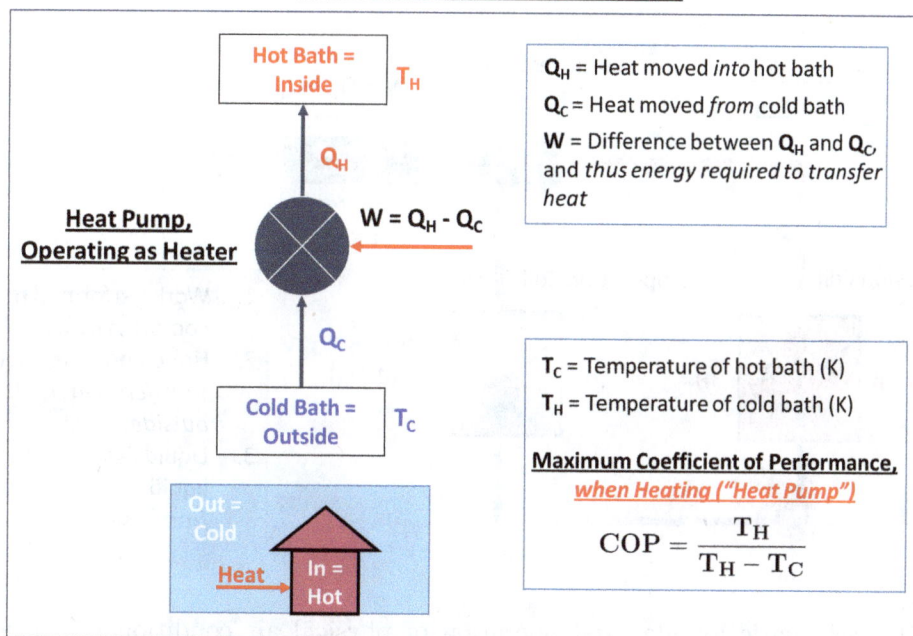

Heat Pump as Heater ("Heat Pump")

Hot Bath = Inside, T_H

Q_H

Heat Pump, Operating as Heater

$W = Q_H - Q_C$

Q_C

Cold Bath = Outside, T_C

Out = Cold
Heat
In = Hot

Q_H = Heat moved *into* hot bath

Q_C = Heat moved *from* cold bath

W = Difference between Q_H and Q_C, and *thus energy required to transfer heat*

T_C = Temperature of hot bath (K)

T_H = Temperature of cold bath (K)

Maximum Coefficient of Performance, *when Heating ("Heat Pump")*

$$COP = \frac{T_H}{T_H - T_C}$$

Figure 12.17: Schematic of a heat pump moving heat from a cold body to a hot one, with the goal to warm the hot body. This is a heater, or what is commercially called a "heat pump." Figure 12.16 similarly shows a heat pump in operation as an air conditioner, where the goal is to cool the cold body.

Idealized Steam Turbine (Rankine Cycle)

Steam Turbine = Engine

$W = Q_H - Q_C$

Electric generator

2.

Q_H

Q_C

Boiler = Hot Bath

1. T_H T_C **3.**   **Condenser = Cold Bath**

1. Heat from hot bath turns liquid to steam, Q_H = heat from hot bath
2. *Steam drives turbine = Work = Q_H - Q_C*
3. Steam converted to liquid in condenser, Q_C = heat into cold bath

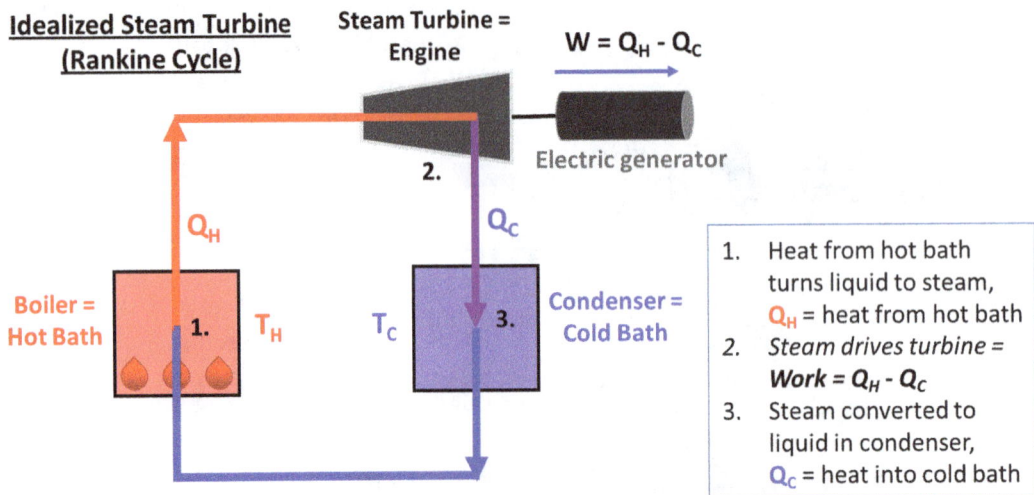

Figure 12.18: Schematic for idealized operation of physical steam turbine. Compare to the idealized AC in Figure 12.19 to see that, both theoretically *and physically*, heat engines and heat pumps are inverses of each other.

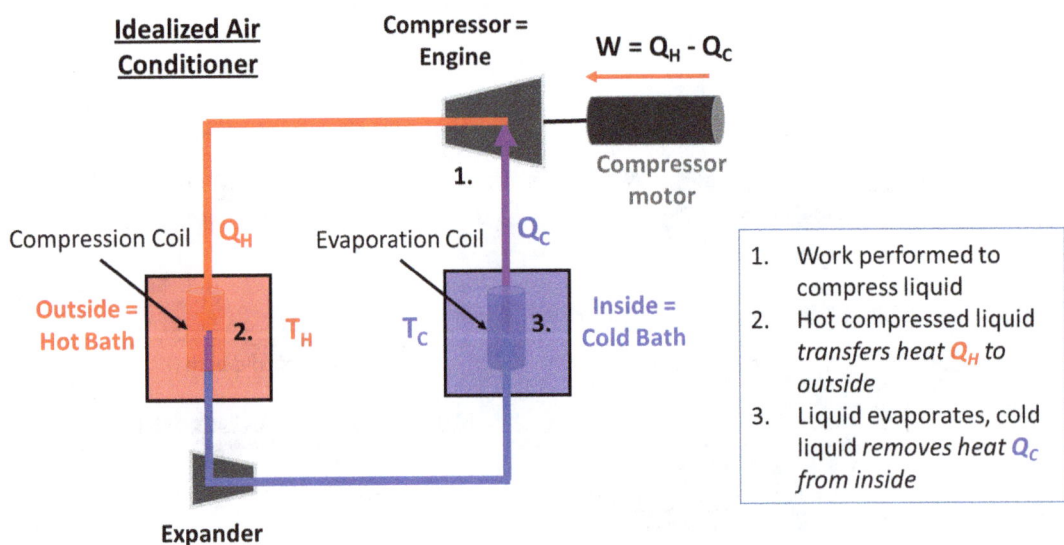

Idealized Air Conditioner

Compressor = Engine

$W = Q_H - Q_C$

Compressor motor

1.

Compression Coil Q_H Evaporation Coil Q_C

Outside = Hot Bath

2. T_H T_C **3.**   **Inside = Cold Bath**

Expander

1. Work performed to compress liquid
2. Hot compressed liquid *transfers heat Q_H to outside*
3. Liquid evaporates, cold liquid *removes heat Q_C from inside*

Figure 12.19: Schematic for idealized operation of physical air conditioner (a heat pump). Compare to the idealized steam engine in Figure 12.18 to see that, both theoretically *and physically*, heat engines and heat pumps are inverses of each other.

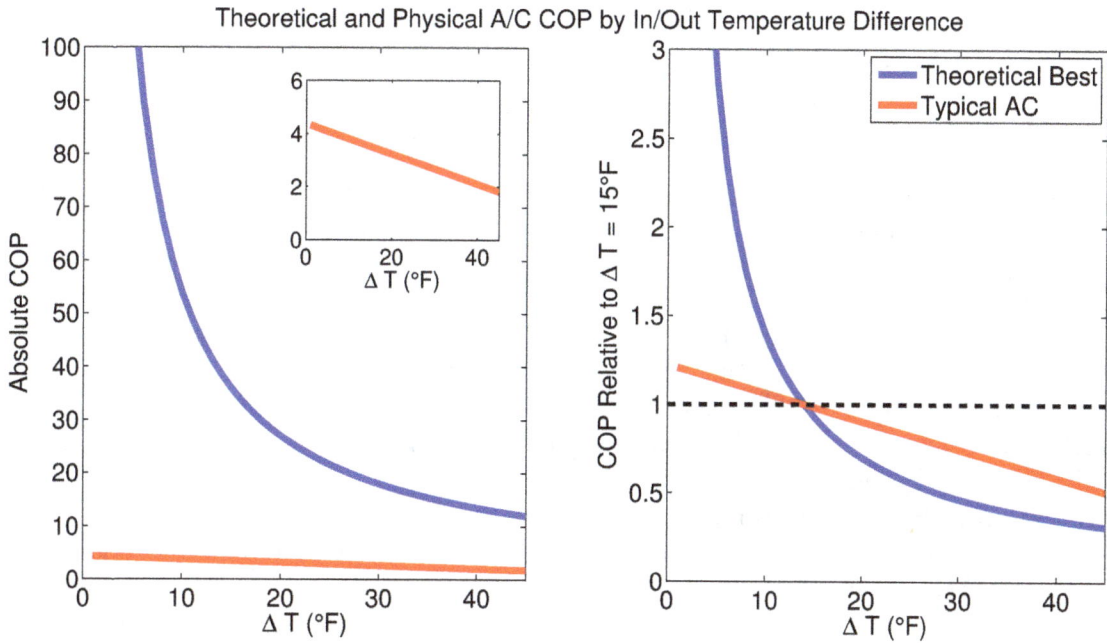

Figure 12.20: Maximum theoretical COP, and approximate COP for typical physical units, as a function of the inside/outside temperature gradient (ΔT), with inside temperature fixed at 80 °F. For physical units, efficiency may vary over twofold over the plausible temperature gradient range.

that this implies an additional advantage to increasing the summer thermostat, beyond reducing the total cooling load.

12.3.2 Efficiency of commercial ACs and heat pumps

On commercial AC labels (in the US), efficiency is typically reported as the energy efficiency ratio (EER) or seasonal EER (SEER). The EER encodes the same information as the coefficient of performance (COP), but it gives the amount of heat energy removed, measured in BTU, per work energy invested measured in watt-hours. This is a somewhat awkward choice of units, and EER can be converted to the more elegant and unit-less COP by a scaling factor,

$$\text{EER} = 3.41214 \times \text{COP}.$$

The EER is derived under standard test conditions of 80 °F indoors and 95 °F outdoors. As just discussed, the COP varies significantly with ambient temperature, and therefore AC units also report a seasonally adjusted EER, or SEER, which is supposed to reflect average operational efficiency over a cooling season. SEER is approximately related to EER by the equation [535]

$$\text{EER} = 0.02 \times \text{SEER}^2 + 1.12 \times \text{SEER}. \tag{12.10}$$

Very loosely, $EER \approx 0.85 \times SEER$, but this is a rather poor approximation.

In 2006, the minimum SEER for central AC units in the US was raised from 10 to 13, and most units must now meet SEER 14 everywhere but the northern US region. A SEER of 13 corresponds to an EER of about 11 (11.18 by Equation 12.10), or a COP of about 3.5. Energy star specifications for central air conditioners require a SEER $\geq$ 14 or $\geq$ 14.5, depending on type, and minimum EER is $\geq$ 12 (COP $\geq$ 3.52) or $\geq$ 11 (COP $\geq$ 3.22). Minimum efficiency

279

metrics for new room air conditioners are somewhat lower, with the minimum EER in the 8.0–9.8 range (COP 2.55–3.12).

12.3.3 Cooling (and heating) capacity

A practical property of physical AC/heat pump units (and furnaces, as well) that should be discussed is the cooling or heating *capacity*. Cooling capacity, which is typically reported in Btu/hour, is a measure of how much heat energy an AC can remove from a living space per hour.

It is important from a consumer standpoint to understand capacity, as AC efficiency decreases with increasing capacity, and unit cost increases. Therefore, one should always choose the AC unit with the lowest practical capacity. Rules of thumb are provided by the US DOE Energy Star program, while detailed heat/cooling load calculations may be performed with commercial software such as Manual J.

For the best Energy Star central air conditioners, the coefficient of performance tends to decrease by 10–20% for a doubling in capacity from about 25,000 to 50,000 Btu/hour. Higher capacity ACs also require greater airflow, which may increase ductal losses (discussed in Section 12.6.3). Smaller living spaces that are better insulated and sealed obviously need less cooling capacity as well, and thus energy-efficiency measures synergize to some degree. In general, oversized AC units are a widespread problem and source of inefficiency (and added expense).

12.4 Evaporative coolers, or, the much-maligned swamp cooler

- Evaporative (swamp) coolers (ECs) are an increasingly uncommon technology that cools air by taking the sensible heat of dry air and storing it in the latent form of water vapor, and thus work well in hot dry climates.

- The effective COP for ECs is on the order of 15–35, compared to 3–5 for ACs. Over the course of a year, ECs also consume several thousand gallons of water, but this is generally only a few percentage points of household water use.

- ECs are used to introduce psychrometric analysis and the notions of sensible and latent cooling loads, which are also applicable to air conditioners.

Growing up in the Southwest, my impression of evaporative coolers, known more widely in the vernacular as the "swamp" cooler, was that it was a sticky, profoundly inferior version of the air conditioner, suitable only for the poorest of abodes. I think this is a common impression, and so I hope here to give a much maligned, but in truth a remarkably effective and simple technology, its proper due. While nearly all new homes now have air conditioners, and the Southwest is now almost irredeemably co-dependent with the technology, modern versions of the humble evaporative cooler can actually cool living spaces in dry environs just as effectively at a fraction of the energy cost. Based on multiple sources [536], modern evaporative coolers are at least three to four times, and likely closer to ten times, as efficient as air conditioners, and thus may achieve energy savings of 75–90% for similar cooling. Furthermore, indirect evaporative coolers (explained in detail in a moment), unlike the traditional swamp coolers the reader may be familiar with, do *not* add water vapor to the conditioned air.

Evaporative coolers are most effective in hot arid climates and are thus are optimal for the American Southwest; they have the further advantage that the hotter the environment, the more efficient they are, contra the air conditioner. However, evaporative coolers do not perform

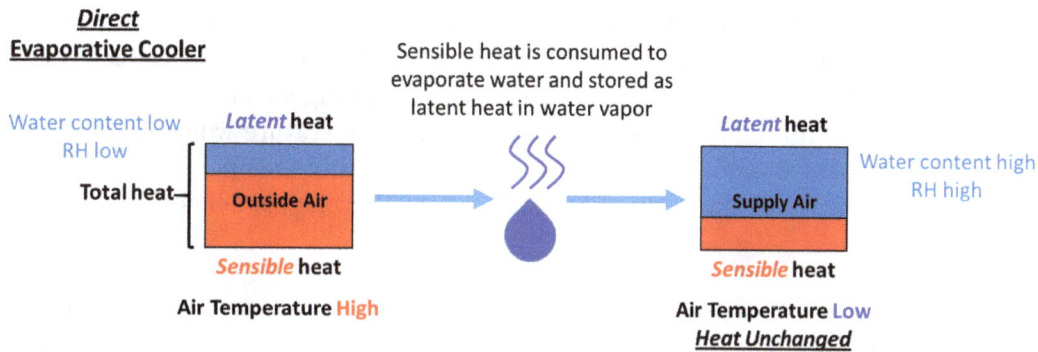

Figure 12.21: Schematic for the operation of a direct evaporative cooler (DEC). Outside air, carrying with it a certain amount of heat, either sensible or latent and in the form of water vapor, is forced through a wet medium. As water evaporates from the medium, sensible heat is converted to latent heat, and the proportion shifts (in this example) from predominantly sensible to predominantly latent heat. Thus, the perceived temperature of the air decreases without moving any heat, while increasing the humidity.

well in humid climates, and are therefore not a practical option in many areas. Many excellent discussions of this technology are available, and I rely in particular upon Amer et al. [536].

12.4.1   Principle of operation and direct evaporative coolers

Unlike the clever AC, the swamp cooler *is not a heat pump*, but operates via the spontaneous evaporation of water into air. The coefficient of performance is not directly connected to ambient temperatures, although evaporative coolers tend to have a *higher* COP at higher temperatures, contra the AC, and the COP may surpass (by a wide margin) the theoretical maximum for ACs. This is because swamp coolers do not directly move heat at all, but rely on the difference between *sensible* and *latent* heat in moist air. Before continuing, note that evaporative coolers act directly on outside air, chilling it, and pumping it into the house, unlike ACs, which cool the air already inside the house.

Sensible heat is the heat that contributes to the temperature of air, and so it is heat that is felt directly and can be measured by a thermometer, i.e. the heat is "sensible." Latent heat, on the other hand, is "hidden" and not directly measurable by a thermometer: it is the heat associated with phase changes. For example, vaporizing liquid water to a gas requires a great deal of heat that is then stored in latent form, and not felt as an increase in temperature. This fact is the basis of the evaporative cooler.

A direct evaporative cooler (DEC) operates by forcing dry air through a wet medium with a fan, in the case of an *active* DEC. Several ancient technologies take advantage of spontaneous air flow to act as *passive* DECs, e.g. the "wind tower" [536]. Now, some of the water in that medium evaporates, and in this process a portion of the sensible heat of air is converted to latent heat stored in the water vapor. Thus, the air becomes more humid and its temperature is lowered, but importantly, no heat is lost or gained: it is merely converted from a sensible to latent form; this process is demonstrated in Figure 12.21. The amount of heat that can be thus stored depends on the relative humidity and temperature of the incoming air. The more humid the air is to start, the less water can be evaporated into it. Hotter air can store more water than colder air, and in general, the relative humidity is least at the height of the day when air is hottest. The latter two facts imply that evaporative coolers are generally more efficient the hotter the ambient temperature, contra the AC.

In practice, we determine how effective an evaporative cooler can be from the dry bulb and wet bulb temperatures. The dry bulb temperature is the usual temperature; wet bulb temperature is, most appropriately, taken using a wet bulb. When the water on the bulb evaporates, sensible heat is lost and the immediately surrounding air has a relative humidity of 100%. Thus, the wet bulb temperature tells us what the temperature would be if the air were to be completely saturated, and thus gives the minimum possible temperature of air exiting a DEC. DECs operate with an efficiency of 70–95%, where efficiency is defined as

$$\frac{T_{\text{drybulb,in}} - T_{\text{drybulb,out}}}{T_{\text{drybulb,in}} - T_{\text{wetbulb,in}}}. \tag{12.11}$$

So, for example, 106 °F air at 15% relative humidity (RH) has an associated wet bulb temperature of about 70 °F, and a DEC operating at 85% efficiency will give 76 °F air with 75% RH. The actual mathematics needed to work this out are presented shortly.

12.4.2 Indirect evaporative coolers

The primary disadvantage of the DEC is that is produces humid air, which may be uncomfortable and possibly lead to mold, etc. A simple innovation is the indirect evaporative cooler (IEC), which is essentially a DEC coupled to a heat exchanger. Working air passes through a DEC, cooling and humidifying it; the cooled working air then passes through a heat exchanger whereby heat is drawn from dry outside air. The dry air is thus cooled without altering the water content, and this dry cool air is injected into the building. Note that, because colder air can hold less water, but the water content of the cooled air is unaltered, there is still a small increase in the *relative* humidity of air exiting an IEC.

A modification of the IEC that increases cooling efficiency is the Maisotsenko-cycle (M-cycle) [536], whereby outside air is drawn in, passing by the heat exchanger and then, instead of all piping into the house, some is piped *into* the DEC, thus being both pre-chilled by the working air and then becoming the working air itself. That is, there is a single air stream that gives rise both to the working and supply air, an innovation that allows the air exiting the M-cycle IEC to be colder than the wet-bulb temperature of outside air, achieving an efficiency greater than 100%, and in the 110–120% range in practice [536].

The efficiency of an evaporative cooling system may also be increased through various couplings of IECs, DECs, solid desiccants for removing moisture, and passive radiative cooling systems [536]. The simplest and most obvious example is a two-stage IEC-DEC chain, which can achieve cooling efficiency over 100%.

12.4.3 Cooling potential at various ambient temperatures and relative humidities

The charts in Figures 12.23 and 12.24 show the temperature and relative humidity of air exiting a given evaporative cooler, as a function of the ambient (dry-bulb) temperature and relative humidity. Results are given for a DEC operating at 85% efficiency, and a high efficiency (115%) modified IEC, and are determined using psychrometric analysis, which is explained next. From Figure 12.24, we see that an IEC can provide sufficient cooling for almost any condition typically encountered in a desert city.

12.4.4 Psychrometric analysis

This section may be skipped by the general reader with little loss of essential content, but I have included it to demonstrate more rigorously the physical principles of evaporative cooler operation, and to give the mathematical background used to derive the results presented in the

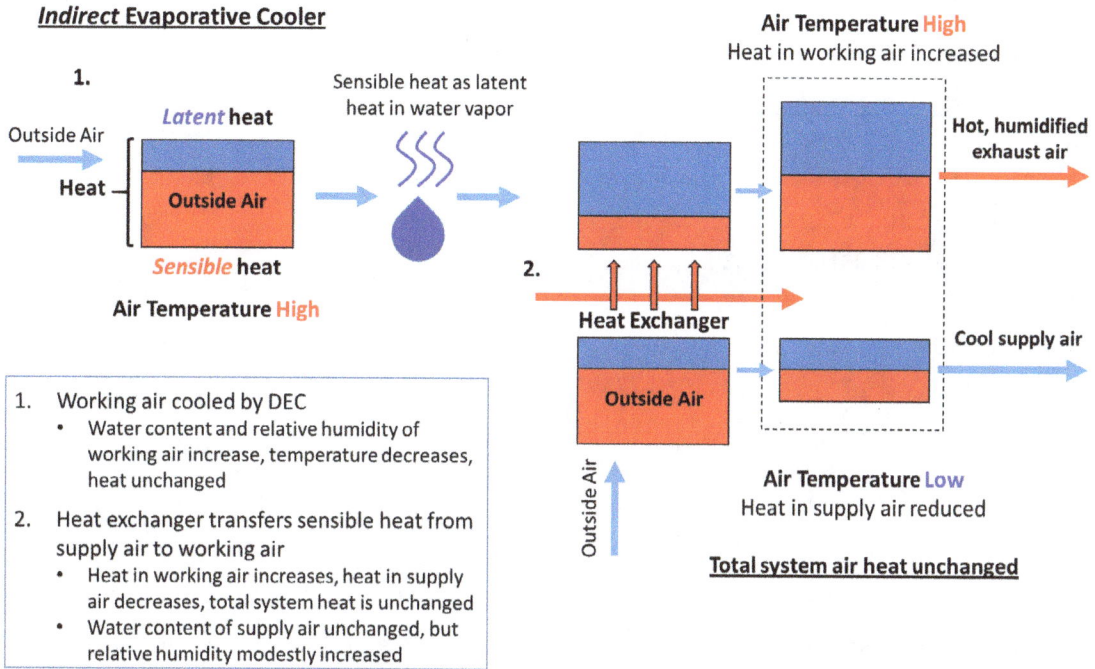

Figure 12.22: Schematic for the operation of an indirect evaporative cooler (IEC), which comprises a DEC coupled to a heat exchanger. Working air passes through a DEC, causing the heat distribution to shift from sensible to latent heat. Outside air then passes through a heat exchanger with this working air, drawing out sensible heat, but leaving the latent component constant. Thus, the working air ends up higher in both sensible and latent heat, while the supply air sees a reduction in sensible heat; the total heat content of both air streams combined remains constant.

RH (%)	DEC at 85% Efficiency								
	Ambient Temp (F)								
	84	88	92	96	100	104	108	112	116
5	57 (73%)	59 (72%)	61 (71%)	63 (71%)	66 (70%)	68 (70%)	70 (69%)	72 (69%)	74 (68%)
12.5	60 (76%)	62 (76%)	65 (76%)	67 (75%)	70 (75%)	72 (75%)	75 (75%)	77 (74%)	79 (74%)
20	62 (80%)	65 (80%)	68 (79%)	71 (79%)	73 (79%)	76 (79%)	79 (79%)	81 (79%)	84 (78%)
27.5	65 (83%)	68 (83%)	71 (83%)	74 (82%)	77 (82%)	80 (82%)	83 (82%)	86 (82%)	89 (82%)
35	67 (85%)	71 (85%)	74 (85%)	77 (85%)	80 (85%)	83 (85%)	86 (85%)	89 (85%)	92 (85%)
42.5	70 (88%)	73 (88%)	76 (88%)	80 (88%)	83 (88%)	86 (88%)	89 (88%)	93 (88%)	96 (88%)
50	72 (90%)	75 (90%)	79 (90%)	82 (90%)	85 (90%)	89 (90%)	92 (90%)	96 (90%)	99 (90%)
57.5	74 (92%)	77 (92%)	81 (92%)	85 (92%)	88 (92%)	92 (92%)	95 (92%)	99 (92%)	102 (92%)
65	76 (93%)	80 (93%)	83 (93%)	87 (93%)	90 (93%)	94 (93%)	98 (93%)	101 (94%)	105 (94%)
72.5	78 (95%)	82 (95%)	85 (95%)	89 (95%)	93 (95%)	96 (95%)	100 (95%)	104 (95%)	108 (95%)
80	80 (96%)	83 (96%)	87 (96%)	91 (96%)	95 (96%)	99 (97%)	102 (97%)	106 (97%)	110 (97%)

Figure 12.23: Temperature and relative humidity (in grayed parentheses) of air exiting an 85% efficient DEC, as a function of ambient air temperature and relative humidity. Temperatures that would generally be considered acceptable cooling are coded blue, borderline are teal, and those too hot are red.

RH (%)	IEC at 115% Efficiency Ambient Temp (F)								
	84	88	92	96	100	104	108	112	116
5	47(18%)	49(19%)	50(21%)	52(22%)	53(23%)	55(25%)	56(27%)	58(29%)	59(31%)
12.5	51(39%)	53(41%)	55(43%)	57(46%)	59(48%)	61(51%)	63(53%)	65(56%)	66(59%)
20	55(54%)	57(57%)	59(59%)	62(62%)	64(64%)	66(67%)	68(70%)	71(73%)	73(75%)
27.5	58(66%)	61(68%)	63(71%)	66(73%)	68(76%)	71(78%)	74(81%)	76(83%)	79(85%)
35	62(74%)	64(77%)	67(79%)	70(81%)	73(83%)	76(85%)	78(88%)	81(90%)	84(91%)
42.5	65(81%)	68(83%)	71(85%)	74(87%)	77(89%)	80(90%)	83(92%)	86(94%)	89(95%)
50	68(86%)	71(88%)	74(89%)	77(91%)	80(92%)	84(94%)	87(95%)	90(97%)	93(96%)
57.5	70(90%)	74(91%)	77(93%)	80(94%)	84(95%)	87(96%)	91(97%)	94(98%)	97(99%)
65	73(93%)	77(94%)	80(95%)	84(96%)	87(97%)	91(98%)	94(99%)	98(99%)	101(100%)
72.5	76(95%)	79(96%)	83(97%)	86(98%)	90(98%)	94(99%)	97(99%)	101(100%)	105(100%)
80	78(97%)	82(98%)	86(98%)	89(99%)	93(99%)	97(99%)	101(100%)	104(100%)	108(100%)

Figure 12.24: Temperature and relative humidity (in grayed parentheses) of air exiting a 115% efficient IEC, as a function of ambient air temperature and relative humidity.

previous sections. A psychrometric chart allows us to determine the wet bulb temperature from a given dry bulb temperature and relative humidity, among other things. It has as its axes the *dry bulb temperature* on the horizontal, and the *humidity ratio* (the mass of water vapor per unit mass of air) on the vertical. Keep in mind that humidity ratio and relative humidity are very different concepts, despite the similar names. The process of constructing such a chart is outlined in Figure 12.25.

Briefly, we begin by creating several curves of humidity ratio vs. dry bulb temperature for a fixed relative humidity. The most important is the 100% RH curve, which tells us how much water can be stored in the air at different temperatures, when the air is fully saturated. The humidity ratio at saturation increases exponentially with dry bulb temperature, and is derived as follows. We first have (in units millibars), the partial pressure of water at saturation (i.e. 100% RH), P_{ws}, given by the equation

$$P_{ws} = 6.11 \times 10^{\frac{7.5T}{237.3+T}}, \tag{12.12}$$

with temperature T in °C. Now, for any given RH, we will have the partial pressure of water, P_w, related to the saturation partial pressure via

$$\text{RH} = \frac{P_w}{P_{ws}}. \tag{12.13}$$

So, starting with temperature and RH given, we can solve for P_w. Having the partial pressure of water and the (known) atmospheric pressure, P_a, we can then calculate the humidity ratio with the equation

$$\text{HR} = 0.62198 \times \frac{P_w}{P_a - P_w}. \tag{12.14}$$

Thus, we are finally able to construct a curve of humidity ratio vs. dry-bulb temperature, for any given relative humidity. Now, the other major step is to construct lines representing air with a constant heat content (constant wet-bulb lines). That is, take some point on the 100% RH curve, say, 70 °F, corresponding to a HR of 0.0158 kg water/kg air. At this point, since RH

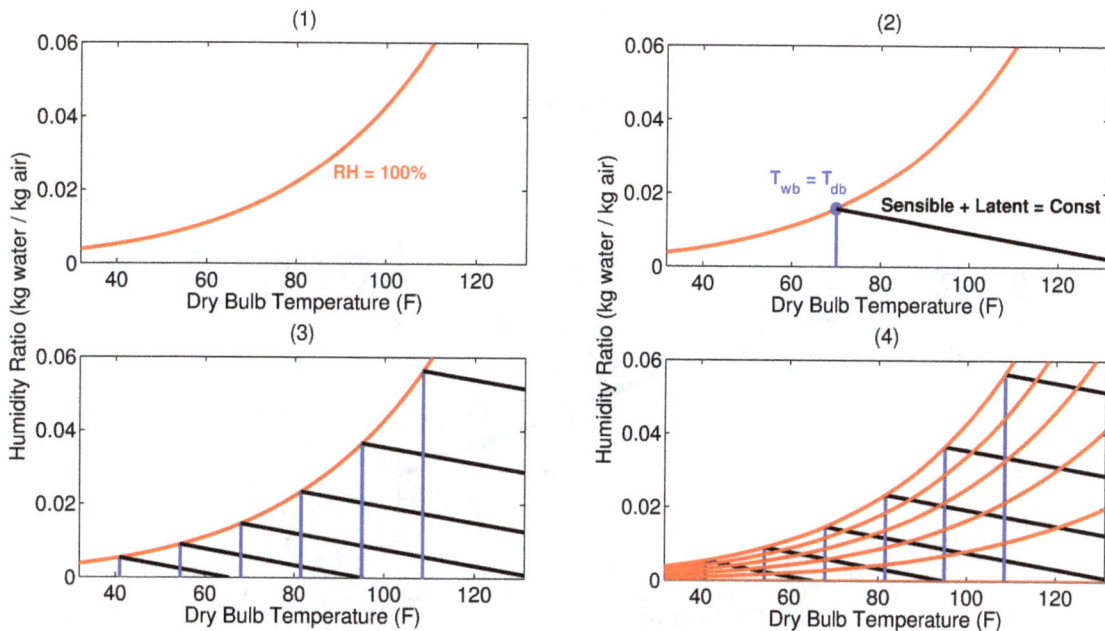

Figure 12.25: Process of constructing a psychrometric chart. As discussed in the text, we first plot the humidity ratio (kg water / kg air) against dry-bulb temperature for 100% relative humidity (RH). We then draw lines of constant total heat from this curve to the x-axis: every point along such a line has the same wet-bulb temperature, which is the temperature where it intersects the 100% RH curve. Finally, we add in several more curves of constant RH.

is 100%, the wet-bulb temperature equals the dry-bulb temperature, and the air contains 40.13 kJ/kg air of latent heat, calculated from

$$\text{latent heat (kJ/kg)} = HR \times (2501 + 1.84T), \tag{12.15}$$

where 2,501 kJ/kg is the enthalpy of evaporation for water (from freezing) and T is in °C. What if this was all converted to sensible heat? Then we would have bone-dry air with a dry-bulb temperature of 142 °F. Connect these two points, (70, 0.0158) and (142, 0), and air at every point along this line contains the same amount of total heat (latent plus sensible) and has the same wet-bulb temperature. We go ahead and do this for a few more points along the 100% RH curve, add in some vertical lines connecting these to the x-axis, and we have a psychrometric chart.

Now, why is this useful? Well, recall that for a DEC, we start with outside air at some arbitrary dry-bulb temperature and relative humidity, and then convert sensible heat to latent. We can easily find the starting point on our chart, and then we simply follow the corresponding line of constant wet-bulb temperature (or constant heat) to the left. Our stopping point will be the new dry-bulb temperature, as determined from the DEC defined efficiency, and an example of this is given in Figure 12.26. We can similarly visualize IECs, where air initially cooled by a DEC engages in sensible heat exchange with the supply air, implying that the supply air line moves horizontally to the left from our starting point. Examples are given in Figures 12.27 and 12.28.

The basic conclusion is that, with a DEC, we follow lines of constant wet-bulb temperature, whereas with an IEC, we follow the lines of constant humidity ratio.

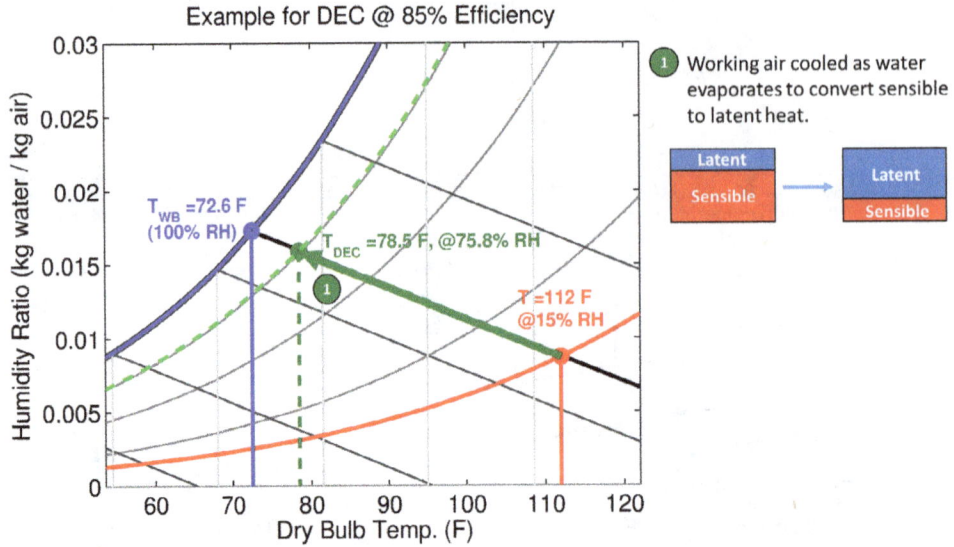

Figure 12.26: Operation of a direct evaporative cooler (DEC), as visualized on a psychrometric chart. Adding moisture to dry air causes us to move along a line of constant heat, where some sensible heat is converted to latent heat (see also Figure 12.21).

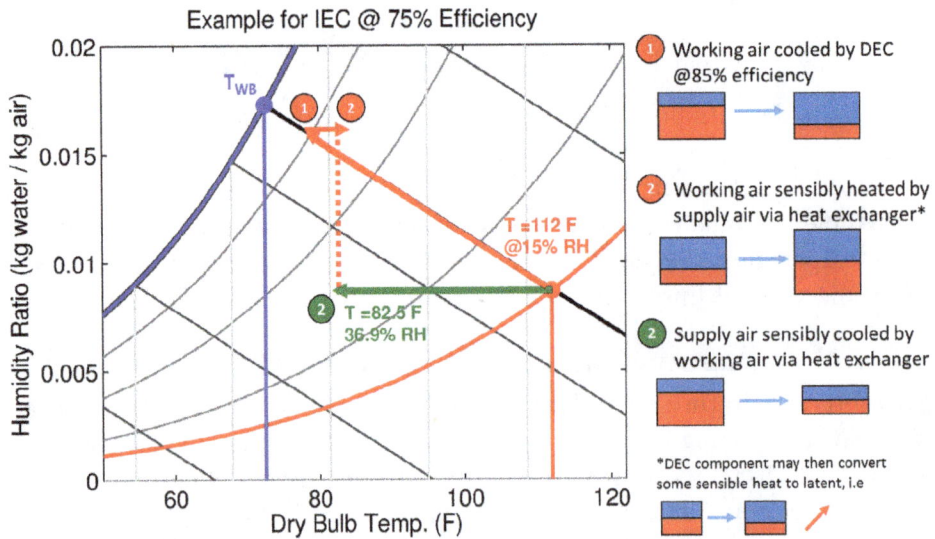

Figure 12.27: Operation of an indirect evaporative cooler (IEC), as visualized on a psychrometric chart. Moisture is added to the working air (red line), and we move along the line of constant heat with some sensible heat converted to latent. Sensible heat exchange then occurs between the working and supply air, moving the supply air along to the left, along the line of constant humidity ratio.

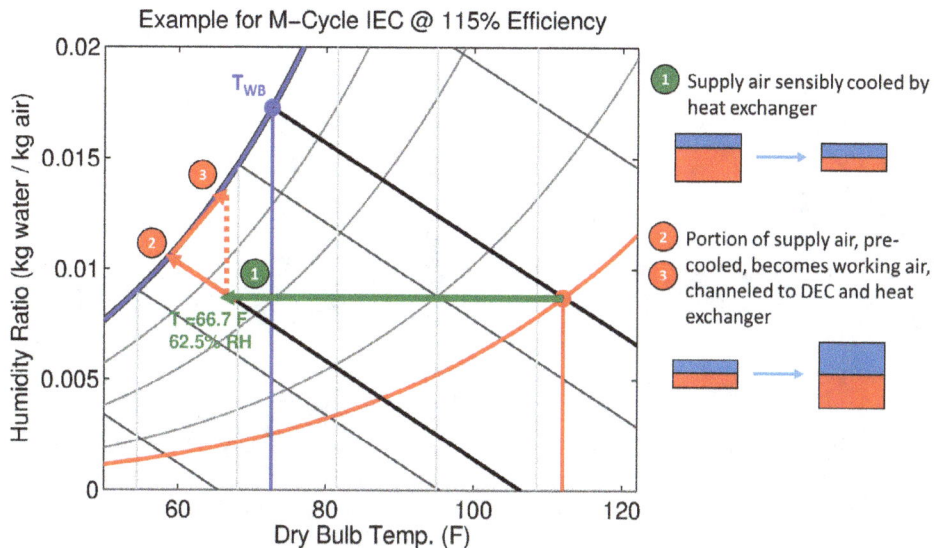

Figure 12.28: Operation of an indirect evaporative cooler (IEC) operating under the M-cycle to achieve >100% efficiency, as visualized on a psychrometric chart.

12.4.5 Psychrometrics, air conditioning, and latent cooling loads

We can also examine air conditioners (and heaters) from a psychrometric perspective. Air conditioners, like IECs, will remove sensible heat from air, moving us across a horizontal line on our psychrometric chart, raising the relative humidity but not the humidity ratio. Unlike IECs, however, an AC can remove an arbitrary amount of heat from the air, and is not "blocked" once it hits the 100% RH curve. To the left of the 100% RH curve, water must condense out of the supersaturated air, and such a phase change takes energy. Thus, to cool further, we must begin to remove both latent and sensible heat, and such removal of latent heat is just what happens when water condenses on an AC coil.

Now, in practice, an AC coil is much colder than the target indoor air temperature, and hence will remove moisture from the air well below 100% RH. Indeed, ACs are explicitly designed to dehumidify as part of their standard operation, and a standard target point is 75 °F at 50% RH. Therefore, there is both a sensible and latent cooling load placed on our equipment. From a psychrometric analysis, as shown in Figure 12.29, we can see how running in AC very hot and dry conditions has essentially no latent load, while latent load can be significantly greater than the sensible load under hot and humid conditions. It follows that more energy consumption for the same level of sensible cooling can be expected in humid climes.

Of course, ACs are generally conditioning *inside*, not outside, air, and therefore interior humidity is partially decoupled from outside humidity, since the majority of heat gain into a house is sensible (as discussed in Section 12.6).

12.4.6 Energy and water use

Evaporative cooling systems generally consume electricity only through air fans and water pumps, processes that are energetically *far* less demanding than the mechanical compression of ACs, and the effective coefficient of performance (kWh of cooling per kWh of electricity consumed) for evaporative coolers used in arid areas may range from about 15 to 35, compared to 2–4 for ACs [536, 537].

To derive this semi-mechanistically, based on [537], we can assume that an EC runs at 0.8 kW (one commercial system used a 0.75 kW blower, plus pumping and other standby electricity),

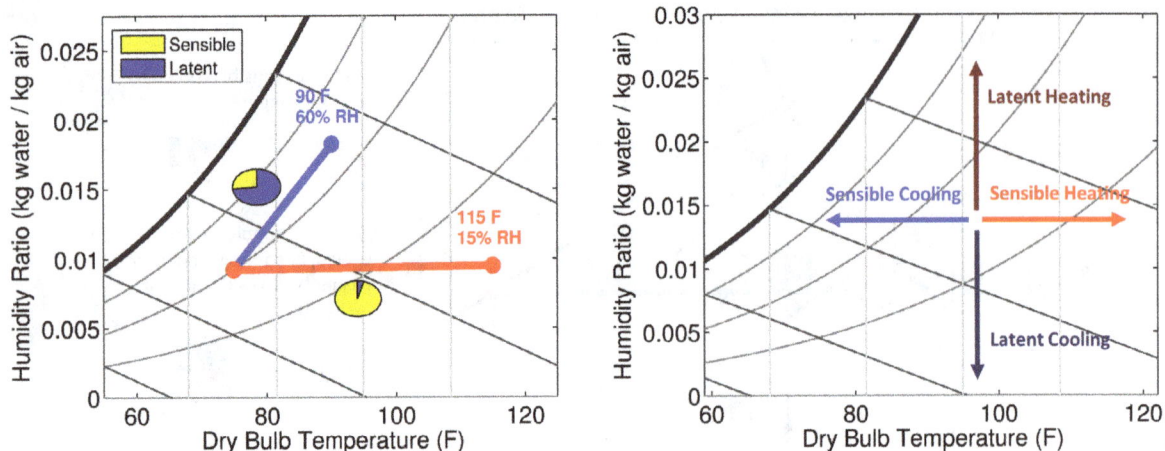

Figure 12.29: The left panel gives sensible and latent cooling loads on an air conditioner acting to shift outside air from a starting point of either 90 °F at 60% RH, or 115 °F at 15% RH. As the inserted pie graphs show, most of the cooling load is latent in the former case, while nearly all (96%) is sensible in the latter. Note that it also takes about 35% more total energy to cool from the 90 °F, 60% RH point (not shown). The right panel simply demonstrates how different directions on the psychrometric chart correspond to sensible or latent heating/cooling.

while evaporating 8–12 gallons of water per hour, depending on conditions. A gallon of water has 2.63 kWh of cooling capacity, and if our EC used water with 85% efficiency, then we should have a COP of 22–34 kWh cooling/kWh electricity. Since evaporation will be more rapid in the hottest, driest regions, this is where the highest COPs are achieved.

Indeed, a 2004 study [537] estimated that cooling energy for a new 1,800 square foot house in Phoenix, AZ, would decrease from 6,043 kWh/yr using AC to just 574 kWh/yr with EC, a 90.5% savings, and equivalent to about 3.75 MgCO2e of avoided emissions. Note, however, that ECs do require a significant amount of water for their operation. Using an average of 4,121 kWh electricity for AC in the hottest climate zone (2009 RECS), and assuming a COP of 3 to yield an annual cooling load of 12,636 kWh, we would require just over 5,500 gallons of water to meet this demand with EC (and likely under 500 kWh of electricity). Compared to overall household water use, this is trivial: residential water consumption in AZ (in 2010) was 147 gallons/day, or about 134,000 gallons per year for a household of 2.5. Furthermore, since electricity production uses a great deal of water as well, once the electricity offset is accounted for, *net* EC water use decreases by about 30% relative to gross [537]. We may conclude that an EC adds no more than 5%, and generally more like 2–3%, to net residential water use.

12.5 Heating: Furnaces, boilers, and fuels

- Among widely used heat sources, gas or propane furnaces and electric heat pumps generate the least carbon, followed closely by fuel-oil furnaces.

- The oldest gas and oil furnaces/boilers may have efficiencies as low as 50%, while modern furnaces have efficiencies in the 80–98.5% range. Upgrading to the highest efficiency technology can therefore save 20–50% of heating energy/emissions.

- Electric resistance heating, used as the primary form of heating in 25% of homes, is very carbon intensive. Replacing all such heating with either electrical heat pumps or gas furnaces (90% AFUE) would reduce total household space heating emissions by roughly 12%, and save about 50 million $MgCO_2e$ overall.

- Wood-burning technologies are relatively inefficient at delivering heat to the home, and a combination of altered forest carbon dynamics and black carbon and methane emissions from wood combustion imply wood burning is usually similar to or worse than fossil-based heating.

Let us now turn to space heating technologies. Most homes in the US are heated by either furnaces or boilers. Furnaces directly heat air which is generally distributed throughout a house via ducts, whereas boilers heat water, either to generate hot water or steam, which is distributed via pipes to radiators. There is little overall difference between furnace and boiler efficiency, although steam boilers are intrinsically less efficient than hot water boilers because of their higher operational temperature.

Furnace/boiler heat is supplied either from direct combustion of fossil fuels (primarily natural gas, followed by fuel oil and propane/LPG) or through electric resistance heating. Under the latter, a current is passed through a resistor, which dissipates the electrical energy as heat, much like a toaster. Electric resistance heating is 2–3 times as carbon-intensive as combustion furnaces. Heat pumps, discussed in Section 12.3, are the other major heating technology, and have a carbon-intensity similar to gas furnaces. Wood fire, contained within a fireplace or wood-burning stove, is a relatively high carbon means of supplying heat, with some exceptions, and it is given special consideration in Section 12.5.6.

Figure 12.30 shows how primary heating sources break down nationally and in the coldest and hottest RECS climate regions. Table 12.1 summarizes the carbon-intensity of major heating technologies, both on an electricity grid-average and marginal emissions-factor basis. A closer look at the major fossil heating fuels and technologies follows.

12.5.1 Lifecycle emissions from fossil heating fuels

- Lifecycle emissions factors for natural gas, propane/LPG, fuel oil, and kerosene are approximately 0.2613, 0.2486, 0.3081, and 0.3156 $kgCO_2e/kWh$, respectively. For comparison, electricity EFs are 0.6820 $kgCO_2e/kWh$ for the US grid-average, and average marginal emissions are 0.8798 $kgCO_2e/kWh$.

This section summarizes the approximate lifecycle emissions from combustion and extraction of fossil fuels. Propane and natural gas (NG), as the lightest hydrocarbons with the highest hydrogen:carbon ratios, generate the least CO_2 per kWh of heat generated at the point of combustion. Upstream emissions are probably somewhat greater for NG, but overall these two fuels are similar. Fuel oil and kerosene are heavier hydrocarbons, and thus generate about 15–

289

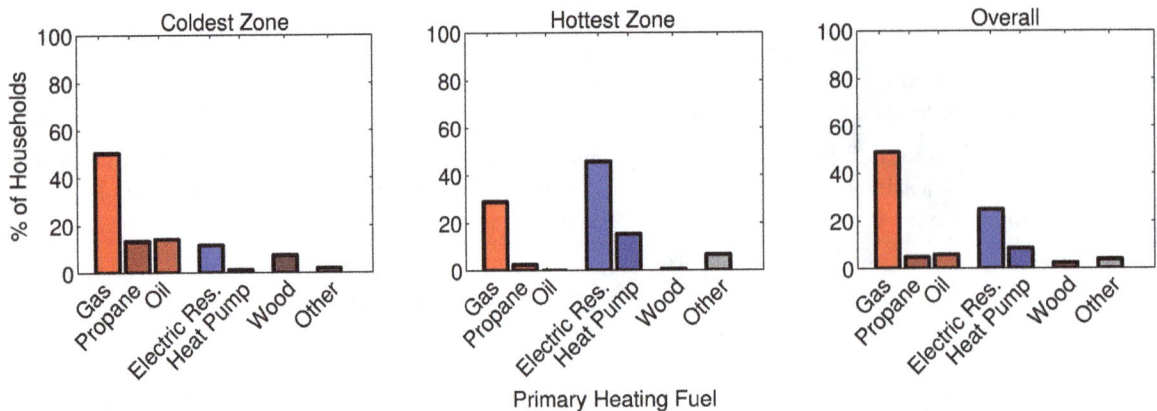

Figure 12.30: Primary heating sources in the 2009 RECS, shown for the coldest and hottest climate zones; the national average is given as well. Note that electric heating is far more prevalent in the hot clime, and there is a consistent trend from relying primarily on fuel combustion to electricity from coldest to hottest zone.

Heat Source	kgCO$_2$e/kWh heat delivered
Gas furnace	
Typical (AFUE 80%)	0.3266
Best (AFUE 98.5%)	0.2653
Oil furnace	
Typical (AFUE 80%)	0.3851
Best (AFUE 98.5%)	0.3128
Electric resistance	
Typical (AFUE 100%)	0.6820 (grid-average) **0.8798 (marginal)**
Heat pump	
Typical (COP 2.5)	0.2728 (grid-average) **0.3519 (marginal)**
Best (COP 3.25)	0.2098 (grid-average) **0.2707 (marginal)**
Wood-burning stove	
Older stove (50% efficiency)	0.5612
Newer stove (70% efficiency)	0.3387
Open fireplace	
Best (10% efficiency)	2.8060

Table 12.1: Lifecycle emissions factors for different heating technologies, given in terms of heat delivered to the house (not including any distribution losses through the ductwork). Those based on electricity include an EF for both grid-average and marginal electricity. Emissions factors are determined from the fuel and electricity source discussion presented in Section 12.5.1, while wood-burning is discussed in Section 12.5.6.

290

25% more CO_2e than propane and NG, per kWh. Note that it is appropriate to use the high heating value (HHV, this includes the latent energy in water vapor generated by combustion) when assessing the energy content of heating fuels, as opposed to the lower heating value (LHV).

Natural gas and upstream leaks (the main source of non-combustion CO_2e) are discussed extensively in Chapter 4 in the context of electricity generation. In addition to combustion emissions of 0.1809 kgCO_2/kWh, I assume an overall NG leak rate of 2.4%, and 13.2 gCO_2e/kWh attributable to extraction, yielding a lifecycle EF of 0.2613 kgCO_2e/kWh.

Upstream emissions for LPG (or propane) are given in the 2014 GREET 2 model as 0.0583 kgCO_2e/kWh on a LHV basis, translating to 0.0537 kgCO_2e/kWh on a HHV basis; using the EPA combustion emissions factor of 5.2214 kgCO_2/gallon, we get propane combustion emissions of 0.1949 kgCO_2e/kWh, and lifecycle emissions summing to 0.2486 kgCO_2e/kWh (HHV basis). Note that this is very similar to natural gas.

Residential fuel/heating oil is essentially identical to diesel fuel, both being subtypes of the broader distillate fuel oil class, one of the fractions obtained in oil refining. Using upstream emissions for diesel reported in GREET 2, we have lifecycle emissions of 0.3081 kgCO_2e/kWh for fuel oil/diesel, on a HHV basis.

Kerosene has an energy density of about 12.833 kWh/kg. Using the EPA combustion emissions factor of 0.2566 kgCO_2e/kWh, and assuming that upstream emissions are similar to diesel, at 23% of combustion emissions, we get 0.059 kgCO_2e/kWh upstream emissions, and a lifecycle emissions factor of 0.3156 kgCO_2e/kWh (HHV basis).

12.5.2   Furnace/boiler efficiency for combustion fuels

- Older furnaces are only 50–70% efficient, the minimum for newer furnaces is around 80%, and the best approach 99% efficiency. Fuel savings from a furnace upgrade can therefore be substantial.
- To achieve better than about 90% efficiency, furnaces must be *condensing*, i.e. they extract the latent heat stored in water vapor generated upon combustion.

The efficiency of furnaces/boilers that operate via direct combustion is typically given by the annual fuel utilization efficiency (AFUE) value, which simply indicates the fraction of energy that is converted to space heat, and accounts for losses to a running pilot light, heat lost in flue gas, etc., but not losses beyond the furnace (e.g. from ductwork, etc.). Older gas and oil furnaces/boilers may have an AFUE of only 50–70%. Until recently the minimum AFUE for a new furnace was 78%, while new standards range 78–83% depending on fuel type; minimum boiler AFUE is 80–84%, depending on type. Energy Star criteria are in the 85–95% range (varying with region), and the most efficient furnaces achieve an AFUE of 97–98.5%. It follows that upgrading an old furnace/boiler could as much as *halve* space heating energy/emissions.

A major difference between very high efficiency combustion furnaces and their less efficient compatriots is that the former are *condensing*. That is, the latent energy of evaporation in water vapor formed during the combustion process, instead of escaping with flue gas, is recovered via condensation, and this heat is then transferred to air via a second heat exchanger. Underlying this, the stoichiometry of natural gas combustion is

$$CH_4 + 2O_2 \rightarrow CO_2 + 2H_2O, \tag{12.16}$$

and this reaction yields 47.141 kJ/kg of heat when the heat contained in the generated water vapor is lost (LHV), and 52.225 kJ/kg when this heat is recovered via condensation (HHV). Thus, non-condensing furnaces have a theoretical maximum AFUE of about 90%.

12.5.3 The vampiric pilot light

- The pilot light from older furnaces and space heaters may consume as much as 5% of household energy when left on outside the cooling season, and could represent around 500-1,500 kWh of wasted energy yearly.

Standing pilot lights, which burn 24/7 to provide a source of ignition for older gas appliances, use a surprising amount of gas energy, and I can personally attest that a gas space heater in standby in summer makes the room noticeably hotter. Many modern appliances use electronic ignition and thus avoid this vampiric loss, and furnaces have been required to use pilot-less ignition since 1980 [539], but pilots are still widely used for gas space heaters (and water heaters as well).

Now, in winter a burning pilot light is of little concern, as it offsets some of the heating load. The pilot for a gas water heater is also generally unimportant yearround, since this "waste" heat goes into the water. Because of these factors, one 1976 study in Ohio [538] estimated that only about one-third of pilot energy is wasted, but nearly all that is, 2.6% of total gas energy, is due to the furnace pilot remaining on in the summer months. Thus, this study suggests that a gas furnace pilot uses about 3 kWh per day (combining the 2.6% figure with RECS data on mean natural gas use), or almost 1,100 kWh per year. On the other hand, one utility company estimates almost 8 kWh per day for a gas furnace pilot [539].

At any rate, turning off furnace pilots when not in use for, say, six months of the year could save in the range of 540–1,440 kWh of natural gas and 0.141–0.376 MgCO$_2$e, not counting any avoided cooling costs (which would actually be nearly as great if all this heat was removed in turn by an air conditioner). Note that these calculations further suggest that, in homes with a furnace pilot, possibly 2–5% of total household energy and 1–3% of household CO$_2$e, is directly attributable to its operation outside the heating season.

12.5.4 Electric resistance heating

- Electric resistance heating has the highest carbon intensity of any furnace fuel, with associated carbon emissions equivalent to those of a <40% efficient gas furnace.
- Electric space heaters can reduce overall heating emissions if they are used to heat less than 30–50% of living space *and* in conjunction with a lower central thermostat set-point.

Electric resistance heating, the second-most popular furnace mode, is essentially 100% efficient in converting current to heat, but the emissions factor for electrical energy is generally much greater than that of gas or propane energy, at about 0.682 kgCO$_2$e/kWh for heat delivered, using grid-average electricity. This is because electricity is generated (mainly) from coal or gas combustion at a power plant at average efficiencies in the 30–45% range, and then transmitted with average losses of about 6.5%. If natural gas was converted to electricity at 42.4% efficiency (national average for gas-fired plants in 2013), transmitted at a 6.5% loss, and converted to heat with 100% efficiency, our *primary* energy efficiency would be only 39.6%, and thus this electric heat would be equivalent to a 39.6% AFUE gas furnace, significantly worse than all but the most medieval of furnaces.

A note on space heaters

Electric space heaters, of course, suffer from the same primary energy efficiency problem as electric furnaces, but generally speaking will offset more primary energy than they consume if they are used as *supplemental* heat for no more than 30–40% of living space, thus allowing the remainder to be kept at a colder temperature. Note that space heaters also do not suffer from heat losses from the duct system, and so, depending upon ductal losses, may even offset carbon emissions when used as supplemental heating for up to around half of living space.

Unvented gas- and propane-powered space heaters are also an option; these are essentially 100% efficient and thus excellent option for reducing overall heating emissions. However, because the water from combustion is directly released into the living space, and the small hazard from other combustion products (especially carbon monoxide, if combustion is incomplete), they must only be used as supplemental heat, professionally installed, and properly maintained.

12.5.5 Heat pumps

- Heat pumps *move* rather than generate heat, moving heat twice or thrice the electrical energy investment. Thus they are approximately equivalent or slightly better than gas furnaces in terms of emissions in most areas.

- Heat pumps are less efficient in colder climates, and these areas also tend to have dirtier electric grids (especially the upper Midwest). In such areas high efficiency furnaces may be slightly preferable over the shorter-term, but heat pumps, not directly dependent upon fossil fuels, are the best long-term solution.

Heat pumps, as discussed previously, do not generate heat, but instead move heat. Thus, the amount of electrical energy invested to move a sum of heat is much smaller for electrical heat pumps than resistive heaters, and heat pumps are roughly comparable to high-efficiency natural gas furnaces in terms of emissions per unit heat delivered when the coefficient of performance is 2.5–3.0. As heat pump COP is strongly dependent upon the inside-outside temperature difference, and a seasonally adjusted COP of 2.5 is sufficient to qualify for the Energy Star program, it follows that in colder climes there may be a small advantage to high efficiency gas furnaces (see e.g. [540]), while heat pumps are generally preferable in warmer cities.

Cold-climate air-source heat pumps typically have a built-in backup resistive heater that engages at colder temperatures, further reducing COP at low temperatures, although some such systems can maintain a COP of 2–3 even at -15 °F [540]. Ground-source heat pumps (which use the earth as a heat source) avoid the problem of fluctuating ambient temperatures and are thus more efficient, but can be very expensive. Since the heat pump COP cannot fall below one, heat pumps are *always* preferable to the much more common electric furnaces. So-called dual fuel systems are also an effective option, where the heat pump is used under relatively warm conditions and at high efficiency, while a backup gas (or other) furnace kicks in at very low temperatures.

The exact COP cutoff point for favoring a heat pump over a gas furnace (the major alternative) will vary with the regional electrical grid and comparator gas furnace AFUE. Compared to an 80% AFUE furnace, under US grid-average electricity the COP cutoff to favor a heat pump is just 2.1, and so typical heat pumps are probably slightly better than typical furnaces; the cutoff is 2.6 versus a 98.5% AFUE furnace, also favoring a high-efficiency heat pump over a high efficiency furnace. Given that electricity emissions factors tend to be higher at night in much of the country (when heating loads are highest) [231], and the fact that in the counterfactual scenarios of a new gas furnace versus a new heat pump, either is essentially a marginal emissions source, it may be more appropriate to compare these technologies on the basis of the marginal electricity EF (0.8798 kgCO$_2$e/kWh). In this case, the COP cutoffs are 2.7 and 3.3, relative to 80% and 98.5% AFUE furnaces, respectively.

Finally, heating loads are also often highest in regions that have the dirtiest regional grids, e.g. the upper Midwest, where marginal EFs can exceed 1 kgCO$_2$e/kWh. Given this, in such areas high efficiency gas furnaces are probably preferable to heat pumps, over the short-term. I emphasize the short-term here, because ultimately, direct use of fossil fuels must be phased out, and heat pumps are thus the best long-term solution to space heating.

12.5.6 Wood burning

> - Wood was the primary heating fuel for 2.8 million households in 2009, but, while a "renewable" energy source, wood burning alters forest and deadwood carbon dynamics, and releases warming black carbon (along with cooling organic carbon) and methane at the point of combustion, such that wood burning is probably as bad or worse than fossil furnaces, in most cases.
>
> - Fireplaces are extremely inefficient, and, via the draft they induce, may even remove more heat from the dwelling than they add when used as supplemental heat.

While wood makes up a small percentage of total heating fuel, it is increasing in popularity and was the primary heating fuel for about 2.8 million households in the 2009 RECS. As I will show in this section, wood burning (using forest products) is most likely comparable to typical fossil fuels in its climate effect, in addition to causing serious locoregional air pollution. Modern pellet-burning stoves that use pellets manufactured from waste wood are the only general exception to this rule, but these represent only a tiny fraction of wood-based heating. There is a limited range of conditions where wood burning is preferable to fossil fuels and may even have an overall cooling effect, but this is likely not the norm.

Overall, wood burning deleteriously affects forest growth and deadwood decay dynamics, and releases a mix of warming and cooling particulate matter at the point of combustion, with the balance likely warming. Wood-burning stoves are generally inefficient compared to fossil fuel furnaces, and, by creating a draft that shunts inside air to the outside, open fireplaces can actually remove more heat from a home than they add (when being used to supplement some other heat source, as is typical).

When wood is burned, the carbon stored in that plant matter is immediately released in the atmosphere. If harvested from living trees, it will take many decades before the carbon is reincorporated into new growth. Furthermore, since forests continue to grow and take up large amounts of carbon, had the forest been left alone, much more carbon would have been sequestered into new growth. Thus, burning living wood leads to a strong warming effect that lasts over decades to centuries, compared to the counterfactual of allowing the forests to grow unmolested. The deleterious effect of biomass harvesting for fuel in the context of electricity generation is discussed extensively in Section 4.10. However, standing or minimally decayed deadwood is a preferred firewood source, and one may respond that dead wood used for fuel would have simply decayed away to CO$_2$ anyway. This is true, but the time-scale of this emission is compressed from decades (for most deadwood) to an instant. The instant bolus of CO$_2$ that results from burning has a significant warming effect over many decades compared to the slow-release of decay.

Setting aside the complexity of growth and decay dynamics, black carbon emitted at the point of combustion also induces powerful warming, and while this is partially counteracted by the cooling effect of co-emitted particulate organic carbon, the balance is likely towards warming for most residential wood burning; methane emitted by wood combustion also contributes to warming.

On the global scale, the harvesting of biomass for cook stoves (and space heating) is an important contributor to global warming via both deforestation and black carbon release. Open cooking fires, widely used in low- and middle-income countries, are extremely inefficient at harnessing the heat from the fire. That is, nearly all the heat energy is simply lost to the winds instead of being deposited into the food. The large amount of wood that is required leads to unsustainable harvesting of forests, subsequent deforestation, and permanent (or semi-permanent) loss of forests as carbon reservoir, carbon sink, and habitat [541].

Wood-burning devices

Fireplaces. Traditional, open-fireplaces are remarkably inefficient at space heating, as nearly all generated heat escapes up the chimney, and the EPA has stated that over 90% of heat is lost this way. Furthermore, when warm room air is lost through the chimney, cold outside air is drawn in, implying that a fireplace can be even worse than useless for space heating (this situation could occur when using the fireplace for secondary heating). Thus, the California EPA has estimated heating efficiencies for traditional fireplaces to range from -10% to +10% [542] (with a negative heating efficiency possible due to the air exchange effect.)

Wood-burning stove. Wood-burning stoves are a much better option than fireplaces, and, among those households that use wood as their primary heating fuel, wood stoves are by far the most common technology. Unfortunately, older wood stoves, which remain in widespread use, have a relatively low heating efficiency of perhaps 50%, and emit large amounts of particulate matter. Newer, EPA-certified stoves are better, with efficiencies in the 60–80% range; 70% is a reasonable rule-of-thumb for newer stoves.

Well-mixed GHG emissions at combustion

> - Wood combustion emissions factors are about 0.33 $kgCO_2$/kWh, 1 gCH_4/kWh (i.e. 0.034 $kgCO_2e$/kWh), plus a trivial amount of N_2O, all adding to about 0.37 $kgCO_2e$/kWh.

Consider the scenario of residential wood burning for heat. At the point of combustion, wood generates 10–80% more carbon emissions than other common fossil fuels (e.g. natural gas, fuel oil) per unit of heat generated, although it is still superior to electric resistance heating.

The combustion properties of wood have been extensively reviewed by Ragland and colleagues [544]. Briefly, green wood has a moisture content of 35–60%, while dry wood is 5–20% water by weight (20% is a commonly assumed value [545]). On a dry weight basis, wood is 47–53% carbon, with softwoods slightly higher (50–53%) than hardwoods (47–50%) in carbon [544]. We further have, for *moisture-free* wood, high heating values (HHVs) of 20–22 MJ kg^{-1} (5.56–6.11 kWh kg^{-1}) and 19–21 MJ kg^{-1} (5.27–5.83 kWh kg^{-1}) for softwoods and hardwoods, respectively.

Assuming 50% carbon by dry weight, complete oxidation of carbon to CO_2, a HHV of 5.56 kWh / (kg dry wood), and moisture content of 15–20% gives a HHV range of 4.45–4.73 kWh / (kg wet wood) and CO_2 emissions of 1.47–1.56 $kgCO_2$/ kg wet wood. This further yields a combustion EF of 0.33 $kgCO_2$/kWh.

Residential wood combustion also generates a small but non-trivial amount of methane: Solli et al. [545] give an estimated 5.8 gCH_4/kg wood for Norwegian wood stoves, which translates into 1.3034 gCH_4/kWh (under HHV 5.56 kWh / kg dry wood and 20% moisture) or 44.31 gCO_2e/kWh. Other works, such as a recent European study [543], show emissions to vary widely with wood-burning technology and wood type, and Johansson et al. [546] observed methane emissions to range from 2.20 to 17.28 gCH_4/kWh for older wood boilers, but were nearly negligible for modern boilers. Based on these works, 1 gCH_4/kWh is a reasonable overall estimate, but this number could be higher for older wood-burning devices and perhaps much smaller for more modern devices.

Wood burning also releases a trivially small amount of nitrous oxide, estimated at 0.0072 gN_2O/kWh in [545], translating to 2.15 gCO_2e/kWh over 100 years.

Black carbon and other aerosol emissions

- Wood combustion emits particulate black carbon (BC) and organic carbon (OC), which are warming and cooling, respectively. Therefore, the OC:BC ratio largely determines the global warming potential of wood burning. A large literature indicates that less efficient combustion conditions favor a higher OC:BC ratio (more cooling), perhaps in the 3:1 to 5:1 range on average, while highly efficient stoves may be more in 1:1 to 2:1 range.

- The net particulate effect is usually warming unless the OC:BC ratio is very high. Even though the OC:BC ratio is higher for low-efficiency combustion, absolute particle emissions are also much larger, often yielding a stronger warming effect than under high-efficiency combustion.

Wood burning releases large amounts of particulate matter, which is generally classified by size as PM_{10}, $PM_{2.5}$, PM_1, where the subscript denotes a particle diameter less than or equal to 10 μm, 2.5 μm, and 1 μm. Smaller particles are better able to infiltrate the small airways of the lung, where they are much more harmful than the larger particles that are cleared in the upper airways.

Particulate emissions vary widely by wood-burning technology and include BC, organic carbon (OC), and inorganic particles. Overall, wood fires emit at least 200 distinct compounds. Black carbon consists of elemental carbon (EC) and while other EC-containing particulates exist, EC and BC are often used synonymously in the literature. While BC absorbs light and is a strong warmer, OC aerosols scatter light and thus have a cooling effect. It must be further clarified that organic carbon is emitted from the flame in either particulate form, and in this case it is referred to as primary organic aerosol (POC), or in the gaseous phase. Later, gaseous organic carbon may condense onto existing atmospheric particles to form secondary organic aerosols (SOA). While it is reasonably-well established that POC has a net cooling effect, great uncertainty surrounds the overall effect of SOA, and I omit it from any further analysis.

In general, higher combustion temperatures lead to more complete combustion and lower particulate emissions. Further, the ratio of OC to BC is an indicator of combustion efficiency, and OC:BC is lower the more efficient the combustion. So, while more efficient combustion leads to an absolute reduction in particulate emissions, a greater proportion is black carbon. The somewhat perverse conclusion is that many more efficient combustion technologies, such as diesel engines and modern residential wood burning technologies, are associated with greater warming attributable to aerosols. This is the conclusion of Bond et al. [41], who determined that diesel and some forms of residential wood use have a warming effect on the planet, while uncontrolled (and hence, highly inefficient) open biomass burning has a cooling effect.

Unfortunately, there is wide variation in literature estimates of the OC:BC ratio from residential wood use, as well as their absolute emissions. Meyer [547], working from multiple literature sources, gave point estimates for black and organic carbon emissions for several wood-burning technologies for Switzerland in 2012: the OC:BC ratio varied from about 5:1 for more inefficient combustion to 1:2 for a standard wood stove, to about 1:1 for a "modern" wood stove. Absolute emissions for both categories fall markedly with newer, more efficient technologies. This overall trend is consistent with other literature. For hot burning wood stoves, Rau [548] found PM to be about 40% BC and 20% OC for softwood, and about 15% BC and OC each for hardwood; numbers from Meyer [547] are generally consistent with BC at 40% and OC at 20–40% of PM for more efficient combustion. Leskinen et al. [549] also indicated that most PM is BC and OC, with an OC:BC ratio of about 1:2.5 under conditions likely representative of typical residential wood burning; the ratio was roughly 1:1 under more efficient burning.

However, other estimates for OC and BC, likely drawn from less efficient combustion conditions, are tilted much more towards OC. In the study by Rao mentioned above [548], the OC:BC ratio was about 10:1 for cool burning wood stoves. A study by McDonald et al.[550] gave OC:BC ratios of 9.0 and 7.9 for hardwood burned in a fireplace and wood stove, respectively. Saud et al. [551] also recently estimated fuel wood emissions in India at 0.95 ± 0.27 g OC kg^{-1} and 0.35 ± 0.07 g BC kg^{-1}, for an average OC:BC ratio of 2.71:1; the observed OC:BC ratio varied from 1.86:1 to 5.11:1 among Indian states. Multiple other sources summarized by this work (see Table 2 of [551]) also give OC:BC ratios for fuel wood combustion ranging from about 1:1 to 5:1.

Numerous other papers have published OC:BC ratios [552, 553, 554, 555, 556] under varying conditions, and my own conclusion is that the balance of the literature suggests that an OC:BC ratio somewhere between 3:1 and 5:1 is likely for fireplaces and older stoves that make up the majority of existing infrastructure, but a ratio somewhere in the range of 2:1 to 1:3 is more likely for newer stoves, newer boilers, and perhaps some new fireplaces. Some older infrastructure, particularly fireplaces, may approach a 10:1 OC:BC ratio, and these devices, while very polluting, could have an atmospheric cooling effect. Absolute emissions are much higher for older devices, and thus they generally still have a stronger warming effect than newer devices, despite the higher OC:BC ratio.

Meyer [547] gives absolute BC and OC emissions on the order of about 0.07–0.16 g/kWh for modern stoves, and in the 0.5–2.3 g/kWh range for inefficient stoves, which is reasonably concordant with numbers for EPA certified stoves. EPA certified wood heaters can emit no more than 4.5 g particulate matter (PM) per hour. Now, examining the heat and PM output of actual EPA certified wood stoves suggests that, on a per kWh basis, particulate emissions may range from 0.1 to 1.5 g/kWh, with the most likely emissions range 0.5–1.0 gPM/kWh. Using 0.3 g PM/kWh would yield 0.120 g BC/kWh and 0.60–0.12 g OC/kWh as a likely low-end estimate for typical use of an EPA-certified wood stove.

AGWP values for BC and OC. From Section 3.4.6, we have an estimated direct absolute GWP of 5.3172×10^{-11} W m^{-2} kg^{-1} yr for BC, based on [43]. Net cloud semi-direct and indirect BC forcing was estimated by Bond et al. [41] to be +0.23 (-0.47 to +1.0) W m^{-2}, while the forcing from snow and ice was +0.13 (+0.04 to 0.33) W m^{-2}. Using total yearly emissions of 14 Tg BC, this yields respective per-kg indirect AGWPs of 1.6429×10^{-11} W m^{-2} kg^{-1} yr and 9.2857×10^{-12} W m^{-2} kg^{-1} yr. These direct and indirect AGWPs sum to 7.8887×10^{-11} W m^{-2} kg^{-1} yr.

Bond et al. [41] give a highly uncertain best estimate for particulate organic aerosols (POA) direct AGWP as $-4 \pm 3 \times 10^{-12}$ W m^{-2} kg^{-1} yr. Indirect RF from cloud aerosol interactions was estimated at -0.74 W m^{-2} for 81 Tg OC yr^{-1}, yielding -9.1358×10^{-12} W m^{-2} kg^{-1} yr for AGWP. However, this is significantly higher than the cited point estimates from numerous models, which ranged from -7.4 to -1.8×10^{-12} W m^{-2} kg^{-1} yr, suggesting that -5×10^{-12} W m^{-2} kg^{-1} yr may be a more reasonable point estimate for OC. Together, these estimates suggest a total direct and indirect AGWP from OC on the order of about 10^{-11} W m^{-2} kg^{-1}

yr, but with significant uncertainty.

Overall aerosol forcing from residential wood use. In general, using 100-year GWPs, aerosol forcing for wood heating may vary widely. For a newer, more efficiently burning EPA woodstove, we might reasonably expect 0.05–0.15 kgCO$_2$e/kWh delivered, while an older inefficient woodstove might yield much higher carbon equivalent emissions, anywhere in the range of 0.2–1.5 kgCO$_2$e/kWh delivered. For very inefficient fireplaces and cooler-burning woodstoves, when very large amounts of organic carbon (OC:BC ratio >8) are produced, the aerosol effect may be become negative. The absolute warming or cooling effect in this case can be very large, on the order of many kgCO$_2$e/kWh, and a switch from warming to cooling is highly sensitive to the OC:BC ratio.

Emissions factors from altered carbon dynamics

- Altered deadwood decay dynamics imply an emissions factor of roughly 0.2–0.25 kgCO$_2$e/kWh of heat delivered for typical woodstoves (over a 100-year time horizon), when comparing the counterfactuals of immediate combustion versus natural decay over several decades. This EF is highly variable, and depends upon wood and stove characteristics.

In the case of living wood, the low carbon argument rests on reincorporation into growing biomass. This is problematic, and I cover this topic extensively when discussing biomass for electricity generation (Section 4.10). The argument translates directly to firewood, and we may infer emissions equivalent to around 0.1–0.5 kgCO$_2$e/kWh(t).

If forest *deadwood*, on the other hand, is harvested for fuel, one can make a low emissions argument on the assumption that the wood would have decayed anyway, leading to equivalent carbon release. Forest deadwood, or coarse woody debris (CWD), is an important carbon and nutrient reservoir in forests of ecological importance that provides habitat for a multitude of wildlife, and aids in tree regeneration as well ("nurse logs") [557]. CWD accounts for 10–20% of above-ground biomass in mature forests, and may make up a much greater proportion in forests that have been degraded by logging activities [558]. CWD is constantly decomposing, and in this process the carbon is metabolized by microorganisms to CO$_2$.

The (exponential) decay rates for CWD for many wood types in different regions have been characterized in numerous works, and a review by Laiho and Prescott [559] reported a range of 0.0025–0.071 yr^{-1}, with a mean and median of about 0.02 yr^{-1}. While a simple exponential decay function is widely used, and apparently reasonable as a first approximation, the overall decay process has been described as tri-phasic [559], with an initial slow phase where the wood is initially colonized by decomposers, a fast phase during which much labile carbon is consumed, and a final slow phase when long-lived lignins (which are relatively resistant to decomposition) decay.

Given a CWD decay rate, it is straightforward to calculate atmospheric carbon and radiative forcing that results over time from burning a set amount of wood versus letting it decay naturally. From the relative radiative forcings, we arrive at a time-dependent CO$_2$e per kWh of heat generated, as shown in Figure 12.31, for several different decay rates. The 100-year GWPs, due to altered decay dynamics, are 0.2880, 0.1292, and 0.0380 kgCO$_2$e/kWh for slow-, average-, and fast-decaying woods, in terms of heat generated. Considering emissions per heat delivered to the living space, we have average 100-year GWPs of 1.29 (0.38–2.88) kgCO$_2$e/kWh delivered for fireplaces (10% efficiency), 0.26 (0.076–0.58) kgCO$_2$e/kWh for older woodstoves (50% efficiency), and 0.18 (0.054–0.41) kgCO$_2$e/kWh for modern woodstoves (70% efficiency), at a 100-year horizon. Compare these numbers to the 100-year GWP for natural gas heating,

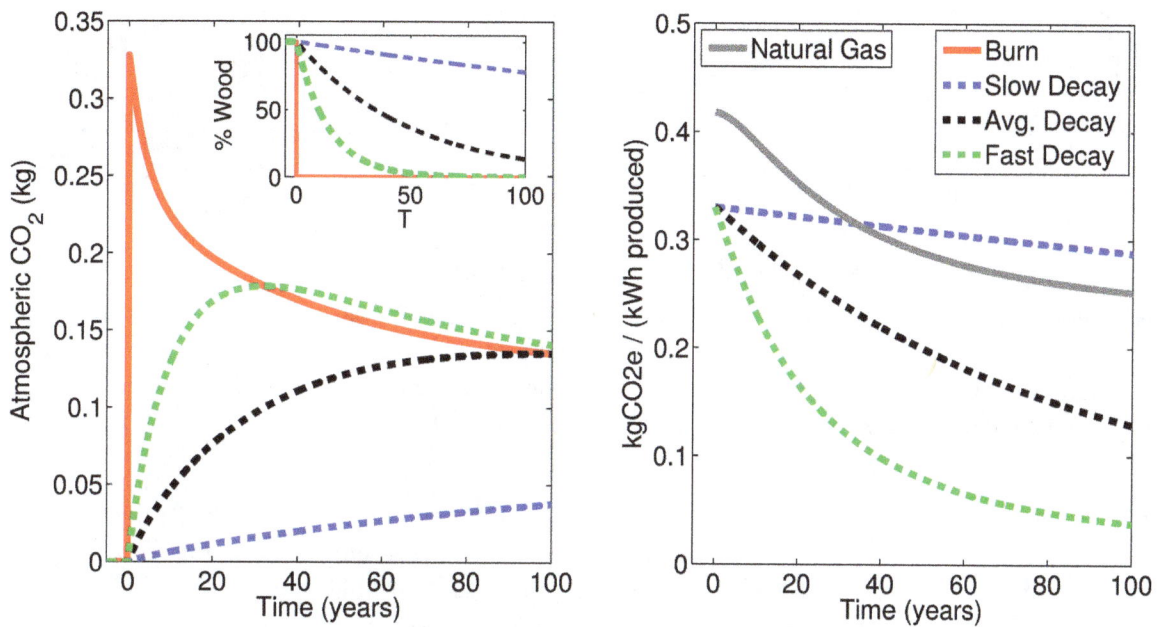

Figure 12.31: The left panel shows the time-course of atmospheric CO_2 concentration either after burning a single kg of wood, or allowing it to decay naturally, at slow, average, and fast decay rates. The inset shows the mass of wood remaining in each case. We see that, when decay would have been slow, most carbon is still locked up in woody mass at the 100-year mark. The right panel then illustrates how this $CO2$ time-course translates into a $kgCO_2/kWh$ heat generated emissions factor, under our three decay scenarios, and with the same calculation performed for natural gas combustion for comparison. Note that the gas calculation accounts for upstream methane released in the provision of the fuel.

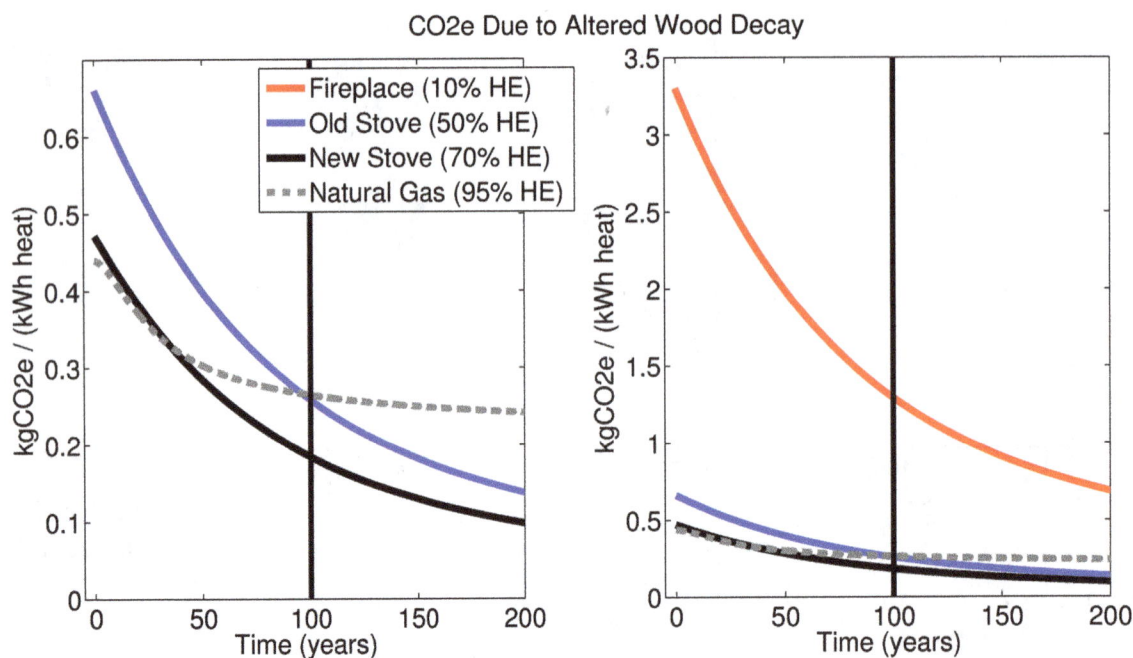

Figure 12.32: Emissions factors in terms of kgCO₂e/kWh *delivered*, for several wood burning technologies of differing efficiencies, over the course of 200 years, attributable solely to altered deadwood decay dynamics (and assuming an average decay rate of 0.02 yr^{-1}). The equivalent emissions factor for a 95% efficient gas furnace is included for comparison. The left panel omits the 10% fireplace to allow a closer comparison of curves, while it is included on the right.

at 0.261 kgCO₂e/kWh generated, and ranging from 0.265 to 0.335 kgCO₂e/kWh delivered for 78–98.5% heating efficiencies. It is apparent that altering the carbon dynamic of deadwood results in lifecycle emissions for wood heat that are highly variable, but on the same order of magnitude as natural gas. These average outcomes are illustrated in Figure 12.32.

Overall wood-burning emissions factor

- Older and newer woodstoves have conservatively estimated emissions factors on the order of 0.56 and 0.34 kgCO₂e/kWh of heat delivered, respectively. About 50% of this emissions factor is due to altered deadwood decay, with black carbon released at the point of combustion the other major factor. Thus, woodstoves are likely no better and as much as twice as bad as gas furnaces.

Deriving single point-estimate for wood-burning emissions factor is fraught, as uncertainty is compounded by uncertainty in this area. Nevertheless, I will give a *conservative* estimate for typical heating emissions in US residences. From the 2009 RECS, about 25% of residential woodstoves were over 20 years old, which would put nearly all at pre-1988 and hence pre-EPA regulations. I assume, and again, these are very rough estimates, that older woodstoves emit OC and BC at a 5:1 ratio, with absolute emissions of 360 mgBC/kWh and 1800 mgOC/kWh. Newer woodstoves are assumed to emit OC:BC at 3:1 with absolute rates of 125 mgBC/kWh and 375 mgOC/kWh. These numbers would suggest aerosol forcings equivalent to 0.1174 kgCO₂e/kWh and 0.0679 kgCO₂e/kWh for older and new woodstoves, respectively, over a 100-year time-horizon (in terms of kWh generated).

300

I assume 1 gCH$_4$/kWh for all stoves, equivalent to 0.034 kgCO$_2$e/kWh over 100 years. Additionally, under the conservative assumption that firewood is harvested from deadwood rather than live trees, and using a decay rate of 0.02 yr^{-1}, we get 0.1292 kgCO$_2$e/kWh. Finally, I use 6 gCO$_2$e/kWh for upstream emissions from extraction and distribution, based on results for a short supply chain in Pierobon et al. [560]. Adding these up, we get 0.2806 kgCO$_2$e/kWh and 0.2371 kgCO$_2$e/kWh for older and new woodstoves, respectively. These numbers are for kWh *generated*: using heating efficiencies of 50% and 70%, we get 0.5612 kgCO$_2$e/kWh and 0.3387 kgCO$_2$e/kWh in terms of kWh *delivered*.

Weighting 25% older and 75% newer woodstoves, one gets overall EFs of 0.2480 kgCO$_2$e/kWh generated and 0.3943 kgCO$_2$e/kWh heating delivered. If we assume open fireplaces are similar in their combustion characteristics to older woodstoves, but deliver heat at only 10% efficiency, then fireplaces would yield 2.8060 kgCO$_2$e/kWh, at best.

I consider these to be conservative, low-end estimates. These factors may be up to several-fold higher or lower, depending upon the particulate emissions and the natural decay rate of the wood. They will be greater if live biomass is harvested for heating. Thus, this final exercise shows that on average residential wood burning is likely as bad as, and possibly worse than, natural gas and petroleum fuels for heating.

12.6 The physical basis for heating/cooling

The temperature of the air within one's dwelling is determined by several basic mechanisms: (1) heat transfer across the walls, floor, windows, and roof that make up the house's thermal envelope, including the heat gained from solar radiation, (2) infiltration of outside air, i.e. exchange of inside and outside air, (3) the "thermal mass" of the house, i.e. the heat stored within the structure itself (and the objects within), (4) internal heating loads, such as cooking, lighting, hot and sweaty humans, etc., and (5) intentional heating via furnaces, etc. and heat removal via air conditioning (or evaporative cooling).

Environmental heat transfer and air infiltration are readily modifiable via modifications such as insulation or weather stripping, and as we shall see, the difference in energy consumption between a well-insulated, tightly sealed house and one drafty and uninsulated is profound. In this section, I present the physical and mathematical basis for mechanisms (1)–(3) above, and then from the resulting mathematical models, explore how different (generally modifiable) residence characteristics affect one's heating and cooling energy use.

12.6.1 Heat transfer: conduction, convection, and radiation

- Conduction, convection, and radiation are the three basic mechanisms by which heat is transferred. The mathematical description for all can be reduced to a single, lumped resistance to heat flow.

- Heat transfer is nearly linearly proportional to the difference between the inside and outside temperatures, but solar irradiance represents an additional heat gain not directly related to ambient temperature. Therefore, the energy cost of heating or cooling is roughly proportional to the inside-outside temperature gradient, providing a physical justification for the degree-day approximation.

Heat is transferred via *only* three basic mechanisms: conduction, convection, and radiation. The first, **conduction**, is the flow of heat via molecular interaction in an solid, gas, or liquid. Since our interest is in conduction across a wall or roof, we can derive from Fourier's Law of

Heat Conduction, that heat transfer across some surface, $\dot{Q}$ (units W), is proportional to the temperature difference across this surface, $T_O - T_I$, where T_O is the outside temperature and T_I is the inside temperature (in units of absolute temperature, K), and surface area, A (m^2), as

$$\dot{Q}_{cond} = \frac{A}{R_C} \left(T_O - T_I \right), \tag{12.17}$$

where R_C, the thermal resistance (m^2 K/W), quantifies how easily heat flows through the given material: the larger this number, the less heat flows. Standard values for thermal resistance are widely available for typical building materials, and in *English* units (ft^2 °F hr/Btu), we call these the "R-values." For example, a single-pane window has an R-value of about 1, while 3.5 inches of wall insulation give an R-value of about 13; in colder climate regions attics should be insulated to R-60, or around 1.5 feet of typical insulation material. We can also define the R-value (or R_C) in terms of thermal resistance per inch of material, with most standard insulation having an R-value of 3–4 per inch.

Convection is a bit more complex, and is heat transfer caused by the movement of a gas or fluid. Roughly speaking, when we have air moving along a surface, it must be that immediately adjacent to the surface, the air is stationary and has the same temperature as the surface. In this "boundary" layer, we will have heat conducting into air from the surface, but then be carried off as it enters the moving bulk air stream. The faster the air flow, the more heat can be carried off, explaining why a freezing wind is more biting than frozen but still air, or conversely, why a convection oven cooks faster than a standard one. Convection is further divided into "natural" convection, which arises spontaneously from heat gradients in a fluid, and "forced" convection, e.g. via a fan or pump, etc.

Now, this rather complex process can fortuitously be reduced to a mathematical form almost identical to Equation 12.17, and we have the transfer of heat from a fluid to a surface as

$$\dot{Q}_{conv} = hA \left(T_\infty - T_S \right), \tag{12.18}$$

where h is the heat-transfer coefficient (W /m^2/K), T_∞ is the temperature within the main fluid stream, and for our purposes will be equal either to the outside or inside air temperature (depending on which side of the wall we're on), and T_S is the temperature of the surface. Fortunately for us, we can easily solve for T_S (if desired) as part of the equivalent circuit below. The major challenge is determining h, towards which much empirical effort has been expended over the years. It enough to know that h increases with wind speed and is in the range of 1–5 for interior walls, floors, and ceilings [561], and 5–10 for exterior walls in the calm, increasing to as much as 40 in higher winds [562].

Consider heat transfer across a wall, where it is hotter outside and cooler inside. In this case we have convective heat transfer from the hot exterior air to the wall surface (defined by the heat transfer coefficient h_o), conductive transfer through the wall, and then convective transfer of heat from the interior surface to the inside air (determined by h_i). This can be modeled, as one often does in engineering, with an analogous electrical circuit, which is simply three resistors in series, with resistances $1/h_i$, R_C, and $1/h_o$, from in to out, as shown in Figure 12.33. This is itself equivalent to a single resistor with resistance

$$R = \frac{1}{h_i} + R_C + \frac{1}{h_o}, \tag{12.19}$$

and our final equation for convective+conductive heat transfer is

$$\dot{Q} = \frac{A}{R} \left(T_O - T_I \right). \tag{12.20}$$

302

Note that the heat flow across every individual element in this three-resistor circuit must be equal, $\dot{Q} = \dot{Q}_{conv,outer} = \dot{Q}_{cond} = \dot{Q}_{conv,inner}$, and so we can easily calculate the inner and outer wall surface temperatures, if we are curious. As an example, suppose we have an insulated wall with $R_C = 2.29$ m^2 K/W (R-value 13 in English units), $h_o = 10$, and $h_i = 3$. Then total effective resistance to heat transfer is 2.72 m^2 K/W (R-value 15.5), and we see that the conductive resistance is generally dominant. If the outside and inside temperatures were 95 °F (308 K) and 75 °F (297 K), then total heat transfer is 4.1 W/m^2, and our outside and inside wall temperatures are 94.26 °F and 77.45 °F, respectively.

We can also infer from the equation that, for uninsulated walls ($R_C \approx 0.7$, R-value 4), convective and conductive resistance will be of roughly equal importance, while for any reasonably well-insulated surface, conductive resistance will always dominate. Further, as outside wind speed becomes arbitrarily high, we will have $h_o \to \infty$, the outside wall surface temperature will approach that of the air stream, and the resistance at our first node, $1/h_o$, goes to zero. Presumably, blasting an industrial fan at the inside wall would have a similar effect on h_i.

Equivalent R-values for composite materials and complex material geometries can be determined using equivalent circuit models, and indeed, we can even use this process to determine a "whole-house R-value" for any specified building geometry, but I defer a detailed exploration of this concept here.

Radiation is the spontaneous emission of energy as electromagnetic radiation (EMR) at a rate proportional to absolute temperature, something which all matter does, and is the basis for the incandescent lamp (as discussed in Section 14.2.2). The rate at which a black-body, a theoretical entity that absorbs all incident radiation, emits energy as EMR, is given (in W) as

$$\dot{Q}_{rad} = \sigma A T^4 \tag{12.21}$$

where σ is the StefanBoltzmann constant, 5.676×10^8 W/m^2/K^4, and A is the body's surface area (m^2). For any *particular* body, the emissivity, ε, gives the ratio of radiation emitted relative to a blackbody of the same temperature, and is greater than 0.9 for almost all materials. Now, it follows from Equation 12.21 that the net radiative heat transfer between two bodies, of temperatures T_1 and T_2, is

$$\dot{Q}_{rad} = \sigma A F \varepsilon (T_2^4 - T_1^4), \tag{12.22}$$

where $F \leq 1$ is a shape factor. Alas that the fourth power of temperature is involved, as we desire nothing more than to put this into a form equivalent to convection or conduction, e.g. $\dot{Q} = h(T_O - T_S)$, and so we perform the mathematical trick of linearization: assuming the difference between T_1 and T_2 is relatively small, we can approximate Equation 12.22 as[3]

$$\dot{Q}_{rad} = 4\sigma A \varepsilon F T_1^3 (T_2 - T_1) = h_{rad} A F (T_2 - T_1), \tag{12.23}$$

where $h_{rad} = 4\sigma\varepsilon T_2^3 \approx 6.1\varepsilon$, using $T_2 = 300$ K (a reasonable assumption for ambient temperature), is the effective radiative heat-transfer coefficient. In the case of the outer wall surface, we have radiative heat transfer between the wall and the sky, as

$$\dot{Q}_{rad} = h_{rad} A F (T_{sky} - T_S), \tag{12.24}$$

where T_{sky} is the effective sky temperature, and $F \approx 0.5$ for an outer wall. The value of T_{sky} varies, but it is generally about 5–15 K below ambient air temperature [563]. Now, there is one final mathematical trick we must do. The heat flux into the outer surface of the wall is given by the sum of convective and radiative heat fluxes,

---

[3]Derived by taking the Taylor series expansion of 12.22 and evaluating at T_1.

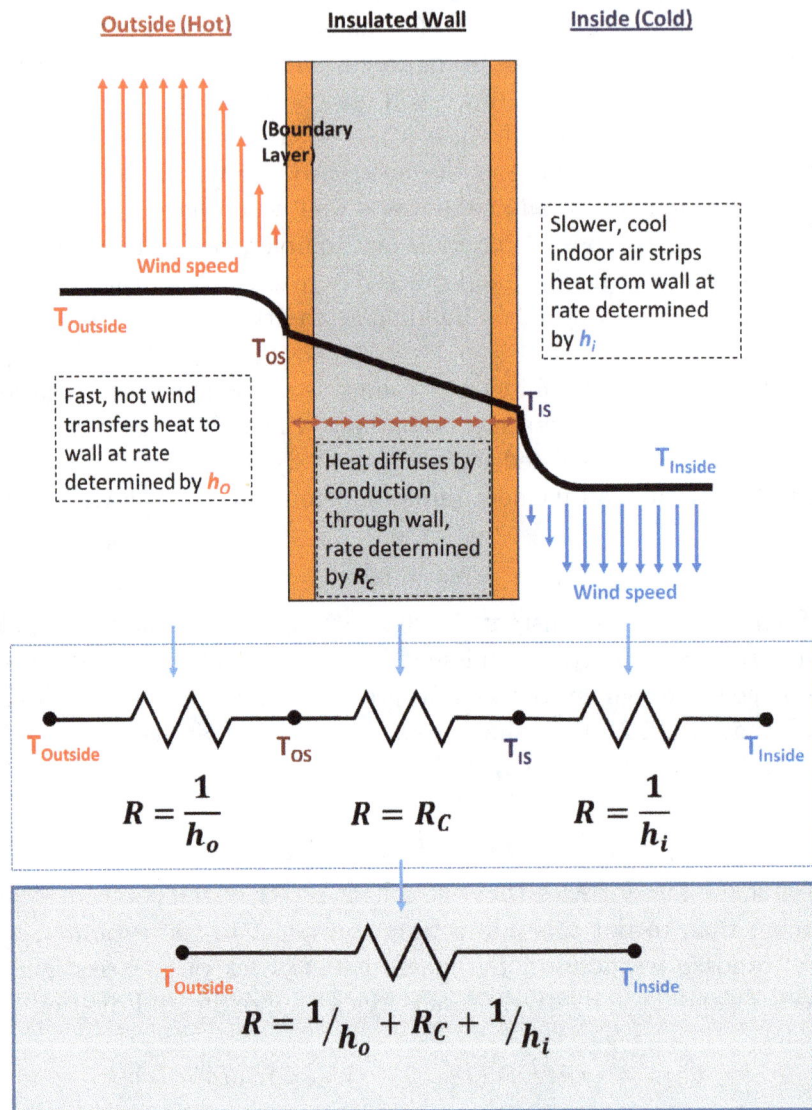

Outside (Hot)   **Insulated Wall**   **Inside (Cold)**

(Boundary Layer)

Wind speed

$T_{Outside}$

T_{OS}

T_{IS}

T_{Inside}

Slower, cool indoor air strips heat from wall at rate determined by h_i

Fast, hot wind transfers heat to wall at rate determined by h_o

Heat diffuses by conduction through wall, rate determined by R_C

Wind speed

$T_{Outside}$ T_{OS} T_{IS} T_{Inside}

$$R = \frac{1}{h_o} \qquad R = R_C \qquad R = \frac{1}{h_i}$$

$T_{Outside}$ T_{Inside}

$$R = {}^1\!/_{h_o} + R_C + {}^1\!/_{h_i}$$

Figure 12.33: The transfer of heat from the hot outdoors to the cooled indoors, by a combination of convection and conduction, given schematically and as an equivalent circuit model.

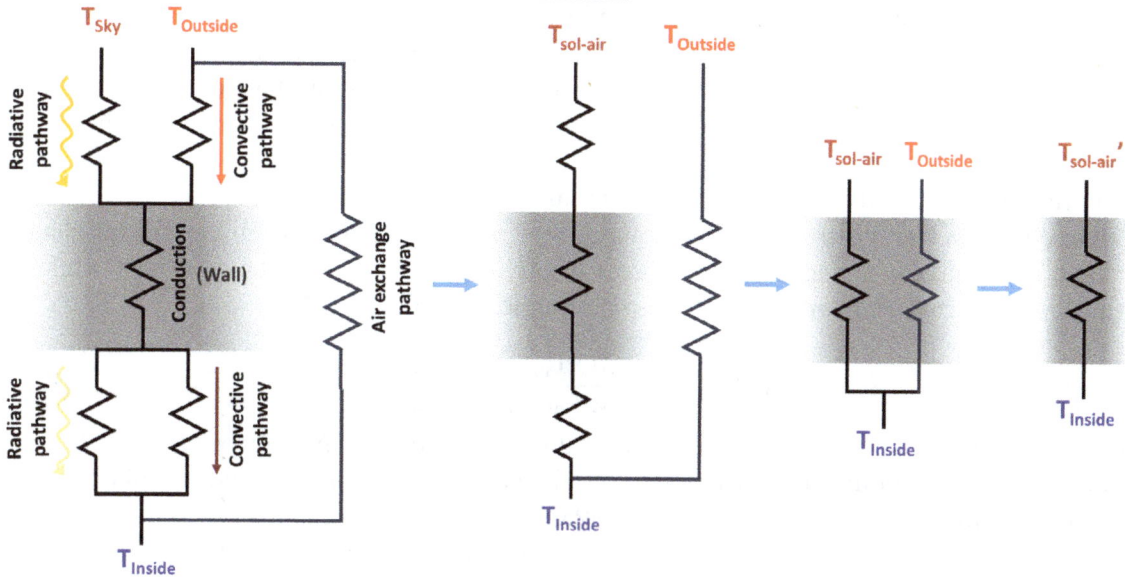

Figure 12.34: The transfer of heat from the outdoors to indoors, via radiative, convective, conductive, and air exchange pathways, with a general schematic for the equivalent circuit model, along with the basic process for reducing all pathways to a single lumped resistance. See the text for details on parameter values, and the $T_{sol-air}$ concept.

$$\dot{Q} = A(h_c(T_O - T_S) + h_{rad}F(T_{sky} - T_S)), \qquad (12.25)$$

but the problem is that $T_O \neq T_{sky}$. However, with a bit of algebraic manipulation, we arrive at

$$\dot{Q} = Ah_c(T_{sol-air} - T_S), \qquad (12.26)$$

where

$$T_{sol-air} = T_O + \frac{h_{rad}F\Delta T_{O-sky}}{h_c}, \qquad (12.27)$$

and $\Delta T_{O-sky} \approx -10$ K is the difference between the ambient and apparent sky temperatures. Now that all heat transfer terms are represented using the same basic linear format, we can represent all three components of heat transfer across a surface using a single equivalent circuit. This, along with the parallel air exchange pathway for heat transfer (discussed in the next section), is demonstrated schematically in Figure 12.34.

Easily incorporated into the above framework, the roof represents a special case, as we have radiant heat via both transfer from the background sky and via direct solar irradiation. Supposing all direct sunlight strikes the roof with irradiance I (W/m^2), and that of this, some fraction a is absorbed and $1 - a$ is reflected away, we have an additional radiative gain of aI, which we simply add to our expression for $T_{sol-air}$ as follows:

$$T_{sol-air} = T_O + \frac{aI + h_{rad}F\Delta T_{O-sky}}{h_c}, \qquad (12.28)$$

and $F \approx 1$ for a roof. In reality, depending upon the incident angle of incoming light, a fraction will strike walls and windows, and only part of the roof will be illuminated, but I take all light falling upon the roof as a first approximation.

305

12.6.2 Air leakage and infiltration

Any air that leaks out of a house must be replaced by entering air, and, depending on how "leaky" a building is, the entire volume of air may be replaced by outside air anywhere from multiple times per hour to less than once per day. Clearly, more air changes per hour (ACH) imply greater energy costs, as all the new air entering must be cooled or heated to the desired level. This background rate of air exchange is referred to as "natural" ventilation, while "active" ventilation entails intentional air exchange via fans, etc.

Air changes per hour. The DOE Energy Star program defines ACH_{50} as the air changes per hour in a building induced by 50 Pascals of pressure during a blower door test, while the natural air changes per hour, ACH_{nat}, is some fraction of ACH_{50},

$$ACH_{nat} = \frac{ACH_{50}}{LBL\ Factor} \approx \frac{ACH_{50}}{20}. \tag{12.29}$$

The LBL factor depends on climate region, number of stories, and how exposed a building is, and varies from about 10–30 depending on all these factors; for single-story buildings with normal exposure, the LBL factor is about 15–25, with 20 a good rule of thumb.

A poorly sealed home may have an ACH_{nat} of one to two, while the Energy Star program requires $ACH_{50} \leq 7.0$, or roughly $ACH_{nat} \leq 0.35$. Even more stringently, the 2012 International Energy Conservation Code requires $ACH_{50} \leq 5.0$ in climate zones 1–2 (essentially southern Arizona, southern Texas, and Florida) and $ACH_{50} \leq 3.0$ in all other climate zones. These requirements correspond to $ACH_{nat} \leq 0.25$ and $ACH_{nat} \leq 0.15$. Finally, the Passive House (or PassivHaus) requirement is $ACH_{50} \leq 0.6$, or $ACH_{nat} \leq 0.03$.

Note, however, that extremely low natural ventilation rates would lead to poor indoor air quality, and so we must have an active ventilation system to provide fresh air. At first glance this would seem to defeat the entire purpose of a low ACH, but some active ventilation systems can pass fresh incoming air and conditioned outgoing air through a heat-exchanger, thus conserving most of the energy invested in heating/cooling. I will not consider such systems in further calculations, and assume that a typical residence is ventilated only by natural ventilation, with ACH referring to ACH_{nat} from here out.

Mathematics. It can be shown that the rate of heat loss or gain, $\dot{Q}$ (in W), due to air exchange only, is

$$\dot{Q} = \dot{V}\rho_{air}c_{air}(T_O - T_I) = \frac{ACH}{3600}V_{air}\rho_{air}c_{air}(T_O - T_I), \tag{12.30}$$

where $\dot{V}$ is the volumetric rate of air movement in ft^3/s or m^3/s, and is equal to $ACH/3600 \times V_{air}$, with the factor 3600 converting from per hour to per second; V_{air} is the total volume of air in the house. The parameter ρ_{air} is the density of air, while c_{air} is the thermal capacity of air, and together they give tell us much heat energy is stored in the air that is exchanged.

This equation confirms one's intuition the energy required to offset infiltration is directly proportional to both the inside-outside temperature gradient, and to the ACH. So, a poorly sealed house, with an ACH of 1–2, will require about 3–6 times as much energy to offset air exchange as a home meeting Energy Star standards (ACH 0.35).

What is especially noteworthy about Equation 12.30 is that is in the same basic format as our final equations for lumped convective, radiative, and conductive heat transfer across a surface. Thus, we can incorporate air exchange into the same basic equivalent circuit, where air exchange represent a *parallel* path for the movement of heat in/out of the house. This is illustrated in Figure 12.34.

12.6.3 Leaky ductwork

- Air distribution systems may reduce HVAC efficiency by 20–50%, for typical houses.

- Leaky ductwork results in both a loss of conditioned supply air and increased infiltration of outside air into the home; sealing ductwork can reduce overall heating/cooling energy by 15–20%.

- Uninsulated ducts also suffer significant conductive heat loss in unconditioned spaces, but this is less important than leakage.

Following the prior section, we should mention one significant but oft overlooked source of heating/cooling inefficiencies: leaking and/or poorly insulated ducts. Central furnaces and air conditioners, which have become the dominant technologies in the US, distribute conditioned air through a network of ducts, and problems with this system decrease HVAC efficiency via several mechanisms: (1) conditioned air leaks out of the ductwork to the environment, (2) outside air is drawn into the return ductwork, (3) whole-house pressure changes due to unbalanced leaks markedly increases overall air infiltration into the house, and (4) conductive heat transfer across ducts passing through unconditioned spaces decreases efficiency.

These problems only manifest themselves when the ductwork is (at least partly) exterior to the conditioned space: a home with all interior ducts will see essentially no efficiency losses from air distribution, but such homes are in the minority [567]. Ductwork located in attics is especially problematic in hot climates, where attic temperatures can reach 130 °F, and thus very effectively transfer heat into the cool air within the pipes, and even for optimum duct sealing and insulation, overall distribution efficiency may not exceed 80% (during the cooling season) in hot Southwestern cities.

Turning first to leakage, we divide this into supply and return leaks, where the return leak fraction (RLF) is the fraction of air returning to the air handler that comes from outside the house, and is typically 10–20% (<5% for well-sealed ducts, and potentially >30% for those that are poorly sealed) [564, 567]. The supply leak fraction (SLF) is the portion of heated or cooled air that escapes from the air distribution, and is probably slightly smaller [564, 567].

Now, the direct effects of supply and return leakage are fairly obvious: if 10% of the supply air leaks into the environment, heating/cooling requirements increase by 10%. Similarly, if 10% of the air entering into an AC return is drawn from the attic, then the AC cooling load is significantly increased. The second-order effect that can be just as important, however, is the increase in whole-house ACH induced by leak imbalances. Suppose, for example, that our HVAC system moves 1,000 cubic feet per minute (cfm), and the RLF is 10% (100 cfm), while SLF is 20% (200 cfm). Thus, on balance, 900 cfm are drawn from the house but only 800 cfm returned, and 100 cfm of additional outside air must enter the house interior to balance this out. This effect may be particularly pernicious in the summer, when much of this additional air is drawn from the hot attic. When RLF > SLF, the imbalance will similarly force conditioned air out of the building interior. Overall, duct leakage increases ACH by roughly 20–30% when averaged over the year [566].

A study of Florida homes [564] found that sealing ductwork reduced the RLF by 73%, from 15.8 to 4.4%, reduced ACH when the air distribution system (ADS) was on from 1.1 to 0.54 (ACH was 0.25 when the ADS was off entirely), and reduced overall cooling energy by 17.2%. Other studies have similarly found ductwork sealing to reduce leakage by 60–70%, and reduce overall HVAC energy by 15–20% (see [565, 566, 567, 568] and references therein).

Conduction losses across ducts in unconditioned attic spaces may also be substantial, with Modera [566], for example, measuring a 23% heat loss across supply ducts in 26 California homes, while Francisco et al. [567] measured a smaller 10–15% loss. Generally speaking, conductive losses are less important than leakage [565]. Finally, note that some heat lost from ducts to unconditioned spaces is regained via conduction, partially mitigating these losses [567, 568].

When conductive losses are combined with the summed effects of air leakage, most authors have found that about 30–40% of all cooling/heating energy is typically lost through the air distribution system, either directly or indirectly (i.e. increased ACH). Aggressive retrofitting, primarily via sealing to reduce leakage, can reduce energy losses to around 15–25% on average, and to perhaps 5–15% for the best houses [567, 568]. Moving ductwork into conditioned spaces is necessary to completely eliminate these losses.

12.6.4   Thermal mass

The thermal mass of a building (and its contents), i.e. its capacity to store heat, affects heating/cooling load by stabilizing interior temperature relative to diurnal variations. A high thermal mass can therefore reduce cooling energy, for example, when there are large swings in temperature such that the height of the day is intolerably warm, but the 24-hour average represents an acceptable temperature. Since the thermal mass maintains temperature at a stable, near 24-average level, no (or at least less) cooling need be done during the day. The opposite applies to heating loads, but, because temperatures tend to fall below an acceptable level at all times during the heating season, the relative benefit of thermal mass is less notable [569].

It should be emphasized that, while larger houses with higher thermal masses may actually require less cooling/heating in those times when diurnal temperature varies about an acceptable mean, this is generally a minority of the year, and overall yearly cooling/heating load is still lower the smaller the abode under most conditions, and especially in the hottest and coldest areas.

12.6.5   A (more or less) complete model

From all the above, we can finally arrive at a reasonably complete mathematical description of heat transfer into our house, and from there make predictions about how different types of dwellings, thermostat strategies, and modifications affect energy use. For the interested reader, there are many professional software packages (including sophisticated freeware, e.g. the Los Alamos National Laboratory DOE-2 Model, which is widely used in academic works) which can model heat transfer in a more sophisticated manner than what I do here, but my approach is geared towards understanding general order of magnitude effects, and so I believe it is appropriate. Now then, I propose the following simple description of a one-story house. I assume that the square footage (SF) is given, while perimeter area is determined as

$$4 \times \sqrt{SF} \times \phi. \tag{12.31}$$

That is, the perimeter area is that of a square-shaped house times some "shape factor," $\phi > 1$ to account for more complicated wall geometries. Now, assuming standard 8-foot high walls, we have total wall surface area, and I also assume roof and floor area are equal, and we have an unfinished, insulated attic. Finally, I suppose about 7.5% of the wall area is window. Heat transfer across walls and windows occurs according to the heat transfer principles discussed above.

As for flooring, we either generally have either a crawl space (essentially a "mini-basement"), basement, or concrete slab foundation. For simplicity, I will assume a 1 m deep, sealed crawl space below the floor, with the temperature of the surrounding soil. As given in [570], among others, the temperature of the soil basically tracks that of the air (with high frequency variations filtered out), approaching the yearly average air temperature at great depth. At 1 m depth, we can expect the amplitude of soil temperature variations to be about 50% that of air (based on [570]).

The attic space is modeled as a separate air-space, to which there is heat transfer across the roof, and then heat transfer from the attic into the house. The reason for this two-compartment design is that it allows us to incorporate air-exchange in a ventilated attic. Fortunately, because of the magic of equivalent circuits, we can ultimately transform this complex process into a single resistance. The only complication that arises is that the effective outer temperature for the convective/radiative/conductive pathway is $T_{sol-air}$, while the outer temperature for air exchange is T_O. Not to worry though, we simply do some algebra to arrive at yet another effective temperature, T_{op}, and have heat transfer into the attic as

$$\dot{Q} = (\alpha + \beta)(T_{op} - T_A) \tag{12.32}$$

where T_A is the attic temperature and

$$T_{op} = \frac{\alpha T_{sol-air} + \beta T_O}{\alpha + \beta}, \tag{12.33}$$

$$\alpha = \frac{A_{roof}}{R_{ccr}}, \tag{12.34}$$

$$\beta = \frac{1}{R_{ACH,attic}}. \tag{12.35}$$

Everything else follows from all the same principles discussed above.

12.7 Upgrades, strategies, and conclusions for reducing heating/cooling energy

- For a single-story house, heating and cooling energy/emissions increase with square footage in an almost 1:1 manner.

- Thermostat set-point is the most immediately modifiable factor: in the warmest climes, cooling energy may vary ten-fold across a plausible thermostat range, while in the coldest areas, heating energy can vary two-fold across a plausible thermostat range.

- Setting back the thermostat 5–10 °F while away from home during the day can save 15–30% of cooling energy in hotter cities, while a similar nighttime thermostat setback in cold cities may reduce heating energy by 5–15%.

- Extremely energy-efficient houses may plausibly use one-tenth (or even less) the heating/cooling energy of extremely low-efficiency houses, for the same thermostat settings, a conclusion compatible with RECS results on actual energy consumption.

- Insulation performs best when all major barriers (attic, walls, windows, floor) are well-insulated: improving one in isolation of the others is only of small benefit. The attic is the single most important location for adequate insulation, especially in cold regions.

- Sealing/insulating ductwork, upgrading furnaces/ACs, and sealing against excessive air exchange all independently affect cooling/heating loads. Reasonable implementations of all, compared to typical baselines, could independently decrease heating/cooling energy 20–50% each, and by perhaps 70% collectively.

- A 50% or better reduction in heating/cooling emissions can reasonably be achieved via just one or two of a variety of modest improvements (upgraded HVAC, insulating, etc.) in conjunction with a conscientious approach to the thermostat (more conservative baseline setting combined with a set-back strategy).

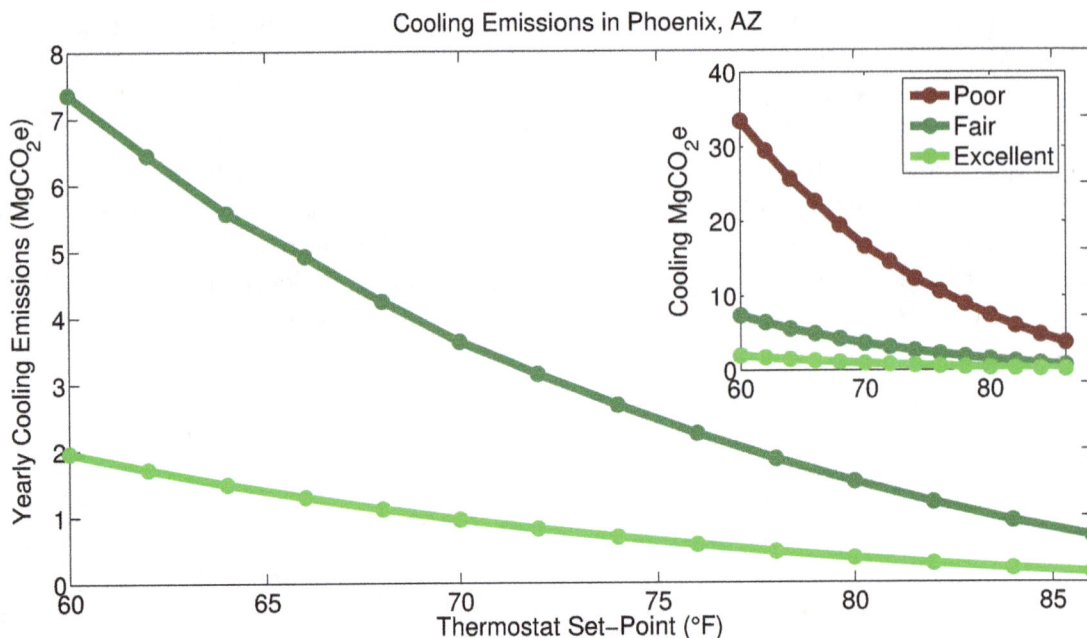

Figure 12.35: Yearly cooling CO_2e emissions for a 1,700 ft^2 house in Phoenix, AZ, under a range of constant thermostat set-points, and for three general levels of household energy efficiency. Moving from a set-point of 60 to 86 °F reduces energy/emissions dramatically, by about 90%, and a single degree change reduces energy by around 7%.

12.7.1 Square footage and energy consumption

Using our heat-transfer model, we can essentially confirm the conclusions drawn from RECS data and a simple analysis based on degree-days: when all else is equal, both heating and cooling energy increase nearly linearly with square footage, and although energy per square foot does fall somewhat with increasing house size, this in no way makes up for the increased total energy demand of a larger house.

12.7.2 Thermostat set-point, and the set-back strategy

Figure 12.35 shows how (model-predicted) yearly cooling emissions change as a function of the thermostat set-point, in Phoenix, AZ, while Figure 12.36 shows how the thermostat affects heating emissions in Minneapolis, MN, with are shown for three basic levels of whole-house energy efficiency: very poor, fair/typical, and excellent. These more detailed simulations confirm the basic conclusions drawn from our analysis based on degree-days. In Phoenix, regardless of overall energy efficiency, each one degree (°F) increase in the thermostat reduces summer cooling emissions by about 7%, while in Minneapolis, each one degree thermostat drop translates into an almost 4% decrease in winter heating emissions. The general level of energy efficiency, however, can be even more dominant than thermostat set-point at the extremes, as is apparent from the figures.

One effective strategy for reducing heating/cooling energy is simply to "setback" the thermostat when one is gone during the day (or vacation, etc.), or during the heating season, turning down the thermostat overnight. Simulations suggest that, under parameters representative of a fairly typical, 1,700 ft^2 house in Phoenix, AZ, increasing the thermostat by just five degrees

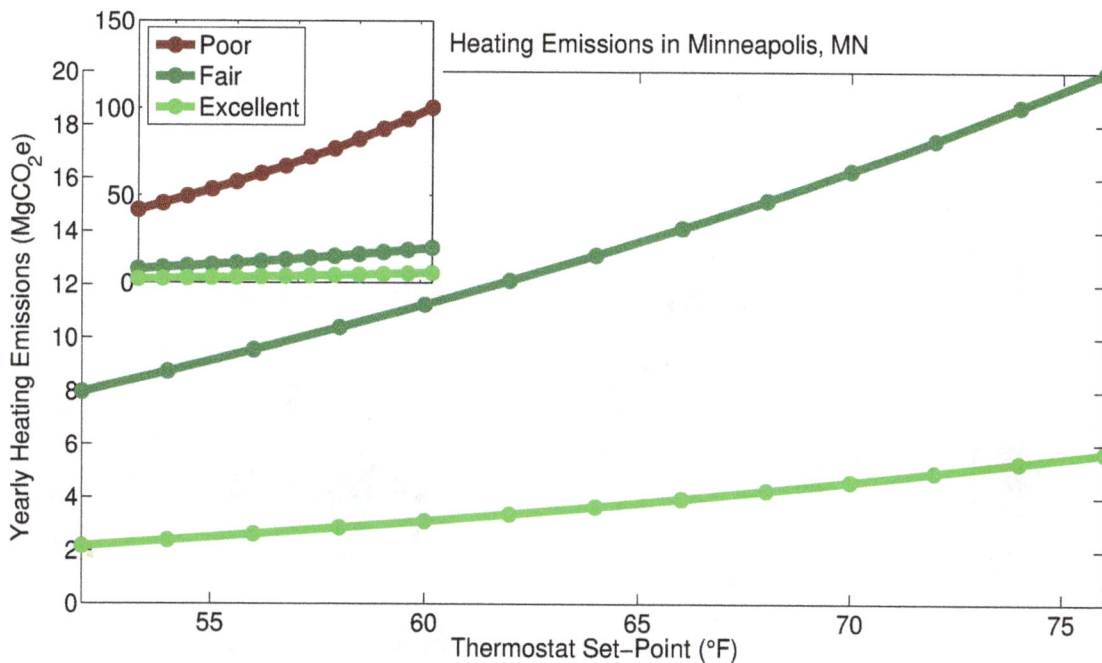

Figure 12.36: Yearly heat CO_2e emissions for a 2,400 ft^2 house Minneapolis, MN, assuming a natural gas furnace, across a range of constant thermostat set-points, and for three general levels of household energy efficiency. Each degree change saves about 4% of heating energy. Shifting from 78 to 52 °F avoids about two-thirds of heating emissions, and a more plausible 70 to 58 °F shift still offsets over one-third of energy.

for eight hours during the day, from 9 AM to 5 PM, may save around 13–14% of annual cooling energy, while bumping the thermostat 10 degrees for 12 hours (7 AM to 7 PM) is likely to offset over 30% of one's annual cooling emissions. The energy savings resulting from permutations of either 5 or 10 °F daytime setbacks for 8 and 12 hours, from either a base thermostat of 73 or 78 °F are demonstrated in Figure 12.37, and Figure 12.38 shows an example simulation (over three days) of cooling a house during the early Phoenix summer to 73 °F, with and without an eight-hour 10 °F daytime setback. As seen, any concerns over the energy required to force the daytime temperature down after the setback period ends are more than offset by the overall energy savings which amount, in this case, to almost 25%. In general, a daytime setback may save anywhere from 13–33%, while a setback combined with a somewhat elevated (e.g. 78 °F) baseline thermostat may save >50% of cooling energy, relative to a more typical baseline (about 73 °F). Energy savings are appreciably greater in colder cities, which includes all other cities in the US, and may be on the order of 20–40% (for the setback strategy alone) outside the hotter southern areas of the US.

The *relative* savings potential of a nighttime setback in the heating season is lower than the potential of a daytime setback in the cooling season (given that winter temperature excursions from the thermostat are typically far higher than those in summer), but the *absolute* savings may still be appreciable. Figure 12.39 shows possible savings under the relatively cold clime of St. Louis., where we may expect about 6–15% energy savings for 5–10 °F overnight setbacks. In St. Louis, compared to a fixed 68 °F, setting the thermostat to 63 °F reduces heating energy by 21%, and adding an overnight setback bumps savings to 26–33%.

311

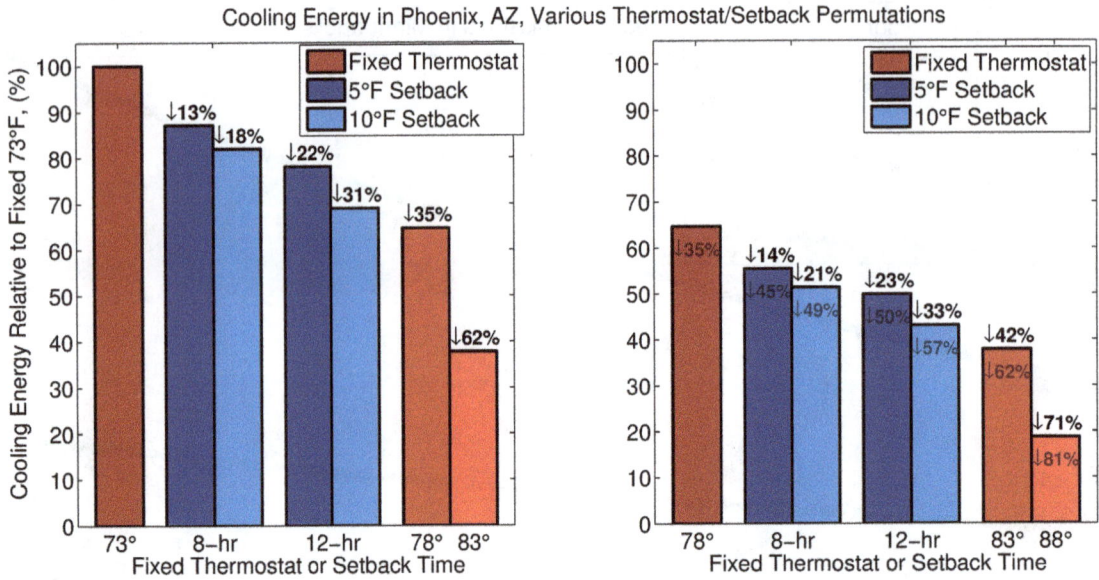

Figure 12.37: Relative annual cooling energy savings for different thermostat and daytime setback settings, under model simulations of a "typical" 1,700 ft² home in Phoenix, AZ. On the left panel, we have energy normalized to the maximum of a fixed thermostat at 73 °F, which is compared to 8- and 12-hour setbacks of either 5 or 10 °F, and fixed thermostat settings of 78 and 83 °F (decreases indicated are relative to fixed 73 °F). On the right, we use a baseline setting of 78 °F, and compare various setbacks and alternative fixed thermostats. On the right, the decreases above the bars are relative to thermostat fixed at 78 °F, while below the bars are relative to 73 °F, so for example, setting the thermostat back 10 °F for 12 hours from a baseline of 78 °F reduces energy use 33% relative to the 78 °F baseline, and 57% relative to a fixed 73° F thermostat.

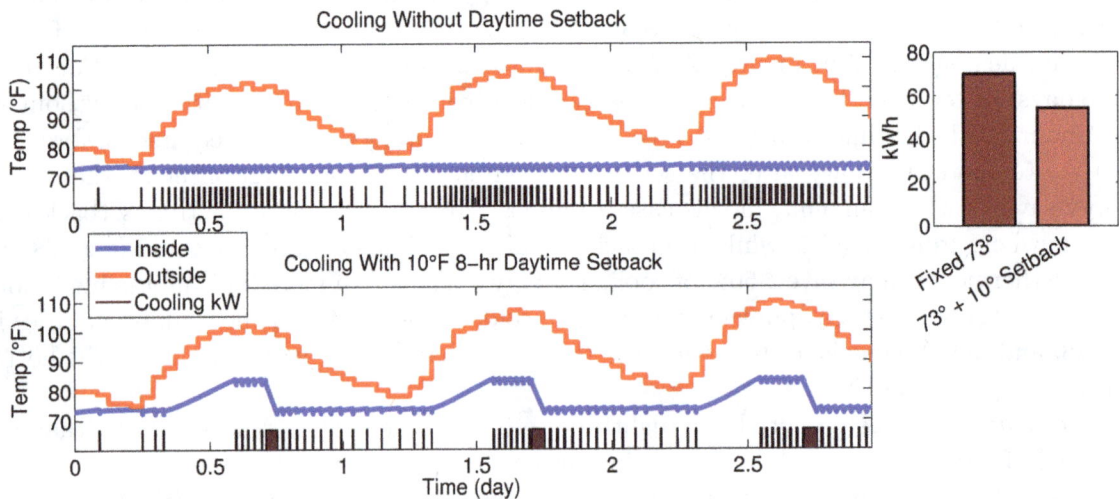

Figure 12.38: Example simulation of cooling a house in Phoenix, AZ, over three days in early summer, to a 73 °F thermostat set-point, with (lower panel) and without (upper panel) a 10 °F setback from 9 AM to 5 PM; inside and outside temperatures are tracked, while the AC energy is also inscribed. As we seen, with the setback, relatively little energy is used during the day, and while it takes a significant amount of energy to pull the indoor temperature back down at the end of the day, the overall energy savings are still quite significant (right graph).

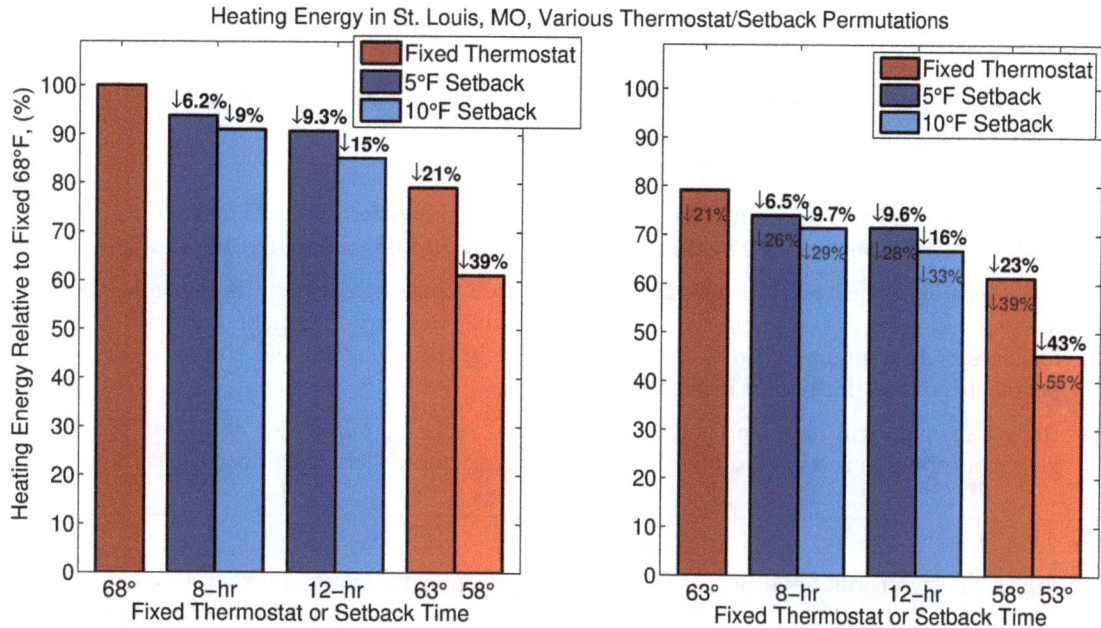

Figure 12.39: Similar to Figure 12.37, relative heating energy over the year in St. Louis, MO, for a simulated 2,400 ft^2 house, with the left panel comparing 8- and 12-hour overnight 5 and 10 °F setbacks to a baseline fixed 68 °F thermostat, while the right gives savings relative to a 63 °F baseline (and decreases below the bars are savings relative to the 68 °F baseline).

12.7.3 Energy consumption and efficiency

We can characterize a house's energy efficiency, with respect to heating and cooling, by the insulation level of the walls, attic, floor, and windows (e.g. single, double-pane, etc.), the air infiltration rate, e.g. the ACH value, and the efficiency of the HVAC equipment, and the efficiency of the air distribution system (i.e. how leaky the ducts are). Now, insulation as a lumped category, ACH, ductal losses, and heater/AC efficiency all affect energy consumption essentially *independently* of each other. Note that insulation is a "lumped category," in the sense that the marginal benefit of increasing any one insulated barrier, say the attic, depends upon the insulation level of the other barriers (walls, windows, floor). In our equivalent circuit analogy, all paths for conductive heat loss (walls, attic, etc.) are resistors in parallel, so any resistor with a low R value behaves something like a short-circuit, providing an easy path for heat (electricity in analog) to escape. Thus, while the attic is clearly the single most important barrier to properly insulate, one must ensure that entire house is well-insulated for optimal results.

Supposing a single house is either poor, fair/typical, good, or excellent in all these categories together, we can compare the yearly cooling and heating energy/emissions. Such comparisons are shown in Figure 12.36 for the cooling emissions associated with a 1,700 ft^2 house in Phoenix, AZ (approximate mean house size in the hottest RECS climate zone), and heating emissions generated by a 2,400 ft^2 home in Minneapolis, MN.

A drafty, completely uninsulated home would require an exorbitant amount of energy to cool and heat, outstripping the capacity of many ACs and heaters. Less dramatic, but still significant, the model suggests that, in Phoenix, AZ, reasonably achievable improvements on a fairly insulated house, one with wall insulation, about six inches of attic insulation, and double pane windows, could still have the potential to roughly *halve* cooling costs. Such improvements would chiefly entail high efficiency HVAC equipment, duct sealing and duct insulation, decreased air exchange, and/or increased attic insulation.

12.7.4 How useful is a white roof? "Cool roofs" and cooling versus heating loads

- Compared to dark shingles, in hotter southern cities highly reflective roofs may reduce the cooling load by 20–30% but increase heating requirements by 10–20%, compared to black shingles.

- Whether this translates into energy and/or carbon emissions savings is dependent upon the heating technology: under electric resistance heating the carbon benefit is minimal, while gas furnaces or heat pumps yield a 10–15% reduction in total heating and cooling emissions. The benefit is greatest for poorly insulated houses.

- Reflective roofs will increase total energy use and emissions in colder climates. Implemented globally, reflective roofs could increase heating energy more than they decrease cooling energy.

Highly reflective roofs have been proposed as a method to lower summertime cooling loads, especially in the sweltering cities of the Southwest and Florida. Standard asphalt shingles, even "white" (asphalt) shingles, reflect very little sunlight, with reflectances ranging from 0.05, for black, to 0.21, for white shingles [572]. Reflective coatings, which are applied onto an existing roof, and some specially designed roofing materials have solar reflectances in the 0.75–0.85 range [572], and thus can rather markedly reduce the solar heat gain into a roof.

It is important to note that highly reflective coatings can generally only applied to low-slope roofs, but the majority of residential roofs are steep-slope and use shingles, for which coatings are not advised. Highly reflective shingles are marketed, but achieve a reflectance of only about 0.26–0.28 when new, based on a review of manufacturer websites. Energy Star criteria for steep-slope roofing is reflectance ≥0.25 when new, and ≥0.15 after three years. For low-slope, criteria are reflectance ≥0.65 initially, and ≥0.50 after three years. Clay tiles may have slightly higher reflectances (0.33 in [572]), while metal roofs are fairly even better: white-coated metal roofing has an initial reflectance of about 0.65–0.75 (based on Energy Star and a review of manufacturer websites), and both materials are appropriate for steep-slope roofs. After aging, metal roof albedos are probably more like 0.50.

Multiple simulation studies and field measurements suggest that retrofitting an existing building with a new "cool roof" (with a reflectance of about 0.80) can reduce cooling energy by 10–50% [574, 575], with most houses seeing about 20% savings under actual field measurements [573, 574]. Furthermore, studies have consistently found that the benefit to a cool roof is much greater in poorly insulated compared to well-insulated homes [573, 574]. This AC reduction is partially offset by an increase in the heating load: with less solar energy heating the house in winter, one must make up the difference, and while the balance is generally favorable in more southerly climes, reflective roofs can be counterproductive in colder, northern cities [575].

Using our simplified heat transfer model, I too find that, for "typical" insulation levels and thermostat set-points in Phoenix, AZ, by increasing roof reflectance from 0.05 to 0.85, cooling load is reduced by about 20–30%, but the heating load is increased 10–15%. Significantly, when the heat source is an inefficient (from a primary energy perspective) electric resistance furnace, overall carbon emissions are reduced only by about 5%, but when the heating source is a heat pump or gas furnace, overall carbon savings are in the 10–15% range.

For the coldest cities, total energy use increases due to the effect on heating. In Minneapolis, for example, a cool roof could increase heating emissions by 300–900 kgCO$_2$e/yr (for a typical 2,400 SF house using a 90% AFUE natural gas furnace), without any cooling benefit. A global-scale modeling study by Oleson et al. [571] similarly predicted that a global increase in the albedo of urban surfaces would increase heating energy use more than decreasing cooling energy.

White roofs as geoengineering

- Compared to a typical asphalt shingle house, highly reflective shingles could reflect solar radiation equivalent to 25–100 $kgCO_2e$/yr, for a 2,000 square foot roof, while a higher albedo metal roof (weathered albedo 0.50) could offset 100–500 $kgCO_2e$/yr, depending on region and the particulars of our calculation.

- This offset would only apply, however to a roof that lasted 100 years. One that lasted a more typical 25–30 years would yield only 25–30% of the benefit (unless, of course, it was replaced thrice in succession).

- This carbon offset will synergize with lowered cooling emissions in hot regions, but will likely be outweighed, or at least neutralized, by the higher heating requirements induced in cold areas.

- As a *global* geoengineering scheme, increasing urban albedo (i.e. highly reflective roofs and pavements) has very limited potential to offset global warming, but can be of small benefit in some hotter regions.

Large-scale increases in the albedo (reflectiveness) of urban surfaces, mainly roofs, have been proposed as a means to reduce the urban heat island effect and partially mitigate global warming via the increased rejection of solar radiation. Some analyses have concluded that the global warming mitigation would be surprisingly substantial, with one widely cited (and somewhat controversial) work [197] concluding that a 0.10 increase in the albedo of all global urban space would be equivalent, over 100 years, to a one-time offset of 44 Gt of CO_2 (or almost one year of global CO_2e emissions). Experimentally, some measurements of heat rejection by white roofs support this notion. However, global climate models show that the cooling effect of a global urban albedo increase would be highly spatially heterogenous and strongly localized, affect the regional hydrologic cycle, and could only mitigate a tiny fraction of the expected warming from CO_2 and other greenhouse gases in the century to come [576].

We can calculate from first principles the first-order cooling effect of any change to a surface's albedo (diffuse reflectance), as follows. Following Akbari [197], as a global average, sunlight reaches the earth at a rate of $I = 172$ W m^{-2}, with roughly $f = 26\%$ absorbed by the atmosphere. Given the surface albedo, α, and assuming that $1 - f$ of the reflected radiation radiates to space, then solar radiation is rejected at a rate (in W m^{-2}) of

$$I \times \alpha \times (1 - f). \tag{12.36}$$

Now, there are several ways we can frame the negative radiative forcing induced by an albedo change in terms of CO_2-equivalents, but, in keeping with the GWP concept, we might integrate this negative radiative forcing over 100 years (so the albedo change must be maintained over all 100 years), and then compare to the warming effect of a 1 kg bolus of CO_2 at year 0. In this case, each 0.01 increase in albedo corresponds to (a *one-time*) CO_2 offset of 2.72 kg, per m^2. This calculation assumes that our green roof would persist for 100 years. If it lasted 30, and was then replaced by a regular ole' roof, we would achieve only 30% of the carbon offset.

Note that this calculation does not take into account that perhaps 10–25% of reflected light will be blocked by trees, other buildings, etc. Also note that the US receives more sunlight (about 205 W m^{-2}, based on 1800 kWh/m^2/yr) than the global average, and so the offset from increased albedo would be slightly higher in much of the US. Taking these factors into account, for a 2,000 square foot roof and compared to typical shingles (albedo 0.10), carbon equivalent savings for white or highly reflective shingles (albedo 0.25) are on the order of 50–100 $kgCO_2e/yr$, while a coated metal roof (albedo 0.50) could conceivably yield 200–400 $kgCO_2e/yr$ savings.

Balance of effects for particular cities. Using TMY datasets, we have both the extraterrestrial radiation (ETR), the amount of radiation reaching the top of earth's atmosphere, and the amount that actually reaches the ground ("global horizontal irradiance," GHI), on an hourly basis. Assuming, as above, that the fraction of energy blocked before reaching the surface is the same as would be blocked leaving, and that about 15% of reflected light is blocked by trees, etc., we can calculate how radiative forcing changes with albedo for specific cities.

So, for example, in Minneapolis, MN, increased heating emissions (based on our heat-transfer model) likely outweigh any direct global warming benefit from increased roof albedo, while in Phoenix, AZ, accounting for the albedo change roughly doubles the efficacy of a cool roof. Such calculations assume, again, that the albedo change persists for a century. Since any given roof will only last about 30 years or so, we should probably discount the albedo benefit (for either city) by 70% when it comes to a personal accounting of one's carbon footprint.

12.7.5   Fans for cooling!

- Individual fans draw about 50–100 W when on, and used liberally, can reduce the apparent air temperature by at least 5 °F, but do not affect actual air temperature (so don't leave them on alone).

- Used wisely in conjunction with elevated thermostat settings, fans can help to significantly reduce cooling energy/emissions.

Standard ceiling fans generally move air at anywhere from 1,000 to over 6,500 cubic feet per minute (cfm), have an airflow efficiency somewhere in the range of 50–100 cfm/W (based on product review and Energy Star criteria), while Energy Star criteria for ceiling fans are 155 cfm/W on low, 100 cfm/W on medium, and 75 cfm/W on high (the best models achieve 500–1,000 cfm/W). Box and circulating fans deliver 1,000–3,000 cfm at about 25–50 cfm/W, for most models. So for example, a modestly efficient ceiling fan moving 3,000 cfm at 50 cfm/W will draw 60 W of power, and one can assume that typical fans draw 25–150 W, depending on settings (compare this to about 1.5 kW for a wall-mounted AC, and at least twice that for central AC). This is as much as a larger incandescent light bulb, and if you would never deign to leave one of those on in an empty room, pray, do the same for a fan.

Now, how much cooling can we expect? Fans do not, of course, reduce air temperature, and in fact, they ever so slightly increase it via heat generation, but recall from Section 12.6.1 that moving air transfers heat via convective heat transfer at a rate that is proportional to wind speed, and so the faster the air, the more heat is removed from the skin of any animal that happens to be in the area (assuming the air is cooler than human skin: if it is hotter then the fan will act as a heater, notwithstanding any enhancement of evaporative cooling from sweat). As a first step, we can quickly calculate the wind speed exiting the fan as the volume of air moved per time divided by the fan blade area, which evaluates to about 0.5–2.5 m/s for typical ceiling fans, and 3–7 m/s for typical box fans (based on an online product review).

Now, for the human body, we will have both convective and radiative heat transfer, and to a first-approximation, the heat transfer from the indoor environment is given by

$$\dot{Q} = A_{cl}(h_r + h_c)(T_{cl} - T_{op}) \tag{12.37}$$

316

where A_{cl} is the surface are of a clothed human (1.73 m^2 is a standard value), $h_r \approx 4.7$ W m^{-2} is the radiative heat-transfer coefficient [577], h_c is the convective heat-transfer coefficient, determined in a moment by air speed, T_{cl} is the surface temperature of our clothed human (which I take to be 34 °C [577]), and T_{op} is the *operative temperature*, which we can assume, without too much error, to simply be the ambient temperature. For much more detail, one may consult [577]. Now, for still air, h_c is about 3.3 W/m^2, and h_c has been found to increase according to a power-law with wind velocity, v, as

$$h_c = Bv^n, \tag{12.38}$$

for $v \geq 0.2$ m/s, with de Dear [577] finding $B = 10.3$, $n = 0.6$, while another estimate is $B = 8.3$ and $n = 0.5$ [577]. Using this relation and Equation 12.37, we can calculate, for any given indoor temperature and air flow rate, the rate at which heat transfers to skin, $\dot{Q}$. From that, we determine what still air temperature would give the same $\dot{Q}$. That is, we determine what temperature it *feels like*.

From Figure 12.40, we see that common fans and settings can reduce the apparent air temperature by at least 5–10 °F, and perhaps over 20 °F for particularly powerful fans. Even if we reduce air speed by 50% to account for the spread and dissipation of moving air, a 5–10 °F drop is still readily achievable. It follows that with the liberal use of fans, one could increase their thermostat by 5 °F or more for the same comfort level, and thus reduce AC-associated emissions by perhaps 40% (based on a 9% reduction per degree °F in thermostat)! Now, fans do use energy, but running three 75 W fans 12 hours a day, for a full six months straight, would use 486 kWh and generate 334 kgCO$_2$e. This is about 12% of the cooling emissions in the hottest RECS climate zone, and combined with a 40% reduction in AC would still yield a net 28% emissions reduction. More efficient fans (see, e.g., the Energy Star website) would obviously help even more.

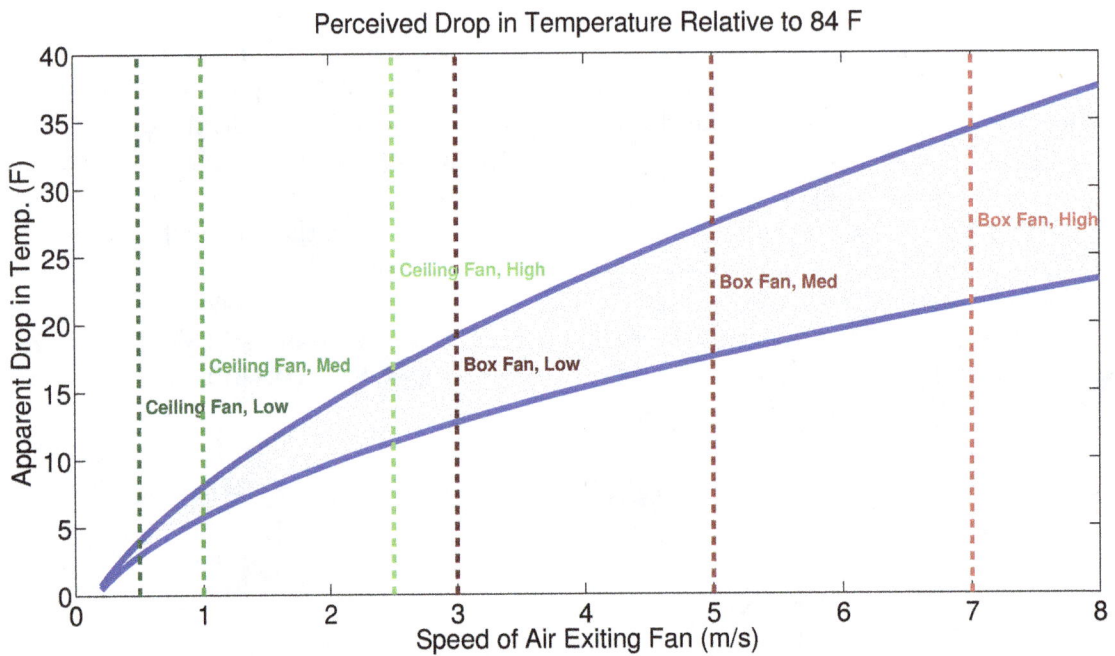

Figure 12.40: The apparent decrease in temperature of air exiting a fan at various speeds, when the actual temperature is 84 °F. The two lines are determined from two different models for h_c as a function of air speed, v, $h_c = 8.3v^{0.5}$ and $h_c = 10.3v^{0.6}$, to give a sense of the uncertainty in the calculations.

Chapter 13

Water: heating and provision

- Most freshwater is used either for thermoelectric power generation or agricultural irrigation, with residential use actually a more minor end-use.

- Most energy/emissions associated with residential water come not from conveyance, but from *heating* the water, equating to about 0.650 $MgCO_2e$/person/year overall, and about 1.125 $MgCO_2e$/person/year when restricting ourselves to single-person households only. Tap water conveyance and treatment adds around 0.1 $MgCO_2e$/person/year.

- Gas water heaters are superior to electric resistance heaters by a factor of almost two, but (still uncommon) electrical heat pump water heaters reduce water heating emissions two to fourfold compared to either option, and are similar to typical solar hot water systems (due to backup electricity use in solar systems).

- Most energy/emissions associated with municipal water delivery come from water pumping, either from the primary source (surface or groundwater) or into secondary pressurized distribution systems.

- Provision of surface water via large-scale aqueduct and canal systems in the Southwest, especially in Southern California and Arizona, is much more energy/emissions intensive than municipal water in other areas, and also has serious ecological consequences for rivers such as the Colorado.

- Simple efficiency measures such as low- or ultra-low flow showerheads (ultra-low flow shower heads are likely the single most effective measure), shorter showers, clothes washing in cold water, and avoiding unnecessary hot tap water use can easily decrease hot water use by 50%; upgrading to a heat pump water heater (likely the best option) or solar heater can reduce hot water emissions by 50–75%.

Water conservation is a major focus of popular environmental education and outreach programs, but it is my experience that such materials tend to focus almost exclusively on reducing water use *per se*, while the environmental impact (certainly in terms of carbon-equivalent emissions) of *residential* water use actually manifests primarily via *hot* water. About 25% of the roughly 90 gallons each person uses per day (as a national average), are heated, while, as a national average, this accounts for >85% of the carbon equivalent emissions associated with residential water, although water provision is more emissions-intensive in much of the Southwest. In this chapter, I review the energy, emissions and technologies associated with water heating at the point of use, as well as the water provision cycle, with a special focus upon Southwestern water projects.

13.1 Water heating

While air conditioning, space heating, and lighting receive a decent amount of "green press," in my experience the humble water heater gets almost none, save the occasional story on solar hot-water heating. But in fact, water heating uses more energy than any other single household activity except space heating. The 2009 RECS suggests that 17.7% of household energy use and 13.9% of emissions are due to water heating. Mean energy use was 4,652 kWh per household (1,809 kWh per capita, and 12.75 kWh/day/HH), corresponding to about 1.69 $MgCO_2e$/HH, or 0.66 $MgCO_2e$/person. As usual, the distribution is skew, with the medians somewhat smaller than the means. The bottom and top quintiles are 1,941 kWh (0.87 $MtCO_2e$) and 6,781 kWh (2.31 $MtCO_2e$) per household, respectively. The deciles and energy/emissions distributions are summarized in Figures 13.1 through 13.3.

There are two major fuels used for water heating: electricity and natural gas (with propane and fuel oil lesser-used fuels). Energy use in households that use natural gas-fired water heaters is more than two times higher than in households that use electric heaters. This can be partly, but not completely, explained by the much lower heating efficiency of gas heaters. Despite this, households with gas heaters actually generate slightly fewer emissions overall, as grid-average electricity has associated emissions more than 2.5 times those of direct gas combustion emissions, per unit of energy delivered.

Again using 2009 RECS data, the bottom decile and quintile (i.e. bottom 10% and 20%) of households have mean water-heating emissions of only 38% and 52% of the overall mean, respectively. These numbers are similar even when controlling for household size, water-heating fuel type, and climate, although there is greater variation for smaller households and for those that use natural gas versus electricity. It follows, as with almost every other realm of consumption, that for an average household, regardless of size, climate, or heat source, a 50–67% reduction in water-heating energy and emissions is very achievable.

Consistent with other household energy uses, smaller households have a larger per capita impact. Single-person households in the 2009 RECS used a mean of 3,096 kWh, with corresponding emissions of 1.127 $MgCO_2e$, and over 70% higher than the per capita average across all households. This also translates to around 33 gallons of hot water per day.

Note that, unlike space heating or cooling, actual energy and emissions for water heating do not vary appreciably between climate zones. This is somewhat surprising, as inlet tap water temperature varies with the clime and season from perhaps as low as about 45 °F, to approaching 80 °F. Heating 40 °F inlet water to 120 °F (80 °F difference) requires twice as much energy as heating 80 °F (40 °F difference). Thus, basic math still indicates that the energy and emissions to heat a gallon of water *can* vary up to two-fold depending on the climate and season.

Back-solving from the energy required to heat a gallon of water and the overall efficiency of water-heaters, the 2009 RECS suggests about 20 gallons of hot per day per person for typical households. Other water-use surveys, e.g. the Residential End Uses of Water Survey (REUWS) [578, 579], suggest closer to 25 gallons per person per day. The largest uses of hot water are showering and general faucet use, at 36% and 31% of hot water use, respectively, in a survey of Denver households [579], with clothes washing (10%) and leaks (10%) the other two major uses.

Back-of-the-envelope calculations, along with survey results, suggest that rather trivial changes in habits could reduce typical hot water use by 50–75%. These changes amount to: (1) use a very low-flow shower head (0.5–1.5 gallons per minute, vs. 2.5 GPM for a standard low-flow head) and/or reduce shower duration, (2) don't use hot water from the tap for no reason, and (3) don't wash clothes in hot water. Fixing leaks is obvious as well, and could save around 10% of hot water. There is rarely any reason to use hot tap water, and clothes can be

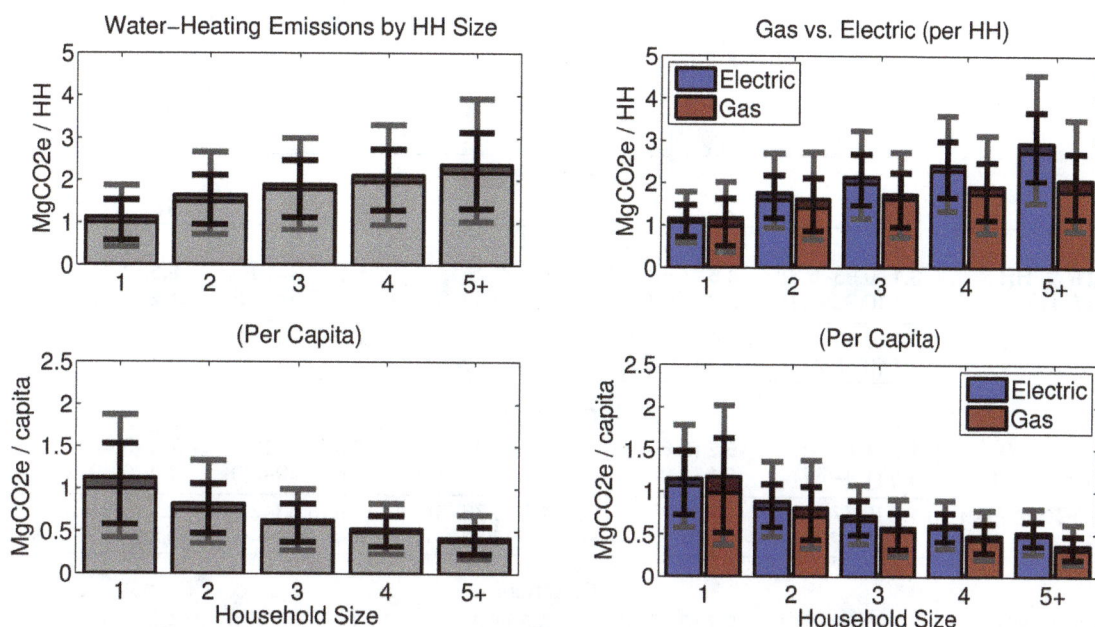

Figure 13.1: 2009 RECS water heating emissions on a per-household and per-capita basis for different household sizes. The median emissions are given by the light bar, while the darker portion of the bar indicates the mean emissions: mean emissions are always slightly greater than the median. The darker error-bars gives the top and bottom quintiles, while the lighter error-bars give the top and bottom deciles. The plots on the left shows overall emissions, while those on the right give emissions when considering households that use natural gas and electricity for heating separately.

cleaned perfectly well in cold water. More detailed calculations are presented presently.

13.1.1   Overall efficiency, emissions factors, and typical gallons per day

Energy factor for efficiency

Not all energy consumed for water heating is converted into usable hot water; inefficiencies are introduced at three points: (1) not all heat generated is transferred into the water, i.e. the heating efficiency is less than unity for gas water heaters, (2) heat is constantly lost from the water tank to the environment as so-called standby losses, and (3) some heat is lost from pipes during distribution. These losses are summarized by a single metric given for commercial water heaters, the *energy factor* (EF) (ratio of actual heating energy to input energy) which is about 0.67 and 0.93 for typical gas and electric heaters, respectively. Tankless heaters (gas or electric) may have an EF of 1.0, while heat pump heaters achieve an EF of 3.25–3.5. Solar water heating systems with electric (or more rarely gas) backup are similarly characterized by a *solar energy factor* (SEF), the ratio of total system energy input to electrical (or gas) energy, and is effectively identical to the EF. Typical solar heating systems have an SEF of about 2–3, with 1.8 (for electric backup) the minimum for Energy Star certification.

	Household Size				
	1	**2**	**3**	**4**	**5+**
	Overall				
MWh / HH	**3.1** (0.93–6.3)	**4.6** (1.6–8.6)	**5.1** (2–9.1)	**5.8** (2.3–10)	**6.6** (2.8–11)
MWh / cap	**3.1** (0.93–6.3)	**2.3** (0.82–4.3)	**1.7** (0.68–3)	**1.4** (0.57–2.6)	**1.2** (0.49–2)
MgCO2e / HH	**1.1** (0.42–1.9)	**1.6** (0.72–2.7)	**1.9** (0.83–3)	**2.1** (0.95–3.3)	**2.4** (1–3.9)
MgCO2e / cap	**1.1** (0.42–1.9)	**0.82** (0.36–1.3)	**0.63** (0.28–1)	**0.53** (0.24–0.83)	**0.42** (0.18–0.69)
	Gas–Fired				
MWh / HH	**4.5** (1.4–7.7)	**6.2** (2.6–10)	**6.6** (2.8–10)	**7.3** (3.2–12)	**7.8** (3.4–13)
MWh / cap	**4.5** (1.4–7.7)	**3.1** (1.3–5.2)	**2.2** (0.93–3.5)	**1.8** (0.79–2.9)	**1.4** (0.59–2.3)
MgCO2e / HH	**1.2** (0.37–1.9)	**1.6** (0.67–2.7)	**1.7** (0.73–3)	**1.9** (0.83–3.3)	**2.1** (0.88–3.9)
MgCO2e / cap	**1.2** (0.37–1.9)	**0.81** (0.34–1.3)	**0.58** (0.24–1)	**0.48** (0.21–0.83)	**0.36** (0.16–0.69)
	Electrical				
MWh / HH	**1.7** (0.58–1.8)	**2.7** (0.95–2.7)	**3.2** (1.2–3.2)	**3.6** (1.4–3.6)	**4.5** (1.5–4.6)
MWh / cap	**1.7** (0.58–1.8)	**1.3** (0.47–1.4)	**1.1** (0.39–1.1)	**0.91** (0.34–0.9)	**0.8** (0.27–0.8)
MgCO2e / HH	**1.1** (0.58–1.8)	**1.8** (0.95–2.7)	**2.1** (1.2–3.2)	**2.4** (1.4–3.6)	**2.9** (1.5–4.6)
MgCO2e / cap	**1.1** (0.58–1.8)	**0.88** (0.47–1.4)	**0.72** (0.39–1.1)	**0.61** (0.34–0.9)	**0.52** (0.27–0.8)

Figure 13.2: Table of per-household and per-capita water-heating energy and emissions, under the 2009 RECS. The primary value in each table entry is the mean, and range given in parentheses is the top and bottom deciles.

	Mean	Median	Std Dev	Deciles								
	–	–	–	10	**20**	30	**40**	50	**60**	70	**80**	90
				Energy (MWh)								
Overall	**4.65**	**3.63**	**3.79**	1.41	**1.94**	2.45	**2.99**	3.63	**4.45**	5.49	**6.78**	8.87
Gas	**6.17**	**5.49**	**4.19**	2.19	**3.09**	3.9	**4.7**	5.49	**6.29**	7.13	**8.34**	10.7
Electrical	2.77	2.39	2	1.15	**1.52**	1.8	**2.1**	2.39	**2.73**	3.14	**3.67**	4.59
				Emissions (MgCO2e)								
Overall	**1.69**	**1.5**	**1.1**	0.62	**0.87**	1.08	**1.29**	1.5	**1.71**	1.96	**2.31**	2.91
Gas	**1.62**	**1.43**	**1.11**	0.57	**0.81**	1.02	**1.23**	1.43	**1.64**	1.87	**2.19**	2.8
Electrical	1.85	1.63	1.08	0.79	**1.04**	1.23	**1.43**	1.63	**1.86**	2.14	**2.49**	3.09

Figure 13.3: Table of water-heating energy and emissions metrics, for households surveyed in the 2009 RECS.

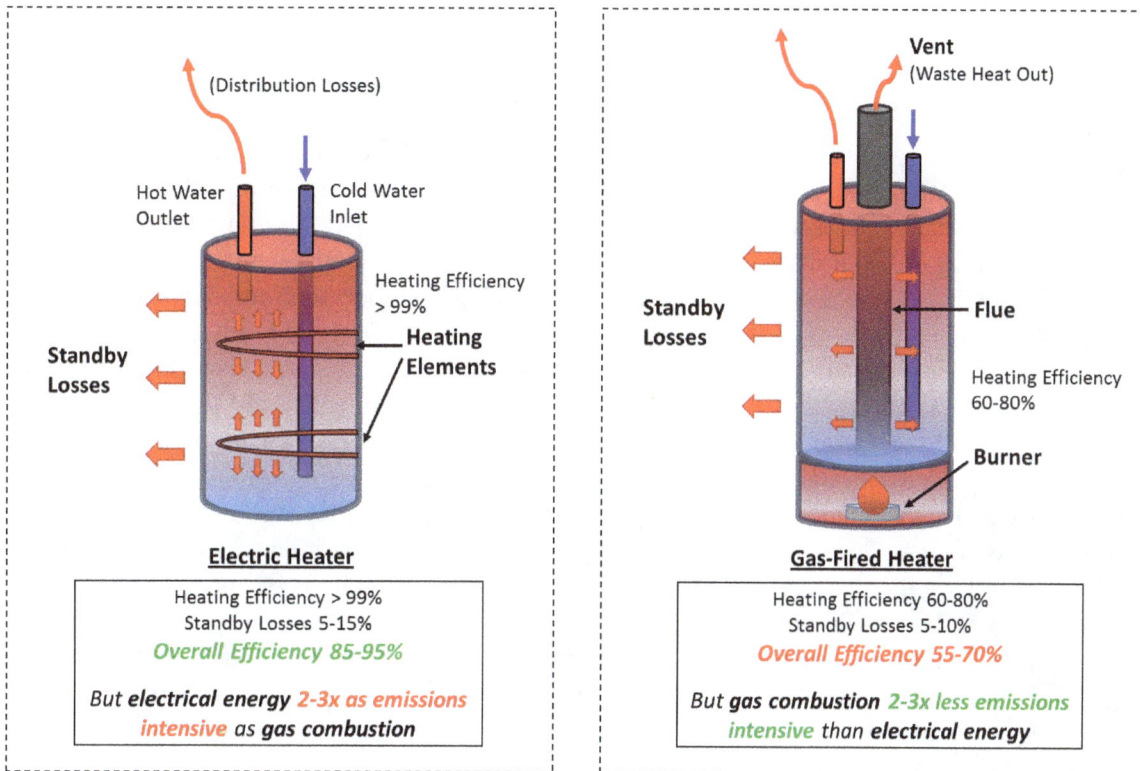

Figure 13.4: Schematic for operation and heat losses for electrical and gas-fired water-heaters. As shown on the left, electric heating elements inject heat into water with near 100% efficiency, and standby and distribution losses account for essentially all heat loss. Thus, overall efficiency is generally greater than 90% (under typical water use). On the other hand, gas-fired heaters waste a significant amount of heat (lost from the vent), and only around 70% of the heat energy generated is actually transferred into the water, lowering overall efficiency to about 60% or so, for a typical heater. Despite the higher efficiency of electric heaters, the greater emissions intensity of electricity means that, overall, emissions are about twice as great when heating with electricity. Not schematized here, newer *heat pump* electrical heaters achieve efficiencies on the order of 250–325%, and are thus by far the best "plug-and-play" option.

Energy/emissions to heat a single gallon

- It takes just under 200 Wh to heat a single gallon of water to typical tank temperatures, about 120–140 °F (135 °F is the DOE test temperature).

- Including efficiency losses, gas heaters generate 1 kgCO$_2$e for every 13.6 gallons of hot water delivered (0.0735 kgCO$_2$e/gallon), while for electric heaters this factor is 1 kgCO$_2$e for every 7.25 gallons (.138 kgCO$_2$e/gallon); the weighted average emissions factor is 0.09 kgCO$_2$e/gallon. Since water used at the tap is 20–40% cooler than at the tank (via mixing with cold water), we have roughly 20 gallons/kgCO$_2$e hot tap water and 10 gallons/kgCO$_2$e hot tap water for NG and electric heaters, respectively.

Let us determine the energy and emissions associated with heating a single gallon of water. First, cold tap water must be heated from an inlet temperature ranging anywhere between 45 and 80 °F, depending on location and season, to the thermostat set-point, which typically ranges between 120 and 140 °F. For water-heater testing purposes, the DOE assumes an average inlet temperature of 58 °F, and a set-point of 135 °F, for a 77 °F temperate difference. Other sources suggest 50 °F as a more typical inlet temperature, but heating from 50 F to 125 °F would give a similar 75 °F temperature change, so I consider the DOE test conditions reasonable for basic emissions factor estimates. Note that one should generally use a relatively low set-point, as a set-point above 120 °F both increases standby energy losses and dramatically increases the risk of hot-water scalding; nevertheless, I use 135 °F for all baseline estimates.

The specific heat of liquid water is 4.187 kJ/kgK (1.1631 Wh/kgK), and assuming a constant density of 1 kg/L, it requires 49.75 Wh to heat a single liter of water from 58 to 135 °F, equating to 188.34 Wh per gallon. Thus, a single kWh of heat yields 5.3 gallons of hot water. Adding in typical efficiency losses, this translates into the following rules of thumb (1) about 3.5 gallons delivered per kWh of gas, and (2) 5 gallons delivered per kWh of electricity. So for example, 20 gallons of hot water would consume either 5.7 kWh of gas or 4 kWh of electricity.

For hot water heated from 58 to 135 °F, we have emissions factors of 0.0735 kgCO$_2$e/gallon *delivered* if using natural gas at a 0.67 EF, and 0.138 kgCO$_2$e/gallon delivered for grid-average electricity under a 0.93 EF. In general, such emissions factors will vary with both the thermostat set point and the temperature of the inlet water. Considering just the two dominant fuels, NG and electricity, and weighting by energy energy use (68.01% and 23.75% per the 2009 RECS, as below), we have about 0.090 kgCO$_2$e/gallon delivered as an overall average EF. And thus as a crude average, we have about 1 kgCO$_2$e per every 11 gallons used.

One complication to note is that these factors are for hot water heated to 135 °F, and so represent gallons of hot water at the *tank* level. Hot water used at the tap will not exceed 120 °F (as higher temperatures will scald) and for example, a typical shower is only around 105 °F. The energy required to yield 105 °F water (from a 58 °F tap) is 114.96 Wh/gallon, just 61% of the energy for 135 °F water. Thus, while 20 gallons of tank-level hot water may yield 2.76 kgCO$_2$e under electric heat, at the shower-head 20 gallons is associated with a lower 1.68 kgCO$_2$e.

Typical gallons per day

In the 2009 RECS the energy sources for water heating were 23.75% electricity, 68.01% natural gas, 3.74% fuel oil, 4.46% liquid petroleum gas (propane), and 0.06% kerosene. Assuming an EF of 0.93 for electricity and 0.67 for other sources, and total energy use as given in the survey, these figures suggests overall hot water use at 49.5 and 19.25 gallons per day per household and per capita, respectively (and 33 gallons per day for single-person households). As an aside, although 68.01% of energy use was gas, a lower percentage of heaters, 51%, were gas-fired, while 42% were electric. This is probably mostly explained by the lower EF of gas heaters.

The REUWS suggests slightly higher hot water use, with the 2012 update [579], for example, giving 60.5 gallons of hot water per day, per household, or 24.2 gallons per person, if we assume 2.5 persons/household. In any case, 20–25 gpd per capita of hot water is a reasonable approximate range.

13.1.2 Standby losses and effect on overall efficiency

Most water heaters are traditional storage units, i.e. there is a large storage tank, typically ranging in volume from 30 to 80 gallons, of water that is continuously kept at the thermostat set-point, and therefore there is always some continuous heat loss to the environment. The basic physical dimensions of the tank and its insulation level, along with the thermostat set-point and ambient temperature, determine this background, or "standby," rate of heat loss. Generally speaking, standby losses are actually relatively small, accounting for 5–15% of energy use under typical operating conditions, although the relative loss becomes greater the less hot water one uses. That is, standby losses are essentially fixed, whereas total heating energy obviously varies with use.

We can infer that standby losses are small from the DOE efficiency factor for electrical heaters, which is generally around 0.93. Since resistive electrical heating has essentially 100% heating efficiency, i.e. no heating energy is lost to the surroundings, that 7% reduction in efficiency must be due to standby (and distribution) losses.

We may also adopt the heat-transfer model from Chapter 12, such that heat transfer from the water tank to surroundings is proportional to the water tank surface area, A, ambient and water temperature difference, $(T_O - T_I)$, and the thermal resistance, or R-value, R, of the heater insulation:

$$\dot{Q} = \frac{A}{R}\left(T_O - T_I\right).$$ (13.1)

For a cylindrical water-heater, surface area is given by the radius, r, and height, h, as

$$A = 2\pi r h + 2\pi r^2.$$ (13.2)

Obviously, for a given tank volume there are an arbitrary number of heights and radii, and thus there is no unique surface area for a tank volume. For simplicity, and since it minimally affects conclusions, in simulations I simply alter height to alter volume. Now, I assume the tank is kept in an unheated space, and use the 2014 average contiguous US temperature of 52.6 °F as the outside temperature; heaters kept in conditioned spaces will generally have somewhat lower standby losses, which may also partly offset heating loads in some cases as well. An inch of foam heater insulation has an R-value of about 8, and most commercial water heaters have overall R values of either 8, 16, or 24 (English units).

As a baseline estimate, for an electric water heater with a heating efficiency of 100%, a 60-gallon water tank with an R-value of 16, and 60 gallons of hot water used per day, we lose 358 kWh of heat from standby loss over a year, or 8.0% of the total hot water energy use. Despite the low relative value, the energy lost to standby *is* equivalent to the energy used in heating about five gallons of water per day, and a poorly insulated water heater with R-value 8 would lose energy equivalent to 10 gallons of hot water per day, which is not insignificant at all. A lower thermostat set-point (e.g. 120 °F) can also reduce standby losses by up to 20% compared to higher values (e.g. 135 °F).

All else being equal, a larger tank will have greater standby losses, although the absolute differences in standby losses are small for reasonably well-insulated tanks, and insulation level is the most important factor. This analysis demonstrates that standby losses are generally fairly minor, but are still nontrivial for poorly insulated tanks, and can be a relatively large component of water heating energy when overall hot water use is small. As an overall conservation strategy, reducing hot water use and shifting to a lower carbon heating technology/fuel source are generally far more effective than lowering standby losses, but of course the two are not mutually exclusive! The latter can be pursued either when purchasing a new heater (higher insulation or tankless models), with fairly inexpensive commercial water-heater insulating blankets, and by lowering the tank set-point.

13.2 Water provision

- Pumping is the major source of energy use in water supply systems, accounting for 2–3% of global electricity use.

- Cooling water for thermoelectric (TE) power generation and irrigation water are the major water withdrawal end-uses in the US; even excluding TE power, domestic water use accounts for only 14% of water withdrawals, at about 89 gallons per day (gpd) per person.

- Domestic/municipal water is, however, far more energy- and emissions-intensive than irrigation water or other uses, mainly due to pumping for pressurized distribution systems and energy consumed by wastewater treatment. Emissions are on the order of 0.003 $kgCO_2e$/gallon as a national average, or around 0.25 $kgCO_2e$/person/day.

- Large-scale water projects in the Southwest are much more energy intensive than most other water supplies. Desalination is a minor but increasing water source that is very energy/emissions-intensive, especially when seawater is the source.

- Overall, domestic water provision (and wastewater treatment) accounts for less than 15% of total residential water emissions (about 100 $kgCO_2e$/person/year), with the rest due to water heating. In the Southwest, however, this percentage may be as much as 20–40% (200–400 $kgCO_2e$/person/year), while in much of the North and East it is probably only 5–10% (about 50 $kgCO_2e$/person/year). Exact energy/emissions vary greatly among individual systems in all areas.

Freshwater is a fundamental limiting resource for agriculture, industry, and human settlement that is under increasing global pressure, with global water consumption trends clearly unsustainable. Moreover, water and energy supplies are *bidirectionally* linked, in that both the provision of water and its end-uses (e.g. washing, showering, cooking, etc.) require large amounts of energy, while energy generation itself is a major user of water; this is the so-called "water-energy nexus" [580]. At just the provision point in this nexus, about 2–3% of all global electricity goes toward water pumping in conventional water supply systems, and pumping is by far the largest use of energy at the supply stage (80–90%) [583].

Now, my principal aim here is to relate how energy is consumed in the provision of water for different purposes, and how this in turn generates greenhouse gas emissions. In addition to the energy used for pumping, and secondarily, fresh- and waste-water treatment, water supply also causes global warming via reservoir creation, as dams are largely constructed to control water flow and divert it for human use; the resulting reservoirs lead to significant carbon emissions as flooded vegetation decays (these emissions are discussed extensively in the context of hydropower in Section 4.9). Not considered in this section is water end-use, which, via water heating, is actually the dominant source of emissions in the water-energy nexus and discussed further in Section 13.3. To properly understand water provision and emissions, we must first understand at least the basics of water supply systems.

13.2.1 Overall water withdrawals

Water *withdrawals* come either from surface water (SW), e.g. lakes, rivers, or man-made reservoirs, or groundwater (GW). Globally, 70% of water withdrawals go toward agriculture, 20% of water use is industrial, including thermoelectric (TE) power generation, while only 10% is domestic [580]. The situation is a bit different in the US and Europe, where the major water use is TE power (about 45% of withdrawals in the US), for which water acts as a coolant to reject waste heat. However, because 94% of such withdrawals are "once-through," where water is withdrawn from local surface waters, used for cooling, and then discharged back, this use category is often excluded in reporting, as it does not represent a permanent withdrawal (although it obviously can detrimentally affect aquatic life). Furthermore, nearly all once-through generation occurs in the Eastern US, where local surface water resources are widely available. In the more arid west, cooling water is usually recycled in closed-loop systems, and withdrawals are comparatively negligible [586].

As illustrated in Figure 13.5, excluding thermoelectric generation, public water supplies and irrigation are the major ends towards which water is withdrawn, with surface water yielding about 60% of each. Domestic water use, that which individuals have direct control over, amounts to 89 gallons per day (gpd) per capita—only 14% of these withdrawals—while irrigation (almost entirely for agriculture) is 60%, or 373 gpd/capita: clearly, most water consumption is indirectly embodied in the food system (and especially meat and dairy, since the greater part of crop production goes to feed animals). In the face of these numbers, domestic water conservation may seem like a fool's errand, but it is important to understand that (1) municipal water is significantly more energy-intensive than irrigation water, mainly due to the energy required to pressurize municipal distribution systems and treat wastewaster, (2) urban water supplies can severely stress local/regional supplies and ecosystems (for example, the Los Angeles Aqueduct), and (3) *most* water-related energy is actually that used for end-use water heating, e.g. for showers, washing, and cooking, so conservation is energetically significant for such hot water uses.

13.2.2 Geography of water withdrawals

Broadly speaking, water withdrawals vary between the wet eastern and arid/semi-arid western United States. As already mentioned, most thermoelectric plants in the west recycle cooling water, while many in the east use large amounts of local surface water for a once-through cycle. Surface water conveyance in parts of the west is highly dependent upon large-scale waterworks that convey water many hundreds of miles and sometimes up to several thousand vertical feet, whereas local water resources are far more prevalent in the east; this makes western water conveyance much more energy-intensive.

Finally, western agriculture is highly dependent upon irrigation, and this is where most irrigation water is consumed: 14 western states[1] use 82% of all US irrigation water. In dry Arizona, where desert cities such Phoenix have been derided as abominations for failing to respect the arid ecology of the region, 75% of all water withdrawals actually go towards irrigation [586].

---

[1]These states are Arizona, California, Colorado, Idaho, Kansas, Montana, Nebraska, Nevada, New Mexico, Oregon, Texas, Utah, Washington, and Wyoming.

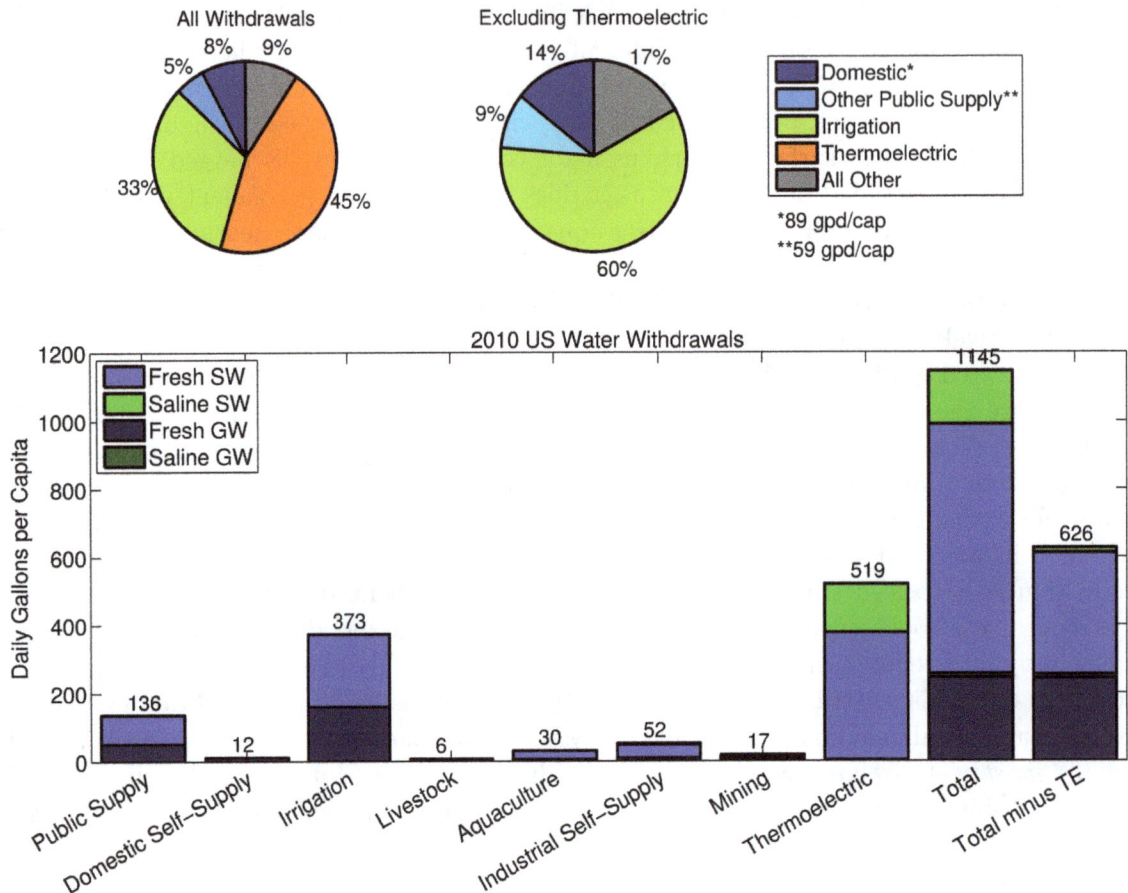

Figure 13.5: Summary of water withdrawals in the US in 2010 [586]. The top pie charts break total withdrawals down by percentage into domestic use, i.e. public supply for domestic customers, which was 23.8 billion gpd, and domestic self-supply, 3.6 billion gpd, other public supply (18.2 billion gpd), irrigation, thermoelectric power, and other, either including or excluding thermoelectric power. The bottom graph gives a more detailed breakdown of all categories, subdivided by whether withdrawals were from surface water (SW) or groundwater (GW), and whether water was saline or fresh (mining and thermoelectric power are the only significant uses of salt water).

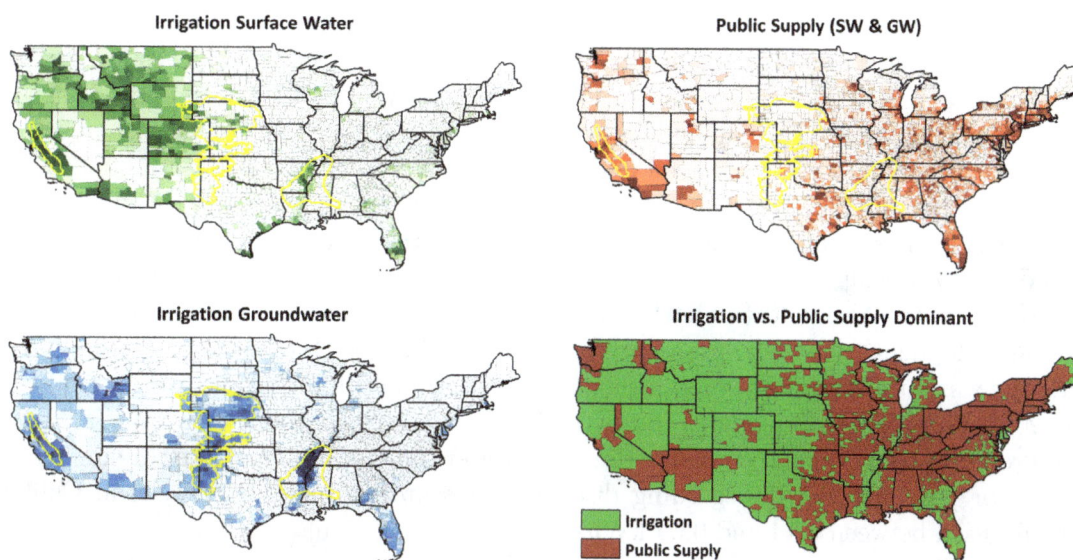

Figure 13.6: The left two maps show the relative intensity of surface water (top) and groundwater (bottom) withdrawals for irrigation at the county level, which are clearly concentrated in the western US. Three major groundwater aquifers under severe stress (from west to east), the Central Valley Aquifer in California, the Ogallala (High Plains) Aquifer in the middle west, and the Mississippi Embayment Aquifer system are outlined in yellow. The top right map gives public supply withdrawals and basically serves to highlight highly urbanized counties, both east and west (note also the partial spatial dissociation between water sources and their urban destinations). The bottom right map has each county coded such that green means more water is used for irrigation, while red implies public supply is dominant. Most of the west is irrigation predominant, while water withdrawals in the east are principally for municipal use. Source: USGS.

While public and other domestic water withdrawals are dwarfed by irrigation overall, across a majority of states—including almost all eastern states—municipal and domestic water withdrawals actually outweigh irrigation withdrawals about four to one (Mississippi, Missouri, Arkansas, and Florida are the major eastern exceptions, which together use 13% of the nation's irrigation water: add these states to the 14 western ones above, and 95% of US irrigation is accounted for) [586]. These geographic patterns are shown in detail in Figure 13.6.

13.2.3 Water provision cycle and energy use

The water provision cycle can be divided into several basic stages: (1) conveyance from the source, (2) treatment, (3) distribution, (4) end-use (domestic, irrigation, etc.), and then, possibly, (5) wastewater collection, (6) wastewater treatment, and finally, (7) wastewater discharge into a receiving body of water or aquifer [580]. We may also have a subcycle wherein wastewater is recycled and re-used. Water *pumping* is the major driver of energy use throughout this cycle.

Water conveyance

First, to convey water, it must either be pumped from a groundwater reservoir or diverted from a surface resource (river, lake, or man-made reservoir). Surface water diversion may either be gravity-driven, or may require pumping up a vertical gradient. Note that it is customary in the literature to express energy use in kWh/m^3 (one m^3 is equal to 264.17 gallons), and I will

329

follow this convention in the following.

To pump a cubic meter of water uphill requires changing the potential energy by $E = mgh$, where the mass m is 1,000 kg, g is acceleration due to gravity, at 9.807 m s^{-2}, and h is height in meters. Thus, the theoretical minimum pumping energy is 0.002724 kWh/m^3 per m, while in practice electrical pumps generally achieve about 0.004 kWh/m^3/m, or an overall 68% efficiency [580]. Greater pumping heights tend to require slightly more energy, as does providing discharge pressure after pumping, and groundwater wells in Southern California, for example, use about 0.006 kWh/m^3/m [580].

Groundwater. About 40% of both municipal and irrigation water is pumped groundwater. Water table depths vary widely, from surface discharge (i.e. depth of zero) to several hundred meters; the water table depth in the southern Ogallala Aquifer, for example, ranges from about 30 to 120 meters [581]. A national study by EPRI [582] suggested about 0.16 kWh/m^3 for municipal groundwater pumping, corresponding to a water table of perhaps 40 m (assuming 0.004 kWh/m^3 for pumping and disregarding discharge pressure); groundwater wells in California reportedly used between 0.14 and 0.69 kWh/m^3 for pumping [580].

Surface water conveyance. Total energy for surface water conveyance varies markedly, from negligible for gravity-driven systems that use local surface waters (e.g. local rivers, a common source in the East) and typically transport water only ones to tens of miles, to several kWh/m^3 for several water projects that transport water hundreds of miles and up thousands of vertical feet in California and Arizona. It is worthwhile to review several such projects, given the longstanding controversies surrounding water in the West. In California, the State Water Project (SWP) entails 701 miles of canals and tunnels, 21 reservoirs, and 4 hydroelectric generating stations, and transports water from Northern California, through the Central Valley via the California Aqueduct, and then up about 3,000 vertical feet to cross the Tehachapi Mountains and into the vast southern cities. As a whole, the system uses a *net* 5.1 billion kWh after discounting 6.5 billion kWh of hydroelectric generation [587], delivering 3.0 billion m^3 of water, and equating to 1.7 kWh/m^3. This average masks significant heterogeneity, as pumping over the Tehachapi Mountains at the southern end makes this water much more energy-intensive than that further upstream: with a gross 3,000–4,000 foot elevation change, we require about 3.7–4.9 kWh/m^3 at this leg. After crediting for hydroelectic generation, a California Energy Commission report estimated maximum conveyance energy to Southern California at 3.7 kWh/m^3 [588].

The Colorado River Project (CRP) also supplies significant water to Southern California, requiring an estimated 1.6 kWh/m^3 to deliver water to Los Angeles [580]. Not all such projects require pumping energy: the Los Angeles Aqueduct is a gravity-driven system that diverts water from the Owens River and creeks that feed into Lake Mono to (surprisingly) Los Angeles. While requiring little energy, these diversions ended agriculture in the Owens Valley, and have seriously harmed the Lake Mono ecology. Similarly, the gravity-driven Central Valley Project (CVP) is a massive conglomeration of canals, reservoirs, and power plants, that generates significant hydropower and provides the better part of the Central Valley's irrigation water. Despite benefits, including low-carbon hydroelectricity, the CVP has also wreaked local/regional environmental havoc on rivers, wetlands, and ecosystems.

As a final example, in Arizona, the Central Arizona Project (CAP), is a 336 mile system of canals and aqueducts that annually transports over 1.4 million acre-feet (1.73 billion m^3, 456 billion gallons) from the Colorado River to the cities and fields of southern Arizona, with its pumps driven by the largest coal-fired power plant in the nation, the Navajo Generating Station (NGS) (24.3% of the plant's output is dedicated to the CAP, while the rest goes to regional utilities). This amounts to 3.86 billion kWh (24.3% of the 15.89 billion kWh generated in 2012, per eGRID), or 2.23 kWh/m^2. CAP water is also ultimately pumped nearly 3,000 vertical feet, implying that at its terminus in Tuscon, AZ, water conveyance requires at least 3.5 kWh/m^3 (based on a minimum of 0.004 kWh/m^3/m). Similarly, Perrone et al. [584] reported about 4.1667 kWh/m^{-3} for delivery of CAP water to Tuscon, AZ, and equivalent to about 0.0172 kgCO$_2$e/gallon under coal-fired electricity, or 0.0108 kgCO$_2$e/gallon under the US grid-average.

Now, these large-scale projects are not the norm for the United States as a whole, and even in the Southwest, local water sources are also used (for example, the Verde River near Phoenix, AZ).

Overall conveyance energy for municipal water. Given that multiple sources generally supply any municipal utility, and these sources vary in their energy intensity, it is difficult to derive any nationwide average. We may assume that many surface water withdrawals require almost no energy, and I will adopt the EPRI estimate [582] of 0.16 kWh/m^3 as a central estimate for groundwater, although this is probably an underestimate for most of the West; weighting groundwater 40% and surface water 60% gives 0.064 kWh/m^3 as a likely lower bound for US municipal water conveyance.

For those living in Southern California and Arizona, I suggest that, after weighting groundwater 40% and imported surface water 60%, 2.0 kWh/m^3 is a very crude estimate of the energy-intensity for water for readers in these regions. Note also, that if the "marginal" water (similar to the concept of marginal electricity, this is the source affected by marginal changes in water demand) is imported water, then this water will have somewhat higher conveyance energy.

Treatment and Distribution

> - Freshwater treatment usually requires a trivial amount of energy; energy demand for booster pumping for distribution is variable, but perhaps 0.372 kWh/m^3 on average.
> - Desalination is an exception, requiring about 3.5–4.0 kWh/m^3 for seawater desalination, but less than half this for brakish groundwater (the major source of desalinated water overall). While a very minor source in most of the US, desalination is growing, and is most prevalent in California, Texas, and Florida.

Following conveyance, municipal water treatment plants treat water and pump it into pressurized distribution systems ("booster pumping"). Not including desalination, the energy required for water treatment is nearly trivial, while distribution pumping is fairly significant (accounting for 85% of plant energy). Reported energy-intensities for booster pumping reviewed by Plappally and Lienhard [580] ranged widely, from nearly zero to 2.27 kWh/m^2, and are likely higher for smaller distribution systems; the EPRI study [582] estimated booster pumping to consume 0.372 kWh/m^3 for a typical system.

Desalination, on the other hand, is quite energy-intensive; desalination processes are either thermal (i.e. distillation) or membrane (e.g. reverse osmosis) based. Either seawater or other brakish waters (e.g. groundwater) may be subject to desalination, with seawater requiring about twice as much energy to purify [585]. Globally, most seawater thermal desalination is carried out at steam power plants, where waste heat drives the desalination process: seawater is used for thermoelectric cooling, heating the seawater and driving the downstream distillation process (such plants are most widely used in several Persian Gulf countries). Since this technology uses waste heat, the net energy/emissions impact is lower than membrane-based technologies, which rely upon electrical energy.

Membrane desalination of seawater requires about 3.5–4.0 kWh/m^3 [585, 580], and is thus more energy intensive than most imported water sources (desalination of other brakish water takes about 0.7–1.6 kWh/m^3 [580]). Desalination also can harm marine life both through seawater intake, and via discharge of highly saline brines [589].

While most global seawater desalination capacity is located in the Middle East [580], US desalination capacity is growing [138], with plans for major expansion in some arid and coastal areas of the US [589]. For example, the nation's largest seawater desalination plant, the Carlsbad Desalination Plant (which operates under membrane desalination), opened in 2015, and will provide about 7% of San Diego County's water [139]. It should be emphasized that, overall, most desalinization plants purify brakish groundwater or briny inland surface water, rather than seawater. Desalination is most prevalent in California, Texas, and Florida [138].

Wastewater

Municipal wastewater passes through primary, secondary, and sometimes tertiary treatment, processes that in sum consume a significant, but highly variable amount of energy (depending upon plant characteristics, quality of discharge water, etc.). Wastewater pumping, aeration, and digestion (to remove organic compounds) are all major points of energy use, and Plappally and Lienhard [580] provide an in-depth review. They give 0.43 kWh/m^3 as an overall average for US municipal wastewater treatment, but again, individual cities and plants may vary markedly. The EPRI report similarly estimated between 0.252 and 0.505 kWh/m^3 as national averages across four treatment plant types [582].

When water is treated further to be suitable for re-use (possibly potable), additional energy is approximately in the 1–3 kWh/m^3 range [580]. Discharge pumping, or pumping of treated wastewater to recharge groundwater supplies, is an additional source of pumping energy.

Adding it up

To sum up, we have, as an overall US average, energy used to convey, distribute, and treat domestic municipal water at about 0.5 kWh/m^3 for conveyance (assuming 2 kWh/m^3 in CA and AZ, 0.064 kWh/m^3 elsewhere, and weighting according to USGS consumption numbers), 0.372 kWh/m^3 for distribution, and 0.43 kWh/m^3 for treatment, summing up to 1.3 kWh/m^3, or 0.005 kWh/gallon. Assuming grid-average US electricity, this translates into 0.0034 kgCO$_2$e/gallon. Since 30% or more of domestic water is used outdoors, this fraction is not treated, and thus such water is associated with only 0.0033 kWh/gallon and 0.0023 kgCO$_2$e/gallon. Our overall weighted average then becomes 1.173 kWh/m^3 = 0.0044 kWh/gallon = 0.003 kgCO$_2$e/gallon.

At 89 gpd per person, the above figures suggest just under 100 kgCO$_2$e/person/year due to domestic water conveyance and municipal treatment. In the Southwest, we will have emissions factors on the order of perhaps 0.005–0.0075 kgCO$_2$e/gallon, while in the East emissions are more like 0.0011–0.0022 kgCO$_2$e/gallon (0.0011 for outdoor use, 0.0022 for indoor use, and 0.0019 a weighted average). It also follows that residents of the Southwest (who also use more water than the national average) may generate 200–400 kgCO$_2$e/year from water provision (this figure may be appreciably lower for areas that rely more on local vs. imported water, and higher where imported water dominates), while those more north and easterly may only be responsible for about 50 kgCO$_2$e/year.

13.3 Major hot and cold water end-uses

There are several major surveys of residential water use. Perhaps the most widely cited is the 1999 Residential End Uses of Water Study (REUWS) [578], which collected detailed end-use data for 1,188 households in 12 metropolitan areas, located primarily in the west and southwest US. Note that this survey was not representative of the US as a whole, but is still a very valuable

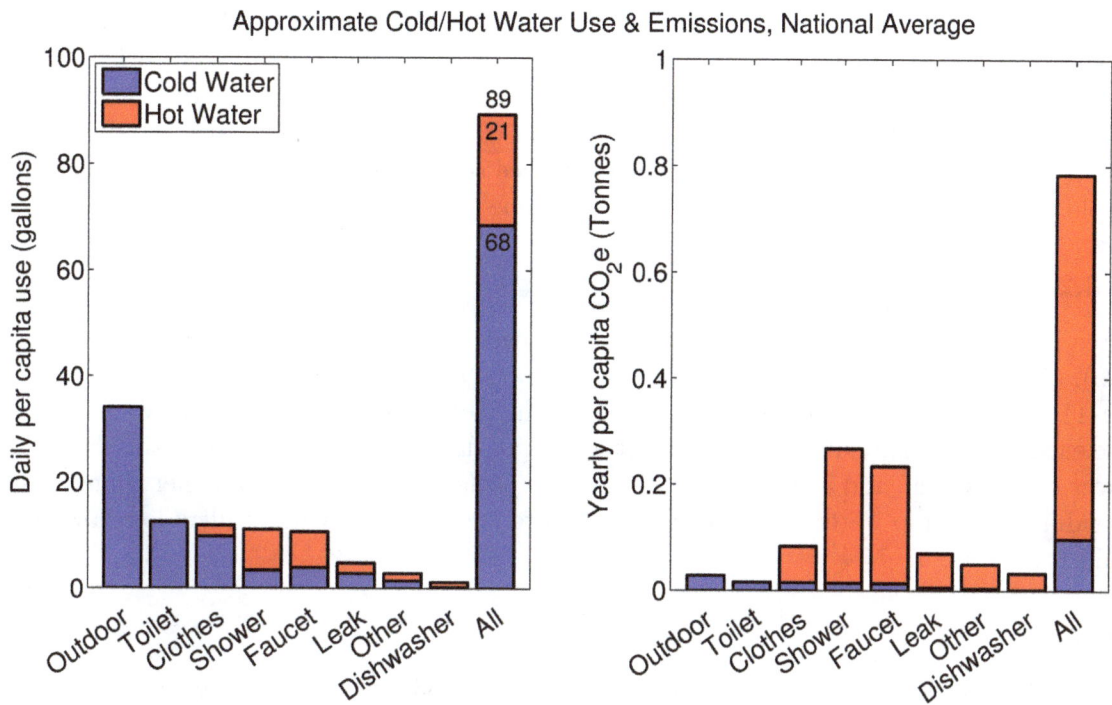

Figure 13.7: *Approximate* daily per capita use of cold and hot water, and associated emissions (yearly basis), considered as a crude national average.

resource. At the time of this writing, some interim results are available from the 2012 REUWS survey as well [579].

Taken as a whole, outdoor watering is the single largest category of use, while of indoor applications the toilet is the single biggest offender. These are both strictly cold-water uses; other indoor uses rely much more on hot water, especially showering, and are thus responsible for the largest share of emissions. Approximate daily per-capita use by category, inferred from the 1999 and 2012 interim REUWS study results [578, 579], along with USGS figures [586] on total use and the sources for individual water uses reviewed in this section, is given in Figure 13.7, along with associated carbon-equivalent emissions. These numbers are not meant to be extremely precise, but to serve as rules-of-thumb accurate enough for order-of-magnitude carbon footprinting. It is clear that hot water for showers and general faucet use are the most important drivers of water-associated emissions at the residential scale.

In the Southwest, cold water conveyance emissions will be somewhat higher (especially because outdoor water use is higher in the West, perhaps twice the national average or more), but still only fraction of the overall carbon footprint. While conveyance requires less energy elsewhere, it is also possible (and likely) that colder ambient (and hence tap water) temperatures increase the energy required for hot water applications.

13.3.1 A note on toilet water

The humble toilet is consistently the largest single user of indoor water, at about 32 gallons per household per day in 2012 [579], or perhaps 12.8 gallons per person per day, and equal to 8 daily flushes each (for a low-flush toilet). This was actually a significant decrease from 1999 (45 gallons/HH/day), presumably due to increased penetrance of low-flush toilets.

It follows that hundreds of miles of aqueduct, canal, and sprawling desalination plants have been constructed in some part because humans simply cannot bring themselves to allow that which is yellow to mellow. Since this is an intolerable proposition to some, low-flush toilets, which are standard and widespread at this point, use ≤1.6 gallons per flush (GPF) and save a great deal of water compared to older models. Further, 1.28 GPF and dual-flush models (typically 1.1/1.6 GPF) can reduce water use by 20–30% compared to standard 1.6 GPF models. Cheap commercial toilet "tank banks" that displace some of the toilet tank water, as well as simple flow diverters, can further reduce GPF by one-third to one-half.

13.3.2 Hot water uses

Three basic measures, (1) using 1.5 gpm showerheads in lieu of standard low flow heads (2.5 gpm), (2) using tap water that is merely warm (70–80 °F) instead of hot (as is typical, at about 100–110 °F) for general faucet uses, and (3) washing clothes only in cold water, would together save each household about 25 gallons of hot water per day (about 10 gpd due to the shower and faucet changes each, and 5 gpd from washing), i.e. 50% of all hot water. Further changes, such as enduring cold tap water, fixing leaks, and shortening showers or using ultra-low-flow heads, could reduce use 67–75%.

Showering

Showering is the single largest residential hot water use (33% in the 2012 REUWS [579]). The 1999 REUWS [578] reports total water use at 11.6 gpd per capita for showering, while the 2012 update gives 28.6 gpd per household, with 21.9 of these gallons hot (for Denver, CO); if households are 2.5 persons, this equates to per capita consumption of 11.44 gpd, 8.76 gallons of which is hot. Assuming 77% of water reaching the showerhead is heated, then 8.88 gpd of hot water is used per person per day to shower under the 1999 REUWS. Typical adults certainly use somewhat more than these averages suggest, since they include younger children who more rarely shower.

An analysis of showering data from the large National Human Activity Pattern Survey (NHAPS) and the 1999 REUWS by Wilkes et al. [590] suggested average shower durations of 13.2 and 7.6 minutes under the NHAPS and REUWS, respectively. Several other smaller surveys, as reviewed in [590] are more consistent with the lower number. Additionally, each shower consumed 18.6 gallons at a mean flow rate of 2.4 gpm, just under the maximum flow rate for low-flow showerheads. In various surveys, people tend to take about 0.8 showers per day [590], suggesting about 15 gpd for showering, with around 10 of these gallons hot.

In sum, typical showering patterns are probably around 7.5–12.5 minutes per shower with water flowing at ≤ 2.5 gpd, and 18–25 gallons of water are consumed per shower, 10–20 of which are hot. Ultra-low shower heads with 1.25–1.5 gpm maximum flow rates are widely available, and using such a device may reduce shower water use by 50–60%, and overall hot water use by about 20%. Reducing shower duration to about five minutes could also save 33–50% of shower water.

General faucet

After bathing, miscellaneous faucet use is the second largest category of hot water use, at 31% in the 2012 REUWS. As with much else, this is likely a largely unnecessary use of energy. For hand-washing, there is actually little basis for using warm or hot water as a sanitary measure. For example the World Health Organization's hand hygiene guidelines for health workers [591] state that water temperature does not meaningfully influence microbial removal from washed hands, and indeed, warmer water temperatures are strongly associated with skin irritation [591].

Similarly, it is unlikely that there is any need for hot water, beyond comfort, for other routine hygiene practices, such as tooth brushing, etc.

Overall, 69% of faucet water was hot (18.9 out of 27.4 gallons/HH/day) in the 2012 REUWS, suggesting 100–110 °F water at the tap, on average (similar to a hot shower). Flipping this proportion would still yield tap water in the 70–85 °F range, which seems more than adequate for comfort, and would save each household just over 10 gallons of hot water per day (and equivalent to about 330 kgCO$_2$e/year overall, more under electric heat, less under gas).

Clothes washers

As also discussed in Section 15.3, there appears to be little reason to use hot water for most clothes washing, according to multiple popular sources. For example, according to a 2011 New York Times article, cold-water detergents perform just as well as those designed for hot water and are no more expensive. Nevertheless, a majority of people continue to use hot or warm water for clothes washing. While a majority of wash cycles are in hot or warm water, overall the majority of water used in a wash is cold, as there is typically a cold rinse cycle (2009 RECS).

From several data sources (summarized below), we can estimate that the average household uses on the order of 5–6 gallons of hot water per day for clothes washing, amounting to about 200 kgCO$_2$e/year per household, or 80 kgCO$_2$e/year/person. I suspect that most people either habitually use either hot or cold water for washing, and thus few households will be average, with most either markedly above theses averages in their use and emissions, or near to zero. Thus, if you, dear reader, make it your custom to always use hot water, your household emissions in this subcategory may be over four times the average. Remember also that electric water heaters have roughly twice the associated emissions of gas heaters, and this will further add to the variability.

Interim results from the 2012 REUWS indicated that the average household in Denver, CO used 6.3 gallons of hot water (and 24.3 gallons of cold water) per day for clothes washing; thus 21% of the water overall was hot. This translates into 164.4 kgCO$_2$e/year for gas heaters and 317.3 kgCO$_2$e/year for (grid-average) electrical heaters. Weighting 42% electrical and 58% gas, this suggests overall an average of 228.6 kgCO$_2$e/year.

The 2009 RECS gives data on the temperature of the wash cycle, and suggests that the shares of hot, warm, and cold cycles are 6%, 48%, and 46%, respectively, yielding 30% of washing cycle (versus rinsing) water as hot. If all rinses cycles were cold and equal in volume, then 15% of water involved in the entire washing process is hot, in rough agreement with the REUWS.

Numbers from the 1999 and 2012 REUWS are roughly consistent with about 300 laundry loads per year, at 40 gallons of water per load. This agrees with the 2005 RECS, which reported 301 loads per year per household. Three hundred yearly 40 gallon loads, at 15% hot water, is equivalent to about 33 gallons per day, 5 of which is hot.

Dish washing

Automated dishwashers uniformly use hot water, but are overall a minor component of household hot water use. A significant portion of general faucet use, however, almost certainly goes towards hand-washing dishes. Given that hot water does not improve hand hygiene over tepid water [591], it seems unlikely that it is necessary to use heated water towards this end beyond the minimum needed for comfort.

Chapter 14

Lighting

14.1 Overview

14.1.1 Residential lighting

Lighting, the poster-child for energy savings. Perhaps no use of energy is more (literally) visible, and high-efficiency lighting, in the form of compact fluorescent lights (CFLs) or LEDs, is widely promoted, and sometimes derided, as an energy savings measure. As we shall see, lighting, based on 2008/2009 numbers, accounts for about 15% of residential electricity use and 9.7% of residential household emissions. It is almost certain, however, that this share has declined with the ongoing phase-out of (most) traditional incandescent bulbs and the broader transition towards high-efficiency lighting sources, which use only 10–25% as much energy as traditional bulbs. Fully converting from the 2008/2009 average lighting profile to high efficiency lighting (a mix of CFLs and LEDs) would reduce household electricity use by about 10.3% and household residential energy emissions by 6.6%. Eliminating any unnecessary lighting would obviously have a further benefit.

In 2012, the US Department of Energy published its Residential Lighting End-Use Consumption Study (RLEUCS) [592], which mainly combined data from the 2008-2009 California Residential Lighting Metering Study, which directly inventoried and monitored lighting in 1,200 California homes, with 2009 EIA RECS data, to obtain multiple state and national estimates of residential lighting use. Overall, this study estimated 1,708 kWh per home per year devoted to lighting, equating to 1.173 $MgCO_2e$ per household, and 0.456 $MgCO_2e$ per individual. The average home had 67.4 lighting fixtures, each of which was on for an average of 1.6 hours (and drawing power at an average rate of 47.7 W). Multiplying gives 107.84 lighting-hours per day, equivalent to having 13.5 lights on for eight hours a day. I rather suspect most households could run perfectly well on (much) less than half. The average masks significant variation in home type, with smaller homes using much less lighting, as shown in Figure 14.1. Overall, these figures translate into 200.15 billion kWh of electricity on a national scale, and 137.4 million $MgCO_2e$ of emissions.

Of note, regional differences in energy use were largely explained by the CFL vs. incandescent household lighting preference; LEDs were not widely commercially available until after 2008, and so were not meaningfully represented in this survey. On the whole, only 25% of lighting-hours were generated by CFLs (but they consumed only 8.8% of lighting electricity), and these bulbs used on average 15 W, about one fourth of the incandescent bulbs (58 W), and less than a fifth of the "other" category (80 W). Overall average lamp power was 47.7 W. Further results from this study are summarized in Figure 14.1.

It follows that, if all residential lighting was replaced by high efficiency bulbs with no change

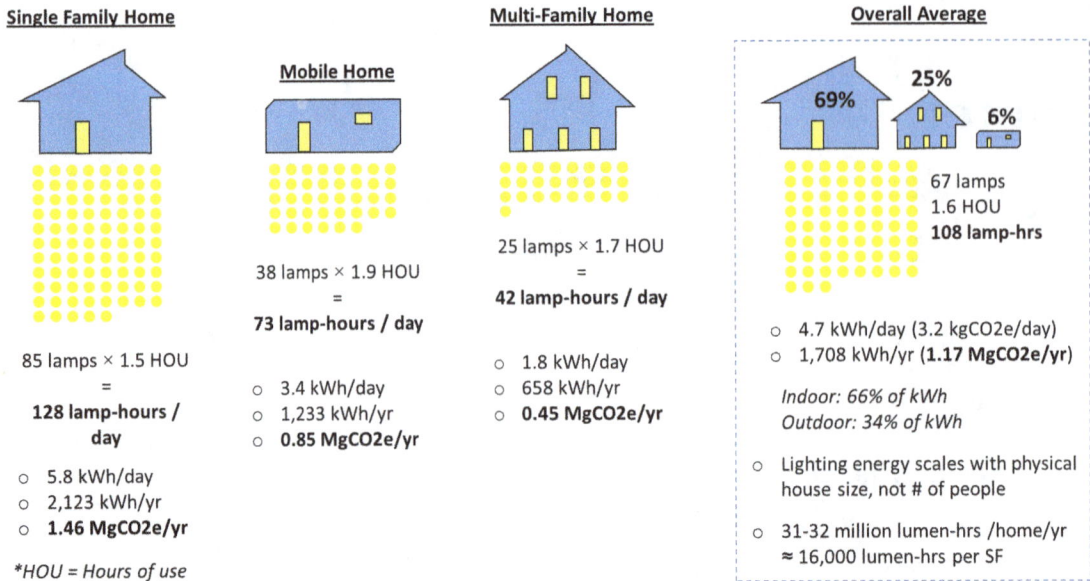

Single Family Home

85 lamps × 1.5 HOU
=
128 lamp-hours / day

- 5.8 kWh/day
- 2,123 kWh/yr
- **1.46 MgCO2e/yr**

HOU = Hours of use

Mobile Home

38 lamps × 1.9 HOU
=
73 lamp-hours / day

- 3.4 kWh/day
- 1,233 kWh/yr
- **0.85 MgCO2e/yr**

Multi-Family Home

25 lamps × 1.7 HOU
=
42 lamp-hours / day

- 1.8 kWh/day
- 658 kWh/yr
- **0.45 MgCO2e/yr**

Overall Average

69%   25%   6%

67 lamps
1.6 HOU
108 lamp-hrs

- 4.7 kWh/day (3.2 kgCO2e/day)
- 1,708 kWh/yr (**1.17 MgCO2e/yr**)

Indoor: 66% of kWh
Outdoor: 34% of kWh

- Lighting energy scales with physical house size, not # of people

- 31-32 million lumen-hrs /home/yr ≈ 16,000 lumen-hrs per SF

Figure 14.1: Summary of 2008/2009 residential lighting profile for different housing types, based on the 2012 Residential Lighting End-Use Consumption Study [592].

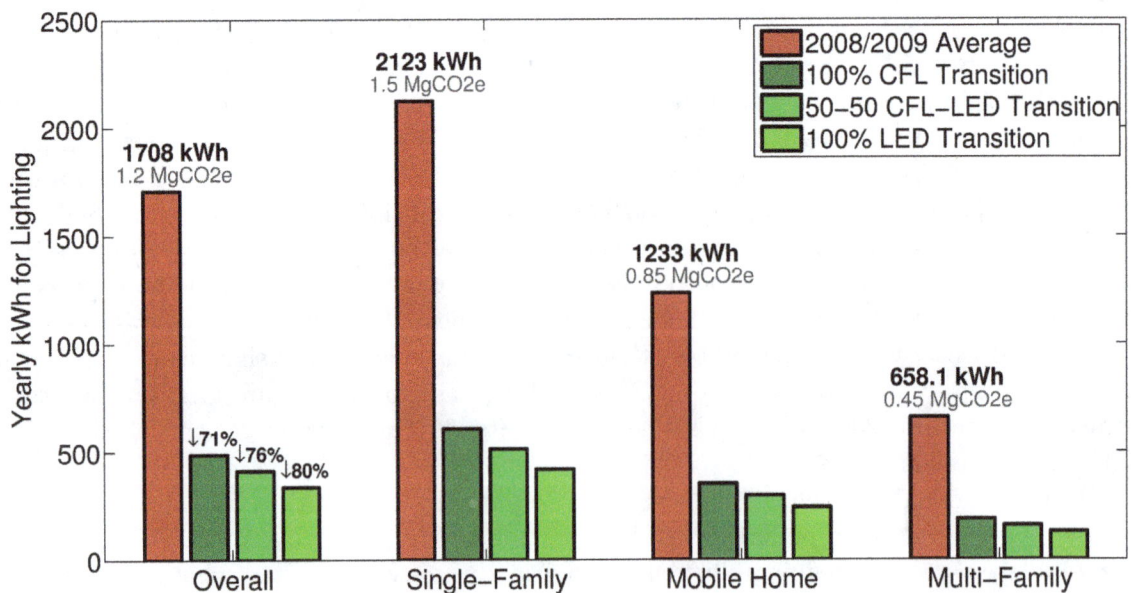

Figure 14.2: Potential household energy savings from CFL and LED transitions, relative to the 2008/2009 baseline.

in lighting use, energy use per home would drop by 71% to 488 kWh under a transition to CLFs (assuming a 60 to 13 W equivalency), or by 80% to 338 kWh under a complete transition to LEDs (assuming an average 60 to 9 W equivalency), and a 50/50 mix would give 413 kWh/home (76% reduction). These results are summarized in Figure 14.2. Transitioning to such a 50/50 mix would represent a household-level savings of 889 kgCO$_2$e (350 kgCO$_2$e/capita), and overall, 151.8 billion kWh of electricity and 104.2 million MgCO$_2$e avoided, under grid-average emissions rates (or 119.3 million MgCO$_2$e at marginal emissions rates) and 2009 household size; this is about 1.6% of the EPA estimated US territorial emissions in 2013.

This transition is already occurring in earnest, due mainly to improved CFL and LED technology and rapidly falling prices, and government mandated lighting efficiency standards that effectively banned most general service traditional incandescent bulbs in 2014. I discuss projected changes in lighting energy use under this transition further in Section 14.1.3.

Lighting energy use seems to be mostly explained by the square footage of the dwelling (and by the efficiency of the lighting technology used), not by the size of the household. While extremely detailed data is not available, lighting energy increases with number of bedrooms. It also is much greater for single family residences compared to multi-family or mobile homes, and furthermore, lighting increases with household head education level. As education increases, lighting per square foot is nearly constant, yet lighting per individual doubles, implying that more educated, and hence wealthier, households are simply lighting larger spaces containing similar numbers of occupants.

It is also important to emphasize that, while the focus of most media, government reports, and analyses has been on the impact of high efficiency lighting, simply reducing lighting use is very much a viable option. While there are no explicit estimates of lighting variability in the RLEUCS, we can infer from the prior discussion that people tend to light a dwelling at a fairly uniform level, regardless of how many individuals inhabit it. It should be a simple matter to only light the space that one is actually using, and I would venture an informed guess that at least half of lighting is not really necessary. My anecdotal experience certainly supports this.

14.1.2 Commercial lighting and overall lighting energy

While my focus is on residential consumption, it is worth noting that the commercial sector is a much larger user of lighting, a fact that may not be surprising to anyone who has had to spend their days bathed in the sickly fluorescent ambience of the largely windowless office buildings, schools, malls, and hospitals that dominate our landscape. As an aside, multiple studies have found that increased natural daylight in hospital rooms can result in faster recovery, less stress, and less pain medication, as can windows that open to a view of nature [594, 593].

The DOE has estimated that 26% of energy use in commercial buildings goes towards lighting, while the EIA's 2003 commercial building energy survey (CBES) reported 20.5% and 37.7% of energy and electricity, respectively, going towards lighting, for a sum total of 392.7 billion kWh of electricity, almost twice what is used at the residential scale. Given this, I would encourage any reader to commit the admittedly subversive and universally frowned upon act of turning off unused lights at work. No doubt one will face fierce resistance over the many seconds lost to the finding and flipping of switches, but the numbers shall vindicate you.

The DOE estimated that lighting consumed 17% of all US electricity in 2013, for a grand total of 609 TWh [595]. Including roadway lighting and industrial lighting, converting all lighting in the US to high-efficiency types could potentially save over 10% of electricity and almost 5% of emissions. As discussed next, such a major transition in lighting is in fact taking place.

14.1.3 Lighting transition

In 2007, the Energy Independence and Security Act of 2007 was signed into law. Included were new incandescent lighting efficiency standards, to be phased in between 2012 and 2014, that effectively banned most *general service* incandescent bulbs. Understand that the general service category, essentially the familiar screw-in light bulbs, does not include track and downlight lighting, which are significant residential light sources, nor does it include decorative lighting. Thus, while general service incandescents are disappearing from US homes, other incandescent sources will still remain on the market for some time.

Although most standard incandescents no longer can be manufactured or imported, halogen incandescents have taken their place to some degree on store shelves. Like other incandescents, these bulbs contain a tungsten filament that is heated by an electric current to high temperature, producing light from the glowing metal, but also causing some tungsten to evaporate. Unlike other incandescents, these bulbs contain a halogen gas which mixes with the evaporated tungsten, allowing it to redeposit on the filament later. This evaporation-deposition cycling allows the filament to be heated to a higher temperature (one that would otherwise result in excessive tungsten evaporation), thus increasing the bulb efficiency and lifespan somewhat (see the subsequent section for a deeper discussion of lighting physics). A 43 W halogen incandescent is equivalent to a 60 W standard incandescent, for an energy savings of about 30%.

Until recently, CFLs have been the only truly high-efficiency lighting source widely used at the residential scale. They have become relatively inexpensive over the last few years, and are leading the short-term lighting transition, with the largest market share of any bulb type since 2014. However, the reign of CLFs may be short lived. LED, or "solid-state" lighting, is being widely promoted by government agencies for clean energy retrofits, and the DOE has projected LEDs to achieve nearly 100% market share for new general service lighting by 2030, with CFLs (and to a lesser degree, halogen incandescents) acting as a bridging technology to an LED future [595]. This is rather remarkable, considering that general service LED lighting only became commercially available in 2007, and was initially less efficient than CFLs and prohibitively costly. LEDs are still more costly, but they have decisively outstripped CFLs in efficiency and will continue to improve in this area; they also have a threefold longer lifespan. It is important to note that, while some currently available LEDs use just 50% as much energy as typical CFLs, the *absolute* difference in energy consumption is not great: the 6 W difference between a 7 W LED and a 13 W CFL is trivial when comparing either to a 60 W incandescent. Therefore, CFLs still represent an acceptable alternative to LEDs, in my view.

Figure 14.3 shows DOE projections for market share for different lighting technologies from 2013 to 2030. It also gives DOE forecasted residential lighting energy consumption from 2013 to 2030. This forecast was determined by modeling the failure of old lighting and replacement according to market-share, lighting installed in new residential construction, as well as projected increases in LED efficiency. Since this forecast includes a 1.31% annual increase in lit residential floor space, we can factor this out to get a rough estimate of how individual household lighting energy consumption is projected to fall over the next 15 years, as also shown.

From Figure 14.3, we see that simply by responding to the overall likely market trend, the average individual will decrease their lighting-associated energy and emissions by around 34% by 2020, and by 63% by 2030. However, one can easily achieve a >70% energy reduction immediately via a complete switch to either CFL- or LED-based lighting, or some combination thereof (see Figure 14.2). As a final note, one should not feel compelled to use old incandescent bulbs until they burn out. Their manufacture uses a trivial amount of energy, and so discarding and replacing even a brand-new incandescent bulb *will save far more energy overall.*

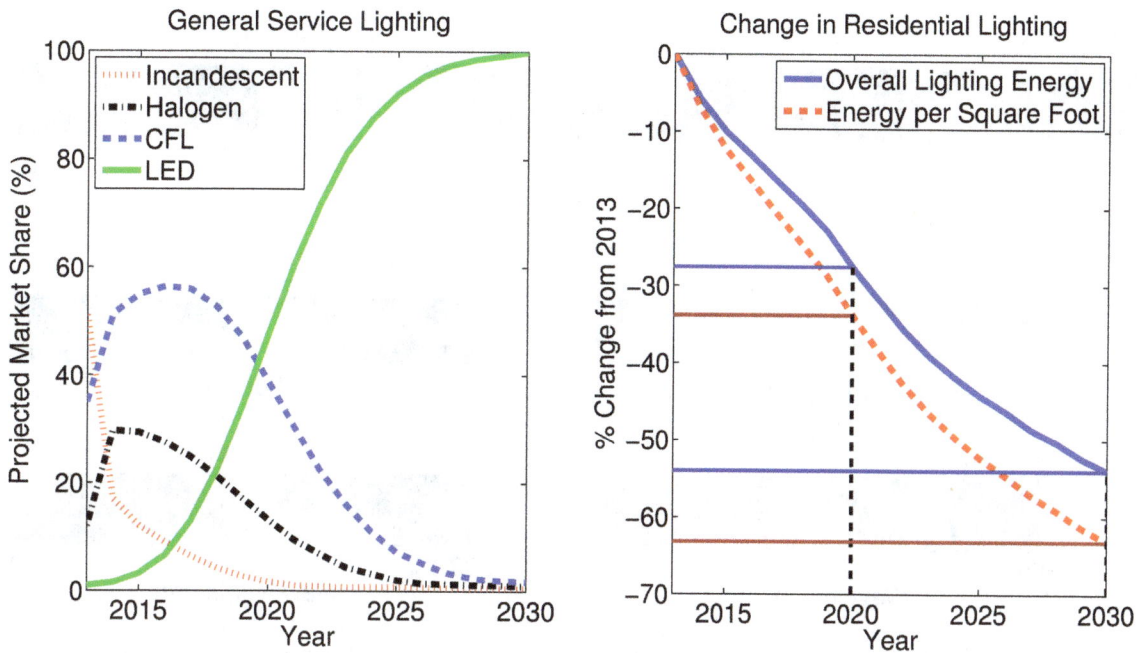

Figure 14.3: DOE projections for changes in general lighting market share (note that this does not include other lighting categories, such as track, downlight, and decorative lighting), and in residential lighting energy use for the year 2013 through 2030 (based on [595]).

14.2 The basic physics of lighting

14.2.1 Electromagnetic spectrum, photopic luminosity, and basic photometric units

As a preamble, recall the electromagnetic spectrum, the range of possible frequencies of electromagnetic (EM) radiation, ranging from long-wave radio waves to very short-wave gamma rays. Note that the product of frequency (v, in Hertz) and wavelength (λ, in m) is equal to the speed of light (c)

$$\lambda v = c, \tag{14.1}$$

and so high frequency implies short wavelength, and inversely, a low frequency implies a long wavelength. A narrow band of frequencies makes up the visible spectrum, i.e. that range of frequencies which is perceptible to the human eye, as shown in Figure 14.4, and the frequency/wavelength defines the color of visible light. Furthermore, the human eye is not uniformly sensitive to visible light: the eye is maximally sensitive to bright green light with a wavelength of 555 nm, and sensitivity varies according to a roughly normal distribution between 380 and 780 nm, as also demonstrated in Figure 14.4. This distribution is called the *photopic luminosity function*[1].

Now, a lamp emits energy as EM radiation, and the *radiant flux* is the amount of energy emitted as EM radiation per second (Joules/s, or W). The key point to consider is that not all of this EM radiation is visible to the eye, as a lamp emits EM radiation across a range of frequencies; the distribution of frequencies is described by the "power spectral density,"

---

[1] At very low light intensities, light sensitivity is mediated by rods and not cones, and under such conditions we have as our sensitivity distribution the *scotopic* luminosity function, with a peak shifted slightly towards the blue, relative to the photopic.

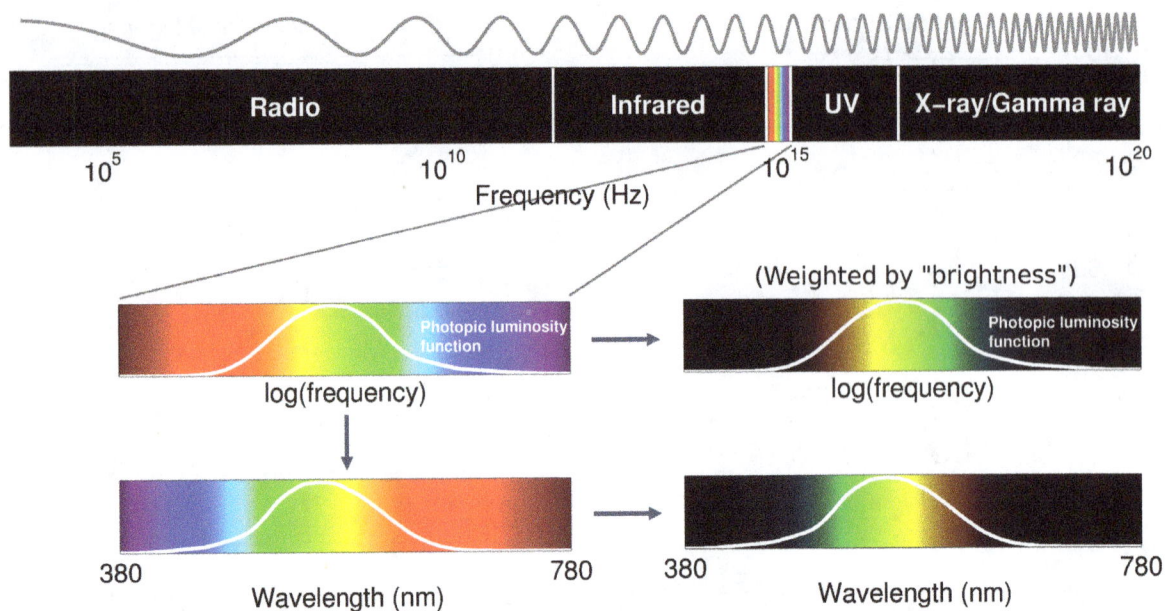

Figure 14.4: The electromagnetic spectrum, consisting of, from lowest frequency (and longest wavelength), radio waves, infrared light, visible light, ultraviolet light, and X-rays/gamma rays. The visible spectrum makes up only a very narrow portion of the overall spectrum, and, as demonstrated in the lower portion of the figure, the human eye's sensitivity to light even within the visible spectrum varies according to a roughly normal distribution. The lower portion also gives the visible spectrum as a function of the logarithm of frequency and as a function of wavelength (plots from here out generally use wavelength).

or "power spectrum." Furthermore, since the eye is maximally sensitive to light with a 555 nm wavelength, a lamp that emitted all energy as light only at 555 nm (i.e. a bright green monochromatic light, defined by a power spectrum consisting of a single peak at 555 nm) would induce the maximum possible visual response. Any other frequency distribution, or spectrum, will necessarily induce less than 100% of the maximum response.

This leads us to the idea of *luminous efficiency*, defined as the fraction of radiant flux which is perceived by the eye as light, relative to the maximum possible fraction perceptible as light. For a "perfect" eye, one that at maximum converted all radiant flux into perceived light, we could define *luminous flux* as luminous efficiency times radiant flux, and luminous flux would thus be the "perceived power of light": since radiant flux is a measure of power (W), any fraction of this (e.g. luminous flux) is also a measure of power.

Real eyes do not perceive 100% of radiant flux even at peak sensitivity, and so the units of luminous flux and radiant flux differ. In the SI system, luminous flux has units of "lumens." Photometric units be quite confusing, but at the risk of repeating myself, just realize that lumens are basically a scaling of watts, i.e. a lumen is a unit of power, and luminous flux is the power of visible light weighted by the degree to which is it *perceived* as visual power. Now, by definition, a point-source monochromatic green light at 555 nm that emits 1/683 watts has a luminous flux of 1 lumen. That is, at best, 1 watt of radiant flux is perceived as 683 lumens. The *luminous efficacy* of a light is the lumens emitted per watt, so a perfect light has a luminous *efficacy* of 683 lumens/watt, and a luminous *efficiency* of 100% (as above, such a light would consist of a single wavelength, 555 nm). A light with luminous efficiency of 50% would yield a luminous flux of 341.5 lumens/W.

Given the photopic luminosity function of an eye, and the power spectrum of a light source,

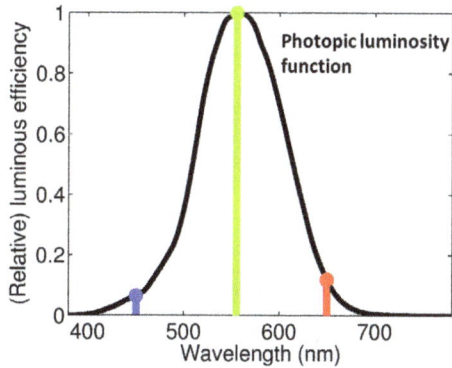

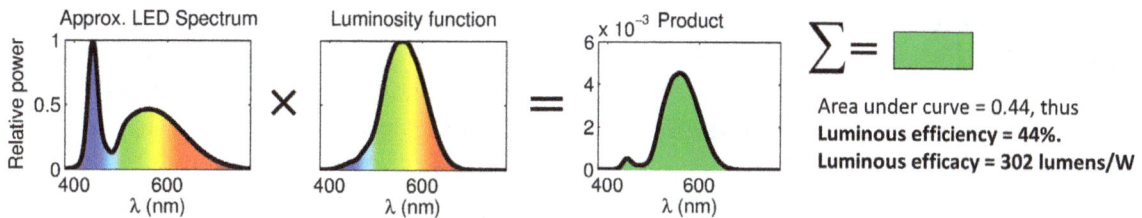

Figure 14.5: The top half of the figure gives the photopic luminosity function, and the luminous efficiencies and efficacies for three example monochromatic (single wavelength) lights. The bottom half shows how the luminous efficiency and efficacy are calculated for a polychromatic light (multiple wavelengths), using an example LED spectrum. We take the product of the luminosity function and the spectral density, and integrate over the resulting curve ("area under the curve") to yield, in this case, 44% luminous efficiency and 302 lumens/W as the the efficacy. Note that real LEDs will have other power losses, and thus achieve appreciably lower wall-plug luminous efficacies.

we can multiply the two distributions and integrate the result to get the luminous efficiency and luminous flux. An example for an LED spectrum is illustrated in Figure 14.5.

For an ideal lamp, luminous efficacy would refer to the lumens generated per watt of EM radiation. For actual physical lamps, luminous efficacy frequently refers, in practice, to the lumens generated per watt of power consumed by the entire lamp apparatus, i.e. the power drawn from the wall plug. For example, if a physical lamp emitted 10 W of radiant power, 3 W of which was perceived as light, i.e. 2,490 lumens, while the lamp drew an additional 10 W of power not converted into radiant flux, the luminous efficacy of radiation would be 2,490 lumens / 10 W = 204.9 lumens/W, while the wall-plug luminous efficacy would be 2,490 lumens / 20 W = 102.5 lumens / W.

Illuminance is the luminous flux per unit area, with SI units of lumens/m², called *lux*. Illuminance is essentially a measure of how brightly lit a surface is. For example, a 10 W LED with a luminous efficacy of 100 lumens/W that is used to light a single square meter will put out 1,000 lumens and give an illuminance ("brightness") of 1,000 lux. If the same LED is used to light ten times the area, 10 m², then the illuminance is 100 lux, and the surface is only one-tenth as brightly lit.

Finally, lumens are officially defined in terms of the SI base unit candela. A candela is a measure of luminous intensity, which is the luminous power per solid angle emitted in a particular direction. A candle has a luminous intensity of roughly 1 candela. For a point light source radiating 1 candela equally in all directions, i.e. a spherical distribution of light, and thus a solid angle of 4π steradians, we would have a luminous flux of $4\pi \times 1 \approx 12.57$ lumens. Do not, gentle reader, overly concern yourself with luminous intensity and the candela. Luminous flux is much more relevant and intuitive.

14.2.2 Black body radiation and the incandescent bulb

From experience, everyone intuitively knows that when an object is heated enough, it begins to glow, e.g. a red-hot stove top. Indeed, all bodies emit electromagnetic radiation, and the spectrum of that radiation shifts as the body grows hotter. Cooler objects emit low-frequency, long-wave radiation that is invisible to the human eye, but once a body becomes hot enough to emit significant light in the visible spectrum, it is said to be incandescing. An incandescent light bulb consists, essentially, of a filament that is heated by an electrical current until it glows.

The theoretical construct used to model incandescence is the blackbody, so called because an ideal blackbody absorbs all electromagnetic radiation hitting it (thus is it black, in the sense that it *reflects* no light at all, although it does *emit* light). The blackbody was a central concept in the development of quantum theory, but I will only quote the results of those towering investigations. At thermal equilibrium with its surroundings, the spectrum of EM radiation emitted by a blackbody is described by Planck's Law, as a function of wavelength, as

$$B_\lambda(\lambda, T) = \frac{2hc^2}{\lambda^5} \frac{1}{e^{\frac{hc}{\lambda k_B T}} - 1}, \tag{14.2}$$

where c is the speed of light, h is Planck's constant, k_B is the Boltzmann constant, and T is the absolute temperature in K; B_λ has units W^{-1} sr^{-1} m^{-3}. The hotter a blackbody, the greater and whiter its emission spectrum. This is salient, as both incandescent bulbs and stars, such as the Sun, are both well-modeled as blackbodies. The temperature at the top of the sun's photosphere is 6,600 K, while the *effective* temperature of the sun's surface is 5,778 K [597], and thus the sun's EM spectrum peaks over the visible range of light, as shown in Figure 14.6. The hotter the filament in an incandescent bulb, the more closely the emission spectrum resembles that of the sun. Tungsten is widely used as a filament material because of its high melting point (3,680 K), and resistance to evaporation. The temperature of incandescent bulbs varies between 2,400 K (standard incandescent lamps) to 3,000 K ("warm white" lamp).

The spectrums of several incandescent bulbs are given in Figure 14.6; this and Figure 14.7 also demonstrate the extremely low luminous efficiency of these bulbs, which is 1%, 1.9%, and 3.2% for 2,400, 2,700, and 3,000 K bulbs, respectively (also shown for comparison is a candle flame, which gives a remarkably low 0.046% luminous efficacy; candles are discussed more in Section 14.5); these figures translate into luminous efficacies of about 7, 13, and 22 lumens/W. As an aside, note that many sources will state something along the lines of, "less than 5% of the energy used in incandescent bulbs is converted to light, while the rest is lost as heat." This is basically correct, but can be confusing: as electrical energy is passed through a filament, it acts to heat that filament; this heat is then converted to EM radiation, with a spectrum determined by the temperature of the filament. When this radiation strikes surrounding matter, it is converted back into heat, and thus all EM radiation from a lamp ultimately goes towards heating the surroundings, via "thermal radiation." However, some fraction of this EMR also induces a photopic response in human eyes ($< 5\%$ for incandescents), which is the logic underpinning statements such as those above.

Incandescent bulbs have been widely used because they are cheap and easy to manufacture, do not require electric current regulation, and work with either AC or DC current. Their major

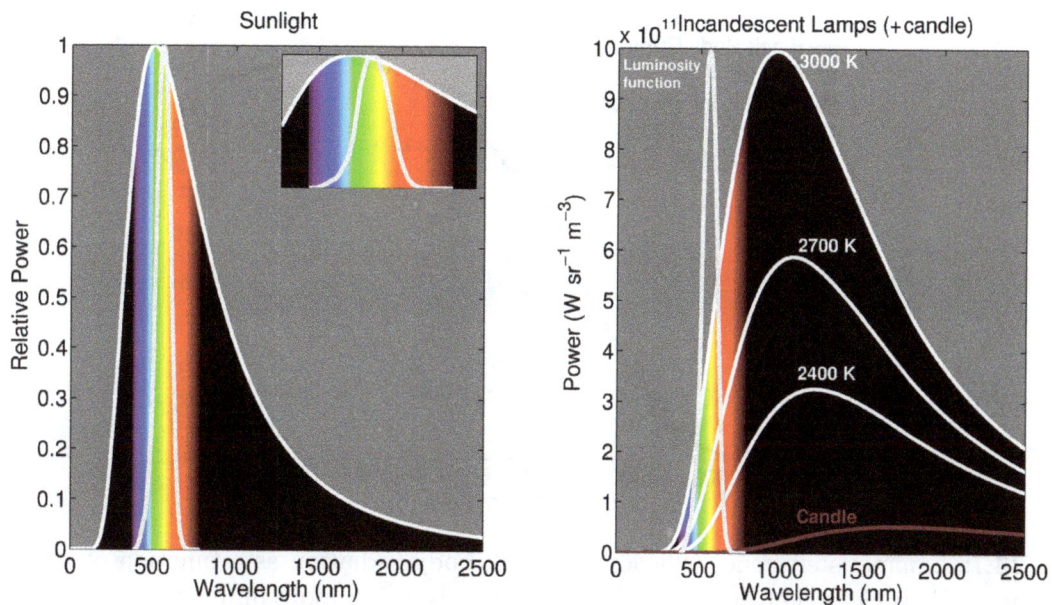

Figure 14.6: Spectrums for sunlight and several incandescent bulbs, as modeled as ideal black bodies. The left panel gives the normalized sunlight spectrum, with the photopic luminosity function inscribed for reference (this is shown in close-up in the inset). The right panel gives the absolute (i.e. not normalized) spectra for several incandescent bulbs, as well as the spectrum of a candle flame (1,673 K); the hotter the bulb, the more efficient it is.

drawback, of course, is their low luminous efficacy. As I discuss in the next section, several alternative technologies achieve much better luminous efficacy, with fluorescent lighting and LEDs by far the most significant high efficiency lighting sources used at the residential scale.

14.3   Major classes of light bulbs

The classical incandescents were just discussed at length in the prior section; here I present the basic principles and characteristics of the two major alternative technologies: gas-discharge lamps (e.g. fluorescents) and LEDs, both of which are far more efficient than incandescents.

14.3.1   Gas-discharge lamps: fluorescents and friends

Fluorescent lights achieve a luminous efficacy in the 50–100 lumen/W range, with 65 lumens/W common for commercial CFLs; straight fluorescent bulbs are similar, although some achieve up to 100 lumens/W. Thus, the average CFL consumes one-fifth the energy of a 2,700 K incandescent bulb for the same luminous flux.

General principle, and fluorescents

In general, gas-discharge lamps operate via electrical discharge through an ionized gas, exciting the atoms in the gas and leading to EM radiation. The gas within these lamps is generally some noble gas, e.g. argon, xenon, or neon, along with vaporized mercury, sodium, or a halide. Fluorescent lights are the most important and widely used of the gas-discharge lamps, and, as the reader may know, contain mercury. These lamps use a noble gas and vaporized mercury mixture contained at very low pressure (0.3% of atmospheric), and electrical discharge induces the emission of high energy UV light, invisible to the eye. The short-wave UV light is absorbed

345

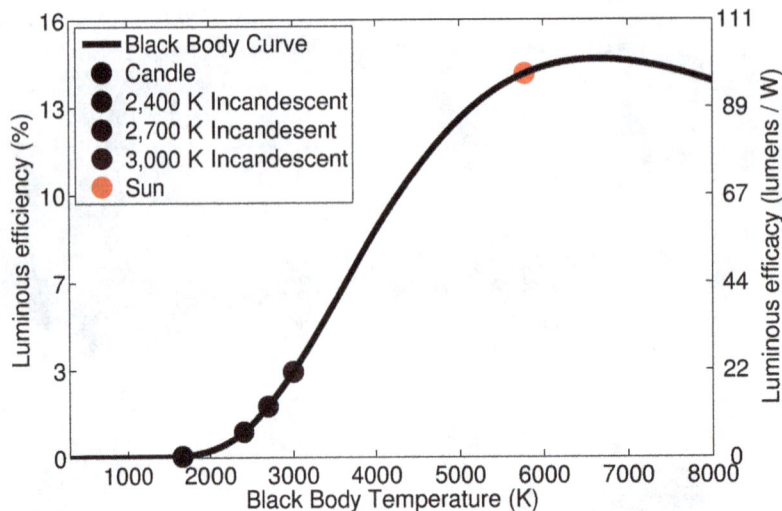

Figure 14.7: Luminous efficiency/efficacy for black body radiation as a function of the body's temperature. It is clear that hotter bodies are generally much more luminous.

by a fluorescent *phosphor* coating the bulb, which then in turn emits radiation as longer-wave visible light ("fluorescence"). This is a form of *luminescence*: as opposed to incandescence, which is the emission of light due to heat, luminescence is the emission of light not due to heat, but to a variety of other mechanisms, and can be much more efficient than incandescence.

Because most of the visible light emitted from a fluorescent light comes from the phosphor coating, visible light is emitted in several narrow wavelength bands, giving an emission spectrum with multiple narrow peaks, much different from the smooth black body spectra. An approximate fluorescent bulb spectrum is shown in Figure 14.9, along with its relation to the photopic luminosity function. Fluorescent lights are more complicated than incandescents, largely because they require a *ballast* to limit electrical current through bulb: because fluorescent bulbs develop a *negative* electrical impedance, plugging directly into the wall would lead to catastrophically high current. Ballasts can be either magnetic (of lower efficiency and responsible for the hum and flicker seen in older bulbs) or electronic.

While the luminous efficiency of a typical fluorescent lamp spectrum may be around 50%, actual fluorescent lamps achieve a luminous efficiency of around 7–15%, i.e. luminous efficacy of 50–100 lumens/W, for several reasons. First, magnetic ballasts can slightly reduce efficiency. More fundamentally, some energy is lost in the electrical discharge, not all UV light emitted from the discharge is absorbed by the phosphors (quantum efficiency), and most importantly, energy is lost when high energy UV light is absorbed and re-emitted as lower energy visible light, a phenomenon known as the Stokes shift. Most energy losses are due to electrical discharge (30–40% loss) and the Stokes shift (roughly 55% loss). Straight fluorescent lights lose slightly less energy to electrical discharge, and thus can be slightly more efficient than compact fluorescent bulbs [598].

Sodium-vapor lamps, etc.

Briefly, sodium-vapor lamps emit light directly, without the use of an intermediate phosphor, and thus are very efficient light sources. Low-pressure sodium lamps are extremely efficient, with a luminous efficacy approaching 200 lumens/W [603]. However, they emit unpleasant, nearly monochromatic yellow light, that can only be (practically speaking) used for street lighting. High-pressure sodium lamps are not quite as efficient, but still achieve an excellent 95–140

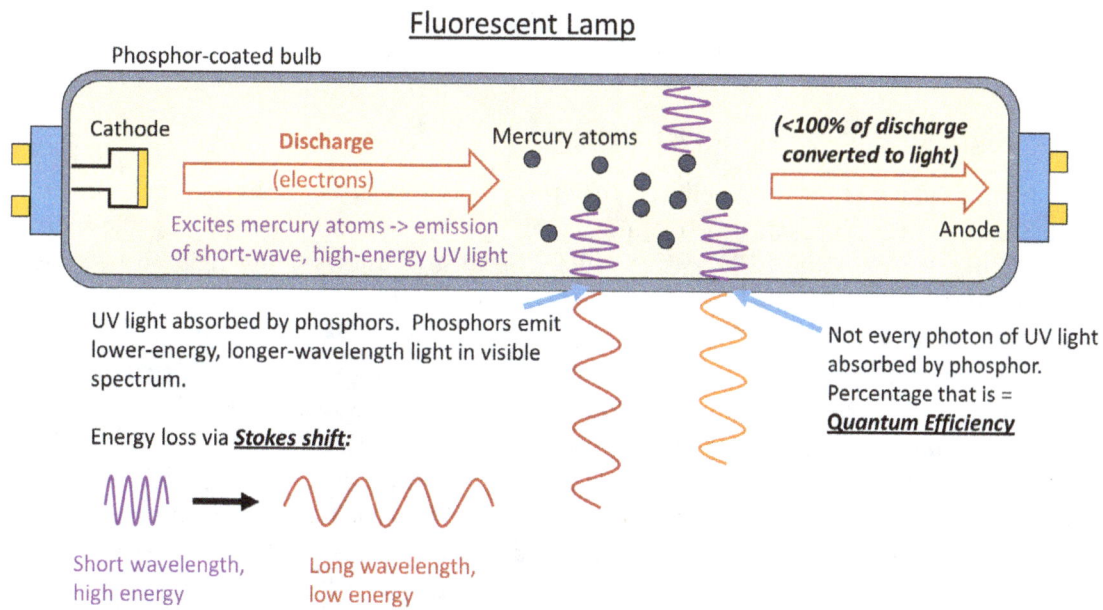

Fluorescent Lamp

Phosphor-coated bulb

Cathode

Discharge

(electrons)

Excites mercury atoms -> emission
of short-wave, high-energy UV light

Mercury atoms

*(<100% of discharge
converted to light)*

Anode

UV light absorbed by phosphors. Phosphors emit
lower-energy, longer-wavelength light in visible
spectrum.

Energy loss via *Stokes shift*:

Short wavelength,
high energy

Long wavelength,
low energy

Not every photon of UV light
absorbed by phosphor.
Percentage that is =
Quantum Efficiency

Figure 14.8: Schematic for the basic operating principle of fluorescent lamps. The bulb contains a mix of argon gas and vaporized mercury, and an electrical discharge generated at the cathode excites vaporized mercury atoms, causing them to emit high energy ultraviolet light. Some fraction (the quantum efficiency) of this UV light is absorbed by phosphors coating the bulb, which re-emit longer-wavelength light in the visible spectrum. Efficiency loss occurs at this step via the Stokes shift; efficiency loss also occurs because not all energy expended in the discharge is converted into UV light.

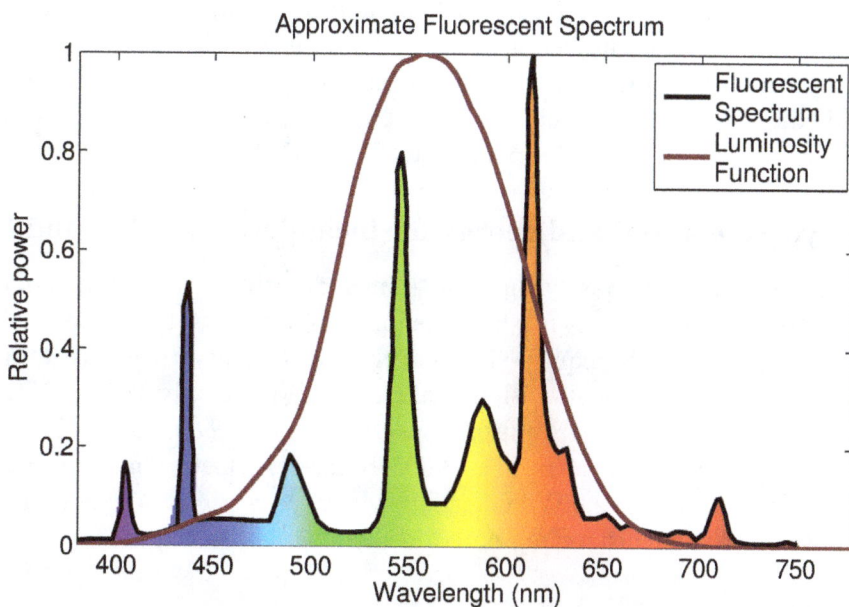

Figure 14.9: Approximate spectrum of a fluorescent lamp, with the photopic luminosity function inscribed. Note that this spectrum gives a theoretical maximum luminous efficiency of over 50%, but actual efficiency is no more than 15%, due to the energy losses elaborated in Figure 14.8.

lumens/W; about 85% of street lighting is high-pressure sodium. Mercury vapor lamps are another light source primarily used in road lighting, but are largely being phased out due to their lower luminous efficacy (36–54 lumens/W [603]).

14.3.2   Light-emitting diodes (LEDs)

Light-emitting diodes are widely believed to represent the future of lighting. Indeed, the 2014 Nobel prize in physics was awarded to three Japanese researchers for developing the blue LED in the early 1990s. The blue LED was so crucial because it allows white light to be produced in combination with previously existing red and green LEDs. The Nobel press release stated "Incandescent light bulbs lit the 20th century; the 21st century will be lit by LED lamps." This choice was not without its controversy, and justifiably. A blue LED had in fact been invented much earlier by researchers at RCA in 1972, and the award disregards those scientists responsible for the far more fundamental advance: the invention of the LED itself. Several researchers in the US independently developed red semiconductor LED lasers in 1962, forming the basis for most work to come, but it was a Russian scientist by the name of Oleg Lasov who truly discovered the LED, as described in a series of papers published between 1924 and 1930. Alas, but Lasov's work languished in obscurity, and in 1942 he perished in the Seige of Leningrad [599].

Now, let us turn to the technology itself. LEDs are based on p-n junction semiconductor technology, and are essentially solar cells acting in reverse; one may consult Section 4.13 for an extensive discussion on semiconductors and solar cells. Very briefly, two wafers of silicon are pressed together and *doped* (impurities are added) such that one side (n-type) has an excess of electrons, while the other (p-type) has an excess of "holes," i.e. silicon atoms with unpaired electrons that can act as charge acceptors. When an external voltage is applied across the semiconductor, electrons from the n-type region are forced into the p-type silicon (and holes migrate from P to N), where they recombine with holes to release photons (in solar cells, incident photons create an electron-hole pair, compared to LEDs where an electron-hole pair combines to create a photon). The light emitted by an LED is nearly monochromatic, and white bulbs are constructed either by combining red, green, and blue LEDs or by coating a single-color LED with a phosphor, similar in principle to the operation of a fluorescent bulb. Typical commercial LEDs are quite efficient, with a luminous efficacy on the order of 90 lumens/W. For example, 9 W bulbs generating 800 lumens are a popular 60 W incandescent replacement model.

14.3.3   Lifecycle emissions and energy for incandescent, CLF, and LED lamps

While CFLs and LEDs are both vastly superior to incandescent lamps in terms of direct energy use, as always we should consider the full lifecycle impacts of these different technologies. While it is true that CFLs and LEDs require significantly more energy to manufacture, use-phase energy/emissions are still preeminent, with manufacturing adding only about 4.5%, and perhaps 10%, to CFL and LED lifetime energy use, respectively, per a DOE analysis [596]. In other words, a 13 W CFL is roughly equivalent to a 13.6 W bulb, when factoring in upstream emissions, and a 9 W LED is equivalent to a 9.9 W bulb. This difference is trivial, and does not appreciably affect any calculations or conclusions presented in this chapter.

14.4   Roadway and parking lighting

14.4.1   How much electricity goes towards roadway lighting?

> - Roadway lighting electricity demand is uncertain, but likely consumes 30–100 billion kWh per year, translating into about 20–70 million $MgCO_2e$, and 30 million $MgCO_2e$ is used as a conservative estimate.

Survey results

It is unclear how much electricity is used for roadway and parking lighting on a national scale. One DOE report [600] estimated 44,882,000 roadway lamps in the US, with an average wattage of 221 W. If on for 12 hours per day, this translates into 43.4 TWh per year, although the report gives a total of 51 TWh for roadway lighting. On the other hand, a 2011 DOE report [601] estimated street lights and highway lights to consume 23.1 and 29.7 TWh, respectively. The street light estimate was based on data from 25 local municipalities, while the highway estimate was determined from the total number of highway miles and average roadway spacing between lamps. However, since most highway lighting is the responsibility of the local municipality, this methodology may have double-counted a great deal of highway lighting [602].

Early results from a DOE survey of 240 municipalities, utilities, DOTs, etc., showed almost exactly 1 lamp per 10 residents [601]; at 221 W per lamp, this would imply 30.9 TWh. Electricity use was reported at 5.7 TWh by 137 respondents; since municipalities and utilities covered about 100 million residents and owned 83% of lamps in the survey, a very crude scaling up would suggest 26.43 TWh for nationwide roadway lighting. Furthermore, a recent report by the New York State Energy Research and Development Authority inventoried street lighting and estimate 0.99 billion kWh for lighting in that state [602], and scaling by population suggests 16 billion kWh for street lighting nationally. Thus, something in the 20–50 billion kWh range for total US roadway lighting is reasonable.

Bottom-up calculations.

As elaborated in Section 9.2.1, we can very roughly estimate that there are about 6 million paved lane-miles equating to 16,000 square miles of paved roadway in the US, and perhaps 9,000 square miles of parking infrastructure. Recommended minimum illumination levels for roadways vary by road classification and surfacing (more reflective concrete surfaces require less lighting than darker asphalt surfaces), but are around 3–9 lux for local roads, i.e. residential and collector roads, 8–17 lux for larger arterial roads, and 6–12 for freeways (there is a general trend for more illumination on larger roads up to freeways, which are a bit lower). Local roads make up about 50.8% of all paved lane-miles (I assume all unpaved roads are local and lack street lighting), and collector roads account for 27.1%, arterials and freeways make up 18.5%, and the final 3.7% is interstate. If we assume 6 lux for local, 8 lux for collector, 12 lux for arterial roads, and 8 lux for interstate, we get a weighted average of 7.72 lux. If we assume that roads are primarily lit by high pressure sodium lamps with a luminous efficacy of 90 lm/W and a downward luminaire efficacy of 70%, we arrive at 3,687 kWh/lane-mile/year (for a lane 12 feet wide with 2 feet of shoulder), and a grand national total of 22.1 billion kWh/year, on the low range of the estimates cited above.

What about parking lighting?

A DOE report [600] estimated parking lighting at 52,168,000 fixtures drawing, on average, 153 W, which would translate, at the reported 16 hours of use per day, into 46.61 billion kWh/yr, while the report gives 52 billion kWh/year (35.5 million MgCO$_2$e/year). A rough bottom-up calculation suggests lower energy use: I assume parking areas are illuminated at an average of around 15 lux (A Minnesota DOT manual [603] recommends 11 lux for parking areas, while the City of Los Angeles recommends 22 lux [604]). We can very roughly estimate 8,880 square miles of parking in the US (see Section 9.2.2), and supposing lamps with luminous efficacy of 90 lm/W and downward efficiency of 70%, this translates into just 11.64 billion kWh (7.9 million MgCO$_2$e). Thus, parking and roadway lighting together probably generate at least 25 million MgCO$_2$e/year, and up to 70 million MgCO$_2$e/year.

14.5   Candles

- Per lumen of light, paraffin wax candles generate 20 times more emissions than incandescent lighting, and 100–150 more emissions than high efficiency (CFL or LED) lights.

- Beeswax candles emit 60–90% less than paraffin candles over their lifecycle, but are still much worse than electric lighting on a per-lumen basis.

- Compared to a single 60 W incandescent bulb, 13 W CFL, and 9 W LED, a single wax candle runs at 40 W and its emissions are 30% of the incandescent, 136% of the CFL, and 197% of the LED.

- Candles, even beeswax candles, are high-carbon lighting sources and are not a feasible alternative, at any meaningful scale, to modern electric lighting.

14.5.1   Paraffin candles

Candles are occasional considered a green alternative to electric lighting. A prominent example has been Earth Hour, a one-hour lights-off event, during which candles are often burned instead. Now, the vast majority of candles are made of paraffin, a petroleum product derived from oil refining and very similar to kerosene, which is widely used as a lighting fuel in the developing world. I have previously estimated the life-cycle emissions of kerosene to be 0.3035 kgCO$_2$/kWh, and I will assume paraffin to be the same. Paraffin has a density of around 900 kg/m^3, and paraffin candle flames burn at a maximum of 1,673 K [605]. As discussed in Section 14.2.2, for lighting purposes this is a very low temperature, and the luminous efficiency of such a flame is less than 0.05%, about 40 times less than a standard incandescent bulb, and around 300 times less than typical LEDs.

Let us analyze the carbon emissions for a functional unit of one lumen-hour. A paraffin candle has a luminous efficacy of 0.3135 lumens/W, and so we require 3.1898 Wh to get a lumen-hour. Using the EF for kerosene above, this brings us to 0.9681 gCO$_2$e/lumen-hr; this does not include the lifecycle emissions from candle manufacture, packaging, shipping, etc., so let us simply round up to 1 gCO$_2$e/lumen-hr as a lower-end estimate. Using US grid-average electricity, a 13 lumen/W incandescent lamp (2,700 K), will yield 0.0527 gCO$_2$e/lumen-hr, while a 65 lumen/W CFL and 95 lumen/W LED would yield 0.0106 gCO$_2$e/lumen-hr and 0.0072 gCO$_2$e/lumen-hr, respectively. Thus, the candle generates at least 19 times the carbon emissions of an incandescent, and 139 times emissions of an LED, on a per-lumen basis.

It follows that even if one were willing to accept only one tenth the illuminance with candle light versus electric light, emissions would still be twice as high compared to inefficient incandescent lighting, and about 10–15 times as high compared to high efficiency electric sources (CFL or LED). Note that the above calculations are nearly the same if we factor in the manufacturing-associated emissions for CFLs and LEDs.

14.5.2 Beeswax candles

The calculations above are for paraffin wax-based candles, but beeswax candles are sometimes promoted as a green alternative that is carbon neutral. However, there are still significant emissions associated with agricultural activities, including commercial honey production. Beeswax is a co-product of honey production, with the two produced in a fixed ratio. A 2010 report by UC Davis researchers estimated carbon emissions associated with honey production to be 0.43–1.39 kgCO$_2$e/kg honey [606]; I assume beeswax is similar. Note that the low-end estimate corresponded to a small "backyard" producer, while higher estimates were obtained for larger commercial operations.

To get emissions per lumen, we have a net energy content of 11.97 kWh/kg for paraffin wax [605], and assuming beeswax is similar, then each lumen-hour of light ultimately corresponds to 0.2665 grams of beeswax or paraffin (using 0.3135 lumens/W). Therefore, we would have in the range of 0.11–0.37 gCO$_2$e/lumen-hr for beeswax candles, about 60–90% lower than paraffin candles, with 0.20 gCO$_2$e/lumen-hour as a reasonable mid-point estimate (80% lower than paraffin wax). Thus, beeswax candles are probably much better than paraffin wax candles, but likely still several times worse than incandescent lighting and 10–50 times worse than CFLs or LEDs (on a per-lumen-hour basis).

14.5.3 A national perspective

Since over 1 billion pounds of candle wax (almost all paraffin) are used yearly in the US (per the National Candle Association), this corresponds to about 5.43 billion kWh of energy and 1.70 million MgCO$_2$e. This is 5.34 kgCO$_2$e per capita, equivalent to almost half a gallon of gasoline. Note that the US produced only 1,600 Mg of beeswax in 2012 (FAO), or 0.35% of candle wax demand, and so beeswax is not a viable large-scale alternative even to this already niche use of petroleum. Global beeswax production (64,688 Mg in 2012, per the FAO) still amounts to only 14% of US candle wax demand.

Finally, the average residence uses on the order of 31.5 million lumen-hours (this is only a rough estimate, calculated assuming 65 lumens/W for fluorescent bulbs, 14 lumens/W for other bulbs, and 1,708 kWh per annum for lighting, 8.8% of which is from fluorescents; this corresponds well to a DOE estimate of 16,300 lumen-hours per square foot, and using 1,971 ft^2 per house based on the 2009 RECS). To generate all these lumens with candles would require 8.4 tonnes of candle wax and generate 31.5 MgCO$_2$e, almost three times the emissions from all residential energy use. Therefore, although candles are a meager source of emissions nationally, using them to displace electric lighting would be a profoundly irrational strategy for carbon reduction, and one that could not possibly be meaningfully scaled up.

14.6 CFLs and mercury

Fluorescent lamps contain a small amount of mercury (Hg), with most bulbs containing less than 4–5 mg of mercury, while some only contain 1–2 mg Hg [608]. This puts into question the environmental efficacy of these bulbs, as a small amount of mercury can be released if a bulb is broken or discarded instead of recycled. However, as coal power plants are currently the major source of atmospheric mercury pollution, the electricity savings of CFLs generally reduce net mercury release.

The mercury content of CFLs varies by brand, and has fallen over time, but we may safely assume no more than 5 mg Hg per bulb. In general, mercury from discarded CFLs can leach or vaporize out of landfills, it may be released into the air from municipal solid waste incinerators, and broken bulbs release significant amounts of mercury vapor.

Li and Jin [608] found only a small fraction of Hg in CFLs to be leachable (<4% in new bulbs, but up to 15% in used bulbs), and it is likely that only a very small fraction of mercury escapes landfills [610]. Of more concern, a significant fraction of CFL mercury, perhaps close to 100% for some bulbs, can vaporize from broken bulbs, although complete vaporization takes several months. Aucott et al. [611] estimated 17–40% of mercury is released from broken bulbs over two weeks, with 25% a best estimate. Johnson et al. [609] found 2–12% of mercury in new bulbs to vaporize within 24 hours, with 30% of a new bulb's mercury released within four days. Much less mercury evaporated from used bulbs compared to new bulbs in several studies [609].

Overall estimates have 6.6 to 30% of mercury in fluorescent lights making it into the environment; one comprehensive estimate is 13% [610]. If we assume that 15% of mercury in every CFL ultimately enters the wider environment at the end of life, and that there are 5 mg Hg per bulb, then 0.75 mg of mercury is released per bulb.

A survey of the 100 largest power producers in 2010 [607] found an average mercury emissions rate of 0.0136 mgHg/kWh for coal power generation. Now, comparing a 13 W CFL with an 8,500 hour lifetime [596] to a 60 W equivalent incandescent bulb, the incandescent bulb results in an additional 399.5 kWh of electricity, and an additional 2.032 mg Hg released into the atmosphere (assuming 37.4% coal generation, using the 2012 eGRID database). Thus, on balance the CFL reduces mercury pollution by over 60%.

An analysis by Eckelman et al. [610] also concluded that CFLs generally reduce mercury emissions appreciably, although they may slightly increase mercury emissions in those few states with very low coal use. As a caveat, these estimates are all dependent on uncertain mercury release from discarded CFLs, and moreover will vary with the actual mercury content of the bulb.

Mercury pollution from CFLs can be eliminated almost entirely by recycling, yet in the US, only about 20% of fluorescent bulbs are properly recycled [610]. CFL recycling is available for free at a number of Big Box chains, and therefore is accessible to the vast majority of consumers. For the individual concerned about mercury and their personal use of CFLs, they need only recycle.

Chapter 15

Other residential energy: refrigerators, cooking, washing, phantom, and miscellaneous loads

A variety of individually minor end-uses of household energy, not already considered separately, sum up to a significant amount energy consumption and carbon emissions. Some of these particular loads, including refrigeration, cooking, clothes washing (most especially drying), phantom loads, and general electric loads such as televisions and other electronics (which are becoming a much more important category of electricity use) are discussed in this chapter. A general theme is that, for each, the numbers involved are not particularly impressive, but collectively such end-uses sum to around one-third of all carbon emissions related to residential energy use, or about 4 $MgCO_2e$ per household, on average. Therefore, taking the effort to achieve relatively small savings (in the absolute sense) across all these categories can yield significant energy/emissions savings in the aggregate. I briefly discuss some of these consumption categories in the remainder of this chapter.

15.1 Refrigerators

- In 2009, the mean household had 1.25 refrigerators that used about 1,250 kWh/year (850 $kgCO_2e$), while the median household had a single refrigerator and used almost 1,000 kWh/year (680 $kgCO_2e$). This translates into about 500 kWh/person/year, or about 340 $kgCO_2e$/person/year.

- From the 1980s through 2001, federal efficiency standards led to a sharp drop in refrigerator energy use. A new full-size fridge uses about 450–700 kWh/year. Smaller fridges use less energy overall, but compact refrigerators, especially, are much less efficient per unit volume.

Refrigerators and freezers, essentially air conditioned boxes, are significant users of electrical energy, responsible for about 11% of US household electricity use in the 2009 RECS (1,249 kWh/HH), and just over 7% of CO_2e emissions (852 $kgCO_2e$/HH) attributable to residential energy use; this is a somewhat declining share compared to earlier years, as refrigerators continue to get more efficient. As usual, there is significant variability in refrigeration electricity, with the top 20% of households using twice the mean, while the bottom 20% consume just under 40% of the mean (this is not attributable to a lack of refrigerator, as virtual no households go without this appliance). It is noteworthy that, at the global scale, domestic refrigeration has expanded dramatically in recent years, especially in China and other urbanizing developing countries [612], with attendant increases in energy use and carbon emissions. However, the impact of this expansion is not wholly negative, as it significantly reduces food spoilage and waste [612], an environmental good at multiple levels (land use, energy, etc.).

In the 2009 RECS, about 80% of households had a single refrigerator (the others had two or more), and these used almost exactly 1,000 kWh/year. Since the 1980s, a series of federal efficiency standards in 1990, 1993, 2001, and most recently, 2014, have more than halved refrigerator energy usage relative to the 1980s [613]. The most recent standards reduce energy consumption by about 25% compared to 2001, depending upon model characteristics, and absolute energy use for a new (full-size) refrigerator is on the order of about 400–750 kWh/year.

All else equal, smaller refrigerators use less energy, and federal efficiency standards are defined as a linear function of adjusted volume in cubic feet, for each refrigerator class. However, the slope of this efficiency line is rather small, and so particular model characteristics are generally more important than size. An efficient compact refrigerator (less than 7.75 ft^3) may use on the order of 200–250 kWh/year, but typical units are 2.4–4.5 ft^3, implying far lower efficiency per unit volume than full-size units, which are about 17–25 ft^3, and very inefficient or older compact units may actually use as much energy as a new full-size fridge.

While the manufacture of a new refrigerator is an energy- and carbon-intensive endeavour, as with most energy-consuming items (vehicles, furnaces, etc.) operational energy/emissions are dominant, at perhaps 90% of lifecycle energy [613], and it is almost always wise to replace an old or (especially) deteriorating refrigerator: deteriorating components, such as seals, can markedly increase energy consumption. A 2006 study by Kim et al. [613], concluded that the optimal time to replacement for an inefficient fridge manufactured in 1985 would be just 2–4 years, on a GWP basis. However, for a fridge manufactured within roughly the last 10 years (i.e. 2007 or later), the optimum time to replacement was estimated at 7–11 years, reflecting the fact that efficiency improvements over the last 15 years are much smaller, in both the relative and absolute senses, than in the preceding years when efficiency standards were first phased in. Given this, it is quite reasonable to operate any relatively new Energy Star fridge over its useful life, but any "legacy" fridges should be immediately disposed of.

15.2   Cooking and kitchen appliances

- Americans eat about two-thirds of meals at home, and preparing a typical meal generates any-where from one-third to several kgCO$_2$e. Overall, this sums to about 500 kgCO$_2$e annually for households that use electricity as their primary cooking fuel, and a slightly lower 400 kgCO$_2$e for those that use gas (such households presumably still use microwaves and other electric accessories); these figures equal about 200 and 160 kgCO$_2$e on a per capita basis.

- Simmering, keeping lids on pots, and other simple strategies to avoid wasting heat can significantly reduce cooking energy.

The direct act of cooking, that is, the putting of fire or microwave to meat, not counting the chain of refrigeration, dishwashing, and the rest of the food chain, is a minor source of residential energy use and emissions, at perhaps 2–5% of either category (with 3–4% likely typical). Sensible cooking strategies, such as appropriate technology selection, and especially simply using a lid on pots and simmering just below boiling, can significantly reduce cooking energy. On a per-meal basis, associated cooking emissions are probably typically in the 0.33–1.0

kgCO$_2$e range. It follows that even one or two miles of driving to eat out (in an average vehicle), would yield equivalent emissions.

15.2.1 Overall cooking energy

First, let us get some idea of how much energy goes towards this end. As reviewed by Parker et al. [614], several household energy monitoring studies carried out by utilities in the late 1980s measured between 300 and 656 kWh for electric stoves/ovens. For gas stoves, which are less efficient at transferring heat, Parker et al. report that 92 CA homes averaged 938 kWh of gas energy. Assuming an average thermal efficiency (i.e. the fraction of energy imparted to food, and not lost to the air) for electric stoves of 74%, compared to 40% for gas ranges [615], 938 kWh of gas is equivalent to about 500 kWh of electric energy, right within the range for electric stoves.

We can also approach the issue with a few back of-the-envelope calculations. Based on a review of commercial products, most stovetop heating elements pull between 1.2 and 3.2 kW (at maximum), while ovens typically draw 2.4 to 4.8 kW. Gas burners are slightly more powerful, running roughly from 1.5 to 6 kW, with about 3 kW typical.

Now, running an electric stovetop element at, say, 2.0 kW for 30 minutes a day, and using the oven for an hour at 2.4 kW once weekly would translate into 489.8 kWh yearly. From this and the above, reasonable rules-of-thumb for stove/oven use might be 500 kWh for electric users, translating into 343 kgCO$_2$e, and 925 kWh for those favoring gas, equal to 242 kgCO$_2$e.

Additionally, one study [616] estimated microwaves to use 131 kWh/year, toasters to use 39 kW (present in 90% of households), toaster ovens to consume 33 kWh (present in 60% of HHs), and coffee machines to use 61 kWh/year (also present in 60% of HHs). Weighting by their penetrance, this gives an additional 222.5 kWh of electrical energy on average, or 152.8 kgCO$_2$e.

To summarize, a typical electric-only household might use 750 kWh for all cooking purposes in a year, equating to a little over 0.5 MgCO$_2$e. Gas-powered households still would use 225 kWh ofelectricity from the microwave, etc., and an additional 1,000 kWh of gas heat. Together, this adds to just over 0.4 MgCO$_2$e. Thus, there is a small (at least in the short-term) advantage to gas-based cooking, despite the fact that gas stoves have a lower thermal efficiency than electric.

15.2.2 Electricity vs. gas

Electricity is the dominant source of cooking heat: in the 2009 RECS, 54.5% of households used an electric stove, while 35.5% used either natural gas (NG) or propane (10% did not use a stove). As with space and water heating, when it comes to stoves and (conventional) ovens, electricity is more efficient at transferring heat at the point of use, but less efficient in terms of primary energy. As a consequence, natural gas has, overall, a somewhat lower carbon emissions-intensity. Note that, compared to conventional ovens, the higher end-use efficiency of electric microwave ovens may be five times that of conventional ovens, in which case they are greatly preferable.

Adjusting for the increased thermal efficiency of electric (74%) vs. gas (40%) [615], we have that, per kWh of heat energy delivered to food, electric ranges emit 0.93 kgCO$_2$e (or 1.25 kgCO$_2$e, using a marginal emissions factor), while gas is 30% (48% marginal basis) lower at 0.65 kgCO$_2$e.

15.2.3 The microwave oven

A report on miscellaneous electrical loads [616] concluded that the average microwave is operated 70 hours a year, drawing 1,500 W when active, and 3 W when idle. This translates into 131 kWh/year (89 kgCO$_2$e), 20% of which is attributable to the idle mode. Thus, anticipating our discussion of phantom loads, unplugging one's microwave when not in use could save about 26 kWh, or 18 kgCO$_2$e (equivalent to about 1.5 gallons of gasoline).

15.2.4 Per-meal emissions

In 2007/2008, Americans derived roughly two-thirds of their calories from their home food supply [617]. If we naively assume, then, that each household cooks two meals a day, we get about 0.3–0.7 kgCO$_2$e per meal, overall. Obviously, meals very widely in their cooking intensity, and I would guess that dinners may be closer to 1–2 kgCO$_2$e, with other meals much lower. For example, a meal that required ten minutes of microwave time (at 1 kW), 30 minutes of electric stove time at medium heat (say, 1.2 kW), and 45 minutes of electric oven (at 2.4 kW), would use 1.97 kWh of electricity, corresponding to 1.35 kgCO$_2$e. A more elaborate holiday dinner that logs many hours of oven and burner time could easily use tens of kilowatt hours.

15.3 Clothes washing and drying

Clothes washing and drying represent a minor, but still significant fraction, portion of household electricity (or gas) use, and hence carbon emissions. Energy for the mechanical operation of washing machines is essentially trivial, although upstream energy to heat washing water may be significant (as discussed in Chapter 13), and hot water for washing should generally be avoided. Most other energy in the washing/drying process is consumed by the dryer, and there is little variation between dryers in their energy use, although (relatively uncommon) gas-powered dryers have appreciably lower lifecycle emissions. Fortunately, drying racks and clotheslines are simple solar-powered alternatives.

Typical household energy use from dryer operation likely amounts to 700–900 kWh/year, based on DOE estimates summarized in a 2011 EPA Energy Star scoping report [618], and estimates from a 2011 report commissioned by the NRDC [619]; the latter report estimated an average of just under 800 kWh/year, and in aggregate about 6% of household electricity is consumed drying laundry. Just under 80% of households have clothes dryers, and I suspect those that don't rely strongly upon commercial laundromats.

Briefly, DOE estimates for a standard electric dryer (4.4 ft^3 or greater capacity) range from 684 kWh/yr for 283 cycles per year at a combined efficiency factor (CEF) of 3.73 lbs clothes/kWh (2.42 kWh/load, 9.0lb load) to 967 kWh/yr for 416 cycles at an efficiency factor of 3.01 lbs clothes/kWh (2.32 kWh/load, 7.0lb load). NRDC test results [619] suggested energy use of 747–778 kWh/year for full-size electric dryers, assuming 260 loads per year, i.e. 5 loads/week, and corresponding to 2.87–2.99 kWh/load. Note that the NRDC report found drying times to be 35% longer when using a "real world" mix of fabrics, compared to standardized test sheets used in DOE tests. A separate study by Keoleian et al. [624] cites 875 kWh per year, based on 1995 EIA data, for clothes drying.

A quick bottom-up calculation that assumes a load size of 7–9 lbs at an energy factor of 3.01 lbs/kWh gives 2.33–2.99 kWh/load, which corresponds well to the above direct estimates. Therefore 2.5–3.0 kWh is a reasonable range for energy use per load, with corresponding emissions factors of 1.71–2.05 kgCO$_2$e/load. If a household dried five loads per week, at 3 kWh/load, then 780 kWh of electricity would be consumed, with corresponding CFs of 0.532 MgCO$_2$e/household and 0.209 MgCO$_2$e/capita.

Gas dryers are generally similar to electric dryers in energy use, although minimum CEFs are about 5–15% lower than for electric dryers. However, because direct combustion of gas is far more efficient than generating heat from electrical resistance, gas dryers still typically generate about 60% less CO_2e.

In sum, line-drying all clothes instead of using an electrical dryer would save the typical household roughly 0.5 $MgCO_2e$ (and the typical individual about 0.2 $MgCO_2e$). Gas dryers are a marked improvement over electrical dryers, and are also expected to save money over the life of the dryer. Avoiding over-drying, and using longer drying times at lower heat, can also decrease dryer energy consumption.

15.4 Standby, or phantom loads

- Standby (a.k.a. phantom or vampire) electricity losses account for around 5% of residential electricity consumption, and perhaps 0.4 $MgCO_2e$ for an average household (with high variability).

- Turning off power strips, unplugging unused devices, and using Energy Star appliances would eliminate most standby losses, and most likely save several hundred $kgCO_2e$ per year per household (and around 100–200 $kgCO_2e$ per person).

A series of studies in the 1990s and early 2000s identified standby loads as significant sources of electricity "leakage," accounting for perhaps 5–10% of all electricity use in developed countries [622], and about 1% (or more) of all global carbon emissions. Standby, or phantom, power has been somewhat variously defined, but it is generally defined as the power that is drawn by appliances and devices plugged into a home or business's electrical main but ostensibly "off" or not performing the device's main function, with standby loads typically in the 0.5–30 W range. One study of 10 California homes by Ross and Meier [620] found total household standby power to range 14-169 W, with 67 W the average, corresponding to 123–1,480 kWh annually (587 kWh average). Although small in absolute terms, multiplied across dozens of devices per household, and across billions of households, phantom loads are non-trivial. The One Watt Initiative, an international program launched in 1999 to reduce standby loads for all appliances to <1 W, was originally proposed by Meier in 1998 [621], and has seen significant success, influencing a number of national efficiency standards.

While standby power has fallen for traditional appliances, this per-unit efficiency gain has almost certainly been (at least partially) offset by two trends. First, an increasing number of appliances now have standby modes that previously did not, e.g. coffee makers, rice cookers, washers and dryers, and the absolute number of plugged-in devices continues to grow, especially internationally. Second, the trend towards equipping appliances with internet access and broadcasting capabilities means that such devices must always be on, and may draw fairly significant amounts of power at all times.

As a general rule, the only ways to eliminate standby losses are either to unplug a device, or, more conveniently, connect it to a power strip with an on/off strip and then turn off the strip. A number of related loads can be connected to a single power strip. Judicious use of power strips and keeping rarely used appliances, chargers, etc. unplugged unless needed certainly has the potential to reduce around-the-clock power consumption by several 10 W, and perhaps 100 W or more for some households, translating into several hundred kWh of avoided electricity per year (and possibly ≥1,000 kWh for households with the highest standby loads).

Chapter 16

The construction footprint and the "building-cycle"

It takes a great deal of energy and materials, with high corresponding carbon emissions, to construct residential buildings (and commercial buildings too, of course!), and to outfit them with major appliances, including HVAC equipment, water heaters, refrigerators, and so on. Maintenance, periodic refurbishment, and major appliance replacements also represent nontrivial emissions sources. Taken as a whole, the construction and outfitting of a residence probably generates on the order of 30–60 kgCO$_2$e/square foot of living space (based on [624, 623] and the discussion below), which equates to 0.6–1.2 kgCO$_2$e/SF when amortized over a 50-year building lifetime, as is commonly done.

However, in 2013, the US Census American Housing Survey reported there were 115.852 million occupied housing units, while 925,000 new housing units were constructed, although the latter number had increased to 1.175 million by 2016. In any case, this suggests newer housing can be expected to last on the order of 100–125 years before complete replacement, and thus it seems likely that a lower amortized construction footprint of 0.3–0.6 kgCO$_2$e/SF of living space may be more appropriate. Even so, the construction of a typical new single-family home can generate well over 100 MgCO$_2$e, and construction emissions are probably equivalent to around 5–10 years of operational emissions for a typical household.

16.0.1 Major emissions sources

By mass, buildings are primarily concrete, gravel, lumber (including dimensional lumber and composite board), gypsum board (drywall), asphalt shingles, and steel [624]. The majority of emissions are attributable to wood and concrete, with steel a significant runner up. PVC pipe, gravel, shingles, insulation, and a variety of other materials are more minor, but still significant emissions sources. Therefore, to get a reasonable order-of-magnitude estimate for the construction footprint, we really need take a detailed look only at wood and concrete. While the concrete lifecycle is reasonably straightforward (and presented in more depth in Section 9.2.1), the picture is more complicated for wood, due to the deleterious influence of logging on forest carbon dynamics.

Lumber

Lumber emissions factor. Many lifecycle inventories for lumber have examined only the energy/emissions involved in extracting and processing trees, yet the carbon footprint (and broader ecological footprint) of lumber depends far more crucially upon how logging affects forest carbon dynamics. This holds true for other uses of wood, including firewood, biomass-fired electricity, and paper and other wood-pulp products. Long-lived wood products, such as furniture or framing, represent large carbon reservoirs with half-lives on the order of 50–100 years, and so logging for lumber has a somewhat different effect on overall carbon dynamics than logging for either short-lived paper products or fuelwood.

The EPA WARM model [522] has estimated that, for every tonne of avoided roundwood harvesting, forest carbon stocks increase by 0.99 tonnes (over a timescale of a few decades). Dry softwood is about 53% carbon by weight, and lumber used in construction is a long-term carbon stock. Since the ratio of roundwood to finished wood product is about 1.1 (i.e. 10% of the virgin roundwood is lost in processing), each tonne of finished wood corresponds to about 0.53 MgC stored long-term, and 1.09 MgC lost from the forest, for a net balance of 0.56 MgC entering the atmosphere as CO_2, i.e. 2.05 $MgCO_2$. Thus, we have an approximate upstream emissions factor of 2 $kgCO_2e$ per kg of (dry) lumber, due to altered forest carbon dynamics, and accounting for the lumber itself as a long-term carbon sink.

These calculations are congruent with observations on forest carbon dynamics in the Pacific Northwest [625] (and are actually comparatively conservative). In the 1960s, intensive logging removing perhaps 90% of stem wood over a 60-year rotation resulted in net forest carbon losses of 3–4 MgC/Ha/year, and since prior to intense harvesting, stem wood mass was on the order of 100 MgC/Ha (assuming half of live biomass is woody stem), we get about 1.5 MgC/Ha/year in timber harvests. We then arrive at a net loss of at least 1.2 MgC per tonne of finished wood, or an emissions factor of 4.4 $kgCO_2e$/kg of lumber. Consistent with the idea that logging strongly depletes forest carbon, carbon stores in federally managed Pacific Northwest forests have rebounded dramatically following the cessation of intense logging in the early 1990s [625].

Per the EPA WARM model, the energy involved in extracting, milling, and transporting lumber adds around 0.20 $kgCO_2e$/kg for dimensional lumber, while more highly processed composite board may generate 0.43 $kgCO_2e$/kg. Other estimates range from around 0.45–1.25 $kgCO_2e$/kg wood for various timber products [624, 623].

Lumber in residential construction and per ft^2 emissions factor. In 2010, new residential construction averaged 5.6 board feet (BF) of wood per ft^2 (a BF is equal to 1/12 cubic feet) [626], and this translates, assuming that lumber has a density of about 500 kg/m^3, into 6.6 kg wood per ft^2, or about 15–18 $kgCO_2e$/ft^2 (assuming 2.25–2.75 $kgCO_2e$/kg wood). Since older construction used more wood (9.0 BF per ft^2 in 1950) [626], older houses probably embody more like 20–25 $kgCO_2e$/ft^2 due to lumber (although this will generally be more than offset by the smaller overall size of such homes). Note that, because I have included the upstream effects of lumber on forest carbon, my estimates for lumber-associated emissions are appreciably higher than in some other works.

Concrete

By weight, concrete is by far the largest single material category used in residential construction: it is the principal component of the foundation, which may weigh up to several hundred tonnes [624]. I discuss the energy- and carbon-intensive process of portland cement concrete (PCC) production in the context of roadway infrastructure, and for standard strength concrete, our EF is on the order of 0.13 $kgCO_2e$/kg. Foundations are also typically reinforced with steel, which (under typical recycled content) has an EF of about 1.75 $kgCO_2e$/kg [623]; if we assume that the reinforced concrete is 2.5–5% steel by weight, then our new EF is perhaps 0.1738–0.2175 $kgCO_2e$/kg, and I will take 0.2 $kgCO_2e$/kg as a central estimate.

The exact amount of concrete per foundation will vary regionally and with foundation style. Poured concrete slab foundations may be on the order of 6–12 inches thick, while foundations with basements will necessarily use much more concrete. If we assumed an overall average of

12 inches of concrete foundation per square foot, then at a concrete density of 2,400 kg/m^3, we should have 68 kg reinforced concrete/ft^2, translating into perhaps 13.6 kgCO$_2$e/ft^2, with about 8–25 kgCO$_2$e/ft^2 a reasonable range for most construction.

Some other materials

Keoleian et al. [624] found steel to be the third-most important contributor to GHG emissions, after lumber and concrete, at around 3.75 kgCO$_2$e/ft^2. This is congruent with an analysis by Hammond et al. [623], which suggested about 3 kgCO$_2$e/ft^2 due to steel. A variety of more minor materials probably add something like 5–10 kgCO$_2$e/ft^2, in total [624, 623]. For example, based on [624] and the EPA WARM model [522], asphalt shingles generate about 1 kgCO$_2$e/ft^2.

Summing it up

To sum up, at the lower range, on a per square-foot basis, we have about 15 kgCO$_2$e/ft^2 from lumber, 8 kgCO$_2$e/ft^2 from reinforced concrete, and maybe 7.5 kgCO$_2$e/ft^2 from all other materials, summing to 30.5 kgCO$_2$e/ft^2. Midrange estimates are 18, 14, and 10 kgCO$_2$e/ft^2 for lumber, concrete, and other materials, respectively, adding up to 42 kgCO$_2$e/ft^2, while an emissions-intensive house could embody $25 + 25 + 12.5 = 62.5$ kgCO$_2$e/ft^2.

Thus, for the average housing unit in 2013 (including multi-family and single-family homes), with a floor area of 1,500 ft^2, per the US Census American Housing Survey (although mean floor area in the 2009 RECS was, 1,971 ft^2), total construction emissions plausibly range from about 45–90 tonnes of CO$_2$e, while for the average *new* single-family home in 2015, at 2,687 ft^2, our range is more like 80–160 tonnes CO$_2$e. Thus, amortized over 100 years, typical emissions profiles are on the order of 0.6–0.8±0.2 MgCO$_2$e/year for typical occupied units, and about 1.1±0.4 MgCO$_2$e/year for new single-family housing, equating to roughly 5–10% of the operational emissions; these footprints would double if housing lasts only 50 years before replacement. Put another way, construction emissions are equivalent to 5–10 years of household energy use.

Chapter 17

The lawn

- On average, turfgrass lawns are probably nearly carbon neutral, with emissions from nitrogen fertilizer, irrigation water conveyance, mower operation, and landfilled grass clippings roughly balanced by soil organic carbon sequestration and municipal composting of some grass clippings.

- As a *very* crude average estimate, per hectare turfgrass emissions are 0.5 $MgCO_2e$ from irrigation, 1 $MgCO_2e$ from nitrogen fertilizer, 0.5–1 $MgCO_2e$ from mowing, and *negative* 2–4 $MgCO_2e$ from soil carbon sequestration (although this sink may be saturated in many lawns, and thus older lawns may tilt toward net positive emissions).

- If grass clippings are landfilled, this adds perhaps 4 $MgCO_2e$/Ha from landfill methane.

- Overall global warming impact will vary widely by region and lawn management: lawns may serve as net carbon sinks through soil carbon storage in some regions, especially in the Northeast and other areas where irrigation is not needed or where the carbon-intensity of municipal water supplies is low, but net sources in others, especially more arid regions where irrigation water is more carbon-intensive and more used, such as the Southwest. If grass clippings are landfilled, then the lawn will almost certainly serve as a net carbon source.

- It is probably better to focus on landscaping for semi-native habitat (for insects, birds, small mammals, etc.), independent of the carbon impact, and monoculture turfgrass tends to provide relatively little.

The well-manicured lawn remains a fixture of suburban life in America, a symbol of prosperity and orderliness. Turfgrass lawn is also prolific throughout urban space as city and commercial parks, athletic fields, and golf courses. There are several estimates of total turfgrass area in the US, but the most widely cited study is a 2005 work by Milesi et al. [627], who estimated 16.38± 3.585 million Ha, or 1.9% of the continental US land area and equivalent to the fifth-most cultivated crop (after corn, soy, wheat, and hay) by land area overall, and equivalent to the single largest irrigated crop. Of this total, we may very roughly estimate that half is residential lawn (based on [627] and references therein).

Lawns provide urban green space, deeply rooted perennial grasses can serve as a significant carbon sink, and prairies dominated by grass species are the native biome of much of the United States. However, the modern turfgrass lawn, as she is played, resembles but little the prairies she has replaced, and instead of representing a wide mixture of grass species, forbes (wildflowers), and shrubs that support a diversity of wildlife, she is all too often a non-native monoculture that poorly supports bird and pollinator life, and where often what insect life *is* supported is injudiciously poisoned. Significant fossil fuel resources are used to support fertilizer inputs (which also result in downstream N_2O emissions), irrigation water delivery (typically municipal drinking water), and frequent gas-powered mowing. What's more, grass clippings are often bagged and landfilled, where their anaerobic decomposition emits methane. On the other hand, fertilization and irrigation do increase biomass production and hence soil carbon sequestration, and balance of global warming harms and benefits is not immediately obvious.

It is my primary purpose here to evaluate the net carbon impact of typical US lawns and how different management practices affect this balance. Note that while well-managed lawns can act as carbon sinks, in general it is likely that they still compare unfavorably to reasonable native landscaping alternatives available to the average consumer, e.g. prairie plant mixes and/or trees in the Midwest, or drought-tolerant xeriscaping in the Southwest. In the former case, native grasses provide more habitat and similar or greater carbon sequestration (and mature trees store a great deal of carbon), while in the latter carbon-intensive irrigation water is avoided, while again, providing some semi-native habitat. As an overall balance, low-input lawns may be carbon neutral or weak carbon sinks in much of the country, where grassland is the native biome, but are likely net carbon sources in dry regions where excessive irrigation water is required; if grass clippings are landfilled, then the balance will almost always tilt toward net positive emissions.

17.1 Carbon sequestration

Multiple studies have reported long-term carbon sequestration in urban lawns on the order of 0.5–1.5 MgC/Ha/yr (1.83–5.5 MgCO$_2$e/Ha/yr), over perhaps 30–50 years, generally attributed to the deep perennial root system of turfgrass and the absence of tilling, and increased turfgrass productivity due to irrigation and fertilization (although one recent modeling study suggested minimal carbon storage [636]). In our assessment, however, we must be careful to keep in mind the prior land management. For example, Townsend-Small and Czimczi [628] inferred seques-tration of 1.40 MgC Ha^{-1} yr^{-1} for ornamental lawns in Irvine, CA between 2 and 33 years old, on land that had, prior to 1970, been under agricultural management for over 100 years. Given that agriculture tends to sorely deplete the soil of carbon, this is not surprising, and conversion of arable lands to wild grassland also generally results in significant soil carbon storage [378]. Note that athletic fields, which are subject to frequent soil disruption and re-seeding with im-ported sod, not too unlike annual cropping systems, demonstrated no sequestration over time [628].

Similarly, a well-done study by Raciti and colleagues [630] compared carbon stocks in 32 residential lawns in Baltimore of varying ages, and found that those planted over prior agricul-tural land sequestered carbon at 0.82±0.39 MgC/Ha/yr, while those on previous forest showed no sequestration trend, and, indeed, young lawns replacing forest already had relatively high soil carbon. Compared to forested reference sites, lawns collectively had higher soil carbon (69.5±6.3 vs. 54.4±3.6 MgC/Ha). In considering the benefit of an urban lawn versus a forest, one must also consider that a carbon debt on the order of 100 MgC is incurred when clearing average US forestland [617], given the massive above-ground carbon stock and clearly outweigh-ing any increase in soil carbon (grasslands have a comparatively trivial amount of above-ground carbon: less than 0.5 MgC/Ha for shortgrass prairie [397]).

Qian and Follet [633] also observed SOC accumulation at an average rate of 0.78 MgC/Ha/yr for golf course fairways (0.54 MgC/Ha/yr if the site was formerly grassland and not agricultural) until plateauing after 31 years, for a total sink capacity of 16.9–24.2 MgC/Ha. They observed even larger accumulation under putting greens, which may have been a consequence of golf course construction: greens are placed over imported sand, and while fairways overlying former grassland had higher initial carbon stores than those atop former agricultural lands, no such difference was observed for greens.

Selhorst and Lal [632] reported far more rapid SOC sequestration in residential turfgrass than any other works of which I'm aware (e.g. [633, 630, 628, 635]), giving an average sequestration rate of 2.8 MgC/Ha/yr and total carbon sink capacity of 45.8 MgC/Ha, across sites in 16 US cities. This may be a valid initial rate, but it suggests saturation of the sink at 16 years, and from Figure 1 of [632], I calculate a sequestration rate of between about 0.5 and 1.1 Mg/Ha/yr for Austin, TX, when amortizing between 30 and 100 years, far lower than the given rate of 2.9 Mg/Ha/yr. Amortizing a total sink of 45.8 MgC/Ha over 30–50 years would give SOC sequestration of 0.9–1.5 MgC/Ha/yr, much more in line with the rest of the literature. In any case, the total sink value and range is comparable to other works.

Pouyat et al. [635], among others, have advanced the hypothesis that anthropogenic inputs (e.g. irrigation, fertilization) overwhelm regional climate in determining soil carbon storage in managed urban soils, and thus carbon sequestration could vary greatly by region, being strongly positive in the semi-arid Denver region (where irrigation can dramatically increase plant growth), while perhaps weakly negative in Baltimore. Finally, it should be noted that remnant urban forests have been observed to have higher soil carbon than similar rural forest [635], which may suggest that other factors specific to the urban environment promote soil carbon sequestration.

In sum, the literature suggests that residential lawns are not unlike other grasslands, in that increased precipitation/irrigation and nitrogen fertilization can both increase biomass productivity and soil carbon stocks (see [397, 396, 398]), although this increase cannot continue indefinitely (perhaps 30–50 years for lawns after initial establishment). Conversion of exhausted agricultural soil to grass/prairie results in a fairly long-lived carbon sink in either case, and while conversion of native grassland or forest to lawn may still sequester soil carbon, any benefit is smaller and, in the case of forests, does not compensate for the loss of above-ground carbon.

Finally, note that median housing unit age in the US was 38 years, in the 2013 US Census American Housing Survey, which would imply, assuming that turfgrass lawn was established at the time of construction, typical lawns may be nearing saturation for carbon storage, and ongoing carbon sequestration beyond 0.5 MgC/Ha/yr in most lawns would be highly unlikely.

17.2 Fertilizer: upstream and nitrous oxide emissions

As discussed in Section 20.1.3, we have about 5 kgCO$_2$e/kgN-fertilizer of upstream manufacturing emissions, translating to 237.6 kgCO$_2$e/Ha/yr, for 47.52 kgN/Ha/yr (overall N lawn application average [636]), and 420.5 kgCO$_2$e/Ha/yr for 84.1 kgN/Ha/yr (average for that half of lawns actually fertilized).

Some fraction (about 1%) of nitrogen fertilizer applied to soils evolves directly to nitrous oxide, N$_2$O, as discussed extensively in Chapter 20. However, newly fixed anthropogenic nitrogen typically undergoes a cascade of transformations as it moves through different nitrogen pools, and thus overall, 3–5% of new anthropogenic nitrogen may ultimately evolve to N$_2$O. I consider it likely that lawns are likely similar in this to other soils under human management.

Raciti et al. [631] observed annual denitrification of 14 kgN/Ha for lawns fertilized at 98 kgN/Ha; if only 10% of denitrified N evolved to N$_2$O (see Section 20.1.3), this would suggest an N$_2$O emission factor of about 1.5%. Townsend-small and Czimczik estimated a lower 0.4–1.0% of applied N evolving to N$_2$O [628]; Groffman et al. [637] did not observe increased N$_2$O flux with increased N application to lawns, but this was a one-time sampling and almost all N$_2$O flux occurred within two weeks of fertilizer application in [628]. For an overall average N application rate of 47.52 kgN/Ha/yr, we would have 334, 668, and 1,113 kgCO$_2$e/Ha/yr, at 1.5, 3, and 5% N$_2$O EFs. Similarly, an N rate of 84.1 kgN/Ha/yr would yield 591, 1,182, and 1,969 kgCO$_2$e/Ha/yr.

17.3 Grass clippings

Estimates for the productivity of *mown* grass (i.e. the fraction that is mown, not the total) range from 77–197 gC/m^2 (see Kaye et al. [638] and references therein), with Kaye et al. [638] giving a roughly midpoints estimate of 133 gC/m^2. If we assume grass is 44.87% carbon by dry weight and 70% water [509], this translates into 9.88 Mg wet grass clippings per Ha. If left in place, these clippings provide carbon and nitrogen, significantly offsetting some fertilizer requirements and increasing soil carbon stores [634], although they will presumably generate some N$_2$O emissions as well.

If, on the other hand, they are bagged and landfilled, they generate significant quantities of methane via anaerobic decomposition within the landfill, with my estimate for carbon equivalent emissions at 0.43 MgCO$_2$e/wet Mg (see Chapter 24 for a much more detailed analysis on landfills). For 9.88 Mg/Ha of clippings, this is a full 4.25 MgCO$_2$e/Ha, or enough to more than offset any soil carbon sequestration. However, since a large proportion of yard waste is now recovered for composting/recycling (about 61% in 2014, per the EPA), grass recovered for this purpose will have a net carbon storage effect, estimated at 0.1764 MgCO$_2$e/wet Mg in the WARM model, or net storage of 1.74 MgCO$_2$e/Ha. Also including combustion as an end fate (7.6% of yard waste in 2014, with an EF of -0.1984 MgCO$_2$e/wet Mg), as an overall average, disposing grass clippings is essentially carbon neutral, with the carbon savings from composting and combusting nearly equal to the emissions from the landfilled fraction (see Chapter 24 for further calculations and references on waste management). It must finally be noted, however, that while composting grass clippings off-site may result in off-site carbon storage, this is also likely to reduce any on-site lawn carbon storage.

17.4 Irrigation

> • The upstream carbon emissions embodied in conveying irrigation watering vary widely, ranging from perhaps 0.4–4.0 MgCO$_2$e/Ha for most irrigated lawns (and zero for non-irrigated, which make up half of lawns). Regionally, irrigated lawn may yield 2–8 MgCO$_2$e/Ha in the Southwest, 0.5 MgCO$_2$e/Ha elsewhere, and 1 MgCO$_2$e/Ha/yr as a national average. Thus, we can probably attribute 0.5 MgCO$_2$e/Ha/yr to lawn irrigation overall (including the unirrigated component).

Irrigation requirements vary widely by region, with rainfall able to supply much water in some regions and all water in some of the rainy Northeast [627], but a standard recommendation is one inch of precipitation per week (during the growing season), and sprinkler systems often operate rain or shine. Supposing half of lawn area is irrigated [636], and, conservatively, that which is watered receives 1 inch per week for 12 weeks of the year (or equivalent to watering all lawns for six weeks), we should need about 402,000 gallons per hectare per year, which, extrapolating to the entire US turfgrass area, is equivalent to 56 gallons per day per capita (under 2015 population). This is quite plausible with respect to 89 gallons/capita/day of domestic water [586], as over 30% of residential water goes toward outdoor uses, as much as three-quarters of water is used outdoors in some arid cities [578], turfgrass is the dominant permeable urban land cover, and only about one-half of turfgrass is residential.

Now, conveying and distributing all this municipal water is energy- and emissions-intensive, generating between 0.001 and 0.01 kgCO$_2$e/gallon, depending on the source and region (see Section 13.2). At the above irrigation rate, this translates into anywhere from 0.4 to 4.0 MgCO$_2$e/Ha/yr. As a national average, I have calculated an outdoor water emissions factor of 0.0023 kgCO$_2$e/gallon, or just under 1 MgCO$_2$e/Ha/yr. In the Southwest, the EF is more like 0.005 kgCO$_2$e/gallon, and water use is greater, so 4 MgCO$_2$e/Ha/yr (and easily up to 8 MgCO$_2$e/Ha/yr if watering is year-round) is probably a reasonable estimate in this region, while 0.5 MgCO$_2$e/Ha/yr may be more likely elsewhere. Since only about half of lawns are irrigated, 0.5 MgCO$_2$e/Ha/yr seems likely as a crude national average.

17.5 Emissions from lawn care

A technical report by Sahu [629] cites gasoline consumption for lawn mowing as 1.0 gallons/acre (2.47 gallons/Ha) for a push-mower, and 0.75 gallons/acre (1.85 gallons/Ha) for a riding mower, according to a "major equipment manufacturer." At this fuel rate, and at an average of 18.1 mowings per year [636], we would require about 45 gallons per Ha of *residential* lawn, or carbon emissions of 0.502 MgCO$_2$e/Ha.

Irvine, CA contractors reported using 32,400 gallons of gasoline annually to maintain 200 Ha of city park [628], including transportation, blowing, and mowing, or 162 gallons per Ha (1.806 MgCO$_2$e/Ha), almost four-fold higher than residential lawns. This may be the more likely figure for other intensively managed greens, such as golf courses and sports fields.

The 2013 ORNL Transportation Energy Data Book reports 1,259.2 and 147.6 million gallons of gasoline and diesel, respectively, used for mowing equipment (by both residential and commercial users), equivalent to 15.88 million MgCO$_2$e overall, or 0.97 MgCO$_2$e/Ha if divided among 16.38 million Ha. However, other soil and turf equipment, leaf blowers, trimmers, and wood cutters were estimated to consume an additional 1,454.3 and 188.8 million gallons of gasoline and diesel, respectively, for a grand total of 34.45 million MgCO$_2$e attributable to lawn and garden equipment. Obviously not all this is attributable to lawns, but omitting wood cutters, blowers, and trimmers from the total would suggest about 80% of the gasoline and 50% of the diesel use, and over 75% of emissions, are still attributable to turfgrass, equivalent to over 1.6 MgCO$_2$e/Ha.

17.5.1 Emissions embodied in mower manufacture

A review of riding mowers available from Home Depot showed weights to range from 393 to 660 pounds, with most models around 500 pounds; push mowers, on the other hand, were only about 50 pounds. Using the 4–5 kgCO$_2$e/kg inferred for passenger vehicle manufacture in Section 6.4 (admittedly something of a stretch, but I am just interested in order-of-magnitude here), we might reasonably guess that a riding mower has embodied emissions of about 1 MgCO$_2$e, while a push mower is tenfold lower, at 100 kgCO$_2$e. Amortizing over an equipment lifetime of 10 years, we would arrive at a very uncertain 100 and 10 kgCO$_2$e/yr for riding and push mowers, respectively. For a 10,000 square foot lawn (about 1/4 acre and < 1/10 Ha), then, riding mower manufacturing could be a major emissions source (equating to about 1 MgCO$_2$e/Ha), while it would be fairly trivial if used on an entire hectare (2.47 acres), at least on a per-hectare basis.

17.6 General strategies for reducing emissions

In general, the two most important practices to follow are avoiding landfilling of grass clippings, best done by leaving the clippings in place as this also reduces fertilizer requirements, and avoiding excessive irrigation in arid areas, especially the Southwest. Minimizing gasoline-powered mowing is also of benefit.

Returning clippings to the lawn has been variously found to reduce N fertilization requirements by 30–75% [634], and as above, methane emissions from landfilled clippings are also very harmful. A simulation study by Qian et al. [634] using the CENTURY model found returning clippings to increase soil carbon and decrease N requirements by 25% initially, and by 60% after lawn establishment. Similarly, Gu et al. [636] recommend using fertilizer mainly to establish lawn, and minimize or eliminate use once it is mature. Note that for a constant fertilizer rate, removing clippings will also reduce soil carbon and nitrogen stores.

Bagging and disposing of grass clippings is extremely inadvisable if their fate is the land-fill, in which case methane from anaerobic decomposition will dominate the lifecycle. If their fate is municipal composting, this could result in a small carbon benefit, but one that is still likely smaller than that associated with simply leaving the clippings in place, and would also necessitate some fertilization for lawn maintenance.

By eliminating irrigation, which is possible mainly in the Northeast and upper Midwest [627], avoiding or minimizing fertilization, and recycling grass clippings (preferably via in-place grass-cycling) or avoiding mowing entirely, lawns could act as a modest overall carbon sink for several decades if established on previous agricultural land, but would likely be carbon neutral if they replaced grassland or forest (not counting the carbon debt from forest clearance). Low input lawns, with minimal fertilization, e.g. 25 kgN/Ha/yr, minimal irrigation, and infrequent mowing without removal of clipping, could also act as weak carbon sinks, depending on carbon sequestration capacity, emissions-intensity of the local water system, and land-use history of the site. Heavy irrigation in the Southwest likely outweighs any carbon sequestration benefit, and intensively managed lawns in general are more likely to be net carbon sources than sinks.

Since the average age of an existing lawn (assuming establishment at the time of house construction) is nearly four decades, most carbon sequestration potential has likely already been met for the typical homeowner, underlining the general advisability of a low-input maintenance strategy.

Finally, the absolute value of lifecycle carbon emissions from most residential lawns is not large, be they sources or sinks, and one may be better advised to focus on creating habitat (e.g. for birds, bees, and/or small mammals) and supporting biodiversity in semi-native urban ecosystems, independent of the carbon impact.

Part IV

Food and Agriculture

Chapter 18

Introduction to food and agricultural emissions

In this part of the book, I review the GHG emissions embedded in food consumption, and determine what aspects of food consumption contribute most to emissions, with my primary focus on agricultural production systems, and not the downstream energy involved in the larger food system, such as that devoted towards refrigeration and cooking. Overall, the indirect emissions associated with diet likely sum to at least 3.0 $MgCO_2e$ per capita, with 4.0–4.5 $MgCO_2e$ a more likely figure (in my accounting), and thus food is responsible for around 15–20% of the average American's total carbon footprint. Via a conservative accounting, by simply avoiding beef and cutting most (or all) food waste, one can decrease their food-related emissions by 40%, and avoid 1.25–2 $MgCO_2e$ in absolute terms.

The production and consumption of animal products, i.e. meat, dairy, and eggs, but especially beef, is the dominant contributor to the negative environmental profile of modern agriculture. The carbon impact manifests mainly through the upstream emissions embodied in feed production (via fertilizer production and application, fuel use, irrigation, etc.), the end towards which a large fraction of crops in the US are actually grown, methane emissions from ruminants (cows, goats, sheep), nitrogen oxide (N_2O) and methane emissions from manure management, and carbon lost from agricultural land-use change. Agriculture is also, by far, the largest use of land in both the US and globally (about 60% of the continental US is devoted either to crops or grazing), and is the principle cause of deforestation and habitat destruction worldwide; a disproportionate amount of this land is devoted either directly or indirectly to livestock. Thus, in addition to their carbon emissions, livestock are the dominant cause of land-use change and habitat loss worldwide.

Beef is uniquely destructive, varying qualitatively in its carbon and land-use impact relative to all other animal products (this holds true for 100% grass-fed, or "grass-finished" beef too, which probably has a *higher* global warming impact than conventional beef), while eggs and poultry are probably the least damaging, with pork also relatively low-impact. Depending upon how we perform our comparisons (i.e. energy- vs. protein-based) dairy is either similar or somewhat worse than monogastric products, i.e. poultry, eggs, and pork, and so beef stands alone as the top dietary environmental offender in the US. Seafood is a small component of US diets, and is generally similar to poultry or pork in terms of per-unit GHG emissions, although deep-water fish harvested via seabed trawling may have a carbon impact more similar to beef (and are highly damaging to the fragile seafloor ecosystem), and some shrimp sources from Southeast Asia are associated with truly astronomical emissions due to the destruction of mangrove forests for shrimp ponds.

It should be noted that different animal production systems also entail issues of animal

371

welfare that do not closely correlate with their carbon and land-use impacts: beef cattle may be treated most humanely on average, typically living half- to two-thirds their lives in unconfined pasture systems with calves staying with their mothers until weaning, and only moving to feedlots, which are certainly less confining than battery egg cages or gestation crates, in the last part of life. While raising cattle for dairy is more efficient overall and less emissions-intensive, calves are removed from their mothers within days, males enter directly into the beef system, most conventional dairies are based on confinement systems, the average lifespan of a lactating dairy cow is actually less than that of a mother beef cow, and the ultimate fate of all dairy cattle is slaughter for beef anyway. Highly confined systems are the norm for most pork, poultry, and egg production systems, but these systems (especially poultry and eggs) also are very efficient, and generate the fewest emissions per unit of protein produced. However, the "carbon premium" for free-range pork and poultry/egg systems is minimal, and so more humane alternatives are entirely feasible at a large scale, especially if overall consumption is concomitantly decreased.

Waste is remarkably prevalent throughout food systems worldwide, and embodies vast amounts of avoidable emissions: between 30 and 50% of all produced food is ultimately wasted, and in developed countries the greater portion of this waste occurs at the post-consumer level, i.e. following purchase at the supermarket, etc.

Transportation-related emissions are actually a more minor component of dietary carbon footprints, and while buying locally may have other, more difficult to quantify advantages, the issue of "food miles," while not entirely trivial, is a more secondary concern. In addition, about three-fourths of food-related transportation occurs upstream of the farm gate (transport of animal feed, etc.), and far more transport miles overall are embodied in animal products, especially, again, beef, than other plant-based foods. Thus, even "local" beef likely embodies more transportation-related emissions than even the most exotic plant faire. It must also be emphasized that demanding local production in many areas that are intrinsically poorly suited to arable crops would require destroying vast tracts of native ecosystems for much lower yields than might be obtained elsewhere.

Organic production, which eschews synthetic fertilizer and pesticides, avoids emissions and other externalities from these sources, but is also less productive, requiring more land and other inputs for the same output, and thus yields per-unit carbon emissions similar to conventional foodstuffs. Additionally, because manure and animal by-products such as bone and blood meal, which may (and usually do) come from conventional animal systems, are the primary fertilizers in organic systems, organic production is largely dependent upon conventional agriculture as an input source, and it is highly unlikely that organic, as a system, is scalable (and less than <1% of cropland is currently under organic management in the US, with this share even smaller globally). The most environmentally beneficial aspect of organic production is likely the avoidance of synthetic pesticides, which are associated with multiple serious ecological harms, including food web disruptions, grassland bird declines, and large-scale pollinator declines. However, even in conventional systems, around 50% of pesticide use is likely avoidable without yield reductions, and the sparing use of pesticides may help to reduce agricultural land requirements.

It follows from all this that reducing meat and dairy, *especially beef*, consumption and food waste are by far the two most important things that consumers can do. The majority of synthetic fertilizer, pesticide, freshwater, fuel, and agricultural land go towards these two ends, and thus dietary shift away from these two categories (meat and waste) is far more impactful than a dietary shift to, say, a local or organic diet otherwise equivalent to that of a typical American. To reiterate once more, *dietary shift away from beef primarily (and all other animal products including dairy, secondarily)* and *food waste reduction* are the most effective strategies for reducing the climate impacts of diet. One need not follow a strictly vegetarian diet, and indeed, dairy is slightly worse overall in terms of carbon and land-use impact than poultry, and the emphasis should be on rational harm reduction over abstention. Organic or other variations on non-conventional food may be preferable in selected cases, but there is not an obvious systemic advantage, and indeed, a broad shift to organic methods, without other dietary shifts, would likely require more land and widespread habitat destruction. Any environmental benefit to local food is context-dependent, and avoiding excessive "food-miles" is generally of minor importance.

The final obviously destructive component of modern agriculture is the production of food crops for fuel, most notably corn ethanol. This, like most other bioenergy and biofuels, is, in my view, an environmental farce. Corn ethanol is predominant in the US, where nearly half the annual corn crop now goes toward this end, which is futile for several reasons. First, the energy return on investment is abysmally low for corn ethanol, hovering around unity, i.e. it takes exactly as much energy to produce ethanol as one gets from it. Second, while optimistic analysis by the EPA projects lifecycle carbon emissions about 20% lower for ethanol than gasoline, proper accounting of nitrogen fertilizer-related emissions and land-use change may imply a carbon footprint as much twice as bad. Third, even under the most optimistic analysis, corn ethanol can only reduce US passenger car emissions by about 1.5%, equivalent to increasing the fleet-average MPG from 21.6 to 21.9 MPG (see Section 22.3). Thus, animal products but mainly beef, waste, and biofuel form a trifecta of excess and avoidable climate and broad environmental harm attributable to the agricultural system. The former two are directly in the hands of the consumer.

18.1   Overall impact

Multiple reviews have estimated that globally, food systems generate on the order of 15–30% of all anthropogenic carbon emissions, including indirect emissions from land-use changes. A recent comprehensive review by Vermeulen and colleagues [316], gave an emissions estimate range of 9.8–16.9 $GtCO_2e$ for 2008, or 19–29% of the global total, with the vast majority, 80–86%, attributable to on-farm agricultural production (including land-use changes). Note however that, in developed countries such as the US, post-farm energy and emissions make up a relatively greater share of food system emissions [317].

The major points in the overall food system include [316] (1) production of *inputs* (fertilizers, seeds, pesticides), (2) agricultural production (crops, livestock, and wild food production, e.g. fisheries), (3) processing, packaging, transportation, and distribution, and (4) consumer-level management and waste disposal. Note that waste occurs at all stages of this system, and while waste disposal is a relatively minor *direct* source of emissions (via landfill methane from anaerobic food breakdown), 30–50% of all food is ultimately wasted, representing a vast quantity of embodied energy, land, water, and carbon. In the subsequent section, I derive an order-of-magnitude estimate for US food system CO_2e, and then give this estimate in disaggregated form.

18.1.1 Top-down estimate for US agricultural and food system emissions

- Major aggregated emissions sources in the US food system include methane from cattle herds (1.0–2.5 MgCO$_2$e/capita); synthetic nitrogen fertilizer production and downstream N$_2$O emissions from soils and manure (perhaps 1.0 MgCO$_2$e/capita); other fertilizer production, pesticide production, on-farm energy use, irrigation, transportation, and food waste landfilling (together around 1.0 MgCO$_2$e/capita); ecosystem carbon loss from conversion to agriculture (0.5–1.5 MgCO$_2$e/capita depending upon accounting), and finally, downstream food processing and handling, not counting residential refrigeration, cooking, etc. (0.75–1.75 MgCO$_2$e/capita).

- From a top-down accounting, total US food system emissions likely sum to about 1.25–2 billion MgCO$_2$e (about 4–6 MgCO$_2$e/capita), with around 75–80% of emissions upstream of the farm gate.

In this section I derive an order of magnitude top-down emissions estimate for the entire US agricultural system, which is likely to be in the 3–4 MgCO$_2$e/capita range when not counting food processing, packaging, retailing, and commercial preparation downstream of the farm gate, and more likely 3.75–5.5 MgCO$_2$e/capita if these activities are included. In the subsequent chapters, I provide much more detailed derivations of the basic numbers presented here.

For cropping systems, we have emissions from (1) fertilizer, mainly nitrogen (N) fertilizer (at both the production and application phases), (2) pesticides, (3) irrigation water, (4) on-farm fuel and electricity use, (5) land-use change emissions, and (6) transportation from the farm gate. Downstream of the farm gate, corn, soy, and forage are fed to animals, where a fraction of feed carbon is converted to methane via enteric fermentation, another fraction becomes methane via manure management, and some manure nitrogen also evolves to N$_2$O. Enteric fermentation, primarily from beef herds, is likely the single largest source of agricultural GHG emissions. Grazing and pasture/rangeland management also affect ecosystem carbon stores, often, but not uniformly, negatively.

First let us consider methane due to livestock. Recent top-down, satellite-based measurements of US methane emissions by Turner and colleagues [96] estimated annual livestock-attributable CH$_4$ at 12.6–17.0 million tonnes, or 29–44% of US methane (over the 2009–2011 period). This translates into 428.4–578.0 million MgCO$_2$e (under a non-fossil methane GWP of 34), or, using the 2011 population, 1.37–1.85 MgCO$_2$e/capita. Similarly, a top-down work by Miller et al. [25] gave 17.0±6.7 TgCH$_4$ due to ruminants and manure, or up to 2.5 MgCO$_2$e/capita, while Wecht et al. [320] gave 12.2±1.3 TgCH$_4$ for the year 2004 (i.e. 1.41±0.15 MgCO$_2$e/capita). These figures significantly exceed the EPA inventory estimate of 9.381 million MgCH$_4$ from livestock (for 2013), which equates to "just" 1.0 MgCO$_2$e/capita. Note that in this inventory, the EPA attributes about 70% to ruminant enteric fermentation (with the remainder from manure), and of this, 71% to beef cattle. The sum of the evidence therefore suggests *at least* 1–1.5 MgCO$_2$e/capita attributable to CH$_4$ from livestock. Note that US rice cultivation (not including international imports), may have added another relatively scant 11.3 MgCO$_2$e of methane (based on EPA inventory).

Fertilizer, mainly nitrogen (N) produced via the industrial Haber-Bosch process, leads to N$_2$O emissions both from fertilized fields and from downstream transformations, e.g. through animal manure. Supposing about 5 kgCO$_2$e/kg N for fertilizer production, and 3–5% ultimately evolving to N$_2$O [348] (see Section 20.1.3), then the 11.648 million Mg of synthetic N used in US agriculture (in 2011), would yield on the order of 222–331 million MgCO$_2$e of N$_2$O.

Nitrous oxide is also released via mineralization of N during the decomposition of soil organic matter, and via free-living bacteria in agricultural soils [10], and the EPA inventory estimated

263.7 million $MgCO_2e$ of direct and indirect N_2O emissions from agricultural soils in 2013 (including grasslands), with only 61 million $MgCO_2e$ directly attributed to synthetic fertilizer (the largest contributor was actually mineralization and asymbiotic N fixation, at 119.2 million $MgCO_2e$). Manure management added just 17.3 million $MgCO_2e$, for a total of 280 million $MgCO_2e$. It is thus unclear exactly how much N_2O is attributable to US agriculture, but it is probably on the order of 1 $MgCO_2e$/capita, and possibly more.

Other fertilizer production, i.e. phosphorus and potassium (potash), together adds a fairly trivial 6 million $MgCO_2e$, and pesticide production may add around 7.5 million $MgCO_2e$. On-farm energy use, excluding pumping energy for irrigation, most likely generates on the order of 100 million $MgCO_2e$ annually, while irrigation may add something like another 50 million $MgCO_2e$. Transportation throughout the food system (including transportation upstream of the farm gate) may generate 80–100 million $MgCO_2e$. These numbers are derived in Chapter 20. Landfilled food waste could add another 30–50 million $MgCO_2e$ through anaerobic decomposition to methane, as discussed in Chapter 24. Summing all these factors gives around 275–315 million $MgCO_2e$, or about 0.85–1 $MgCO_2e$/capita.

Bringing pristine land under cultivation tends to release large amounts of carbon, but how and whether to include this as a dietary carbon emission is uncertain, as most historical prairies fell to the plow long ago. As discussed in Section 20.3, long-term prairie soil organic carbon (SOC) losses are about 2–3 $MgCO_2e$/hectare[1]/yr after conversion to cropland, suggesting emissions equivalent to 330–495 million $MgCO_2e$/yr. Since taking cropland out of production restores soil carbon, but at a lower rate, 180 million $MgCO_2e$/yr is probably a reasonable minimum SOC penalty to US agriculture, not counting the uncertain effects of grazing on pasture and rangeland.

Summing up all production emissions suggests slightly over 1 billion $MgCO_2e$/year due to US agricultural production, but possibly >1.5 billion $MgCO_2e$ at the extreme upper end, especially if we use a high estimate for land-use change emissions. In other words, per capita agricultural emissions are likely in the 3–4.75 $MgCO_2e$ range.

Considering the extended food system of food processing and purveyance downstream of the farm gate, but upstream of the consumer, is a major end-use of energy in American society. A USDA study [318] concluded that food processing consumed 0.79 trillion kWh of *primary* energy in 2002, while packaging and the wholesale, retail, and food services together used 1.4 trillion kWh of primary. Cuellar et al. [319], on the other hand, gave just under 1 trillion kWh of primary energy for all such processes. In either case, supposing an emissions factor of about 0.2 $kgCO_2e$/kWh for primary energy, these numbers translate into anywhere from about 200 to 450 million $MgCO_2e$ per year overall, and thus, downstream of the farm gate, an additional 0.7–1.75 $MgCO_2e$/capita are added by various food handling and processing processes, aside from residential energy use. We then arrive at a grand total of 3.6–6.5 $MgCO_2e$/capita as a plausible range for total upstream US food system emissions, with 4.5–5.0 $MgCO_2e$/capita perhaps a reasonable best guess. This estimate is slightly higher than that of most existing bottom-up lifecycle analyses, largely due to the inclusion of land-use changes, downstream food processing, and higher ruminant methane emissions. Figure 18.1 shows low, "best," and high estimates of pre-consumer food system emissions by mechanism.

---

[1] A hectare (Ha) is 10,000 m^2, and equal to 2.47 acres; this is the primary unit for land area used in this book.

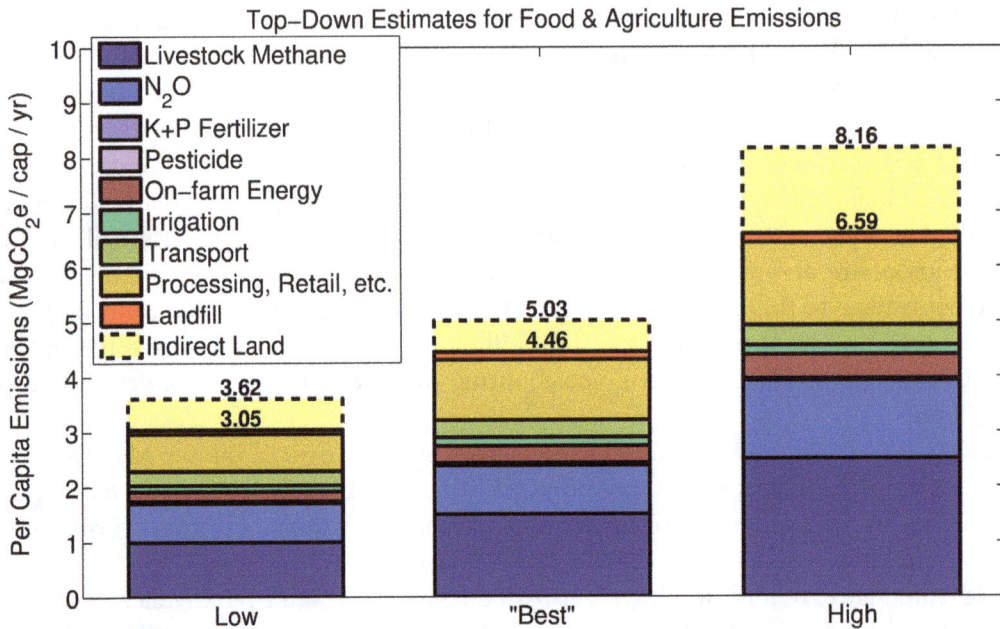

Figure 18.1: Low, best, and high top-down estimates for the US food and agriculture system (not including consumer-level energy use related to cooking, refrigeration, washing, etc.), disaggregated by major mechanism, and given in per-capita terms. Ruminant methane includes enteric fermentation from ruminants and manure emissions from all livestock; N_2O is the sum of N_2O from new reactive nitrogen (mainly synthetic fertilizer), agricultural soils, and manure. Note that phosphorus (P), potassium (K), and pesticide production emissions are too minor to be clearly seen. Indirect land use emissions related to carbon loss from agricultural soils is less certain, and the numbers above each bar give per-capita emissions with and without its inclusion.

18.2 Bottom-up emissions factors for major foods, and total bottom-up emissions

Overall, as a general guideline, bottom-up estimates suggest *production* emissions factors (by weight) are around 0.25–2 kgCO$_2$e/kg for various plant products, including high-protein legumes, grains, tubers such as potatoes, fruits and vegetables, etc. (see, e.g. [321]), around 2.5–7.5 kgCO$_2$e/kg for monogastric animal products, i.e. eggs, poultry meat, and pork, as well as much seafood, and on the order of 1.5–2.0 kgCO$_2$e/kg milk but around 10–15 kgCO$_2$e/kg for solid cheeses. Beef is qualitatively worse, with an EF on the order of 30–50 kgCO$_2$e/kg. These factors are derived in detail in the following chapters.

In other words, as one moves from plants to eggs, poultry, pork, fish, and dairy, production emissions factors increase by a factor of 5–10. Moving from this latter category of moderate-impact animal foods to beef (and other ruminant meat, such as lamb and goat), emissions again increase five to tenfold. Thus, per unit weight, CO$_2$e impact varies across two orders of magnitude, from plant to cow. Once post-farm processing, packaging, and distribution are accounted for, we likely add 0.75–2.5 kgCO$_2$e/kg for any given food, depending upon its post-farm gate lifecycle.

Note that these bottom-up estimates, when applied to retail-level food availability, suggest a total diet carbon footprint on the order of 2.25–3.5 MgCO$_2$e/capita/year, appreciably lower than the top-down range derived above. Part of the gap is probably explained by higher ruminant methane emissions in top-down studies: bottom-up analyses are more consistent with just about 750 million MgCO$_2$e from beef and dairy enteric fermentation, which is probably too low by at least 25% and quite possibly by more than a factor of two. Crudely correcting adds around 5–10 kgCO$_2$e/kg beef and 0.5–0.67 kgCO$_2$e/kg milk, and bumps our overall bottom-up estimate range to about 2.6–3.85 MgCO$_2$e/capita/yr.

Additionally, land carbon losses are, in the main, disregarded in bottom-up estimates. Taking a reasonable minimum total, 180 million MgCO$_2$e (see Section 20.3), and our bottom-up estimate corrects to about 3.15–4.4 MgCO$_2$e/capita/yr, with at least 0.25 kgCO$_2$e/kg added to plant products (derived from a crude division of about 750 million tonnes of primary productivity across 180 million MgCO$_2$e), perhaps 1 kgCO$_2$e/kg added to monogastric products, and >3 kgCO$_2$e/kg added to beef (assuming feed conversion factors in [322], and that 67% of beef cattle dry matter intake is pasture, based on [439]).

Finally, adding in transportation emissions on the order of 100 million MgCO$_2$e, and we should add at least 0.133 kgCO$_2$e/kg to plant products, about 0.5 kgCO$_2$e/kg to monogastric products, and 1.75 kgCO$_2$e/kg to beef (also based on 750 million tonnes of primary productivity and feed conversion efficiencies as above), and our final bottom-up emissions estimate corrects to 3.45–4.7 MgCO$_2$e/capita/year, mostly within the lower half of our top-down estimate above, and a reasonable bottom-up midpoint estimate might be a somewhat lower 4.0 MgCO$_2$e/capita/year, or 3.5 MgCO$_2$e/capita/year if we disregard indirect land use. Note now, that even with production emissions on the order of 0.5 kgCO$_2$e/kg, plant-based products may be expected to have a net impact anywhere from 1 to 3.5 kgCO$_2$e/kg, with half or more of the impact, on average, post-farm. This is still anywhere from 2–10 times better than typical monogastric and dairy products, and 10–50 times better than beef. Emission factors for animal products, on the other hand, are still uniformly dominated by on-farm production.

Estimated food system emissions, using per-unit emissions factors as above and with adjustments for post-farm processing, packaging, retail, transport and indirect soil carbon losses are given in Figure 18.2; results are given for representative low, "best," and high emissions factors, both with and without indirect land use losses. Reasonably consistent with other lifecycle analyses (e.g. [321]), beef alone is responsible for one-third or more of all food system emissions,

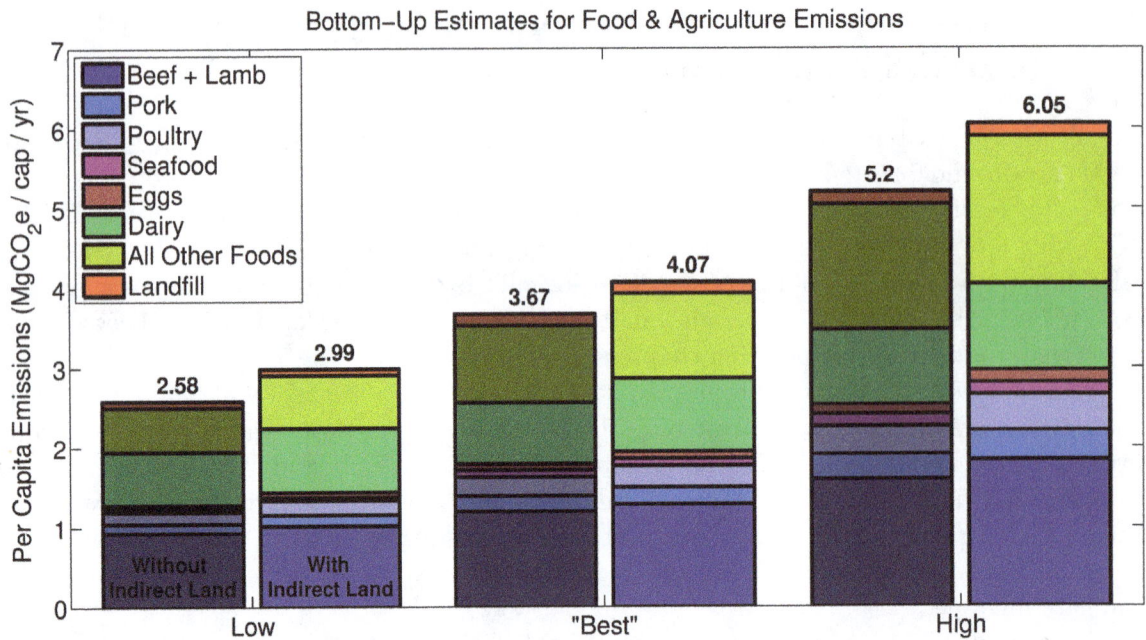

Figure 18.2: Low, best, and high bottom-up estimates for the US food and agriculture system CO_2e emissions, on a per-capita basis (not including consumer-level energy use), either with or without approximate corrections for indirect soil carbon losses (left and right bars for each estimate, respectively), using food-specific emissions factors (including transport and post-farm processing, retail, etc.) for major food categories (and using USDA LAFA database for major food retail availability). Comparing to Figure 18.1, even with some adjustments as discussed in the text, a bottom-up analysis gives a dietary carbon footprint about 20% lower than a top-down approach.

while dairy contributes on the order of 20–25%, other meat and eggs together yield 15–18%, and all other foods taken together account for just 25% of food-related CO_2e; landfilling of food waste and the resulting methane from anaerobic breakdown gives the final 3–4%.

18.3 Carbon footprint for typical dietary patterns

Several lifecycle analyses have addressed the carbon footprint of typical whole diet patterns, the most common being typical omnivorous, lacto-ovo vegetarian, and vegan, but we may also compare lower and higher meat diets, pescetarianism, etc. Unsurprisingly, there is generally an association between less dietary meat and dairy and lower emissions.

18.3.1 Some existing whole-diet studies

Multiple studies have attempted to quantify the GHG emissions associated with typical diets. The majority of estimates are specific to the UK and several other western European countries, although several US-specific estimates exist. There is significant uncertainty in the precise EFs for various food types, post farm-gate emissions are often unaccounted for in these works, food wastage is variably considered, and the effects of upstream land use are also generally disregarded, so we must be cautious in drawing conclusions concerning the overall CF of a typical western diet, but several findings are robust, mainly that waste and (red) meat are large contributors to the diet CF, with lower meat diets less impactful, and that vegan diets generally hold a small advantage over vegetarian diets (given the relatively high overall and per-unit impacts of dairy products).

In brief, to estimate average dietary CF, the commodity components of a typical diet are estimated from national inventories of consumption (e.g. LAFA as used by Heller and Keoleian [321]) or from diet questionnaires, e.g. [323], and then matched to CO_2e emissions factor estimates abstracted from the larger literature. In addition to the omissions noted above, the use of diet questionnaires almost certainly leads to underestimation of dietary CF, as they are well-known to systematically underreport energy intake, do not account for food that is purchased and not consumed, and further ignore food waste upstream of the consumer [324]. Given that roughly 30% of food is wasted at the consumer level, we may safely increase estimates derived from diet questionnaires by about 40%. Furthermore, in some studies consumption is normalized to a 2,000 kcal diet, a practice that clearly reduces estimated emissions substantially, and we may safely double the estimated impact from such studies (about 4,000 kcals per capita reach the retail stage in the US).

Turning to results, Berners-Lee et al. [324], using a combination of FAO data on per capita calorie availability, combined with survey results on actual diets, estimated the average UK diet embodies 2.7 $MgCO_2e$/year (scaling from 3,548 kcal/capita in the UK to about 4,000 kcal/capita in the US suggests 3.0 $MgCO_2e$/yr, likely an underestimate), and moreover concluded that vegetarian and vegan diets might, on average, offset 22% and 26% of this total, respectively, although they suggest this offset may be higher if land use changes are accounted for.

Scarborough et al. [325] also recently estimated dietary GHG emissions for UK diets with different levels of meat consumption, using data from a large survey on dietary habits to inform diet habits. Since diets are likely to vary pervasively depending upon the degree of meat consumption, this methodology may yield results that correspond better to the GHG impact of broad classes of diets, as actually practiced. In short, yearly emissions for normalized 2,000 kcal diets were 2.65, 2.16, 1.70, 1.41, and 1.07 $MgCO_2e$/year for high-meat, average-meat, low-meat, vegetarian, and vegan diets, respectively. Note that we should roughly double the numbers to reflect the actual 4,000 kcal US diet for more accurate absolute emissions estimates. It is very notable that a low-meat diet reduced CO_2e by 36% and 21% relative to high and average-meat diets, respectively, and thus abstinence is not an absolute requirement for a relatively low-carbon diet. It is also notable that meat consumption was actually relatively low in this study, with "high-meat" diet defined as >100 g/day, yet average per-capita meat consumption in the US is about 240 g/day. Along similar lines, Heller and Keoleian [321] found beef alone to account for 36% of dietary CO_2e, in the US.

It should finally be noted that vegetarian and vegan diets tend to be slightly lower in calories than typical omnivorous diets [325], and incorporating this improves the relative benefit of meat-free diets by perhaps 3–5%.

18.3.2 Some general conclusions

From the prior sections, it should be apparent that a diet that simply cuts out beef, without any other change, will avoid 30–40% of diet-associated CO_2e. If one were to additionally reduce other meat and dairy by, say one-third, with protein replaced by legumes or other high-protein plant foods, then overall diet CF would fall to just 50–60% of the typical diets. The bottom line is that, although vegetarian and vegan diets are fine choices, a more nuanced harm-reduction approach that gives beef avoidance top billing can be just as or even more effective than vegetarianism, and likely far easier to implement in practice.

379

18.4 Carbon footprint of food waste

Food loss and waste consumes vast quantities of resources. Note that we distinguish between *loss*, which includes all types of loss, including cooking losses, shrinkage, spoilage, etc., and *waste*, which refers more strictly to otherwise edible food that is discarded. USDA researchers have estimated food loss at the retail and consumer levels to amount to 60.45 million tonnes per year, or 31% of the entire (195.45 million tonne) US retail food supply [326]. The methodology is based on the LAFA database, and excludes inedible commodity portions, e.g. bones and skins. While the method cannot distinguish between loss and waste, given the vast disparity in consumer level wastage between developed and developing nations as estimated by the FAO [327], these losses are probably overwhelmingly waste *per se*.

Heller and Keoleian [321] recently applied LCA to study the CF of food waste at the retail/consumer level, and estimated that, by weight, 31% of the US food supply is lost at the retail/consumer level, with two-thirds of this waste occurring at the consumer level. They further estimated the GHG emissions of a typical diet to be 5.0 kgCO$_2$e/day, with 1.4 kgCO$_2$e due to wastage, i.e. 28% of the total CF. Only a very small portion of waste is inedible, e.g. fruit peels, egg shells, with the majority simply food that goes bad in storage. Furthermore, a 2011 FAO report [327] estimated that while European and North American consumers directly waste 95–115 kg of food per year on a per capita basis, consumers in sub-Saharan Africa and South and Southeast Asia waste only 6–11 kg/year. These facts together suggest that a 90% reduction in typical consumer-level food waste is achievable via behavioral change alone.

Under my estimates of dietary CF, as derived above, yearly upstream GHG emissions attributable to waste are likely around 1–1.5 MgCO$_2$e per capita, assuming about 30% of diet-related emissions are due to waste. Including the GHG impact of landfilling food scraps (due to anaerobic decomposition yielding methane, as discussed in Chapter 24) increases waste-related emissions by 10% or so, at perhaps 100–200 kgCO$_2$e/capita.

18.5 Eating locally: food miles and land suitability

The idea of reducing "food miles" and thus emissions related to burning fossil fuels for transport has been widely promoted as the environmental basis for eating locally. There are any number of calculations that give staggering miles counts for a meal, with total meal miles numbering in the 10,000s. As a general caveat, there is no general agreement on what counts as local, although the "100-mile diet" is a popular rule-of-thumb.

An economic input-output lifecycle analysis by Weber and Matthews [328] found "food miles," i.e. transportation from producer to retail, to account for only about 4.4% of food system emissions, or 0.36 MgCO$_2$e/household/yr, translating into about 0.144 MgCO$_2$e/capita/yr. Transportation in general was estimated at about 11% of US agricultural emissions, and thus about two-thirds of transportation emissions in the food system actually involve upstream activities, such as feed and fertilizer transport, and not final delivery to retail. Of note, total embodied miles in the supply chain and transportation-related emissions were, unsurprisingly, highest for red meat, with the supply chain entailing a total of 12,680 miles (of which just 9%, about 1,200 miles, was final delivery to retail), thrice the average of 4,200 miles (with only 1,020 of this being final delivery).

Obviously, eating "locally" does not eliminate all food miles, and we can do some simple back-of-the-envelope calculations to compare local farmers' market products to the commercial food chain. Suppose a (gasoline-powered) pickup truck transports 1 tonne of produce 20–100 miles from a farm, and gets 17.2 MPG. Then (including both legs of the trip) we would have 25.92–129.60 kgCO$_2$e/tonne food and an emissions factor of 0.648 kgCO$_2$e/tonne-mile.

Average final delivery in the food system is 1,020 miles, Weber and Matthews used 0.2897 $kgCO_2e$/tonne-mile for truck freight, which would give 295.5 $kgCO_2e$/tonne produce. Thus, eating locally would indeed save 56–92% of the food-mile emissions, in this case. Note that food-mile emissions could plausibly be worse for the farmers' market: a 40-mile round-trip would yield equivalent food-mile emissions if 87.7 kg (193.3 lbs) of produce were sold, while a 200-mile round-trip would need to move 438.6 kg (966.9 lbs) of produce to break even.

Also noteworthy, a more rigorous analysis by Cleveland and colleagues [329] concluded that if all produce consumed in Santa Barbara County was grown in the county, then food-related greenhouse gas emissions would fall by a scant <1%.

Commonly neglected in popular discourse is the fact that certain areas are intrinsically better suited to agriculture, and that agricultural extensification into many areas, such as tropical forests, carries very high costs in carbon, biodiversity, and other ecosystem services. For example, a model by Johnson et al. [330] found that, globally, there are a few geographic hot spots where selective extensification preserves far more ecosystem carbon than a business as usual scenario that expands production in all areas proportionally. That is, by expanding farming into a few concentrated regions where ecosystem carbons losses are comparatively low and productivity high, overall environmental impact is minimized, despite the fact that such expansion is necessarily non-local to most of the populace. In the US, carbon stores were maximized by expanding agriculture primarily at the edges of the US corn belt, while crop expansion into much of the American West, and especially Southwest, carried a very high carbon cost and was best avoided [330].

Overall then, there may be good reasons to eat locally, but reducing food-miles may not be a primary one, as it is only a marginal contributor to total emissions, and depending upon the product, certain local foods (mainly meats) may embody far more supply chain miles than non-local alternatives. Agricultural production practices, the particular foods being produced, and the intrinsic suitability of the land to farming are of far greater importance to the agricultural footprint. Therefore, consuming a local product may or may not be more environmentally friendly than a more globalized alternative.

Chapter 19

US and global land use, and major crops

19.1 US land use

In the United States, just over 50% of the entire landbase is used for agricultural production, with about 60% of land in the contiguous 48 states involved in agriculture [331]. Major sources on agricultural and overall land use and land cover include the USDA Major Land Use (MLU) survey, the USDA agricultural census, and the National Land Cover Database from the US Geological Survey. I primarily rely on the MLU survey in this section.

Restricting ourselves to the lower 48 states, the 2007 MLU [331] survey reports that 22% of land is "cropland," which represents (generally) intensively managed arable land, while 39% is some type of grazing land: either pasture or rangeland (32%), or grazed forest (7%). Rangelands represent a variety of semi-natural grasslands, savannas, deserts, marshes, meadows, and tundras that are naturally dominated by native grasses, shrubs, or other grass-like vegetation. They are distinguished from pastures in that they are quasi-natural, but still managed, ecosystems that are also used for domesticated animal grazing, whereas pastures are more intensively managed with practices such as introduced forage plant species, irrigation, and fertilization. Note that many native species are intentionally suppressed on both range and pasture.

We see then, that much agricultural land overlaps with semi-natural ecosystems and wildlife, the greater part is used for domestic animal grazing, and only about one third of all agricultural land is intensively managed cropland that represents a more complete appropriation of nature for human use (although non-native pasture also often differs markedly from the native ecosystem). It is finally instructive to compare the agricultural footprint to urban land use, which accounts for a comparatively small 3.18% of the contiguous 48 states, and only 2.68% of the entire US.

19.2 Overall US and global agricultural production

Considering first the primary production of various crops (i.e. excluding animal production), the US agricultural system is dominated by just a few commodity crops, namely, maize (corn[1]), soybeans, wheat, and hay (with various hays considered as one), which together make up about 75% of all agricultural production by weight, 90% of all food energy and protein production, and nearly 90% of all cropland area. Most of this output goes not to feeding humans directly, as it might, but to livestock, which are either used to convert feed to dairy or egg products or

---

[1]Curiously, the term "corn" traditionally refers to the dominant cereal crop in a region, and only in the US does the term unambiguously refer to some variety of maize.

are themselves slaughtered and consumed in turn. Indeed, in recent years the US agricultural system has produced around 16,000–18,000 kcals per capita per day, at the primary crop level, of human-consumable crops (based on FAO data), and yet less than 4,000 kcals/capita/day make it to retail (per the USDA). This production total does *not* include the roughly 2,000–2,500 kcals of hay produced per day, on a per capita basis. Not being directly consumable by humans, hay is the odd man out here, and it is not tracked by the FAO, which I primarily rely on for data in this section (for hay, I supplement with the USDA's 2012 Census of Agriculture). Note at this time that all calculations and data, unless otherwise specified, are based on FAO crop production data through 2013, and on standard nutritional information for foodstuffs.

These numbers already suggest that the fresh fruit and vegetable produce typically evoked when one thinks of farming and gardening amount to little more than a rounding error when it comes to assessing the impact of our agricultural system. This sentiment is validated once we see that the remaining crops produced at anything approaching a large scale, at least as compared the Big Four above, are mainly sugar crops, i.e. sugar cane and sugar beet, and other staples such as potatoes, rice, barley, and beans; tomatoes and oranges are the only fruits/vegetables to break the top ten in terms of production by weight (and these two are also used mainly in a commodity-like fashion, i.e. as tomato pastes/sauces and orange juice).

Now then, there are several metrics by which we can measure agricultural output. While weight is most obvious, even more salient are crop energy content (typically measured in kcal) and the three basic food macronutrients, carbohydrates, protein, and fat, with protein being the most important to consider separately from carbohydrates and fat, which are essentially pure energy sources. So for example, maize is quite calorie-dense, and while it dominates at 45% of all US primary crop weight (excluding hay, based on 2012 FAO data), it accounts for an even larger 56% of all primary kcals, but a somewhat lower 35% of protein. Soybeans account for 13% of crop weight, 20% of crop energy, but 42% of crop protein (again, excluding hay). Thus, we see that maize is primarily grown to provide animal feed energy (and energy for ethanol fuel), while soybeans are mainly grown as a protein source for animal feed, and for oilseed.

Total production, in terms of mass, energy, and protein, along with harvested crop area, for the top ten US crops is summarized in Figure 19.1. Figure 19.2 also shows yields for top-yielding crop varieties, both in terms of mass/Ha and kcals/Ha. Finally, for comparison, the somewhat more diverse array of leading global crops is shown in Figure 19.3.

As already mentioned, in the US, close to 20,000 kcals/capita/day are grown at the primary crop, nearly 10 times the average daily calorie requirement, but only 4,000 kcal/capita/day reaches the retail level, largely because most commodity crops are fed either to animals, with a minuscule energy return on investment (EROI) that is far less than unity, or to vehicle engines in the form of corn ethanol (40–45% of the maize crop), likely with *no* net energy return on investment (see Section 22.3 for an exhaustive discussion).

19.2.1 Major crops

The geography of major crop production is summarized graphically in the maps in Figure 19.4; similarly, Figure 19.10 gives the distribution of the major farm animals. Time-series for harvested acreage, annual production, yield, and production on a kcal/capita basis for the four major US crops, corn, soy, wheat, and hay, are given in Figure 19.5. Yield time-series for these crops are also given separately in Figure 19.6. I now devote a brief discussion to the two largest commodity crops, maize and soybean.

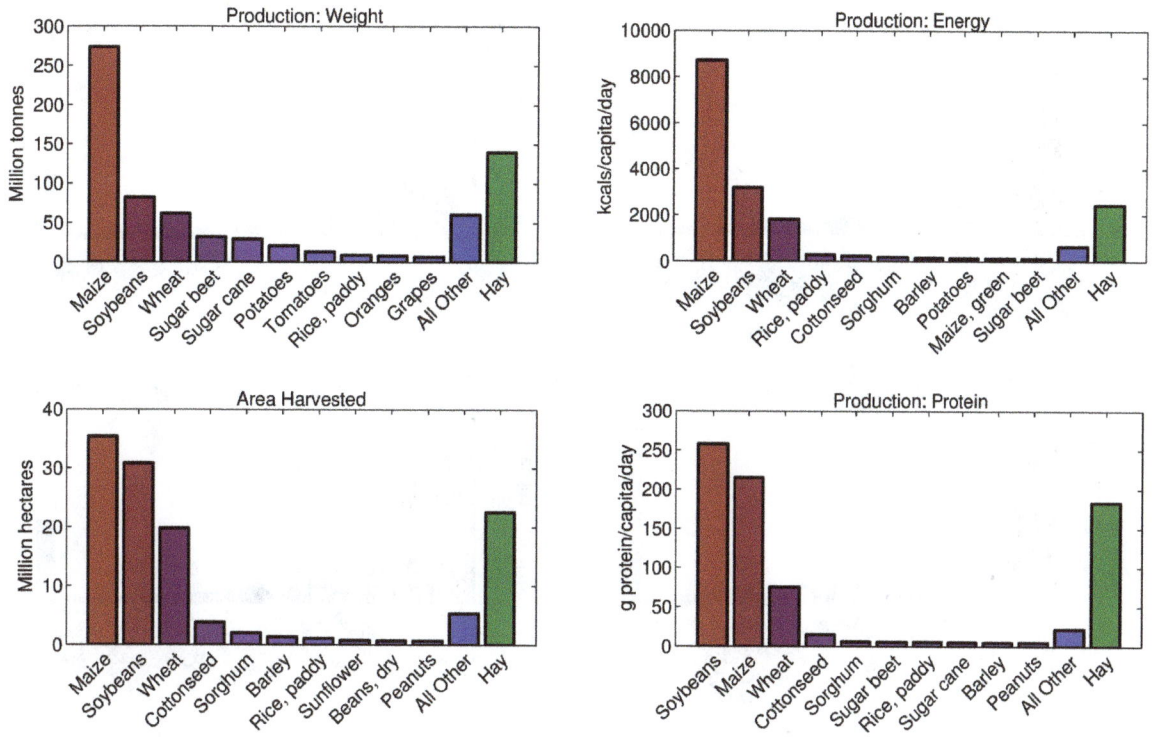

Figure 19.1: US crop productivity and land area harvested in 2012, based on FAO data. Shown are the top crops in terms of tonnage produced, energy (displayed as kcals/capita/day equivalent, for the US population), and protein (similarly normalized to per capita protein supply), and the principal crops by area harvested. Note that hay is displayed separately, as it is not consumable by humans, and data for this is from the 2012 US agricultural census (tonnage is converted to energy and protein under the rough approximation of 2,000 kcals/kg hay and 150 g protein/kg hay). Finally, note that, among recent years, production in 2012 was relatively low, with per-capita maize production often exceeding 10,000 kcals/capita/day in other years, and passing 11,000 kcals/capita/day in 2014 and 2015 (USDA ERS).

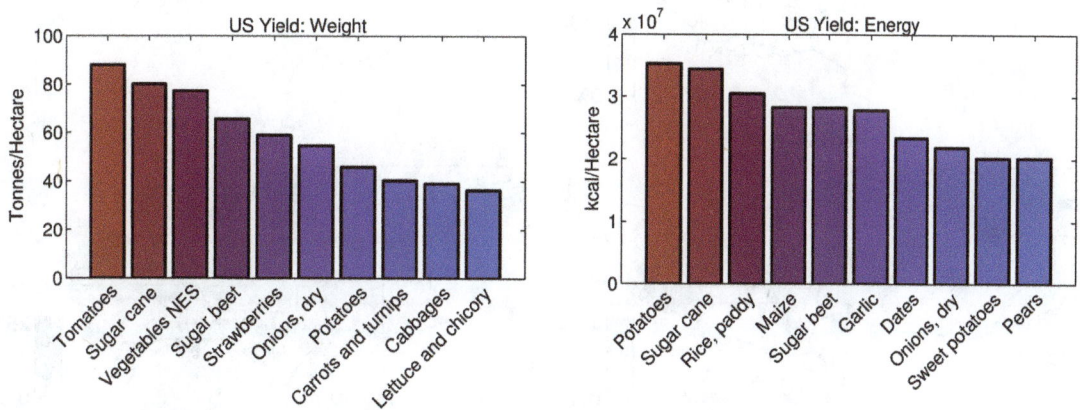

Figure 19.2: Top-yielding US crops in 2012 (FAO), given both in terms of tonnage per hectare and energy per hectare (excluding green maize). Note that while many horticultural crops, such as tomatoes, are extremely high yielding by weight, they have extremely low caloric and protein density, and so staple grains, tubers, and sugar crops are still higher energy yielders.

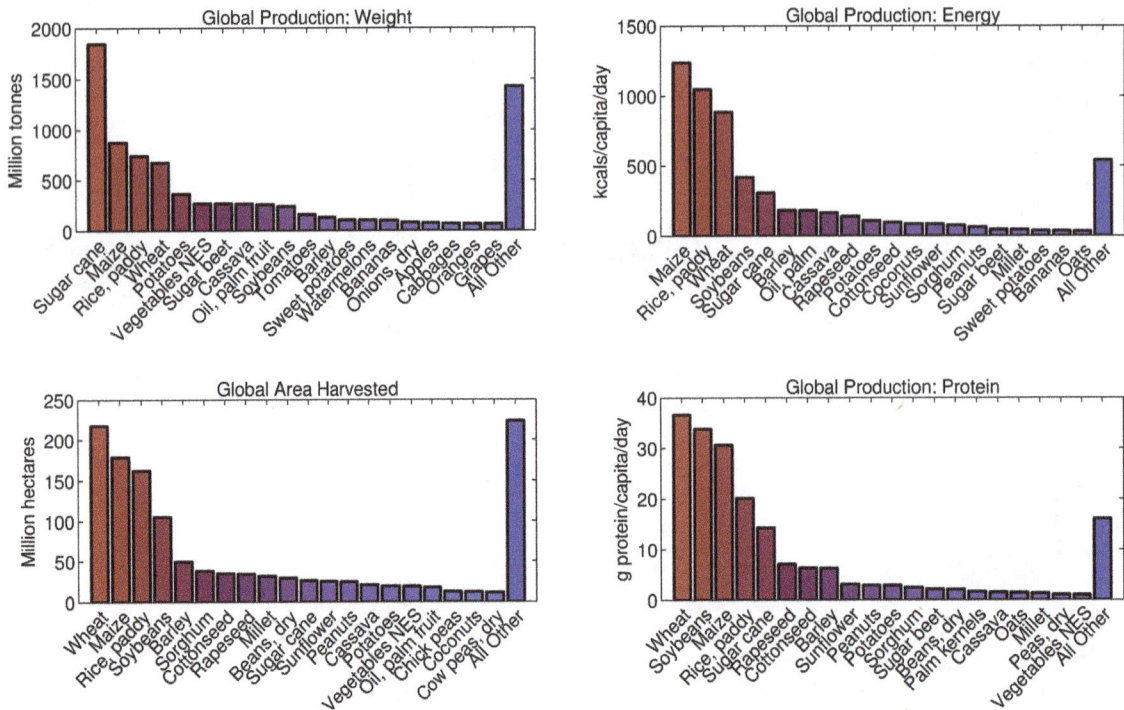

Figure 19.3: Top global crops by tonnage, energy, and protein production, and by land area harvested (FAO data). As can be seen, global agriculture is more diverse than that in the US, but it is still dominated by staple and oilseed crops, especially wheat, maize, rice, sugar cane, and soy.

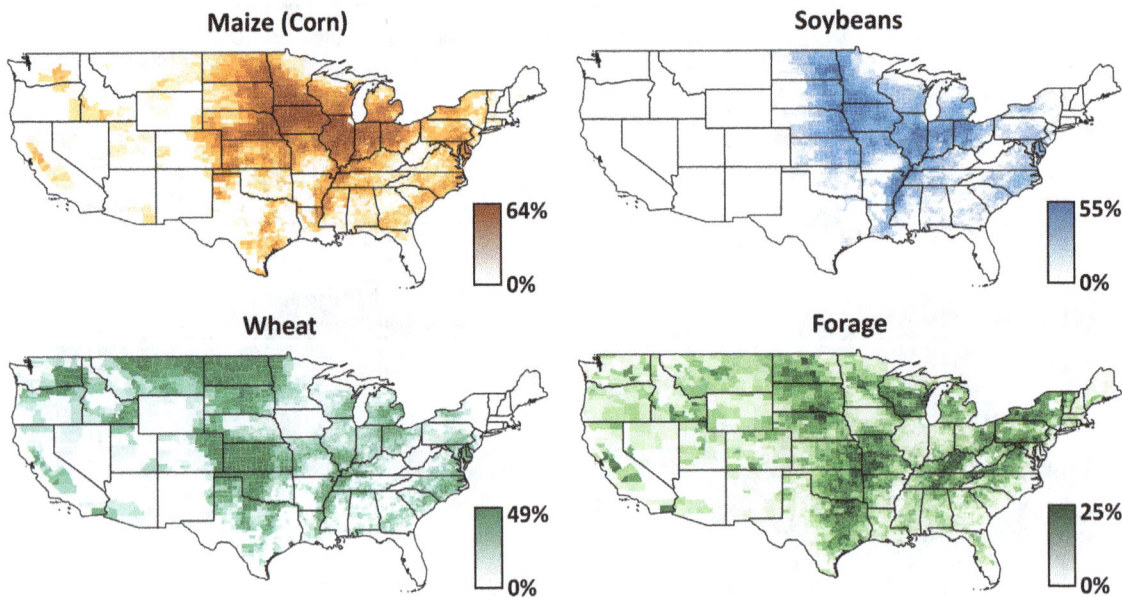

Figure 19.4: Relative geographic distribution (at the county scale) of corn, soy, wheat, and forage acreage, given as a percentage of total land area devoted to each crop; the colorbars indicate the maximum percentage of land area that is harvested crop, reaching a maximum of 64% for corn. Maps based on the 2012 USDA Census of Agriculture.

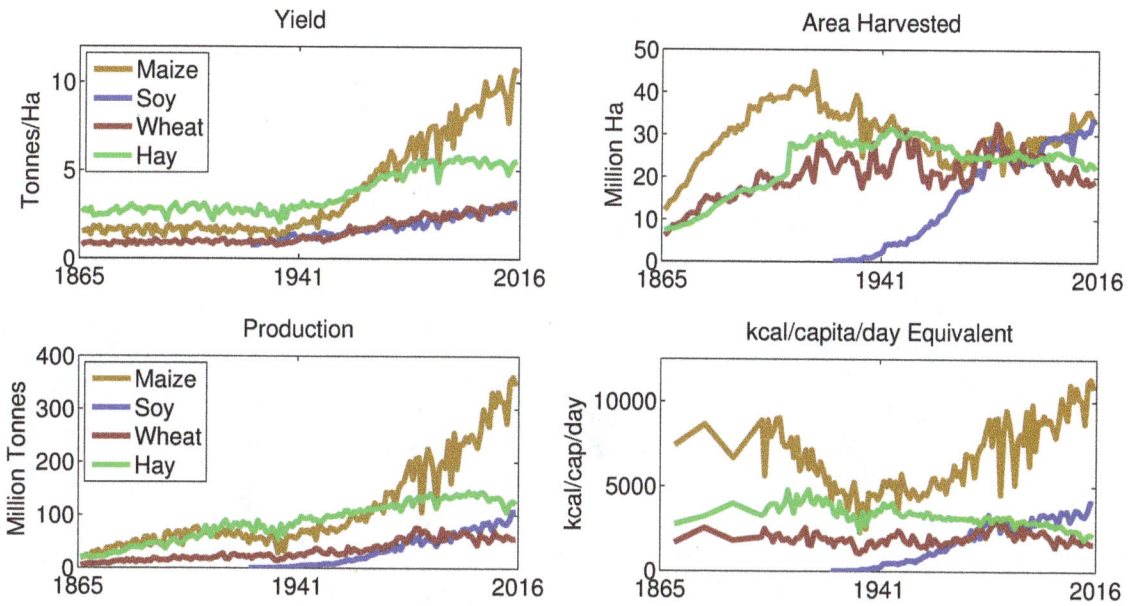

Figure 19.5: Production trends for the top four US crops, since 1866. We see that yields for all were steady until about World War II, when the Green Revolution and modern industrialized agriculture began in earnest. Overall calorie availability declined steadily until that point, but now exceeds any time in the past.

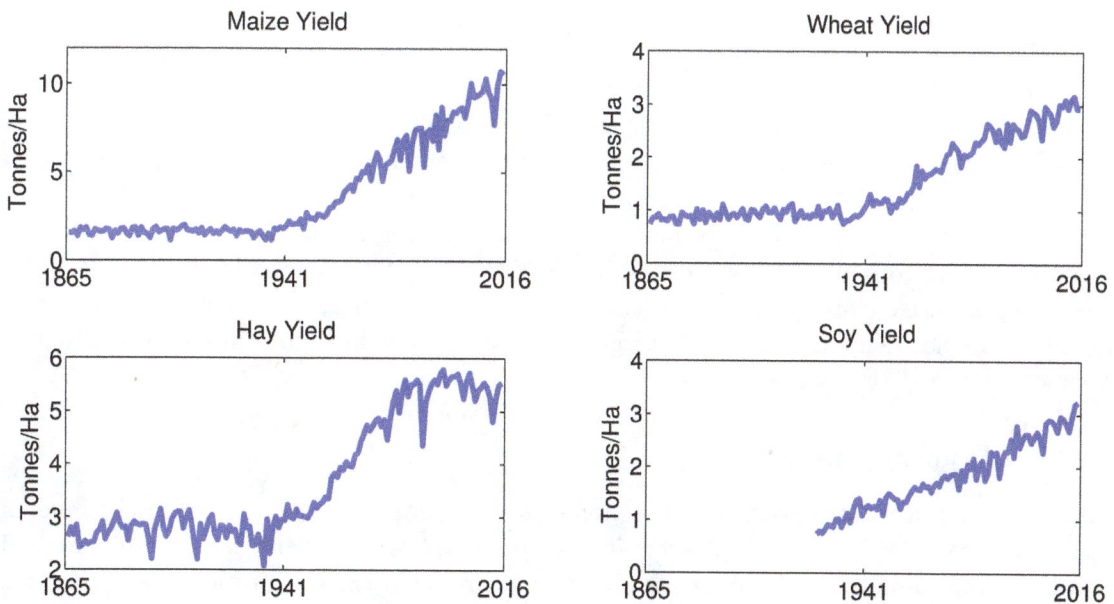

Figure 19.6: Historical yield (tonne per hectare) trends for the top four US crops, maize, soy, wheat, and hay, since 1866, which uniformly increased markedly with the advent of World War II and the Green Revolution.

Maize (corn)

Maize dominates US agriculture, and has always been the dominant staple of the Americas, both during pre-Columbian times and through the expansion of the US across the continent. Corn yields were stable from the nineteenth century to the 1930s, and then increased steadily as agriculture transitioned from a low-input to high-input model (the Green Revolution), beginning with the wide adoption of hybrid varieties, synthetic nitrogen fertilizer, and pesticides (see [332] for an overview of this trend). Figure 19.7 gives historical production, yield, and land area harvested dating to 1866. Notably, per capita maize energy in the late 1800s was near modern levels, and similar to now, most was "processed" through livestock into meat. Significant amounts were also processed into whiskey, another value-added transformation. Corn was mainly used as animal feed until very recently: 40–45% of the corn crop has been diverted to ethanol fuel production in the last few years, although absolute feed corn supply has remained relatively constant. This recent trend is summarized in Figure 19.8. For an informative book-length treatment of the history of maize in US history and especially in the Midwest, the reader may consult Clampitt [333].

US maize production yields food energy equivalent to over 10,000 kcals/capita/day and protein greater than 250 g protein/capita/day in most years (FAO, ERS). Thus, both the crude protein and caloric requirements of over 1.5 billion people could be met by the US maize crop alone, although of course they are not. Overall, in a typical year about 15% of corn produced is exported, 40% goes to ethanol, 35–40% goes to animal feed, perhaps 7–8% is processed by *wet milling* to high fructose corn syrup, corn syrup, and cornstarch mainly for use in processed foods, about 1% is fermented to beverage alcohol (ethanol for the people), and under 1.5% is consumed directly (FAO, USDA ERS). Thus, about 10% of the corn crop ultimately is eaten somewhat directly by Americans, but most of this is in processed form (per capita, wet milling yields over 800 kcals/day, alcohol is equivalent to 100 kcals/day, and direct consumption also amounts to about 100 kcals/day).

Soybeans

Soybeans are relatively new to America, and were not grown at meaningful scale until the 1920s. Corn and soy now form the backbone of US agriculture, and the corn-soy crop rotation dominates throughout the country. These two crops are used mainly as complementary animal feeds: corn provides energy, while soy provides protein.

After harvesting, nearly all soybeans are processed to separate a protein-rich meal from the oil; the beans are roughly 40% protein and 20% oil on a dry matter basis [334], with soybean meal used almost exclusively as animal feed. Note that as a legume, soy fixes its own nitrogen, and thus does not generally require nitrogen fertilizers, even in highly intensive conventional systems.

19.2.2   Major animal products

American and European agriculture has, for centuries, been heavily focused on animal production relative to other historical societies. Cattle have traditionally been dominant in US agriculture, but in recent years, increasingly efficient broiler chicken operations have led chicken to displace beef as the number one meat consumed by Americans. Time-series of American beef, milk, pork, and egg production since 1899 are given in Figure 19.9. The geography of major animal product production is illustrated in Figure 19.10.

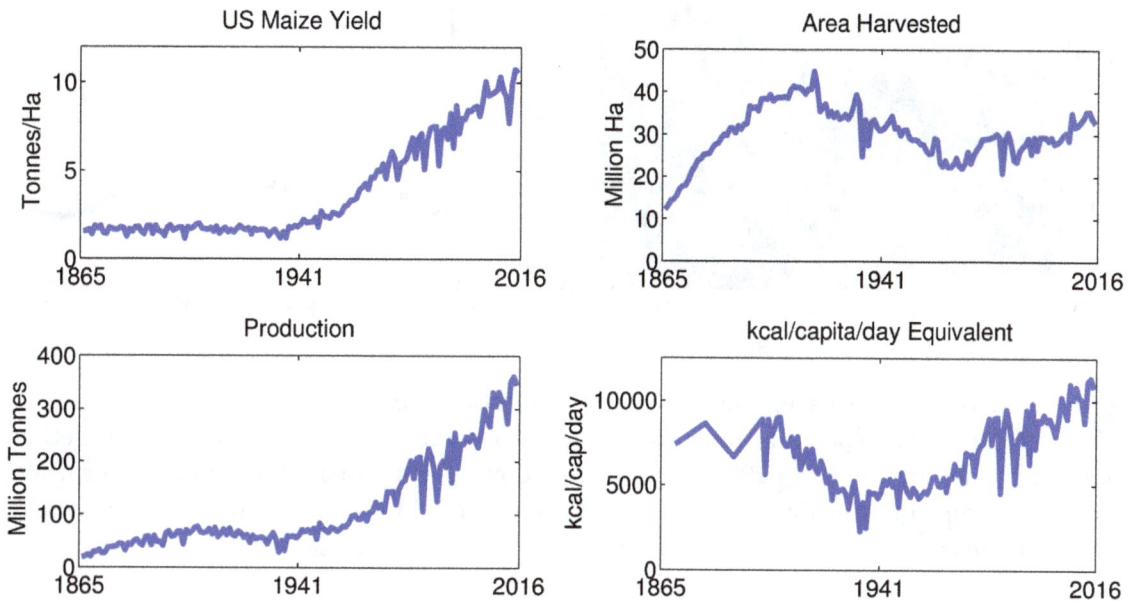

Figure 19.7: Historical changes in US maize yield, harvested crop area, absolute production in terms of mass and per capita energy (in kcals/capita/day), since 1866. As can be seen, from 1866 to 1900, the area under cultivation more than tripled with commensurate gains in productivity, while yield stayed flat until roughly World War II. Maize energy yield on a per capita basis was actually quite high in the late 1800s, supporting high levels of beef, pork, and milk production, but as harvested area plateaued and the US population continued to grow, per capita energy supply dropped until the Green Revolution beginning in the 1930s. With increased yields, but little change in overall harvested area, the US population, which is now over 10 times larger than it was in 1866, sees even higher per capita maize energy and corresponding meat supplies. Source: USDA NASS.

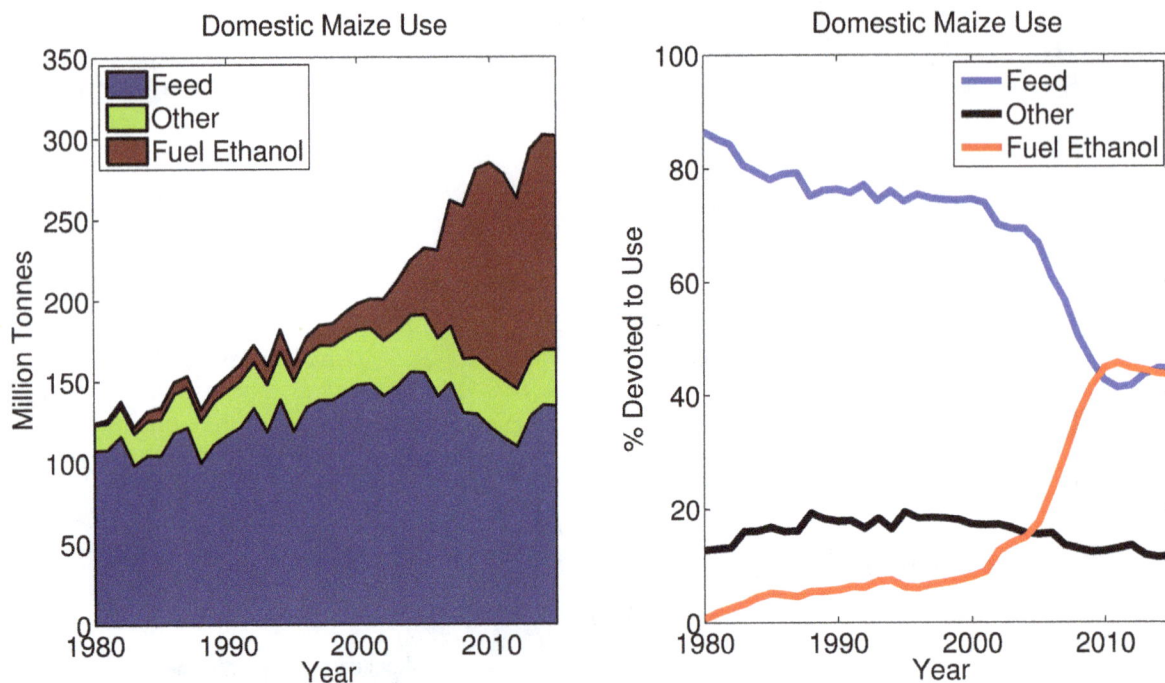

Figure 19.8: Recent trends in the end-use of domestic corn (i.e., excluding corn for export, 10–20% of the total depending on year) from 1980 to 2015. As can be seen, fuel ethanol rapidly expanded by the early 2000s, and now accounts for about half of domestic corn. With increasing production, the absolute amounts of corn devoted to animal feed and other uses have not changed markedly. Source: USDA ERS.

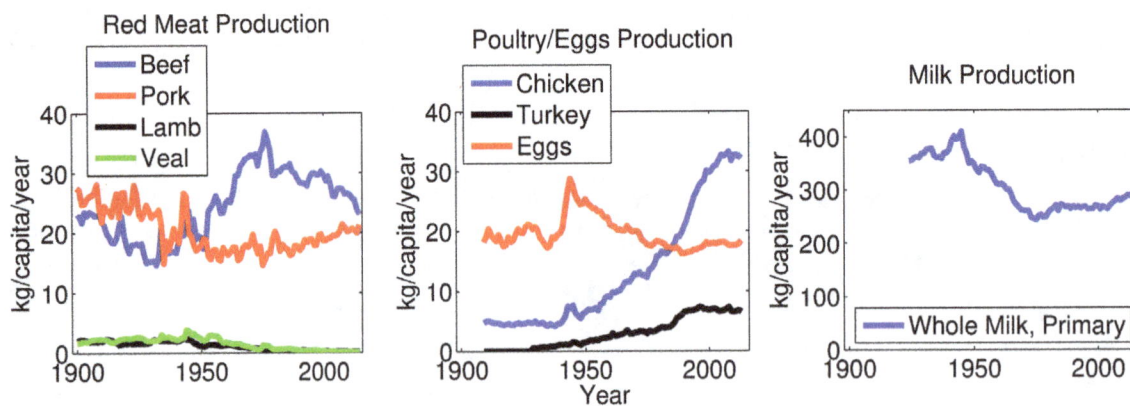

Figure 19.9: US production trends for major meat products, eggs, and whole milk, on a per capita basis. As can be seen, broiler chicken production has greatly expanded since the second world war, while, after peaking in the late 1970s, beef production has slowly fallen. Also of note, the veal and lamb markets have largely collapsed in recent decades in the face of decreasing consumer demand. Source: USDA ERS.

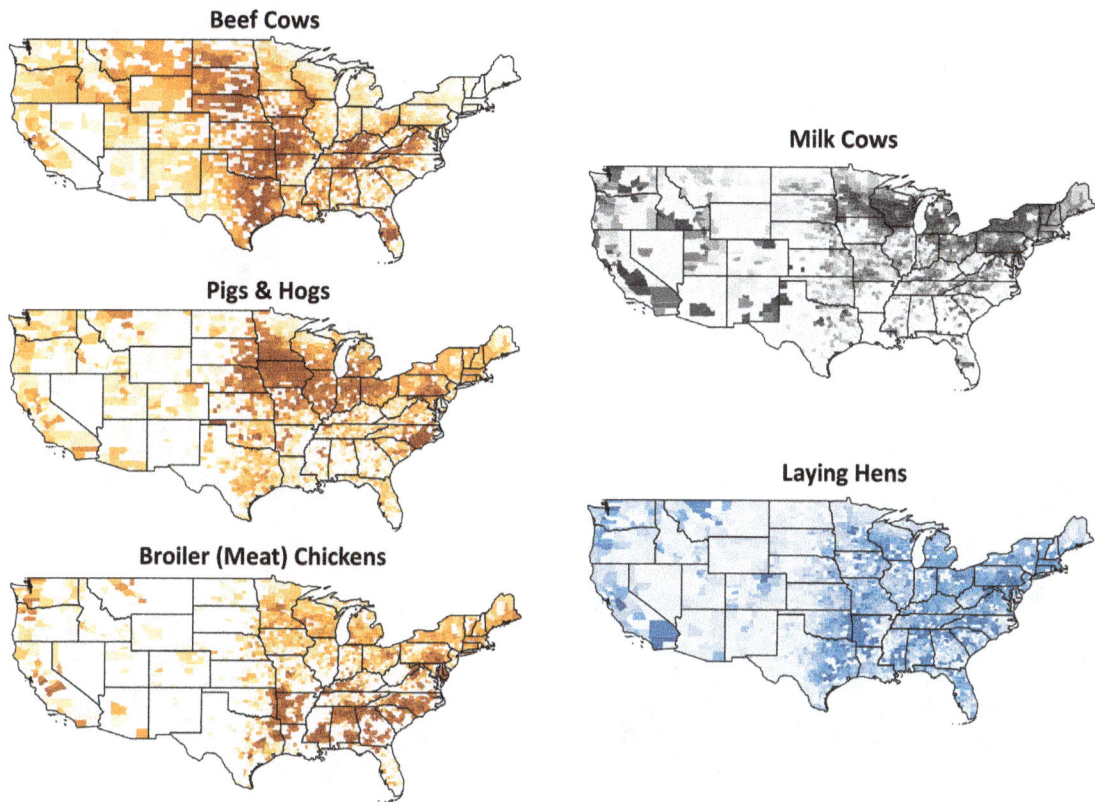

Beef Cows

Milk Cows

Pigs & Hogs

Laying Hens

Broiler (Meat) Chickens

Figure 19.10: Relative geographic distribution (at the county scale) of beef cattle, milk cows, pigs and hogs, broiler (meat) chickens, and layers, based on the 2012 USDA Census of Agriculture. Maps are scaled to the number of animals per county; note that some empty counties represent missing or withheld data.

Chapter 20

Cropping systems and associated emissions

In this chapter, I provide a detailed discussion of the inputs, processes, and land-use changes associated with cropping systems, and the associated global warming emissions. I discuss, in turn, fertilizer, pesticides, land use change, on-farm energy use, irrigation, transportation, and final downstream processing, packaging, and delivery of agricultural products.

20.1 Fertilizer and limiting nutrients

Via photosynthesis, plants store solar energy by converting water (H_2O) and CO_2 into hydrocarbon chains (carbohydrates), with oxygen (O_2) a "waste" product. Reversing the reaction yields useful energy, and these carbohydrate energy stores provide for all the plant's needs, including synthesis of new proteins, DNA, RNA, etc., and ultimately support all animal life. Clearly, sunlight and water are limiting factors on growth, and atmospheric CO_2 levels also can affect growth (although not necessarily positively). Several other elements are ubiquitous in biological molecules, and their availability also limits growth.

The concept of limiting nutrients and the vital role of nitrogen was recognized in the early 1800s by Justus von Liebig (among others), who studied the "matters which supply the nutriment of plants," [335] and in 1863 he described his Law of the Minimum [336]:

> Every field contains...a minimum of one or several other nutritive substances. It is by the minimum that the crops are governed...Where lime or magnesia, for instance, is the minimum constituent, the produce of corn and straw, turnips, potatoes, or clover, will not be increased by a supply of even a hundred times the actual store of potash, phosphoric acid, silicic acid, etc., in the ground.

Nitrogen is a component of both the amino acids that are the basic building blocks of all proteins and the nitrogenous bases of DNA and RNA. Phosphorus forms part of the backbone of DNA and RNA, acts as an energy carrier (e.g. in ATP and ADP), and is involved in myriad cell signalling pathways. These two elements are the most important limiting nutrients in plant growth, and nitrogen and phosphorus fertilizers are fundamental to agriculture. Nitrogen in particular is the fertilizer and nutrient of foremost importance, as it is generally *the* limiting nutrient, has by far the greatest environmental impact of the common fertilizers, and organic agriculture is largely defined by a rejection of synthetic nitrogen. Potassium, in the form of potash (K_2O), is the third major plant fertilizer.

Phosphorus and potash are produced from mining operations that generally exploit *finite* ancient seabed deposits of element-rich organic matter (not unlike the fossil fuels mined from energy-rich deposits of ancient organic matter), whereas nitrogen is mainly produced from the effectively *infinite* atmospheric nitrogen via the industrial Haber-Bosch process, but is also produced naturally by legumes (e.g. beans, peas, etc.) and other "nitrogen-fixing" crops.

Before continuing, it is important to clarify the difference between "organic" and "synthetic" fertilizers, as the distinction is only indirectly related to organic agriculture. Organic fertilizers are directly derived from plant or animal biomass, and include compost and manure. Simply enough, synthetic fertilizers are *synthesized* by industrial processes, mainly the Haber-Bosch process for synthesizing ammonia, a nitrogen source, and via processing of phosphate and potassium rock.

20.1.1 Man's open-loop nutrient cycle

A *fundamental* aspect of modern agriculture is the open-loop nature of the nutrient cycle. That is, nutrients are effectively mined from the soil to feed humans and are never returned. While much nutrient loss from agricultural soils is attributable to erosion or inefficient application practices, even under the best agricultural management, very large amounts of inorganic nutrients are, by necessity, removed from the local system with crop harvesting. Further nutrient losses occur at the animal stage (when crops are used as feed), and then there is another nutrient drain via sewage systems that, in the main, dump these nutrients into waterways, where they are effectively lost. To a much lesser extent, burial and cremation practices also remove the nutrients contained in human bodies from the larger agricultural ecology.

Therefore, *all* agricultural systems, both conventional and organic, lose large amounts of soil nutrients that must be replenished, generally via large fertilizer inputs. While some organic systems (and conventional ones as well) may replace nitrogen by incorporating nitrogen-fixing legumes into the crop rotation, phosphorus, potassium, and other nutrients are permanently lost and must eventually be replaced, lest the soil be exhausted. Organic systems may employ "natural" sources such as bonemeal and manure, but the mineral nutrients in these products must also be sourced from somewhere, and will generally represent a mining of other soils or simply a transfer of synthetic nutrients from conventional to organic fields via an animal intermediary.

20.1.2 Nitrogen, part I: the cycle(s)

As already mentioned, nitrogen is the most important limiting nutrient in agricultural soils, and anthropogenic alteration of the global nitrogen cycle has markedly increased greenhouse gas emissions in the form of nitrous oxide (N_2O), perhaps the third-most important single greenhouse gas. In this section I review both the local soil-plant nitrogen cycle and the global cycle, while the following section discusses the direct and indirect greenhouse gas impacts of nitrogen fertilizer synthesis and application.

Basic soil-plant nitrogen cycle

Nitrogen is of profound importance in the global ecology, as it cycles through atmospheric and various biological and soil pools. Primarily through fossil fuel combustion and synthetic nitrogen production and (mis-) application, Man has profoundly altered the global nitrogen cycle. A basic understanding of how nitrogen moves between soil and vegetable pools will serve us well in understanding the effects of nitrogen fertilizers on agricultural lands and beyond.

Nitrogen, as N_2, is the major component of the atmosphere, and it composes 78% of dry air by volume. However, N_2 is fairly inert, and cannot be utilized by plants for growth. First, it must be "fixed" into a reactive inorganic or "mineral" form, the chief forms being ammonia/ammonium (NH_3/NH_4^+) and nitrate (NO_3^-). Overall, nitrogen may be fixed via lightning strikes, biological nitrogen fixation, or via industrial processes and fossil fuel combustion. Fossil fuel combustion generates gaseous nitrogen oxides, mainly NO_2, and these and other nitrogen aerosols deposit onto soils and waterways globally. The Haber-Bosch cycle, discussed further in Section 20.1.3, uses large amounts of heat to fix atmospheric nitrogen to NH_3.

Lightning strikes provide enough heat energy to spontaneously convert some atmospheric N_2 and O_2 to NO_3^-, but this accounts for only a small fraction of naturally-occurring nitrogen fixation. The major natural mechanism is nitrogen fixation by bacteria known as *diazotrophs*, which may either exist as free living organisms in soil or water, or associate in symbiotic relationships with plants or fungi. The most famous and agriculturally relevant example of these symbiotes are the *rhizobia* that fix nitrogen within root nodules of legumes (e.g. beans, peas, clover, soy). While legume/rhizobia systems are by far the most studied, a much wider array of microbes is capable of nitrogen fixation, and this is an area of active discovery [344]. The reaction scheme for all biological nitrogen fixation (BNF) is:

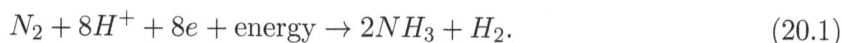

$$N_2 + 8H^+ + 8e + \text{energy} \rightarrow 2NH_3 + H_2. \tag{20.1}$$

Within diazotrophs (and other microbes), this reaction is catalyzed by biological enzymes. Via the industrial Haber-Bosch process, the source of essentially all synthetic nitrogen fertilizer, the exact same reaction occurs under high pressure and heat, catalyzed by iron-containing particles.

Upon fixation, NH_3 is either incorporated directly into plant matter, in the case of fixation by root nodule rhizobia, or enters into the soil nitrogen pool. Within the soil, nitrogen exists either in inorganic ("mineral") form (generally as NH_4^+ or NO_3^-) or in organic form as either a component of decaying plant or animal residues, living soil microorganisms, or some other part of the soil organic matter (SOM), e.g. humus. While NH_4^+ and NO_3^- can be taken up by plant roots and used for growth, as much as 99% of all soil nitrogen is in organic form [337]. This organic nitrogen *is not available* to plants, and thus it is referred to as "immobilized" nitrogen.

Immobilized (organic) nitrogen is made available to plants via the action of soil microbes, who metabolize decaying leaf litter and SOM. Some of the nitrogen in these sources is incorporated into microbial biomass, but some is converted to NH_4^+, a process called *mineralization* (minerals are inorganic, hence the name). Soil microbes themselves grow, die, and decompose, and nitrogen also shuttles between the mineral and organic pool as part of this process.

Organic nitrogen exists in a variety of forms, but mineral nitrogen exists mainly as either NH_4^+ and NO_3^-. NH_3 is the product of biological nitrogen fixation, but as a fairly basic molecule, it readily accepts a proton to become the NH_4^+ cation. *Nitrification* is an aerobic, two-step process by which NH_4^+ is oxidized to NO_3^- by bacteria, first to NO_2^- by nitrosomas, and then to NO_3^- by nitrobacters:

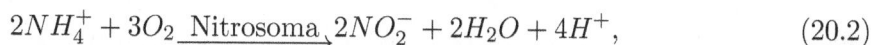

$$2NH_4^+ + 3O_2 \xrightarrow{\text{Nitrosoma}} 2NO_2^- + 2H_2O + 4H^+, \tag{20.2}$$

$$NO_2^- + H_2O \xrightarrow{\text{Nitrobacter}} 2NO_3^-. \tag{20.3}$$

The nitrate molecule, NO_3^-, is soluble in water, and as such, may leach out of the soil with flowing water. The ammonium molecule, NH_4^+, on the other hand, as a positively charged ion, reversibly binds with the soil's cation exchange complex (CEC), a collection of negative binding sites on soil clays and organic humus. It may also permanently bind with clays, and NH_4^+-bound clays also occasionally serve as a soil nitrogen source when eroded.

Finally, nitrogen is lost from the system in several basic ways: (1) leaching, (2) denitrification, (3) volatilization, and (4) crop removal. The water-soluble NO_3^- can leach out with water, especially under very wet conditions. This nitrate can contaminate ground and surface waters, causing eutrophication and other toxicities. In low-oxygen conditions, such as water-logged soil, some bacteria catalyze *denitrification*, a series of reactions by which oxygen is stripped from NO_3^-, converting it back to N_2 and, when the reaction is incomplete, N_2O and NO, thus serving as the major source of N_2O emissions from agricultural soils. The reaction series follows:

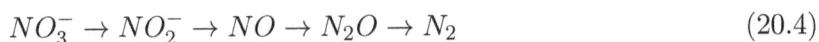

$$NO_3^- \rightarrow NO_2^- \rightarrow NO \rightarrow N_2O \rightarrow N_2 \tag{20.4}$$

Ammonia, NH_3, can also volatilize to be lost as gas. Volatilization is more likely in basic (versus acidic), warmer, and drier, soils. Finally, in agricultural systems large amounts of nitrogen are lost via crop removal, and must eventually be replenished by an outside source to maintain fertility. The basic plant-soil nitrogen cycle is illustrated in Figure 20.1.

With this basic understanding, we can see why nitrogen fertilizers are valuable in supporting plant productivity, and how organic and synthetic fertilizers differ. Organic fertilizers, as plant and animal residues, contain immobilized nitrogen that must be mineralized by microbes before they support plant growth. Therefore, organic fertilizers release their nitrogen relatively slowly, and a basic challenge when using organic fertilizers is synchronizing mineral N supply with peak plant N demand.

Synthetic nitrogen, on the other hand, traditionally was produced primarily as anhydrous ammonia (NH_3) or ammonium nitrate, $(NH_4^+)(NO_3^-)$, and thus is immediately available to plant roots upon application. In recent years, urea, with chemical formula $CO(NH_2)_2$ and manufactured from NH_3, has been the most popular form of synthetic fertilizer. Urea is water soluble and rapidly converted (within 48 hours) to ammonia and carbonic acid (H_2CO_3) by the enzyme urease, present in soil microbes, as follows:

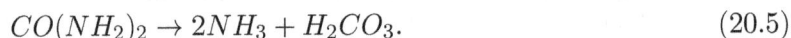

$$CO(NH_2)_2 \rightarrow 2NH_3 + H_2CO_3. \tag{20.5}$$

Thus, we see mechanistically why organic fertilizers are relatively slow-release and synthetic fertilizers are rapid- or immediate-release. Note that organic fertilizers also contain large amounts of organic carbon, whereas synthetic fertilizers are pure nitrogen sources. Furthermore, urea is the main form of nitrogen found in urine, and most nitrogen consumed by animals, including humans, is excreted as highly concentrated urine urea. Urea rapidly breaks down to ammonia which will volatilize and be lost from manure within a matter of days, if not applied to soils.

Understanding the basic mechanics of the system informs our discussion of the environmental externalities of nitrogen fertilizer and manure management, as well as comparisons of conventional and organic production systems, as discussed extensively in Section 22.1.

Quantifying nitrogen use efficiency, yield from fertilizer, and nitrogen losses

Generally speaking, 50% or less, and often much less, of the nitrogen fertilizer applied in a given year will make it into crop biomass. Typically, 15–25% may be immobilized in soil organic matter, while 15–20% is lost to ammonia volatilization, 5–15% is lost via nitrate leaching, and 5–10% is lost to denitrification, with wide variability in these losses depending on climate, soil type, and management practices. Losses and nitrogen use efficiencies can also vary between agricultural system type, e.g. animal production on grasslands vs. field crops. Practices that

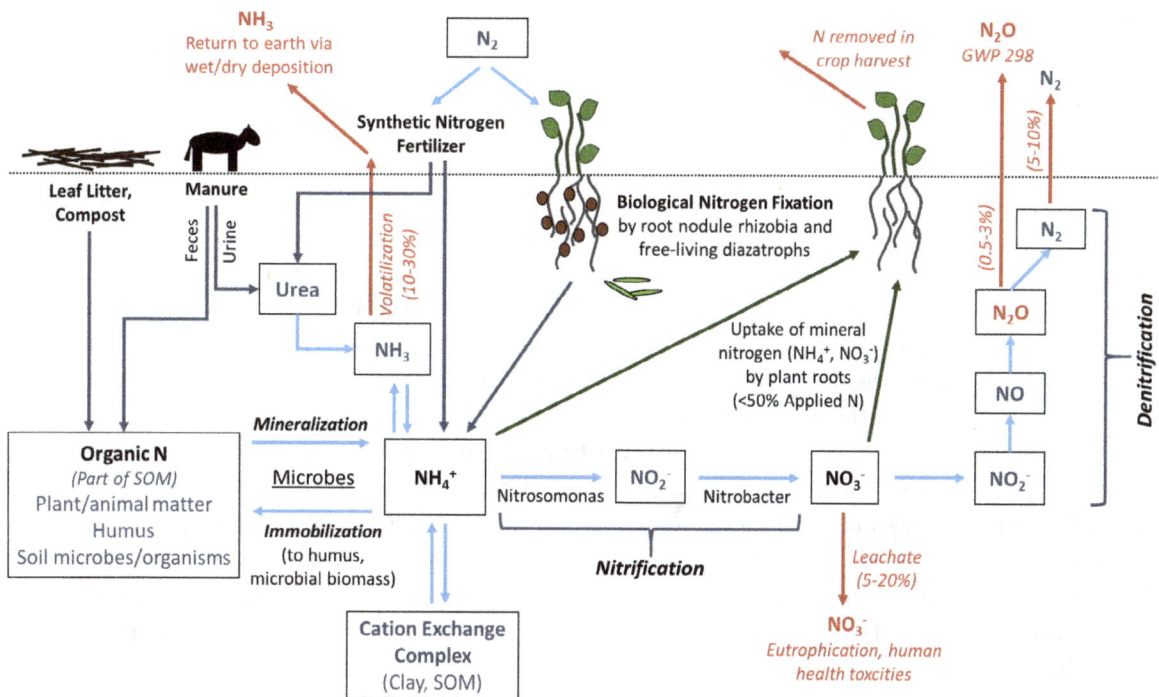

Figure 20.1: Schematic for the basic soil/vegetation nitrogen cycle.

minimize any one loss tend to minimize all losses, as well as improve nitrogen use efficiency, and Cameron and colleagues provide an excellent quantitative review [338].

Volatilization. While large amounts of soil ammonia nitrogen may volatilize to the air, atmospheric ammonia rapidly deposits back onto soil [344], and thus re-enters the plant-soil cycle until meeting some other ultimate fate. The location of re-entry, however, may not be one that is desired.

Leachate. Nitrogen that leaches into groundwater represents both a loss of soil fertility, a threat to human health, and may cause eutrophication with attendant algae blooms and fish kills when it enters local waterways [338].

Denitrification. Denitrification, either to N_2 when the reaction is complete, or to N_2O (or other intermediates) when incomplete, is the ultimate fate of all new reactive nitrogen not lost to the seas. The fraction of N entering the denitrification that becomes N_2O is variable; for example, 16–26% of nitrogen applied to cabbage fields evolving to N_2 or N_2O by denitrification ended up as N_2O in [340]. The ratio of N_2:N_2O varied dramatically with moisture, soil, carbon, and nitrogen applied in [341], and N_2O may make up 20–50% of denitrification products depending on moisture [342]. If we suppose that 5–10% of applied nitrogen is initially lost via the denitrification pathway, then about 10–20% of this nitrogen is N_2O, if overall N_2O emissions are 1% of applied N.

Global nitrogen cycle

Having reviewed the basic plant-soil nitrogen cycle, and the basic process by which inert N_2 is "fixed" to become new reactive nitrogen, we can scale our conversation up to the global nitrogen cycle, which has been dramatically altered by new anthropogenic nitrogen. Via biological nitrogen fixation, the Haber-Bosch process, and fossil fuel combustion (plus a few other minor processes, such as lightning), on the order of 400–450 Tg of nitrogen is fixed each year to become reactive nitrogen (N_r). Reactive nitrogen is a catch-all that includes NH_3, NH_4^+, NO, NO_2,

NO_3^-, and organic N compounds [343], i.e. those nitrogenous molecules that readily participate in biological and ecological processes, but all reactive nitrogen begins as NH_3/NH_4^+ or NO_x. As reviewed by multiple authors, approximately half of all newly fixed reactive nitrogen is now directly attributable to human activities [343].

A general model for understanding the nitrogen cycle is to compare the current, rapidly changing cycle to the pre-industrial nitrogen cycle, which we assume to have been, essentially, at steady-state (at least at the global scale). Under pre-industrial conditions, all reactive nitrogen inputs to land must be balanced by equal losses to denitrification (conversion to gaseous N_2 and N_2O) and hydrologic losses, i.e. transport through groundwater and rivers to ultimately reach the ocean. Terrestrial inputs include biological nitrogen fixation (BNF), lightning, transport of nitrogen from the oceans to land, and weathering of nitrogen fixed within rocks. Current anthropogenic inputs entail new reactive nitrogen from fossil fuel combustion, additional BNF via the cultivation of legumes, and most importantly, the Haber-Bosch process.

At the present day, it is impossible to separate a human-driven anthropogenic nitrogen cycle from a background pre-industrial nitrogen cycle, for several reasons. First, the global lands and ecosystems responsible for BNF in the pre-industrial era have obviously been radically altered since, and second, anthropogenic nitrogen is ubiquitous in the global ecosystem, modifying plant growth, ecosystem function, *and* ecological BNF. In particular, BNF by many legumes increases when nitrogen is scare, and is suppressed when nitrogen is abundant [344].

The global nitrogen cycle is depicted schematically in Figure 20.2. This, and our model of the soil-plant nitrogen cycle, informs a simplified, lumped representation of the global N cycle, and its basic coupling with the carbon cycle, depicted in Figure 20.3. That is, new reactive nitrogen enters the cycle as NH_3/NH_4^+, *reversibly* combining with carbon to form organic matter, thus acting as a coupled nitrogen/carbon sink. This reactive NH_3/NH_4^+ ultimately evolves to NO_3^- via nitrification (which may irreversibly be incorporated into organic matter), and then to N_2 or N_2O via, denitrification, the ultimate fate of all new nitrogen (note that N_2O itself eventually decays to N_2 by photolysis). There also exist long-term N_r sinks in the form of deep ocean sediments, burial in peatlands, etc. Furthermore, while N_r can enhance carbon storage, it can also enhance microbial breakdown of organic substances, and the net effect of increasing global N_r on carbon stocks is not fully understood [345].

20.1.3 Nitrogen, part II: global warming impacts

- Synthesis of reactive nitrogen via the Haber-Bosch process is emissions-intensive, at around 5 $kgCO_2e/kgN$ of fertilizer, translating into about 50 million $MgCO_2e$ in the US.

- Nitrogen fertilizer evolves to N_2O via a variety of mechanisms as it cascades through the ecosystem, yielding perhaps 14–23 $kgCO_2e/kgN$, or about 150–300 million $MgCO_2e$ for the US. Overall, a best guess is that US nitrogen fertilizer emissions sum to about 275 million $MgCO_2e$.

Reactive nitrogen production and embodied energy/emissions

In 2014, global nitrogen *fertilizer* demand (in terms of elemental nitrogen, N), was 113.1 million Mg, and the total demand for nitrogen (i.e. fertilizer + non-fertilizer) was 147.293 million Mg; supply was near demand, at 152.769 million Mg, per the FAO. According to the USDA, the US consumed about 11.648 million Mg of N fertilizer in 2011, or around 10% of global use.

It is commonly stated that 1–2% of global energy (Wood and Cowie [347] cite a 1.2% figure) and 3–5% of global natural gas is directly consumed in the Haber-Bosch process, corresponding

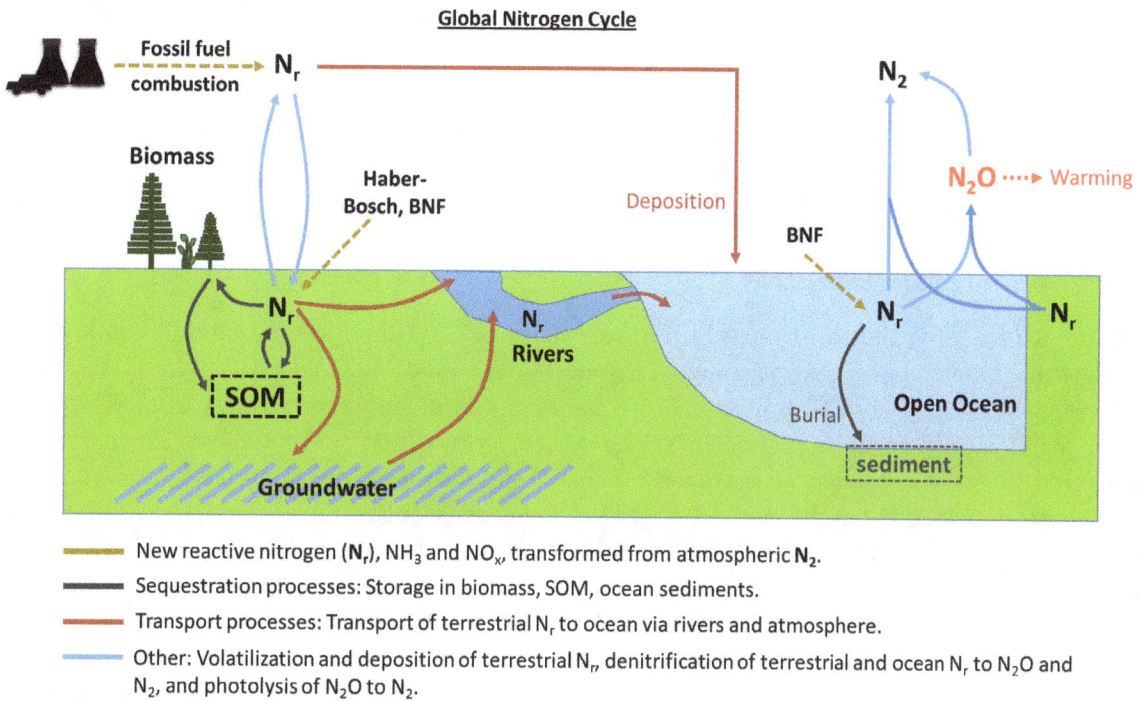

Figure 20.2: Major processes in the global reactive nitrogen cycle. Fossil fuel combustion, the Haber-Bosch process, and BNF are all major sources of new N_r from N_2, which then is subject to various transport, sequestration, and transformation processes that ultimately result in long-term sequestration or evolution back to N_2.

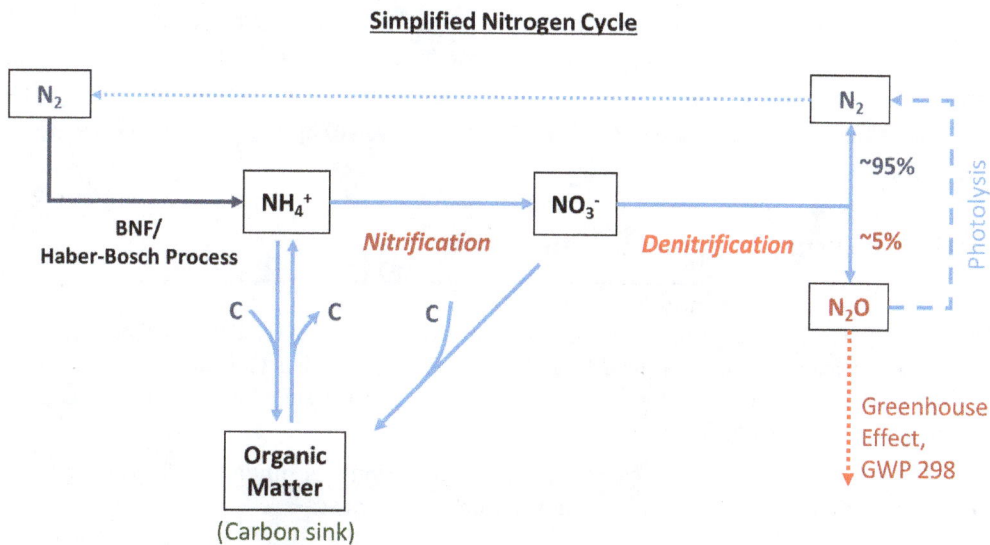

Figure 20.3: Simplified representation of the global nitrogen cycle.

to 1–2% of global carbon emissions, but actual energy and emissions estimates for the process are rather hard to come by. Based on numbers reviewed by Wood and Cowie [347], we can assume an absolute minimum of 2 $MgCO_2e/MgN$ of fertilizer, with at least 4 or 5 $MgCO_2e/MgN$ much more likely. This suggests a ballpark estimate of 50 million $MgCO_2e$ attributable to the production of synthetic nitrogen fertilizer used in the US, roughly equivalent to 10 million cars. Note that this is just about 0.1% of the estimated 49 $GtCO_2e$ global anthropogenic emissions in 2010 [8], about what we would expect for 10% of 1% of global energy use, and it is less than 1% of US territorial emissions.

N_2O from nitrogen fertilizer

Nitrous oxide, N_2O, the third-most important well-mixed greenhouse gas overall, is fairly long-lived (atmospheric perturbation time 121 years), with an IPCC AR5 100-year GWP of 298 (20-year GWP 268). Globally, agriculture is by far the most important source of anthropogenic N_2O. A quick note on units and calculations before proceeding. Synthetic fertilizer is typically measured in mass of elemental nitrogen (N), and suppose 5 kg out of 100 kg of elemental N applied to a field ultimately evolved to N_2O. Then we would write 5 kgN_2O-N, but should actually have $44/28 \times 5 = 7.86$ kg N_2O, yielding a global warming potential of $44/28 \times 5 \times 298 = 2,341$ $kgCO_2e$.

Globally, overall N_2O is fairly well-understood, and multiple authors have derived top-down estimates of the fraction of synthetic fertilizer N that evolves into N_2O. As reviewed by Crutzen et al. [348], top-down estimates suggest that 3–5% of "new" anthropogenic N is ultimately converted to N_2O (see also the more recent review by Reay et al. [349]). Note that this is in contrast to the AR4 IPCC bottom-up estimate of 1% direct emissions factor of N from agricultural fields, plus 0.35–0.45% indirect emissions. However, there is wide uncertainty in bottom-up emissions, and it seems likely that the complex transformation of N as it flows through agriculture systems accounts for the disparity. For example, N in fertilized crops may be converted into animal feed, some of which is then emitted as N_2O from animal dung and urine (in AR4 the IPCC estimated 2% of excreted N is emitted as N_2O). Thus, a bottom-up inventory expanded in scope can be consistent with the 3–5% top-down estimate, as pointed out in [348], and thus I consider this to be a credible range.

It follows from the above numbers that on the order of 150–300 million $MgCO_2e$ are attributable to N_2O evolving from US synthetic nitrogen fertilizer, with the majority of emissions those stemming from the nitrogen cascade downstream of direct application to agricultural fields. The per-unit emissions factor is about 14–23 $kgCO_2e/kgN$ attributable to downstream N_2O production.

20.1.4 Phosphorus: yet another peak?

- The direct global warming impact of synthetic phosphate (P_2O_5) fertilizer derived from phosphate rock is comparatively minor, around 1 $kgCO_2e/kg$ P_2O_5, and thus the 3.92 million tonnes of P_2O_5 applied to US fields in 2011 was equivalent to abound 4 million $MgCO_2e$, a tiny fraction of the warming due to nitrogen fertilizer.

- Phosphate rock is a finite resource, and resource depletion is a fundamental concern, with "peak phosphorus" likely to occur within a century, or two at most.

Phosphorus is an essential mineral nutrient for all life, and, since World War II, humanity has vastly increased the rate at which phosphorus enters the environment via mining of phosphate rocks and application as fertilizer. Phosphate is a finite, mineral resource, for which there is no substitute, and recent interest in a potential "peak phosphate" was largely initiated by a 2009 paper by Cordell and colleagues [350]. Peak phosphate, and the potential to exhaust the physical phosphate resource base, is on one level far more concerning than even fossil fuel peaks, for while substitutes may be found for such fuels, none exist for phosphate: as related in [351], the venerable Isaac Asimov once described phosphate as "life's bottleneck." More immediately, phosphate fertilizer runoff (like nitrogen runoff) causes eutrophication and nutrient pollution across many bodies of water. Phosphorus is highly heterogenous spatially: excessive use in many areas, especially in the developed counties, causes locoregional pollution and surplus soil phosphorus, while many poorer countries suffer from serious deficiencies that limit agricultural productivity [351].

From a global warming perspective, phosphate fertilizer is relatively benign, at least when compared to nitrogen fertilizer: as reviewed by several authors [501, 500], mining and production of phosphate fertilizer takes only about a fifth the energy as nitrogen fertilizer (using kg N and kg P_2O_5 as our respective functional units), with estimates range from roughly 5–25 MJ/kg P_2O_5. Using 10–15 MJ/kg P_2O_5 a reasonable midpoint, this equates to perhaps 1 $MgCO_2e$/Mg P_2O_5 on average. Therefore, after accounting for downstream N_2O emissions, P_2O_5 has only 4–10% the global warming potential as synthetic N, on a per-kg basis. Since, moreover, phosphate fertilizer application rates are less than one-fifth nitrogen application rates worldwide [352], and about 33% the nitrogen application rate in the US (USDA), phosphate is only a minor source of agricultural CO_2e emissions.

Phosphate cycle and anthropogenic alteration

The natural phosphorus cycle occurs on a timescale of several million years. Reactive phosphorus makes its way to the ocean, where it is buried in deep sediments with dead sea life, forming sedimentary rocks that, eventually, make their way back to the surface via tectonic activity and uplift, where they weather and return phosphorus to the biosphere [353]. This cyclical process is the source of all phosphate rock reserves, which thus are nonrenewable resources over a time-scale of millions of years (similar to the fossil fuels formed via somewhat similar geologic processes) [353].

Since antiquity, agricultural phosphorus came only from recycled local sources, chiefly animal manure, and, especially in China and Japan, human waste. By the nineteenth century it was discovered that bone meal could provide nutriment to plants, and vast deposits of bat guano discovered in several islands off the coast of South America were mined at scale, with England importing large quantities of both bone and guano for their phosphorus [351]. Guano reserves were depleted in relatively short order, and the world turned to synthetic phosphorus extracted from phosphate rock, with exploitation of this resource exploding after the second World War, altering the global phosphorus cycle from a slow, closed loop cycle with more rapid local cycling, to one largely characterized by rapid one-way flow from rock to field and then onto (essentially permanent) dispersal into the oceans, with a portion (perhaps 50%) retained in soils [351].

Phosphates reserves and peak phosphate

The actual magnitude of global mineral phosphate reserves is unclear, but it obvious that the geologic distribution of these resources is extremely uneven, with Morocco alone the location of over 70% of global phosphate [350]. In 2009, Cordell and colleagues [350] predicted, using 2009 USGS reserve data and a Hubbert-style curve for resource extraction (the same basic curve that formed the basis for Hubbert's famous Peak Oil theory, as discussed in Section 3.7.3), that phosphorus production would peak by 2033. In 2010, the International Fertilizer Development

Center (IFDC) published updated phosphate reserve estimates nearly four times higher than previous estimates, and these were adopted by the USGS to more than quadruple more recent phosphate reserve estimates from 16,000 Mt phosphate rock in 2009 to 65,000 Mt in 2010. However, this update was not based on any new data, with Edixhoven et al. [354], for example, very critical of the methodology; these authors concluded these new estimates are likely inflated and act to confuse the phosphate scarcity debate.

In any case, using higher phosphate reserve estimates shifts the time of peak phosphorus to the latter part of the twenty-first century. While more optimistic estimates give 300–400 years before phosphorus depletion when assuming constant use rates, estimates that include increased demand and any "Hubbert-like" curve are more pessimistic (see [351] for a summary of such estimates), and it seems likely that phosphorus will peak before 2100. Efficient and appropriate use, agricultural practices that minimize erosion, and recycling of waste, including human waste, can extend phosphorus supplies [352], and will almost certainly *eventually* become mandatory to preserve the long-term fertility of global agricultural soils.

20.1.5 Potash (potassium)

> - Potash is a source of soluble potassium (K) salts, and derived mainly from salt deposits left by ancient seas. The global warming impact of potash is relatively low, perhaps 0.5 ± 0.25 kgCO$_2$e/kg, and use rates are similar to phosphorus (about 4.16 million tonnes of K$_2$O in the US in 2011). Potash is a finite, nonrenewable resource that, like phosphorus, will last only a few centuries.

Potash is a general term referring any of several compounds containing soluble potassium (K), the third major plant fertilizer and a non-substitutable electrolyte needed by all organisms. In soil, nearly all K is unavailable to plants, and thus soluble additions support growth. Potash was initially derived mainly from wood ash (hence the term), a source still occasionally used, but, especially since World War II, nearly all K is mined from evaporite deposits that derive from ancient evaporated seas. For an in-depth history of potash, one may consult Ciceri et al. [356].

Producing potash fertilizer takes relatively little energy, perhaps about half that required for phosphate [501, 500], and thus 0.5 kgCO$_2$e/kg K$_2$O is a reasonable guess for the associated carbon emissions. Potash (as K$_2$O) is applied to agricultural soils at roughly the same rates as phosphate fertilizer, both globally and in the US. As with phosphate rock, the geographical distribution of potash reserves is highly uneven, and over 50% of reserves are found in Canada. Potash reserves are likely sufficient for several centuries [355], but, as with phosphate, without recycling and improved agricultural practice, we will eventually face "peak potash" as well.

20.2 Pesticides: herbicides and insecticides

- Synthetic pesticides came into widespread use following World War II, another technology of the Green Revolution. Since 1960, absolute pesticide use in the US dramatically increased through about 1980 but has since declined slightly, with most current pesticide use *herb*icide, rather than *insect*icide.

- Insecticides are generally far more toxic to all animal life than herbicides, and there also is greater potential to reduce insecticide use without compromising agricultural productivity. Pesticide use, overall, can probably be reduced by as much 50% with no or minimal effect on productivity. Complete abstinence may lead to significant crop losses.

- Pesticide production and application generates few carbon-equivalent emissions directly, perhaps 7.5 million $MgCO_2e$ per annum in the US, but pesticides do have multiple negative effects upon ecosystem health and productivity, as well as human health. These include direct wildlife kills, chronic toxicity, loss of both vegetable and insect food sources, widespread loss of pollinators, and increased cancer and immunologic disorders in humans exposed to pesticides.

- The effect of pesticides upon crop yields is probably smaller than that of fertilizers, but generally does increase productivity (at least short-term), and also enables highly productive cropping systems that are relatively vulnerable to pests and weed competition. It is therefore possible that, by increasing yield, pesticides reduce agricultural land-use and associated habitat loss and other agricultural input use. Thus, determining the balance of effects upon both general ecological health and CO_2e emissions is not a wholly straightforward calculus.

- Conventional fruits and vegetables are exposed to much higher pesticide spraying intensities than commodity crops such as corn and soy.

- Household pesticide use for purely aesthetic reasons ("healthy" lawns) harms both urban wildlife and humans, and is therefore generally an inexcusable practice.

20.2.1 Overview and history

Crop protection, *broadly speaking*, is profoundly important to agricultural production, and we may broadly divide crop "pests" into four major groups: pathogens (fungi such as powdery mildew, bacteria, etc.), viruses (also a pathogen, but considered as a separate category in some literature), animal pests (various insects, mites, and other arthropods, nematodes, snails/slugs, mammals, and birds), and weeds, which compete for space, light, and nutrients. Of these, weeds are by far the most important, accounting for over half of *potential* pest losses [357]. A variety of non-pesticide means exist to counter pests, such as crop rotation and mechanical weeding/tilling, but the era since World War II has seen a massive expansion in the use of synthetic pesticides targeting all categories of crop pest (mainly insecticides, fungicides, and herbicides), both globally and within the US. This expansion helped shape modern cropping systems by enabling simpler crop rotations, the use of higher-yielding but more vulnerable crop strains [357], decreased tillage for weed control, and, most recently, the widespread use of herbicide-resistant (mainly glyphosate-resistant) transgenic (GMO) crops [361].

Herbicides, rather than insecticides, are the primary pesticide class in use today (about 85% of US pesticides by mass), and, especially in the US, this category is now dominated by a single agent, glyphosate (the active ingredient in Roundup, the most common commercial version), which came into widespread use following the almost universal adoption of glyphosate-resistant corn, soy, and cotton GMO seeds since the mid-1990s [361]. Glyphosate and its various commercial formulations (which contain many poorly studied "adjuvants" that enhance the toxicity of glyphosate) may be less toxic to animal life than many other agents, but this molecule, and its intimate connection to GMOs, has been of great controversy, especially in popular environmental literature.

While global crop losses to pests, as a *percentage* of potential yield, from 1960 to about 2000 actually remained *constant*, the more modern cropping systems which are (at least partially) enabled by chemical suppression of pests are also intrinsically higher yielding, and so pesticide use has likely increased *absolute* agricultural productivity [357]. From a global warming perspective, the emissions directly attributable to pesticide production and application are fairly trivial, but their larger effects upon the agriculture system, e.g. possible land-sparing from higher yields, and modified cropping (simpler crop rotations with higher yielding varieties) and tilling patterns (decreased dependence upon mechanical tilling of weeds), are almost certainly significant. Of course, pesticides also have numerous ecosystem toxicities, and thus the overall balance of benefits and harms remains unclear. Pesticide intensity and agents also vary by both crop and farm, and it is probably possible to appreciably reduce use while maintaining yield. Finally, it should be noted that organic farming does not eschew all pesticides, just *synthetic* pesticides, and several "natural" pesticides can have significant toxicities, especially those containing copper (e.g. copper sulfate).

Early history

From the late 1800s, a variety of mainly copper, lime, and sulfur-based compounds were used in small amounts as general fungicides [358], but the modern era of crop protection via industrially produced pesticides largely began in the early 1940s with the rapid advance of the chemical industry during the War Years. A variety of antifungals were introduced in this period, while that most famous and controversial of insecticides, DDT, was first synthesized in 1939 by Paul Müller, a feat which would earn him the 1948 Nobel Prize in Physiology or Medicine; DDT entered commercial production in 1942 [359]. During the war, it was used on the European continent to control potato beetle plagues, while the American armed forces widely deployed the agent for malaria control in the Pacific theater [359]. Entering into civilian use in late 1945, DDT was soon hailed as a miracle compound, yet toxicity to fish and birds was recognized as early as 1946 [359]. A wide variety of new fungicides, herbicides, and insecticides would be synthesized and brought into use over the following decades.

In 1962, Rachel Carson published Silent Spring, a fierce critique of pesticide overuse and the chemical industry, focusing particularly upon DDT. The book sparked national and international debate, helped usher in the fledgling environmental movement, and has been widely credited as the major impetus for the formation of the Environmental Protection Agency (EPA), in 1970, and the subsequent banning of DDT for agricultural use in 1972 (in the US). It is notable that Carson herself explicitly *did not* call for an outright ban even on DDT, but for a more limited and rational use of these agents that takes into account their harms.

A common misconception (and one, apparently it seems, intentionally promoted by pesticide advocates) is that Rachel Carson and the DDT ban caused the deaths of perhaps millions of (mainly African) children by undermining malaria control efforts. In fact, DDT was never (and has never been) banned for malaria control, and the World Health Organization's (WHO's) Global Malaria Eradication Programme, which was indeed based largely spraying highly persistent pesticides indoors, such as DDT ("indoor residual spraying"), was disbanded due to futility in 1969. In subsequent years, newly independent African governments abandoned pesticide-based strategies largely because of evolving pesticide resistance in mosquitoes, among other reasons [360].

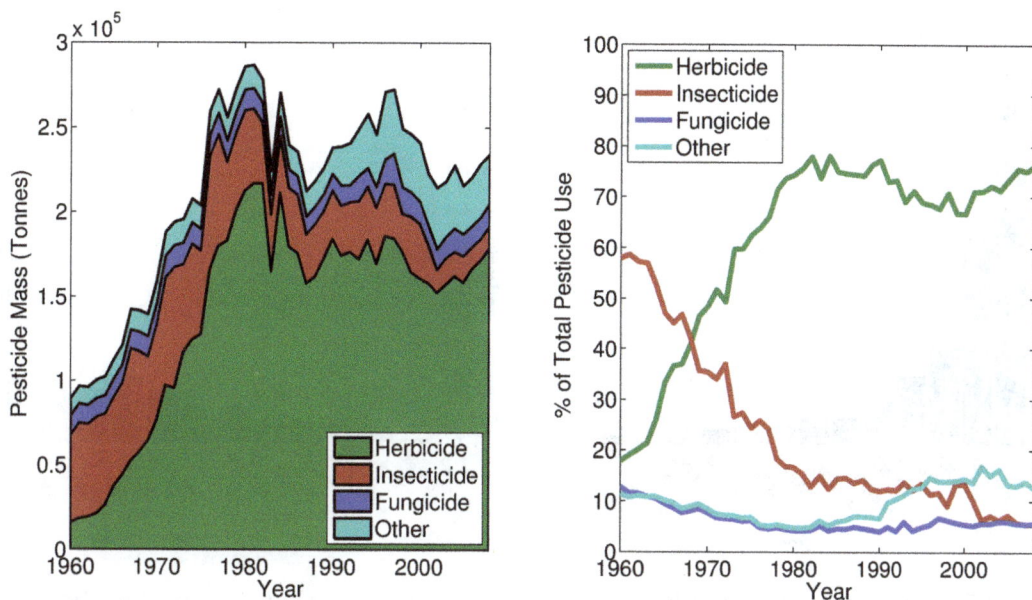

Figure 20.4: Broad trends in pesticide use from 1960 to 2008, based on 21 selected crops that represent about 70% of US pesticide use [361] (note then, that total pesticide use for the US as a whole is about 40% higher than the numbers given here would indicate).

Recent trends in US pesticide use

Since 1960, overall pesticide use in the US, as measured by total pesticide mass, increased dramatically until peaking in 1981, with the subsequent decades showing a slow but uneven overall decline [361]. Between 1960 and 1981, the overall increase in pesticide use was driven entirely by increased herbicide use, which peaked in the early 1980s and has since remained relatively constant, but with a recent uptick, especially in glyphosate (the active ingredient in Monsanto's Roundup), which now makes up at least 50% of all herbicide applications. Insecticide use, on the other hand, has declined appreciably over the last few decades.

Broadly speaking, we can divide 49 years spanning from 1960 to 2008 (the most recent year for which comprehensive USDA data is available) into three pesticide eras, following Fernandez-Cornejo et al. [361]. The expansion of herbicide use in the 1960 to 1981 period represented a broad shift away from tillage and cultivation as the primary weed control strategy and towards chemical suppression. From 1982 to 1995 was a quasi-stable era, while the period from 1996 onward has been strongly influenced by the adoption of genetically engineered corn, soy, and cotton varieties that either express the naturally occurring Bt insecticide (Bt trait) or are resistant to the broad-spectrum herbicide glyphosate. This era has seen a slow *decline* in total applied pesticide mass, while glyphosate has rapidly displaced other herbicides to become by far the most used single agent. The overall toxicity, and not just mass, of applied pesticides has declined as well, as the most toxic insecticides have declined in use or been banned [361].

Pesticide intensity on major crops today

As shown in Figure 20.5, most pesticides are applied to several commodity crops, especially corn and soy. However, this is mainly attributable to the massive land area devoted to these crops, not use intensity. Indeed, as also shown in Figure 20.5, use intensity (i.e kg of pesticide per hectare) is generally much higher for fruits and vegetables, with potatoes by far the most intensely sprayed crop (this particular crop is fairly unique in that it is subject to extensive soil

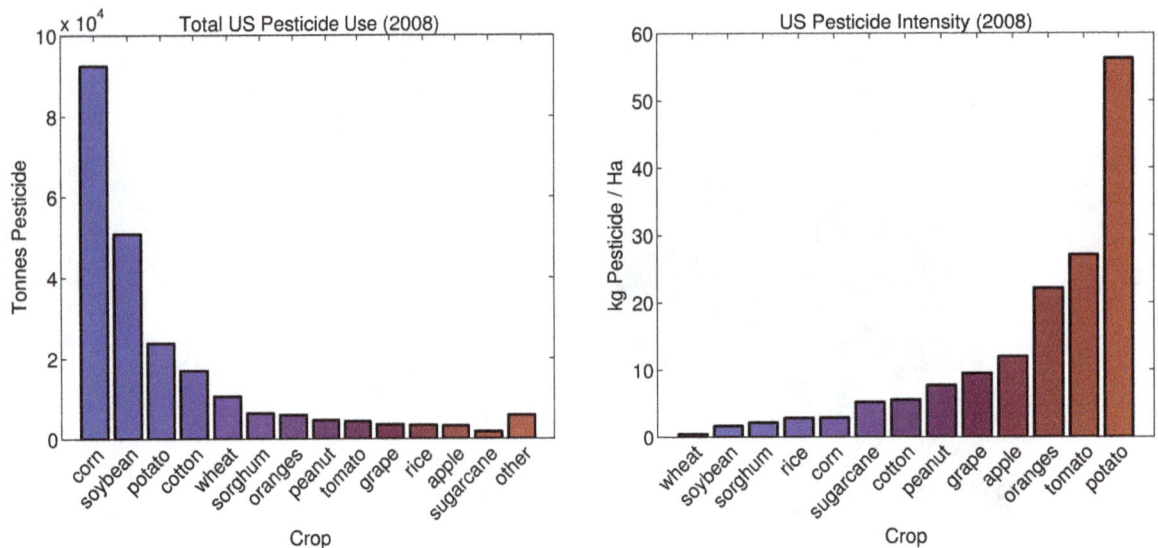

Figure 20.5: Total mass of pesticides used on 21 selected crops that account for about 70% of US pesticide use, along with use intensity for the top 13 crops; data is from Fernandex-Cornejo et al. [361]. As seen in the left panel, corn and soy account for most pesticide use overall, but pesticides are used at relatively low intensity on these crops (right panel). Fruits, vegetables, and especially potatoes, are subject to more spraying. Note that, dissimilar to other crops, fumigants accounted for 83% of all pesticide application, by weight, on potatoes.

fumigation).

20.2.2 Direct production emissions

On a per-kg basis, the production of various pesticides is much more energy/emissions-intensive than fertilizer production. However, since the total mass of pesticide applied in the US (241,000 tonnes in 2008) is about two orders of magnitude lower than the fertilizer mass, the net direct global warming impact of these substances is relatively small. Lal [501] compiled a number of emissions factors for these substances, and gave ranges of 6.23–46.2, 4.4–29.7, and 4.4–29.3 $kgCO_2e/kg$ for herbicides, pesticides, and fungicides, respectively. The respective un-weighted means (i.e. mean across many different particular substances, not weighted according to use) were 23.1 (herbicide), 18.7 (pesticide), and 14.3 (fungicides) $kgCO_2e/kg$. However, as glyphosate accounts for 50% of herbicide use, and its production is relatively carbon-intensive (33.4 $kgCO_2e/kg$ in [501]), when weighting by actual herbicide use we get about 25.6 $kgCO_2e/kg$ for herbicides, a slight increase over the unweighted average.

Now, using the above mean estimates for carbon intensity, and pesticide use-rates in 2008 (per the USDA), we arrive at about 7.6 million $MgCO_2e$ attributable to all pesticide production, with about 85% of this due to herbicides. For major crops, then, as an overall average, about 2.5 kg of pesticide is applied per hectare, equating to around 50 $kgCO_2e$. On a per-kg basis, pesticide production is a trivial component of the carbon footprint for major commodity crops (corn, soy, wheat), adding, using corn as an example, less than <0.01 $kgCO_2e/kg$ corn. The impact is somewhat larger for the highly fumigated potatoes, at perhaps 0.02–0.035 $kgCO_2e/kg$ potato, or around 5–10% of this tuber's production carbon footprint; for highly sprayed fruits, such as oranges, apples, and grapes, pesticides may account for about 3–5% of the production carbon impact (which is generally on the order of 0.5 $kgCO_2e/kg$ produce).

It follows in conclusion that most significant global warming effects attributable to pesticide will be downstream, mainly in the form of altered yields and/or land management (e.g. no-till or reduced-till systems). Increased yields for similar levels of non-pesticide inputs would obviously decrease the carbon impact of food production (at least on a direct per-kg basis) and may decrease the agricultural land-base, while reduced-till systems can reduce soil carbon losses. On the other hand, ecosystem toxicities are an independent axis of harm that must be considered. This is further complicated by the fact that it may be better to poison the forest than to cut it down outright for crops (a fate possibly avoided if pesticides do indeed decrease land-use).

20.2.3 Effectiveness and ecological effects

A basic question that must be answered is, simply, how effective are pesticides in increasing agricultural productivity? The answer is not straightforward, as pesticides help shape the overall cropping system, and there are also secondary effects that may undermine pesticide effectiveness. Quantifying overall effectiveness and detrimental ecological effects is difficult, but I review many of the qualitative patterns and arguments here.

Yield, crop protection, and cropping systems

Pesticides do appear to increase the productivity of conventional cropping systems, although the magnitude of benefit is unclear, and there is likely the potential to significantly reduce, if not completely eliminate, pesticide application without negatively impacting yields. We must also be clear that while "crop protection" is absolutely essential to adequate agricultural productivity, this term is *not* synonymous with pesticide use. Various diseases and pathogens, animal pests, and weeds all affect crop productivity, and may be subject to different control measures.

A basic method for quantifying the impact of different pest categories, and the potential and actual impact of various crop protection strategies, is to quantify a crop's *potential* yield in the absence of any pests, and then estimate the degree to which different pest categories reduce potential yield. Perhaps the most widely cited such work, a review by Oerke [357], concluded that there was no change at all in potential yield reduction as a *percentage* from 1960 to 2000. However, because pesticide-based crop protection enabled newer and simpler high-yielding, input-responsive agricultural systems, they still were of net benefit to productivity.

More complex cropping systems (i.e. greater crop variety, longer crop rotations) tend to better suppress weeds. Widespread herbicide use thus goes hand in hand with the broad shift to simpler cropping rotations and large scale monocultures [362]. On the other hand, herbicides have also facilitated reduced-till and no-till systems, which may decrease soil organic carbon losses.

Potential for reducing use

Multiple authors have examined the relationship between pesticide intensity and productivity, with an eye towards the potential for significant pesticide reduction. In general, it seems likely that, for conventional farms, appreciable reductions in use, probably on the order of 50% for major cereal and commodity crops, are possible with only a small or even no effect upon productivity. However, complete abstinence may lead to much more significant crop losses.

A review of earlier studies conducted mainly in experimental farms examining herbicides applied at below-label rates by Zhang et al. [363] concluded that low herbicide application rates generally yield good weed control, often at as little as 20–40% of the label rate, and especially if herbicide was combined with inter-row cultivation (mechanical weeding). While there was generally a trend towards improved weed control at higher application rates, the benefit was uniformly very small beyond about 60–80% of the label rate. A more recent meta-analysis [364] comparing organic, low-input, and conventional farming found no yield difference between the low-input and conventional systems for corn (mean 50% reduction in pesticides for the low-input systems) and only a small difference for wheat (70% reduction in pesticides), while organic systems that eschewed pesticide entirely were clearly less productive than either the low-input or conventional ones.

Several recent studies have focused on wheat yields in France, and similarly suggest 50% pesticide reductions are feasible. Most recently, Lechenet and colleagues [365] performed a regression analysis upon data from 946 farms in France relating treatment frequency index (TFI), various other farm parameters, and farm productivity, as measured by gross energy output per unit area (GJ Ha^{-1} yr^{-1}). The TFI is a lumped metric that summarizes the intensity of all pesticide use, and is defined as the sum of the ratio of applied to recommended dose across pesticide treatments, or more formally,

$$\sum_T \frac{\mathrm{AD}_T}{\mathrm{RD}_T} \tag{20.6}$$

where AD_T and RD_T are the applied dose and recommended doses of treatment T, respectively. These authors' results suggested that a majority of (but not all) farms could decrease pesticides without affecting productivity and that, overall, pesticide use could be reduced by 42% without affecting productivity. Herbicides had the least potential for reduction (37% overall), and the generally more harmful insecticides the greatest (60%).

Gaba and colleagues [366] were unable to detect any relationship between herbicide intensity and yield or overall weed control in 150 winter wheat fields divided between 30 farms in France. They did find, however, that herbicides suppressed rarer plants not the focus of suppression efforts, leading to the conclusion that reducing herbicide on the order of 50% would not affect yield and increase weed diversity. On the other hand, a regression analysis of 176 experimental wheat plots in France by Hossard et al. [367], which used the TFI index as an explanatory variable, concluded that a 50% drop in pesticide use from the mean would reduce wheat yields 5–12%, while complete avoidance would lead to more a dramatic 24–33% yield drop. However, beyond a TFI of about 6, there was no (or very little) apparent benefit to increased spraying, and thus this study would be consistent with the idea that heavier pesticide users, at least, could safely decrease use.

While the analyses above suggest that very significant pesticide sparing is possible without penalties to yield, it is important to note that variations in pesticide use at the farm level occur within a *landscape* that, as a whole, is subject to relatively intense pesticide application. Therefore, if a single farm or experimental plot experiences only a small or nonexistent reduction in yield upon reducing pesticides, this may not apply at a landscape scale, as our reduced pesticide plot may be indirectly benefitting from pest suppression in surrounding farms and fields. Several analyses also used data from French farms, where GMOs are generally prohibited, and thus may not be wholly generalizable to the US, where most commodity crops (except wheat) are GMO varieties that either produce insecticide (e.g. Bt corn) or are resistant to herbicide.

Some general systemic drawbacks

Wilson and Tisdell [368], among others, have described how insecticide use may initially increase agricultural productivity, but eventually lead to overall decreased productivity and/or increased pest epidemics later in time. Initially, target pests are destroyed by insecticide application.

However, this has the side-effect of destroying many natural predators that normally check pest populations, as well as other insect competitor species. Furthermore, as resistance evolves, ever more pesticide must be used, with ever diminished agricultural and economic returns. Eventually, it becomes uneconomical to use the pesticide, and pest populations *rebound* beyond their initial population size, now that natural controls are reduced.

This pattern whereby the natural predator-prey is undermined can lead to pest epidemics that cannot be controlled by predation, nor, once sufficient pesticide resistance is evolved, can they be controlled by chemical means. A related phenomenon, termed *secondary pest outbreak*, occurs when pesticide use targeting one pest species leads to outbreaks of other pest species that were previously not problematic. Similar to rebound, potential mechanisms include elimination of pest predators and/or other competing insect species, as well as induced changes in non-target species.

Even if they increase the productivity of the fields to which they are applied, pesticides can decrease productivity in other food production systems. For example, aquatic ecosystems are sensitive to pesticide exposure, which may lead to both outright large-scale fish kills, as well as reduced fishery productivity. An older estimate gives 6–14 million fish kills annually attributable to direct pesticide exposure [368], and one study concluded that common pesticide exposure is sufficient to reduce wild salmon productivity [369].

20.2.4 (Some) ecosystem toxicities

Pesticides affect ecosystems via at least three major mechanism: (1) direct, acute poisonings causing either death or impaired development and/or reproductive success, (2) chronic toxicity, also decreasing survival, development, or reproduction, and (3) direct or indirect elimination of food resources (i.e. direct and indirect plant and insect declines).

There is no such thing as a truly selective insecticide: all insecticides affect both the target insect as well as non-target insects and other invertebrates, and every class of vertebrate animal, i.e. fish, amphibians, reptiles, birds, and mammals. Herbicides, which target an entirely different kingdom of life, are generally much less toxic but can still directly affect a broad range of animal species, and can have important indirect effects upon food sources and ecologies. Acute insecticide exposure can result in mass deaths; this is best documented in fish and birds, and was largely the focus of Silent Spring and early attempts at regulating pesticide use. With increased regulation and broad shifts in pesticide use patterns (decreasing insecticide use and a withdrawal from the market of some of the most toxic and persistent insecticides), acute poisonings have probably declined in the developed world, but may still be quite substantial.

Quantifying direct avian mortality from insecticides is challenging, although at their peak, insecticides almost certainly killed at least several tens of millions of birds annually and perhaps hundreds of millions (in the US) [371]. Mineau [372] estimated that a single pesticide, carbofuran, killed between 17–91 million birds in US cornfields at its peak of popularity, although this particular agent is now effectively banned in the US. Rather remarkably, Mineau and Whiteside [371] found that the risk of lethal insecticide exposure was the best predictor of grassland bird species decline from 1980–2003, more so even than changes in land area under crop cultivation. With the phasing out of more toxic insecticides, there has been some signal of a concomitant reduction in bird declines [371].

Even if they do not directly poison certain animals, pesticides can affect their populations by altering species interactions. For example, herbicide application reduces plant cover and "weed" species that are important food sources for some birds and many insects. In turn, other birds rely on the herbivorous insects and their invertebrate predators, e.g. spiders, as a food source. Furthermore, intensive herbicide helps facilitate large-scale monocultures, and the loss of diversity in plant species also reduces food sources. Therefore, one sees an overall decline in bird populations and diversity as an indirect consequence of herbicide use via multiple causal cascades [370]. Insecticides also directly reduce insect food resources as well. Space limits a more thorough discussion, but other orders of animal life, such as fish and amphibians, also likely suffer greatly from man's pesticidal activity.

Pollinator decline and pesticides

Pesticides, particularly the neonicotinoid insecticides, have been widely implicated in global declines in pollinator insects, especially bees. This has been best studied in the (semi-) domesticated European honey bee (*Apis mellifera*), colonies of which are commercially raised at large scales for honey production and orchard pollination services, and subject to the colony collapse disorder (CCD) featured so prominently in the media in the last few years. One must, however, understand that outside of Eurasia and Africa, the honey bee is an introduced, domesticated species (including in the US), and (usually solitary, non-honey producing) wild bee populations are far more important to both natural ecologies and human agriculture [377].

Pollinator insects play a pervasive role in supporting most natural ecosystems, and are thus of profound, if somewhat indirect, importance to human civilization as well. Angiosperms, or flowering plants, are by far the dominant form of plant life on planet Earth, with nearly 90% of extant land species belonging to this phylum [373], and Ollerton and colleagues [374] have calculated that 87.5% of all angiosperm species are pollinated by animals (although not all necessarily *require* biotic pollination). Similarly, a majority of the world's food crop *species* depend at least partly upon animal pollination, and one of the most cited sources has been interpreted as stating that humans depend upon pollinators for about one-third of all their food: Klein et al. [375] reported that, of 115 leading global crops, 87 of these species (76% of the total) rely upon animal pollination, with these species accounting for 35% of all global food production.

Note however, that a disproportionate amount of agriculture production comes from crops that depend upon pollinators not at all. The world's major staple grains, including maize, wheat, soy, rice, and barely, are all wind-pollinated, while tubers (e.g. potatoes), also major staples, do not require pollination (as tubers are not fruits, but specialized roots, and these plants are typically propagated from the root). Furthermore, while animal pollinators *enhance* the productivity of many food crops, only a small minority of these are absolutely dependent upon animals (less than 10%), and thus Aizen et al. [376] calculated that a complete loss of pollinators would reduce global agricultural output (in terms of weight) by "just" 5–8%; significant to be sure, but much less than one-third. Nevertheless, because insect pollinated crops tend to have lower yields than others, compensating for pollinator losses could disproportionately increase agricultural land requirements [376]. Moreover, the share of pollinator-dependent crops has increased dramatically in recent decades, with demand for pollinators outstripping supply. Finally, even if the effects on grain-based agriculture are comparatively minor or manageable (at a global scale), pollinator decline has potentially dire ramifications for the larger global ecology.

Returning to pesticides and pollinators, Europe and North American have seen severe declines in managed honey bee populations in recent decades, although increased bee-keeping in Asia has led to a net global honey bee increase [377]. Data is far sparser for wild bee populations, but it is clear that both Europe and North America have also suffered severe bumblebee (which form small colonies) and other wild bee (which are generally solitary) losses over the last century or so. This is actually even more worrisome than honey bee loss, as wild pollinators perform most crop pollination globally, and wild bees support ecosystem services more generally.

Multiple anthropogenic stressors, including habitat loss, introduced pathogens, and chronic agrochemical exposure all undoubtedly interact to drive bee declines [377]. Massive losses of flower-rich grasslands to farmland historically drove major bee declines, and agricultural inten-sification continues to eliminate habitat, while monocultures provide very limited diets. Long-distance transport of commercial bees has contributed to the spread of pathogens and disease, especially the parasitic mite, *Varroa destructor*, a major cause of colony collapse; commercially raised bees at high densities can also introduce devastating diseases into wild bee populations. Finally, pesticides play a significant role: broad herbicide applications kill many food sources, but most focus has been on the more direct effects of insecticides, particularly the neonicoti-noids. These agents are applied as seed treatments, and are present throughout the mature plant. They can directly kill bees, and have a variety of sublethal effects, including decreased learning, foraging, and reproduction, and they likely increase vulnerability to disease and other stressors. Such effects occur at very low doses, and bees living in farmed areas are likely rou-tinely exposed to doses sufficient for harm. An extensive scientific literature, partially reviewed in [377], confirms the hazard of insecticides to bees, although there remains great uncertainty.

20.2.5 Household pesticide use

Pesticides are widely used by residential households and on commercial properties, largely for purely aesthetic landscaping and gardening. This, I believe, is inexcusable. We may have a serious debate over the merits of pesticide use in agriculture, given the potential environmen-tal benefit of improved yields, but to use these products that clearly affect both human and ecosystem health towards no end other than a "nice" lawn is, again, inexcusable.

20.3 Land use change and carbon emissions

- Converting US prairie and grassland to arable cropland likely results in long-term soil organic carbon (SOC) losses of around 50%, or 150–225 $MgCO_2e$/Ha.

- In the first few years of cultivation, carbon losses may be a staggering 20–30 $MgCO_2e$/Ha/yr, but carbon content eventually stabilizes at a much lower equilibrium after about half a century. Long-term losses average to 2–3 $MgCO_2e$/Ha/yr.

- These figures naively suggest amortized emissions of 330–495 million $MgCO_2e$/year due to con-version of US prairie to cropland.

- Taking cropland out of production tends to store 1.1–2.2 $MgCO_2e$/Ha/yr for 55–75 years; thus the (hypothetical) marginal benefit of taking all cropland out of production would be to store 182–363 million $MgCO_2e$/year.

Converting native lands to arable cropland almost always results in massive and relatively rapid losses of carbon stores, due to lost biomass and soil organic carbon (SOC). The magnitude and type of losses vary by native biome, with live biomass dominating in forests, while SOC is the major loss when grasslands are put to the plow. In tropical rainforests, carbon losses exceed 800 $MgCO_2e$/Ha, with above-ground biomass (trees, etc.) dominant [70]. In the US, while much forest has been cleared for agriculture, crops are widely grown on former prairie and grassland, where most carbon stores are below-ground, principally in the form of SOC and secondarily as root biomass; it is on this biome that I now focus.

20.3.1 Soil carbon loss in US grasslands

Fargione et al. [70] estimated that conversion of native central US prairie to annual crops would incur a carbon debt of 134 $MgCO_2e$/Ha, with the vast majority of the debt attributable to losses of soil organic carbon and below-ground biomass (as opposed to above-ground biomass) such as perennial root structures. This estimate was based on several works comparing SOC change after conversion of prairie to cropland and above-ground biomass carbon, and the root:shoot ratio was used to derive below-ground biomass carbon. My own review of the literature suggests that this is a reasonable but perhaps low estimate, and long-term losses may be 150–225 $MgCO_2e$/Ha.

Almost all data appears to be from experiments comparing wheat-fallow rotations to native prairie in the Great Plains regions, many of which date from the early 1900s, and which have consistently found SOC losses in the upper layers of soil to ultimately reach 50% by mass, and exceed 50% by concentration, in the decades following conversion to cultivation, with the greatest losses in the first few years. We distinguish between SOC *mass* and *concentration* because, while many studies focus on organic carbon concentration as a percentage of the soil matter, cultivated soils are denser near the surface than prairie sods, implying that concentration reductions of around 50% translate to mass reductions on the order of 30–40% [381]. I am more interested in overall carbon mass loss, and focus on this metric in the remainder of the discussion. The reader may consult the excellent review by Peterson et al. [381] for details on many of these studies.

Briefly, data from Bowman et al. [379] suggests that, in the first 30 cm of soil, only three years after converting prairie to agriculture a total of 85.8 $MgCO_2e$/Ha are lost from the SOC pool, and 145.2 $MgCO_2e$/Ha are lost by 60 years, or 50% of the initial carbon mass. Note that this is may be an underestimate of total carbon loss, as while soil carbon losses occur predominantly in the upper layers, one study in forests converted to plantations found losses down to 1 to 2 meters in older plantations [49]. Some other studies suggest long-term cultivation-induced mass losses of 221.1 $MgCO_2e$/Ha in the first 50 cm of central Missouri prairie (35% of carbon mass) [380], and >73.3 $MgCO_2e$/Ha in the first 20 cm (>50% of carbon mass; number derived by assuming a total of 36–40 MgC/Ha in first 20 cm) [382, 381]. The general pattern of carbon loss, again, is rapid loss for the first few years followed by a slower gradual decline. Overall, CO_2e losses probably average around 2.5 $MgCO_2e$/Ha/yr long-term, but initial losses are likely extremely high, on order of tens of $MgCO_2e$/Ha/yr (say 20–30 $MgCO_2e$/Ha/yr).

20.3.2 Soil carbon gain after reversion to grassland

The natural follow-up question is, what happens to SOC stores when cultivated land is allowed to revert to grassland (or other native ecosystem)? Is this carbon simply lost, or does it return whence it came? The evidence suggests that SOC does indeed recover at a relatively constant rate in the years and decades following the end of cultivation, at a rate of perhaps 0.30–0.60 MgC/Ha/yr overall (or 1.1–2.2 $MgCO_2e$/Ha/yr; see [378] and references therein).

McLauchlan et al. [378] found SOC in the first 10 cm to increase by about 0.62 MgC/Ha/yr (2.27 $MgCO_2e$/Ha/yr) at a linear rate after conversion of formerly tilled sites in Minnesota to one of several grassland types (sites which in turn were formerly tallgrass prairie). Note that these sites were not grazed, but did infrequently experience fire (fire does not appreciably affect SOC). Carbon continued to accumulate up to 35 years (the maximum time of the study), and a simple mathematical model suggested that carbon stocks would saturate at about 60 MgC/Ha (220 $MgCO_2e$/Ha) within the first 10 cm, comparable to native prairie, and therefore SOC accumulation could be expected to continue for 55–75 years, at which time a full recovery, more or less, would be achieved. As an aside, these numbers suggest that, if 25% of US cropland were converted to native grassland, about 94 million $MgCO_2e$ could be sequestered yearly for at least several decades, or about 1.4% of US annual emissions.

20.4   On-farm fuel and energy use

- Excluding irrigation, on-farm fuel and energy use, including field operations (plowing, fertilizer application, etc.), and grain drying and storage, is probably equivalent to somewhere between 100 and 250 L of diesel per hectare (on a GWP basis), or 55–136 million $MgCO_2e$ across all US cropland.

- I suggest 100 million $MgCO_2e$ as a very crude midpoint estimate for US on-farm energy use (excluding irrigation and transportation).

On-farm fuel and electricity use is non-trivial, being needed for field operations such as plowing, cultivating, fertilizer and pesticide application, and harvesting, as well as grain drying. When irrigation is used, this is usually by far the biggest consumer of on-farm energy [384, 383], but I consider irrigation separately, as only a small minority of US cropland is irrigated.

Grassini and Cassman [384] reported on-farm energy use of 124 liters of diesel-equivalent, per hectare, for irrigated Nebraskan maize, divided nearly evenly between field operations and grain drying, and roughly equivalent to 400 $kgCO_2e$/Ha. On-farm fuel/electricity figures compiled by Murphy et al. [72] suggest nearly 600 $kgCO_2e$/Ha for maize cultivation, or about 175 L diesel-equivalent/Ha. Diesel inputs for South American soybean systems ranged from 35 to 94 L/Ha, and it seems probable that the energy/emissions-intensity of soybean drying is roughly comparable to that of maize drying (about 60 L/Ha in [384]).

Direct diesel consumption for field operations for wheat, barley, and sorghum under a range of farming practices and regions ranged from as little as 10.5 to as much as 63.85 L/Ha in a study by Maraseni et al. [383]; under conventional tillage, wheat and barley fuel requirements averaged about 40 L/Ha. Note that study did not include post-harvest grain drying, storage, or transport. Field operations and post-harvest cold storage of potatoes, another major crop, required 41 L diesel/Ha and about 2,300 kWh electricity/Ha for cold storage (assuming a yield of 45 tonnes potato/Ha) in [385]. This electricity requirement is roughly equivalent to 475 L diesel (on a carbon-equivalent basis), and we thus sum to a total 516 L diesel-equivalent/Ha for potatoes. Tomato farming also uses a great deal of on-farm fuel, with requirements tabulated at 529 and 286 L diesel/Ha for California and Michigan tomatoes, respectively (equivalent to about 0.048–0.074 $kgCO_2e$/kg tomato) [386].

In sum, an exhaustive review of all crops is not possible here, but 100–250 L diesel-equivalent/Ha (GWP basis, i.e. 330–826 $kgCO_2e$/Ha) is probably reasonable as a crude overall average for US on-farm operations. Supposing such a range for farming operations across 165 million hectares of US cropland, this would yield 55–136 million $MgCO_2e$ for all arable land, and so 100 million $MgCO_2e$ is a reasonable midpoint estimate. These calculations do not include fuel use for pasture and rangeland operations, but if we very conservatively assume that such lands require just 10% the fuel of cropland (i.e. 10–25 L diesel/Ha), then across all rangeland and grazed forest (300 million Ha), we should generate an additional 5.5–13.6 million $MgCO_2e$.

20.5 Irrigation water

- Less than 20% of US cropland is irrigated, but this land consumes about 42 trillion gallons of irrigation water per year, generating, as a very rough estimate, on the order of 36–54 million $MgCO_2e$.

- For irrigated crops, the average emissions factor may be about 0.25–0.35 $kgCO_2e$/kg produce due to irrigation, but this will vary widely with irrigation level and yield.

In the US, most cropland is exclusively rainfed and not irrigated. Most irrigation occurs in the more arid western half of the country, relying upon both surface and ground water withdrawals in roughly equal measure. The energy and emissions embodied in water conveyance systems are discussed more thoroughly in Chapter 13, mainly in the context of municipal water conveyance, distribution, and treatment. Briefly, I estimate energy requirements to be, as an overall national average, about 0.33–0.5 kWh/m^3 for water pumping and conveyance. Assuming grid-average electricity, and 41.975 trillion gallons of irrigation water yearly (per the USGS [586]), this translates into 36–54 million $MgCO_2e$ attributable to irrigation water annually.

Based upon irrigated acreage (25.3 million Ha in the US), these figures are roughly comparable to per-hectare irrigation emissions estimates given by Lal [501], and amount to 1.42–2.15 $MgCO_2e$/Ha. For irrigated vegetable and fruit crops with yields on the order of tens of tonnes per hectare, this translates to < 0.1 $kgCO_2e$/kg produce, while it may approach 1 $kgCO_2e$/kg for some lower-yielding crops. Based on primary US crop productivity of around 750–800 million tonnes (including hay, based on FAO and USDA numbers), and assuming about 19% of US cropland is irrigated (based upon World Bank and USGS figures), this gives us an overall average of perhaps 0.25–0.35 $kgCO_2e$/kg produce at the primary crop level attributable to irrigation alone, for irrigated crops.

20.6 Transportation

- Transportation in the US food system embodies on the order of 100 million $MgCO_2e$/year, with just under 40% related to final delivery, i.e. food miles, or about 0.3125 $MgCO_2e$/capita for all transport and 0.125 $MgCO_2e$/capita associated with food miles.

While "food-miles," the distance that food must travel from the farm gate to reach the consumer are prominently billed in the popular discourse, about three-fourths of all tonne-miles in the agricultural system actually occur upstream of the farm gate, as already discussed in the introductory chapter. Weber and Matthews [328] determined transportation in the US food system to embody, in 1997, roughly 92 million $MgCO_2e$ generated across 7.46 trillion tonne-miles, or 11% of food-system emissions (as calculated by these authors). Major transportation modes for the overall system, by tonne-mile, were truck (28%), rail (29%), international water (29%), domestic water (10%), pipeline (3%), and air (<1%), while final delivery, i.e. transportation to retail ("food miles") was dominated by truck, at 62%, with the balance divided between rail and water.

Calculation of total transportation emissions is somewhat sensitive to the truck and air freight emissions factors. Using either the freight emissions factors developed in Section 9.4 or those reported in [328], and adjusting for population growth from 1997 to 2015, we have total transport emissions in the 80–120 million $MgCO_2e$ range, with 100 million $MgCO_2e$ a central estimate; emissions associated with final delivery are about 30–40 million $MgCO_2e$, or just under 40% of all transportation emissions. In per-capita terms, a best estimate is about 0.3125 $MgCO_2e$/capita for all food transportation, and 0.125 $MgCO_2e$/capita related to final delivery.

20.7 Downstream processing, packaging, distribution, etc.

- Beyond the agricultural system, processing and packaging food upstream of the consumer, and the energy consumed by upstream purveyors of foodstuffs, e.g. restaurants, likely add emissions on the order of 0.75–1.5 $MgCO_2e$/capita, with 1 $MgCO_2e$/capita a rough best guess. Note that this does not include the roughly 0.5–0.6 $MgCO_2e$/capita of refrigeration and cooking-related emissions at the residential scale.

Considering the extended food system, food processing downstream of the farm gate, but upstream of the consumer, appears to be a surprisingly large and increasing user of energy in the food system, with a USDA [318] study finding food processing to consume 0.79 trillion kWh of *primary* energy in 2002. Supposing an emissions factor of about 0.2 $kgCO_2e$/kWh for primary energy, this would translate into 158 million $MgCO_2e$, or just over 0.5 $MgCO_2e$ per capita, based on the 2002 US population. This further suggests that processing adds about 0.75 $kgCO_2e$ per kg of food at the retail level. Packaging and the wholesale, retail, and food services systems (not counting residential energy use for cooking, refrigeration, dish washing, etc.) may together have used about 1.4 trillion kWh of primary energy [318], or about 275 million $MgCO_2e$, and nearly 1 $MgCO_2e$/capita.

These numbers, however, seem quite uncertain, and Cuellar et al. [319] gave markedly lower numbers, at 0.33 trillion kWh and 0.65 trillion kWh for processing alone and packaging, retail, and food services together, respectively, and translating to about 66 million $MgCO_2e$ (0.23 $MgCO_2e$/capita) and 130 million $MgCO_2e$ (0.45 $MgCO_2e$/capita), respectively.

In any case, it seems likely from these system-scale estimates that, downstream of the farm gate, an additional 0.75–1.5 $MgCO_2e$/capita are added by various food handling and processing processes, not counting residential energy use. Given that about 633 kg per capita of food reaches the retail stage [321], in the US, this suggests about 1–2.5 $kgCO_2e$/kg food attributable to these processes. This is reasonably consistent with several lifecycle analyses focused on particular products, commercial building energy use, and with residential refrigeration and cooking energy, as follows.

An LCA on potatoes [385] suggested that processing and packaging of potato chips could generate on the order of 2 $kgCO_2e$/kg chips, from cradle to retail, compared to <0.5 $kgCO_2e$/kg for straight potatoes, and about 1 $kgCO_2e$/kg for frozen potato chips. A study on tomatoes [386] also indicated that processing and packaging added around 0.5 $kgCO_2e$/kg for diced tomatoes, and >1 $kgCO_2e$/kg tomato paste. From the EIA's 2012 Commercial Buildings Energy Consumption Survey, I calculate that food sales and service buildings together generated about 118.8 million $MgCO_2e$ (with about 80% due to electricity and the remainder from natural gas). If we suppose that 32% of food is consumed outside the home (based on the USDA ERS finding that 32% of calories are so consumed), i.e. about 200 kg food per capita per year, then the commercial provision of foodstuffs adds just under 2 $kgCO_2e$/kg food. This is in addition to any upstream processing and packaging of food delivered to the restaurant or supermarket.

The above numbers suggest processing, packaging, distribution, and commercial preparation of foodstuffs most likely add about 1 $MgCO_2e$/capita. As discussed in Chapter 15, residential refrigeration and cooking emissions are, together, on the order of 0.5–0.6 $MgCO_2e$/capita (or about 1.25 $kgCO_2e$/kg of food consumed at home), and while I have apportioned this to the residential energy use emissions category, it should be recognized that such emissions are also diet-related. Hot faucet water and dishwashers related to cooking also add a bit to the residential energy toll.

Avoiding highly processed and packaged foods may reduce total diet-related emissions, although fresher, less processed foods may in turn require more refrigeration and cooking energy at the residential stage. Consuming food at home is also likely less emissions-intensive than eating out, although the absolute savings from eating only at home may be relatively small. If we suppose 1.25 $kgCO_2e$/food of final prep energy at home versus 2–2.5 $kgCO_2e$/food eating out, then we should save 150–250 $kgCO_2e$/year by consuming 200 kg food inside vs. outside the home.

Chapter 21

Animal products and livestock

As already discussed in the introductory chapter to this part of the book, animal agriculture, which entails the processing of grain and forage inputs at a marked energy and mineral nutrient loss to animal outputs, dominates the environmental footprint of agriculture, including its global warming impact. In addition to the feed conversion efficiency issue, methane from enteric fermentation in ruminants (cows, sheep, and goats) and both methane and nitrous oxide from manure management are major greenhouse gas sources. Furthermore, land degradation and land-use change also lead to carbon emissions. Not all animal products, however, are created equal, with beef (and other ruminant meat such as sheep and goat) qualitatively far worse than all other major animal products: while accounting for less than 4% of the American diet (on both a weight and energy basis), beef is responsible for 30–35% of the dietary greenhouse gas impact (and a *majority* of the land-use impact). An iso-caloric shift to any other major non-ruminant meat animal product (dairy, poultry, pork, or egg) would result a 75–90% reduction in GHGs, while a shift to any plant source of protein would likely reduce emissions by >90%. Dairy, poultry, pork, and eggs are all relatively similar in their GHG impact, although dairy is the worst when measured on a $kgCO_2e$ per protein basis, with poultry and eggs the best.

Overall, animal products together probably account for as much as 70–75% of the greenhouse gas impact (and broader environmental impact) of American diets. Vegetarian diets are likely about 30–35% less emitting than omnivorous, while vegan diets may be as much as 50% less emitting. Again, a majority of the climate benefit to a vegetarian diet stems from eliminating beef in particular, rather than meat in general. One need not abstain completely, as low meat diets are much more benign than either typical or high meat diets; limiting, even if not eliminating, dairy is also quite beneficial. I also believe that the somewhat all-or-nothing mental model of complete abstinence (vegetarian or vegan) versus "anything goes" can be unhelpful, and it bears emphasizing that (meaningful) harm *reduction*, not harm elimination (which is impossible) or moral purity should be our primary goal.

It should further be noted that while vegetarian diets are generally superior to non-vegetarian, when these diets substitute large amounts of dairy for meat this is not necessarily a boon, as dairy systems are relatively high carbon, have many associated animal welfare issues, and in fact supply a nontrivial fraction of meat to the beef system. Therefore, the omnivorous/vegetarian dichotomy is not necessarily the best diet distinction, and some meat products, e.g. chicken, are as good or better than dairy, at least on a global warming (and land-use) basis. A vegetarian diet that substituted dairy for all meat would be similar to and possibly worse than an omnivorous diet that simply cut out beef.

The goals of the remainder of this chapter are to review, in detail, some of the major US animal production systems (especially beef and dairy), their greenhouse gas impacts—both magnitude and mechanism (to wit, enteric fermentation, manure management, and feed production)—and some associated controversies, including the effect of grazing systems on soil carbon stores.

21.1 Major prior assessments

Two major reports are frequently cited in the popular media, and it is worth reviewing both. The first, "Livestock's long shadow" by the FAO, is a masterful work that reveals the gravity of the problem, while the second, a Worldwatch Institute report, is perhaps well-meaning, but profoundly exaggerates the impact of livestock and is an example of results-driven research that must be rejected.

21.1.1 Livestock's long shadow: The FAO report

In 2006, the Food and Agriculture Organization (FAO) of the United Nations published a report entitled "Livestock's long shadow" [389]. The report stated in no uncertain terms that the global livestock sector is one of the top contributors to almost every serious environmental problem, including land change and degradation, loss of biodiversity, water use and pollution, and global climate change. The report estimated that livestock accounts for 18% of global greenhouse gas emissions overall, and in particular, livestock were estimated to be responsible for 9% of CO_2, 34% of all methane emissions, mainly from enteric fermentation, and 65% of nitrous oxide emissions, largely from manure.

Tilling of arable lands to grow animal feed, deforestation for pasture, and rangeland degradation from overgrazing are all also significant sources of carbon emissions. Indeed, the report related that, on a global scale, livestock grazing uses 26% of the earth's ice-free land, 33% of cropland goes towards producing animal feed, and in sum, 70% of agricultural land and about 30% of *all* land goes towards raising livestock. Note that these are global figures, and in the US the situation is even worse, with closer to 50% of the nation's entire landbase directly or indirectly devoted to livestock. Conversion of Amazonian forest to pasture and feedcrop production (e.g. soybeans) is the major driver of rain forest deforestation, and a significant portion of grazing lands have been degraded by overgrazing. These vast tracts of agricultural land were once wildlife habitat. This, combined with the depletion and pollution of water sources and the stressing effect of climate change, imply that agriculture, especially livestock, is the greatest global threat to biodiversity. Thus, the environmental impact of livestock extends far beyond carbon emissions alone.

21.1.2 Worldwatch report: 51%? An implausible estimate.

While the FAO gave 18% as their estimate of the global warming share attributable to agriculture, a widely cited report by the Worldwatch Institute, published essentially as a response to *Livestock's long shadow*, claimed that livestock alone (not even including the remainder of the agricultural system) accounts for a staggering 51% of all anthropogenic carbon emissions. Note that this calculation is derived *after* the authors claim that existing inventories undercount emissions by 25.0 $GtCO_2e$ per year (all additional emissions attributable—supposedly—to livestock), raising the cited global total emissions from 41.8 $GtCO_2e$ to 66.8 $GtCO_2e$. That is, the authors assert that the FAO accounting was almost fivefold too low (as the FAO 18% figure adjusts to 11.8% of emissions under the new grand total), and that existing carbon inventories have ignored almost 40% of all emissions. This claim approaches absurdity almost *prima facie*, and a closer look validates my skepticism. Some of the errors are worth examining in detail.

Respiration is not a carbon source

The most obvious error, and it is *unambiguously* an error, is the claim that livestock respiration counts as an anthropogenic carbon source, to the tune of 8.8 $GtCO_2$/yr (the largest "correction" to the FAO figure, and more than the US's entire annual emissions). Respiration is the conversion of food energy to CO_2 and H_2O, and (nearly) every living thing does this. Respiration by any animal, including humans, that does not act on some long-term carbon store is *never* counted as an anthropogenic emission: it is a natural background process that entails rapid cycling between carbon pools, and does not add to long-term CO_2 atmospheric concentrations. It does not matter if the cows are a creation of Man, or that they are consuming food grown on deforested land with an attendant carbon debt. It is correct to account for the carbon debt incurred when one switches to a forest system to a cropping system, but one cannot count the carbon from metabolizing food grown on the cropping system.

It is an important point that respiration generally should not be counted towards emissions totals, as it occasionally is in various contexts. I recall a GOP politician once claiming that bike lanes are bad because cyclists emit CO_2 when exercising. While the objection is obvious nonsense at any level, it also fails to appreciate that respiration, even by humans, should not be counted as an anthropogenic emission at all. The major exception is when long-term carbon stocks (e.g. soil carbon, peatland) are altered by human intervention such that they are oxidized to CO_2 via respiration, and thus yield *new* atmospheric CO_2. By the same token, CH_4 produced via enteric fermentation is *new* to the atmospheric system in a way that oxidized crops are not, and is thus tabulated as a new emission.

Other claims

Of other dubious claims, the most important is that the "appropriate" time-frame for methane GWP is 20 years, not 100, and using 20 years inflates the effect of methane from livestock approximately threefold compared to FAO estimates. Indeed, it is not at all obvious what the proper time-horizon for evaluating global warming potentials is, and there is no general method to decide what is appropriate, but the international standard has been 100 years, and in any case, all other greenhouse gases must be recalibrated to 20 year GWPs for a consistent inventory.

Some of the other claims are not necessarily unreasonable. For example, land-use changes are posited to be undercounted by the FAO, as the use of agricultural land is not compared against a counterfactual of either abandonment or cultivation of food for direct human consumption or biofuels. Additionally, increases in livestock head compared to the time FAO statistics were compiled would imply increased emissions. Finally, meat cooking, waste disposal, livestock-related infrastructure, animal-borne illnesses, and several other more indirect activities are claimed to account for 13% of global carbon emissions (5.6 $GtCO_2$e, close to the US's annual emissions from all sectors). This latter sum also strikes me as implausible on its face, but I will not take the time to thoroughly dispute it. The inclusion of respiration alone as a carbon source discredits any conclusions, and it strikes me as a fine example of results-driven research. We must reject all such work, even if it is well-meaning and seeks to address a real problem of global significance, in favor of a clear and honest assessment, regardless of what conclusions it should leads us to.

21.2   Beef and other ruminant meat production

It is universally accepted in the scientific literature that beef production has the highest overall and per-unit environmental impact of any large-scale agricultural activity, being both extremely emissions-intensive and land-intensive per kg of production (see, e.g. [322, 457, 390, 467]). Cows are ruminants, a class of animals that includes a variety of hoofed mammals, and that are unique in possessing a four-compartment stomach wherein coarse plant matter is broken down by symbiotic microbes via *enteric fermentation*, a process that releases methane and is thus largely responsible for the high global warming impact of cattle production. Ruminants' major evolutionary advantage is the ability to digest cellulose via fermentation, a feat many animals are incapable of. In addition to methane release, ruminants also very inefficiently convert feed into growth compared to monogastrics (i.e. single-stomached animals, including humans, pigs, and chickens) [322], and thus both grass- and grain-fed ruminants are associated with excessive embodied emissions, fertilizer, pesticide, water, and land-use.

It follows that, compared to most other animal production or direct consumption of crops (vs. feeding them to animals), large-scale cattle and other ruminant, e.g. goat or lamb, production is intrinsically more harmful, thanks to enteric fermentation and feed conversion inefficiency. This applies not only to conventionally produced beef, which are "grain-finished" in feedlots, but to purely grass-fed beef as well. Indeed, the relative impact of grass-fed vs. conventional (grain-finished) beef is a recent controversy, but the literature does seem to support the conclusion that grain-fed beef actually has a lower carbon footprint and requires less land to produce [439, 407]. This is true for two basic reasons: (1) grain fattens cows up more efficiently, and (2) less methane is released per unit of grain than per unit grass.

The perhaps disturbing conclusion is that simply purchasing grass-fed beef in lieu of standard fare is no solution at all: if we demanded the same level of beef in our diets as we have now, then converting to an all grass-fed system would in fact generate more carbon emissions, use and degrade more land, and be wholly unsustainable. The proper solution then must simply to reduce consumption of what is properly understood as an intrinsically harmful luxury product, but that may be acceptable as a *small* part of a larger system.

21.2.1 Survey of LCAs

- Lifecycle analyses suggest carbon emissions on the order of 30–50 $kgCO_2e$/kg boneless beef, for meat produced in American beef-only herds, with higher figures, e.g. 40–50 $kgCO_2e$/kg boneless beef more consistent with top-down estimates of livestock methane emissions.

- More intensive management systems generally have lower CO_2e and land requirements per kg of beef; lifecycle emissions for grass-finished beef are probably 15–30% higher than for conventional systems.

Lifecycle analyses have generally reported carbon emissions on the order of 20–50 $kgCO_2e$/kg of beef, for the US, Canada, and other Western Countries (Canadian and US production systems are fairly similar). For example, 15 estimates compiled by Desjardins et al. [407] for mainly US, Canadian, and Western European conventional beef systems averaged to around 30–35 $kgCO_2e$/kg beef (converted from about 12–14 $kgCO_2e$/kg live weight), although these studies varied significantly in how they accounted for carbon sequestration or loss from land-use change and land-management, the interaction with the dairy system, the role of co-products (e.g. leather hides), and overall system boundaries. However, other estimates are closer to 50 $kgCO_2e$/kg beef [439, 388]. Analyses have also fairly consistently concluded that "grass-fed" (a term that, in popular use, really refers to "grass-finished," as explains below) actually have higher greenhouse gas emissions than conventional systems [439, 392].

Globally, emissions estimates from life-cycle analyses have ranged from 9–129 $kgCO_2e$/kg beef, with more intensive production systems generally lower in their carbon impact [390].

Desjardins [407] and colleagues have suggested that carbon emissions are likely lowest in the West, where production systems are much more efficient than those in the developing world and indeed, analysis by the FAO [388] suggested that more industrialized North American beef is less than half as emissions-intensive as that in sub-Saharan Africa and several other developing areas. Along these lines, Capper et al. [391] concluded that with increased efficiency, the carbon footprint of beef in the US decreased by 16% from 1977 to 2007. Note also that, since the methane GWP potential estimates have been recently revised upwards, most existing analyses likely underestimate CO_2e emissions by 12–17% (calculated assuming 40% of lifecycle emissions are from methane, and correcting with a 34 to 21 or 25 ratio).

Before continuing, it is important to note that the functional unit of analysis in studies varies between carbon footprint per live weight (LW), carcass weight (CW), and boneless meat. Live weight is obvious, whereas carcass weight is the weight following slaughter and removal of the head, hide, feet, and viscera, and is only about 50–65% of live weight [407, 408]. Further, only 55-75% of the carcass weight ultimately yields boneless, trimmed retail cuts of meat [408]. Therefore, generally speaking only about 40% of an animal's live weight ends up as retail meat. Thus, if one estimated, for example, 20 $kgCO_2e$/kg live weight, this should translate into about 50 $kgCO_2e$/kg retail boneless beef.

Multiple studies have examined the emissions from model herd systems. Indeed, we must understand that the beef cattle system entails a whole herd with several different roles. Only a fraction of cows are brought to slaughter each year, but we must sum the emissions from the entire herd system to get a valid per product emissions factor. Further complicating the matter is the fact that dairy and beef herds are interconnected, making emissions allocations trickier. I discuss correcting for the dairy system in Section 21.4.3, but for now, let us examine isolated beef production, which is the dominant model in the US. Worldwide, a slight majority of beef actually comes from the dairy system, in the form of fatted calves and culled milk cows [432].

A prototypical beef-only herd consists of cow/calf, stocker, and feedlot components [439]. In the cow/calf herd, there are, say, 100 mother cows who give birth each year with a success rate of about 90%, along with 15 heifers (young female cows that have yet to give birth), and perhaps three bulls for breeding (alternatively, artificial insemination may be used), one of which is culled yearly. The mother/calf pairs (and heifers) graze on pasture or range, with the calf fed on its mother's milk until weaning at about six months of age. Shortly after weaning, the calf undergoes one of several fates. First, perhaps 15 female calves are kept in the cowherd as replacement heifers, to ultimately replace 15 older cows that are culled (i.e. sent to slaughter for beef). Note that this implies mother cows last about seven years, on average, before culling.

Now, the remaining calves (both male and female), are used for beef. Some fraction is sent to a stocker system ("yearling" system), which entails a further 6–10 months of grazing (also called backgrounding), and then to a feedlot for "finishing" on a high-grain diet (additional 6 months). Thus, these calves take about 450 days beyond weaning to reach slaughter. The other calves are sent directly to feedlot ("weanling" system), where they finish in about 300 days [439]. A tiny minority (<1%) of calves are finished on pasture instead of feedlot. This is what labels such as "100% grass-fed" or "grass-finished" refer to: all beef cattle spend the first portion of their life grass-fed. At every stage of this process there are direct emissions, primarily CH_4 from enteric fermentation and N_2O from manure, along with indirect emissions from feed production; the process is illustrated in Figure 21.1. Note that even during the grass-fed stages, cattle require significant outside inputs in the form of hay (especially over winter) and other supplemental feed (e.g. oilseed cakes) grown on arable land.

A fairly comprehensive analysis by Pelletier and colleagues [439] compared stocker/feedlot, feedlot-only, and pasture finishing systems for Upper Midwestern beef production systems. This system is geographically notable in that the pasture systems are intensively managed, i.e.

fertilized, seeded, etc., as opposed to unimproved rangeland, stocker cattle are maintained on wheat fields, and a significant portion of the forage fed to pastured cows is, in fact, alfalfa hay that is grown on cropland (although this is common). Thus these pasture systems are more energy intensive than in many parts of the US. At any rate, including emissions from enteric fermentation, manure, and feed production, we arrive at about 40, 37, and 47 $kgCO_2e/kg$ beef for the stocker/feedlot, feedlot-only, and pasture systems, respectively. Note that the pasture finishing system also required 25–50% more land than the conventional systems. By far, most emissions are generated by maintaining the cows and heifers in the cow/calf system.

I have done my own calculations using Pelletier et al.'s figures on feed, fertilizer and fuel inputs for feed production and have obtained (also using updated CH_4 GWP values) about 20 $kgCO_2e/kg$ boneless beef attributable to enteric fermentation in the two conventional systems, and 27 $kgCO_2e/kg$ in the pasture-finished system. Pelletier et al.'s analysis also indicated 9–10 $kgCO_2e/kg$ from manure management in the conventional systems, and about 10 $kgCO_2e/kg$ in the pasture system, with almost all this due to N_2O. Adding in upstream feed production adds about 10–15 $kgCO_2e/kg$ beef, and thus, we sum to about 40–45 $kgCO_2e/kg$ for conventional beef, and 45–50 $kgCO_2e/kg$ for grass-finished.

Numerous other studies (e.g. [392, 393, 390, 388]) have been published in the last few years with all coming to essentially the same conclusions, and while there is some variation in exact emissions estimates, there is no real disagreement concerning the order of magnitude of beef's impact.

21.2.2 Carbon sequestration from grazing?

A central issue in lifecycle analysis for the overall impact of beef (and dairy) in general, and in determining the comparative impact of grass- and grain-finished beef, is how grazing and other land management strategies affect grassland (and other rangeland biome) carbon stores. Overgrazing can lead to land degradation and desertification, characterized by a shift in dominant vegetation and loss of soil organic carbon. This is especially true in the more arid areas of the world [389]. On the other hand, well-managed grazing may beneficially affect vegetation and increase soil carbon, mainly in biomes that were subject to grazing by large generalist herbivores during their evolutionary history. It is sometimes claimed that, because of such carbon sequestration, grass-fed beef has a lower, nil, or even negative greenhouse gas impact.

Despite decades of work, overall conclusions on the effect of grazing on SOC stocks are unclear, with studies variously showing a positive, neutral, or negative effect; it is doubtful that grazing has a globally uniform effect, implying that grazing strategies must be tailored, at the least, to grassland biome. Conclusions drawn from studies of the American Great Plains, for example, which historically evolved under grazing pressure from large bison herds, are not generalizable to the more arid lands of the far American West and Southwest, which evolved largely in the absence of such pressures. It clear that climate, precipitation, soil type, and vegetation type interact to determine carbon sequestration and the net effect of grazing. In any case, grazing is *not* a dominant factor in determining soil carbon, except when it induces desertification, with rainfall and climate far more important.

Great Plains

Let us first focus primarily on studies of the Great Plains region which, again, was subject to heavy bison grazing in recent evolutionary history (although it should be noted that the exact size of historical bison herds is remarkably uncertain, with widely cited estimates based upon almost nothing, and it can only be stated with any certainty that at least several million and possibly several ten million bison once roamed the Great Plains, as reviewed by Shaw

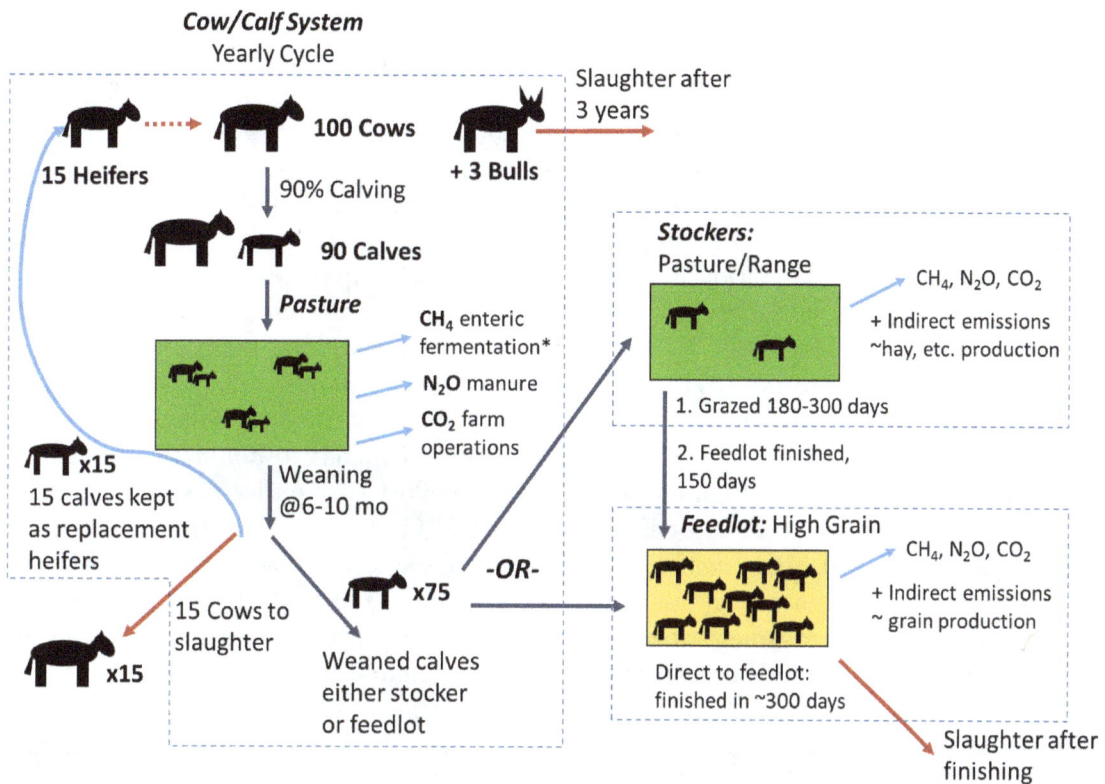

Figure 21.1: Schematic for the beef-only production system that dominates in the US. At the first stage, we have cow/calf systems where out of 100 cows, perhaps 90 will calve each year. These calves are kept with their mothers on pasture until weaning, at which point they are sent either to a stocker (or "backgrounding") system where they continue to graze for 6–10 months before feedlot finishing, or older calves may go directly to feedlot. Note that about one in seven cows are culled yearly, and so 15 heifers are retained. Most emissions in the beef system are attributable to the cow/calf stage.

[394]). This region is characterized by a general increase in precipitation from west to east, with a corresponding shift in grassland type from shortgrass to mixed-grass (or "midgrass") to tallgrass prairie. As one moves west to east the productivity of the land increases, the soil carbon stores increase markedly, and the root:shoot grass ratio decreases (i.e. there is comparatively more above-ground than root biomass) [397, 396]. These factors all shift in proportion to the precipitation level [397, 398]. It makes sense that grasses in the less productive, drier west devote more mass and energy to their root system (higher root:shoot ratio), the better to acquire water and avoid transevaporation. With less carbon input from plant growth and decay, it is also logical that soil carbon stores are much lower in the drier west.

Now, it has been fairly consistently found that annual precipitation determines not only prairie composition, but also (at least indirectly) carbon sequestration in response to grazing, with *lower* average precipitation predicting increased SOC under grazing, while higher precipitation may actually lead to decreased SOC under grazing. Derner and Schuman [396], reviewing Great Plains studies, observed the SOC grazing effect to shift from positive to negative at about 440 and 600 mm of precipitation for the 0–10 cm and 0–30 cm soil horizons, respectively. Note, however, that *drought* (i.e. a negative departure from average precipitation levels) may conversely interact with grazing to *reduce* SOC [399].

Short-term carbon sequestration in grazed versus ungrazed patches has been measured on the order of 0.07–0.12 MgC Ha^{-1} yr^{-1} in shortgrass prairie, and around 0.3–0.4 MgC Ha^{-1} yr^{-1} in mixed-grass prairie ([396, 401]); the effect may be negative in tallgrass prairie as well as in some mixed-grass prairies [397]. While sequestration has been observed over multiple decades, e.g. [401], *overall*, the influence of grazing on soil carbon stores tends not to increase, as might be expected if additional carbon is stored every year, but to *diminish* in longer-term trials [395, 396, 399] (see, e.g., Figure 3 of [395], where effect size, either positive or negative, clearly diminishes with trial length, or Figure 1 of [396]). Given that, with time, the difference in soil carbon between grazing treatments tends to regress towards zero, it is clear that any grazing-induced SOC increases will not continue indefinitely, and that at best grazing induces a (likely small) shift toward a new steady-state SOC level. I discuss the likely mechanism for this pattern below. These general patterns are shown in Figure 21.2.

In general, grazing increases nutrient *cycling* in a prairie system, reduces the carbon stored in above-ground plant matter (as it is being eaten), may or may not reduce root biomass, and can shift carbon into the soil with a net increase in total ecosystem carbon, but a soil carbon loss is also possible. To get a clearer theoretical understanding of these disparate grazing effects, it is worth taking a closer look at one study, by Derner et al. [397] on the effect of grazing on shortgrass, midgrass, and tallgrass grasslands in the US Great Plains. In this study, all prairie types saw a reduction in above-ground biomass, but while root mass in the midgrass and tallgrass biomes was reduced, it was stable in the shortgrass system; although root mass was stable, root *distribution* was not. This is because the essential effect of grazing in shortgrass system was to reduce *C3 grasses* in favor of *C4 grasses*, particularly the C4 grass "blue grama" (*Bouteloua gracilis*). This C4 grass is characterized by a fine root system concentrated in shallower soil (83% of root mass in the first 0–15 cm [399]), contra the root systems of C3 grasses, which tend to be deeper and coarser (more uniform distribution up to 60–90 cm deep).

Before continuing, I should briefly clarify that grasses are broadly divided into C3 and C4 types, with the two having different photosynthetic pathways. The C3 grasses are "cool season" grasses, adapted to wetter and colder environs, while "warm season" C4 grasses are better adapted to hotter, drier conditions.

Now, we have that grazing, likely through preferential consumption, can cause a shift in grass type from C3 to C4, potentially causing an overall drop in root biomass, but shifting the bulk of the roots into shallower soil, and thus more root carbon is brought near the surface.

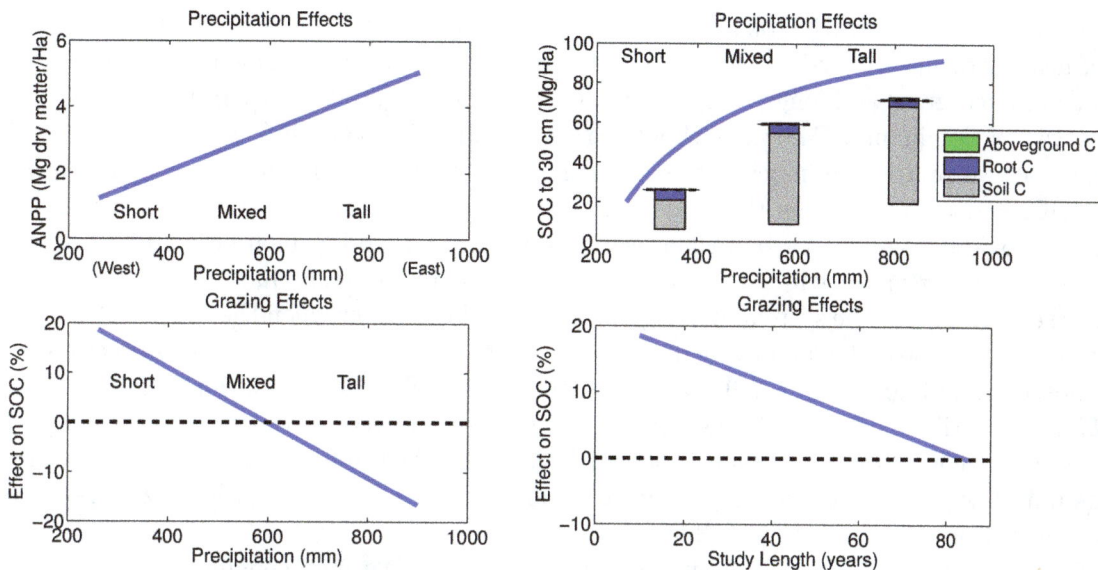

Figure 21.2: General trends in precipitation and grazing effects on the Great Plains. The top left shows how above-ground net primary productivity (ANPP) increases linearly with precipitation as one moves geographically west to east, and through shortgrass, mixed-grass, and tallgrass prairie types (data from [398]). The top right shows the hyperbolic increase in SOC with precipitation, again over this west-to-east gradient (SOC curve adapted from [396]). Inscribed is the distribution of ecosystem carbon for example (ungrazed) shortgrass, mixed-grass, and tallgrass prairie sites (from [397]), and we see that absolute carbon stores increase from west to east, while root mass decreases. Above-ground biomass is barely perceptible, but also increases west to east. The bottom panels show (left) that grazing may have a positive impact on SOC in drier western shortgrass and (possible) mixed-grass prairie, but generally shifts to a negative effect in the wetter east, while the absolute impact (positive or negative) tends to diminish towards zero (right panel) with study duration, implying that any short-term effect is generally not sustained.

In addition to this vertical carbon shift, C4 grasses also innately devote more photosynthate to the roots, and root matter nearer the surface decays faster than deeper roots [396, 399]. Thus we can see an overall increase in carbon entering the soil, with the increase concentrated in the upper horizon. With heavy grazing, this C3 to C4 shift may occur faster [399]. High precipitation, relative to the biome average, will lead to more overall plant growth and more carbon entering the soil as a consequence, and thus heavy rains and heavy grazing may synergize to store carbon beyond either light or no grazing. I say high precipitation *relative* to biome average because this cascade seems mainly to occur in the semi-arid shortgrass biome.

On the other hand, if grazing reduces overall plant biomass (especially below-ground) without a change in the vertical root distribution (or possibly even with such a change) then ultimately less carbon enters the soil as decaying plant matter, and SOC stocks will decrease. In this case, heavy grazing will reduce SOC stocks more than light grazing. This is what appeared to occur in mixed-grass and longgrass prairie studied by Derner et al. [397]. In these systems, the grass was predominantly C4 even without grazing, and no C3 to C4 shift occurred with it. Note that these two prairies are much wetter than the shortgrass system, with much higher baseline soil SOC, and it may also be that increased nutrient cycling under grazing leads to excessive microbial oxidation of soil carbon when water is more available for microbe metabolism.

Finally, heavy grazing in shortgrass or mixed-grass biomes that has a short-term carbon storage effect may set the system up for greater carbon losses later when the system is perturbed by drought. This would explain why little long-term grazing effect is seen, and is precisely what was observed by Ingram et al. [399] who reported marked SOC accumulation in heavily grazed (50% forage utilization) mixed-grass prairie after 12 relatively wet years (see also [400]), but even greater SOC losses in the subsequent drier decade. At the last count, the heavily grazed fields were lowest in soil carbon and above-ground biomass, while the lightly grazed fields (10% forage utilization) were healthiest [399].

These effects of grazing are illustrated in Figure 21.3. We may conclude that, in general, light grazing is probably sustainable throughout grasslands that evolved under large herbivore grazing pressure, with little change in range productivity and possibly small increases in soil carbon in the western Great Plains. While certain grasslands may experience large short-term soil carbon gain with heavy grazing, this reduces range productivity [399, 422], is unlikely to be sustained long-term in most cases, and is most likely ultimately counterproductive. The C3 to C4 grass shift is likely the major mechanism for increased SOC under grazing, and this tends not to occur in more easterly, wetter prairies. Climatic factors, mainly precipitation, are clearly far more important than grazing in determining SOC (for example, Derner and Schuman [396] found precipitation to explain 83% of the SOC variance across the Great Plains), except when severe overgrazing degrades the land.

Grazing in the far American West: carbon loss and widespread ecological damage

- Much cattle ranching occurs west of the Great Plains, where it almost certainly causes widespread ecologic harm and dramatically reduces soil carbon stores.

- Widespread predator and wild herbivore suppression and eradication is another ongoing consequence of livestock ranching.

The biogeography of the far American West, and particularly the American Southwest, is distinct from that of the Great Plains. These arid and semi-arid lands are characterized by low vegetation cover, low levels of soil organic matter, vulnerability to erosion, and very low grazing

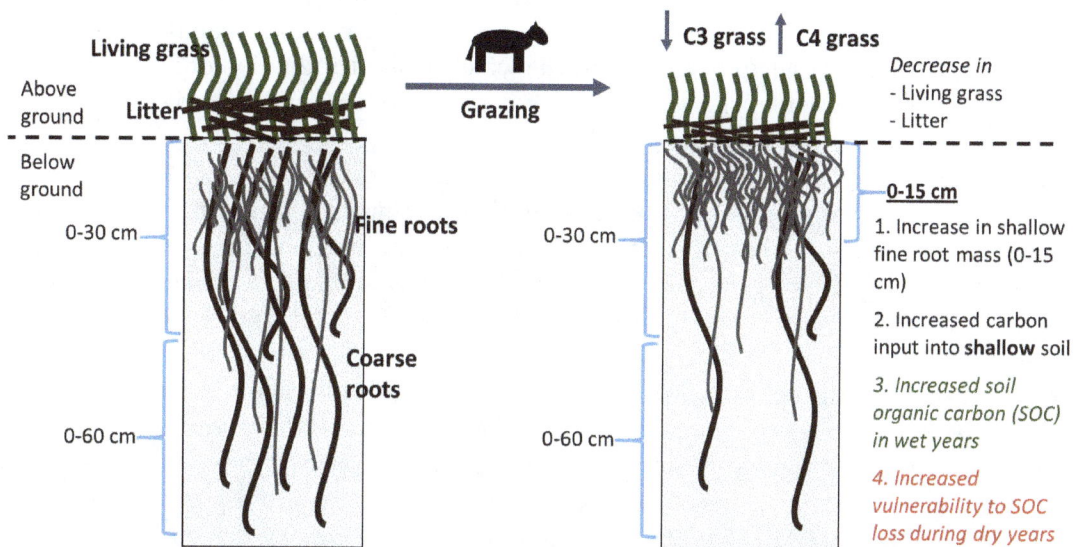

Figure 21.3: Schematic for how grazing might increase SOC via a shift in C3 to C4 grass species, mainly blue grama (*B. gracilis*), and a corresponding shift of root biomass to shallow fine roots. Note that this shift does not occur in all biomes, e.g. tallgrass prairie and some mixed-grass prairie, and in such cases grazing probably has a negative effect on SOC regardless. In shortgrass biomes where this shift does occur, heavy grazing can make the shallow SOC vulnerable to erosion and loss during drought years, and thus may be counterproductive overall. Light grazing may sustain small increases in SOC longer-term.

pressures in their evolutionary history [419]. Thus, grazing by human-imposed large herbivores can severely degrade these lands. For example, Neff et al. [402] compared a never-grazed site in the Canyonlands National Park of southeastern Utah to two very similar sites grazed from the 1880s through 1974, and found, even 30 years following grazing cessation, markedly reduced mineral nutrients, SOC (40–70% lower), and soil nitrogen (40–85% lower) in the grazed sites. This was likely attributable to a combination of increased decomposition of carbon in disturbed soils, increased wind erosion, changes in vegetation, and destruction of biological soil crusts, which stabilize soil and fix nitrogen.

Indeed, livestock use of largely public land in the American West is generally understood to be an ecological disaster [403]. While land and ecologies in the American West have been dramatically affected by numerous human activities, including widespread fire suppression, large-scale re-engineering of the region's hydrology (e.g. massive damming and canal projets), logging, and near-eradication of large apex predators, the widespread introduction of non-native grazing animals is perhaps greatest in the scale of effects, and there is very little wild land remaining in the West never subjected to grazing [403].

With global climate change, temperatures in the American West are increasing, and the region is becoming increasingly vulnerable to drought, and thus also increasing soil vulnerability to livestock, which may increase erosion via several mechanisms. As mentioned above, introduced livestock trample biological soil crusts, which stabilize soil and can take centuries to fully regenerate. Trampling also compacts soil and decreases water infiltration, causing increased runoff and erosion.

Feral horses and burros, living primarily on BLM land, also have deleterious ecological effects. Native wild ungulates can cause problems as well, mainly because human control and eradication programs of large predators have disrupted the basic predator-prey ecology [403]. Indeed, suppression of wild animals, both predator and prey, is also a grievous byproduct of livestock ranching. As the Midwest and West were settled by European-descended families, large prey such as bison and elk were nearly eradicated by over-hunting. As their former predators increasingly turned to introduced domestic livestock for sustenance, the US government carried out multiple predator eradication campaigns. Furthermore, multiple extermination campaigns were waged against native herbivores perceived as competitors to livestock, none so much as the beleaguered prairie dog. These campaigns are not merely historical, but continue to this day: In 2014 Bergstrom et al. [404] documented that Wildlife Services, an agency of the USDA, since the year 2000 directly killed (either intentionally or unintentionally, e.g. via indiscriminate trapping) two million native mammals and 15 million native birds. Moreover, these campaigns were largely unscientific and often ineffective or even counterproductive at achieving even their stated goals. These culls largely stem from a desire to promote domestic livestock herding (and to promote certain game species hunted by humans).

Global desertification

Desertification, the conversion of arid grassland to desert shrubland, has been estimated to affect a full 25% of the world's land, and is primarily driven by overgrazing and drought, with the former likely the more important factor [405]. While traditionally viewed as an irreversible process, there is evidence that, with the removal of livestock, desertified lands may slowly recover [405]. Given the magnitude of the problem, and the clear role of livestock grazing, it seems obvious that grazing is clearly deleterious to soils on a global scale.

21.3 Alternative grazing systems and meat sustainability

I would argue strongly that red meat must be considered a luxury, as its production requires vast amounts of land, directly generates a substantial portion of all greenhouse gases, is the leading cause of rangeland degradation and desertification, and is also a major global driver of deforestation. However, the idea has recently come into vogue that this is all a consequence of altering natural herbivore-grassland interactions and that, when *properly* managed, cattle grazing, at even higher intensities than currently done, can not only restore the earth's rangelands to their former vigor, but could actually sequester *all* anthropogenic carbon emissions. This idea was rather famously popularized in a 2013 TED Talk by Allan Savory, who promotes so-called holistic management (HM), also known as short-duration grazing or intensive rotational grazing (IRG). Some of the major claims follow [420]: (1) all grasslands evolved in the presence of large herbivores that grouped into large herds, due to predation, (2) such herds exposed lands to brief, intensive periods of grazing, manuring, and trampling, (3) intensive "hoof action" is necessary to break up soil and incorporate organic matter, (4) rangelands degrade in the absence of intensive grazing, and (5) therefore, by very frequently rotating cattle, (ostensibly) the modern equivalent of ancient wild herds, through small paddocks of land which are then subject to intense but brief grazing, we can improve rangeland health, produce vastly more meat on the same land, and sequester *all* anthropogenic carbon emissions.

While benefit to IRG has a somewhat plausible theoretical basis, it has been criticized in the scientific literature, with very few studies finding any favor to this strategy [420, 419]. Indeed, from my own review of the subject I conclude that there is little evidence to support the idea that a rotational grazing strategy is of any benefit over continuous grazing, increasing stocking densities by the amounts proposed would almost certainly harm all existing rangelands, and it is pure fantasy to suppose that it could sequester any more than a tiny fraction of humanity's carbon emissions,as elaborated below.

IRG was popularized in the 1960s in sub-saharan Africa, and the idea has been aggressively promoted by Allan Savory since then. Surprisingly, governmental agencies in North America and Africa have encouraged the idea, despite dubious evidence of benefit. As reviewed below, above-average rainfall, which dramatically increases the productivity of arid and semi-arid rangelands, coincided with early phases of adoption in both regions, but in either case the parting of the rains put the lie to the notion that IRG had anything to do with increased yields [422], and multiple reviews of dozens of studies [423, 425, 426] show absolutely no experimental evidence of benefit to IRG over simpler continuous grazing (despite theoretical arguments advanced by some authors), and even suggest that continuous grazing is the more productive strategy [426]. Moreover, many rangelands in the US, especially west of the Rockies, did *not* evolve under heavy herd grazing pressures, undermining the evolutionary logic supporting IRG [419], and the claim that all grasslands degrade in the absence of grazing is patently false, with the opposite more often observed [419, 420].

Finally, the idea that mostly arid and semi-arid rangelands, which are characterized by intrinsically low productively, could ever absorb even a fraction of humanities carbon emissions is nonsense, *regardless* of grazing strategy [409, 413, 419] and there is little evidence that IRG results in greater soil organic carbon stores than a continuous grazing strategy [413]: while one study by Teague et al. [418] observed slightly higher SOC concentrations under IRG than continuous grazing, Allen et al. [415] conversely found SOC to be lower under IRG across 18 Australian properties, and most studies have found no significant differences between grazing strategies [416, 414, 417, 411, 412, 410]. Since range productivity appears to be similar between IRG and continuous grazing [425, 410] there is also little theoretical reason to think that IRG would improve carbon stores. I now further discuss the historical literature comparing rotational and continuous grazing strategies, where the focus has primarily been on productivity, not carbon stores.

21.3.1 Comparisons of grazing strategies

There are two basic grazing *strategies*: (1) continuous grazing (CG), and (2) rotational grazing (RG). Continuous grazing is the traditional method, and it is just what it sounds like: cattle are allowed season-long access to a single paddock. Rotational grazing was first described over two hundred years ago, and rotational systems were widely implemented beginning in the 1950s, in hopes that they could represent a solution to the severe overgrazing of the previous century [425]. There are many versions of rotational grazing, but all involve dividing a range into multiple paddocks, and then rotating the herd through these paddocks. This increases the intensity of grazing in the active paddock, but then provides a rest period for the vegetation to recover. The more paddocks, the higher the grazing intensity.

Multiple reviews since the 1960s have concluded that there is little difference between the two systems in terms of either range condition or livestock productivity when the stocking rates, i.e. the number of animals per hectare, are similar [425]. Indeed, the literature overwhelming indicates that stocking rate is the key management variable [422, 425]: In general, the more livestock on a piece of land (within the range of commercial stocking densities), the more it is degraded. Increasing the stocking rate also increases animal productivity (as measured in kg beef/hectare) up to a point, beyond which the resource base is overwhelmed and productivity begins to decline. Overstocking can yield short-term gains, but is not sustainable longer-term, and it is well-understood that overgrazing has been the major driver of pasture and rangeland degradation over the last century. How many cattle is what matters; it matters little how they are grazed, and much of the literature comparing CG and RG is confounded by different stocking rates between plots.

This conclusion is also supported by a critical review by Joseph and colleagues [421] of the so-called "Charter Grazing Trials" conducted in Zimbabwe between 1969 and 1975, the major research foundation for Savory's IRG system. Overall, these trials included two test (IRG) and two control plots, each about 2,000 acres in size. The tests happened to occur during a period of above average rainfall (24% higher than long-term average), and IRG plots were stocked at about 50% higher density than the controls. The IRG plots did indeed yield 30–40% higher beef production (from 50% more cows), but at the cost of poorer forage, lower individual cow weight, and an increase in supplemental feeding from outside (an extra nutritional input, confounding the test further). Moreover, the control plots were periodically burned (IRG plots were not) to eliminate excess shrub and vegetation growth, strongly suggesting that they were comparatively under-stocked. There was no significant difference in grass or vegetation cover type between the plots. Joseph et al. concluded [421], quite validly in my opinion, that the increased stocking density in the test plots explained the higher beef yield, not the grazing strategy, and that the above average rainfall during that period was the factor that allowed higher stocking densities in the first place (as reviewed, other ranches that increased stocking and productivity at this time using IRG were later forced to drastically decrease or even eliminate grazing in the face of severe land damage when rainfall returned to historical norms).

This was not a new critique, and in a very worthwhile 1987 paper, Jon Skovlin [424] expressed astonishment at the acceptance of Savory's IRG system in North America, following years of apparent failure in sub-Saharan Africa. Even at this time Skovlin noted that increased stocking in Savory's experiments was likely enabled by very high rainfall, with dramatic cutbacks in stocking necessary once drought set in. Skovlin also documented decreased weight gain and increased stress in cattle subjected to IRG, and cited a 1982 World Bank study that found no justification for claims of long-term doubling or tripling of stocking rates under IRG, and furthermore, that most rancher clients who had adopted Savory's IRG method had since reverted to traditional, lower-stocking grazing.

Holechek and colleagues reviewed a number of classical long-term studies on grazing systems, including rotational systems, in [422], as well as 13 studies of short-term IRG grazing systems at 13 US locations [423]. Contra Savory's claim that the hoof action, i.e. trampling, of a large number of animals will improve soil health and increase water infiltration, it was very consistently found across studies that trampling compacted the soil, reduced water infiltration, greatly increased erosion, decreased soil fungus, decreased soil organic carbon, and furthermore did not affect incorporation of organic litter into the soil. Compaction and erosion both increased with stocking density, again contradicting one of Savory's central claims. As with other reviews and individual studies, there was little difference in vegetation or forage production, and beef production was similar for similar stocking rates regardless of grazing strategy (except for one study which found lower cattle weights under IRG).

The most significant recent publication was a 2008 synthesis paper of 41 experiments comparing CG and RG [425]. Across all studies, in terms of animal production per unit of land, 50% showed no difference between the strategies, 34% favored CG, and 16% favored RG. Plant productivity was comparable in 83% of studies, while 13% favored RG and 4% CG. From this, it would seem to be a wash. However, this "vote-counting" methodology is overly simplistic, and recently a more sophisticated meta-analysis of the same data-set was performed [426], which found that *continuous* grazing yielded 7% higher animal production in terms of kg/head ($p < .0001$), and 5% higher animal production in terms of kg/hectare ($p < .0001$); both differences were highly statistically significant. There was possibly a very slight advantage to RG in terms of plant productivity (2%), but this may have been an artifact of lower overall foraging in the RG group, and was not considered reliable by the authors.

It is notable that there was no difference between the two strategies in studies with larger land areas (> 1,000 hectares) [426], which may be more representative of large commercial ranches, and the analysis suggested a *possible* benefit to RG with longer rotational periods, in more arid environments, and with more seasonal weather variability. In any case, the difference between the strategies was small, and the preponderance of the evidence actually seems to favor continuous grazing.

Several other authors, most notably the group of Teague and colleagues (see e.g. [418]), have promoted rotational grazing in the literature almost entirely on theoretical grounds, and have strongly criticized and dismissed the empirical literature. Nevertheless, at some point theory must yield to facts, and over 60 years of study have failed to show any meaningful benefit to rotational grazing systems, intensive or not, and dozens of trials have failed to support Savory's method, including studies overseen by Savory himself.

21.3.2 Final word

I have spent the time discussing this controversy largely as a warning against "magic bullet" thinking: it is possible that intensive rotational grazing may have very modest benefits in some circumstances, but it is a dangerous delusion to believe it can single-handedly reverse global warming and feed the world. Much like half-baked geoengineering schemes, it represents an appeal to the hope that there really is a *deux es machina* that can save us, that the problem of global warming is not one fundamental to the Western lifestyle. The claim itself makes no sense with respect to basic ecological constraints: rangelands can only produce so much vegetation, and they can only store a limited amount of carbon. Cattle produce methane regardless, and overstocking degrades the land no matter what. Consuming cows for food as anything other than a relatively rare luxury item is too environmentally destructive to be sustained. This includes grass-fed cattle, it clearly includes grain-finished cattle, and it includes "holistically managed" cattle.

21.4 Dairy

21.4.1 Overview of production system

Dairy systems are the most complex of modern livestock systems to analyze, chiefly because they yield both milk and meat, with meat coming from the calves born yearly to each milking cow and from spent cows culled from the heard. The dairy production system is somewhat similar to the cow/calf beef production system, but with important differences. Typically, dairy cows are impregnated yearly, and after a gestation time of nine months (similar to humans), they give birth to a calf and begin lactating. Recall that mammals *do not* lactate unless they have recently given birth, and yearly impregnation with subsequent calving is necessary to maintain milk production. Several straightforward handbooks for dairy operation are available through the FAO, e.g. [428] and [427], which I have partially relied upon for the following discussion.

The calf is typically removed from the mother no later than two or three days after birth (and sometimes within hours). Similar to humans, mammals that we are, the first milk produced is known as *colostrum*, a thick, antibody-rich substance that is essential to imparting passive immunity against infection for the first few months of the calf's (or baby's) life [427]. Thus, calves are allowed to stay with their mother at the beginning chiefly that they may suckle of the all-important colostrum, which is not marketable for human consumption anyway. These calves are then raised individually in hutches until weaning, and ultimately enter the beef system or are retained as replacement milk cows.

With the taking of the calf, we now have a new mother cow producing milk, and one that has been bred to produce far in excess of what a calf needs, and she, along with the rest of the milking herd, is milked two or three times daily. When not being milked, the cows may feed at pasture or indoors. About three months after calving, the lactating cows are again impregnated. Milk production peaks about 6–8 weeks after calving, and then slowly decreases until ceasing about two months before the next calving [427]; this two month interval is known as the dry period, and the lactation, calving, and milk production cycle is illustrated in Figure 21.4.

As with the cow/calf beef systems, some fraction of the cowherd is regularly culled. In the US, cows first calve at about 2 years of age [322, 432], and then last, on average, about 2.5–3 more years (i.e. three total lactations, with one cut short) before culling[1]. Thus, out of 100 milking cows, as many as 30–40 may be culled for meat yearly, and 30–40 female calves will be retained as replacement heifers. The remaining female and male calves will be sold into the beef production system. While a few unfortunates become veal, the veal market has collapsed in recent decades (as a consequence, one should note, of collapsing consumer demand), and most dairy calves will enter the beef system, generally going first to a stocker system and then ultimately to feedlot for finishing and slaughter. A very small number of bulls are also maintained for breeding purposes (or artificial insemination may be used). The dairy system and its connection to the beef system are illustrated in Figure 21.5.

---

[1] The FAO gives a per-annum 34% adult cow replacement rate for North America, suggesting 3 years before culling. A shorter milking lifetime of 2.5 years is calculated using the total US dairy herd size of 17,515,149, with 9,762,171 either milk cows or heifers that calved (2012 Agricultural Census, Table 17), and assuming that all other cattle are replacement heifers that take two years to first calving.

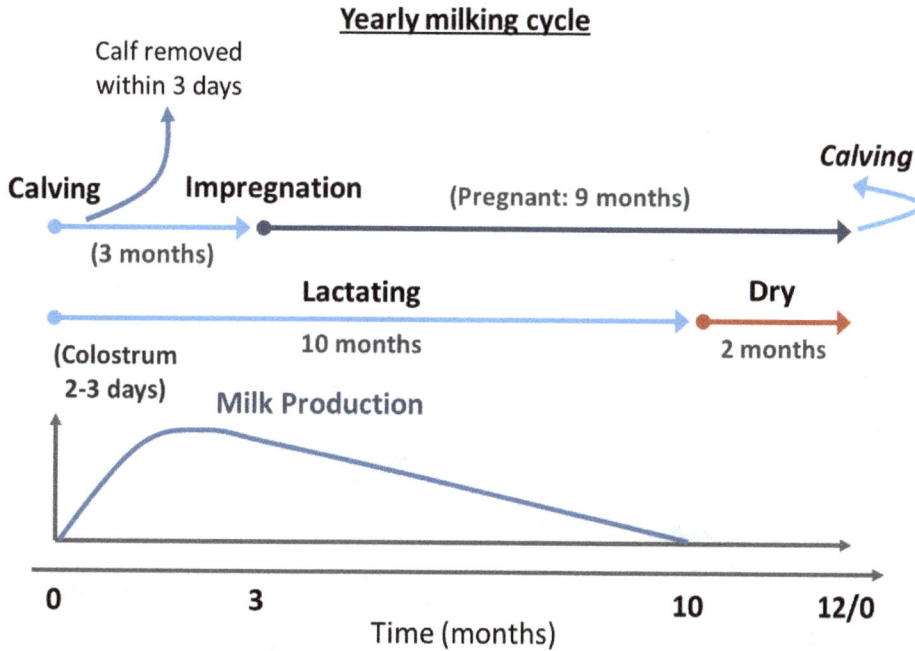

Figure 21.4: Ideal yearly milking cycle for a dairy cow. A cow gives birth at month 0, and after producing antibody rich colostrum, essential to the newborn calf, for two or three days, she lactates for about 10 months, with peak milk supply occurring at 6–8 weeks, as shown at the bottom graph. She is impregnated at the third month to yield another calf at the end of the year, following a two month "dry" period, thus restarting lactation for another yearly cycle.

21.4.2 Survey of lifecycle analyses and milk/dairy emissions factors

- On a per-kg basis, production of fat and protein corrected milk (FPCM, 4% fat and 3.3% protein) in the US likely generates 1.5–2.0 kgCO$_2$e/kg, with 1.75 kgCO$_2$e/kg a reasonable best estimate.

- Using energy or protein as the basis of conversion, common solid cheeses have mass-based emissions factors around 10–15 kgCO$_2$e/kg, or about one-fourth to one-fifth the EF of retail beef cuts, but two or three times higher than the (mass-based) EFs for poultry and pork.

Emissions sources from milk production are, quite naturally, very similar to those for beef production, and mainly entail CH$_4$ from enteric fermentation (the largest single source, at 52% in an FAO analysis [432]), N$_2$O from manure, upstream emissions from feed production, and a small amount of CO$_2$ from on-farm fuel and electricity use. In the US, these on-farm factors account for about 75–85% of all lifecycle emissions [432, 431]. The remainder, beyond the farm gate, are attributable to further milk processing, refrigeration, and transport.

A large number of lifecycle analyses have examined the dairy system, generally arriving at an emissions factor on the order of 0.75–2.0 kgCO$_2$e/kg milk (with most around 1.5 kgCO$_2$e/kg) for European (the majority of studies) and American systems (see [430, 431] and references therein), where the functional unit is fat and protein corrected milk (FPCM), assumed to be 4% fat and 3.3% protein, also known as energy-corrected milk (ECM) [430, 432][2] That is,

---

[2]The energy contained in any particular milk can be estimated from percentage fat and percentage protein, according to empirically derived regression equations. The IFCN formula for normalizing to 4% fat, 3.3% protein

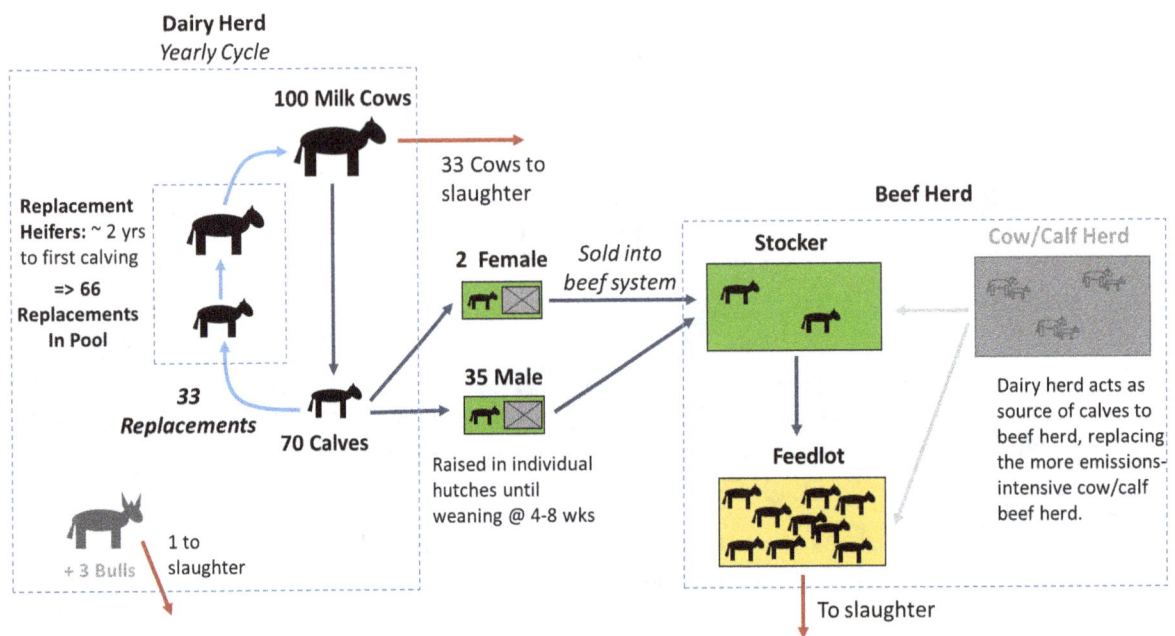

Figure 21.5: Schematic for the (ideally) yearly dairy herd cycle. From 100 milk cows, perhaps 70 surviving calves are born. Since milk cows last less than three years before culling, this translates into 33 cows sent to slaughter, and 33 female calves retained as replacements. Since it takes at least two years before first calving and "graduation" to milk cow status, the replacement pool is 66+ heifers strong. All calves are raised until weaning in individual hutches, and those not retained enter the beef system, typically to a stocker system, and then to feedlot and slaughter. Therefore, the beef cow/calf herd is cut out when dairy calves enter beef production. Note that the most beef cattle in the US come from a cow/calf, beef-only production system.

emissions are normalized based on the *energy* content of the milk produced, a necessity for any coherent analysis, given the wide array of both liquid milk products (e.g. 2%, 1%, skin milk), and other dairy, chiefly cheese, that raw milk is processed into. Further, since the dairy system produces both milk products and meat, any analysis must allocate emissions between these two products. This is commonly done on the basis of either protein or energy content of these two products, with 75–90% of emissions attributed to milk [430, 432].

An extensive analysis by the FAO [432], produced as a follow-up to *Livestock's Long Shadow*, examined the global dairy system, and found it to be responsible for 4.0% of global anthropogenic GHG emissions. Per unit emissions ranged from 1.3–7.4 $kgCO_2e$/kg FPCM at the farm gate, with developing countries having much higher emissions than Europe and North America. Further processing emissions where about 0.225 $kgCO_2e$/kg FPCM for the USA (with emissions to process different dairy products similar except for fermented milk/yogurt, which are about twice those of fluid milk or cheese), suggesting about 1.525 $kgCO_2e$/kg FPCM at the retail level.

A slightly more recent series of US-specific analyses, performed by multiple authors but summarized by Thoma et al. [431], similarly suggested about 1.63 $kgCO_2e$/kg of milk at the retail level (including 12% waste upstream of retail), and, including both consumer-level energy use for refrigeration, etc. and 20% consumer waste, 2.05 $kgCO_2e$/kg milk consumed. In sum, disregarding consumer-level waste, one may reasonably assume that emissions are on the order of 1.5–2.0 $kgCO_2e$/kg FPCM purchased at retail, with 1.75 $kgCO_2e$/kg FPCM a reasonable point-estimate (I use this slightly higher number than the point estimates of either [432] or [431] to correct for an updated methane GWP).

Given that both milk and beef are products of our friend the cow, it is salient to compare the emissions intensities of these two foods. While beef is over 20-times as emissions-intensive on a mass-basis (approximately 40–50 $kgCO_2e$ vs 1.75–2.0 $kgCO_2e$ per kg), this is not the appropriate comparison, since a much higher fraction of milk is water. On either an energy basis (i.e. CO_2e per kcal) or protein basis ($kgCO_2e$ per kg protein), beef is about four to five times as emissions-intensive (using FAO numbers for beef carcass, we have 3,230 kcal and 165 g protein per kg beef); Figure 21.6 graphically summarizes these comparisons.

Emissions factors for non-milk dairy

Given an emissions factor for FPCM, and estimates for additional processing energy for other dairy products (per the FAO), we can derive approximate emissions factors for commonly consumed items such as cheese and yogurt on a mass-basis (as mass is most readily discovered at the retail level), converting from FPCM either on the basis of energy or protein content. Figure 21.7 gives such emissions factors for several common cheeses and yogurt, assuming 1.75 $kgCO_2e$/kg FPCM. Note that the EF is slightly higher on a protein basis, and 10–15 $kgCO_2e$/kg is a reasonable rule (or range) of thumb for common solid cheeses, i.e. about one-third the impact of beef on a mass-basis.

---

is given by the FAO [432]. I use the regression given by [430]; one may also consult [429]. Standard energy content of FPCM is about 748 kcal/kg [429], but this appears to be combustion energy. Based on published milk nutrition information, I calculate that digestible energy is about 89% of combustion energy, or about 665 kcal/kg of FPCM.

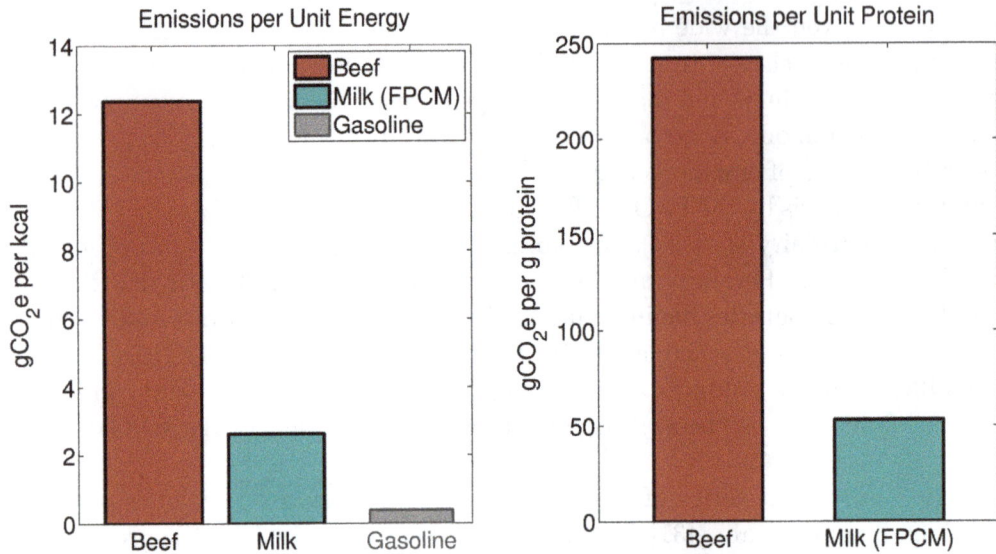

Figure 21.6: Approximate carbon emissions of milk (and most milk products) versus beef, on caloric and protein bases. The energetic comparison is given on the left; as a curiosity, the emissions per kcal of energy in gasoline is included for comparison, showing that these food products actually embody much higher emissions than fossil fuels (although the comparison is not exactly fair, as men cannot eat oil). The protein-based comparison is shown on the right, and as can be seen, the relative impact is similar on either basis.

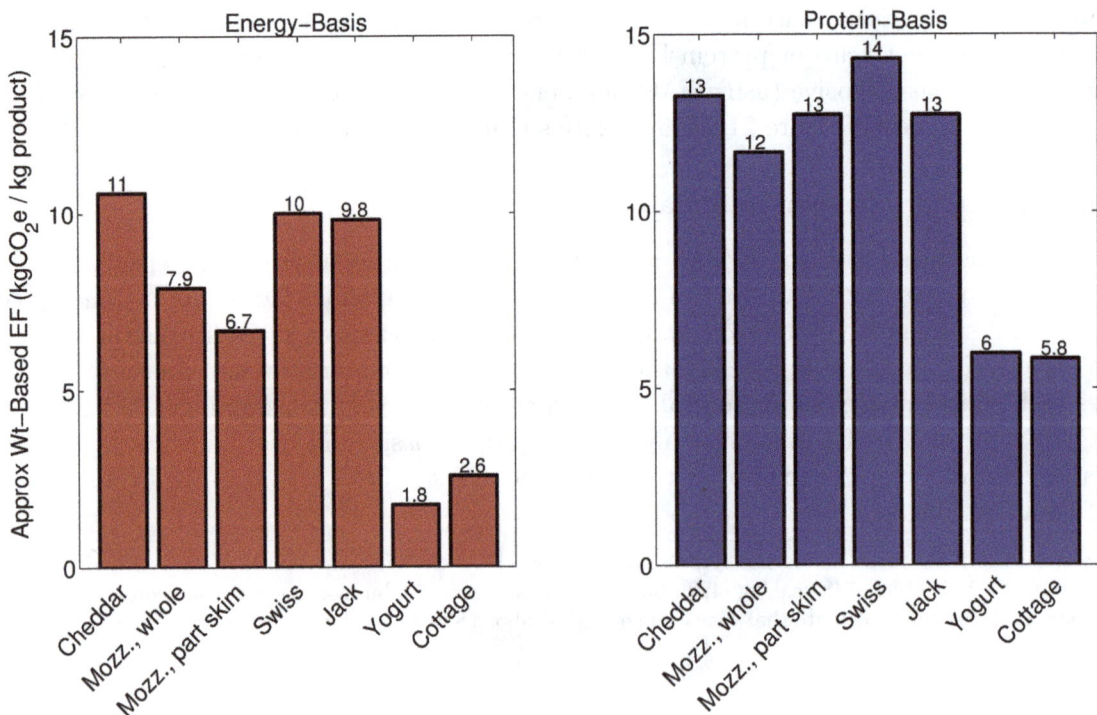

Figure 21.7: Approximate carbon emissions factor for common cheeses and yogurt, on the basis of weight, converting from an EF of 1.75 kgCO$_2$e/kg FPCM either on the basis of food energy (left panel) or protein (right panel).

21.4.3 Dairy-Beef system connection and emissions corrections for beef

- Meat ultimately sourced from dairy herds has a 33–50% lower carbon footprint than beef-only herds, and perhaps 20% of US beef is so sourced. Correcting for this, the likely carbon footprint of retail beef decreases slightly, from about 35–50 kgCO$_2$e/kg to 32–46 kgCO$_2$e/kg.

The carbon footprint of beef produced as a side-product of the dairy system has a somewhat lower carbon footprint than beef produced in designated beef-only production systems, e.g. the cow/calf-stocker-feedlot process discussed in Section 21.2. This is for two reasons: (1) milk cows culled have spent much of their life producing milk, and thus not all emissions from feed, etc. are attributable to biomass accumulation, and (2) calves that exit dairy farms to enter beef production essentially skip the cow/calf stage of beef production, the most emissions-intensive phase in beef-only herds.

Using data from Pelletier et al. [439], I calculate that, if a calf was raised entirely in a stocker to feedlot system (i.e. excluding the calf-cow phase), emissions per kg boneless beef would fall from about 40.5 kgCO$_2$e/kg beef to perhaps 27 kgCO$_2$e/kg beef, a 33% decrease. A more detailed analysis of Irish beef production [433] suggested emissions savings of 33–36% using calves from a dairy system rather than a designated beef system, very similar to my rougher estimate.

The FAO analysis [432] gave, as a global average, emissions intensities of 15.6 kgCO$_2$e/kg carcass weight and 20.2 kgCO$_2$e/kg carcass weight for culled dairy cows and fatted calves, respectively. If we assume that culls from the US system are similarly about 25% lower than fatted dairy calves, then we can reasonably estimate that meat from dairy culls has only half the associated emissions of meat from beef-only system.

So, the question now is: how do these lower emissions factors for dairy-sourced meat affect our estimate of the carbon footprint of the average supermarket beef purchase? There is no data that I am aware of tracking exactly how many cattle slaughtered for meat come from either dairy culls or dairy calves sold into meat production, but one can reverse-engineer a reasonable guess. Per the FAO, in the US about 34 million cattle were slaughtered for beef in 2012, while the 2012 USDA Agricultural Census gives 9.25 million milk cows. Assuming 77% fertility and an 8% calf death rate [432], we have about 6.5 million surviving calves. If 34% of milk cows are culled yearly, that means 3.15 million cows sent to slaughter, 3.15 million calves kept as replacements, and 3.35 million calves entering the beef system, with an equal number of (former) calves therefore slaughtered from past years. Thus, in total, around 10% of beef may come from culls (given that adult cows are slightly heavier than beef steers and heifers at slaughter [432]) and 10% from former dairy calves. Supposing, as above, that fatted dairy calves are 33% lower in their impact, and culls 50% lower, then the average retail beef emissions factor would decrease by 8.33%, from around 30–50 kgCO$_2$e/kg beef to 28–46 kgCO$_2$e/kg beef.

21.5 Monogastrics: pig, chicken, and egg

Monogastric ("single stomach") animals, such as the pig, chicken, and horse, are much more efficient at converting feed to biomass than ruminants, and methane from enteric fermentation in monogastrics is negligible. Thus, the (per-unit) carbon emissions attributable to monograstric production are nearly an order of magnitude lower than those due to beef production, and are comparable to or slightly lower than diary emissions. While Americans consume almost three times as much pork, chicken, and eggs (on a weight-basis) as beef, the former systems all together probably only generate 30–50% the greenhouse gases, and require a fraction of the land as the beef system. Monogastric production is roughly comparable in scale to dairy, with the monogastric system as a whole generating about 78% of the calories as the dairy system (using FAO numbers for fresh carcass nutritional content), while likely yielding on the order of 60–80% of the greenhouse gases. However, the monogastric system yields slightly more protein than dairy (perhaps 13% more), and so is somewhat less emissions-intensive on this basis. Land requirements (mainly in the form of pasture) are also at least two-fold higher for dairy [457].

Now, although monogastric systems are relatively efficient at converting feed to edible products, their per-unit impact is still perhaps two to tenfold greater than comparable plant sources of protein, and due their sheer scale, these production systems still represent a non-trivial environmental impact, and modern animal confinement systems entail very serious animal welfare issues as well.

As with beef, monogastric production systems generally entail a stock of breeding animals, which produce a string of offspring that are then "finished" for slaughter (or raised as laying hens), with a portion retained in the breeding population as replacement for culls. Carbon emissions (as well as land-use, etc.) must be calculated across the entire system to obtain accurate per-unit emissions factors.

Egg and chicken meat ("broiler") production systems are typically separate in the US. In broiler systems, day-old chicks from the breeding flock are raised in confined housing (but not in individual cages, and the "cage-free" label when applied to chicken *meat* is utterly meaningless marketing), fed a soy and maize-based diet, and generally brought to slaughter within six weeks. In egg systems, day-old chicks become "pullets," i.e. young birds not yet ready to lay, who then graduate to the laying hen flock, the majority of whom (in the US) are housed in battery cages and then culled after a few years of laying (the meat from such spent hens does not generally reach the retail market).

Exact emissions factors are uncertain, but overall my reading of the literature suggests a likely range of about 5–7 $kgCO_2e$/kg pork, 3–6 $kgCO_2e$/kg poultry, and 2–4 $kgCO_2e$/kg egg, at the production level. Similarly, Heller and Keolian [321] averaged results from several studies, giving 6.87 $kgCO_2e$/kg meat for pork, 5.05 $kgCO_2e$/kg meat for poultry, and 3.54 $kgCO_2e$/kg for eggs. However, these products vary in their energy and protein content, and beef and pork carcass meat (pork especially) is relatively high in calories and lower in protein, compared to poultry meat, which complicates slightly any weight-based comparison of lean retail cuts. On a caloric basis, eggs and pork may have the lower per-unit emissions, while on a per-protein basis pork is worst; eggs and chicken meat are similar, likely with a slight advantage to chicken.

Alternative production systems focusing on improved animal welfare, e.g. free range chicken and eggs and deep-bedded swine systems, tend to require slightly more feed but less on-farm energy, and may typically generate around 10–15% more greenhouse gas emissions (and require <10% more land) than conventional confinement systems. However, these relative differences usually amount to a trivial <0.5 $kgCO_2e$/kg meat or egg in absolute terms, a minor and highly defensible trade-off for improved animal welfare, at least in my view. The following sections review some of the literature supporting my emissions factor estimates.

21.5.1 Poultry

Before the Second World War per capita US egg consumption was similar to the present day, but chicken meat was a very minor component of American diets, with annual consumption less than five kg, about one-tenth of red meat consumption. The post-war years, however, saw Americans learn to love the chicken, as cheap chicken accompanied a revolution in poultry production. Compared to ruminant meat, chicken is low in carbon emissions for two major reasons: (1) chicken has a very high feed conversion ratio (FCR) and nitrogen retention fraction, i.e. a larger proportion of feed is converted to bird mass and protein (note that reduced feed requirements reduce both upstream feed production emissions and downstream manure emissions), and (2) birds, as monograstrics, emit essentially no methane from enteric fermentation.

As with pork, carbon emissions from chicken production come chiefly from feed production, manure management, and farm operations. A 2013 FAO [387] analysis estimated North American broilers to generate about 4.5 $kgCO_2e$/kg carcass weight, or 7.5 $kgCO_2e$/kg boneless meat (assuming boneless retail weight is 60% of carcass weight, per USDA data), but this estimate is high compared to other US- and UK-specific estimates. Several slightly older analyses, as reviewed by Nijdam et al. [390], gave estimates in the 2–6 $kgCO_2e$/kg meat range, and averaged just about 3 $kgCO_2e$/kg. Pelletier [434] similarly estimated US broilers to cost 1.395 $kgCO_2e$/kg live weight, or about 2.8 $kgCO_2e$/kg boneless meat. A study of the UK broiler systems [435] gave 4.41, 5.13, and 5.66 $kgCO_2e$/kg meat for standard, free range, and organic systems, respectively. Analyses have consistently concluded that a majority of emissions are attributable to feed production (almost entirely maize- and soy-based feed in the US), with on-farm energy generally coming in second, and manure management usually a very small contributor.

Several other studies suggest the greenhouse gas impact of both free range broiler [435] and egg [437, 442, 436] production to be only 10–20% higher than for conventional confinement systems, mainly due to modestly increased feed requirements. Almost all land associated with poultry/eggs is upstream, at the level of feed production, and the increased on-farm land requirement for free-range flocks is likely well under 10% [437]. Even a 20% increase in carbon footprint, from a baseline of 3–6 $kgCO_2e$/kg, corrects to just 3.6–7.2 $kgCO_2e$/kg, with a midpoint estimate of 5 $kgCO_2e$/kg meat increasing to only 6 $kgCO_2e$/kg meat. Thus, the improved animal welfare in free-range systems is, in my view, worth the small (absolute) land and carbon cost, and can (and should) be offset at the individual level by modestly decreased overall meat consumption. Finally, "organic" is not synonymous with "free range," and organic broiler systems may carry a somewhat higher cost than either conventional or free range systems [435].

21.5.2 Pork

On balance, the literature suggests somewhat higher emissions for pork than poultry, with feed production similarly dominating as the number one source of emissions, but with manure emissions also fairly significant, and more important than for poultry. The FAO analysis [387] gives about 4.75 $kgCO_2e$/kg carcass weight for North American pig systems, translating into 6.6 $kgCO_2e$/kg boneless meat (boneless weight is about 72% of carcass weight per the USDA). Pelletier and colleagues [439] estimated swine production emissions in the Upper Midwest to be about 4.9–6.1 $kgCO_2e$/kg meat for conventional industrial systems (assuming 50% of live weight is converted to boneless meat), and 5.0–6.7 $kgCO_2e$/kg for small-scale, deep-bedded "niche" systems. In a report commissioned by the National Pork Board, Thoma et al. [438] gave 5.9 $kgCO_2e$/kg boneless meat as an average across US swine. Finally, eight studies reviewed by Nijdam et al. [390] gave a range of 4–11 $kgCO_2e$/kg meat for pork (with 5 $kgCO_2e$/kg typical), and Heller and Keoleian [321] gave 6.87 $kgCO_2e$/kg as their study average.

21.5.3 Eggs

A recent and quite thorough analysis by Pelletier et al. [440], gave 2.1 $kgCO_2e$/kg egg for US production systems in 2010. Assuming an average egg weight of about 60 g, this translates into 0.1260 $kgCO_2e$/egg, or just about 1.5 $kgCO_2e$ for a dozen eggs. This study followed a more theoretical one by the same authors [441], which suggested emissions in the 2–5 $kgCO_2e$/kg egg range. Emissions sources are dominated by feed production, with manure management the only other major contributor. This is fairly consistent with the FAO analysis [387], which gave about 2.8 kgCO$_2$e/kg egg for North American flocks. Other emissions estimates, as reviewed in [440] and [390], have also generally ranged from <2 to about 5 kgCO$_2$e/kg egg, and the balance of the literature seems to suggest an emissions factor closer to 2 kgCO$_2$e/kg egg as more likely.

Of note, the FAO analysis found emissions from backyard chicken flocks to be comparable to those from industrial systems, mainly because, while backyard flocks were less efficient at converting feed to growth and had increased manure emissions, they could subsist on more marginal feedstock and their feed did not have the associated land use changes of some industrial systems. Leinonen and colleagues [436] calculated UK emissions factors of 2.92, 3.45, 3.38, and 3.42 kgCO$_2$e/kg egg for caged, barn (i.e. cage-free), free range, and organic egg production, suggesting only a slight carbon cost to improved welfare. Along the same lines, Dekker et al. [442] gave 2.24 kgCO$_2$e/kg egg using battery cages and 2.74 kgCO$_2$e/kg under free range conditions (organic and barn systems were intermediate), for Dutch egg systems. Meta-analysis of this and several other comparisons found no significant difference between the GWP for conventional and organic eggs, although lower feed conversion efficiency in organic systems does increase the land footprint [442, 467].

21.6   Seafood

- The carbon impact of fish ranges from <1 kgCO$_2$e/kg meat to over 30 kgCO$_2$e/kg meat, with mussels and herring examples of low-impact fish, while deep-trawled shrimp and various rockfish are high-impact. Salmon and tuna are middling in their impact (3–10 kgCO$_2$e/kg) and similar to chicken or pork.

- The carbon impact of wild-caught seafood is dominated by diesel fuel used to operate fishing gear, with deep trawling most energy-intensive, in addition to the terrible damage it does to the seafloor.

- Shrimp is the top seafood in the US, and the carbon impact may range from about 7–38 kgCO$_2$e/kg for wild-caught shrimp, while some farmed southeast Asian shrimp could have emissions of well over 1,500 kgCO$_2$e/kg, due to the destruction of mangrove forests. Shrimp should generally be avoided unless one is sure of a relatively benign source.

Americans consumed 7.0 kg (15.5 lbs) of seafood per capita, in 2015, with a somewhat uncertain impact. Seafood is often viewed as a more benign alternative to other animal products ("pescetarianism," etc.), but in fact the environmental impact of this broad category of food varies widely, with some quite sustainable, much of it deeply harmful, and most typical seafoods are probably similar, in terms of carbon emissions, to chicken and pork. The emissions factor for shrimp, the number one seafood in the US, may vary from anywhere from <10 to *thousands* of kgCO$_2$e/kg edible meat [445, 447], depending upon the source, and so should generally be avoided. Salmon, the number two species, may have an impact in the 3–8 kgCO$_2$e/kg range [390], while number three tuna likely generates around 3–10 kgCO$_2$e/kg [444].

Carbon emissions vary widely with fishery, fishing method, and freight shipping mode, but overall the emissions factor may generally range from <1 to 35+ kgCO$_2$e/kg of edible fish, but likely averages around 3–10 kgCO$_2$e/kg fish [390]. For wild-caught fish, the carbon footprint is dominated by on-ship diesel use, and it is generally far more efficient to catch fish that dwell in the shallower levels of the open ocean ("pelagic" species), e.g. via purse seine nets, than via deep-sea methods such as bottom trawling and longline fishing; at the extreme end, bottom trawling for Norwegian lobster required just over seven gallons of diesel fuel for a single kg of meat, equivalent to a remarkable 86 kgCO$_2$e/kg [390], about twice the carbon footprint of beef.

However, anywhere from 5 to 30 kgCO$_2$e/kg may be more typical for deep trawling [443, 445].

Bottom trawling is also widely destructive to the seafloor and generates large amounts of by-catch. Thus, fish caught via deep trawling, which generally includes rockfish (redfish, orange roughy, Chilean sea bass, and others), some cod, halibut, and shrimp and other crustaceans, should be avoided for multiple reasons. Longline methods, while also energy-intensive, are considerably less destructive to the ocean ecology than deep trawling. Small pelagic species, such as herring, may have carbon footprints well under 1 kgCO$_2$e/kg [445, 390].

Shrimp is worth a special mention, as it is the number one seafood product consumed in the US by a good stretch, accounting for over a quarter of all fish consumed in this country. At industrial scale, it is fished by deep trawling, and some estimates for North Atlantic shrimp come in at about 7 kgCO$_2$e/kg meat (based on [445], and assuming 45% of live weight is edible). On the other hand, Ziegler et al. [446] calculated a much higher emissions factor of 38 kgCO$_2$e/kg edible shrimp for industrial Senegalese pink shrimp. Furthermore, in tropical Asia, mangrove forests have been widely converted to extensively managed shrimp ponds, resulting in a massive loss of ecosystem carbon (and the ponds are abandoned after just 3–9 years due to acidification and contamination), with Kauffman et al. [447] estimating that such systems yield a staggering 1,603 kgCO$_2$e/kg shrimp on average, about 40 times worse than beef and equivalent to almost 150 gallons of gasoline.

Finally, in addition to fishing energy, cold storage and downstream transportation also add to the carbon impact of fish, although these tend to be minor components, with the exception of transport via air freight [444] (and so any seafood known to be flown in for freshness should be shunned, and not coveted).

21.7   Emissions by mechanism

Not counting direct and indirect land-use effects, three major mechanisms account for most of the greenhouse gas impact of animal agriculture: enteric fermentation, manure management, and feed production. While these have already been extensively addressed, at least in passing, I give each a dedicated treatment here.

21.7.1   Enteric fermentation

Some fraction of the carbohydrates consumed by ruminant animals is converted into methane within the rumen of the animal. It is standard to define the CH$_4$ conversion factor (Y_M) as a percentage of the animal's *gross energy intake*, and standard IPCC values are 3% for feedlot cattle and 6.5% for grass-fed cattle [450], although the value may vary between 2–12% overall [451]; several studies of Canadian beef cattle have found values of 4% for feedlot and 6% for pasture cattle [451], and Todd et al. [449] measured a value of about 3.0% in US feedlot cattle. A negligible amount of methane is produced by milk-fed calves.

The Y_M factor tells us the energy content (not mass) of methane produced for a given amount of food energy fed; since the energy content of methane is 55.65 MJ/kg (15.46 kWh/kg, HHV) [450], we can then convert to kgCH$_4$ and the CO$_2$ equivalent. For example, suppose daily energy intake for one feedlot cow is 30,000 kcal, equal to 125 MJ. If 4% of this energy turns into methane, we arrive at 90 gCH$_4$ per day, which on a yearly basis amounts to 32.85 kgCH$_4$/yr, and 1.12 metric tons CO$_2$e on a 100-yr GWP basis.

Several factors affect the CH$_4$ conversion factor, with the major factor the digestibility of the feed. Grasses and other forage (and low-quality agricultural byproducts) are less efficiently digested, and digestion depends more heavily on the rumen for forage than grains [448]. Indeed, the major evolutionary advantage of a rumen is that it allows some energy to be extracted from

441

cellulosic carbohydrates that are otherwise indigestible, but this process is less efficient than absorbing simpler carbohydrates from grain. It is generally found that the more ruminally digestible carbohydrate consumed, the higher the methane emissions [448].

It therefore follows, and has been consistently observed, that cattle fed a high-forage diet (e.g. grass-fed) convert about 50–115% more of their feed to methane than those fed diets very high (> 90%) in grain, especially corn-based diets [448, 451, 449]. A second factor observed to affect methane production is feeding rate [448]. Heavy feeding seems to reduce the CH_4 conversion factor (perhaps because food passes more quickly through the digestive tract, spending less time in the rumen). Additionally, this leads to more rapid weight gain, shorter time to slaughter, and thus decreased CH_4 emissions per kg of meat due to the combined effects of less time spent alive and emitting, and a lower per unit feed CH_4 emission rate during life. The depressing conclusion is that purely grass-fed beef is no solution at all to the environmental harms of cattle ranching, and the more difficult task of meaningful dietary shift, rather than simply a shift in labelling, is needed to overcome them.

21.7.2 Manure management

Manure, a mixture of both urine and dung, is a valuable source of all three major crop nutrients, nitrogen, phosphorus, and potassium, although it is particularly valued as a nitrogen source, and it is a major external source of fertilizer in organic farming systems (the extent to which this represents unsustainable mining of other soils and/or a masking of conventional synthetic nitrogen input via processing through an animal's digestive tract is discussed in Section 22.1.4). However, manure management is also one of the primary agricultural sources of greenhouse gases, mainly nitrous oxide formed through nitrification/de-nitrification of manure nitrogen and methane resulting from anaerobic manure storage systems. For the first part of this discussion, I focus on nitrogen and nitrous oxide, and defer a brief discussion of methane emissions to Section 21.7.3.

Manure may be deposited either on pasture by grazing animals or within housing. For theoretical analysis, overall manure N may be straightforwardly calculated as the balance between feed N input and N retained either in animal growth or products, e.g. milk and eggs. To a first-approximation, some fraction of this excreted N, which we may label F_{N2O}, then ultimately evolves via the nitrification/de-nitrification pathway to N_2O, with the exact fraction varying with climate, animal species and diet, manure storage system, etc. Our basic equations quantifying N_2O emissions from an animal manure system are therefore

$$N_{\text{Manure}} = N_{\text{Feed}} - N_{\text{Retained}} = N_{\text{Feed}}(1 - F_{\text{Retained}}), \tag{21.1}$$

$$N_2O_{\text{Manure}} = \frac{44}{28}F_{N2O}N_{\text{Manure}}, \tag{21.2}$$

where N_x denotes mass of some nitrogen pool and F_x denotes a fraction, e.g. F_{Retained} is the nitrogen fraction retained in animal products. From the first of the above equations, two obvious methods for reducing warming potential from animal manure are (1) reduce nitrogen in feed, and (2) increase the fraction of nitrogen retained. The first is achieved by balancing protein requirements with intake, i.e. avoiding excessive protein in the diet [452]. As reviewed by Rotz [452], various lower protein feeding strategies reduced N excretion by roughly 10–35% in swine and poultry, while in cattle low protein diets may reduce N excretion by up to 70%, relative to high protein diets.

Now, the fraction of nitrogen retained varies markedly by animal system, with ruminant meat production by far the lowest: N retention is 10% or less for beef production [452], and the IPCC default value is 7% [450]. Dairy is more efficient, typically about 20–30% [452, 450, 453], while poultry and swine raised for meat retain 30–40% of feed nitrogen [452, 450]. Near the theoretical limit, 40–45% of feed nitrogen may be retained in eggs [455, 456]. It follows that manure N_2O emissions scale somewhat similarly.

A more detailed hierarchy of equations for estimating N content of feed, animal mass, and products is provided by IPCC GHG Inventory Guidelines [450], along with reasonable parameter values, but now let us turn to the fate of nitrogen once excreted and the second of our guiding equations above.

Fate of manure and manure nitrogen

It is the ultimate fate of essentially all manure to be returned to fields or pastures, although not necessarily those fields from whence it originally came. However, while the bulk matter may return, *most* nitrogen is either lost en route or within days of application to fields. Indeed, although the greater part of nitrogen contained in US crops becomes animal feed, a USDA study [459] estimated only 520,000 tonnes of manure nitrogen are applied to fields (based on Figure 2 of [459]), or less than 5% of synthetic nitrogen fertilizer initially applied (this low figure is even more impressive when one realizes that little nitrogen fertilizer is applied to soy, the major protein/nitrogen feed crop, since it is a nitrogen-fixing legume).

The most basic reason for nitrogen loss from manure is that ammonia, NH_3, is extremely prone to volatilization if not rapidly incorporated into soil, and across the various stages of manure management, i.e. housing, storage, and spreading onto fields, most nitrogen simply evaporates away [458, 452]. This is probably a good point to clarify that, in mammals, nitrogen that is absorbed into the body and not incorporated into tissues, etc. is excreted in *inorganic* (or mineral) form, mainly as urea, which, as mentioned earlier rapidly degrades into ammonia and may then volatilize. Nitrogen lost in the feces is that fraction that is not absorbed (along with some excreted nitrogen); it remains primarily in organic form, and must be mobilized to mineral form before becoming either available to plants or vulnerable to loss.

Now, we can see how a high protein/nitrogen diet not only leads to increased manure nitrogen and hence N_2O emissions as a straightforward consequence of the mass-balance described in Equation 21.2, but also increases the fraction of nitrogen excreted in urea/ammonia form (as excess nitrogen is absorbed but not utilized), and thus the fraction of manure nitrogen subsequently lost [453]. In dairy cows, for example, as one moves from a low to high protein diet, the share of nitrogen excreted in urine (vs. the feces) increases from 50% to around 67–75% [453, 454].

Returning our focus to manure management, let us briefly review the stages and losses. Overall losses are reviewed expertly by Rotz [452], while a detailed overview of nitrogen loss once manure reaches the field is given be Meisinger and Jokela [458]; the IPCC, as always, is also a very valuable resource [450]. Unless otherwise clarified, one may assume that all (or nearly all) N losses referenced are due to ammonia volatilization. At most stages, small amounts of N_2O emissions occur as well.

Housing

Housed animals obviously produce manure, and the longer the manure stays within the housing apparatus, the more time the contained ammonia has to volatilize. The amount of total manure nitrogen lost at this stage varies widely, from as little as 5% for systems where manure is removed from housing daily, to as much as 90% for cattle feedlot systems where manure is not removed for many months [452]. Note that feedlots, in addition to volatilization losses, also suffer some nitrogen loss from runoff and leaching into soil. Furthermore, some decomposition and nitrification/denitrification occurs under longer-term storage, and so, depending on the system, between 0.1–4.0% of nitrogen may evolve to N_2O [450]. Typically, one may expect nitrogen losses of about 50% across various poultry, swine, and cattle systems [452].

Storage

Once removed from animal housing, manure is usually stored long-term, up to one year. This allows manure application to coincide with the very narrow peak fertilizer demand of annual crops [452]. The most significant exception to this rule occurs in some dairy systems, where manure may be spread daily (or nearly daily) on surrounding pasture, and with little time for it, nitrogen loss can be much lower [450]. Cow/calf and pasture beef systems also see most manure deposited directly on pasture.

Since housing facilities sometimes also act as long-term storage, the line may blur a bit between these stages. In any case, manure is stored in one of three forms: (1) "solid," where partially dried manure mixed with litter/bedding is stored in stacks, and 10–40% of N is lost as the manure decomposes, with losses higher under active composting of manure (also producing N_2O); (2) "slurry," where wet manure is stored in largely static ponds or tanks, with very little N lost due to the stability of the tank; or (3) "liquid" form, common in large-scale operations, where manure is stored in lagoons that are >95% water, and recycled lagoon effluent is used to flush out manure, resulting in constant mixing and near total N loss (70 to 99%) [452].

Solid and liquid lagoon storage likely leads to some N_2O emissions from nitrification/denitrification, while nitrogen can also be lost through leachate and runoff at this stage [450].

Application

At the final step, manure is applied to agricultural fields. These fields are often part of the same farm as the animals are raised in, but increasingly there are large spatial separations between animal feeding operations and crop production. This is especially true for cattle feedlots, with 68% of such operations having no crop acreage at all, and 45% of poultry is also raised on farms lacking crops [459]. Given its bulk and high water content, it is generally impractical to transport manure long distances, resulting in a quasi-open-loop nutrient cycle, where much manure goes to crops and lands not associated with its genesis, some fields may be over-manured, and ultimately only 5% of US cropland receives manure [459]. Note that poultry litter, being drier and higher in nitrogen, is more valued and may be sold and transported further than other manures.

Volatilization, the bugbear of both the housing and storage systems, is prominent at the application point as well. Irrigating (to help work the manure into soil) or direct incorporation into soil may decrease this loss, but without some kind of immediate incorporation, essentially all remaining NH_3 nitrogen may be lost [458]. Like ammonia/ammonium introduced by grazing animals, legumes, or synthetic fertilizers, applied manure NH_3/NH_4^+ enters into the nitrification/denitrification pathway to become N_2 and warming N_2O.

Now then, since typically 25–75% of nitrogen is already lost prior to field application (during housing and storage), an additional 50% loss at the field stage implies at least two-thirds of all manure nitrogen is lost before incorporation into soil, and that therefore the remaining fraction is likely to be mainly more stable organic nitrogen, rather than volatile urea and ammonia.

Overall nitrous oxide from the manure management chain

While most nitrogen is lost from the manure management chain as ammonia, as discussed in Sections 20.1.2 and 20.1.3, detailing the nitrogen cycle, volatilized ammonia does not travel far before depositing onto soils, where it re-enters the local soil-plant nitrogen cycle, and ultimately about 1% (more or less depending on nitrogen loading and plant demand) evolves to N_2O as a direct soil emission. Furthermore, some fraction will re-volatilize, re-deposit, and partially evolve to N_2O, and thus, by iteration, we should ultimately have about 1.5% of volatilized ammonia evolving to N_2O through this cycle. This is slightly higher than the IPCC value of 1.35%. Now, of manure excreted, the fraction that becomes N_2O varies markedly with storage system, but likely averages about 2% overall. Thus, of new nitrogen that is incorporated into plant matter and subsequently passes through an animal, on the first pass perhaps 3.5% evolves to N_2O.

21.7.3 Methane from manure management

Undigested organic matter that makes up the bulk of the dry matter of manure can decompose anaerobically to produce large quantities of methane. Liquid and slurry storage systems can be largely anaerobic and thus result in very high methane emissions, while dry systems are reasonably aerobic and emit smaller amounts [450]. Manure storage systems are discussed above, and I refer the reader to the IPCC for a more detailed discussion of methane emissions [450].

21.7.4 Feed production and conversion efficiency

Feed production is a significant source of emissions for beef production, and the dominant emissions source for monogastric animal products (pork, chicken, egg). All the ways in which raising crops leads to GHGs are discussed in Chapter 20, and here I restrict discussion to a brief mention of the feed conversion factor. Peters et al. [322] derived dry matter (DM) to edible weight conversion factors of 35.5 for beef. Factors for swine, chicken, and turkey were 4.91, 4.32, and 4.25, respectively, while eggs had a DM to edible weight conversion factor of 2.62.

In terms of DM to energy and protein conversion, beef took about an order of magnitude more DM to yield the same amount of edible energy and protein, compared to all other animal products (all other animal products were quite similar in these regards). In terms of energy and protein conversion efficiency, milk was similar to the monogastric animal products. These results were qualitatively similar to those reported by Eshel and colleagues [457].

Chapter 22

Other issues: Organic vs. conventional, the problem of productivity, and corn ethanol

22.1 Organic vs. conventional

22.1.1 Overview

As a consumer one often has a choice between a product labelled "organic," and one lacking such a label, with little or no further information. So, which is better, and why? While their animating philosophies may vary, the two major practical differences between organic and conventional crops are organic prohibitions on (1) synthetic nitrogen fertilizer, and (2) synthetic pesticides. Genetically modified organisms are also prohibited, but this is far less fundamental (in my view); sewage sludge for fertilizer is, somewhat curiously, prohibited as well (this strikes me as a beneficial reuse in keeping with the organic ethos). Other prohibitions also apply for organic livestock, with hormones and antibiotics in particular prohibited (animals requiring antibiotics for acute illness must be sold into the conventional system). Thus, "organic" can encompass a wide range of farms and agricultural practices, some of which may, more or less, be "organic in name only." While it is widely supposed that these prohibitions make organic production more environmentally friendly, this is not at all obvious, either from first principles or from the data.

Organic agriculture (OA) is *clearly* less productive than conventional alternatives and thus requires appreciably more land for the same output, and is likely similar to conventional in its global warming impact (on a per-product basis) [468, 464, 465, 466, 364, 467]. On-farm, organic farms do tend to support higher levels of biodiversity per unit area, but upon adjustment for lower yields, conventional and organic farms may be similar [467]. Furthermore, organic management may also result in somewhat higher soil carbon stores, but this must be weighed against the increased land requirements, as conversion of pristine land to agriculture releases far more carbon than even the best managed farm could store [467]. Additionally, this finding could also be an artifact resulting from the transfer of organic matter, via manure, from other source fields (OA is far more reliant upon manure as a fertilizer than is conventional).

While fruits and vegetables make up a greater portion of organic sales than conventional sales, overall the organic sector (at least in the US) is reasonably similar to the larger food system in its output: livestock and poultry products remain the top single category of organic sales, and the top organic crops (by land area) are the commodities hay, wheat, corn, and soybeans. Organic animal agriculture is characterized by relatively high levels of milk and egg production, and relatively low levels of beef and chicken production, but is, overall, similar to the general system in the ratio of livestock to land base. The environmental and global warming impact of organic meat and dairy is similar to or slightly worse than the conventional analogs and thus, as a sub-system of US agriculture, OA has likely done little to mitigate the environmental harms of a food system focused on animal products.

Given the magnitude of the harms associated with pesticides (discussed in detail in Section 20.2), the most beneficial aspect of organic agriculture is likely pesticide avoidance, but it is unclear if, at a systems scale, this outweighs the costs in land. It follows that simply substituting the components of a typical conventional diet for organic alternatives is unlikely to be of much efficacy, and indeed, the effect of a broad shift to organic production methods without a concomitant shift in dietary habits (e.g. less waste and meat, especially ruminant meat) could be one of net environmental harm, mainly through increased land conversion and habitat destruction.

22.1.2 Scale of the US organic system

While organic has become far more mainstream in the last few years, global area under organic management remains minuscule, at only about 0.33% of total cultivated land area, with most concentrated in the developed world, where price premiums and, as in Europe, government subsidies support this mode of production. In the US, USDA numbers indicate that, in 2011, just 0.83% of cropland and 0.49% of rangeland (and 0.64% of all agriculture land taken together) was under organic management. As already mentioned, despite the strong association between organic and produce, the top individual organic crops are commodities, mainly hay, wheat, corn, and soy, as seen in Figure 22.1.

Eggs and dairy dominated organic animal production, with 2.78% of US milk cows, and 1.97% of laying hens raised organically, while meat animals were under-represented: 0.34% of beef cattle, 0.33% of broiler chickens, and 0.20% of turkeys were organic in 2011 (USDA ERS). While < 1% of production is organic, price premiums are such that >4% of retail sales by value are organic in the US.

22.1.3 Comparative yields

For organic agriculture to move beyond a niche product for wealthy westerners to a viable large-scale alternative system, it is necessary that it be productive enough to "feed the world." Several meta-analyses performed in the last few years have attempted to quantify the yield gap between organic and conventional crops, and further, to determine how the yield gap varies among crop types, e.g. legumes, grains, perennials, etc. Overall, the weight of the evidence suggests that yields are appreciably lower under organic management, with the difference likely greatest for cereal grains. It must also be emphasized that these comparisons are limited to plot- and field-level comparisons, and extrapolations to higher systems levels are fraught, with multiple challenges inherent in scaling organic beyond isolated fields, as discussed presently, but for now, let us focus upon the organic:conventional *yield gap (or ratio)* in this more limited setting. Perhaps the earliest systematic review addressing the yield gap was that of Stanhill [461], who in 1990 compared the two systems using several lines of evidence and arrived at an organic:conventional yield ratio of 0.91 (i.e. organic production was 91% of conventional), averaged across 26 crop types and two animal products, but there was significant variability, and in this review, organic milk and beans actually tended to outperform conventional. In any case, this analysis is now of mostly historical interest, as the comparisons included are now many decades out of date, with some dating to the 1930s.

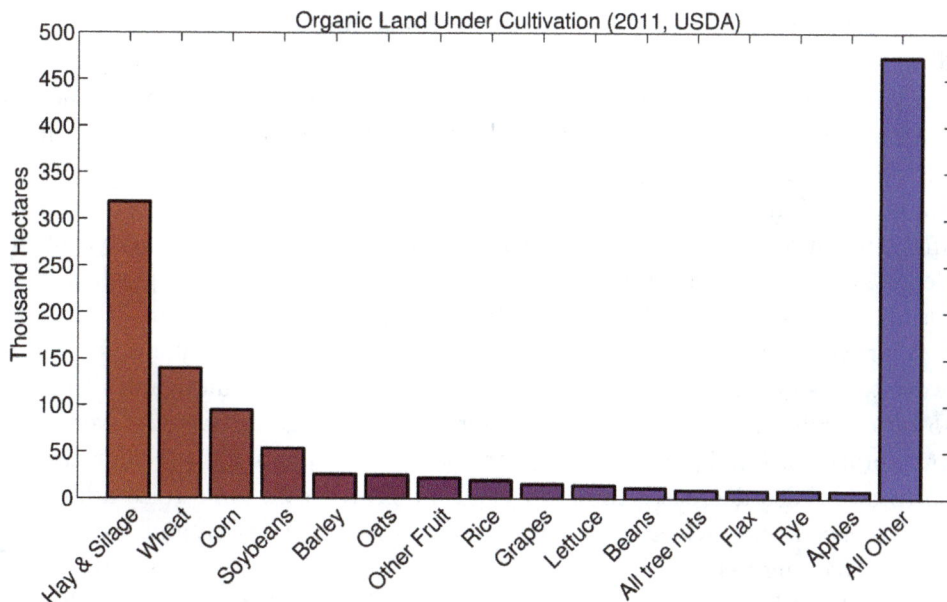

Figure 22.1: Top 15 organic crops by cultivated area, in 2011 (based on USDA ERS). Hay and other commodities clearly dominate, although the share of other crops is appreciably higher in the organic system than in the conventional one. Note that area harvested for any given crop is measured in the thousands of hectares, compared to millions under the conventional counterparts.

Later, a highly controversial 2007 analysis by Badgley et al. [460] of 293 yield comparisons concluded that, on a global basis, organic is actually the more productive mode, with an average yield ratio of 1.32 across all agricultural products, and modeling by this group further suggested a broad shift to organic could increase the global food supply. However, results varied markedly between developed countries, where the yield ratio was 0.92, and developing countries, where this ratio was a remarkably high 1.80. Applying such ratios to FAO food production statistics is the origin of the authors' conclusion that global adoption of OA could increase the food supply by >50%, but such a conclusion is deeply suspect: the favorable yield ratio in developing countries is best explained not by any intrinsic advantage to OA, but by the fact that the comparisons cited largely compared poorly productive subsistence agriculture with little or no access to inputs to optimized organic systems supplemented with large amounts of external organic inputs, e.g. off-farm manure, and could not reasonably be considered fair comparisons [462, 464, 465]. The finding that organic inputs are better than none at all is unsurprisingly and uncontroversial, and does little to inform the debate concerning the relative efficacy of OA in the developed world (or in the developing world, for that matter).

More recent meta-analyses are consistent in that, when comparing relatively comparable farming systems, the yield ratio for OA is almost uniformly <1 [468, 464, 465, 466, 364, 467]. In 2012, a systematic review by De Ponti et al. [464] excluded comparisons prior to 1985 and, notably, found only 14% of the comparisons considered by Badgley et al. met pre-defined quality criteria. Across 362 organic:conventional comparisons (mostly in North America and Europe), these authors arrived at an overall yield ratio of 0.80. They further speculated that, as many organic plots relied on very high levels of external manure input, the yield gap would likely increase when scaling up to higher system scales. Further, while they allowed that longer organic crop rotations including non-food legumes may (negatively) affect total system yields and should be adjusted for, no systematic effort was made in this respect.

Published just months after De Ponti et al, a similar work by Seufert and colleagues [465] also found organic to be less productive, with a lumped yield gap of -25% (yield ratio 0.75). Notably, the gap varied with context, and was smallest for perennials and legumes, consistent with the notion that OA is nitrogen-limited: legumes fix their own nitrogen, while perennials have larger root systems and longer growth periods that are more robust to slow or variable nitrogen availability. The gap was largest for vegetables (-33%) and cereals (-26%) and, perhaps discouragingly, was also large when the conventional and organic systems were considered most comparable, at -34%. A much larger 2015 study by Ponisio et al. [466], which used largely the same study criteria as [465], but incorporated over 1,000 comparisons, about three times more than prior works, gave a lumped yield gap of -19.2±3.7% for organic. Unlike in Seufert et al., however, no significant differences between perennials, legumes, and other crops were seen. Also noteworthy, a smaller gap was observed when organic systems were polycultures or had more rotations, but is does not appear that this analysis adjusted for the fact that a longer crop rotation also decreases the crop yield integrated over time (also addressed below). Meta-analysis focused upon wheat and maize yields in North America and Western Europe by Hossard et al. [364] suggested an average yield gap of around -30% (closer to -20% for corn, but nearly -40% for wheat), while a very recent work by Clark and Tilman also found land requirements for organic crops to be >25% higher [467].

Note that most of the above analyses are focused upon crop yields, but comparative meat and dairy yields are as or more important, given their disproportionate environmental impact. Meta-analysis by Clark and Tilman [467] suggested organic milk, dairy, and meat may require around twice the land for production, organic dairy cattle in the US produce only about 61% as much milk as their conventional counterparts [476], and my own review of individual studies on dairy and meat systems (see Sections 21.5 and 22.1.6) is also consistent with higher land requirements, largely due to lower feed conversion efficiencies in organic systems, as well as decreased yields in organic feed production [479].

In sum, the most recent and highest quality evidence suggests that, under experimental conditions, OA is about 20 to 33% less productive than conventional, on the basis of yield [468, 464, 465, 466, 364]. However, these estimates are generally plot/field-based comparisons where the organic plot often received large external inputs (mainly manure), not whole system comparisons, and Connor [463] has therefore stated, with some justice, that these are better understood as studies comparing organic and synthetic *fertilizer*, not organic and conventional *systems*, and the yield gap for an organic system could be markedly worse. These studies also do not typically adjust for sometimes longer organic crop rotations that may incorporate non-food crops for green manure (adjusting would tend to increase the yield gap), and the lower yield gap observed for longer rotations [466] may thus be essentially artifactual.

22.1.4 External inputs and organic: can an organic system be self-sustaining?

Organic systems tend to be nitrogen-limited, for several reasons [472, 465]. The most basic is the need for alternative N sources when synthetic fertilizer is eschewed, either in the form of "green manures," typically nitrogen-fixing cover crops (generally legumes) that are plowed under at the end of the season to provide nutriment for the following crop, or imported animal byproducts and manures (note that the N in manure itself must ultimately come from some soil or other). Also important is the fact there tends to be a temporal mismatch between nitrogen availability from the mineralization of organic N sources and peak demand by fast-growing annual crops. Unlike most organic amendments, the timing of soluble synthetic N applications can be more directly matched with plant demand.

The need for *external* nitrogen inputs, mainly manures and meat industry by-products (e.g. feathers, meat and bone meals [469]), is of fundamental importance and may severely limit the viability of organic agriculture as a large-scale alternative system. Indeed, in most experimental studies comparing conventional and organic yield, overall nitrogen inputs are similar in magnitude (and sometimes *greater* in the organic system), with large amounts (typically all) coming from off-farm manure. For example, Clark et al. [474] supplied organic corn and tomato beans with 150–200 kgN/Ha from composted poultry manure (comparable to conventional synthetic N application rates), and Delate et al. [475] similarly applied 157 kgN/Ha of swine manure to organic corn, identical to the conventional 157 kgN/Ha of urea.

As has been pointed out by Connor [463] and others, much of this manure and other inputs may come from conventional sources, and hence simply represent covertly processed synthetic fertilizer. Indeed, Nowak et al. [470] found that nearly all N input to 63 French farms across three agricultural districts came either from conventional sources or atmospheric N deposition (actually the major N source, which itself partially originates from volatilization of synthetic N applied to neighboring conventional farms [463]): of the N imported through manure and animal by-product fertilizers, >95% of byproduct N came from conventional sources, and 82% of manure N was conventional in origin. Once considering other nutrient inputs in the form of feedstuffs, fodders, and straw, a smaller but still hefty 66% of imported N came from conventional farms. Organic farms were also heavily reliant upon conventional agriculture for P and K fertilizers, at 73% and 53% of inputs, respectively [470].

Despite the problem of mineral nutrients possibly (or even probably) being sourced from conventional farms, we also must confront the basic fact that manure cannot hope to replace synthetic N at anything remotely approaching current use rates. Per [459], about 520,000 tonnes of manure N are applied to US fields annually, a figure wholly dwarfed by the >11 million tonnes of synthetic N spread o'er these same fields. Indeed, at the corn application rate employed by Clark et al. [474], the entire manure output of this nation's bloated animal sector would supply a scant 3.3 million hectares, just 10% of all harvested corn area and 2% of all cropland, leaving the other 98% without any inputs at all. Clearly, manure application rates in experimental organic systems are not scalable to a national system, and we must be skeptical of extrapolating such yields to any alternative hypothetical food system. Finally, it should go without saying that manure nitrogen is not some kind of "free" N source, as it must ultimately be supplied either by synthetic N fixation, or biological N fixation in some soil or other.

"Green manures," (GMs) generally either nitrogen-fixing legumes or nitrogen-scavenging grasses, are crops grown specifically to serve as a soil amendment and nutrient source for subsequent crops [471], and represent a far more viable replacement for synthetic N than manure or by-product fertilizers, although they are not without their limitations. Green manures are most typically cover crops (CC), planted during the cool season between main-season crops, especially in grain-based systems, that are then plowed under before planting the next production crop. In warmer areas, leguminous CCs can fix approximately 100–200 kgN/Ha in a season, and may obviate much or all nitrogen fertilizer need. In colder areas, however, the growth and N-fixing potential of CCs is lower, although they may still be beneficial [472]. Grass species may also be used as CCs, as they produce large amounts of biomass and also scavenge residual soil N, hence decreasing N leaching. Co-planting of grassses and legumes is a particularly beneficial practice that both effectively retains existing soil N while fixing new N. In addition to providing N, CCs have other benefits: CC biomass augments soil organic matter, and the practice of cover cropping can build soil organic carbon stores over many years [473]. Cover crops also provide habitat and suppress weed growth in subsequent main-season crops [472, 473].

While generally broadly beneficial as cover crops, green manures do have some important limitations. First, only about 10–50% of N fixed by a cover crop is available to the next crop [472]. Second, there is an asynchrony between N supplied by a CC and peak crop demand: upon incorporation into soil, CCs degrade rapidly, providing a short-lived burst of N (perhaps 6–8 weeks), which mainly predates peak demand by the subsequent main crop, although grasses tend to decay slower than legumes [472]. Third, as already mentioned, CC productivity is limited in colder areas. Fourth, in some organic rotations GMs are grown during a regular growing season, thus displacing a food crop and lowering the time-integrated productivity of the overall system. Fifth, while valuable N and C sources, green manures do not generally supply new P and K.

Finally, it should also be emphasized that green manures and cover cropping are not exclusive to organic systems, and may be (and often are) beneficially incorporated into conventional farming practices.

22.1.5 Organic livestock and animal welfare

USDA organic standards do set minimum animal welfare standards that exceed those for conventionally raised animals, and, given their philosophical leanings, many smaller organic producers appear to make animal welfare a special focus, e.g. in small pastured laying flocks. Organic livestock are required by the USDA to have access to the outdoors and direct sunlight year-round, and must not be confined in such a way that prevents free movement. In the case of poultry and eggs, the "access to outdoors" requirement has been followed by some (typically larger) egg producers more to the letter than spirit, where producers build large hen houses (with several 10,000 hens) with an adjoined outdoor screened-in "porch," thus technically meeting the requirement, but clearly not providing any meaningful pasture; close to half of organic eggs may be produced in such systems, while a new regulation, on hold at the time of this writing and with a five year phase-in period, would require a more meaningful two square feet of outdoor space for hens and do away with porches [477].

Organic ruminant production is based on pasture, with the USDA explicitly requiring that animals have access to pasture during the grazing season (no less than 120 days), and meet at least 30% of dietary dry matter intake from pasture. Therefore, organic milk systems are generally pasture-based, whereas the majority of dairy cattle raised in the US live in large confinement systems: in 2010, 73%, 17%, 6%, and 5% of milk came from conventional confinement, nonorganic semipasture-based, nonorganic pasture-based, and organic operations, respectively [476].

22.1.6 Comparative global warming impacts

In terms of global warming impact, there is no apparent benefit to organic agriculture, and for some products, the organic option may even be slightly worse. The fairly consistent conclusion across multiple studies reviewed by Mondelaers et al. [468] is that organic cropping systems tend to have lower CO_2e emissions per unit area, but, due to the generally lower productivity of OA, there is little difference between systems when CO_2e is expressed per unit product, the far more relevant metric (in my view). Recent meta-analysis by Clark and Tyler [467] similarly found a very slight, but not statistically significant trend towards increased emissions from organic systems.

Several individual analyses of animal products have found the global warming impact of organic options to be slightly higher than non-organic, mainly through somewhat lower feed conversion efficiencies. For example, Leinonen and colleagues found organic chicken and eggs to be 28% [435] and 17% [436] higher in GHG impact, respectively, and Dekker et al. [442] similarly found organic eggs to be 13–14% higher in CO_2e emissions. Multiple publications also suggest a carbon premium for organic pork production, with studies reporting increased CO_2e per kg pork on the order of 7-22% [483], 14–35% [484], or 73% [485]; Kumm [486] also reported higher CO_2e for organic vs. conventional pork, but did not provide a precise number.

Organic milk production is relatively inefficient and requires appreciably more land per unit product, but it is uncertain if its global warming impact differs significantly from conventional systems. More intensive milk production has been associated with lower $kgCO_2e/kg$ emissions factors in some studies [479], but not in others [480, 481] (although land use more consistently decreases with intensity), emissions factors for organic and conventional milk were similar in several works [478, 480, 482], and while globally, milk production emissions are much higher outside industrialized areas, meta-analysis of mainly US and European studies [467] also showed no appreciable difference between the two methods.

As discussed extensively in Section 21.2, 100% grass-fed beef likely has a higher ecological impact, both in terms of land use and greenhouse gases, than grain-finished beef. While grass-fed or finished is not synonymous with organic, the USDA mandates access to pasture and a minimum 30% pasture feed intake for organic cows (although this requirement is waived in the final 120 days of life), so there is significant overlap.

22.1.7 Conclusions

Overall, it seems to me that minimizing pesticide, especially insecticide, applications is the most beneficial aspect of organic agriculture over conventional, although a significant body of work supports the notion that pesticide use can be markedly reduced (but not wholly eliminated) with little to no effect on yields (see Section 20.2.3). Cover crops planted between growing seasons can clearly provide significant nitrogen as well as organic matter input, but these are not limited to organic systems, and while they are unlikely to wholly replace synthetic inputs in much of the world, they can at least offset some requirements. The most environmentally friendly production system is likely a "conventional" one that does not abandon the benefits of Green Revolution technologies, but that seeks to minimize the impacts of their overuse. Moreover, such a system would be far more scalable than organic, which would quickly run into problems of organic nutrient availability if deployed at a truly global (or even national) scale. "Organic," then, defined in negative terms that (perhaps) arbitrarily prohibit potentially useful synthetic fertilizers, pesticides, antibiotics, and other inputs is ultimately more of an ideology than a scientific paradigm for environmentally friendly agriculture (see also Trewavas [487] for a discussion along these lines).

Finally, at the point of purchase, there are several products for which it may be reasonable to choose the organic option. First, produce items including grapes, apples, oranges, and tomatoes are subject to very high pesticide spraying intensities, and thus total pesticide avoided may be maximized by choosing organic versions of such products. On the other hand, while field crops such as corn and soy are sprayed at relatively low intensities, given that feed conversion ratios are on the order of 4–5 for monogastric meat [322], total upstream pesticide embodied in such meats is likely similar to heavily treated horticultural crops (and beef, with a conversion ratio approaching 40 [322], likely embodies more far pesticide than any other common food). The yield gap for legumes, such as beans and soy, may be relatively small (as observed in [465], but not in the larger analysis by Ponisio et al. [466]), and so these might also be more reasonable organic choices.

One may purchase organic animal products in the hopes of somewhat improved animal welfare, but probably at some cost to the larger environment, and the best answer is to reduce consumption, period. Indeed, unless accompanied by an absolute reduction in consumption (at least relative to a typical diet), the consumption of even organic animal products seems difficult to justify. The absolute carbon and land premium for organic monogastric products, e.g., free range organic eggs, is relatively small, and again, so long as these products are *minimized* overall, organic eggs and dairy products are probably reasonable selections. Beef is best avoided regardless, and as already discussed extensively, grass-fed beef likely carries a higher environmental cost than the grain-finished alternative. Overall, a conventionally produced diet that minimizes or eliminates animal products and waste is almost certainly vastly superior to a completely organic diet that is otherwise typically American, and organic dietary substitutions are far less meaningful than an overall dietary shift (but may be done to a limited extent as one component of a larger shift).

22.2 The problem of productivity: are higher yields truly an environmental good?

In the prior section I compared organic and conventional agriculture, and concluded that organic yields are probably lower, with a yield gap of at least 20%. This leads to the central argument in favor of conventional high-input agriculture, namely that is an efficient use of land, a finite resource. To lower the efficiency of agriculture would require the expansion of the land-base, so the argument goes, and thus conventional agriculture is a great good, preserving wild land and ecosystems that would otherwise be appropriated for Man's use. This argument is particularly salient, given that global food demand is variously projected to increase by 70% to over 100% by 2050, in the face of both global population growth and increasing worldwide demand for meat and high calorie diets, and the problem of meeting the twin demands of increased agricultural productivity and minimizing its environmental impact is clearly a fundamental one [465]. The general strategy of increasing yields via intensification with an eye toward minimizing environmental impact has been termed *sustainable intensification*, and one may consult, e.g., Garnett and colleagues [490] for a thoughtful commentary on the concept.

Indeed, it is largely true that the *per-unit* environmental impact, both in terms of land required and carbon emissions, has fallen over the latter half of the twentieth century with agricultural intensification [488, 391], and emissions factors, for animal products especially, are much lower in Western industrialized systems. For example, both GHG emissions and land use are higher in extensive pastoral beef production systems compared to intensive industrial systems [390], and FAO analyses have concluded that beef and milk emissions factors are both several times higher in the developed world compared to North America and Western Europe [387, 432]. The trend towards more intensive systems having a lower impact is likely true within the US as well: For US maize production, Grassini and Cassman [384] observed higher agricultural intensity and higher corresponding yields to result in lower GHG emissions-intensities, and as reviewed previously, more intensive animal production systems within the US also tend to save carbon on a per-product basis.

Burney and colleagues [488] constructed counterfactual scenarios to estimate how land use and GHG emissions would have differed in hypothetical worlds without the historically observed agricultural intensification between 1961 and 2005, and concluded that intensification avoided 317–590 GtCO$_2$e over that period, or as much as one-third of humanity's historical CO$_2$e emissions, and suggested that improving yield is an excellent harm mitigation strategy. Forward-looking projections by Tilman et al. [489] similarly suggested that, to meet a rough doubling in global calorie and protein demand by 2050, moderate agricultural intensification in mainly developing countries could avoid 80% and 67% of the land clearing and carbon emissions, respectively, that would otherwise occur.

A related debate is that of "land-sparing" vs. "land-sharing." The notion of land-sharing is to promote farming practices that increase on-farm biodiversity and, at least to some extent, share the land with wildlife. The central problem is that such an approach tends to give lower yields, and requires more farming land overall. Land-sparing entails higher-intensity agriculture that, while less friendly to wildlife on the area actively farmed, frees, at least in principle, more land to be wholly untroubled wilderness. On the whole, so long as areas spared by high-yield agriculture are actually protected from other development, land-sparing may be the better strategy [491]. This point is key, because as discussed in just a moment, decoupled from a larger policy framework, intensification could also have the perverse effect of even greater land appropriation for farming [492].

The evidence would thus seem to fairly clearly come down upon the side of agricultural intensification and land-sparing over land-sharing. However, one possible fly in this ointment is the famous Jevons paradox, which comes from the observation of Jevons, in 1865, that increased efficiency of coal-use led to increased coal consumption. That is, as efficiency goes up, the cost of using a resource goes down, and so overall demand increases. It is worth quoting the original passages, from Chapter 7 of The Coal Question [19], at some length (emphasis in original):

> It is very commonly urged, that the failing supply of coal will be met by new modes of using it efficiently and economically. The amount of useful work got out of coal may be made to increase manifold, while the amount of coal consumed is stationary or diminishing. We have thus, it is supposed, the means of completely neutralizing the evils of scarce and costly fuel...

> *It is wholly a confusion of ideas to suppose that the economical use of fuel is equivalent to a diminished consumption. The very contrary is the truth...*It is the very economy of its use which leads to its extensive consumption. It has been so in the past, and it will be so in the future. Nor is it difficult to see how this paradox arises....

> It needs but little reflection to see that the whole of our present vast industrial system, and its consequent consumption of coal, has chiefly arisen from successive measures of economy.

The implications translate naturally to agricultural land, and I am not the first to make this comparison: Lambin and Meyfroidt [493] have reviewed the at least partially flawed of notion land-sparing through agricultural intensification expertly. As they point out, demand for staple grains (the basic provisioners of calories and protein) in a society is largely inelastic, but demand for meat and biofuels, which actually consume the majority of calories produced at the primary crop level, at least in the West, are elastic. Thus, efficiency increases may primarily act to increase production of and demand for the latter products, with little change in the area under cultivation. Furthermore, increased efficiency increases profitability, giving an incentive for expansion into more marginal lands, and therefore agricultural intensification can increase rather than decrease cropland expansion, especially within the context of a globalized agricultural trade.

The effect of agricultural intensification varies between developed and developing countries, and between agricultural systems that grow food primarily for local consumption and systems that grow cash crops for export. Increasing efficiency of local food production can indeed reduce pressure on the land, whereas increased efficiency of cash crop systems has led instead to agricultural expansion [493]. Another problem with agricultural intensification is that it may open new lands to cultivation. For example, cropland expansion into the Amazon has been facilitated by new soy varieties and heavy fertilizer and pesticide use [492]. Clearly, high yields are probably necessary, but are by no means sufficient for a relatively "green" agriculture [492].

Returning to the problem statement above, namely that agriculture must evolve to meet the demands of increasingly high-calorie and meat diets of roughly nine billion individuals by 2050 while minimizing environmental impact, the problem statement itself would seem to suggest two possible solutions beyond agricultural intensification: reduce demand for meat and high-calorie diets, or stop growing the population. Indeed, my reading of history suggests a basic problem in agriculture throughout world history that one might term the "productivity trap" [494], a seemingly never-ending ratcheting up of agricultural productivity, only to have it undermined by increasing demand and population growth. Population growth projections are always treated as *exogenous*, i.e. imposed from without, in these discussions of productivity, which simply must, it seems, expand to meet rising demand. But if productivity cannot support meat-heavy, calorie-rich diets throughout the world, then it will not; if it cannot support a population of nine billion, then it will not. The point here is that we should think of demand and supply in an integrated sense, and the answer to whether industrialized high-yielding agriculture is an environmental good is subtler than per-unit emissions or land-use factors. It is beyond me to provide a complete answer here; it remains true that, *all else equal*, a more intensive agricultural system is likely of benefit, although in reality this probably must be coupled with other policies and/or dietary shifts to be a true environmental boon.

22.3 Corn Ethanol

Analysis by the EPA concluded that corn ethanol, as produced in 2022, would generate 21–23% fewer lifecycle GHG emissions than gasoline [502]. Such conclusions motivated the 2007 US Renewable Fuels Standard (RFS2), which mandates the blending of renewable fuels into the transportation fuel supply, an ends towards which nearly half the US corn crop now goes. As we shall see, many other analyses have concluded that ethanol is actually *worse* than gasoline when it comes to greenhouse gas emissions, it saves little if any energy, and has likely driven recent large-scale agricultural expansion at the expense of natural grasslands and habitat.

Even if ethanol does achieve a 20% reduction in GHG on a per-gallon basis compared to gasoline, this would represent only a very small carbon savings overall, but at the expense of much habitat destruction and biodiversity loss. Indeed, if, hypothetically, *all* US cropland was converted to corn ethanol production, we would achieve only about a 15% reduction in light-duty vehicle emissions (i.e. I am not even counting heavy duty vehicles and freight transportation), or roughly the equivalent of increasing the average passenger vehicle fuel efficiency by 4 MPG. This cannot be emphasized enough: if ethanol were somehow scaled to the absolute theoretical maximum, it would in the best case be equivalent to increasing average fuel economy from about 22 to 26 MPG, and would require a truly vast (and likely impossible) expansion of agriculture into all remaining wild lands.

Given this, the question that remains then is, is corn ethanol, on the scale it is currently employed, valid as a minor component of a larger clean energy portfolio? Well, corn ethanol likely takes just as much fossil energy to produce as it provides, and, once the nitrous oxide emissions from agricultural soil and indirect land-use changes are properly accounted for, probably generates as many if not more greenhouse gas emissions as it offsets. Thus, no matter how you look at it, corn ethanol is disastrous. From an energetics perspective, corn ethanol is simply spinning the wheels or worse: it provides no additional energy and is probably actually a net drain on the existing fossil-based energy system. From a climate perspective, it is probably a wash or worse, and even the most optimistic assessment yields minimal overall climate benefit.

22.3.1 Energetic analysis: return on energy investment

Multiple studies have analysed corn ethanol on the basis of energy return on energy investment (EROI), a standard measure of net energy return that can be calculated for an array of energy sources, defined as the ratio,

$$EROI = \frac{\text{Energy Out}}{\text{Energy In}}. \tag{22.1}$$

An EROI of 1.0 implies that, for every unit of energy invested, a single unit of energy was extracted and thus the project was a wash; an EROI > 1 implies a net energy return on investment. The EROI for historical oil fields was around 50 (i.e. it took only 1 unit of energy to obtain 50 units), but EROI for fossil fuels tends to decrease over time as those reserves of highest quality and easiest access are spent, and extraction shifts towards lower quality resources. Tar sands, in particular, have an extremely low EROI, at perhaps 4, and oil shales have an EROI of just 7 [496]. Solar and wind have reasonably good EROIs (perhaps 10 or so for photovoltaics and closer to 20 for wind [496]) that, while lower than some fossil sources, are not subject to the same law of diminishing returns and indeed, are likely to improve with time, especially in the case of solar PV, with improving manufacturing technologies that use less raw material and energy. Note also that the energy produced by solar and wind is in the form of electricity, a higher quality form than the thermal energy contained in fossil sources.

The EROI for corn ethanol, on the other hand, hovers around 1.0, and edges above or below this magic number depending on the particular study [72]. While one may be tempted to conclude that if the EROI is greater than 1.0 then ethanol is a good idea, this is false, as emphatically pointed out by Murphy et al. [72]. Consider, at the civilization scale, EROI for a society's energy source. If it is only slightly above 1.0, then nearly all of society's energy must be used to obtain more energy, with little left over for other use. The amount of energy required to deliver a single unit of net, usable energy to society is given as

$$\frac{EROI}{EROI - 1} - 1, \tag{22.2}$$

so for an energy source with an EROI of 10, we require 0.11 units of energy to extract a single unit. The total (or gross) energy use by society then sums to 1.11 (also given as EROI/(EROI - 1)), with 90% usable. As EROI decreases, we begin to fall of the "net energy cliff," where most of society's total energy is devoted to energy extraction. If corn ethanol has an EROI of 1.3, this implies that 77% of all energy goes toward energy extraction, leaving little to support the basic infrastructure of society. Hall and colleagues [495] suggested that, as any society must gain appreciably more energy than it expends, the minimum EROI for energy sources used in support of an industrial society must be about 3; any sources with an EROI < 3 are thus subsidized by the fossil energy system.

Murphy and colleagues [72] summarized energetic inputs for corn production from five previous studies on ethanol, and under meta-analysis found an EROI of 1.07±0.2. A county-level analysis by the same group across 1,287 counties suggested a national average EROI of just 1.01. A reasonable best-case EROI estimate for corn ethanol is 1.3, still far below the approximate minimum useful EROI of 3.

22.3.2 Some global warming effects of corn ethanol

- The EPA suggests an ethanol emissions factor (EF) 21–23% lower than that of gasoline, under a new gas-fired plant in 2022. My own calculations suggest an EF anywhere from 11% lower to 77% higher, using current technology, and depending upon our accounting of land use, ethanol could well be twice as bad as gasoline.
- Indirect land-use changes and N_2O from N fertilizer are major, if uncertain, factors, that undermine any benefit to ethanol.

Aside from the energetic calculus above, it is unclear if corn ethanol has a lower carbon impact than gasoline, and N_2O emissions from synthetic fertilizer and land uses changes are the major areas of controversy, as briefly discussed here. Disregarding land-use changes entirely, and using energy inputs as tabulated in [72] and [500], my own calculations suggest that, assuming a maize yield of 10 Mg/Ha and N fertilizer inputs of 157 kgN/Ha (with 1.35% evolving to N_2O), in the best case ethanol has an emissions factor of about 0.297 kgCO$_2$e/kWh, about 11% less than that of gasoline (EF 0.333 kgCO$_2$e/kWh on LHV basis). However, a more realistic 3% conversion factor of N to N_2O gives an ethanol emissions factor of 0.3615, 9% worse than gasoline. Alternatively, even a conservative estimate of 0.05 kgCO$_2$e/kWh due to land-use change gives an EF of 0.347 kgCO$_2$e/kWh, 4% worse than gasoline. Under a worst-case scenario, with a 5% N to N_2O factor and 0.15 kgCO$_2$e/kWh due to land use changes, we have an ethanol EF of 0.5896 kgCO$_2$e/kWh, 77% higher than the gasoline EF.

Nitrogen fertilizer and nitrous oxide

The global warming impacts of nitrogen fertilizer in general are discussed extensively in Section 20.1.3. Crutzen et al. [348] have argued that, once extra N_2O emissions from fertilizer are accounted for, at a "proper" 3–5% conversion factor of new reactive N to N_2O, then the global warming impact of biofuels is worse than the fossil-based alternatives, even disregarding all other lifecycle factors (fertilizer, on-farm energy use, etc.). While my own calculations are not quite as dramatic, if the N to N_2O factor is indeed in the 3–5% range, then ethanol will be definitively worse than gasoline.

Land-use changes

There have been concerns for years that the expansion of biofuels may lead to clearing of carbon-rich ecosystems for new cropland, thus incurring a massive "carbon debt" that could take up to centuries to repay by biofuel use [70], as discussed already in Section 20.3. Land use change can also occur indirectly: if land already under cultivation is converted to biofuel production, new agricultural land may be cleared for those displaced crops. One of the most pessimistic, and famous, conclusions was that of Searchinger et al. [498], who calculated that the inclusion of land-use changes gave an ethanol global warming impact 93% higher than gasoline, over 30 years. Even less pessimistic projections (see below) negate any benefit to ethanol.

An additional concern, when new land is either cleared for corn ethanol production or even when other crops are displaced to produce corn, is that this new land will almost invariably be of lower quality. That is, the best and most fertile land in the optimal environment is used for production first, where the best return on investment can be expected. As production expands, marginal lands are cultivated, where yields will be lower for the same (or even greater) energy inputs. This applies at a local scale, i.e. the best fields in an area are used first, and at a regional scale, e.g., corn yields in Iowa, the prototypical corn belt state, are about 50% higher than yields in Texas [72]. Thus, even if there is some advantage to ethanol under optimal growing conditions, if its use drives the cultivation of marginal lands, the overall EROI will decrease and associated emissions will increase.

Such effects on land-use occur within a complicated economic system, and so directly quantifying the influence of biofuel production is difficult and controversial. However, several recent studies [497] make it clear that high commodity prices for corn and soy, largely attributable to biofuel demand, have driven a dramatic expansion of corn/soy cropland into formerly uncultivated grasslands, and that moreover, these lands are of marginal quality, highly vulnerable to drought and erosion. In addition to the carbon debts incurred, this obviously represents a massive loss of natural ecosystems and biodiversity.

Lark and colleagues [497] recently performed an analysis, using multiple land cover databases and satellite data, of continental US cropland changes from 2008–2012, those years immediately following the passage of the 2007 US RFS2, (again, mandating the blending of "renewable" fuels into the transportation fuel supply). They found that 2.97 million hectacres of land uncultivated since at least 2001 were converted to cropland, while 1.76 million Ha were taken out of production, for a net expansion of 1.21 million Ha. Most of the newly converted land was of marginal quality, and included significant expansion into the hilly landscapes of southern Illinois and northern Missouri, land formerly used primarily for grazing, and expansion into the panhandles of Oklahoma and Texas, land that is irrigated by the rapidly depleting Ogallala aquifer. Corn was responsible for 51% of increase in cropped area, expanded more than any other crop, with a net increase of 3.48 million Ha under cultivation, and was also the most planted on newly converted land, including 0.65 million Ha that had been grassland for at least 20 years. While other crops also were planted on new land, especially wheat, this was largely due to displacement from other areas.

Lark et al. [497] estimated perhaps 94–186 million $MgCO_2e$ attributable clearing new lands for corn and soy between 2008 and 2012, across 1.38 million Ha. Even attributing just half these emissions to corn, and supposing a very optimistic yield of 10 Mg maize per Ha over 50 years (yielding 18,850 kWh per Ha [500]), this gives at least 36.1–71.5 gCO_2e/kWh ethanol, enough to offset any GHG benefit to ethanol. Attributing all emissions to corn ethanol and amortizing over 20 years and we would have 181–358 gCO_2e/kWh ethanol, enough to nearly double the GHG impact of ethanol relative to gasoline. Qin et al. [499] also modeled soil organic carbon changes resulting from land-use changes (LUC) attributable to corn ethanol, and gave a much lower overall estimate of 7.56–33.48 gCO_2e/kWh of ethanol from LUC. However, if we suppose a yield of about 20,000 kWh/Ha/yr of ethanol, and long-term soil carbon losses equivalent to 2–3 $MgCO_2e$/yr (see Section 20.3), then we would have an additional 100–150 gCO_2e/kWh, more consistent with my inferences from [497], and with the conclusions of Searchinger et al. [498].

22.3.3   Ethanol is not scalable

While I argue that it is unlikely that ethanol provides *any* emissions benefit, even if we assume that the 21% carbon reduction estimated by the EPA in 2022 (for a *new* natural gas plant) is accurate, ethanol still provides very little net climate benefit, nor can it possibly be scaled up to provide more significant benefit. At best, with about 45% of the corn crop going to ethanol to yield almost exactly 300 billion KWh of fuel energy, we would reduce carbon emissions by 21.0 million $MgCO_2e$, equivalent to about 1.52% of the emissions attributable to gasoline consumption by personal vehicles in the US, or 0.32% of US territorial emissions. That is, if literally every square inch of Iowa were devoted to corn ethanol production, overall US emissions would decrease by barely one third of one percent!

On the other hand, suppose this crop land was simply taken out of production and converted back to native grassland. As discussed in Section 20.3.2, we could then *conservatively* expect soil carbon sequestration on the order of 0.30–0.60 MgC/Ha/yr [378], implying 17.6–35.1 million $MgCO_2e$ sequestered per year; this calculation does not include other increases in root mass, etc. Therefore, even a conservative estimate of the carbon savings from simply allowing the corn fields to revert to grassland is greater than the most optimistic estimate of the carbon offset of using this land for corn ethanol. Both carbon offsets are relatively low, but the grassland scenario would likely have other enormous ecological benefits.

While clearly not plausible, suppose the entirety of US cropland (165 million Ha) was devoted to corn ethanol, and the very optimistic 21% CO_2 reduction per gallon of ethanol was achieved. Then we should hypothetically save 217.43 million $MgCO_2e$, still less than 3.3% of overall US emissions (and likely much less, as much cropland is not well-suited to corn). Compare this to solar. A similar area of land covered in modestly efficient panels (obviously a purely hypothetical, and supposing 15% efficiency and a 75% performance factor) would generate about 17 times global electricity consumption in 2012, and over twice the global primary energy consumption, thus sending global, not just US, fossil fuel emissions to zero. Wind farming this area would give only about 4% the energy (assuming 1 W/m^2), but still yield over thrice the US's electricity consumption. Clearly, solar and wind are vastly superior to even an optimistic assessment of ethanol in particular, and biofuels in general.

Part V

Goods, Services, Waste and Recycling

Chapter 23

Goods and services

- The provision of various goods and services, mainly clothing, household goods, entertainment, healthcare, and education, likely generates anywhere from 12 to 16 $MgCO_2e$ per household, based on input-output analyses, and including internationally imported emissions. A reasonable midrange estimate is 14 $MgCO_2e$/household, or about 5.6 $MgCO_2e$/person.

- Overall, household spending has a crude emissions per dollar factor of around 0.4–0.5 $kgCO_2e$/\$. However, excluding housing (where most spending is debt-servicing), transportation, food, and savings, the emissions factor for most direct expenditures is likely closer to 1 $kgCO_2e$/\$.

- In terms of total attributable emissions, the single most carbon-intensive manufactured good is likely clothing, with almost all emissions imported from abroad, while the single most carbon-intensive service is almost certainly healthcare, accounting for as much as one-third of the emissions attributable to goods/services.

23.1 Overview

Several input-output analyses of the carbon footprint of consumption have generally found the general consumption categories of goods and services (i.e. household consumption excluding electricity, direct fossil fuel use, personal transportation, and food) to together account for about 12–20 $MgCO_2e$ per household (with all emissions occurring upstream of the consumer) [1, 2, 4], or about 4.8–8 $MgCO_2e$ per person, with emissions fairly evenly divided between the goods and services categories, although these broad designations are disaggregated in slightly different ways, depending upon the analysis [1, 2]. Of these, the most important subcategories are healthcare, clothing, and entertainment (broadly defined) [2], with impact generally correlating strongly, and approximately linearly, with consumer expenditure [2]. Of note, while not the focus of this chapter, food expenditures are likely the only major type of expense for which associated CO_2e emissions do not clearly scale with expenditures and household income: while wealthier households spend much more on food than poorer ones, it would appear that they are (in general) simply buying more expensive versions of food products that all have similar carbon impacts [2]. Interestingly, this trend seems to mostly hold at the global scale as well: Hertwich and Peters [4] found the correlation between national income and food-associated carbon emissions, while positive, to be much weaker than for all other categories of consumption, where the correlation was clear and strong (although this analysis did not include emissions from land use changes).

A global-scale trade-linked input-output study by Hertwich and Peters [4] estimated US per capita emissions at 28.6 MgCO$_2$e/capita, with 3, 12, and 16% attributable to clothing, manufactured goods, and services, respectively, summing to 8.87 MgCO$_2$e/capita. Supposing that only 72% of this total is attributable to household spending [4] still gives 6.38 MgCO$_2$e/capita (or about 16 MgCO$_2$e per household). This total also does not include emissions due to trade, i.e. the energy required to move products from producer to consumer, which accounted for an additional 8% of the footprint.

As all consumer spending carries an indirect carbon cost, this may be quantified in terms of a kgCO$_2$e/\$ emissions factor, and a very crude estimate of the emissions intensity of economic activity may be obtained by dividing US GDP by territorial emissions: Using the EPA's estimate of 2013 US emissions (6.673 billion MgCO$_2$e) and 2013 US GDP of \$ 16.7681 trillion (World Bank), implies a crude, economy-wide emissions intensity of 0.40 kgCO$_2$e/\$, not including emissions imported from abroad. However, once we consider that household consumption emissions are on the same order as US territorial emissions, generally estimated to be at least 50 MgCO$_2$e/yr [1, 2], and that average household expenditures have been on the order of 50–55,000 \$/year the last few years, then we have a emissions factor closer to 1 kgCO$_2$e/\$ for direct expenses. Jones and Kammen [2] gave lumped emissions factors of 0.565 and 0.507 kgCO$_2$e/\$ for generic goods and services, respectively, but this was weighted down by very low emissions categories such as personal savings and charitable contributions, and most other goods and services (including entertainment, healthcare, apparel, and personal care) had emissions factors roughly within 0.75–1.25 kgCO2e/\$. Caron et al. [503] somewhat similarly gave 0.484 kgCO$_2$e/\$ across all consumer consumption.

Based upon the Bureau of Labor Statistics' annual Consumer Expenditure Survey (CES) [504], and using kgCO$_2$e/\$ emissions factors (based mainly upon [2, 4] and the brief discussions below on apparel and healthcare), we can obtain a rough approximation of the carbon-equivalent emissions attributable to direct expenditures upon goods and services, as summarized graphically in Figures 23.1 and 23.2. As can be seen, household wealth dramatically affects goods/services emissions, with a nearly fivefold difference between the poorest and wealthiest quintiles. Given the greater number of persons in wealthier households, this difference is not as great on a per capita basis, but even under this accounting the richest and poorest vary by a factor of almost 2.5. Unsurprisingly, single-person households have a relatively high per capita footprint, at around 8.4 MgCO$_2$e/capita/year, compared to an overall mean of 5.7 MgCO$_2$e/capita/year. It should again be noted that these calculations are *approximate*, as there is appreciable uncertainty as the exact emissions factors involved, but I nevertheless consider them grossly reliable in terms of overall magnitude.

23.2   Health care

An economic input-output analysis by Chung and Melzter [505], in 2009, concluded that the American health care sector, responsible for 16% of GDP, was also directly and indirectly responsible for 8% of US GHG emissions, at 546 million tonnes CO$_2$e, or almost 1.8 MgCO$_2$e/capita, and nearly 4.7 MgCO$_2$e per household. Most of the emissions were attributable to hospitals and the building sector, although pharmaceutical drugs were the second largest single category. It's also worth noting that salaries in the health care sector, especially of doctors, are high, and thus this spending also supports many high-carbon lifestyles. Finally, using direct healthcare expenditures of \$3,126 per household (i.e. not counting payments made by insurance and governmental healthcare spending) in 2009 (per the CES), this implies an adjusted emissions factor of 1.49 kgCO2e/\$, while if we more conservatively use 2015 numbers for expenditure (\$4,342 per household) and suppose no change in healthcare-associate emissions, we get 1.07 kgCO2e/\$.

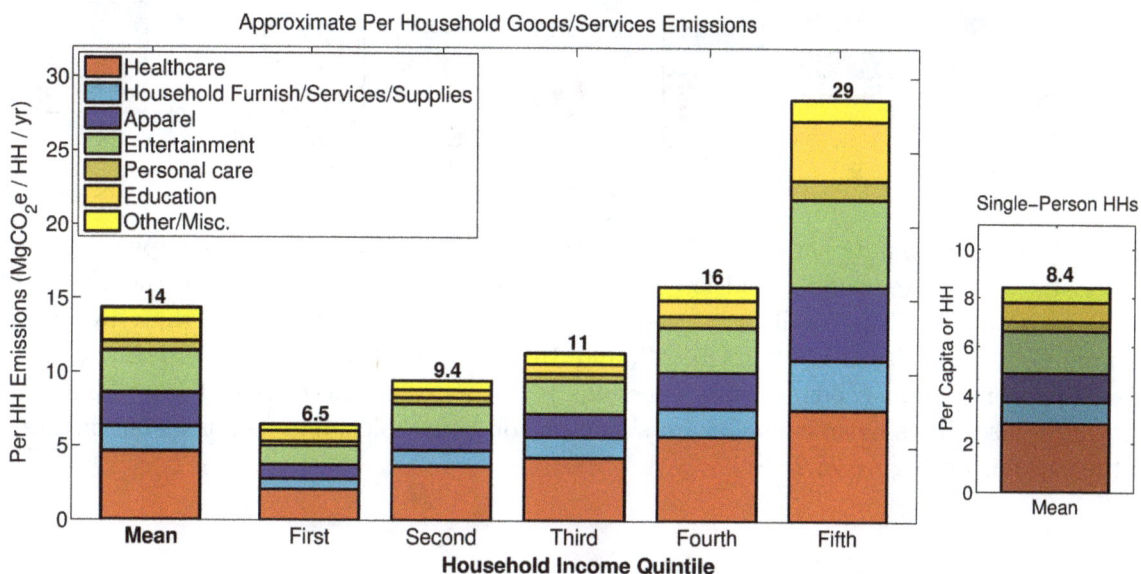

Figure 23.1: *Approximate* household CO_2e emissions related to direct consumer spending upon goods and services (excluding utilities, transportation, and food), based upon the 2015 CES [504], and stratified by overall mean and household income quintile. The right inset also gives the mean for single-person households. Emissions factors are only approximate, and, based upon [2] and the discussion in this chapter, I use 0.134 $kgCO_2e/\$$ for household service and operations, 0.614 $kgCO_2e/\$$ for household supplies and furnishings, 1.23 $kgCO_2e/\$$ for apparel, 1.07 $kgCO_2e/\$$ for healthcare, 0.95 $kgCO_2e/\$$ for personal care products and services, and 1.07 $kgCO_2e/\$$ for education. The miscellaneous/other category includes miscellaneous goods/services at 0.5 $kgCO_2e/\$$, tobacco at 0.5 $kgCO_2e/\$$, and reading at 2 $kgCO_2e/\$$.

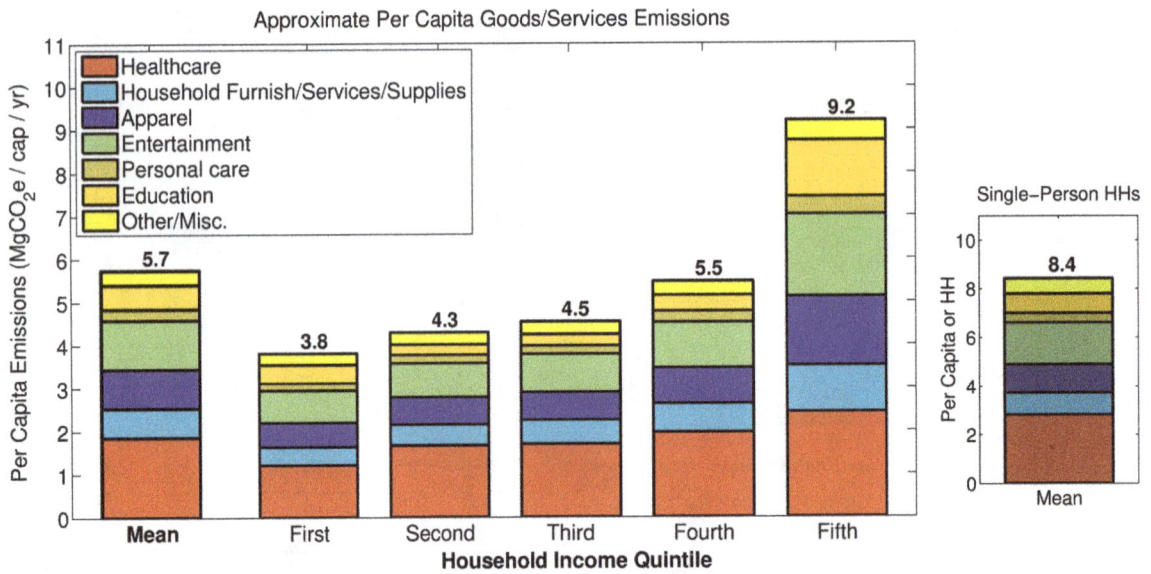

Figure 23.2: *Approximate* per capita CO_2e emissions related to direct consumer spending upon goods and services (excluding utilities, transportation, and food), based upon the 2015 CES [504], and stratified by overall mean and household income quintile. Note that, by income quintile, household size is 1.7, 2.2, 2.5, 2.9, and 3.1 persons/HH in ascending order, and so the effect of income upon goods/services emissions is not as pronounced as in Figure 23.2. The right inset also shows mean emissions for single-person households, and emissions factors are as in the caption for Figure 23.2.

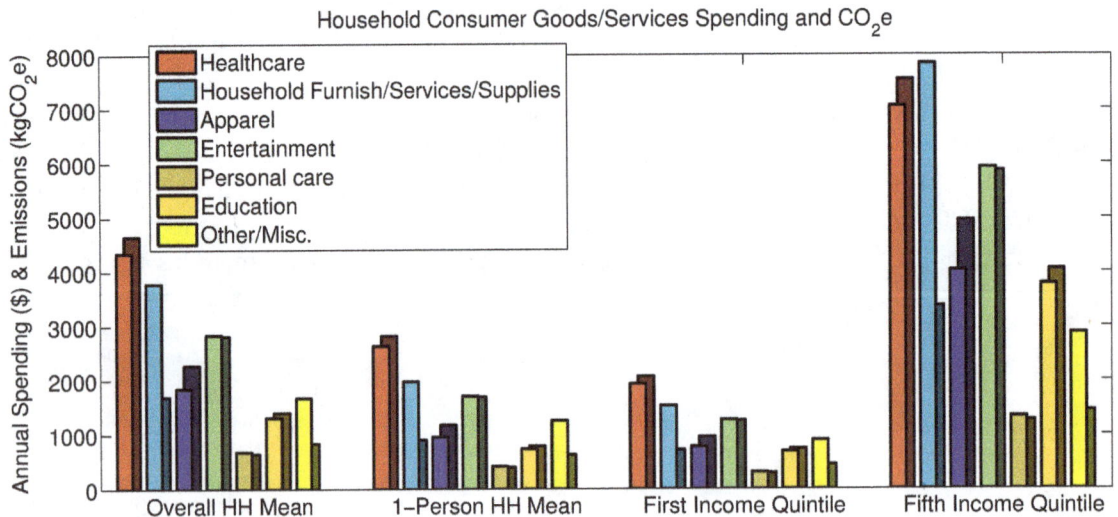

Figure 23.3: Major consumer spending, per the 2015 CES [504] (light bars in foreground), and approximate associated emissions (darker background bars), for several major household categories, namely the overall household mean, single-person household mean, and first and fifth household income quintiles.

466

23.3 Clothing

Hertwich and Peters [4] estimated that 3% of US emissions were attributable to clothing in 2001, out of per-capita emissions of 28.6 $MgCO_2e$/year, equating to 0.86 $MgCO_2e$/capita and about 2.1 $MgCO_2e$/household. Using the $1,743 of annual household spending on "apparel and services" given by the 2001 Consumer Expenditure Survey, this translates into about 1.23 $kgCO_2e$/$, and is reasonably consistent with the results of Weber and Matthews [1].

Chapter 24

Waste and Recycling

- Household recycling has a nontrivial but limited actual and potential impact on carbon emissions (and, more broadly speaking, societal energy and material use), with paper the most important material to recycle (annual CO_2e offset potential of about 250 kgCO_2e per capita), followed by plastic (50 kgCO_2e potential offset) and aluminum (35 kgCO_2e potential offset).

- Generally speaking, source reduction results in roughly double the carbon offset of recycling, with aluminum the major exception (recycling is nearly as good as source reduction in this case).

- Landfilling of carbon-containing materials, mainly food waste, yard waste, and paper products, leads to anaerobic methane emissions (with some captured and flared or used for energy generation). Thus, landfilling food waste likely yields a net 100–150 kgCO_2e/capita/year (most emissions and other resource use associated with food waste come from the production stage).

- Plastic bags are only a tiny fraction of plastic waste, disposable alternatives (paper or biodegradable bags) are worse, and while a complete ban might save about 1 million MgCO_2e ($< 0.05\%$ of US emissions), increased use of paper etc. might result in net positive emissions. In my view, plastic bag bans are largely meaningless and a waste of political and organizational capital.

Direct waste management, i.e. collection, transport, and disposal, is a very minor source of carbon emissions overall. However, recycling (and composting), by reducing the extraction of virgin materials, does represent a modest carbon offset, with paper and cardboard recycling by far the most significant offset (on an absolute basis), followed by plastic and aluminum. Re-use and source reduction, i.e. not using stuff you'll throw away in the first place, are obviously superior still. Furthermore, organic matter that is landfilled is anaerobically degraded to methane, and this represents a significant new GHG emission resulting from (primarily) food, yard waste, and paper disposal, and indeed, landfills rival oil and gas exploration as an anthropogenic methane source [95].

While recycling is often the almost exclusive focus of environmentally conscious behavior, compared to the other major categories of carbon-intensive consumption, the potential to offset CO_2e via recycling is very limited, with paper recycling a minor exception. In general, at typical use rates, the carbon offsets under maximal recycling of plastic (about 50 kgCO_2e), glass (8 kgCO_2e), steel packaging (mainly food cans, 12 kgCO_2e), and aluminum cans (35–40 kgCO_2e) would collectively amount to about 110 kgCO_2e per person, yearly, or about 0.5% of a typical American's carbon footprint. Mainly by increasing forest carbon stores, but also through avoided landfill methane emissions, maximal recycling of residentially generated paper waste might avoid 250 kgCO_2e per person, and, if we include commercial sources of paper waste (shipping boxes, office paper, etc.), this increases to 600 kgCO_2e. Source reduction has a greater potential offset in general, but this is particularly true for paper as well, and roughly doubles the avoided emissions across waste categories (per kg), with the major exception being aluminum: aluminum recycling is extremely efficient, and so in this case recycling is almost as good as source reduction.

All this is not to say I view recycling as pointless, and indeed, the carbon offset, while relatively small, is another piece of the silver buckshot (to borrow Bill McKibben's phrase) needed to address the climate crisis, nontrivial with respect to a lower-carbon lifestyle, and, moreover, it is important from a materials conservation perspective. The fact remains, however, that transportation habits, residential energy use, diet, etc. are far more impactful: one cannot drive an SUV to the recycling center and think oneself environmentally responsible.

While landfilling, *per se*, of inert materials such as metals, plastic, etc. has little net climate effect, landfilling organic wastes, especially food waste and some paper types, does have a nontrivial effect. The EPA estimated that 34.8 million tonnes of food waste entered the waste stream in 2014 [506], or 109 kg/capita, although based on other studies of food waste, this may be a dramatic underestimate. For example, based on USDA figures, Heller and Keoleian [321] gave 195.7 kg of food wasted yearly, per capita, at the retail and consumer levels, translating into 62.4 million tonnes of waste in 2014, and almost 80% higher than the EPA estimate. These latter numbers are far more consistent with about 30% of all food being wasted at the consumer and retail levels. About 20% of waste is combusted, with nearly all the rest landfilled (only about 5% of food waste is recovered for composting or other recycling, e.g. animal feed) [506], and therefore, since I estimate net emissions on the order of 1 kgCO$_2$e per kg food waste landfilled (see Section 24.1.2), 100–200 kgCO$_2$e is a reasonable per capita estimate for food waste disposal emissions. Such emissions are best avoided simply by avoiding food waste in general, and indeed, the upstream production emissions far outweigh those from landfilling food, but those unavoidable food scraps, such as coffee grounds, egg shells, peels, etc., may be composted instead of discarded, promoting soil fertility as well as storing a small amount of organic carbon.

For this chapter, I rely strongly (but not exclusively) upon the EPA's WARM model, which extensively documents emissions sources associated with various disposal and recycling options across all major waste categories, although my own results on net emissions from organic landfilling disagree to some degree. In the following sections I first characterize landfill methane emissions, as well as the carbon sink that occurs when carbon-containing materials are buried long-term, and I then discuss each commonly recycled material.

24.1 Landfills and landfill emissions

When organic matter (food waste, yard waste, paper, wood) is buried in landfills, the carbon ultimately exits as CO$_2$, methane, volatile organic compounds (VOCs), or leachate, or it remains trapped. Organic carbon decomposition in landfills passes through a brief aerobic state, an anaerobic acid state, and finally a *methanogenic* state, in which CO$_2$ and CH$_4$ are generated by anaerobic bacterial digestion.

Thus, we have two basic effects on carbon dynamics: (1) anaerobic digestion yielding methane has a strong warming effect, and (2) undigested carbon remains sequestered long-term, and is thus removed from the carbon cycle as a cooling anthropogenic carbon sink. Note that the CO$_2$ resulting from decomposition, but not the CH$_4$, is generally considered a biogenic or non-additional emission, as oxidation to CO$_2$ is the expected fate of carbon in rapidly cycling carbon pools such as food, while for the same reason long-term sequestration is a sink: again, the natural fate of food carbon would be rapid oxidation to CO$_2$. Long-lived wood products, woody waste, and some paper products (e.g. books) are at least partial exceptions to this rule, as it may naturally be years to decades before such substances oxidize to CO$_2$; woody waste also takes much longer to decay in landfills compared to food.

Landfill methane can be (partially) captured and either flared (i.e. simply burned, converting the potent methane to much weaker CO_2), or used to run a gas turbine for energy, offsetting some fossil energy. Even with landfill gas capture and energy generation, the balance of effects is generally net warming. The main exception is woody matter that strongly resists degradation to act as a carbon sink with net cooling. Wetter conditions also favor anaerobic digestion and methane production. I characterize these phenomena more precisely using a relatively simple mathematical model in the following section, largely based upon the EPA WARM model and IPCC guidelines [522, 507].

24.1.1 Mathematical model for landfill emissions

Upon anaerobic breakdown, organic carbon "'dissimilates" to landfill gas (LFG), a roughly 50:50 mix of CO_2 and CH_4, with only the CH_4 component counted as an anthropogenic GHG. For any given waste type i, we can characterize the methane yield, Y_i, in terms of Mg per *wet* tonne, as

$$Y_i = (1 - MC_i)CF_i DF_i GF \left(\frac{16}{12}\right) \tag{24.1}$$

where MC_i is moisture content, CF_i is carbon fraction on a dry mass basis, DF_i is the fraction of carbon that is dissimilated to LFG, GF is the methane fraction of LFG, and the 16/12 factor converts C to CH_4. Now, a fraction of this methane is captured at the landfill, with the absolute amount, YC_i (Mg CH_4/wet tonne), given as

$$YC_i = RY_i, \tag{24.2}$$

where R is the methane capture efficiency, while methane released, YR_i (Mg CH_4/wet tonne), is given by

$$YR_i = (Y_i - YC_i)(1 - OX), \tag{24.3}$$

with OX being the fraction of uncaptured methane that is oxidized by bacteria present in the landfill soil. Let us characterize each parameter somewhat further.

Dissimilation fraction and carbon sequestration

The fraction of carbon dissimilated to LFG, DF_i, may be estimated using one of several methods, but the simplest is simply to assume that all carbon not stored long-term in the landfill evolves to LFG, giving

$$DF_i = \left(1 - \frac{CSF_i}{CF_i}\right), \tag{24.4}$$

where CSF_i is the *carbon storage factor* (CSF) for waste i, and is generally given in units of kg C (kg dry waste)$^{-1}$, and thus we divide by the dry carbon fraction, CF_i to convert to the fraction of total carbon sequestered within the landfill. Table 24.1 gives some estimates of the CSF and other parameters for different waste types. The carbon footprint of landfilling depends strongly upon DF_i (or CSF_i), and I have found that newspaper, in particular, may shift from being strong net CO_2e sink to a weak CO_2e source when using different published CSF estimates for this paper product.

471

Table 24.1: Characteristics for major organic wastes.

Waste type	C (% DM)	MC (%)	CSF (kg C (kg DM)$^{-1}$)
Food scraps	50.8 [509]	70 [509], 65–75 [508]	.08±.04 [509], .08 [511]
Yard trimmings	47.14 [509]	45 [509], 40–60 [508]	.39 [509], .31 [511]
Grass	44.87 [509]	70 [509]	.32±.02 [509], .24 [511]
Leaves	49.4 [509]	30 [509]	.54±.06 [509], .385 [511]
Branches	49.4 [509]	10 [509]	.38±.02 [509], .38 [511]
Office Paper	38.6–39.1 [512], 40.3 [509]	5–6 [509, 510]	.05 [509], .02–.12 [512]
Newspaper	35.7–42.6 [512], 49.2 [509]	5–6 [509, 510]	.42±.02 [509], .09–0.17, .18–.22 [512]
Corrugated boxes	43.5–50.8 [512], 46.9 [509]	5–6 [509, 510]	.26±.01 [509], .18–.27 [512]
Glossy paper	34.3 [509]	6 [510]	.27 [510]

Methane gas fraction (GF)

It general, the CH_4:CO_2 ratio of generated landfill gas is nearly 50:50, implying GF = 0.5. Measurements of LFG suggest a ratio of 55:45 is more common, but this has been attributed to CO_2 being absorbed in seepage water [507]. Themelis and Ulloa [513] calculated a theoretical GF of 0.54 for anaerobic methanogenesis, assuming an average molecular composition of $C_6H_{10}O_4$ for organic wastes.

Methane oxidation fraction

Some fraction, OX, of un-captured methane is oxidized by methanotrophic bacteria in the landfill cover material. Estimates of the oxidation fractions vary widely and have been found to depend strongly upon soil type and temperature. A model by Bogner et al. [514] estimated oxidation from 1–23%, Czepiel et al. [519] estimated yearly average oxidation rate at 10%, and Liptay et al. [515] estimated 25-34% in the warm season but 10% for the year in a New England site; Chanton and Liptey [516] similarly estimated a yearly average of 20% with marked variation between summer and winter at a Florida site.

However, other estimates are appreciably higher. A literature review by Chanton et al. [517] gave percent oxidation for cover types as organic-34%, clayey-18%, sandy-53%, and other-28%, with an overall mean of 36±6%. Chanton et al. [518] recently estimated oxidation fraction to be 37.5±3.5%, using samples from 20 landfills in five climate types. It was found that a greater methane loading resulted in lower oxidation rates (perhaps through enzyme saturation), arid climates had the highest oxidation rates, and there was marked variation between individual landfills even within a fixed climate type.

Methane capture efficiency, and methane generation kinetics

Methane capture efficiency, R, is a somewhat controversial parameter. The IPCC 2006 Guidelines [507] cite estimates in the 10–85% range, and recommend a default value of only 20%. However, this is a global average estimate and not applicable to landfills in the U.S, where most landfills now actively capture LFG. Further complicating matters, R changes with time, as gas capture is minimal for initially buried refuse in an active landfill cell, but may approach 100% once a final cover is installed [510]. Along these lines, Spokas et al. [520], for example, measured recovery efficiencies from 41–94%, and conservatively estimated recovery efficiencies of 35% for an operating cell (with an active LFG recovery system), 65% for a temporary covered cell, 85% for a final clay covered cell, and 90% for a final geomembrane covered cell.

If we use a time-dependent approach for methane capture efficiency, we must also then consider the kinetics of methane generation. I assume first-order kinetics, as is widely done, and use component-specific estimates for the decay rate, k (yr^{-1}). Most studies use a fixed value for k to represent all municipal solid waste (MSW), and estimates range widely, from 0.01 to as much as 0.50 yr^{-1} [521], with 0.04 yr^{-1} being a common default value. Amini et al. [521] recently estimated k to vary from 0.04 to 0.13 yr^{-1} in five Florida landfills. The decay rate is strongly dependent upon moisture content, with wet and bioreactor landfills having much larger rates than traditional or dry landfills. The EPA WARM model uses decay rates, for undifferentiated MSW, of 0.02, 0.04, 0.06, and 0.12 yr^{-1} for dry, average, wet, and bioreactor conditions, respectively, with $k = 0.052$ yr^{-1} given as a national average. Food and yard wastes have much higher decay rates than most paper and wood products, and substance-specific decay rates used by the EPA WARM model are 0.19, 0.26, and 0.39 yr^{-1} for food, yard trimmings, and grass, respectively [522].

It is important to point out that, in general, the faster the decay rate, the more methane escapes from the landfill. This should be kept in mind, as popular authors are frequently aghast at the eons it takes waste to decay in landfills, failing to realize that, to avoid harmful methane emissions, we *want* slow decay rates. It follows that there is, in fact, little reason to prefer disposable products marketed as biodegradable, assuming their ultimate fate is a landfill.

Finally, returning to the model construction, I also impose a six month delay on methane generation, as the methanogenic phase takes several months to begin [507, 510].

Flaring and energy generation

Methane that is captured may either be flared to CO_2, or it may be used to generate electricity, usually on-site. Following the EPA WARM model [522], I assume a thermal efficiency of 29% for landfill gas generators, and a capacity factor of 85% (i.e. only 85% of captured methane is used for electricity, the rest is flared). Finally, I conservatively assume that landfill generators displace marginal electricity generation (the EPA WARM model assumes non-baseload energy is offset), and so the emissions factor is -0.880 $kgCO_2e$/kWh (note that this EF is negative, reflecting the energy offset). Also following the EPA WARM model (version 14), I assume that 18%, 38%, and 44% of all landfill methane is generated at landfills with no LFG capture, LFG capture and flaring, and LFG capture with energy generation, respectively.

24.1.2 Net landfilling emissions for food, yard waste, and paper products

Figure 24.1 demonstrates approximate 100-year emissions factors for landfill disposal of different organic substances, assuming that gas capture is phased in over eight years for each landfill cell, as (1) 0% capture in years 0–1, (2) 50% capture in years 2–4, (3) 75% capture in years 5–7, and(4) 90% capture thereafter; this is a slightly more aggressive version of the EPA's "typical" collection scenario. Note that my results differ from the EPA's chiefly in that mine suggest newspaper burial to be a net carbon source, while the EPA considers it a net sink. This is due to differing assumptions for newspaper's CSF: I consider the lower values reported by Wang and colleagues [512], based upon field studies at two landfill sites, to be more reliable than the older laboratory study by Barlaz [509] (see Table 24.1).

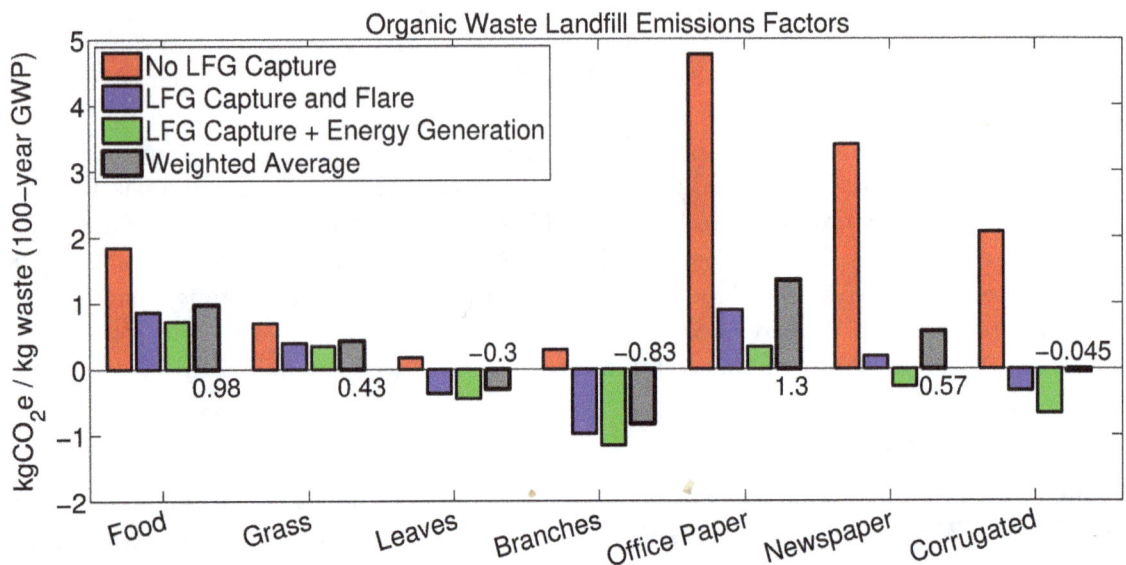

Figure 24.1: Approximate emissions factors for landfilling of major organic wastes that consumers typically discard, i.e. food waste, several yard waste types, and several paper products, under different landfill conditions (the text in the figure gives the numeric value of the weighted average for each waste substance). Positive EFs indicate that methane emissions outweigh carbon sequestration in the landfill, for net warming, while negative EFs imply the opposite. Unsurprisingly, woody waste, which decays slowy and is the most refractory to decay overall, acts as the strongest carbon sink, while food waste and office paper have the highest warming effect. Note that emissions factors are in terms of *wet* weight, and so paper products, which contain almost no moisture and thus have the highest carbon content on a wet basis, also have the highest warming potential when LFG is not captured.

24.2 Commonly recycled materials

Municipal recycling programs generally accept paper, cardboard and paperboard, glass bottles and containers, plastic bottles and containers, aluminum cans and (sometimes) foil, and steel cans. Let us examine each in turn and the materials/emissions reductions potential for recycling each, at typical consumer use rates.

24.2.1 Paper products

- Paper is by far the most recycled material, and, mainly by increasing forest carbon stores, all paper recycling offsets perhaps 221 million $MgCO_2e$, with over 90% of this due to increased forest carbon, and most of the remainder from avoided landfill emissions (process energy/emissions for virgin and recycled paper are actually quite similar overall).

- At the residential scale, average paper waste generation is about 78 kg/capita/year; recycling 90% (100% recycling is not plausible, given that napkins, paper cups, etc. are not recyclable) could offset just under 250 $kgCO_2e$/year, while 90% source reduction (highly unlikely) could save 500 $kgCO_2e$/year.

- Including paper generated at the commercial scale, recycling could offset over 600 $kgCO_2e$/capita/year.

According to the EPA [506], about 62.2 million tonnes of paper and cardboard enter the municipal waste stream yearly, with, according to a recent EPA study, about 40% attributable to the residential sector, and 60% derives from commercial use. About half of this waste is corrugated boxes, of which nearly 90% are recycled, while for all other paper waste, the recycling rate is 45%, and lumped together, 64.7% of all paper is recycled. This actually represents a fairly significant carbon emissions offset, compared to the counterfactual of discarding all this paper to landfill or incineration, almost entirely through avoided pulpwood logging, and hence increased forest carbon storage. To simplify presentation somewhat, I lump paper waste categories into (1) newspapers; (2) tissue, paper towels, paper plates, etc.; (3) corrugated boxes; (4) other containers; (5) magazines and mail, and (6) office paper, books, and other.

As I have already discussed at length in the context of biomass for energy, firewood, and lumber harvesting, the harvesting of forest products deleteriously affects the carbon dynamics of forests, as harvesting undercuts ongoing growth, disturbs soil carbon pools, and it can take many decades for carbon to be reincorporated into new tree growth. This obviously holds for paper as well, and while producing paper from pulpwood does take a significant amount of energy, altered forest carbon is still the most important impact of paper production and the major offset from recycling.

Paper may be produced either via mechanical (newspapers and phonebooks) or chemical pulping (most other paper), requiring 1.11 and 2.11 tonnes of pulp per tonne of finished paper, respectively. Each tonne of avoided pulpwood harvest increases forest carbon stocks by about 1.04 tonnes C [522], and so source reduction saves 4.23 $MgCO_2e$/Mg mechanical paper and 8.05 $MgCO_2e$/Mg chemical paper, in terms of CO_2-equivalents.

Considering paper recycling, only a fraction of paper, the "retention rate," can be recovered to make new paper, with retention about 95% at the collection phase, while manufacturing phase retention rates are 65–95% (at this stage retention rates are nearly 95% for corrugated containers and newspaper, and about 70% for other paper types) [522]. Supposing 87.5% and 70% retention rates for mechanical and chemical pulp papers, respectively, and we have that recycling would save 3.70 $MgCO_2e$/Mg mechanical paper and 5.63 $MgCO_2e$/Mg chemical paper

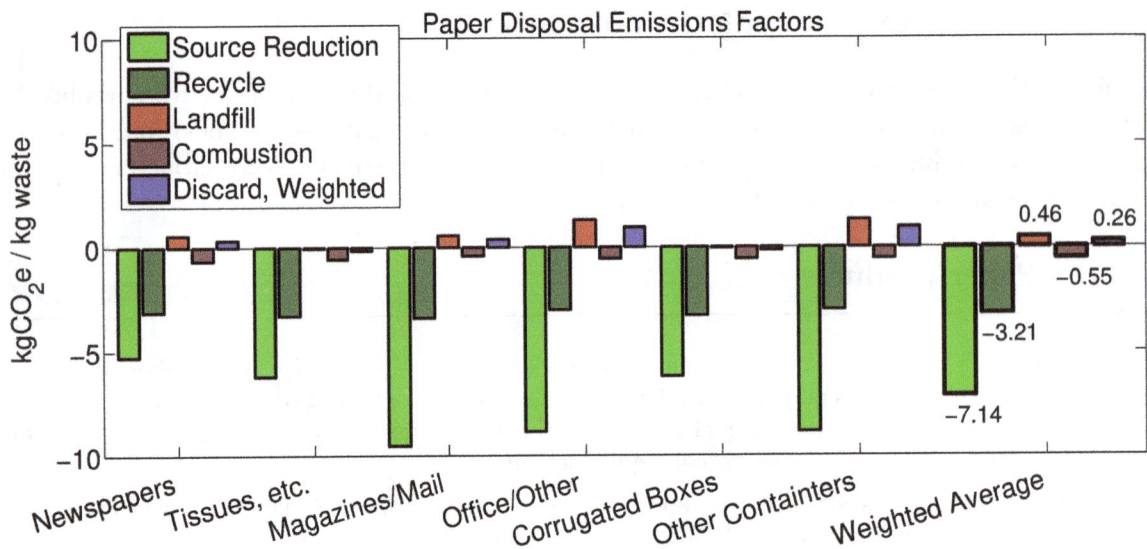

Figure 24.2: Approximate emissions factors (kgCO$_2$e/kg waste) for paper waste management options: source reduction, recycling, landfill, incineration, and the weighted outcome of discarding 80% to landfill and 20% to incineration. Negative values indicate carbon is offset, while positive values imply net carbon emissions. Results are given for major paper types, and as a weighted average of all paper waste. Note that source reduction figures account for the recycled content of avoided paper products.

of forest carbon. However, since about 40% of paper recovered for recycling is exported, the EPA conservatively assumes no offset from this fraction, reducing our forest carbon savings to 2.22 and 3.38 MgCO$_2$e/Mg paper.

To fully compare different disposal options, we must consider that the process energy for recycling paper is reasonably similar to the energy required to process virgin inputs, but is somewhat higher for some paper types (e.g. office paper), and somewhat less for others (such as newspaper), and I adopt process energy figures from the EPA WARM model [522]. Furthermore, almost all paper that is not recycled meets one of two fates, landfilling or incineration with energy recovery, with the former dominant at just over 80% of disposal. Now, as discussed already, the organic matter in landfilled paper partially evolves to CH$_4$ under anaerobic digestion, while undigested carbon acts as a long-term sink, with the balance of effects generally one of net warming. Using landfill emissions factors from the prior section, using EPA values for the carbon offset that occurs from waste-to-energy incineration of paper, and the differences in recycling vs. virgin material processes, we get the net GWP for each possible fate, as shown in Figure 24.2. Figure 24.2 also includes source reduction as an option (i.e. not using paper in the first place), and this takes into account the fraction of recycled paper in the existing paper mix.

Finally, with Figure 24.2, we are equipped to compare, for an individual consumer, the potential global warming impact of source reduction and recycling of paper products. For 2014, per-capita paper waste generation amounted to 195.1 kg, but only about 40% of this, 78 kg, was generated at residences (this amounts to just under half a pound a day, and certainly a plausible figure from personal experience). From this baseline, if one completely eliminated paper products from one's life, this would translate into carbon savings of 557 kgCO$_2$e, while recycling all paper instead of discarding it equates to 271 kgCO$_2$e (including avoided landfill emissions). Of course, not all paper is recyclable, but one might reasonably reduce paper consumption by 30%, and recycle 70% of the remainder, in which case one could save about 300 kgCO$_2$e.

24.2.2 Aluminum

- The per-kg carbon benefit to recycling is greatest for aluminum, out of any commonly recycled material. Overall, aluminum can recycling in the US saves on the order of 6.5 million $MgCO_2e$.

- For the average consumer, recycling all drink cans would save about 3.6 kg of aluminum and 36.5 $kgCO_2e$.

Aluminum can be indefinitely recycled in a closed loop process, and avoids extremely energy-intensive aluminum ore mining (generally mined as bauxite in strip mines) and refining. Overall, producing aluminum from recycled material uses about 5% of the energy required to produce the metal from ore, and generates less than 5% of the CO_2e emissions, with Das [523] reporting aluminum from ore and recycling to embody 12 $MgCO_2e$/Mg metal and 0.5 $MgCO_2e$/Mg metal, respectively. However, for the manufacture of aluminum cans in particular, the difference is a bit less dramatic: the EPA WARM model has estimated 12.225 $MgCO_2e$/Mg metal and 2.161 $MgCO_2e$/Mg metal for cans from virgin and recycled material, respectively. Cans are both the major source and fate of recycled aluminum in the US [522].

The Aluminum Association reports that the average can weighed 12.99 grams in 2014, and contained about 70% recycled aluminum. If one recycles their can, it is reasonable to ascribe emissions for a 100% recycled can to the user. Conversely, if you discard the can, an equal amount of material will presumably be processed from virgin material; in this case we also add 0.04 $MgCO_2e$/Mg metal from landfilling (EPA WARM). Thus, we have 28.07 gCO_2e/can and 159.32 gCO_2e/can under recycling and discarding, respectively, for a difference of 131.25 gCO_2e/can.

Per the EPA, 3.2 million tonnes of aluminum waste were generated in 2014 [506], about half in durable goods (none recycled), while 1.152 million tonnes were soft drink and beer cans. This translates, at 12.99 grams/can, to 278 cans/person/year. Recycling all would offset 11.64 million $MgCO_2e$; the actual recycling rate, 55.1%, saves 6.42 million $MgCO_2e$. At the individual scale, recycling 278 cans saves 3.6 kg of aluminum and 36.5 $kgCO_2e$; recycling 365 cans saves 4.75 kg aluminum and 47.9 $kgCO_2e$. These carbon savings are roughly equivalent to three or four gallons of gasoline.

24.2.3 Steel

- Perhaps 3 million $MgCO_2e$ are avoided by consumer recycling of steel in food containers (about 73% of such steel is recycled), or about 9 $kgCO_2e$/capita.

The vast majority of discarded steel is contained in durable goods. The portion that consumers control, mainly steel food cans (and other packaging), amounted to 1.97 million tonnes of metal, or 6.2 kg steel/capita/year [506]. Following, again, the EPA WARM model, recycling steel offsets 2.0 $kgCO_2e$/kg metal, and thus the *potential* steel packaging recycling offset amounts to 12.4 $kgCO_2e$/year (about 9 $kgCO_2e$/capita is realized at the current recycling rate of 72.8%). Ensuring that durable goods, e.g. old appliances are recycled for scrap would represent a much more significant offset.

24.2.4 Plastic overall

- Less than 10% of roughly 30 million tonnes of plastic waste is recycled, representing an actual carbon offset of about 3 million $MgCO_2e$, out of a theoretical potential offset of 33 million $MgCO_2e$.
- Maximum feasible recycling of non-durable plastic waste would save around 50 $kgCO_2e$/capita/year, and *complete* source reduction could avoid 150 $kgCO_2e$/capita/year.

In 2014, about 30.2 million tonnes of plastic waste entered the municipal waste stream, and only 9.5%, overall, was recycled [506]. Various containers, bags, and packages accounted for 43% of this waste, while other nondurable goods, including trash bags and plastic cups, accounted for another 20% (and durable goods made up the remainder); note that single-use grocery bags, which get so much negative press, were an almost trivial component of plastic waste. The production of various plastics generates on the order of 2.2 $kgCO_2e$/kg (EPA average for mixed plastics), while plastic recycling offsets about 50% of this burden [522]. It follows from these numbers that complete recycling of all plastic not in durable goods, and excluding trash bags, could offset about 78 $kgCO_2e$/capita.

However, only high-density polyethylene (HDPE), low-density polyethylene (LDPE), and polyethylene terephthalate (PET) are recycled in appreciable quantities, and since these plastics made up about 65% of all non-durable plastic waste, 65% recycling is a practical upper bound for consumers, and therefore the maximum recycling offset is more like 50 $kgCO_2e$/capita/year. Complete avoidance of all plastic-containing containers, packaging, etc. would be more appreciable, offsetting around 150 $kgCO_2e$/capita/year.

24.2.5 Plastic bags and alternatives

- Single-use plastic bags are a trivial source of plastic waste, disposable alternative (biodegradable bags or paper bags) are generally *more* carbon-intensive, and reusable alternatives are superior only if they offset around 20–40 plastic bags over their lifetime.

Given that "responsible consumerism" seems almost to begin and end at reusable bags in the minds of many, I spend a decent amount of time examining the issue of single-use plastic bags and alternatives. Generally speaking, we will see that this is a trivial non-issue: it matters a great deal how one gets to the store (a typical two-way trip to the supermarket generates emissions equivalent to 741 plastic bags) and what one purchases there (a single kg of beef equates to around 4,000–5,000 plastic bags, yes, *thousand*), but hardly at all how one bags the purchase.

Scale of the problem

- Single-use HDPE bags typically targeted by bag bans account for only about 2% of all plastic waste, and yearly CO_2e emissions associated with these bags amount to about 3.25 $kgCO_2e$/capita, equivalent to less than a third of a gallon of gasoline.

When it comes to assessing the wisdom of plastic bag bans as a major environmental initiative, the first question we must address is the scale of problem. First note that there are two major classes of single-use bag that consumers regularly encounter. "Singlet" bags are the thinner, flimsier bags generally used at grocery and box stores and are made of high-density polyethylene (HDPE); these are the bags typically targeted by "ban the bag" initiatives. "Boutique" bags are thicker and heavier, made of low-density polyethylene (LDPE), are often branded and are encountered in, for example, department stores, malls, etc.

The EPA estimated that 3.4292 million metric tons of plastic were used for the overall category "bags, sacks, and wraps" in 2013, with only 13.5% recovered for recycling, and the majority of the plastic used was LDPE (used for boutique bags) [526]. In the subcategory of HDPE singlet plastic bags, again, the type usually encountered at grocery check-outs, the EPA estimated 635,000 metric tons of plastic used, corresponding to about 115 billion individual bags (at 5.5 grams each) and 365 bags per person per year (using 2013 numbers), exactly one bag per day. Even fewer of these bags, just 5.7%, are recycled. Note that the bag *number* is a very rough estimate, as cited bag weights vary from 5.5 grams to well over 10 grams. While 115 billion bags is a dramatic number, HPDE plastic bags account for only about 2% of all plastic waste.

The EPA WARM model [527] estimates carbon emissions of 1.6204 $MgCO_2e/Mg$ of HDPE plastic using the current mix of virgin and recycled materials (1.7306 $MgCO_2e/Mg$ using entirely virgin materials), with the majority of emissions generated by manufacturing. This implies yearly emissions of about 1.03 million $MgCO_2e$ attributable to HDPE plastic bag production overall, 8.91 gCO_2e per bag, and 3.25 $kgCO_2e$ per capita (i.e. per 365 bags). In other words, the carbon emissions from the average individual's entire yearly HDPE plastic bag use is less than a third of the emissions from a single gallon of gasoline (11.146 $kgCO_2e/gallon$). Since the average trip length for shopping was 6.4 miles in the 2009 NHTS, based on an average 21.6 MPG, a one-way trip to the store generates just as much carbon as all HDPE plastic bags consumed in a year.

Including all bag plastics, and using EPA emissions factors for production, we have 6.5215 million metric tons of CO_2e overall and 20.61 $kgCO_2e/capita$ attributable to the general category "bags, sacks, and wraps," which excludes trash bags. Trash bags add an additional 1.54 million Mg CO_2e (4.88 $kgCO_2e/capita$), and our grand total hits 25.47 $kgCO_2e/capita$. These results are summarized in Table 24.2.

Plastic vs. paper vs. reusable

- Most reusable bags must displace at least several dozen singlet bags before achieving net benefit. Polypropylene, or "green fiber," bags are the best reusable alternative.

- Paper bags are 2–4 times as carbon-intensive as singlet HDPE bags on the basis of manufacturing energy. If we include upstream forest carbon effects, paper bags may be *twenty to forty* times worse than plastic, and even recycled paper bags are markedly worse than plastic.

- Biodegradable bags are slightly more carbon-intensive than HDPE plastic bags.

While the above analysis would suggest that singlet use bags are too minor a problem for further study, they are indeed the focus of many environmental campaigns, reusable bags have become prominent in many supermarkets, and reusable bags have achieved a special prominence among many of the environmentally minded. Thus, we confront a second question: how do alternatives, e.g. compostable plastic bags, paper bags, and reusable bags, compare in their

Plastic Type	1,000 Mg	MgCO$_2$e/Mg	Total CO$_2$e (10^6 Mg)	kgCO$_2$e/cap
Bags, Wraps, Sacks				
HDPE	635	1.6204	1.0290	3.2512
LDPE	2,050.2	1.9842	4.0680	12.8531
PVC	45.4	2.1605	0.0980	0.3096
PP	571.5	1.7086	0.9765	3.0853
PS	127	2.7558	0.3500	1.1058
Subtotal	3,429.2	-	6.5215	20.6051
Trash Bags				
HDPE	181.4	1.6204	0.2667	0.8427
LDPE	707.6	1.9842	1.2737	4.0243
Subtotal	889.0	-	1.5404	4.8670
Total	4,318.2	-	8.0619	25.4720

Table 24.2: EPA estimated plastic bag production by plastic type and associated emissions, as estimated from the EPA WARM model. From left to right, we have total production in 1,000 metric tons, production emission factor, and total and per capita CO$_2$e emissions. Note that the HPDE subcategory of the bags, wraps, and sacks category (red row) represent typical "singlet" grocery bags, while LDPE bags in this category are typical "botique" bags. Overall per capita emissions from both the generic bag and trash bag categories combined (about 25 kgCO$_2$e) are equivalent to less than 2.5 gallons of gasoline.

impact to those most loathed plastic vessels? Several lifecycle analyses [524, 525, 528] have addressed this question, and they have reached variable conclusions, with much variability due to different assumptions on bag disposal, e.g. landfilling versus incineration with energy recovery, rates of recycling for paper and plastic bags, and credits for re-use of "single-use" bags as trash bags, pet waste bags, etc.

We may summarize the major alternatives as follows. **Paper bags** are the obvious single-use alternative to most American consumers. However, they are heavier (and thus require more material to construct), obviously require upstream tree logging with its attendant effects on forests and biodiversity, and their manufacture is energy-intensive. There is some disagreement over whether paper bags are more emissions-intensive than plastic, but most studies suggest paper is significantly worse on both carbon emissions and several other environmental axes, such as water use, except when paper recycling rates are very high [525]. Note that these studies generally do not take into account the effect of wood harvesting on forest carbon storage, which is most likely by far the dominant global warming effect. I discuss this further in the Paper section, but applying results here suggest that, if all virgin materials are used, then a 42.6 g paper bag has associated lifecycle emissions of 377 gCO_2e, perhaps 30–40 times worse than a single plastic bag. Assuming a maximum systemic recycling retention rate of 70–90%, then recycled paper bags would yield 77–145 gCO_2e/bag.

Biodegradable bags using "bio-plastics" derived from corn (or other plant) starch are occasionally used as single-use bags, but embody upstream agricultural emissions and land-use. Furthermore, once landfilled, if these bags do degrade then they decompose into carbon dioxide or methane, hardly a desirable outcome. Furthermore, simply on the basis of manufacturing energy, these bags are also likely worse for the atmosphere than the plastic ones they displace, with studies consistently finding them to be worse than plastic bags, e.g. [524, 525].

Polypropylene (PP) "green fiber" bags are a popular inexpensive reusable bag, now often found for sale for, say, $1.00, at checkout stands. They are manufactured from petroleum byproducts and have similar upstream emissions and energy use as single-use bags. However, they are superior to single-use bags after only a few uses, and are likely the best option overall.

Canvas (cotton) bags are a heavy-duty reusable alternative, but cotton farming is a land, water, pesticide, fertilizer, and emissions-intensive activity, and so these bags have a heavy environmental impact on a per-bag basis. Even so, once they have displaced around 100 (more or less, depending on the source) single-use bags, they begin to pay a carbon dividend.

Per-bag emissions estimates for different classes of bags from several lifecycle analyses are summarized in Table 24.3.

A note on biodegradability

A common objection to plastic bags is that they take millennia to break down in landfills. However, while its persistence is extremely problematic when plastic disperses into ecosystems, it is in fact *desirable* that plastic remain intact when landfilled. Plastics are *carbon* polymers, and when biologically degraded or burned yield *carbon dioxide*. That is, the last thing we want after landfilling is for carbon sources to decompose. Biodegradable or compostable alternative single-use bags are thus arguably even worse. If a 7 gram biodegradable bag is 85% carbon by weight and oxidizes completely to CO_2, this yields 22 gCO_2, about twice the carbon emissions from the production of a singlet HPDE bag.

Conclusions

While eliminating singlet HPDE bags in complete isolation could avoid about one million metric tons of CO_2e per year across the US, which is not entirely trivial, it is unlikely that under current

Bag type	gCO$_2$e/bag (bag wt.)	Study
Single-use HDPE	8.9 gCO$_2$e/bag (5.5 g)	EPA (derived)
	11.7 gCO$_2$e/bag (6 g)	[528]
Single-use LDPE	57.3 gCO$_2$e/bag (18.1 g)	[528]
Paper bag	22.7 gCO$_2$e/bag (42.6 g)	[528]
	377 (virgin), 77–145 (recycled) gCO$_2$e/bag	This work
Biodegradable	12.7 gCO$_2$e/bag (7 g)	[528]
	Additional 21.8 gCO$_2$/bag if degraded	
	to CO$_2$ and 85% C by weight	This work
Reusable PP "green fiber"	472.3 gCO$_2$e/bag (65.6 g)	[528]
Reusable canvas (cotton)	276.9 gCO$_2$e/bag (125.4 g)	[528]

Table 24.3: Selected estimates of the carbon impact and typical weight for various bag types.

consumer behaviors that appreciable carbon savings would actually be realized: any beneficial reuses of HPDE bags would be eliminated (e.g. reuse as a trash bag), disposable alternatives (especially paper) are worse, and reusable bags still have some impact. Therefore, plastic bag bans as a *policy* measure will do little, even in a best case scenario, and it is my view that it is unwise to expend political and organizational capital on such endeavors. Furthermore, a broad shift to disposable paper bags, which seems the most likely short-term consequence, would actually cause net harm even under high recycling rates. Nevertheless, as an individual it is still clearly a waste to use plastic bags unnecessarily, and it is likely best to use (polypropylene) reusable bags as an alternative.

24.2.6 Glass

Most glass entering the waste stream comes from glass containers (8.35 million tonnes, out of 10.4 million tonnes in 2014), with 32.5% of such container glass recycled [506]. Glass manufacture, mainly entailing melting down sand with some additives and casting them to form, is not especially carbon-intensive compared to most other products, with 0.58 kgCO$_2$e/kg virgin glass estimated by the EPA WARM model. Recycling glass is less energy-intensive, and the EPA gives a glass recycling emissions offset of -0.31 kgCO$_2$/kg glass recycled. Complete recycling of all 26 kg container glass per capita would thus equal carbon savings of just 8.1 kgCO$_2$e/capita, while the current 32.5% rate offsets 2.6 kgCO$_2$e/capita yearly.

Chapter 25

Postscript

I should like to close with a few passages and ideas from other authors, beginning with a passage by the Roman statesman and philosopher Seneca the Younger, drawn from Chapter XI of *Of a Happy Life*, written around the 58th year of the Common Era (and as translated into English by Sir Roger L'Estrange in 1685):

> How long shall we covet and oppress, enlarge and account that too little for one which was formerly enough for a nation? And our luxury is as insatiable as our avarice. Where is that sea, that forest, that spot of land; that is not ransacked to gratify our palate? The very earth burdened with our buildings; not a river, not a mountain, escapes us. Oh, that there should be such boundless desires in our little bodies! Would not fewer lodgings serve us? We lie in but one, and where we are not, that is not properly. What with our hooks, snares, nets, dogs, etc., we are at war with all living creatures; and nothing comes amiss but that which is either too cheap, or too common; and all this is to gratify a fantastical palate. Our avarice, our ambition, our lusts, are insatiable; we enlarge our possessions, swell our families, we rifle sea and land for matter of ornament and luxury. A bull contents himself with one meadow, and one forest is enough for a thousand elephants; but the little body of a man devours more than all other living creatures. We do not eat to satisfy hunger, but ambition; we are dead while we are alive and our houses are so much our tombs that a man might write our *epitaphs* upon our very doors.

The avarice of man and the heaviness of his presence on the earth is not such a new notion then, after all. Given this, what can one do? Well, it is my hope that the contents of this book show that to change one's own life is not merely a futile gesture, and even though on the scale of a single person all such acts are necessarily insignificant, two more stories may provide solace and encouragement.

The Myth of Sisyphus, an essay by Albert Camus, relates the Greek myth of Sisyphus, a mortal king who dared to put Death himself in chains, and in the underworld was condemned by the Gods to forever roll a great rock up a mountain, only to have it forever tumble back down to the plain. Camus concluded that Sisyphus was the embodiment of the absurd hero, and that "the struggle itself toward the heights is enough to fill a man's heart. One must imagine Sisyphus happy." I have always admired David Simon's particular interpretation of this tale and essay: that to commit to a just cause almost certain to fail is absurd, yet not to commit is also absurd, *but only one choice offers any change at dignity.*

Finally, Wangari Maathai, founder of the Green Belt Movement and Nobel laureate, in the film *Dirt! The Movie*, related the widely-told children's story of the hummingbird, wherein a forest is being consumed by a huge fire, and all the animals stand by, amazed and overwhelmed.

All except a little hummingbird, who, fast as he can, takes a drop of water from a nearby stream into his beak and drops it on the fire, and back and forth he goes. The other animals, many of whom are much bigger and could help much more, all say "what do you think you can do? You are too little and this fire is too big. Your wings are too little, and your beak too small!" To this, the hummingbird replies, "I am doing the best I can." And so, Wangari Maathai concludes,

> I may feel insignificant, but I certainly don't want to be like the animals watching as the planet goes down the drain. I will be a hummingbird. I will do the best I can.

Bibliography

[1] Weber, C. L., & Matthews, H. S. (2008). Quantifying the global and distributional aspects of American household carbon footprint. Ecological Economics, 66(2), 379–391.

[2] Jones, C. M., & Kammen, D. M. (2011). Quantifying carbon footprint reduction opportunities for US households and communities. Environmental Science & Technology, 45(9), 4088–4095.

[3] Jones, C., & Kammen, D. M. (2014). Spatial distribution of US household carbon footprints reveals suburbanization undermines greenhouse gas benefits of urban population density. Environmental science & technology, 48(2), 895–902.

[4] Hertwich, E. G., & Peters, G. P. (2009). Carbon footprint of nations: A global, trade-linked analysis. Environmental science & technology, 43(16), 6414–6420.

[5] Davis, S. J., & Caldeira, K. (2010). Consumption-based accounting of CO2 emissions. Proceedings of the National Academy of Sciences, 107(12), 5687–5692.

[6] Kanemoto, K., Moran, D., Lenzen, M., & Geschke, A. (2014). International trade undermines national emission reduction targets: New evidence from air pollution. Global Environmental Change, 24, 52–59.

[7] IPCC. (2013). Climate Change 2013: The Physical Science Basis. Contribution of Working Group I to the Fifth Assessment Report of the Intergovern mental Panel on Climate Change [Stocker, T. F., Qin, D., Plattner, G. K., Tignor, M., Allen, S. K., Boschung, J., Nauels, A., Xia, Y., Bex, V., & Midgley, P. M. (eds.)]. Cambridge University Press, Cambridge, United Kingdom and New York, NY, USA, 1535 pp.

[8] IPCC. (2014). Climate Change 2014: Mitigation of Climate Change. Contribution of Working Group III to the Fifth Assessment Report of the Intergovernmental Panel on Climate Change [Edenhofer, O., Pichs-Madruga, R., Sokona, Y., Farahani, E., Kadner, S., Seyboth, K., Adler, A., Baum, I., Brunner, S., Eickemeier, P., Kriemann, B., Savolainen, J., Schlmer, S., von Stechow, C., Zwickel, T., & Minx, J. C. (eds.)]. Cambridge University Press, Cambridge, United Kingdom and New York, NY, USA.

[9] IPCC. (2007). Climate Change 2007: The Physical Science Basis. Contribution of Working Group I to the Fourth Assessment Report of the Intergovernmental Panel on Climate Change [Solomon, S., Qin, D. Manning, M., Chen, Z., Marquis, M., Averyt, K. B., Tignor, M., & Miller, H. L. (eds.)]. Cambridge University Press, Cambridge, United Kingdom and New York, NY, USA, 996 pp.

[10] U.S. Environmental Protection Agency. (2015). Inventory of U.S. Greenhouse Gas Emissions and Sinks: 1990–2013. EPA 430-R-15-004, April 15, 2015.

[11] United Nations. (2017). World Population Prospects: The 2017 Revision. June 21, 2017.

[12] Hubbert, M. K. (1949). Energy from Fossil Fuels. Science, 109(2823), 103–109.

[13] Azevedo, I. M. (2014). Consumer end-use energy efficiency and rebound effects. Annual Review of Environment and Resources, 39, 393–418.

[14] Small, K. A., & Van Dender, K. (2007). Fuel efficiency and motor vehicle travel: the declining rebound effect. The Energy Journal, 25–51.

[204] Geller, H., Harrington, P., Rosenfeld, A. H., Tanishima, S., & Unander, F. (2006). Polices for increasing energy efficiency: Thirty years of experience in OECD countries. Energy policy, 34(5), 556–573.

[18] Dixon, R. K., McGowan, E., Onysko, G., & Scheer, R. M. (2010). US energy conservation and efficiency policies: Challenges and opportunities. Energy Policy, 38(11), 6398–6408.

[17] Gillingham, K., Rapson, D., & Wagner, G. (2016). The rebound effect and energy efficiency policy. Review of Environmental Economics and Policy, 10(1), 68–88.

[18] Dixon, R. K., McGowan, E., Onysko, G., & Scheer, R. M. (2010). US energy conservation and efficiency policies: Challenges and opportunities. Energy Policy, 38(11), 6398–6408.

[19] Jevons, W.S. (1866). The Coal Question (Macmillan, London), 2nd Ed. Available online: http://www.econlib.org/library/YPDBooks/Jevons/jvnCQ7.html#Chapter%207

[20] Joos, F., Roth, R., Fuglestvedt, J. S., Peters, G. P., Enting, I. G., Bloh, W. V., ... & Friedrich, T. (2013). Carbon dioxide and climate impulse response functions for the computation of greenhouse gas metrics: a multi-model analysis. Atmospheric Chemistry and Physics, 13(5), 2793–2825.

[21] Van den Heede, P., & De Belie, N. (2012). Environmental impact and life cycle assessment (LCA) of traditional and greenconcretes: literature review and theoretical calculations. Cement and Concrete Composites, 34(4), 431–442.

[22] Miller, S. M., & Michalak, A. M. (2017). Constraining sector-specific CO2 and CH4 emissions in the US. Atmospheric Chemistry and Physics, 17(6), 3963–3985.

[23] Shindell, D. T., Faluvegi, G., Bell, N., & Schmidt, G. A. (2005). An emissionsbased view of climate forcing by methane and tropospheric ozone. Geophysical Research Letters, 32(4), L04803.

[24] Shindell, D. T., Faluvegi, G., Koch, D. M., Schmidt, G. A., Unger, N., & Bauer, S. E. (2009). Improved attribution of climate forcing to emissions. Science, 326(5953), 716–718.

[25] Miller, S. M., Wofsy, S. C., Michalak, A. M., Kort, E. A., Andrews, A. E., Biraud, S. C., ... & Miller, B. R. (2013). Anthropogenic emissions of methane in the United States. Proceedings of the National Academy of Sciences, 110(50), 20018–20022.

[26] Velders, G. J., Ravishankara, A. R., Miller, M. K., Molina, M. J., Alcamo, J., Daniel, J. S., ... & Reimann, S. (2012). Preserving Montreal Protocol climate benefits by limiting HFCs. Science, 335(6071), 922–923.

[27] Molina, M. J., & Rowland, F. S. (1974). Stratospheric sink for chlorouoromethanes: chlorine atom-catalysed destruction of ozone. Nature, 249(28), 810–812.

[28] Solomon, S. (1988). The mystery of the Antarctic ozone "hole". Reviews of Geophysics, 26(1), 131–148.

[29] Douglass, A. R., Newman, P. A., & Solomon, S. (2014). The Antarctic ozone hole: An update. Physics Today, 67(7), 42–48.

[30] Morrisette, P. M. (1989). The evolution of policy responses to stratospheric ozone depletion. Natural Resources Journal, 29, 793–820.

[31] Farman, J. C., Gardiner, B. G., & Shanklin, J. D. (1985). Large losses of total ozone in Antarctica reveal seasonal ClOx/NOx interaction. Nature, 315(6016), 207–210.

[32] Velders, G. J., Andersen, S. O., Daniel, J. S., Fahey, D. W., & McFarland, M. (2007). The importance of the Montreal Protocol in protecting climate. Proceedings of the National Academy of Sciences, 104(12), 4814–4819.

[33] White House Fact Sheet. (2016). Nearly 200 Countries Reach a Global Deal to Phase Down Potent Greenhouse Gases and Avoid Up to 0.5C of Warming. October 15, 2016. https://www.whitehouse.gov/the-press-office/2016/10/15/fact-sheet-nearly-200-countries-reach-global-deal -phase-down-potent

[34] Xu, Y., Zaelke, D., Velders, G. J., & Ramanathan, V. (2013). The role of HFCs in mitigating 21st century climate change. Atmospheric Chemistry and Physics, 13(12), 6083–6089.

[35] Minnesota Pollution Control Agency. Accessed 11/26/2016. https://www.pca.state.mn.us/quick-links/climate-ch ange-mobile-air-conditioners

[36] State of California Air Resources Board. (2009). Emissions of HFC-134a in Auto Dismantling and Recycling. October, 2009.

[37] Xiang, B., Patra, P. K., Montzka, S. A., Miller, S. M., Elkins, J. W., Moore, F. L., ... & Wofsy, S. C. (2014). Global emissions of refrigerants HCFC-22 and HFC-134a: unforeseen seasonal contributions. Proceedings of the National Academy of Sciences, 111(49), 17379–17384.

[38] Young, P. J., Archibald, A. T., Bowman, K. W., Lamarque, J. F., Naik, V., Stevenson, D. S., ... & Cameron-Smith, P. (2013). Pre-industrial to end 21st century projections of tropospheric ozone from the Atmospheric Chemistry and Climate Model Intercomparison Project (ACCMIP). Atmospheric Chemistry and Physics, 13(4), 2063–2090.

[39] Fuglestvedt, J. S., Shine, K. P., Berntsen, T., Cook, J., Lee, D. S., Stenke, A., ... & Waitz, I. A. (2010). Transport impacts on atmosphere and climate: Metrics. Atmospheric Environment, 44(37), 4648–4677.

[40] Eyring, V., Stevenson, D. S., Lauer, A., Dentener, F. J., Butler, T., Collins, W. J., ... & Lawrence, M. G. (2007). Multi-model simulations of the impact of international shipping on atmospheric chemistry and climate in 2000 and 2030. Atmospheric Chemistry and Physics, 7(3), 757–780.

[41] Bond, T. C., Doherty, S. J., Fahey, D. W., Forster, P. M., Berntsen, T., DeAngelo, B. J., ... & Kinne, S. (2013). Bounding the role of black carbon in the climate system: A scientific assessment. Journal of Geophysical Research: Atmospheres, 118(11), 5380–5552.

[42] Boucher, O., & Reddy, M. S. (2008). Climate trade-off between black carbon and carbon dioxide emissions. Energy Policy, 36(1), 193–200.

[43] Bond, T. C., & Sun, H. (2005). Can reducing black carbon emissions counteract global warming?. Environmental science & technology, 39(16), 5921–5926.

[44] Jobbgy, E. G., & Jackson, R. B. (2000). The vertical distribution of soil organic carbon and its relation to climate and vegetation. Ecological applications, 10(2), 423–436.

[45] Stockmann, U., Adams, M. A., Crawford, J. W., Field, D. J., Henakaarchchi, N., Jenkins, M., ... & Wheeler, I. (2013). The knowns, known unknowns and unknowns of sequestration of soil organic carbon. Agriculture, Ecosystems & Environment, 164, 80–99.

[46] Pan, Y., Birdsey, R. A., Fang, J., Houghton, R., Kauppi, P. E., Kurz, W. A., ... & Ciais, P. (2011). A large and persistent carbon sink in the worlds forests. Science, 333(6045), 988–993.

[47] Schulze, E. D., Krner, C., Law, B. E., Haberl, H., & Luyssaert, S. (2012). Largescale bioenergy from additional harvest of forest biomass is neither sustainable nor greenhouse gas neutral. Gcb Bioenergy, 4(6), 611–616.

[48] Guillaume, T., Damris, M., & Kuzyakov, Y. (2015). Losses of soil carbon by converting tropical forest to plantations: erosion and decomposition estimated by 13C. Global change biology, 21(9), 3548–3560.

[49] van Straaten, O., Corre, M. D., Wolf, K., Tchienkoua, M., Cuellar, E., Matthews, R. B., & Veldkamp, E. (2015). Conversion of lowland tropical forests to tree cash crop plantations loses up to one-half of stored soil organic carbon. Proceedings of the National Academy of Sciences, 112(32), 9956–9960.

[617] Smith, J. E., Heath, L. S., & Hoover, C. M. (2013). Carbon factors and models for forest carbon estimates for the 20052011 National Greenhouse Gas Inventories of the United States. Forest Ecology and Management, 307, 7–19.

[51] Hong, B. D., & Slatick, E. R. (1994). Energy Information Administration, Quarterly Coal Report, January-April 1994, DOE/EIA-0121(94/Q1) (Washington, DC, August 1994), pp. 1–8. https://www.eia.gov/coal/production/quarterly/co2_article/co2.html#N_5_

[114] U.S. Energy Information Administration. (2017). Monthly Energy Review, November, 2017. http://www.eia.gov/totalenergy/data/monthly/pdf/sec13_5.pdf

[53] Zavitsas, A. A., Matsunaga, N., & Rogers, D. W. (2008). Enthalpies of formation of hydrocarbons by hydrogen atom counting. Theoretical implications. The Journal of Physical Chemistry A, 112(25), 5734–5741.

[54] Weston, K. C. (1992). Energy Conversion. West Publishing Company, 633 p. Online edition: http://www.personal.utulsa.edu/~kenneth-weston/

[55] Hyne, N. J. (2012). Nontechnical Guide to Petroleum Geology, Exploration, Drilling, and Production. PennWell Books, 698 p.

[56] Hubbert, M. K. (1956). Nuclear energy and the fossil fuels. In: Meeting of the Southern District, Division of Production, American Petroleum Institute. Shell Development Company, San Antonio, Texas.

[57] Hubbert, M. K., (1959). Techniques of prediction with application to the petroleum industry. In: 44th Annual Meeting of the American Association of Petroleum Geologists. Shell Development Company, Dallas, TX.

[58] Bardi, U. (2009). Peak oil: The four stages of a new idea. Energy, 34(3), 323–326.

[59] Chapman, I. (2014). The end of Peak Oil? Why this topic is still relevant despite recent denials. Energy Policy, 64, 93–101.

[60] Swart, N. C., & Weaver, A. J. (2012). The Alberta oil sands and climate. Nature Climate Change, 2(3), 134–136.

[61] Food and Agriculture Organization of the United Nations. (2016). Wood Energy. http://www.fao.org/forestry/energy/en/

[62] International Energy Agency. (2017). Bioenergy. https://www.iea.org/topics/renewables/subtopics/bioenergy/

[63] Masera, O. R., Bailis, R., Drigo, R., Ghilardi, A., & Ruiz-Mercado, I. (2015). Environmental burden of traditional bioenergy use. Annual Review of Environment and Resources, 40, 121–150.

[334] Berk, Z. (1992). Technology Of Production Of Edible Flours And Protein Products From Soybeans. Technion, Israel Institute of Technology. Haifa, Israel. FAO Agricultural Services Bulletin No. 97. http://www.fao.org/docrep/t0532e/t0532e00.htm#con

[65] Weyer, K. M., Bush, D. R., Darzins, A., & Willson, B. D. (2010). Theoretical maximum algal oil production. Bioenergy Research, 3(2), 204–213.

[66] Pradhan, A., Shrestha, D. S., McAloon, A., Yee, W., Haas, M., & Duffield, J. A. (2011). Energy life-cycle assessment of soybean biodiesel revisited. Trans. ASABE, 54(3), 1031–1039.

[67] Luyssaert, S., Inglima, I., Jung, M., Richardson, A. D., Reichstein, M., Papale, D., ... & Aragao, L. E. O. C. (2007). CO2 balance of boreal, temperate, and tropical forests derived from a global database. Global change biology, 13(12), 2509–2537.

[625] Krankina, O. N., Harmon, M. E., Schnekenburger, F., & Sierra, C. A. (2012). Carbon balance on federal forest lands of Western Oregon and Washington: the impact of the Northwest Forest Plan. Forest ecology and management, 286, 171–182.

[69] Liao, C., Luo, Y., Fang, C., & Li, B. (2010). Ecosystem carbon stock influenced by plantation practice: implications for planting forests as a measure of climate change mitigation. PloS one, 5(5), e10867.

[70] Fargione, J., Hill, J., Tilman, D., Polasky, S., & Hawthorne, P. (2008). Land clearing and the biofuel carbon debt. Science, 319(5867), 1235–1238.

[71] McKinley, G. A., Pilcher, D. J., Fay, A. R., Lindsay, K., Long, M. C., & Lovenduski, N. S. (2016). Timescales for detection of trends in the ocean carbon sink. Nature, 530(7591), 469-472.

[72] Murphy, D. J., Hall, C. A., & Powers, B. (2011). New perspectives on the energy return on (energy) investment (EROI) of corn ethanol. Environment, development and sustainability, 13(1), 179–202.

[73] U.S. Energy Information Administration. (2011). History of energy consumption in the United States, 17752009. Today in Energy, February 9, 2011,

[74] Siler-Evans, K., Azevedo, I. L., & Morgan, M. G. (2012). Marginal emissions factors for the US electricity system. Environmental science & technology, 46(9), 4742–4748.

[75] Podobnik, B. (2006). Global energy shifts: fostering sustainability in a turbulent age. Temple University Press: Philadelphia, PA.

[76] Mather, A. (2001). The transition from deforestation to reforestation in Europe. In: Agricultural technologies and tropical deforestation, 35–52.

[77] Kaplan, J. O., Krumhardt, K. M., & Zimmermann, N. (2009). The prehistoric and preindustrial deforestation of Europe. Quaternary Science Reviews, 28(27), 3016–3034.

[78] Jaramillo, P., Griffin, W. M., & Matthews, H. S. (2007). Comparative life-cycle air emissions of coal, domestic natural gas, LNG, and SNG for electricity generation. Environmental Science & Technology, 41(17), 6290–6296.

[79] Whitaker, M., Heath, G. A., ODonoughue, P., & Vorum, M. (2012). Life cycle greenhouse gas emissions of coalfired electricity generation. Journal of Industrial Ecology, 16(S1), S53–S72.

[80] Venkatesh, A., Jaramillo, P., Griffin, W. M., & Matthews, H. S. (2012). Uncertainty in life cycle greenhouse gas emissions from United States coal. Energy & Fuels, 26(8), 4917–4923.

[81] Steinmann, Z. J., Hauck, M., Karuppiah, R., Laurenzi, I. J., & Huijbregts, M. A. (2014). A methodology for separating uncertainty and variability in the life cycle greenhouse gas emissions of coal-fueled power generation in the USA. The International Journal of Life Cycle Assessment, 19(5), 1146–1155.

[82] Palmer, M. A., Bernhardt, E. S., Schlesinger, W. H., Eshleman, K. N., Foufoula-Georgiou, E., Hendryx, M. S., ... & White, P. S. (2010). Mountaintop mining consequences. Science, 327(5962), 148–149.

[83] Bernhardt, E. S., & Palmer, M. A. (2011). The environmental costs of mountaintop mining valley fill operations for aquatic ecosystems of the Central Appalachians. Annals of the New york Academy of Sciences, 1223(1), 39–57.

[84] Bernhardt, E. S., Lutz, B. D., King, R. S., Fay, J. P., Carter, C. E., Helton, A. M., ... & Amos, J. (2012). How many mountains can we mine? Assessing the regional degradation of central Appalachian rivers by surface coal mining. Environmental science & technology, 46(15), 8115–8122.

[85] Howarth, R. W., Santoro, R., & Ingraffea, A. (2011). Methane and the greenhouse-gas footprint of natural gas from shale formations. Climatic Change, 106(4), 679–690.

[86] Alvarez, R. A., Pacala, S. W., Winebrake, J. J., Chameides, W. L., & Hamburg, S. P. (2012). Greater focus needed on methane leakage from natural gas infrastructure. Proceedings of the National Academy of Sciences, 109(17), 6435–6440.

[87] Caulton, D. R., Shepson, P. B., Santoro, R. L., Sparks, J. P., Howarth, R. W., Ingraffea, A. R., ... & Miller, B. R. (2014). Toward a better understanding and quantification of methane emissions from shale gas development. Proceedings of the National Academy of Sciences, 111(17), 6237–6242.

[88] Burnham, A., Han, J., Clark, C. E., Wang, M., Dunn, J. B., & Palou-Rivera, I. (2011). Life-cycle greenhouse gas emissions of shale gas, natural gas, coal, and petroleum. Environmental science & technology, 46(2), 619–627.

[89] Weber, C. L., & Clavin, C. (2012). Life cycle carbon footprint of shale gas: Review of evidence and implications. Environmental science & technology, 46(11), 5688–5695.

[90] Karion, A., Sweeney, C., Ptron, G., Frost, G., Michael Hardesty, R., Kofler, J., ... & Brewer, A. (2013). Methane emissions estimate from airborne measurements over a western United States natural gas field. Geophysical Research Letters, 40(16), 4393–4397.

[91] Allen, D. T., Torres, V. M., Thomas, J., Sullivan, D. W., Harrison, M., Hendler, A., ... & Lamb, B. K. (2013). Measurements of methane emissions at natural gas production sites in the United States. Proceedings of the National Academy of Sciences, 110(44), 17768–17773.

[92] Brandt, A. R., Heath, G. A., Kort, E. A., O'sullivan, F., Ptron, G., Jordaan, S. M., ... & Wofsy, S. (2014). Methane leaks from North American natural gas systems. Science, 343(6172), 733–735.

[93] Schneising, O., Burrows, J. P., Dickerson, R. R., Buchwitz, M., Reuter, M., & Bovensmann, H. (2014). Remote sensing of fugitive methane emissions from oil and gas production in North American tight geologic formations. Earth's Future, 2(10), 548–558.

[94] Peischl, J., Ryerson, T. B., Aikin, K. C., Gouw, J. A., Gilman, J. B., Holloway, J. S., ... & Trainer, M. (2015). Quantifying atmospheric methane emissions from the Haynesville, Fayetteville, and northeastern Marcellus shale gas production regions. Journal of Geophysical Research: Atmospheres, 120(5), 2119-2139.

[95] Turner, A. J., Jacob, D. J., Benmergui, J., Wofsy, S. C., Maasakkers, J. D., Butz, A., ... & Dlugokencky, E. (2016). A large increase in US methane emissions over the past decade inferred from satellite data and surface observations. Geophysical Research Letters, 43, 2218-2224.

[96] Turner, A. J., Jacob, D. J., Wecht, K. J., Maasakkers, J. D., Lundgren, E., Andrews, A. E., ... & Dubey, M. K. (2015). Estimating global and North American methane emissions with high spatial resolution using GOSAT satellite data. Atmos. Chem. Phys, 15(12), 7049–7069.

[97] OSullivan, F., & Paltsev, S. (2012). Shale gas production: potential versus actual greenhouse gas emissions. Environ-

mental Research Letters, 7(4), 044030.

[98] Van den Bergh, K., & Delarue, E. (2015). Cycling of conventional power plants: technical limits and actual costs. Energy Conversion and Management, 97, 70–77.

[99] Zhang, X., Myhrvold, N. P., Hausfather, Z., & Caldeira, K. (2016). Climate benefits of natural gas as a bridge fuel and potential delay of near-zero energy systems. Applied Energy, 167, 317–322.

[100] McJeon, H., Edmonds, J., Bauer, N., Clarke, L., Fisher, B., Flannery, B. P., ... & Riahi, K. (2014). Limited impact on decadal-scale climate change from increased use of natural gas. Nature, 514(7523), 482–485.

[101] Lenox, C., & Kaplan, P. O. (2016). Role of natural gas in meeting an electric sector emissions reduction strategy and effects on greenhouse gas emissions. Energy Economics, in press.

[102] Davis, S. J., & Shearer, C. (2014). Climate change: a crack in the natural-gas bridge. Nature, 514(7523), 436–437.

[103] Newell, R. G., & Raimi, D. (2014). Implications of shale gas development for climate change. Environmental science & technology, 48(15), 8360–8368.

[104] Shearer, C., Bistline, J., Inman, M., & Davis, S. J. (2014). The effect of natural gas supply on US renewable energy and CO_2 emissions. Environmental Research Letters, 9(9), 094008.

[105] Vengosh, A., Jackson, R. B., Warner, N., Darrah, T. H., & Kondash, A. (2014). A critical review of the risks to water resources from unconventional shale gas development and hydraulic fracturing in the United States. Environmental science & technology, 48(15), 8334–8348.

[106] Brantley, S. L., Yoxtheimer, D., Arjmand, S., Grieve, P., Vidic, R., Pollak, J., ... & Simon, C. (2014). Water resource impacts during unconventional shale gas development: The Pennsylvania experience. International Journal of Coal Geology, 126, 140–156.

[107] Nicot, J. P., Scanlon, B. R., Reedy, R. C., & Costley, R. A. (2014). Source and fate of hydraulic fracturing water in the Barnett Shale: a historical perspective. Environmental science & technology, 48(4), 2464–2471.

[108] Vidic, R. D., Brantley, S. L., Vandenbossche, J. M., Yoxtheimer, D., & Abad, J. D. (2013). Impact of shale gas development on regional water quality. Science, 340(6134), 1235009.

[109] Rubinstein, J. L., & Mahani, A. B. (2015). Myths and Facts on Wastewater Injection, Hydraulic Fracturing, Enhanced Oil Recovery, and Induced Seismicity. Seismological Research Letters, 86(4), 1060–1067.

[110] McGarr, A., Bekins, B., Burkardt, N., Dewey, J., Earle, P., Ellsworth, W., ... & Rubinstein, J. (2015). Coping with earthquakes induced by fluid injection. Science, 347(6224), 830–831.

[111] Kannan, R., Tso, C. P., Osman, R., & Ho, H. K. (2004). LCALCCA of oil fired steam turbine power plant in Singapore. Energy Conversion and Management, 45(18), 3093–3107.

[112] Weisser, D. (2007). A guide to life-cycle greenhouse gas (GHG) emissions from electric supply technologies. Energy, 32(9), 1543–1559.

[113] Murray, R., & Holbert, K. E. (2014). Nuclear energy: an introduction to the concepts, systems, and applications of nuclear processes. Elsevier.

[114] U.S. Energy Information Administration. (2016). International Energy Outlook 2016. DOE/EIA-0484. May, 2016.

[115] Mudd, G. M. (2014). The future of yellowcake: A global assessment of uranium resources and mining. Science of the Total Environment, 472, 590–607.

[116] International Atomic Energy Agency. (2008). Spent Fuel Reprocessing Options. IAEA-TECDOC-1587. IAEA, Vienna.

[117] Sovacool, B. K. (2008). Valuing the greenhouse gas emissions from nuclear power: A critical survey. Energy Policy, 36(8), 2950–2963.

[118] Beerten, J., Laes, E., Meskens, G., & Dhaeseleer, W. (2009). Greenhouse gas emissions in the nuclear life cycle: A balanced appraisal. Energy Policy, 37(12), 5056–5068.

[119] Warner, E. S., & Heath, G. A. (2012). Life cycle greenhouse gas emissions of nuclear electricity generation. Journal of Industrial Ecology, 16(s1), S73–S92.

[120] Lenzen, M. (2008). Life cycle energy and greenhouse gas emissions of nuclear energy: A review. Energy conversion and management, 49(8), 2178–2199.

[121] Norgate, T., Haque, N., & Koltun, P. (2014). The impact of uranium ore grade on the greenhouse gas footprint of nuclear power. Journal of Cleaner Production, 84, 360–367.

[122] Markandya, A., & Wilkinson, P. (2007). Electricity generation and health. The Lancet, 370(9591), 979–990.

[123] Kharecha, P. A., & Hansen, J. E. (2013). Prevented mortality and greenhouse gas emissions from historical and projected nuclear power. Environmental science & technology, 47(9), 4889–4895.

[124] Caiazzo, F., Ashok, A., Waitz, I. A., Yim, S. H., & Barrett, S. R. (2013). Air pollution and early deaths in the United States. Part I: Quantifying the impact of major sectors in 2005. Atmospheric Environment, 79, 198–208.

[125] Rashad, S. M., & Hammad, F. H. (2000). Nuclear power and the environment: comparative assessment of environ-

mental and health impacts of electricity-generating systems. Applied Energy, 65(1), 211–229.

[126] Fearnside, P. M. (2015). Emissions from tropical hydropower and the IPCC. Environmental Science & Policy, 50, 225–239.

[127] Fearnside, P. M. (2015). Tropical hydropower in the clean development mechanism: Brazils Santo Antnio Dam as an example of the need for change. Climatic Change, 131(4), 575–589.

[128] Cole, J. J., Prairie, Y. T., Caraco, N. F., McDowell, W. H., Tranvik, L. J., Striegl, R. G., ... & Melack, J. (2007). Plumbing the global carbon cycle: integrating inland waters into the terrestrial carbon budget. Ecosystems, 10(1), 172–185.

[129] Hertwich, E. G. (2013). Addressing biogenic greenhouse gas emissions from hydropower in LCA. Environmental science & technology, 47(17), 9604–9611.

[130] Zarfl, C., Lumsdon, A. E., Berlekamp, J., Tydecks, L., & Tockner, K. (2015). A global boom in hydropower dam construction. Aquatic Sciences, 77(1), 161–170.

[131] Barros, N., Cole, J. J., Tranvik, L. J., Prairie, Y. T., Bastviken, D., Huszar, V. L., ... & Roland, F. (2011). Carbon emission from hydroelectric reservoirs linked to reservoir age and latitude. Nature Geoscience, 4(9), 593–596.

[132] Demarty, M., & Bastien, J. (2011). GHG emissions from hydroelectric reservoirs in tropical and equatorial regions: Review of 20 years of CH 4 emission measurements. Energy Policy, 39(7), 4197–4206.

[133] Abril, G., Gurin, F., Richard, S., Delmas, R., GalyLacaux, C., Gosse, P., ... & Matvienko, B. (2005). Carbon dioxide and methane emissions and the carbon budget of a 10year old tropical reservoir (Petit Saut, French Guiana). Global biogeochemical cycles, 19, GB4007.

[134] Abril, G., Parize, M., Prez, M. A., & Filizola, N. (2013). Wood decomposition in Amazonian hydropower reservoirs: An additional source of greenhouse gases. Journal of South American Earth Sciences, 44, 104–107.

[135] Teodoru, C. R., Bastien, J., Bonneville, M. C., Giorgio, P. A., Demarty, M., Garneau, M., ... & Strachan, I. B. (2012). The net carbon footprint of a newly created boreal hydroelectric reservoir. Global Biogeochemical Cycles, 26, GB2016.

[136] de Faria, F. A., Jaramillo, P., Sawakuchi, H. O., Richey, J. E., & Barros, N. (2015). Estimating greenhouse gas emissions from future Amazonian hydroelectric reservoirs. Environmental Research Letters, 10(12), 124019.

[137] Pacca, S. (2007). Impacts from decommissioning of hydroelectric dams: a life cycle perspective. Climatic Change, 84(3-4), 281–294.

[138] Ziolkowska, J. R., & Reyes, R. (2016). Geospatial analysis of desalination in the USAn interactive tool for socio-economic evaluations and decision support. Applied Geography, 71, 115–122.

[139] Perry, T. Backers of desalination hope Carlsbad plant will disarm critics. Los Angeles Times, June 4, 2015.

[140] Vrsmarty, C. J., & Sahagian, D. (2000). Anthropogenic disturbance of the terrestrial water cycle. BioScience, 50(9), 753–765.

[141] Nilsson, C., & Berggren, K. (2000). Alterations of riparian ecosystems caused by river regulation. BioScience, 50(9), 783–792.

[142] Nilsson, C., Reidy, C. A., Dynesius, M., & Revenga, C. (2005). Fragmentation and flow regulation of the world's large river systems. Science, 308(5720), 405–408.

[143] Hudiburg, T. W., Law, B. E., Wirth, C., & Luyssaert, S. (2011). Regional carbon dioxide implications of forest bioenergy production. Nature Climate Change, 1(8), 419–423.

[144] Hudiburg, T. W., Luyssaert, S., Thornton, P. E., & Law, B. E. (2013). Interactive effects of environmental change and management strategies on regional forest carbon emissions. Environmental science & technology, 47(22), 13132–13140.

[145] Repo, A., Tuovinen, J. P., & Liski, J. (2015). Can we produce carbon and climate neutral forest bioenergy?. Gcb Bioenergy, 7(2), 253–262.

[146] Cherubini, F., Peters, G. P., Berntsen, T., Strmman, A. H., & Hertwich, E. (2011). CO_2 emissions from biomass combustion for bioenergy: atmospheric decay and contribution to global warming. Gcb Bioenergy, 3(5), 413–426.

[147] Cherubini, F., Huijbregts, M., Kindermann, G., Van Zelm, R., Van Der Velde, M., Stadler, K., & Strmman, A. H. (2016). Global spatially explicit CO_2 emission metrics for forest bioenergy. Scientific reports, 6, 20186.

[148] Keith, H., Lindenmayer, D., Mackey, B., Blair, D., Carter, L., McBurney, L., ...& Konishi-Nagano, T. (2014). Managing temperate forests for carbon storage: impacts of logging versus forest protection on carbon stocks. Ecosphere, 5(6), 1–34.

[149] Holtsmark, B. (2012). Harvesting in boreal forests and the biofuel carbon debt. Climatic change, 112(2), 415–428.

[150] Holtsmark, B. (2015). Quantifying the global warming potential of CO2 emissions from wood fuels. Gcb Bioenergy, 7(2), 195–206.

[151] Ter-Mikaelian, M. T., Colombo, S. J., & Chen, J. (2015). The burning question: Does forest bioenergy reduce carbon emissions? A review of common misconceptions about forest carbon accounting. Journal of Forestry, 113(1), 57–68.

[152] Johnston, C. M., & van Kooten, G. C. (2015). Back to the past: Burning wood to save the globe. Ecological

Economics, 120, 185–193.

[153] Wiltsee, G; National Renewable Energy Laboratory. (2000). Lessons Learned from Existing Biomass Power Plants, NREL/SR-570-26946.

[154] Bayer, P., Rybach, L., Blum, P., & Brauchler, R. (2013). Review on life cycle environmental effects of geothermal power generation. Renewable and Sustainable Energy Reviews, 26, 446–463.

[155] Bertani, R. and Thain, I. (2002). Geothermal power generating plant CO2 emission survey. International Geothermal Association (IGA) News 2002:3.

[156] Sullivan, J. L., Clark, C., Han, J., Harto, C., & Wang, M. (2013). Cumulative energy, emissions, and water consumption for geothermal electric power production. Journal of Renewable and Sustainable Energy, 5(2), 023127.

[157] Bloomfield, K. K., Moore, J. N., & Nielsen, R. N. (2003). Geothermal energy reduces greenhouse gas emissions. Climate Change Research, GRC Bulletin, March/April, 77-79.

[158] Fitch, A. C. (2015). Climate impacts of large-scale wind farms as parameterized in a global climate model. Journal of Climate, 28(15), 6160–6180.

[159] Miller, L. M., Brunsell, N. A., Mechem, D. B., Gans, F., Monaghan, A. J., Vautard, R., ... & Kleidon, A. (2015). Two methods for estimating limits to large-scale wind power generation. Proceedings of the National Academy of Sciences, 112(36), 11169–11174.

[160] Arvesen, A., & Hertwich, E. G. (2012). Assessing the life cycle environmental impacts of wind power: A review of present knowledge and research needs. Renewable and Sustainable Energy Reviews, 16(8), 5994–6006.

[161] Kabir, M. R., Rooke, B., Dassanayake, G. M., & Fleck, B. A. (2012). Comparative life cycle energy, emission, and economic analysis of 100 kW nameplate wind power generation. Renewable Energy, 37(1), 133–141.

[162] Grogg, K. (2005). Harvesting the wind: the physics of wind turbines. Physics and Astronomy Comps Papers, 7.

[163] Dolan, S. L., & Heath, G. A. (2012). Life cycle greenhouse gas emissions of utilityscale wind power. Journal of Industrial Ecology, 16(s1), S136–S154.

[164] Fleck, B., & Huot, M. (2009). Comparative life-cycle assessment of a small wind turbine for residential off-grid use. Renewable Energy, 34(12), 2688–2696.

[165] Nugent, D., & Sovacool, B. K. (2014). Assessing the lifecycle greenhouse gas emissions from solar PV and wind energy: a critical meta-survey. Energy Policy, 65, 229–244.

[166] Guezuraga, B., Zauner, R., & Plz, W. (2012). Life cycle assessment of two different 2 MW class wind turbines. Renewable Energy, 37(1), 37–44.

[167] Wang, S., Wang, S., & Smith, P. (2015). Ecological impacts of wind farms on birds: Questions, hypotheses, and research needs. Renewable and Sustainable Energy Reviews, 44, 599–607.

[168] Sovacool, B. K. (2009). Contextualizing avian mortality: A preliminary appraisal of bird and bat fatalities from wind, fossil-fuel, and nuclear electricity. Energy Policy, 37(6), 2241–2248.

[169] Sovacool, B. K. (2013). The avian benefits of wind energy: A 2009 update. Renewable Energy, 49, 19–24.

[170] Waldien, D. L., & Reichard, J. (2010). Bats are not birds and other problems with Sovacools (2009) analysis of animal fatalities due to electricity generation. Energy Policy, 38, 2067–2069.

[171] Loss, S. R., Will, T., & Marra, P. P. (2013). Estimates of bird collision mortality at wind facilities in the contiguous United States. Biological Conservation, 168, 201–209.

[172] Kunz, T. H., Arnett, E. B., Erickson, W. P., Hoar, A. R., Johnson, G. D., Larkin, R. P., ... & Tuttle, M. D. (2007). Ecological impacts of wind energy development on bats: questions, research needs, and hypotheses. Frontiers in Ecology and the Environment, 5(6), 315–324.

[173] Marvel, K., Kravitz, B., & Caldeira, K. (2013). Geophysical limits to global wind power. Nature Climate Change, 3(2), 118–121.

[174] Vautard, R., Thais, F., Tobin, I., Bron, F. M., Devezeaux, D. L. J., Colette, A., ... & Ruti, P. M. (2014). Regional climate model simulations indicate limited climatic impacts by operational and planned European wind farms. Nature communications, 5, 3196.

[175] Shockley, W., & Queisser, H. J. (1961). Detailed balance limit of efficiency of pn junction solar cells. Journal of applied physics, 32(3), 510–519.

[176] Rühle, S. (2016). Tabulated values of the ShockleyQueisser limit for single junction solar cells. Solar Energy, 130, 139–147.

[177] Fthenakis, V. M., & Kim, H. C. (2011). Photovoltaics: Life-cycle analyses. Solar Energy, 85(8), 1609–1628.

[178] Wong, J. H., Royapoor, M., & Chan, C. W. (2016). Review of life cycle analyses and embodied energy requirements of single-crystalline and multi-crystalline silicon photovoltaic systems. Renewable and Sustainable Energy Reviews, 58, 608–618.

[179] Kim, H. C., Fthenakis, V., Choi, J. K., & Turney, D. E. (2012). Life cycle greenhouse gas emissions of thinfilm photovoltaic electricity generation. Journal of Industrial Ecology, 16(s1), S110–S121.

491

[180] Safarian, J., Tranell, G., & Tangstad, M. (2012). Processes for upgrading metallurgical grade silicon to solar grade silicon. Energy Procedia, 20, 88–97.

[181] Fthenakis, V., & Alsema, E. (2006). Photovoltaics energy payback times, greenhouse gas emissions and external costs: 2004early 2005 status. Progress in Photovoltaics: research and applications, 14(3), 275–280.

[182] Xakalashe, B. S., & Tangstad, M. (2012). Silicon processing: from quartz to crystalline silicon solar cells. Chem Technol, 32–37.

[183] U.S. Department of Energy. (1982). Basic photovoltaic principles and methods. SERI/SP-290-1448, February 1982.

[184] Archer, C. L., & Jacobson, M. Z. (2005). Evaluation of global wind power. Journal of Geophysical Research, 110, D12110.

[185] Lu, X., McElroy, M. B., & Kiviluoma, J. (2009). Global potential for wind-generated electricity. Proceedings of the National Academy of Sciences, 106(27), 10933–10938.

[186] Adams, A. S., & Keith, D. W. (2013). Are global wind power resource estimates overstated?. Environmental Research Letters, 8(1), 015021.

[187] Jacobson, M. Z. (2009). Review of solutions to global warming, air pollution, and energy security. Energy & Environmental Science, 2(2), 148–173.

[188] Jacobson, M. Z., & Delucchi, M. A. (2011). Providing all global energy with wind, water, and solar power, Part I: Technologies, energy resources, quantities and areas of infrastructure, and materials. Energy Policy, 39(3), 1154–1169.

[189] Alonso, E., Sherman, A. M., Wallington, T. J., Everson, M. P., Field, F. R., Roth, R., & Kirchain, R. E. (2012). Evaluating rare earth element availability: A case with revolutionary demand from clean technologies. Environmental science & technology, 46(6), 3406–3414.

[190] Habib, K., & Wenzel, H. (2014). Exploring rare earths supply constraints for the emerging clean energy technologies and the role of recycling. Journal of Cleaner Production, 84, 348–359.

[191] Lacal-Arntegui, R. (2015). Materials use in electricity generators in wind turbinesstate-of-the-art and future specifications. Journal of Cleaner Production, 87, 275–283.

[192] Kavlak, G., McNerney, J., Jaffe, R. L., & Trancik, J. E. (2015). Metal production requirements for rapid photovoltaics deployment. Energy & Environmental Science, 8(6), 1651–1659.

[193] Feltrin, A., & Freundlich, A. (2008). Material considerations for terawatt level deployment of photovoltaics. Renewable energy, 33(2), 180–185.

[194] Grandell, L., & Thorenz, A. (2014). Silver supply risk analysis for the solar sector. Renewable energy, 69, 157–165.

[195] U.S. Geological Survey. (2016). Mineral Commodity Summaries. January 2016. 152–153.

[196] International Technology Roadmap for Photovoltaic (ITRPV). (2016). 2015 Results including maturity reports. Seventh Edition, October 2016.

[197] Akbari, H., Menon, S., & Rosenfeld, A. (2009). Global cooling: increasing world-wide urban albedos to offset CO 2. Climatic change, 94(3), 275–286.

[198] MacDonald, A. E., Clack, C. T., Alexander, A., Dunbar, A., Wilczak, J., & Xie, Y. (2016). Future cost-competitive electricity systems and their impact on US CO2 emissions. Nature Climate Change, 6(5), 526–531.

[199] Bogdanov, D., & Breyer, C. (2016). North-East Asian Super Grid for 100% Renewable Energy supply: Optimal mix of energy technologies for electricity, gas and heat supply options. Energy Conversion and Management, 112, 176–190.

[200] Jacobson, M. Z., Delucchi, M. A., Cameron, M. A., & Frew, B. A. (2015). Low-cost solution to the grid reliability problem with 100% penetration of intermittent wind, water, and solar for all purposes. Proceedings of the National Academy of Sciences, 112(49), 15060–15065.

[201] Energy and Environmental Economics, Inc. (2014). Investigating a Higher Renewables Portfolio Standard in California. January, 2014.

[202] Weitemeyer, S., Kleinhans, D., Vogt, T., & Agert, C. (2015). Integration of Renewable Energy Sources in future power systems: The role of storage. Renewable Energy, 75, 14–20.

[203] Sivak, M., & Schoettle, B. (2015). On-road fuel economy of vehicles in the United States: 19232013. University of Michigan Transportation Research Institute.

[204] Geller, H., Harrington, P., Rosenfeld, A. H., Tanishima, S., & Unander, F. (2006). Polices for increasing energy efficiency: Thirty years of experience in OECD countries. Energy policy, 34(5), 556–573.

[205] Environmental Protection Agency. (2014). Light-Duty Automotive Technology, Carbon Dioxide Emissions, and Fuel Economy Trends: 1975 Through 2014.

[206] Union of Concerned Scientists. A Brief History of U.S. Fuel Efficiency Standards. http://www.ucsusa.org/clean-vehicles/fuel-efficiency/fuel-economy-basics.html#.WJwfexsrJqM

[207] US Department of Energy. National Energy Technology Laboratory. (2009). An Evaluation of the Extraction, Transport and Refining of Imported Crude Oils and the Impact on Life Cycle Greenhouse Gas Emissions. DOE/NETL-2009/1362.

[208] U.S. Environmental Protection Agency. (2014). Greenhouse Gas Emissions from a Typical Passenger Vehicle. EPA-420-F-14-040, May 2014.

[209] Palou-Rivera, I., Han, J., & Wang, M. (2011). Updates to Petroleum Refining and Upstream Emissions. Center for Transportation Research. Argonne National Laboratory.

[210] Bergerson, J. A., Kofoworola, O., Charpentier, A. D., Sleep, S., & MacLean, H. L. (2012). Life cycle greenhouse gas emissions of current oil sands technologies: surface mining and in situ applications. Environmental science & technology, 46(14), 7865–7874.

[211] Charpentier, A. D., Bergerson, J. A., & MacLean, H. L. (2009). Understanding the Canadian oil sands industrys greenhouse gas emissions. Environmental research letters, 4(1), 014005.

[212] United States Department of State (2014). Final Supplemental Environmental Impact Statement for the Keystone XL Project. Applicant for Presidential Permit: TransCanada Keystone Pipeline, LP. Executive Summary. January 2014.

[213] Ramseur, J. L., Lattanzio, R. K., Luther, L., Parfomak, P. W., & Carter, N. T.; Congressional Research Service (2014). Oil Sands and the Keystone XL Pipeline: Background and Selected Environmental Issues. Congressional Research Service, 7-5700.

[214] JACOBS Consultancy. (2009). Life Cycle Assessment Comparison of North American and Imported Crudes. AERI 1747.

[215] TIAX LLC; MathPro Inc. (2009). Comparison of North American and Imported Crude Oil Lifecycle GHG Emissions. TIAX Case No. D5595.

[216] Lee, P,. Cheng, R. (2009). Bitumen and Biocarbon: Land Use Conversions and Loss of Biological Carbon Due to Bitumen Operations in the Boreal Forests of Alberta, Canada. Global Forest Watch Canada, 2009.

[217] Weiss, M. A., Heywood, J. B., Drake, E. M., Schafer, A., & AuYeung, F. F. (2000). On the Road in 2020. Energy Laboratory Report# MIT EL 00-003. Cambridge, MA: Energy Laboratory, Massachusetts Institute of Technology.

[218] U.S. Environmental Protection Agency. Accessed 2/16/2015, http://www.epa.gov/epawaste/conserve/materials/auto.htm

[219] Lewis, A. M., Kelly, J. C., & Keoleian, G. A. (2014). Vehicle lightweighting vs. electrification: life cycle energy and GHG emissions results for diverse powertrain vehicles. Applied Energy, 126, 13–20.

[220] Samaras, C., & Meisterling, K. (2008). Life cycle assessment of greenhouse gas emissions from plug-in hybrid vehicles: implications for policy. Environmental science & technology, 42(9), 3170–3176.

[221] Schweimer, G. W., & Levin, M. (2000). Life Cycle Inventory for the Golf A4; Volkswagen: Wolfsburg, Germany, 2000.

[222] Sullivan, J. L.,Gaines, L. (2010). A Review of Battery Life-Cycle Analysis: State of Knowledge and Critical Needs. Argonne National Laboratory.

[223] SAE International. (2010). Leaf to be sold with battery pack at C-segment price. March 11, 2010. http://articles.sae.org/7714/

[224] Notter, D. A., Gauch, M., Widmer, R., Wger, P., Stamp, A., Zah, R., & Althaus, H. J. (2010). Contribution of Li-ion batteries to the environmental impact of electric vehicles. Environmental science & technology, 44(17), 6550–6.

[225] Dunn, J. B., Gaines, L., Sullivan, J., & Wang, M. Q. (2012). Impact of recycling on cradle-to-gate energy consumption and greenhouse gas emissions of automotive lithium-ion batteries. Environmental science & technology, 46(22), 12704–12710.

[226] Zackrisson, M., Avelln, L., & Orlenius, J. (2010). Life cycle assessment of lithium-ion batteries for plug-in hybrid electric vehiclesCritical issues. Journal of Cleaner Production, 18(15), 1519–1529.

[227] Ellingsen, L. A. W., MajeauBettez, G., Singh, B., Srivastava, A. K., Valen, L. O., & Strmman, A. H. (2014). Life cycle assessment of a lithiumion battery vehicle pack. Journal of Industrial Ecology, 18(1), 113–124.

[228] Kim, H. C., Wallington, T. J., Arsenault, R., Bae, C., Ahn, S., & Lee, J. (2016). Cradle-to-Gate Emissions from a Commercial Electric Vehicle Li-Ion Battery: A Comparative Analysis. Environmental Science & Technology, 50(14), 7715–7722.

[229] U.S. Environmental Protection Agency. (2013). Application of Life-Cycle Assessment to Nanoscale Technology: Lithium-ion Batteries for Electric Vehicles. EPA 744-R-12-001.

[230] Zivin, J. S. G., Kotchen, M. J., & Mansur, E. T. (2014). Spatial and temporal heterogeneity of marginal emissions: Implications for electric cars and other electricity-shifting policies. Journal of Economic Behavior & Organization, 107, 248–268.

[231] Tamayao, M. A. M., Michalek, J. J., Hendrickson, C., & Azevedo, I. M. (2015). Regional variability and uncertainty of electric vehicle life cycle CO_2 emissions across the United States. Environmental Science & Technology, 49(14), 8844–8855.

[232] Yuksel, T., & Michalek, J. J. (2015). Effects of regional temperature on electric vehicle efficiency, range, and emissions in the United States. Environmental science & technology, 49(6), 3974–3980.

[233] Axsen, J., Kurani, K. S., McCarthy, R., & Yang, C. (2011). Plug-in hybrid vehicle GHG impacts in California:

Integrating consumer-informed recharge profiles with an electricity-dispatch model. Energy Policy, 39(3), 1617–1629.

[234] Elgowainy, A., Burnham, A., Wang, M., Molburg, J., & Rousseau, A. (2009). Well-to-wheels energy use and greenhouse gas emissions of plug-in hybrid electric vehicles. No. ANL/ESD/09-2. Energy Systems Division, Argonne National Laboratory. 70 p.

[235] LeBlanc, D. J, Sivak, M., & Bogard, S. (2010). Using naturalistic driving data to assess variations in fuel efficiency among individual drivers. University of Michigan, Transportation Research Institute. Report No. UMTRI-2010-34.

[236] Barkenbus, J. N. (2010). Eco-driving: An overlooked climate change initiative. Energy Policy, 38(2), 762–769.

[237] Greene, D. L. (1986). Driver energy conservation awareness training: Review and recommendations for a national program. Oak Ridge National Laboratory, ORNL/TM-9897.

[238] Kenneth S. Kurani, K. S., Stillwater, T., Jones, M, & Caperello, N. Ecodrive I-80: a large sample fuel economy feedback field test. Final report. Institute of Transportation Studies University of California, Davis. Report: ITS-RR-13-15

[239] Ho, S. H., Wong, Y. D., & Chang, V. W. C. (2015). What can eco-driving do for sustainable road transport? Perspectives from a city (Singapore) eco-driving programme. Sustainable Cities and Society, 14, 82–88.

[240] Beusen, B., Broekx, S., Denys, T., Beckx, C., Degraeuwe, B., Gijsbers, M., ... & Panis, L. I. (2009). Using on-board logging devices to study the longer-term impact of an eco-driving course. Transportation research part D: transport and environment, 14(7), 514–520.

[241] Alam, M. S., & McNabola, A. (2014). A critical review and assessment of Eco-Driving policy & technology: Benefits & limitations. Transport Policy, 35, 42–49.

[242] Thomas, J., Hwang, H., West, B., & Huff, S. (2013). Predicting Light-Duty Vehicle Fuel Economy as a Function of Highway Speed. SAE Int. J. Passeng. Cars - Mech. Syst., 6(2):859–875.

[243] Gaines, L., Rask, E.,& Keller, G. (2012). Which Is Greener: Idle, or Stop and Restart? Comparing Fuel Use and Emissions for Short Passenger-Car Stop. Argonne National Laboratory, Energy Systems Division. November, 2012.

[244] Weyand, P. G., Smith, B. R., Puyau, M. R., & Butte, N. F. (2010). The mass-specific energy cost of human walking is set by stature. Journal of Experimental Biology, 213(23), 3972–3979.

[245] Ross, M. (1997). Fuel efficiency and the physics of automobiles. Contemporary Physics, 38(6), 381–394.

[246] Farrington, R., & Rugh, J. (2000). Impact of Vehicle Air-Conditioning on Fuel Economy, Tailpipe Emissions, and Electric Vehicle Range: Preprint (No. NREL/CP-540-28960). National Renewable Energy Lab., Golden, CO (US).

[247] Clean Air Vehicle Technology Center. (1999). Effect of Air-conditioning on Regulated Emissions for In-Use Vehicles, Phase I, Final Report. Prepared for Coordinating Research Council, Inc. CRC Project E-37, October 1999.

[248] Weilenmann, M. F., Alvarez, R., & Keller, M. (2010). Fuel Consumption and CO2/Pollutant Emissions of Mobile Air Conditioning at Fleet Level-New Data and Model Comparison. Environmental science & technology, 44(13), 5277–5282.

[249] Multerer, B., & Burton, R. L. (1991). Alternative Technologies for Automobile Air Conditioning. Air Conditioning and Refrigeration Center. College of Engineering. University of Illinois at Urbana-Champaign.

[250] Fuhs, A. (2008). Hybrid Vehicles: and the Future of Personal Transportation. CRC Press.

[251] Parker, D. S. (2005). Energy Efficient Transportation for Florida. Available online: http://www.fsec.ucf.edu/en/Publications/html/fsec-en-19/#2.

[252] Bettes, W. H. (1982). The Aerodynamic Drag of Road Vehicles-Past, Present, and Future. Engineering and Science, 45(3), 4–10.

[253] Kurtz, D. W. (1980). Aerodynamic design of electric and hybrid vehicles: a guidebook (No. JPL-PUB-80-91). Jet Propulsion Lab., Pasadena, CA (USA).

[254] Knibbs, L. D., De Dear, R. J., & Atkinson, S. E. (2009). Field study of air change and flow rate in six automobiles. Indoor air, 19(4), 303–313.

[255] Fruin, S. A., Hudda, N., Sioutas, C., & Delfino, R. J. (2011). Predictive model for vehicle air exchange rates based on a large, representative sample. Environmental science & technology, 45(8), 3569–3575.

[256] Hill, W., Lebut, D., Major, G., & Schenkel, F. (2004). Affect [sic] of windows down on vehicle fuel economy as compared to AC load. In: Proceedings of SAE Alternate Refrigerant Systems Symposium, Scottsdale, Arizona.

[257] Hucho, W. H., & Sovran, G. (1993). Aerodynamics of road vehicles. Annual review of fluid mechanics, 25(1), 485–537.

[258] U.S. Environmental Protection Agency; Nam, E. (2004). Advanced Technology Vehicle Modeling in PERE. March, 2004, EPA420-D-04-002.

[259] U.S. Environmental Protection Agency; Nam, E. (2005). Fuel Consumption Modeling of Conventional and Advanced Technology Vehicles in the Physical Emission Rate Estimator (PERE). February, 2005, EPA420-P-05-001.

[260] Åhman, M. (2001). Primary energy efficiency of alternative powertrains in vehicles. Energy, 26(11), 973–989.

[261] Van Vliet, O., Brouwer, A. S., Kuramochi, T., van Den Broek, M., & Faaij, A. (2011). Energy use, cost and CO2 emissions of electric cars. Journal of Power Sources, 196(4), 2298–2310.

[262] Campanari, S., Manzolini, G., & De la Iglesia, F. G. (2009). Energy analysis of electric vehicles using batteries or fuel cells through well-to-wheel driving cycle simulations. Journal of Power Sources, 186(2), 464–477.

[263] Penner, J. E., Lister, D. H., Griggs, D. J., Dokken, D. J., & McFarland, M. (1999). IPCC special report. Aviation and the global atmosphere. Intergovernmental Panel on Climate Change.

[264] Elgowainy, A, Han, J., Wang, M., Carter, N., Stratton, R., Hileman, J., Malwitz, A., & Balasubramanian, S. (2012). Life-cycle analysis of alternative aviation fuels in GREET. Argonne National Laboratory, June 2012, ANL/ESD/12-8.

[265] Bureau of Tranportation Statistics. (2014). Summary 2014 U.S.-Based Airline Traffic Data. https://www.bts.gov/bts/newsroom/summary-2014-us-based-airline-traffic-data.

[266] Stuber, N., Forster, P., Rdel, G., & Shine, K. (2006). The importance of the diurnal and annual cycle of air traffic for contrail radiative forcing. Nature, 441(7095), 864–867.

[267] igojet (2010). Understanding Business Aviations Costs. http://www.igojet.com/downloads/wp2-understandingbusinessaviationscosts_final.pdf

[268] Dessens, O., Khler, M. O., Rogers, H. L., Jones, R. L., & Pyle, J. A. (2014). Aviation and climate change. Transport Policy, 34, 14–20.

[269] Lee, D. S., Fahey, D. W., Forster, P. M., Newton, P. J., Wit, R. C., Lim, L. L., ... & Sausen, R. (2009). Aviation and global climate change in the 21st century. Atmospheric Environment, 43(22), 3520–3537.

[270] Lee, D. S., Pitari, G., Grewe, V., Gierens, K., Penner, J. E., Petzold, A., ... & Iachetti, D. (2010). Transport impacts on atmosphere and climate: Aviation. Atmospheric Environment, 44(37), 4678–4734.

[271] Schumann, U., Penner, J. E., Chen, Y., Zhou, C., & Graf, K. (2015). Dehydration effects from contrails in a coupled contrailclimate model. Atmospheric Chemistry and Physics, 15(19), 11179–11199.

[272] Burkhardt, U., & Krcher, B. (2011). Global radiative forcing from contrail cirrus. Nature climate change, 1(1), 54–58.

[273] Chen, C. C., & Gettelman, A. (2013). Simulated radiative forcing from contrails and contrail cirrus. Atmospheric Chemistry and Physics, 13(24), 12525–12536.

[274] Chen, C. C., & Gettelman, A. (2016). Simulated 2050 aviation radiative forcing from contrails and aerosols. Atmospheric Chemistry and Physics, 16(11), 7317–7333.

[275] Gettelman, A., & Chen, C. (2013). The climate impact of aviation aerosols. Geophysical Research Letters, 40(11), 2785–2789.

[276] Righi, M., Hendricks, J., & Sausen, R. (2013). The global impact of the transport sectors on atmospheric aerosol: simulations for year 2000 emissions. Atmospheric Chemistry and Physics, 13(19), 9939–9970.

[277] Righi, M., Hendricks, J., & Sausen, R. (2016). The global impact of the transport sectors on atmospheric aerosol in 2030Part 2: Aviation. Atmospheric Chemistry and Physics, 16(7), 4481–4495.

[278] Kapadia, Z. Z., Spracklen, D. V., Arnold, S. R., Borman, D. J., Mann, G. W., Pringle, K. J., ... & Scott, C. E. (2015). Impacts of aviation fuel sulfur content on climate and human health. Atmos. Chem. Phys. Discuss, 15, 18921–18961.

[279] Airlines for America. (2017). Air Travelers in America: Survey Highlights. February 28, 2017

[280] Sausen, R., & Schumann, U. (2000). Estimates of the Climate Response to Aircraft CO_2 and NO x Emissions Scenarios. Climatic Change, 44(1-2), 27–58.

[281] Asselin, M. (2000). Introduction to Aircraft Performance. Reston, US: American Institute of Aeronautics and Astronautics.

[282] Federal Transit Administration, U.S. Department of Transportation. (2010). Public Transportations Role in Responding to Climate Change. January, 2010.

[283] Fischer, L. A., & Schwieterman, J. P. (2011). The Decline and Recovery of Intercity Bus Service in the United States: A Comeback for an Environmentally Friendly Transportation Mode?. Environmental Practice, 13(01), 7–15.

[284] American Bus Association. (2015). Motorcoach Census: A Study of the Size and Activity of the Motorcoach Industry in the United States and Canada in 2013. March 12, 2015.

[285] Federal Transit Administration. (2016). National Transit Database: 2015 National Transit Summary and Trends. October, 2016.

[286] Woldeamanuel, M. (2012). Evaluating the Competitiveness of Intercity Buses in Terms of Sustainability Indicators. Journal of Public Transportation, 15(3), 77–96.

[287] Hallmark, S., Sperry, B., & Mudgal, A. (2011). In-use fuel economy of hybrid-electric school buses in iowa. Journal of the Air & Waste Management Association, 61(5), 504–510.

[288] Zhang, S., Wu, Y., Liu, H., Huang, R., Yang, L., Li, Z., ... & Hao, J. (2014). Real-world fuel consumption and CO2 emissions of urban public buses in Beijing. Applied Energy, 113, 1645–1655.

[289] Barnitt, R.A. (2008). In-use performance comparison of hybrid electric, CNG, and diesel buses at New York City transit. SAE international technical paper, [200801-1556].

[290] Chong, U., Yim, S. H., Barrett, S. R., & Boies, A. M. (2014). Air quality and climate impacts of alternative bus

technologies in greater London. Environmental science & technology, 48(8), 4613–4622.

[291] Network Rail. New Lines Programme. Comparing environmental impact of conventional and high speed rail.

[292] Lane, B. W. (2008). Significant characteristics of the urban rail renaissance in the United States: a discriminant analysis. Transportation Research Part A: Policy and Practice, 42(2), 279–295.

[293] Bhattacharjee, S., & Goetz, A. R. (2012). Impact of light rail on traffic congestion in Denver. Journal of Transport Geography, 22, 262–270.

[294] Ewing, R., & Hamidi, S. (2014). Longitudinal analysis of transit's land use multiplier in Portland (OR). Journal of the American Planning Association, 80(2), 123–137.

[295] White, P., Golden, J. S., Biligiri, K. P., & Kaloush, K. (2010). Modeling climate change impacts of pavement production and construction. Resources, Conservation and Recycling, 54(11), 776–782.

[296] Ang, B. W., Fwa, T. F., & Ng, T. T. (1993). Analysis of process energy use of asphalt-mixing plants. Energy, 18(7), 769–777.

[297] Zapata, P., & Gambatese, J. A. (2005). Energy consumption of asphalt and reinforced concrete pavement materials and construction. Journal of Infrastructure Systems, 11(1), 9–20.

[298] Santero, N. J., & Horvath, A. (2009). Global warming potential of pavements. Environmental Research Letters, 4(3), 034011.

[299] Santero, N. J., Masanet, E., & Horvath, A. (2011). Life-cycle assessment of pavements. Part I: Critical review. Resources, Conservation and Recycling, 55(9), 801–809.

[300] Athena Institute. (2006). A Life Cycle Perspective on Concrete and Asphalt Roadways: Embodied Primary Energy and Global Warming Potential. September 2006. 68 p. http://www.athenasmi.org/resources/publications/

[301] Blissett, R. S., & Rowson, N. A. (2012). A review of the multi-component utilisation of coal fly ash. Fuel, 97, 1–23.

[302] Shoup, D. C. (1999). The trouble with minimum parking requirements. Transportation Research Part A: Policy and Practice, 33(7), 549–574.

[303] Duany, A., Plater-Zyberk, E., & Speck, J. (2001). Suburban nation: The rise of sprawl and the decline of the American dream. Macmillan.

[304] Marsden, G. (2006). The evidence base for parking policiesa review. Transport policy, 13(6), 447–457.

[305] Chester, M., Horvath, A., & Madanat, S. (2010). Parking infrastructure: energy, emissions, and automobile life-cycle environmental accounting. Environmental Research Letters, 5(3), 034001.

[306] Chester, M. V., & Horvath, A. (2009). Environmental assessment of passenger transportation should include infrastructure and supply chains. Environmental research letters, 4(2), 024008.

[307] Chester, M., & Horvath, A. (2010). Life-cycle assessment of high-speed rail: the case of California. Environmental Research Letters, 5(1), 014003.

[308] Chang, B., & Kendall, A. (2011). Life cycle greenhouse gas assessment of infrastructure construction for Californias high-speed rail system. Transportation Research Part D: Transport and Environment, 16(6), 429–434.

[309] Mohamed-Kassim, Z., & Filippone, A. (2010). Fuel savings on a heavy vehicle via aerodynamic drag reduction. Transportation Research Part D: Transport and Environment, 15(5), 275–284.

[310] Giannelli, R. A., Nam, E. K., Helmer, K., Younglove, T., Scora, G., & Barth, M. (2005). Heavy-duty diesel vehicle fuel consumption modeling based on road load and power train parameters. SAE World Congress.

[311] Chowdhury, H., Moria, H., Ali, A., Khan, I., Alam, F., & Watkins, S. (2013). A study on aerodynamic drag of a semi-trailer truck. Procedia Engineering, 56, 201–205.

[312] Bhattacharya, S. (2017). Main deck or belly: who leads the race? The STAT Trade Times. http://www.stattimes.com/main-deck-or-belly-who-leads-the-race-air-cargo

[313] International Maritime Organization (IMO); Smith, T. W. P., Jalkanen, J. P., Anderson, B. A., Corbett, J. J., Faber, J., Hanayama, S., ... & Raucci, C. (2015). Third IMO GHG Study 2014.

[314] Psaraftis, H. N., & Kontovas, C. A. (2009). CO2 emission statistics for the world commercial fleet. WMU Journal of Maritime Affairs, 8(1), 1–25.

[315] Eyring, V., Isaksen, I. S., Berntsen, T., Collins, W. J., Corbett, J. J., Endresen, O., ... & Stevenson, D. S. (2010). Transport impacts on atmosphere and climate: Shipping. Atmospheric Environment, 44(37), 4735–4771.

[316] Vermeulen, S. J., Campbell, B. M., & Ingram, J. S. (2012). Climate change and food systems. Annual Review of Environment and Resources, 37(1), 195–222.

[317] Pelletier, N., Audsley, E., Brodt, S., Garnett, T., Henriksson, P., Kendall, A., ... & Troell10, M. (2011). Energy Intensity of Agriculture and Food Systems. Annual Review of Environment and Resources, 36, 223–46.

[318] Canning, P., Charles, A., Huang, S., Polenske, K. R., & Waters, A. (2010). Energy Use in the US Food System. United States Department of Agriculture, Economic Research Service, Economic Research Report Number 94, March 2010.

[319] Cuéllar, A. D., & Webber, M. E. (2010). Wasted food, wasted energy: the embedded energy in food waste in the United States. Environmental science & technology, 44(16), 6464–6469.

[320] Wecht, K. J., Jacob, D. J., Frankenberg, C., Jiang, Z., & Blake, D. R. (2014). Mapping of North American methane emissions with high spatial resolution by inversion of SCIAMACHY satellite data. Journal of Geophysical Research: Atmospheres, 119(12), 7741–7756.

[321] Heller, M. C., & Keoleian, G. A. (2015). Greenhouse gas emission estimates of US dietary choices and food loss. Journal of Industrial Ecology, 19(3), 391–401.

[322] Peters, C. J., Picardy, J. A., Darrouzet-Nardi, A., & Griffin, T. S. (2014). Feed conversions, ration compositions, and land use efficiencies of major livestock products in US agricultural systems. Agricultural Systems, 130, 35–43.

[323] Aston, L. M., Smith, J. N., & Powles, J. W. (2012). Impact of a reduced red and processed meat dietary pattern on disease risks and greenhouse gas emissions in the UK: a modelling study. BMJ Open, 2(5), e001072.

[324] Berners-Lee, M., Hoolohan, C., Cammack, H., & Hewitt, C. N. (2012). The relative greenhouse gas impacts of realistic dietary choices. Energy Policy, 43, 184–190.

[325] Scarborough, P., Appleby, P. N., Mizdrak, A., Briggs, A. D., Travis, R. C., Bradbury, K. E., & Key, T. J. (2014). Dietary greenhouse gas emissions of meat-eaters, fish-eaters, vegetarians and vegans in the UK. Climatic change, 125(2), 179–192.

[326] Buzby, J. C., Wells, H. F.,& Hyman,J. (2014). The Estimated Amount, Value, and Calories of Postharvest Food Losses at the Retail and Consumer Levels in the United States. USDA Economic Research Service, Economic Information Bulletin Number 121, February 2014.

[327] Food and Agriculture Organization of the United Nations. (2011). Global food losses and food waste Extent, causes and prevention. Rome.

[328] Weber, C. L., & Matthews, H. S. (2008). Food-miles and the relative climate impacts of food choices in the United States. Environmental science & technology, 42(10), 3508–3513.

[329] Cleveland, D. A., Radka, C. N., Müller, N. M., Watson, T. D., Rekstein, N. J., Van M. Wright, H., & Hollingshead, S. E. (2011). Effect of localizing fruit and vegetable consumption on greenhouse gas emissions and nutrition, Santa Barbara County. Environmental science & technology, 45(10), 4555–4562.

[330] Johnson, J. A., Runge, C. F., Senauer, B., Foley, J., & Polasky, S. (2014). Global agriculture and carbon trade-offs. Proceedings of the National Academy of Sciences, 111(34), 12342–12347.

[331] Nickerson, C., Ebel, R., Borchers, A., & Carriazo, F. (2011). Major Uses of Land in the United States, 2007. USDA Economic Research Service, Economic Information Bulletin Number 89, December 2011.

[332] Egli, D. B. (2008). Comparison of corn and soybean yields in the United States: Historical trends and future prospects. Agronomy journal, 100 (S3), S79–88.

[333] Clampitt, C. (2015). Midwest Maize: How Corn Shaped the U.S. Heartland. University of Illinois Press.

[334] Berk, Z. (1992). Technology Of Production Of Edible Flours And Protein Products From Soybeans. Technion, Israel Institute of Technology. Haifa, Israel. FAO Agricultural Services Bulletin No. 97.

[335] Liebig J: Organic Chemistry in Its Application to Agriculture and Physiology. London: Taylor Walton, 1840.

[336] Liebig J: The Natural Laws of Husbandry. London: Walton Maberly, 1863.

[337] Rosswall, T. (1976). The internal nitrogen cycle between microorganisms, vegetation and soil. Ecological Bulletins, 157–167.

[338] Cameron, K. C., Di, H. J., & Moir, J. L. (2013). Nitrogen losses from the soil/plant system: a review. Annals of Applied Biology, 162(2), 145–173.

[339] Rheinbaben, W. V. (1990). Nitrogen losses from agricultural soils through denitrificationa critical evaluation. Zeitschrift fr Pflanzenernhrung und Bodenkunde, 153(3), 157–166.

[340] Bing, C., He, F. Y., Xu, Q. M., Yin, B., & Gui-Xin, C. A. I. (2006). Denitrification losses and N 2 O emissions from nitrogen fertilizer applied to a vegetable field. Pedosphere, 16(3), 390–397.

[341] Weier, K. L., Doran, J. W., Power, J. F., & Walters, D. T. (1993). Denitrification and the dinitrogen/nitrous oxide ratio as affected by soil water, available carbon, and nitrate. Soil Science Society of America Journal, 57(1), 66-72.

[342] Elmi, A. A., Madramootoo, C., Hamel, C., & Liu, A. (2003). Denitrification and nitrous oxide to nitrous oxide plus dinitrogen ratios in the soil profile under three tillage systems. Biology and Fertility of Soils, 38(6), 340–348.

[343] Fowler, D., Coyle, M., Skiba, U., Sutton, M. A., Cape, J. N., Reis, S., ... & Vitousek, P. (2013). The global nitrogen cycle in the twenty-first century. Philosophical Transactions of the Royal Society of London B: Biological Sciences, 368(1621), 20130164.

[344] Vitousek, P. M., Menge, D. N., Reed, S. C., & Cleveland, C. C. (2013). Biological nitrogen fixation: rates, patterns and ecological controls in terrestrial ecosystems. Philosophical Transactions of the Royal Society of London B: Biological Sciences, 368(1621), 20130119.

[345] Reay, D. S., Dentener, F., Smith, P., Grace, J., & Feely, R. A. (2008). Global nitrogen deposition and carbon sinks. Nature Geoscience, 1(7), 430–437.

497

[346] Zhang, W. F., Dou, Z. X., He, P., Ju, X. T., Powlson, D., Chadwick, D., ... & Chen, X. P. (2013). New technologies reduce greenhouse gas emissions from nitrogenous fertilizer in China. PNAS, 110(21), 8375–8380.

[347] Wood, S., & Cowie, A. (2004). A review of greenhouse gas emission factors for fertiliser production. For IEA bioenergy task 38.

[348] Crutzen, P. J., Mosier, A. R., Smith, K. A., & Winiwarter, W. (2008). N 2 O release from agro-biofuel production negates global warming reduction by replacing fossil fuels. Atmospheric chemistry and physics, 8(2), 389–395.

[349] Reay, D. S., Davidson, E. A., Smith, K. A., Smith, P., Melillo, J. M., Dentener, F., & Crutzen, P. J. (2012). Global agriculture and nitrous oxide emissions. Nature climate change, 2(6), 410–416.

[350] Cordell, D., Drangert, J. O., & White, S. (2009). The story of phosphorus: global food security and food for thought. Global environmental change, 19(2), 292–305.

[351] Cordell, D., & White, S. (2014). Life's bottleneck: sustaining the world's phosphorus for a food secure future. Annual Review of Environment and Resources, 39, 161–188.

[352] Koppelaar, R. H. E. M., & Weikard, H. P. (2013). Assessing phosphate rock depletion and phosphorus recycling options. Global Environmental Change, 23(6), 1454–1466.

[353] Filippelli, G. M. (2011). Phosphate rock formation and marine phosphorus geochemistry: the deep time perspective. Chemosphere, 84(6), 759–766.

[354] Edixhoven, J. D., Gupta, J., & Savenije, H. H. G. (2014). Recent revisions of phosphate rock reserves and resources: a critique. Earth System Dynamics, 5(2), 491–507.

[355] Fixen, P. E., & Johnston, A. M. (2012). World fertilizer nutrient reserves: a view to the future. Journal of the Science of Food and Agriculture, 92(5), 1001–1005.

[356] Ciceri, D., Manning, D. A., & Allanore, A. (2015). Historical and technical developments of potassium resources. Science of The Total Environment, 502, 590–601.

[357] Oerke, E. C. (2006). Crop losses to pests. The Journal of Agricultural Science, 144(1), 31–43.

[358] Russell, P. E. (2005). A century of fungicide evolution. The Journal of Agricultural Science, 143(1), 11–25.

[359] Jarman, W. M., & Ballschmiter, K. (2012). From coal to DDT: the history of the development of the pesticide DDT from synthetic dyes till Silent Spring. Endeavour, 36(4), 131–142.

[360] Webb, J., L., A., Jr. (2014). The Long Struggle Against Malaria in Tropical Africa. New York, NY: Cambridge University Press.

[361] Fernandez-Cornejo, J., Nehring, R., Osteen, C., Wechsler, S., Martin, A., & Vialou, A. (2014). Pesticide Use in U.S. Agriculture: 21 Selected Crops, 1960-2008, EIB-124, U.S. Department of Agriculture, Economic Research Service, May 2014.

[362] Chikowo, R., Faloya, V., Petit, S., & Munier-Jolain, N. M. (2009). Integrated Weed Management systems allow reduced reliance on herbicides and long-term weed control. Agriculture, ecosystems & environment, 132(3), 237–242.

[363] Zhang, J., Weaver, S. E., & Hamill, A. S. (2000). Risks and reliability of using herbicides at below-labeled rates. Weed Technology, 14(1), 106–115.

[364] Hossard, L., Archer, D. W., Bertrand, M., Colnenne-David, C., Debaeke, P., Ernfors, M., ... & Snapp, S. S. (2016). A meta-analysis of maize and wheat yields in low-input vs. conventional and organic systems. Agronomy Journal, 108(3), 1155–1167.

[365] Lechenet, M., Dessaint, F., Py, G., Makowski, D., & Munier-Jolain, N. (2017). Reducing pesticide use while preserving crop productivity and profitability on arable farms. Nature Plants, 3, 17008.

[366] Gaba, S., Gabriel, E., Chaduf, J., Bonneu, F., & Bretagnolle, V. (2016). Herbicides do not ensure for higher wheat yield, but eliminate rare plant species. Scientific reports, 6, 30112.

[367] Hossard, L., Philibert, A., Bertrand, M., Colnenne-David, C., Debaeke, P., Munier-Jolain, N., ... & Makowski, D. (2014). Effects of halving pesticide use on wheat production. Scientific reports, 4, 4405.

[368] Wilson, C., & Tisdell, C. (2001). Why farmers continue to use pesticides despite environmental, health and sustainability costs. Ecological economics, 39(3), 449–462.

[369] Baldwin, D. H., Spromberg, J. A., Collier, T. K., & Scholz, N. L. (2009). A fish of many scales: extrapolating sublethal pesticide exposures to the productivity of wild salmon populations. Ecological Applications, 19(8), 2004–2015.

[370] Khler, H. R., & Triebskorn, R. (2013). Wildlife ecotoxicology of pesticides: can we track effects to the population level and beyond?. Science, 341(6147), 759–765.

[371] Mineau, P., & Whiteside, M. (2013). Pesticide acute toxicity is a better correlate of US grassland bird declines than agricultural intensification. PLoS One, 8(2), e57457.

[372] Mineau, P. (2005). Direct losses of birds to pesticides-Beginnings of a quantification. USDA Forest Service Gen. Tech. Rep. PSW-GTR-191.

[373] Crepet, W. L., & Niklas, K. J. (2009). Darwins second abominable mystery: Why are there so many angiosperm species?. American Journal of Botany, 96(1), 366–381.

[374] Ollerton, J., Winfree, R., & Tarrant, S. (2011). How many flowering plants are pollinated by animals?. Oikos, 120(3), 321–326.

[375] Klein, A. M., Vaissiere, B. E., Cane, J. H., Steffan-Dewenter, I., Cunningham, S. A., Kremen, C., & Tscharntke, T. (2007). Importance of pollinators in changing landscapes for world crops. Proceedings of the Royal Society of London B: Biological Sciences, 274(1608), 303–313.

[376] Aizen, M. A., Garibaldi, L. A., Cunningham, S. A., & Klein, A. M. (2009). How much does agriculture depend on pollinators? Lessons from long-term trends in crop production. Annals of botany, 103(9), 1579–1588.

[377] Goulson, D., Nicholls, E., Botas, C., & Rotheray, E. L. (2015). Bee declines driven by combined stress from parasites, pesticides, and lack of flowers. Science, 347(6229), 1255957.

[378] McLauchlan, K. K., Hobbie, S. E., & Post, W. M. (2006). Conversion from agriculture to grassland builds soil organic matter on decadal timescales. Ecological applications, 16(1), 143–153.

[379] Bowman, R. A., Reeder, J. D., & Lober, R. W. (1990). Changes in soil properties in a central plains rangeland soil after 3, 20, and 60 years of cultivation. Soil Science, 150(6), 851–857.

[380] Buyanovsky, G. A., Kucera, C. L., & Wagner, G. H. (1987). Comparative analyses of carbon dynamics in native and cultivated ecosystems. Ecology, 2023–2031.

[381] Peterson, G. A., Halvorson, A. D., Havlin, J. L., Jones, O., Lyon, D. J., & Tanaka, D. L. (1998). Reduced tillage and increasing cropping intensity in the Great Plains conserves soil C. Soil and Tillage Research, 47(3), 207–218.

[382] Halvorson, A. D., Vigil, M. F., Peterson, G. A., & Elliott, E. T. (1996). Long-term tillage and crop residue management study at Akron, Colorado. Soil Organic Matter in Temperate Agroecosystems. Lewis Publishers, Boca Raton, FL, 361–370.

[383] Maraseni, T., Chen, G., Banhazi, T., Bundschuh, J., & Yusaf, T. (2015). An assessment of direct on-farm energy use for high value grain crops grown under different farming practices in Australia. Energies, 8(11), 13033–13046.

[384] Grassini, P.,& Cassman, K. G. (2012). High-yield maize with large net energy yield and small global warming intensity. Proceedings of the National Academy of Sciences, 109(4), 1074–1079.

[385] Ponsioen, T, & Blonk, H.; Blonk Environmental Consultants. (2011). Case studies for more insight into the methodology and composition of carbon footprints of table potatoes and chips. July, 2011.

[386] Brodt, S., Kramer, K. J., Kendall, A., & Feenstra, G. (2013). Comparing environmental impacts of regional and national-scale food supply chains: A case study of processed tomatoes. Food Policy, 42, 106–114.

[387] MacLeod, M., Gerber, P., Mottet, A., Tempio, G., Falcucci, A., Opio, C., Vellinga, T., Henderson, B. & Steinfeld, H. 2013. Greenhouse gas emissions from pig and chicken supply chains - A global life cycle assessment. Food and Agriculture Organization of the United Nations (FAO), Rome.

[388] Gerber, P. J., Steinfeld, H., Henderson, B., Mottet, A., Opio, C., Dijkman, J., Falcucci, A. & Tempio, G. (2013). Tackling climate change through livestock - A global assessment of emissions and mitigation opportunities. Food and Agriculture Organization of the United Nations (FAO), Rome.

[389] Steinfeld, H., Gerber, P., Wassenaar, T., Castel, V. Rosales, M., de Haan, C. (2006). Livestocks' long shadow: environmental issues and options. Food and Agriculture Organization of the United Nations (FAO), Rome.

[390] Nijdam, D., Rood, T., & Westhoek, H. (2012). The price of protein: Review of land use and carbon footprints from life cycle assessments of animal food products and their substitutes. Food Policy, 37(6), 760–770.

[391] Capper, J. L. (2011). The environmental impact of beef production in the United States: 1977 compared with 2007. Journal of animal science, 89(12), 4249–4261.

[392] Lupo, C. D., Clay, D. E., Benning, J. L., & Stone, J. J. (2013). Life-cycle assessment of the beef cattle production system for the Northern Great Plains, USA. Journal of environmental quality, 42(5), 1386–1394.

[393] Basarab, J., Baron, V., Lpez-Campos, ., Aalhus, J., Haugen-Kozyra, K., & Okine, E. (2012). Greenhouse gas emissions from calf-and yearling-fed beef production systems, with and without the use of growth promotants. Animals, 2(2), 195–220.

[394] Shaw, J. H. (1995). How many bison originally populated western rangelands?. Rangelands, 17(5), 148–150.

[395] McSherry, M. E., & Ritchie, M. E. (2013). Effects of grazing on grassland soil carbon: a global review. Global Change Biology, 19(5), 1347–1357.

[396] Derner, J. D., & Schuman, G. E. (2007). Carbon sequestration and rangelands: a synthesis of land management and precipitation effects. Journal of Soil and Water Conservation, 62(2), 77–85.

[397] Derner, J. D., Boutton, T. W., & Briske, D. D. (2006). Grazing and ecosystem carbon storage in the North American Great Plains. Plant and Soil, 280(1-2), 77–90.

[398] Sala, O. E., Parton, W. J., Joyce, L. A., & Lauenroth, W. K. (1988). Primary production of the central grassland region of the United States. Ecology, 69(1), 40–45.

[399] Ingram, L. J., Stahl, P. D., Schuman, G. E., Buyer, J. S., Vance, G. F., Ganjegunte, G. K., ... & Derner, J. D. (2008). Grazing impacts on soil carbon and microbial communities in a mixed-grass ecosystem. Soil Science Society of America Journal, 72(4), 939–948.

[400] Schuman, G. E., Reeder, J. D., Manley, J. T., Hart, R. H., & Manley, W. A. (1999). Impact of grazing management on the carbon and nitrogen balance of a mixedgrass rangeland. Ecological applications, 9(1), 65–71.

[401] Liebig, M. A., Gross, J. R., Kronberg, S. L., & Phillips, R. L. (2010). Grazing management contributions to net global warming potential: A long-term evaluation in the Northern Great Plains. Journal of Environmental Quality, 39(3), 799–809.

[402] Neff, J. C., Reynolds, R. L., Belnap, J., & Lamothe, P. (2005). Multidecadal impacts of grazing on soil physical and biogeochemical properties in southeast Utah. Ecological Applications, 15(1), 87–95.

[403] Beschta, R. L., Donahue, D. L., DellaSala, D. A., Rhodes, J. J., Karr, J. R., OBrien, M. H., ... & Williams, C. D. (2013). Adapting to climate change on western public lands: addressing the ecological effects of domestic, wild, and feral ungulates. Environmental Management, 51(2), 474–491.

[404] Bergstrom, B. J., Arias, L. C., Davidson, A. D., Ferguson, A. W., Randa, L. A., & Sheffield, S. R. (2014). License to kill: reforming federal wildlife control to restore biodiversity and ecosystem function. Conservation Letters, 7(2), 131–142.

[405] Allington, G. R. H., & Valone, T. J. (2010). Reversal of desertification: the role of physical and chemical soil properties. Journal of Arid Environments, 74(8), 973–977.

[439] Pelletier, N., Pirog, R., & Rasmussen, R. (2010). Comparative life cycle environmental impacts of three beef production strategies in the Upper Midwestern United States. Agricultural Systems, 103(6), 380–389.

[407] Desjardins, R. L., Worth, D. E., Verg, X. P., Maxime, D., Dyer, J., & Cerkowniak, D. (2012). Carbon footprint of beef cattle. Sustainability, 4(12), 3279–3301.

[408] Nold, R. (2013). How Much Meat Can You Expect from a Fed Steer? iGrow, SDSU Extension.

[409] Booker, K., Huntsinger, L., Bartolome, J. W., Sayre, N. F., & Stewart, W. (2013). What can ecological science tell us about opportunities for carbon sequestration on arid rangelands in the United States?. Global Environmental Change, 23(1), 240–251.

[410] de la Motte, L. G., Mamadou, O., Beckers, Y., Bodson, B., Heinesch, B., & Aubinet, M. (2018). Rotational and continuous grazing does not affect the total net ecosystem exchange of a pasture grazed by cattle but modifies CO 2 exchange dynamics. Agriculture, Ecosystems & Environment, 253, 157–165.

[411] Sanderman, J., Reseigh, J., Wurst, M., Young, M. A., & Austin, J. (2015). Impacts of rotational grazing on soil carbon in native grass-based pastures in southern Australia. PloS One, 10(8), e0136157.

[412] rgill, S. E., Waters, C. M., Melville, G., Toole, I., Alemseged, Y., & Smith, W. (2017). Sensitivity of soil organic carbon to grazing management in the semi-arid rangelands of south-eastern Australia. The Rangeland Journal, 39(2), 153–167.

[413] Briske, D. D., Ash, A. J., Derner, J. D., & Huntsinger, L. (2014). Commentary: A critical assessment of the policy endorsement for holistic management. Agricultural Systems, 125, 50–53.

[414] Cowie, A. L., Lonergan, V. E., Rabbi, S. F., Fornasier, F., Macdonald, C., Harden, S., ... & Singh, B. K. (2014). Impact of carbon farming practices on soil carbon in northern New South Wales. Soil Research, 51(8), 707–718.

[415] Allen, D. E., Pringle, M. J., Bray, S., Hall, T. J., OReagain, P. O., Phelps, D., ... & Dalal, R. C. (2014). What determines soil organic carbon stocks in the grazing lands of north-eastern Australia?. Soil Research, 51(8), 695–706.

[416] Badgery, W., King, H., Simmons, A., Murphy, B., Rawson, A., & Warden, E. (2013). The effects of management and vegetation on soil carbon stocks in temperate Australian grazing systems. In: Proceedings of the 22nd International Grasslands Congress. Revitalising grasslands to sustain our communities, Sydney, Australia, pp. 1223–1226.

[417] Chan, K. Y., Oates, A., Li, G. D., Conyers, M. K., Prangnell, R. J., Poile, G., ... & Barchia, I. M. (2010). Soil carbon stocks under different pastures and pasture management in the higher rainfall areas of south-eastern Australia. Soil Research, 48(1), 7–15.

[418] Teague, W. R., Dowhower, S. L., Baker, S. A., Haile, N., DeLaune, P. B., & Conover, D. M. (2011). Grazing management impacts on vegetation, soil biota and soil chemical, physical and hydrological properties in tall grass prairie. Agriculture, Ecosystems & Environment, 141(3), 310–322.

[419] Carter, J., Jones, A., OBrien, M., Ratner, J., & Wuerthner, G. (2014). Holistic management: misinformation on the science of grazed ecosystems. International Journal of Biodiversity.

[420] Briske, D. D., Bestelmeyer, B. T., Brown, J. R., Fuhlendorf, S. D., & Polley, H. W. (2013). The Savory method can not green deserts or reverse climate change. Rangelands, 35(5), 72–74.

[421] Joseph, J., Molinar, F., Galt, D., Valdez, R., & Holechek, J. (2002). Short duration grazing research in Africa. Rangelands, 24(4), 9–12.

[422] Holechek, J. L., Gomez, H., Molinar, F., & Galt, D. (1999). Grazing studies: what we've learned. Rangelands, 21(2), 12–16.

[423] Holechek, J. L., Gomes, H., Molinar, F., Galt, D., & Valdez, R. (2000). Short-duration grazing: the facts in 1999. Rangelands, 22(1), 18–22.

[424] Skovlin, J. (1987). Southern Africa's experience with intensive short duration grazing. Rangelands, 9(4), 162–167.

[425] Briske, D. D., Derner, J. D., Brown, J. R., Fuhlendorf, S. D., Teague, W. R., Havstad, K. M., ... & Willms, W. D. (2008). Rotational grazing on rangelands: reconciliation of perception and experimental evidence. Rangeland Ecology & Management, 61(1), 3–17.

[426] Wolf, K., & Horney, M. (2015). Continuous versus rotational grazing, again: Another perspective from meta-analysis. Conference: Society for Range Management, At Sacramento, California. Apr 4, 2015.

[427] Blauw, H., den Hertog, G., & Koeslag, J. (2008). Dairy cattle husbandry: More milk through better management. Agromisa Foundation and CTA, Wageningen, 2008.

[428] Food and Agriculture Organization of the United Nations. (1993). Small-scale dairy farming manual. Rome, Italy: Food and Agriculture Organization.

[429] Tyrrell, H. F., & Reid, J. T. (1965). Prediction of the Energy Value of Cow's Milk 1, 2. Journal of Dairy Science, 48(9), 1215–1223.

[430] Thoma, G., Popp, J., Shonnard, D., Nutter, D., Matlock, M., Ulrich, R., ... & Adom, F. (2013). Regional analysis of greenhouse gas emissions from USA dairy farms: A cradle to farm-gate assessment of the American dairy industry circa 2008. International Dairy Journal, 31, S29–S40.

[431] Thoma, G., Popp, J., Nutter, D., Shonnard, D., Ulrich, R., Matlock, M., ... & Adom, F. (2013). Greenhouse gas emissions from milk production and consumption in the United States: A cradle-to-grave life cycle assessment circa 2008. International Dairy Journal, 31, S3–S14.

[432] Gerber, P., Vellinga, T., Dietze, K., Falcucci, A., Gianni, G., Mounsey, J., et al. (2010). Greenhouse gas emissions from the dairy sector: A life cycle assessment. Rome, Italy: Food and Agriculture Organization of the United Nations.

[433] Casey, J. W., & Holden, N. M. (2006). Quantification of GHG emissions from sucker-beef production in Ireland. Agricultural Systems, 90(1), 79–98.

[434] Pelletier, N. (2008). Environmental performance in the US broiler poultry sector: Life cycle energy use and greenhouse gas, ozone depleting, acidifying and eutrophying emissions. Agricultural Systems, 98(2), 67–73.

[435] Leinonen, I., Williams, A. G., Wiseman, J., Guy, J., & Kyriazakis, I. (2012). Predicting the environmental impacts of chicken systems in the United Kingdom through a life cycle assessment: Broiler production systems. Poultry Science, 91(1), 8–25.

[436] Leinonen, I., Williams, A. G., Wiseman, J., Guy, J., & Kyriazakis, I. (2012). Predicting the environmental impacts of chicken systems in the United Kingdom through a life cycle assessment: Egg production systems. Poultry Science, 91(1), 26–40.

[437] Mollenhorst, H., Berentsen, P. B. M., & De Boer, I. J. M. (2006). On-farm quantification of sustainability indicators: an application to egg production systems. British poultry science, 47(4), 405–417.

[438] Thoma, G., Nutter, D., Ulrich, R., Maxwell, C., Frank, J., & East, C. (2011). National life cycle carbon footprint study for production of US swine. National Pork Board, Des Moines, IA.

[439] Pelletier, N., Lammers, P., Stender, D., & Pirog, R. (2010). Life cycle assessment of high-and low-profitability commodity and deep-bedded niche swine production systems in the Upper Midwestern United States. Agricultural Systems, 103(9), 599–608.

[440] Pelletier, N., Ibarburu, M., & Xin, H. (2014). Comparison of the environmental footprint of the egg industry in the United States in 1960 and 2010. Poultry science, 93(2), 241–255.

[441] Pelletier, N., Ibarburu, M., & Xin, H. (2013). A carbon footprint analysis of egg production and processing supply chains in the Midwestern United States. Journal of cleaner production, 54, 108–114.

[442] Dekker, S. E. M., De Boer, I. J. M., Vermeij, I., Aarnink, A. J. A., & Koerkamp, P. G. (2011). Ecological and economic evaluation of Dutch egg production systems. Livestock Science, 139(1), 109–121.

[443] Madin, E. M., & Macreadie, P. I. (2015). Incorporating carbon footprints into seafood sustainability certification and eco-labels. Marine Policy, 57, 178–181.

[444] Tyedmers, P., & Parker, R. (2012). Fuel consumption and greenhouse gas emissions from global tuna fisheries: a preliminary assessment. International Seafood Sustainability Foundation, McLean, Virginia, USA (ISSF Technical Report 201203).

[445] Tyedmers, P. (2001). Energy consumed by North Atlantic fisheries. Fisheries Impacts on North Atlantic Ecosystems: Catch, Effort, and National/Regional Data Sets. Fisheries Centre Research Reports, 9, 12–34.

[446] Ziegler, F., Emanuelsson, A., Eichelsheim, J. L., Flysj, A., Ndiaye, V., & Thrane, M. (2011). Extended life cycle assessment of southern pink shrimp products originating in Senegalese artisanal and industrial fisheries for export to Europe. Journal of Industrial Ecology, 15(4), 527–538.

[447] Kauffman, J. B., Arifanti, V. B., Trejo, H. H., del Carmen Jess Garca, M., Norfolk, J., Cifuentes, M., ... & Murdiyarso, D. (2017). The jumbo carbon footprint of a shrimp: carbon losses from mangrove deforestation. Frontiers in Ecology and the Environment, 15(4), 183–188.

[448] Beauchemin, K. A., & McGinn, S. M. (2006). Enteric methane emissions from growing beef cattle as affected by diet and level of intake. Canadian journal of animal science, 86(3), 401–408.

[449] Todd, R. W., Altman, M. B., Cole, N. A., & Waldrip, H. M. (2014). Methane emissions from a beef cattle feedyard

during winter and summer on the southern high plains of Texas. Journal of environmental quality, 43(4), 1125–1130.

[450] Dong, H., Mangino, J. McAllister, T. A., Hatfield, J. L., Johnson, D. E., Lassey, K. R., Aparecida de Lima, M., & Romanovskaya, A. (2006). 2006 IPCC Guidelines for National Greenhouse Gas Inventories: Chapter 10: Emissions from Livestock and Manure Management.

[451] Van Haarlem, R. P., Desjardins, R. L., Gao, Z., Flesch, T. K., & Li, X. (2008). Methane and ammonia emissions from a beef feedlot in western Canada for a twelve-day period in the fall. Canadian Journal of Animal Science, 88(4), 641–649.

[452] Rotz, C. A. (2004). Management to reduce nitrogen losses in animal production. Journal of animal science, 82(13S), E119–E137.

[453] Dijkstra, J., Oenema, O., Van Groenigen, J. W., Spek, J. W., Van Vuuren, A. M., & Bannink, A. (2013). Diet effects on urine composition of cattle and N 2 O emissions. Animal, 7(s2), 292–302.

[454] van der Weerden, T. J., Luo, J., de Klein, C. A., Hoogendoorn, C. J., Littlejohn, R. P., & Rys, G. J. (2011). Disaggregating nitrous oxide emission factors for ruminant urine and dung deposited onto pastoral soils. Agriculture, Ecosystems & Environment, 141(3-4), 426–436.

[455] Chi, M. S., & Speers, G. M. (1976). Effects of dietary protein and lysine levels on plasma amino acids, nitrogen retention and egg production in laying hens. The Journal of nutrition, 106(8), 1192–1201.

[456] Liang, Y., Xin, H., Wheeler, E. F., Gates, R. S., Li, H., Zajaczkowski, J. S., ... & Zajaczkowski, F. J. (2005). Ammonia emissions from US laying hen houses in Iowa and Pennsylvania. Transactions of the ASAE, 48(5), 1927–1941.

[457] Eshel, G., Shepon, A., Makov, T., & Milo, R. (2014). Land, irrigation water, greenhouse gas, and reactive nitrogen burdens of meat, eggs, and dairy production in the United States. Proceedings of the National Academy of Sciences, 111(33), 11996–12001.

[458] Meisinger, J. J., & Jokela, W. E. (2000). Ammonia volatilization from dairy and poultry manure. Managing nutrients and pathogens from animal agriculture. NRAES-130. Natural Resource, Agriculture, and Engineering Service, Ithaca, NY, 334–354.

[459] MacDonald, J. M., Ribaudo, M. O., Livingston, M., Beckman, J., & Huang, W. (2009). Manure use for fertilizer and for energy: Report to Congress (Administrative Publication No. AP-037). Washington, D.C., USA: US Department of Agriculture Economic Research Service.

[460] Badgley, C., Moghtader, J., Quintero, E., Zakem, E., Chappell, M. J., Aviles-Vazquez, K., ... & Perfecto, I. (2007). Organic agriculture and the global food supply. Renewable agriculture and food systems, 22(2), 86–108.

[461] Stanhill, G. (1990). The comparative productivity of organic agriculture. Agriculture, ecosystems & environment, 30(1-2), 1–26.

[462] Connor, D. J. (2008). Organic agriculture cannot feed the world. Field Crops Research, 106(2), 187–190.

[463] Connor, D. J. (2013). Organically grown crops do not a cropping system make and nor can organic agriculture nearly feed the world. Field Crops Research, 144, 145–147.

[464] De Ponti, T., Rijk, B., & Van Ittersum, M. K. (2012). The crop yield gap between organic and conventional agriculture. Agricultural systems, 108, 1–9.

[465] Seufert, V., Ramankutty, N., & Foley, J. A. (2012). Comparing the yields of organic and conventional agriculture. Nature, 485(7397), 229–232.

[466] Ponisio, L. C., M'Gonigle, L. K., Mace, K. C., Palomino, J., de Valpine, P., & Kremen, C. (2015). Diversification practices reduce organic to conventional yield gap. Proceedings of the Royal Society B: Biological Sciences, 282(1799).

[467] Clark, M., & Tilman, D. (2017). Comparative analysis of environmental impacts of agricultural production systems, agricultural input efficiency, and food choice, Environmental Research Letters, 12, 064016.

[468] Mondelaers, K., Aertsens, J., & Van Huylenbroeck, G. (2009). A meta-analysis of the differences in environmental impacts between organic and conventional farming. British food journal, 111(10), 1098–1119.

[469] Nowak, B., Nesme, T., David, C., & Pellerin, S. (2013). Disentangling the drivers of fertilising material inflows in organic farming. Nutrient cycling in agroecosystems, 96(1), 79–91.

[470] Nowak, B., Nesme, T., David, C., & Pellerin, S. (2013). To what extent does organic farming rely on nutrient inflows from conventional farming?. Environmental Research Letters, 8(4), 044045.

[471] Cherr, C. M., Scholberg, J. M. S., & McSorley, R. (2006). Green manure approaches to crop production. Agronomy Journal, 98(2), 302–319.

[472] Gaskell, M., & Smith, R. (2007). Nitrogen sources for organic vegetable crops. HortTechnology, 17(4), 431–441.

[473] Schipanski, M. E., Barbercheck, M., Douglas, M. R., Finney, D. M., Haider, K., Kaye, J. P., ... & White, C. (2014). A framework for evaluating ecosystem services provided by cover crops in agroecosystems. Agricultural Systems, 125, 12–22.

[474] Clark, S., Klonsky, K., Livingston, P., & Temple, S. (1999). Crop-yield and economic comparisons of organic, low-input, and conventional farming systems in California's Sacramento Valley. American journal of alternative agriculture, 14(3), 109–121.

[475] Delate, K., Cambardella, C., Chase, C., Johanns, A., & Turnbull, R. (2013). The Long-Term Agroecological Research (LTAR) experiment supports organic yields, soil quality, and economic performance in Iowa. Crop Management, 12(1).

[476] Gillespie, J., & Nehring, R. (2014). Pasture-based versus conventional milk production: Where is the profit?. Journal of Agricultural and Applied Economics, 46(4), 543–558.

[477] Charles, D.; National Public Radio. (2017). Organic Industry Sues USDA To Push For Animal Welfare Rules. September 13, 2017.

[478] Cederberg, C., & Mattsson, B. (2000). Life cycle assessment of milk productiona comparison of conventional and organic farming. Journal of Cleaner production, 8(1), 49–60.

[479] Guerci, M., Bava, L., Zucali, M., Sandrucci, A., Penati, C., & Tamburini, A. (2013). Effect of farming strategies on environmental impact of intensive dairy farms in Italy. Journal of dairy research, 80(3), 300–308.

[480] Guerci, M., Knudsen, M. T., Bava, L., Zucali, M., Schnbach, P., & Kristensen, T. (2013). Parameters affecting the environmental impact of a range of dairy farming systems in Denmark, Germany and Italy. Journal of cleaner production, 54, 133-1-41.

[481] Bava, L., Sandrucci, A., Zucali, M., Guerci, M., & Tamburini, A. (2014). How can farming intensification affect the environmental impact of milk production?. Journal of dairy science, 97(7), 4579–4593.

[482] Thomassen, M. A., van Calker, K. J., Smits, M. C., Iepema, G. L., & de Boer, I. J. (2008). Life cycle assessment of conventional and organic milk production in the Netherlands. Agricultural systems, 96(1), 95–107.

[483] Halberg, N., Hermansen, J. E., Kristensen, I. S., Eriksen, J., Tvedegaard, N., & Petersen, B. M. (2010). Impact of organic pig production systems on CO2 emission, C sequestration and nitrate pollution. Agronomy for Sustainable Development, 30(4), 721–731.

[484] Kool, A., Blonk, H., Ponsioen, T., Sukkel, W., Vermeer, H., De Vries, J., & Hoste, R. (2010). Carbon footprints of conventional and organic pork: Assessment of typical production systems in the Netherlands, Denmark, England and Germany. Blonk Milieu Advies.

[485] Basset-Mens, C., & Van der Werf, H. M. (2005). Scenario-based environmental assessment of farming systems: the case of pig production in France. Agriculture, Ecosystems & Environment, 105(1), 127–144.

[486] Kumm, K. I. (2002). Sustainability of organic meat production under Swedish conditions. Agriculture, Ecosystems & Environment, 88(1), 95–101.

[487] Trewavas, A. (2001). Urban myths of organic farming. Nature, 410(6827), 409–410.

[488] Burney, J. A., Davis, S. J., & Lobell, D. B. (2010). Greenhouse gas mitigation by agricultural intensification. Proceedings of the national Academy of Sciences, 107(26), 12052–12057.

[489] Tilman, D., Balzer, C., Hill, J., & Befort, B. L. (2011). Global food demand and the sustainable intensification of agriculture. Proceedings of the National Academy of Sciences, 108(50), 20260–20264.

[490] Garnett, T., Appleby, M. C., Balmford, A., Bateman, I. J., Benton, T. G., Bloomer, P., ... & Herrero, M. (2013). Sustainable intensification in agriculture: premises and policies. Science, 341(6141), 33–34.

[491] Phalan, B., Onial, M., Balmford, A., & Green, R. E. (2011). Reconciling food production and biodiversity conservation: land sharing and land sparing compared. Science, 333(6047), 1289–1291.

[492] Laurance, W. F., Sayer, J., & Cassman, K. G. (2014). Agricultural expansion and its impacts on tropical nature. Trends in ecology & evolution, 29(2), 107–116.

[493] Lambin, E. F., & Meyfroidt, P. (2011). Global land use change, economic globalization, and the looming land scarcity. Proceedings of the National Academy of Sciences, 108(9), 3465–3472.

[494] Tauger, M. B. (2010). Agriculture in World History. Routledge.

[495] Hall, C. A., Balogh, S., & Murphy, D. J. (2009). What is the minimum EROI that a sustainable society must have?. Energies, 2(1), 25–47.

[496] Hall, C. A., Lambert, J. G., & Balogh, S. B. (2014). EROI of different fuels and the implications for society. Energy policy, 64, 141–152.

[497] Lark, T. J., Salmon, J. M., & Gibbs, H. K. (2015). Cropland expansion outpaces agricultural and biofuel policies in the United States. Environmental Research Letters, 10(4), 044003.

[498] Searchinger, T., Heimlich, R., Houghton, R. A., Dong, F., Elobeid, A., Fabiosa, J., ... & Yu, T. H. (2008). Use of US croplands for biofuels increases greenhouse gases through emissions from land-use change. Science, 319(5867), 1238–1240.

[499] Qin, Z., Dunn, J. B., Kwon, H., Mueller, S., & Wander, M. M. (2016). Influence of spatially dependent, modeled soil carbon emission factors on lifecycle greenhouse gas emissions of corn and cellulosic ethanol. GCB Bioenergy, 8(6), 1136–1149.

[500] Patzek, T. W. (2004). Thermodynamics of the corn-ethanol biofuel cycle. Critical Reviews in Plant Sciences, 23(6), 519–567.

[501] Lal, R. (2004). Carbon emission from farm operations. Environment international, 30(7), 981–990.

[502] U.S. Environmental Protection Agency. (2010). Renewable Fuel Standard Program (RFS2) Regulatory Impact Analysis. EPA-420-R-10-006, February 2010.

[503] Caron, J., Metcalf, G., & Reilly, J. (2014). The CO2 Content of Consumption Across US Regions: A Multi-Regional Input-Output (MRIO) Approach. MIT Joint Program on the Science and Policy of Global Change.

[504] Bureau of Labor Statistics. (2017). Consumer Expenditures in 2015. Report 1066, April, 2017.

[505] Chung, J. W., & Meltzer, D. O. (2009). Estimate of the carbon footprint of the US health care sector. JAMA, 302(18), 1970–1972.

[506] U.S. Environmental Protection Agency. (2016). Advancing Sustainable Materials Management: 2014 Tables and Figures. November, 2016.

[507] Pipatti, R., Svardal, P., Alves, J. W. S., Gao, Q., Cabrera, C. L., Mareckova, K., Oonk, H., Scheehle, E., Sharma, C., Smith, A., & Yamada, M. (2006). 2006 IPCC Guidelines for National Greenhouse Gas Inventories: Volume 5, Chapter 3: Solid Waste Disposal.

[508] El Hanandeh, A., & El-Zein, A. (2010). Life-cycle assessment of municipal solid waste management alternatives with consideration of uncertainty: SIWMS development and application. Waste Management, 30(5), 902–911.

[509] Barlaz, M. A. (1998). Carbon storage during biodegradation of municipal solid waste components in laboratoryscale landfills. Global biogeochemical cycles, 12(2), 373–380.

[510] Barlaz, M. A., Chanton, J. P., & Green, R. B. (2009). Controls on landfill gas collection efficiency: instantaneous and lifetime performance. Journal of the Air & Waste Management Association, 59(12), 1399–1404.

[511] Staley, B. F., & Barlaz, M. A. (2009). Composition of municipal solid waste in the united states and implications for carbon sequestration and methane yield. Journal of Environmental Engineering, 135(10), 901–909.

[512] Wang, X., Padgett, J. M., Powell, J. S., & Barlaz, M. A. (2013). Decomposition of forest products buried in landfills. Waste management, 33(11), 2267–2276.

[513] Themelis, N. J., & Ulloa, P. A. (2007). Methane generation in landfills. Renewable Energy, 32(7), 1243–1257.

[514] Bogner, J. E., Sass, R. L., & Walter, B. P. (2000). Model comparisons of methane oxidation across a management gradient: Wetlands, rice production systems, and landfill. Global Biogeochemical Cycles, 14(4), 1021–1033.

[515] Liptay, K., Chanton, J., Czepiel, P., & Mosher, B. (1998). Use of stable isotopes to determine methane oxidation in landfill cover soils. Journal of Geophysical Research, 103(D7), 8243–8250.

[516] Chanton, J., & Liptay, K. (2000). Seasonal variation in methane oxidation in a landfill cover soil as determined by an in situ stable isotope technique. Global Biogeochemical Cycles, 14(1), 51–60.

[517] Chanton, J. P., Powelson, D. K., & Green, R. B. (2009). Methane oxidation in landfill cover soils, is a 10% default value reasonable?. Journal of environmental quality, 38(2), 654–663.

[518] Chanton, J., Abichou, T., Langford, C., Hater, G., Green, R., Goldsmith, D., & Swan, N. (2011). Landfill methane oxidation across climate types in the US. Environmental science & technology, 45(1), 313.

[519] Czepiel, P. M., Mosher, B., Crill, P. M., & Harriss, R. C. (1996). Quantifying the effect of oxidation on landfill methane emissions. Journal of Geophysical Research, 101, 16721–16729.

[520] Spokas, K., J. Bogner, J. P. Chanton, M. Morcet, C. Aran, C. Graff, Y. Golvan, & I. Hebe. (2006). Methane mass balance at three landfill sites: What is the efficiency of capture by gas collection systems?. Waste Management, 26(5), 516–525.

[521] Amini, H. R., Reinhart, D. R., & Mackie, K. R. (2012). Determination of first-order landfill gas modeling parameters and uncertainties. Waste management, 32(2), 305–316.

[522] U.S. Environmental Protection Agency. (2016). Documentation for Greenhouse Gas Emission and Energy Factors Used in the Waste Reduction Model (WARM). February, 2016.

[523] Das, S. K. (2011). Aluminum recycling in a carbon constrained world: Observations and opportunities. JOM, 63(8), 137–140.

[524] Khoo, H. H., & Tan, R. B. (2010). Environmental impacts of conventional plastic and bio-based carrier bags. The international journal of life cycle assessment, 15(4), 338–345.

[525] Mattila, T., Kujanp, M., Dahlbo, H., Soukka, R., & Myllymaa, T. (2011). Uncertainty and sensitivity in the carbon footprint of shopping bags. Journal of Industrial Ecology, 15(2), 217–227.

[526] U.S. Environmental Protection Agency. (2015). Advancing Sustainable Materials Management: Facts and Figures 2013.

[527] U.S. Environmental Protection Agency. (2015). WARM Version 13: Plastics. March, 2015.

[528] NOLAN ITU. (2002). Plastic shopping bags: Analysis of levies and environmental impacts. Everton Park, Australia: Environment Australia.

[529] U.S. Energy Information Administration. (2015). Annual Energy Outlook 2015. DOE/EIA-0383(2015). April, 2015

[530] U.S. Energy Information Administration. (2009). 2009 Residential Energy Consumption Survey.

[531] National Association of Home Builders. (2006). Housing Facts, Figures and Trends. March, 2006.

[532] National Renewable Energy Laboratory. (2015). National Solar Radiation Data Base.

[533] Motta, S. F. Y., & Domanski, P. A. (2000). Impact of Elevated Ambient Temperatures on Capacity and Energy Input to a Vapor Compression System–Literature Review. Letter report for Air-Conditioning and Refrigeration Institute 21-CR Research Project: 605-50010/605-50015.

[534] Payne, W. V., & Domanski, P. A. (2002). A comparison of an R22 and an R410a air conditioner operating at high ambient temperatures. International Refrigeration and Air Conditioning Conference, Paper 532.

[535] Wassmer, M. R. (2003). A Component-Based Model for Residential Air Conditioner and Heat Pump Energy Calculations (Doctoral dissertation, University of Colorado).

[536] Amer, O., Boukhanouf, R., & Ibrahim, H. G. (2015). A Review of Evaporative Cooling Technologies. International Journal of Environmental Science and Development, 6(2), 111–117.

[537] Southwest Energy Efficiency Project (2004). New Evaporative Cooling Systems: An Emerging Solution for Homes in Hot Dry Climates with Modest Cooling Loads.

[538] Macriss, R. A., & Elkins, R. H. (1976). Energy conservation: standing pilot gas consumption. ASHRAE J., 18(6).

[539] Wisconsin Public Service. Natural Gas Appliance Calculator. Accessed 6/25/2016. http://www.wisconsinpublicservice.com/home/gas_calculator.aspx

[540] Brown, M., Burke-Scoll, M, Stebnicki, J.; Franklin Energy Services. (2011). Air Source Heat Pump Efficiency Gains from Low Ambient Temperature Operation Using Supplemental Electric Heating.

[541] Somanathan, E., Bluffstone, R., & Toman, M. (2014). Biogas Replacement of Fuelwood: Clean Energy Access with Low-Cost Mitigation of Climate Change.

[542] Cal/EPA Air Resources Board. (2005). Wood Burning Handbook.

[543] Evtyugina, M., Alves, C., Calvo, A., Nunes, T., Tarelho, L., Duarte, M., ... & Pio, C. (2014). VOC emissions from residential combustion of Southern and mid-European woods. Atmospheric environment, 83, 90–98.

[544] Ragland, K. W., Aerts, D. J., & Baker, A. J. (1991). Properties of wood for combustion analysis. Bioresource technology, 37(2), 161–168.

[545] Solli, C., Reenaas, M., Strmman, A. H., & Hertwich, E. G. (2009). Life cycle assessment of wood-based heating in Norway. The International Journal of Life Cycle Assessment, 14(6), 517–528.

[546] Johansson, L. S., Leckner, B., Gustavsson, L., Cooper, D., Tullin, C., & Potter, A. (2004). Emission characteristics of modern and old-type residential boilers fired with wood logs and wood pellets. Atmospheric environment, 38(25), 4183–4195.

[547] Meyer, N. K. (2012). Particulate, black carbon and organic emissions from small-scale residential wood combustion appliances in Switzerland. Biomass and Bioenergy, 36, 31–42.

[548] Rau, J. A. (1989). Composition and size distribution of residential wood smoke particles. Aerosol Science and Technology, 10(1), 181–192.

[549] Leskinen, J., Tissari, J., Uski, O., Virn, A., Torvela, T., Kaivosoja, T., ... & Jalava, P. I. (2014). Fine particle emissions in three different combustion conditions of a wood chip-fired applianceParticulate physico-chemical properties and induced cell death. Atmospheric environment, 86, 129–139.

[550] McDonald, J. D., Zielinska, B., Fujita, E. M., Sagebiel, J. C., Chow, J. C., & Watson, J. G. (2000). Fine particle and gaseous emission rates from residential wood combustion. Environmental Science & Technology, 34(11), 2080–2091.

[551] Saud, T., Gautam, R., Mandal, T. K., Gadi, R., Singh, D. P., Sharma, S. K., ... & Saxena, M. (2012). Emission estimates of organic and elemental carbon from household biomass fuel used over the Indo-Gangetic Plain (IGP), India. Atmospheric environment, 61, 212–220.

[552] Caserini, S., Galante, S., Ozgen, S., Cucco, S., de Gregorio, K., & Moretti, M. (2013). A methodology for elemental and organic carbon emission inventory and results for Lombardy region, Italy. Science of the Total Environment, 450, 22–30.

[553] Heringa, M. F., DeCarlo, P. F., Chirico, R., Tritscher, T., Dommen, J., Weingartner, E., ... & Baltensperger, U. (2011). Investigations of primary and secondary particulate matter of different wood combustion appliances with a high-resolution time-of-flight aerosol mass spectrometer. Atmospheric Chemistry and Physics, 11(12), 5945–5957.

[554] Heringa, M. F., DeCarlo, P. F., Chirico, R., Lauber, A., Doberer, A., Good, J., ... & Miljevic, B. (2012). Time-resolved characterization of primary emissions from residential wood combustion appliances. Environmental science & technology, 46(20), 11418–11425.

[555] Bruns, E. A., Krapf, M., Orasche, J., Huang, Y., Zimmermann, R., Drinovec, L., ... & Baltensperger, U. (2015). Characterization of primary and secondary wood combustion products generated under different burner loads. Atmospheric Chemistry and Physics, 15(5), 2825–2841.

[556] Grieshop, A. P., Logue, J. M., Donahue, N. M., & Robinson, A. L. (2009). Laboratory investigation of photochemical oxidation of organic aerosol from wood fires 1: measurement and simulation of organic aerosol evolution. Atmospheric Chemistry and Physics, 9(4), 1263–1277.

[557] Pfeifer, M., Lefebvre, V., Turner, E., Cusack, J., Khoo, M., Chey, V. K., ... & Ewers, R. M. (2015). Deadwood biomass: an underestimated carbon stock in degraded tropical forests?. Environmental Research Letters, 10(4), 044019.

[558] Rock, J., Badeck, F. W., & Harmon, M. E. (2008). Estimating decomposition rate constants for European tree species from literature sources. European Journal of Forest Research, 127(4), 301–313.

[559] Laiho, R., & Prescott, C. E. (2004). Decay and nutrient dynamics of coarse woody debris in northern coniferous forests: a synthesis. Canadian Journal of Forest Research, 34(4), 763–777.

[560] Pierobon, F., Zanetti, M., Grigolato, S., Sgarbossa, A., Anfodillo, T., & Cavalli, R. (2015). Life cycle environmental impact of firewood productionA case study in Italy. Applied Energy, 150, 185–195.

[561] Awbi, H. B. (1998). Calculation of convective heat transfer coefficients of room surfaces for natural convection. Energy and Buildings, 28(2), 219–227.

[562] Emmel, M. G., Abadie, M. O., & Mendes, N. (2007). New external convective heat transfer coefficient correlations for isolated low-rise buildings. Energy and Buildings, 39(3), 335–342.

[563] Adelard, L., Pignolet-Tardan, F., Mara, T., Lauret, P., Garde, F., & Boyer, H. (1998). Sky temperature modelisation and applications in building simulation. Renewable Energy, 15(1), 418–430.

[564] Cummings, J. B., Tooley, J. Jr., & Moyer, N. (1991). Investigation of Air Distribution System Leakage and Its Impacts in Central Florida Homes, Prepared for the Governor's Energy Office, FSEC-CR-397-91, January 31, 1991.

[565] Parker, D., Fairey, P., & Gu, L. (1993). Simulation of the effects of duct leakage and heat transfer on residential space-cooling energy use. Energy and Buildings, 20(2), 97–113.

[566] Modera, M. (1993). Characterizing the performance of residential air distribution systems. Energy and Buildings, 20(1), 65–75.

[567] Francisco, P. W., Palmiter, L., & Davis, B. (1998). Modeling the thermal distribution efficiency of ducts: comparisons to measured results. Energy and Buildings, 28(3), 287–297.

[568] Francisco, P. W., Siegel, J., Palmiter, L., & Davis, B. (2006). Measuring residential duct efficiency with the short-term coheat test methodology. Energy and buildings, 38(9), 1076–1083.

[569] Kalema, T., J´hannesson, G., Pylsy, P., & Hagengran, P. (2008). Accuracy of energy analysis of buildings: a comparison of a monthly energy balance method and simulation methods in calculating the energy consumption and the effect of thermal mass. Journal of Building Physics, 32(2), 101–130.

[570] van Lier, J., & Durigon, A. (2013). Soil thermal diffusivity estimated from data of soil temperature and single soil component properties. Revista Brasileira de Cincia do Solo, 37(1), 106–112.

[571] Oleson, K. W., Bonan, G. B., & Feddema, J. (2010). Effects of white roofs on urban temperature in a global climate model. Geophysical Research Letters, 37, L03701.

[572] Berdahl, P., & Bretz, S. E. (1997). Preliminary survey of the solar reflectance of cool roofing materials. Energy and Buildings, 25(2), 149–158.

[573] Akbari, H., Bretz, S., Kurn, D. M., & Hanford, J. (1997). Peak power and cooling energy savings of high-albedo roofs. Energy and Buildings, 25(2), 117–126.

[574] Parker, D. S., Huang, Y. J., Konopacki, S. J., & Gartland, L. M. (1998). Measured and simulated performance of reflective roofing systems in residential buildings. ASHRAE Transactions, 104, 963–975.

[575] Synnefa, A., Santamouris, M., & Akbari, H. (2007). Estimating the effect of using cool coatings on energy loads and thermal comfort in residential buildings in various climatic conditions. Energy and Buildings, 39(11), 1167–1174.

[576] Irvine, P. J., Ridgwell, A., & Lunt, D. J. (2011). Climatic effects of surface albedo geoengineering. Journal of Geophysical Research: Atmospheres, 116, D24112.

[577] de Dear, R. J., Arens, E., Hui, Z., & Oguro, M. (1997). Convective and radiative heat transfer coefficients for individual human body segments. International Journal of Biometeorology, 40(3), 141–156.

[578] Mayer, P. W., DeOreo, W. B., ... & AWWA Research Foundation. (1999). Residential End Uses of Water Survey. AWWA Research Foundation, ISBN 1-58321-016-4.

[579] DeOreo, B., Mayer, P.; Aquacraft Inc. (2012). Residential End Uses of Water: Progress Report and Interim Results. Drinking Water Research, 22(3), 14–22.

[580] Plappally, A. K., & Lienhard, V. (2012). Energy requirements for water production, treatment, end use, reclamation, and disposal. Renewable and Sustainable Energy Reviews, 16(7), 4818–4848.

[581] Cruse, R. M., Devlin, D. L., Parker, D., & Waskom, R. M. (2016). Irrigation aquifer depletion: the nexus linchpin. Journal of Environmental Studies and Sciences, 6(1), 149–160.

[582] EPRI. (2002). Water and Sustainability: U.S. Electricity Consumption for Water Supply & TreatmentThe Next Half Century, EPRI, Palo Alto, CA: 2000. 1006787.

[583] Vilanova, M. R. N., & Balestieri, J. A. P. (2014). Energy and hydraulic efficiency in conventional water supply systems. Renewable and Sustainable Energy Reviews, 30, 701–714.

[584] Perrone, D., Murphy, J., & Hornberger, G. M. (2011). Gaining perspective on the water energy nexus at the com-

munity scale. Environmental science & technology, 45(10), 4228-4234.

[585] Stokes, J. R., & Horvath, A. (2009). Energy and air emission effects of water supply. Environmental science & technology, 43(8), 2680–2687.

[586] Maupin, M.A., Kenny, J.F., Hutson, S.S., Lovelace, J.K., Barber, N.L., & Linsey, K.S. (2014). Estimated use of water in the United States in 2010: U.S. Geological Survey Circular 1405, 56 p.

[587] California Dept. of Water Resources. California State Water Project Today.

[588] Klein, G., Krebs, M., Hall, V., O'Brien, T., Blevins, B. B.; California Energy Commission. (2005). Californias Water-Energy Relationship. November 2005, CEC-700-2005-011-SF.

[589] Cooley, H, Gleick, P, & Wolff, G. (2006). Desalination, With a Grain of Salt: A California Perspective. Pacific Institute, June 2006.

[590] Wilkes, C. R., Mason, A. D., & Hern, S. C. (2005). Probability Distributions for Showering and Bathing WaterUse Behavior for Various US Subpopulations. Risk analysis, 25(2), 317–337.

[591] Pittet, D., Allegranzi, B., & Boyce, J. (2009). The World Health Organization guidelines on hand hygiene in health care and their consensus recommendations. Infection Control & Hospital Epidemiology, 30(7), 611–622.

[592] Gifford, W. R., Goldberg, M. L., Tanimoto, P. M., Celnicker, D. R., & Poplawski, M. E. (2012). Residential lighting end-use consumption study: Estimation framework and initial estimates. Prepared for US DOE by DNV KEMA Energy and Sustainability and Pacific Northwest National Laboratory (PNNL). December, 2012.

[593] Ulrich, R. S., Zimring, C., Zhu, X., DuBose, J., Seo, H. B., Choi, Y. S., ... & Joseph, A. (2008). A review of the research literature on evidence-based healthcare design. HERD: Health Environments Research & Design Journal, 1(3), 61–125.

[594] Ulrich R. S., Zimring C., Joseph A., Quan X., & Choudhary R. (2004). The role of the physical environment in the hospital of the 21st century: A once-in-a-lifetime opportunity. Concord, CA: The Center for Health Design.

[595] U.S. Department of Energy. (2014). Energy Savings Forecast of Solid-State Lighting in General Illumination Applications. Prepared by Navigant Consulting, Inc.

[596] U.S. Department of Energy. (2012). Life-Cycle Assessment of Energy and Environmental Impacts of LED Lighting Products Part I: Review of the Life-Cycle Energy Consumption of Incandescent, Compact Fluorescent, and LED Lamps. Prepared by Navigant Consulting, Inc.

[597] Williams, D. R.; NASA Space Science Data Coordinated Archive. (2018). Sun Fact Sheet. http://nssdc.gsfc.nasa.gov/planetary/factsheet/sunfact.html

[598] Ronda, C. R. (2008). Luminescence: From Theory to Applications. John Wiley & Sons, 2008.

[599] Zheludev, N. (2007). The life and times of the LEDa 100-year history. Nature Photonics, 1(4), 189–92.

[600] U.S. Department of Energy. (2012). 2010 U.S. Lighting Market Characterization. Prepared by Navigant Consulting, Inc.

[601] U.S. Department of Energy. (2011). Energy Savings Estimates of Light Emitting Diodes in Niche Lighting Applications. Prepared by Navigant Consulting, Inc.

[602] New York State Energy Research and Development Authority. (2014). Street Lighting in New York State: Opportunities and Challenges. Prepared by Energy and Resource Solutions, and Optimal Energy.

[603] Mn/DOT. (2010). Roadway Lighting Design Manual.

[604] City of Los Angeles, Bureau of Street Lighting. (2007). Design Standards and Guidelines.

[605] Hamins, A., Bundy, M., & Dillon, S. E. (2005). Characterization of candle flames. Journal of Fire Protection Engineering, 15(4), 265–285.

[606] Kendalla, A., Yuanb, J., Brodtc, S., & Kramerd, K. J. (2010). Carbon Footprint of US Honey Production and Packing. UC Davis (CA).

[607] M. J. Bradley & Associates. (2012). Benchmarking Air Emissions of the 100 Largest Electric Power Producers in the United States.

[608] Li, Y., & Jin, L. (2011). Environmental release of mercury from broken compact fluorescent lamps. Environmental Engineering Science, 28(10), 687–691.

[609] Johnson, N. C., Manchester, S., Sarin, L., Gao, Y., Kulaots, I., & Hurt, R. H. (2008). Mercury vapor release from broken compact fluorescent lamps and in situ capture by new nanomaterial sorbents. Environmental science & technology, 42(15), 5772–5778.

[610] Eckelman, M. J., Anastas, P. T., & Zimmerman, J. B. (2008). Spatial assessment of net mercury emissions from the use of fluorescent bulbs. Environmental science & technology, 42(22), 8564–8570.

[611] Aucott, M., McLinden, M., & Winka, M. (2003). Release of mercury from broken fluorescent bulbs. Journal of the Air & Waste Management Association, 53(2), 143–151.

[612] James, C., Onarinde, B. A., & James, S. J. (2017). The use and performance of household refrigerators: a review.

Comprehensive Reviews in Food Science and Food Safety, 16(1), 160–179.

[613] Kim, H. C., Keoleian, G. A., & Horie, Y. A. (2006). Optimal household refrigerator replacement policy for life cycle energy, greenhouse gas emissions, and cost. Energy Policy, 34(15), 2310–2323.

[614] Parker, D., Fairey, P., & Hendron, R. (2010). Updated Miscellaneous Electricity Loads and Appliance Energy Usage Profiles for Use in Home Energy Ratings, the Building America Benchmark Procedures and Related Calculations. Florida Solar Energy Center, FSECCR-1837-10.

[615] Hager, T. J., & Morawicki, R. (2013). Energy consumption during cooking in the residential sector of developed nations: A review. Food Policy, 40, 54–63.

[616] Roth, K., Mckenney, K., Paetsch, C., & Ponoum, R.; TIAX LLC (2008). U.S. Residential Miscellaneous Electric Loads Electricity Consumption.

[617] Smith, L. P., Ng, S. W., & Popkin, B. M. (2013). Trends in US home food preparation and consumption: analysis of national nutrition surveys and time use studies from 19651966 to 20072008. Nutrition Journal, 12, 45.

[618] U.S. Environmental Protection Agency. (2011). ENERGY STAR Market & Industry Scoping Report: Dryers. November, 2011.

[619] Denkenberger, D., Mau, S., Calwell, C, & Wanless, E. (2011). Residential Clothes Dryers: A Closer Look at Energy Efficiency Test Procedures and Savings Opportunities. Natural Resources Defense Council, November 9, 2011.

[620] Ross, J. P., Meier, A. (2000). Whole-House Measurements of Standby Power Consumption. Lawrence Berkeley National Laboratory, LBNL-45967.

[621] Meier, A., LeBot, B. (1999). One watt initiative: A global effort to reduce leaking electricity. Lawrence Berkeley National Laboratory.

[622] Meier, A. (2001). A worldwide review of standby power use in homes. Lawrence Berkeley National Laboratory.

[623] Hammond, G. P., & Jones, C. I. (2008). Embodied energy and carbon in construction materials. Proceedings of the Institution of Civil Engineers-Energy, 161(2), 87–98.

[624] Keoleian, G. A., Blanchard, S., & Reppe, P. (2000). Lifecycle energy, costs, and strategies for improving a singlefamily house. Journal of Industrial Ecology, 4(2), 135–156.

[625] Krankina, O. N., Harmon, M. E., Schnekenburger, F., & Sierra, C. A. (2012). Carbon balance on federal forest lands of Western Oregon and Washington: the impact of the Northwest Forest Plan. Forest Ecology and Management, 286, 171–182.

[626] Lutz, J. (2016). Forest Research Notes, 13(1), 1–4.

[627] Milesi, C., Running, S. W., Elvidge, C. D., Dietz, J. B., Tuttle, B. T., & Nemani, R. R. (2005). Mapping and modeling the biogeochemical cycling of turf grasses in the United States. Environmental Management, 36(3), 426–438.

[628] Townsend-Small, A., & Czimczik, C. I. (2010). Carbon sequestration and greenhouse gas emissions in urban turf. Geophysical Research Letters, 37(2), L02707.

[629] Sahu, R. Technical Assessment of the Carbon Sequestration Potential of Managed Turfgrass in the United States.

[630] Raciti, S. M., Groffman, P. M., Jenkins, J. C., Pouyat, R. V., Fahey, T. J., Pickett, S. T., & Cadenasso, M. L. (2011). Accumulation of carbon and nitrogen in residential soils with different land-use histories. Ecosystems, 14(2), 287–297.

[631] Raciti, S. M., Burgin, A. J., Groffman, P. M., Lewis, D. N., & Fahey, T. J. (2011). Denitrification in suburban lawn soils. Journal of environmental quality, 40(6), 1932–1940.

[632] Selhorst, A., & Lal, R. (2013). Net carbon sequestration potential and emissions in home lawn turfgrasses of the United States. Environmental management, 51(1), 198–208.

[633] Qian, Y., & Follett, R. F. (2002). Assessing soil carbon sequestration in turfgrass systems using long-term soil testing data. Agronomy Journal, 94(4), 930–935.

[634] Qian, Y. L., Bandaranayake, W., Parton, W. J., Mecham, B., Harivandi, M. A., & Mosier, A. R. (2003). Long-term effects of clipping and nitrogen management in turfgrass on soil organic carbon and nitrogen dynamics. Journal of Environmental Quality, 32(5), 1694–1700.

[635] Pouyat, R. V., Yesilonis, I. D., & Golubiewski, N. E. (2009). A comparison of soil organic carbon stocks between residential turf grass and native soil. Urban Ecosystems, 12(1), 45–62.

[636] Gu, C., Crane II, J., Hornberger, G., & Carrico, A. (2015). The effects of household management practices on the global warming potential of urban lawns. Journal of environmental management, 151, 233–242.

[637] Groffman, P. M., Williams, C. O., Pouyat, R. V., Band, L. E., & Yesilonis, I. D. (2009). Nitrate leaching and nitrous oxide flux in urban forests and grasslands. Journal of environmental quality, 38(5), 1848–1860.

[638] Kaye, J. P., McCulley, R. L., & Burke, I. C. (2005). Carbon fluxes, nitrogen cycling, and soil microbial communities in adjacent urban, native and agricultural ecosystems. Global Change Biology, 11(4), 575–587.

[639] U.S. Departmentof Agriculture. (2001). U.S. Forest Facts and Historical Trends. FS-696-M, September, 2001.

[640] U.S. Environmental Protection Agency. (2014). Inventory of U.S. Greenhouse Gas Emissions and Sinks: 1990–2012.

EPA 430-R-14-003, April 15, 2014.

[641] American Petroleum Institute, America's Natural Gas Alliance. (2012). Characterizing Pivotal Sources of Methane Emissions from Natural Gas Production: Summary and Analysis of API and ANGA Survey Responses. September 21, 2012.

[642] Santos, A., McGuckin, N., Nakamoto, H. Y., Gray, D., & Liss, S; Federal Highway Administration. (2011). Summary of Travel Trends: 2009 National Household Travel Survey. FHWA-PL-ll-022, June, 2011.

[643] Hu, P. S., & Reuscher, T. R.; Federal Highway Administration. (2004). Summary of Travel Trends: 2001 National Household Travel Survey. December, 2004.

www.ingramcontent.com/pod-product-compliance
Lightning Source LLC
Chambersburg PA
CBHW080902030426
42336CB00017B/2981